Bacterial Adhesion to Cells and Tissues

Bacterial Adhesion to Cells and Tissues

Itzhak Ofek and Ronald J. Doyle

CHAPMAN & HALL
New York • London

First published in 1994 by
Chapman & Hall
One Penn Plaza
New York, NY 10119

Published in Great Britain by
Chapman & Hall
2-6 Boundary Row
London SE1 8HN

Printed in the United States of America

Library of Congress Cataloging in Publication Data

Ofek, Itzhak
Bacterial adhesion to cells & tissues / Itzhak Ofek and Ronald J. Doyle.
p. cm.
Includes bibliographical references and index.
ISBN 0-412-03011-X
1. Bacteria—Adhesion. I. Doyle, Ronald J. II. Title. III. Title: Bacterial adhesion to cells and tissues.
[DNLM: 1. Bacterial Adhesion. 2. Cell Adhesion. QW 52 D754b 1993]
QR96.8.D68 1993
589.9′087′5—dc20
DNLM/DLC
for Library of Congress 93-7370
CIP

British Library Cataloguing in Publication Data

Please send your order for this or any **Chapman & Hall book to Chapman & Hall, 29 West 35th Street, New York, NY 10001, Attn: Customer Service Department.** You may also call our Order Department at 1-212-244-3336 or fax your purchase order to 1-800-248-4724.

For a complete listing of Chapman & Hall's titles, send your requests to **Chapman & Hall, Dept. BC, One Penn Plaza, New York, NY 10119.**

Dedication

In Memory of a Friend: Edwin H. Beachey, M.D.

We dedicate this book to the memory of our friend, Ed Beachey. It seems almost impossible that we live in a world without Ed, who was a victim of cancer in 1989 at a relatively young age. Our dedication of this book to Ed attests to our personal respect for and friendship with him.

When Ed was a medical student at Northwestern University, he became ill with rheumatic fever, a sequellae to a streptococcal infection. His attending physician, Gene Stollerman, proved to be such an influence on Ed that he dedicated his career to the study of streptococcal diseases. Most individuals who study this book will immediately recognize the contributions Ed made to streptococcal adhesion. They will also recognize the name of Ed as a leader in microbial adhesion mechanisms of other bacterial pathogens. In fact, in a review article he wrote in the early 1980s for the *Journal of Infectious Diseases,* Ed captured the most important principles of bacterial adhesion to soft and hard tissues, including the role of adhesion in pathogenesis. The article was especially important in its presentation of basic molecular methodology to study bacterial adhesion. It is not surprising that this review became a Citation Classic and continues to be well-cited.

Ed Beachey grew in stature in the scientific world from the early 1970s to the late 1980s, and his remarkable abilities seemed to emerge more and more from year to year. Young scientists from around the world found a way to visit Ed in Memphis. In spite of demands for teaching, clinical rounds, grants, paper writing, writing of reviews, refereeing, scientific study sections and other responsibilities, Ed always was a gracious host. We do not know of other individuals who have so assiduously devoted themselves to the course of science.

Although Ed has been gone for more than three years, we retain a fond and respectful memory of him. His fundamental work on bacterial adhesion will continue to inspire us and others for generations to come. Long live the legacy of Ed Beachey.

Contents

Preface

Systematic studies on bacterial adhesion began about three decades ago. By the mid 1980s, there were approximately 500 publications per year on adhesion and its relationship to pathogenesis, a level that has been sustained. The emergence of recombinant DNA technology, gene cloning, and rapid DNA sequencing methods and the development of reliable assays for bacterial adhesion have resulted in a shift of research efforts aimed toward an understanding of the molecular mechanisms of adhesin regulation. Concomitantly, new techniques have been developed to identify and isolate adhesin receptors.

This volume emphasizes three general areas of adhesion of pathogenic bacteria: (1) A review of the most extensively studied bacteria, especially *Escherichia coli,* describing the molecular mechanisms of adhesion. Chapters dealing with pyogenic Gram-positive cocci and oral bacteria as well as detailed chapters on the biogenesis and regulation of adhesins are also included. (2) The recognition of common themes in adhesion is made clear at several locations in the text. One chapter devotes considerable space to the common themes of bacterial adhesion. (3) The final area is concerned with data treatment, methods in adhesion, principles of adhesion, and composition of surfaces of bacteria and eucaryotic cell receptors. Additional chapters deal with lectins as adhesins and with adhesion in nonopsonic phagocytosis.

In general, we have attempted to present views dealing with mechanisms of bacterial adhesion and its regulation, focusing on bacteria that have been studied in detail at the molecular level. We wish to express our gratitude to the late Ed Beachey for stimulating discussions and for support and encouragement in convincing us to write this book. We also thank the numerous granting agencies that have supported our research, including the NIDR-NIH, March of Dimes (Ohio Valley), US-Israel Bi-Nat Science Foundation, the US Army R & D Command, and the VA hospital system. Finally, much credit is due to Administrative Assistant Atha Carter and typist Kim Hayes.

I. Ofek, Tel-Aviv
R. J. Doyle, Louisville, KY

Bacterial Adhesion to Cells and Tissues

1

Principles of Bacterial Adhesion

Certain fundamental aspects of bacterial adhesion have been known for years (reviewed in Marshall, 1976; Ellwood et al., 1979; Beachey, 1980a, b; Berkeley et al., 1980; Bitton and Marshall, 1980; Beachey et al., 1982; Schlessinger, 1982; Jones and Isaacson, 1984; Marshall, 1984; Mergenhagen and Rosan, 1985; Savage and Fletcher, 1985; Lark et al., 1986). More discussions on various fundamental aspects can also be found in other selected books (Boedeker, 1984; Mirelman, 1986; Switalski et al., 1989; Doyle and Rosenberg, 1990; Hook and Switalski, 1992). During the first decade of intense research on the adhesion of microorganisms to various substrata a number of points had become clear. One, there is little doubt that the survival of microorganisms in various niches is dependent on their ability to adhere to surfaces or substrata. Second, the adhesion process involves an interaction between complementary molecules on the respective surfaces of the microbe and the substratum. Third, the expression by the organisms of the macromolecules that participate in the adhesion process is under a number of regulatory control mechanisms. During the second decade of research most of the efforts have been focused on molecular mechanisms (fine specificity and genetic control) and consequences of the adhesion phenomenon, with the premise that the process of microbial adhesion may be specifically manipulated. In this first chapter the fundamental principles that have emerged from studies performed during the last two decades are reviewed, with an emphasis on how bacteria adhere to biological substrata, the importance of adhesion in infectious processes, and the basic genetic and phenotypic variables affecting adhesion. A summary of common terms that are now widely used and have become prominent in the concepts developed in the study of bacterial adhesion is given in Table 1–1. The chapters that follow provide more in-depth discussion of the molecular nature of adhesins and their receptors. In addition, thorough reviews of the genetic regulation of adhesins are presented, along with an appraisal of the

Table 1–1. Explanation of terms in bacterial adhesion

Adhesion (or adherence)	The measurable union between a bacterium and a substratum. A bacterium is said to have adhered to a substratum when energy is required to separate the bacterium from the substratum.
Adhesin	A surface molecule of a bacterium capable of binding to a receptor on a substratum. Most bacterial species express more than one type of adhesin. An adhesin site is a region of the adhesin molecule engaged in stereospecific interaction with a receptor. In some cases, an adhesin may possess two or more distinct adhesin sites specific for two distinct receptors.
Amorphin	Amorphous appearing materials observed on the surfaces of some bacteria. Amorphin does not appear to possess regular repeating structures and may be composed of protein and/or polysaccharide.
Association	An encounter between a bacterium and a substratum that may lead to adhesion of the bacterium without specification of mechanism.
Capsule	Secreted polymers, usually acidic polysaccharide, that have a tendency to remain cell associated and may participate in adhesive events. Capsule-like proteins have been implicated as adhesins in a few bacteria.
Colonization	Bacterial growth at a site following adhesion of bacteria on the site.
Cryptitope	Latent receptor for a bacterial adhesin. The cryptitope may be made available as a receptor by proteolysis, heat, or other treatments.
Deposition	Bacteria adhering to a substratum during fluid movement.
Fibrillae	Appendages, consisting of protein subunits, emanating from the wall or membrane compartments of Gram-positive or Gram-negative bacteria. The fibrillae are usually shorter than fimbriae and are usually of an indeterminant diameter. Fibrillae carry adhesins in a not well understood manner.
Fimbriae or pili	Appendages, consisting of protein subunits that are anchored in the outer membrane of Gram-negative bacteria or the cytoplasmic membrane or cell wall of Gram-positive bacteria. Fimbriae may be tubular (rigid) or flexible. Rigid fimbriae are usually 6–10 nm in diameter and carry adhesins at their tips and in some cases along their sides as well. Flexible fimbriae are thin (usually < 5 mm in diamater) and usually carry adhesins along their entire length.
Glycocalyx	The outer coating of polysaccharide possessed by some bacteria. Some refer to capsule or cell-associated slime as a glycocalyx.
Hydrophobic effect	A name given for the tendency of apolar molecules to adhere or bind to each other. A hydrophobic molecule has a greater tendency to adhere to another apolar molecule than to water.
Hydrophobins	Apolar molecules on bacterial surfaces involved in adhesion.
Isolectins	Proteins with identical function and specificity but differing slightly in amino acid composition.
Isoreceptor	A form of an adhesin receptor that differs in composition from another receptor with similar function and stereochemical arrangements.
Lectin	A carbohydrate binding protein of nonimmune origin. Many bacteria employ surface lectins as adhesins.

continued

Table 1–1. continued

Ligand	Molecules complementary to adhesin or receptor.
MATH	Acronym for microbial adhesion to hydrocarbons
Microbioadh	Association between microorganisms leading to aggregation or cohesion. Small cocci adhering with rod-shaped bacteria form aggregating pairs referred to as "corn cobs," a term also employed for other interacting bacteria. Microbioadh is a replacement for corn cobs and includes all coaggregating microorganisms.
Nonspecific adhesion	An association between the bacterium and a substatum that does not require a precise stereochemical fit. The same forces involved in the specific adhesion are also involved in nonspecific adhesion.
Pellicle	A coated surface. Salivary pellicles consist of saliva-coated surfaces. The term "conditioned" surface is a synonym for pellicle that is most often used in environmental adhesion.
RGD	An amino acid sequence, arginine–glycine–aspartic acid, found in the primary structure of proteins or glycoproteins of host (e.g., fibronectin) or bacterial origin and recognized by integrin molecules on many types of animal cell surfaces. Some bacteria (invasin-bearing *Yersinia*) adhere to animal cells via an RGD-dependent interaction.
Receptor	A substratum molecule that is complementary to an adhesin. In some cases the distinction between "receptor" and "adhesin" cannot be made and the terms may be used interchangeably. A receptor site (attachment site) is a defined region of the substratum molecule engaged in stereospecific interaction with the adhesin. In some cases a receptor may have two distinct receptor sites for two distinct adhesins.
Slimectin	Soluble secreted polymers, usually acidic polysaccharide, involved in adhesion; to be distinguished from "slime," a name reserved for noncapsular secreted polymers of similar structures not involved in adhesion phenomena.
Specific adhesion	An association between a bacterium and a substratum that requires rigid stereochemical constraints. Specific adhesion may require hydrogen bond formation, ion–ion pairing, or the hydrophobic effect.
Substratum	A surface bearing receptors for bacterial adhesins.
Tropism	The apparent tendency of bacterium to colonize a specific tissue. Some bacteria are found in high numbers in buccal mucosal cells, so the bacteria are said to have a tropism for those cells. Tropism is a result of many host and bacterial factors, one of which is adhesion.

biological significance of the adhesion process. Finally, recent advances on adhesion of the most widely studied pathogens are provided.

How Bacteria Adhere

Bacteria adhere only to complementary substrata. They adhere by ionic or Coulombic interactions, by hydrogen bonding (Pimental and McClellan, 1990), by the hydrophobic effect (Duncan-Hewitt, 1990), and by coordination complexes

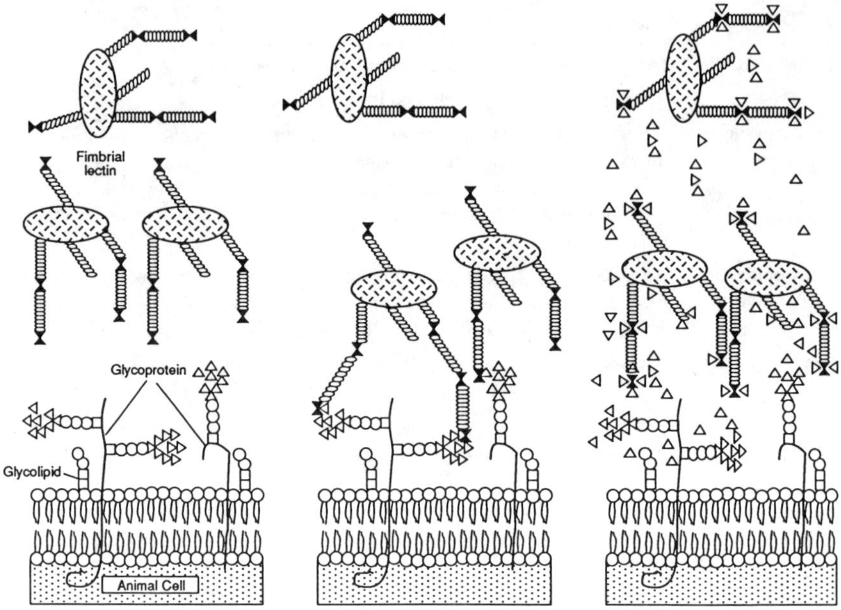

Figure 1–1. A diagram showing carbohydrate-mediated adhesion of fimbriated bacteria to an animal cell. The bacterial lectin that functions as an adhesin is shown on the tips of the fimbriae and binds the organisms to complementary carbohydrate structures. The carbohydrate structures are presented as glycoproteins or glycolipids originating from the animal cell membrane. Excess soluble carbohydrate will occupy available lectin sites on the bacterial surface and competitively inhibit adhesion of the bacteria. Similarly, excess soluble bacterial lectin could also block carbohydrate structures on animal cells and inhibit adhesion.

involving multivalent metal ions. There are no other forces known that are able to participate in adhesion interactions. Frequently, adhesion is highly stereospecific in that a bacterium will bind to a substratum only if the substratum possesses a certain kind of receptor. A typical example for the stereospecific interaction in bacterial adhesion is that between a lectin and a carbohydrate (Ofek et al., 1977, 1978; Mirelman, 1986). Some bacteria bearing carbohydrate-binding lectin adhesins will adhere only to substrata possessing the requisite carbohydrate structure (Figure 1–1). In this sense, the adhesion is highly stereospecific. In contrast, a much less specific adhesion is the case of a hydrophobic bacterium that will bind to any hydrophobic surface. Adhesion by hydrophobic interactions is often referred to as nonspecific (Busscher and Weerkamp, 1987). The designation of nonspecificity for adhesion that is not dependent on stereospecific interactions, but rather on the hydrophobic effect or on ionic interactions, is probably an oversimplification. In fact, evolution has selected some bacteria with hydrophobic surfaces or with surface structures containing hydrophobic domains. It is possible that adhesion to some substrata is dependent on the surface expression of these hydrophobins. The mere potential of bacteria to adhere to a particular substratum is not sufficient to initiate colonization of the substratum because factors such as nutritional requirements and resistance to deleterious physical and chemical pressures are also important. A frequent argument is the issue of how the degree of specificity of adhesion determines the ability of the bacterium to colonize a limited range of substrata (tropism) permissive to the survival and/or growth of the bacteria. If tissue tropism, for example, is due only to stereospecific interactions then some hydrophobic bacteria would be expected to colonize any hospitable hydrophobic surface. The group A streptococci are hydrophobic (Courtney et al., 1990) and adhere to hydrophobic surfaces but are unable to colonize all these surfaces (Rosenberg et al., 1981). Similarly, *Escherichia coli* bearing mannose-specific adhesins do not colonize all mannose-containing substrata. The adhesion process is probably a composite of factors that include presentation, orientation, and accessibility of both the receptors and the adhesins on the surfaces of bacteria and substrata, respectively. In addition, the adhesion process may require the participation of a number of distinct surface constituents that interact in a sequential manner to overcome repulsive forces as discussed below.

The process of adhesion by a bacterium requires a strategy by the cell. The common strategy for all bacteria is to position their adhesins on their surfaces. Almost all adhesive bacteria have devised a number of means to scaffold or present their adhesins. The surface appendages and structures that bear adhesins include fimbriae, fibrillae, flagella, capsule, outer membrane, and other appendages (a description is found in Chapter 4). In some cases, surface appendages can be clearly identified, such as fimbriae or flagella (Figure 1–2). Completely

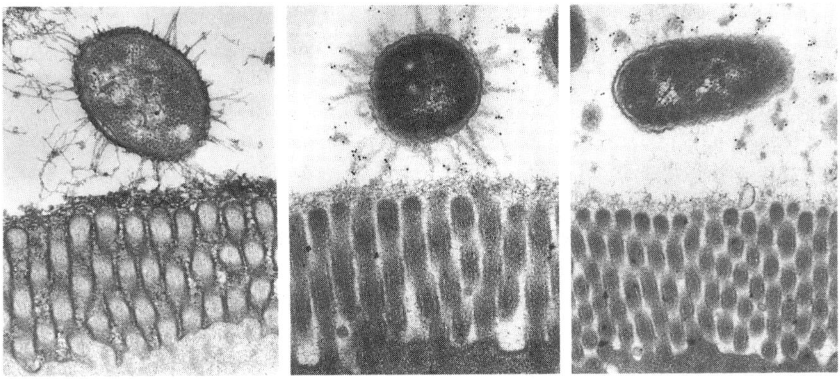

Figure 1–2. Electron micrograph showing the adhesion of a fimbriated *Escherichia coli* strain to cultured human duodenal mucosa. The fimbriae were labeled with gold antibodies and are seen to interact with the brush border of the mucosa to promote adhesion of the bacteria. Left, *E. coli* carrying CS4 fimbriae (stained with ruthenium red); center, same strain, but revealed by gold-labeled antibodies; right, a strain carrying CS6 fimbriae does not appear to adhere. (Reprinted by permission, S. Knutton and the American Society for Microbiology; Knutton et al., 1989.)

bald bacteria capable of adhesion are rare. The use of surface appendages as schleppers of adhesins is logical. Bacteria-bearing adhesins can approach receptors at a distance, form complexes, and ultimately settle onto the substratum. The use of adhesins at a distant site remote from the cell sacculus enables the bacterium to overcome an energy barrier to adhesion. In what has become known as DLVO theory (Derjaguin and Landau, 1941; Verwey and Overbeek, 1984), it is assumed that bacteria and most biological substrata cannot adhere because both are negatively charged and repulse each other. The theory states that repulsion is a result of the electrical double layer existing on any substratum suspended in an electrolyte. The electrostatic layer surrounding bacteria is composed of the negatively charged surface macromolecules and positively charged counterions. There is an electrical potential across the double layer. For the bacteria, the electrical potential is a function of the electrophoretic mobility and the net surface charge, expressed as the zeta potential of the bacterium. The nature of the electrolytes surrounding the bacteria also affect the electrical potential of the surface double layer. For example, cells suspended in a Mg^{2+} salts medium would have a lower electrical potential than cells suspended in an equal ionic strength medium containing potassium salts. Multivalent ions, therefore, tend to reduce the surface potential. The diameter of the electrical double layer is usually of the order of 1–5 nm in physiological buffers, so bacteria are generally repulsed from substrata at distances of $<$ 8–12 nm (Figures 1–3 and 1–4). Once repulsive forces have been overcome, adhesins may now interact with their receptors to lead to a firm adhesion. The bacterium–substratum interaction may involve the simultaneous binding of a large number of adhesin molecules with a large number of complementary receptors to form many individual bonds, rendering the adhesion resistant to detachment and making it virtually irreversible. Such adhesin–receptor interaction depends on spatial arrangement and accessibility of the adhesin and its receptors on the substratum. Surface components other than adhesins may interfere with the accessibility of the adhesin and therefore successful adhesion may depend on the expression and regulation of such interfering molecules. A typical example is formation of polysaccharide capsule on the bacterial surface. In contrast, presence of nonadhesin molecules on the bacterial surface may stabilize the adhesin–receptor interaction. The hydrophobic effect may stabilize polar interactions (Doyle and Rosenberg, 1990). Structural components on the bacterium or on the substratum that tend to promote or to impede adhesion may complicate the identification of the actual adhesin and its receptor.

Adhesion and Infection

Marine microbiologists recognized the importance of bacterial adhesion in the colonization of moving streams of water (Fletcher, 1985; Kjelleberg, 1985; Marshall, 1985) long before adhesion was established as a factor in the bacterial

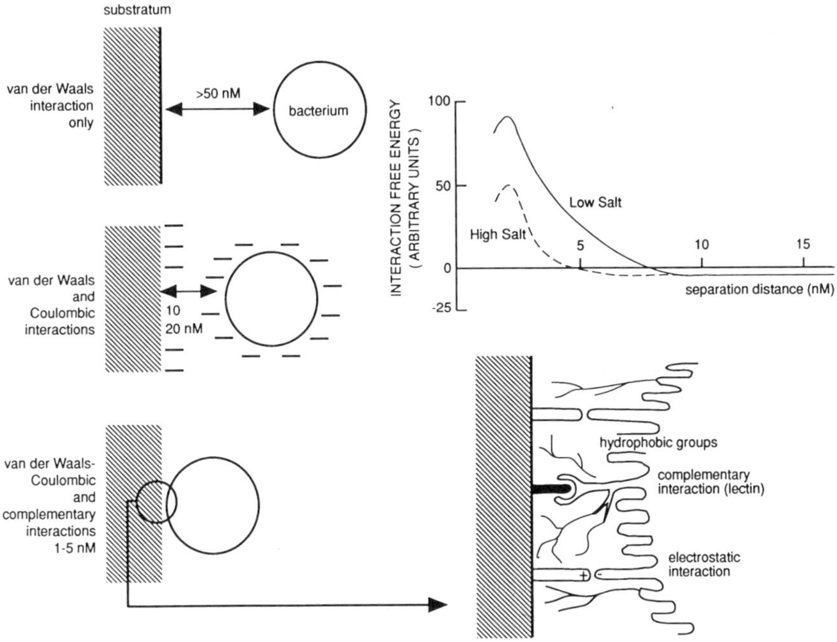

Figure 1–3. Description of the electrical double layer and bacterial adhesion. In this diagram three distinct interaction regions are depicted. One region (>50 nm) reflects van der Waals attractions. A closer region (10–20 nm) involves both van der Waals attractions and Coulombic forces. A third, even closer region (<2 nm), requires complementary binding sites which may involve hydrophobin-hydrophobin, lectin–carbohydrate, and charge–charge interactions. The presence of hydrophobic sites may stabilize other interacting sites. Hydrophobic sites and/or numerous charge–charge sites contribute to form a virtually irreversible adhesion. (Adapted from Busscher and Weerkamp, 1987.)

colonization on animal tissues. It is likely the major reason for this lag is that medical bacteriologists were primarily interested in identifying other virulence factors, such as toxins and capsules and in devising new antimicrobial agents. However, once it became accepted that bacterial adhesion to tissues was a prerequisite step in the infectious process, the dissection of the molecular mechanisms underlying the adhesion process by far surpassed that of marine bacteria. The aim was and remains the same: to design new approaches to the prevention of serious bacterial infections by interfering with the adhesion process. It is important, therefore, to outline why adhesion is important to infection. A number of approaches have been used to examine the relationship between adhesion and infectivity (Table 1–2). One way was to compare the infectivity phenotypes or genotypes expressing or not expressing adhesins. In the early studies, such comparisons were made on nonisogenic pairs of variants. Several in vitro adhesion assays were developed as a means to study the expression of the adhesin (reviewed

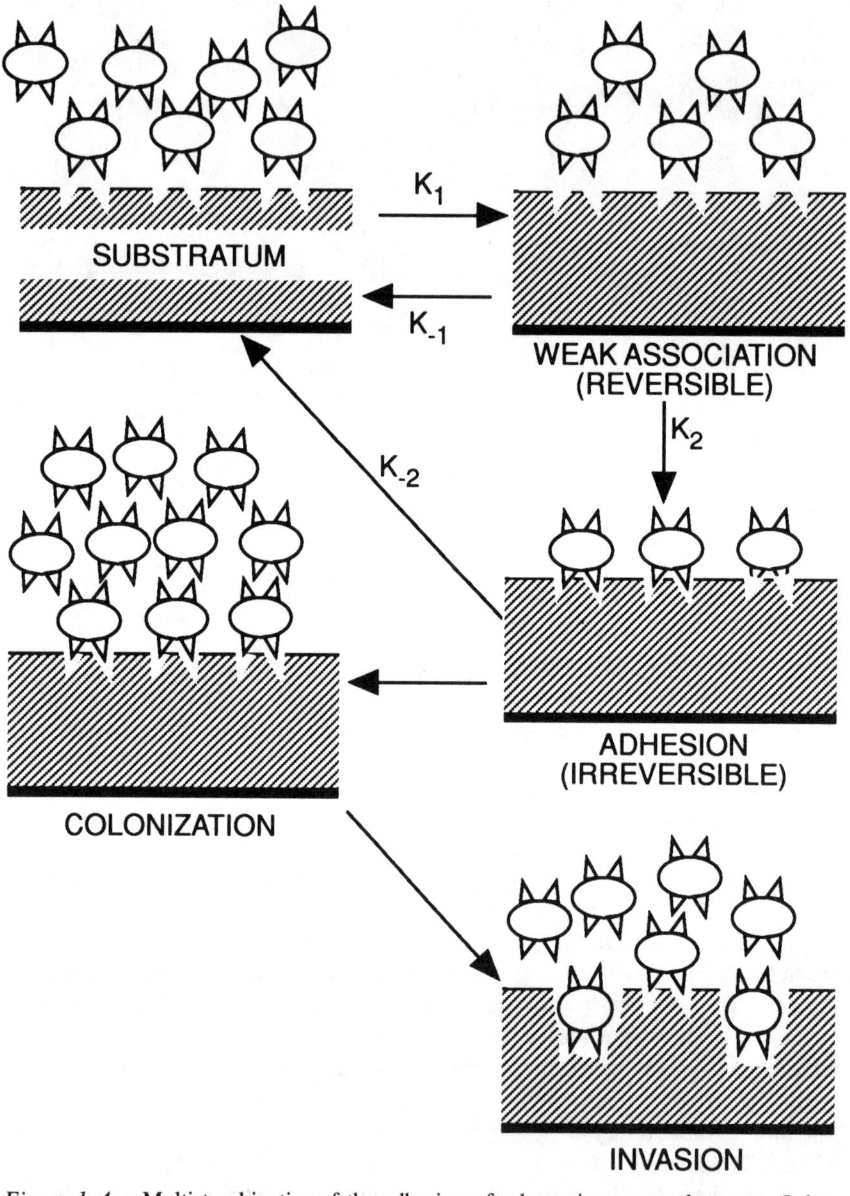

Figure 1–4. Multistep kinetics of the adhesion of a bacterium to a substratum. It has now become clear that bacterial adhesion may involve two (or more) kinetic steps. This figure shows a two-step adhesion leading to a firm ("irreversible") complex. The bacterium may approach the substratum and become loosely bound (K_1 rate). Complementary adhesin–receptor interactions then lead to an "irreversible" complex (K_2 rate). A dissociation rate constant (K_2) may occur, but does not seem to involve an intermediate. Colonization and subsequent invasion may be the consequences of adhesion.

Table 1–2. Relationship between adhesion and infection

Reduced infectivity of phenotypic or genotypic variants that do not express adhesins.
Resistance to infection of hosts that genetically lack the ability to express adhesin receptors.
Sensitivity to infection of hosts possessing tissues that overexpress adhesin receptors.
Ability of animal cells normally colonized by the bacteria and/or its adhesin to bind the organisms in vitro.
Antiadhesin immunity protects against infection.
Specific adhesion inhibitors of nonimmune origin prevent infection.

More complete details describing the role of adhesion in the infectious process are given in Chapters 5, 6, 7, 8, 10, and 11.

in Ofek and Beachey, 1980a). Later, when the adhesin genes were identified and cloned it was possible to compare the infectivity of isogenic mutants, one of which lacked only the genes for a specific adhesin. A classic study is that of Keith et al (1986), who showed that a type 1 fimbriated mutant lacking the *fim*H gene encoding for the mannose-specific adhesin carried by type 1 fimbriae was less infective in mice than its isogenic Fim H^+ type 1 fimbriated parent.

Another way for assessing the role of adhesion in the infectious process is to compare the sensitivity or resistance to infection of hosts that either overexpress or do not express, respectively, adhesin receptors. In general, there is a positive correlation between the ability of the tissue cells to bind the bacterial pathogen and susceptibility of the host to develop infection caused by the pathogen. For example, the bacterial binding capacity of epithelial cells from infection-prone persons is greater than that from similar cells of persons not prone to infection (reviewed in Ofek and Beachey, 1980a). Two of the best examples illustrating this positive correlation between the capacity of the host cells to bind bacteria and susceptibility or resistance to infection are illustrated by studies on infection-prone females and on piglets that are genetically resistant to infection. The epithelial cells of females who have recurrent urinary tract infections bind up to five times the normal number of uropathogenic *E. coli* (Figure 1–5). One of the most convincing lines of evidence describing the relationship between resistance to infection and the capacity of host cells to bind pathogenic bacteria deals with *E. coli* K88-induced diarrhea in piglets (Sellwood et al., 1975). Intestinal epithelial-cell brush borders isolated from piglets that were resistant to the diarrheal disease failed to bind *E. coli* K88 bacteria in vitro, whereas those from susceptible animals were able to bind the organisms in high numbers. Cross-breeding of resistant and susceptible pigs revealed that susceptibility to the disease as well as the ability of the brush border of the epithelial cells to bind the organisms was an autosomal trait. Thus, the lack of epithelial-cell receptor sites for the binding of *E. coli* organisms rendered the piglets genetically immune.

Once the identity of the receptors of adhesins of a particular pathogen becomes known, it is possible to screen large populations for those individuals who are genetically incapable of producing the cellular adhesin receptors. For example,

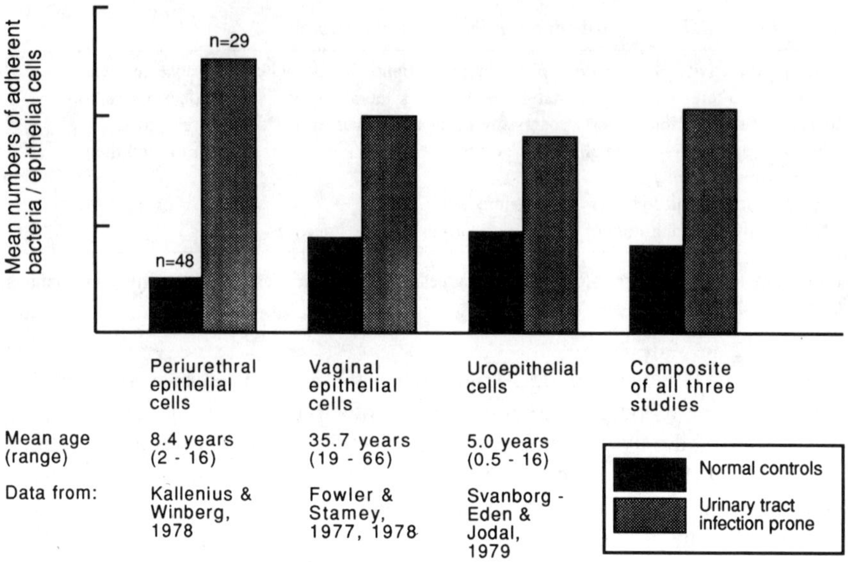

Figure 1–5. Relationship between susceptibility to infection and capacity of epithelial cells to bind *Escherichia coli* bacteria. Note the increased binding of *E. coli* by genitourinary epithelial cells from infection-prone females. (Adapted from Beachey et al., 1982.)

uropathogenic strains of *E. coli* expressing the globo-A specific adhesin bind only to uroepithelial cells of individuals of blood group A_1 with a positive secretor state (Lindstedt et al., 1991). Screening of 1473 children with urinary tract infections revealed that globo-A-bearing individuals were more susceptible to urinary tract infection caused by strains possessing the globo-A specific lectin. Pyelonephritogenic strains of *E. coli* express the P fimbrial adhesin specific for the galabiose part of P blood group substance (see Chapters 5 and 10). Children with blood group P_1 are more susceptible to recurrent pyelonephritis as compared to individuals with blood group P_2 (Lomberg et al., 1983, 1986). It is possible that density or accessibility or affinity of the receptors for the P galabiose specific adhesin may also determine susceptibility to infection caused by P fimbriated *E. coli*. The relationship between the ability to produce adhesin receptors and susceptibility to infection may also determine the host range that contracts infection. Such host selectivity was shown for the *E. coli* expressing the K99 adhesin. Ono et al. (1989) showed that the K99 fimbriae bind to *N*-glycolylneuraminyllactosyl ceramide, but not to the *N*-acetylated derivative of the sialic acid. Interestingly, the host range of *E. coli* bearing K99 fimbriae is limited to animals that carry the *N*-glycolyl linkage on their cells. Humans are deficient in *N*-glycolyl-containing receptors and are not infected by $K99^+$ *E. coli*.

Specific inhibitors of adhesion have not only provided the best evidence that adhesion is important for infection but also enabled the design of new approaches to the prevention of infection. The best examples for such inhibition were bacteria expressing carbohydrate-specific adhesins (more thoroughly discussed in Chapter 5). The bacterial lectin that serves as an adhesin can be inhibited either by specific antibodies or by relatively high concentrations of soluble carbohydrates specific for the lectin. In all cases, infection in animals is prevented by these adhesin inhibitors. Mouricout et al. (1990) isolated glycolipids from the nonimmunoglobulin fraction of bovine plasma and administered the glycoconjugates in drinking water to colostrum-deprived calves. It was observed that the animals drinking glycoconjugate became more resistant to infection caused by K99^{+} *E. coli*. Furthermore, adhesion of the bacteria to the intestines of the calves was reduced by two orders of magnitude. A striking feature of these findings is that relatively low concentrations of adhesin-blocking glycoconjugates were sufficient to prevent infection.

There are a number of reasons to support the view that adhesion is important in infectivity. One of the most convincing and probably the most widely accepted is that bacterial adhesion endows the pathogen with the ability to withstand cleansing mechanisms operating on mucosal and endothelial surfaces (Figure 1–6) (Ofek and Beachey, 1980a,b; Beachey, 1981). In good health status, normal

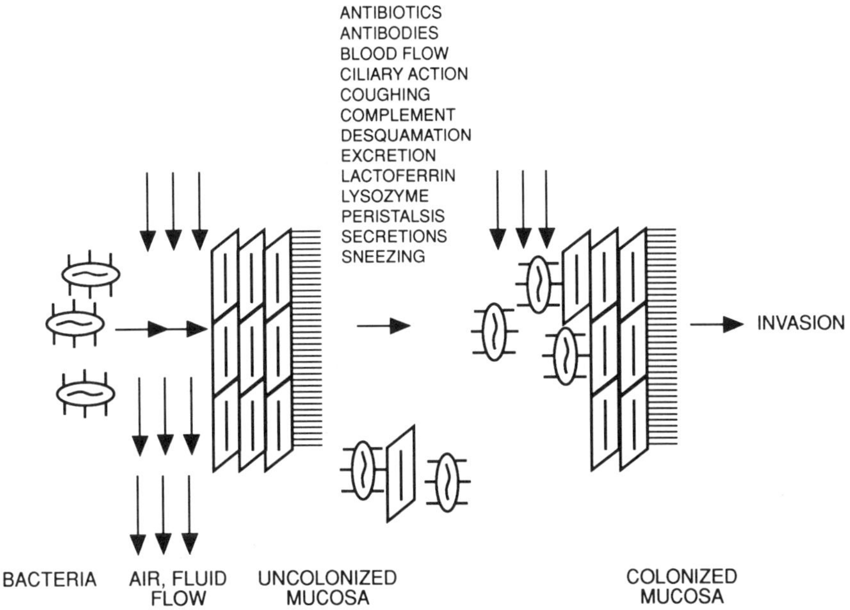

Figure 1–6. A diagram describing the effects of normal cleansing mechanisms on the adhesion of bacteria to mucosal surfaces. Mucosal surfaces are the most common port of entry for most bacterial pathogens. A successful pathogen must overcome many defense mechanisms operating on these surfaces to proliferate and colonize a particular site. Nonadhering phenotypes and progenies of adherent and proliferating bacteria are cleansed away to either a more hospitable site or a deleterious site. Adhesion may lead to tissue invasion.

Table 1–3. Advantages gained by bacteria resulting from adhesion to nonphagocytic animal cells

Advantage	Explanation
Withstand mechanical cleansing mechanisms	Figure 1–6
Growth advantages	Figures 11–3, 11–4
Enhanced toxicity to host	Figure 11–6
Increased resistance to deleterious agents	Figure 11–7

cleansing mechanisms such as secretions, ciliary action, salivation, swallowing, voiding peristalsis, desquamation, excretion, coughing, sneezing, and blood flow, in addition to deleterious agents (e.g., antibodies), prevent the organisms from making contact with binding sites on mucosal or endothelial surfaces. Because the normal fluid flow is more rapid than the rate of multiplication of bacteria in mucosal secretion, unattached organisms are simply eliminated, along with the luminal contents, by the mechanical cleansing mechanisms. Thus, the successful colonizer must adhere to the tissues in order to withstand these cleansing mechanisms. It has been pointed out (Freter, 1978, 1980; Freter et al., 1978) that the successful association of the pathogen with mucosal surfaces such as intestinal cells is a complex process requiring interplay among a variety of virulence factors, including bacterial motility, chemotaxis of the bacteria to the mucous gel overlying the microvilli of the intestinal cells, penetration of the mucous gel, and finally attachment to the epithelial cells and elaboration of injurious toxins. Even then, the organisms must multiply on the cell surface because the cells are constantly shed (desquamation) and washed away along with any attached organisms. Adhesion to animal cells seems to confer at least three main advantages in addition to the ability to withstand mechanical cleansing mechanisms, including growth advantages, toxin targeting, and escape from antimicrobial or antitoxin activities (Table 1–3). It seems likely that even more advantages gained by bacteria as a result of adhesion will be discovered in the next few years.

References

Beachey, E.H. (ed.). 1980a. *Bacterial Adherence*. (*Receptors and Recognition, Vol. 6*). Chapman and Hall, London.

Beachey, E.H. 1980b. Preface. In: Beachey E.H. (ed.), *Bacterial Adherence (Receptors and Recognition, Vol. 6*). Chapman and Hall, London, pp. xi–xii.

Beachey, E.H. 1981. Bacterial adherence: Adhesin receptor interactions mediating the attachment of bacteria to mucosal surfaces. *J. Infect. Dis.* **143**:325–345.

Beachey, E.H., B. Eisenstein, and I. Ofek. 1982. *Adherence and Infectious Diseases. Current Concepts*. The Upjohn Company, Kalamazoo.

Berkeley, R.C.W., J.M. Lynch, J. Melling, P.R. Rutter, and B. Vincent (eds.). 1980. *Microbial Adhesion to Surfaces*. Chichester, Ellis Horwood.

Bitton, G. and K.C. Marshall. (eds.). 1980. *Adsorption of Microorganisms to Surfaces*. John Wiley & Sons, New York.

Boedeker, E.C. (ed.). 1984. *Attachment of Organisms to the Gut Mucosa*, Vols. I and II. CRC Press, Boca Raton, FL.

Busscher, H.J. and A.H. Weerkamp. 1987. Specific and non-specific interactions in bacterial adhesion to solid substrata. *FEMS Microbiol. Rev.* **46**:165–173.

Courtney, H.S., D. Hasty, and I. Ofek. 1990. Hydrophobicity of group A streptococci and its relationship to adhesion of streptococci to host cells. In: Doyle, R.J. and M. Rosenberg (eds.), *Microbial Cell Surface Hydrophobicity*. American Society for Microbiology, Washington, pp. 361–386.

Derjaguin, B.V. and L. Landau. 1941. Theory of the stability of strongly charged lyophobic soils and of the adhesion of strongly charged particles in solutions of electrolytes. *Acta Physicochim.* (*USSR*) **14**:633–662.

Doyle, R.J. and M. Rosenberg (eds.). 1990. *Microbial Cell Surface Hydrophobicity*. American Society for Microbiology, Washington.

Duncan-Hewitt, W. 1990. Nature of the hydrophobic effect. In Doyle, R.J. and M. Rosenberg (eds.), *Microbial Cell Surface Hydrophobicity*. American Society of Microbiology, Washington, pp. 39–73.

Ellwood, D.C., J. Melling, and P. Rutler (eds.). 1979. *Adhesion of Microorganisms to Surfaces*. Academic Press, London.

Fletcher, M. 1985. Effect of solid surfaces on the activity of attached bacteria. In: Savage, D.C. and M. Fletcher (eds.), *Bacterial Adhesion: Mechanisms and Physiological Significance*. Plenum Press, New York, pp. 339–362.

Fowler, J.E. and T.A. Stamey. 1977. Studies of introital colonization in women with recurrent urinary infections. VII. The role of bacterial adherence. *J. Urol.* **117**:472–476.

Fowler, J.E. and T.A. Stamey. 1978. Studies of introital colonization in women with recurrent urinary infections: X. Adhesive properties of *Escherichia coli* and *Proteus mirabilis:* Lack of correlation with urinary pathogenicity. *J. Urol.* **120**:315–318.

Freter, R. 1978. Association of enterotoxigenic bacteria with the mucosa of the small intestine—mechanisms and pathogenic implications. In: Ouchterlony O. and J. Holmgren. (eds.), *Nobel Symposium, vol. 43: Cholera and Related Diarrheas: Molecular Aspects of a Global Health Problem*. S. Karger, Basel, pp. 155–170.

Freter, R. 1980. Prospects for preventing the association of harmful bacteria with host mucosal surfaces. In: Beachey, E.H. (ed.), Bacterial Adherence (Receptors and Recognition, *Vol.* 6). Chapman and Hall, London, pp. 439–458.

Freter, R., P.C.M. O'Brien, and S.A. Halstead. 1978. Adhesion and chemotaxis as determinants of bacterial association with mucosal surfaces. In: McGhee, J.R., J. Mestecky, and J.L. Babb (eds.), *Secretory Immunity and Infection*. Plenum Press, New York, pp. 429–438.

Hook, M. and L.M. Switalski (eds). 1992. *Microbial Adhesion and Invasion*. Springer-Verlag, New York.

Jones, G.W. and R.E. Isaacson. 1984. Proteinaceous bacterial adhesins and their receptors. *CRC Crit. Rev. Microbiol.* **10**:229–260.

Kallenius, G., and J. Winberg. 1978. Bacterial adherence to periurethral epithelial cells in girls prone to urinary-tract infections. *Lancet* **2**:540–543.

Keith, B.R., L. Maurer, P.A. Spears, and P.E. Orndorff. 1986. Receptor binding function of type 1 pili effects bladder colonization by a clinical isolate of *Escherichia coli*. *Infect. Immun.* **53**:693–696.

Kjelleberg, S. 1985. Mechanisms of bacterial adhesion at gas–liquid interfaces: In: Savage, D.C. and M. Fletcher (eds.), *Bacterial Adhesion: Mechanisms and Physiological Significance*. Plenum Press, New York, pp. 163–194.

Knutton, S., M.M. McConnell, B.E. Rowe, and A.S. McNeish. 1989. Adhesion and ultrastructural properties of human enterotoxigenic *Escherichia coli* producing colonization factor antigens III and IV. *Infect. Immun.* **57**:3364–3371.

Lark, D.L., S. Normark, B.E. Uhlin, and H. Wolf-Watz (eds.). 1986. *Protein–Carbohydrate Interactions in Biological Systems. The Molecular Biology of Microbial Pathogenicity*. Academic Press, London.

Lindstedt, R., G. Larson, P. Falk, U. Jodal, H. Leffler, and C. Svanborg. 1991. The receptor repertoire defines the host range for attaching *Escherichia coli* strains that recognize globo-A. *Infect. Immun.* **59**:1086–1082.

Lomberg, H., L.A. Hansson, B.J. Jacobsson, U. Jodal, H. Leffler, and C. Svanborg-Eden. 1983. Correlation of P blood group, vesicoureteral reflux and bacterial attachment in patients with recurrent pyelonephritis. *N. Engl. J. Med.* **308**:1189–1192.

Lomberg, H., B. Cedergren, H. Leffler, B. Nilsson, A.S. Carlstrom, and C. Svanborg-Eden. 1986. Influence of blood group on the availability of receptors for attachment of uropathogenic *Escherichia coli*. *Infect. Immun.* **51**:919–926.

Marshall, K. 1976. *Interfaces in Microbial Ecology*. Harvard University Press, Cambridge, MA.

Marshall, K.C. (ed.). 1984. *Microbial Adhesion and Aggregation* (Dahlem Workshop). Springer-Verlag, Berlin.

Marshall, K.C. 1985. Mechanisms of bacterial adhesion at solid–water interfaces. In: Savage, D.C. and M. Fletcher (eds.), *Bacterial Adhesion: Mechanisms and Physiological Significance*. Plenum Press, New York, pp. 133–161.

Mergenhagen, S.E. and B. Rosan (eds.). 1985. *Molecular Basis of Oral Microbial Adhesion*. American Society for Microbiology, Washington.

Mirelman, D. (ed.). 1986. *Microbial Lectins and Agglutinins*. John Wiley & Sons, New York.

Mouricout, M., J.M. Petit, J.R. Carias, and R. Julien. 1990. Glycoprotein glycans that inhibit adhesion of *Escherichia coli* mediated by K99 fimbriae: treatment of experimental colibacillosis. *Infect. Immun.* **58**:98–106.

Ofek, I. and E.H. Beachey. 1980a. Bacterial adherence. *Adv. Intern. Med.* **25**:503–532.

Ofek, I. and E.H. Beachey. 1980b. General concepts and principles of bacterial adherence. In: Beachey, E.H. (ed.), *Bacterial Adherence* (Receptors and Recognition, *Vol.* 6). Chapman and Hall, London, pp. 1–29.

Ofek, I., D. Mirelman, and N. Sharon. 1977. Adherence of *Escherichia coli* to human mucosal cells mediated by mannose receptors. *Nature* **265**:623–625.

Ofek, I., E.H. Beachey, and N. Sharon. 1978. Surface sugars of animal cells as determinants of recognition in bacterial adherence. *Trends Biochem. Sci.* **3**:159–160.

Ono, E., K. Abe, M. Nakazawa, and M. Naiki. 1989. Ganglioside epitope recognized by K99 fimbriae from enterotoxigenic *Escherichia coli. Infect. Immun.* **57**:907–911.

Pimentel, G.C. and A.L. McClellan. 1960. *The Hydrogen Bond.* W.H. Freeman, San Francisco.

Rosenberg, M., A. Perry, E.A. Bayer, D.L. Gutnick, E. Rosenberg, and I. Ofek. 1981. Adherence of *Acinetobacter calcoaceticus* RAG-1 to human epithelial cells and to hexadecane. *Infect. Immun.* **33**:29–33.

Savage, D.C. and M. Fletcher (eds.). 1985. *Bacterial Adhesion. Mechanisms and Physiological Significance.* Plenum Press, New York.

Schlessinger, D. (ed.). 1982. *Microbiology—1982.* American Society for Microbiology, Washington.

Sellwood, R., R.A. Gibbons, G.W. Jones, and J.M. Rutter. 1975. Adhesion of enteropathogenic *Escherichia coli* to pig intestinal brush borders: The existence of two pig phenotypes. *J. Med. Microbiol.* **8**:405–411.

Svanborg-Eden C. and U. Jodal. 1979. Attachment of *Escherichia coli* to urinary sediment epithelial cells from urinary tract infection-prone and healthy children. *Infect. Immun.* **26**:837–840.

Switalski, L., M. Höök, and E. Beachey (eds.). 1989. *Molecular Mechanisms of Microbial Adhesion.* Springer-Verlag, New York.

Verwey, E.J.W. and J.T.G. Overbeek. 1948. *Theory of the Stability of Lyophobic Colloids.* Elsevier, Amsterdam.

2

Methods, Models, and Analysis of Bacterial Adhesion

It is axiomatic to consider that most living and nonliving surfaces have a tendency to be colonized by microorganisms. The importance of microbial adhesion and colonization to surfaces was not appreciated until molecular techniques were applied to analyze modes and mechanisms of cell–substratum interactions. As more and more techniques became available, new knowledge was gained that made it possible to understand the modulation of the adhesion and subsequent colonization of many microorganisms. To date, no single experimental system has been developed that can be used to adequately characterize all aspects of microbe–substratum interactions. It is therefore essential that the reliabilities, advantages, and limitations of the existing techniques be understood. Most techniques employed in the study of adhesion yield restricted amounts of information, usually about defined events in a complicated series of interactions. This chapter considers methods for the study of adhesion. Consideration is given to model systems, methods for separating adherent from nonadherent cells, controlled and uncontrolled variables in experimental design, and approaches used in analyzing adhesion data. Finally, methods related to the identification and regulation of expression of adhesins and their receptors are reviewed.

Model Systems

The establishment of any model system for the study of bacterial adhesion has three primary goals: (1) development of a means to characterize a related natural system; (2) determination of the properties of adhesins and/or receptors; and (3) examination of the role of molecules that contribute to the adhesion process, but do not directly participate as adhesins or receptors. Figure 2–1 illustrates an arbitrary hierarchy of model systems, as depicted by a gradient. At one extreme the model is used for monitoring the presence of surface molecules involved in

MODEL

Partitioning at hydrocarbon-aqueous interfaces

Binding to solid surfaces coated with
- -buffers, solvents and salts
- -constituents derived from the environment or a hydrophobic probe

Aggregation of particles naturally or artificially coated with constituents bearing structural similarities with those of the natural environment
- -hemagglutination
- -aggregation of yeasts
- -aggregation of latex beads

Binding to
- -tissue culture cells
- -exfoliated (or scraped) cells
- -excised tissue (soft tissue, teeth, plants)

Tissues excised after *in vivo* inoculation

NATURAL

Figure 2–1. Model systems commonly employed to study the adhesion of microorganisms to natural environments.

adhesion, and at the other extreme, for studying adhesion of bacteria onto natural target surfaces. The monitoring system assumes that the molecules participating in the interaction play a role in adhesion of bacteria to natural target surfaces. This assumption, once validated for a particular bacterial species by appropriate correlation tests, becomes valuable in the study of adhesion of bacteria to target surfaces. For example, partition by hexadecane seems to be proportional to the ability of many members of the genus *Streptococcus* to adhere to saliva-coated surfaces or to epithelia (Figure 2–2). The relationships shown in Figure 2–2 imply that the complementary ligands and receptors possess lipophilic characteristics. This may be the case for certain bacteria, but it is possible to account for the relationships in other ways, as discussed in the preceding chapter. The hydrophobic residues may facilitate the approach of hydrophilic residues on

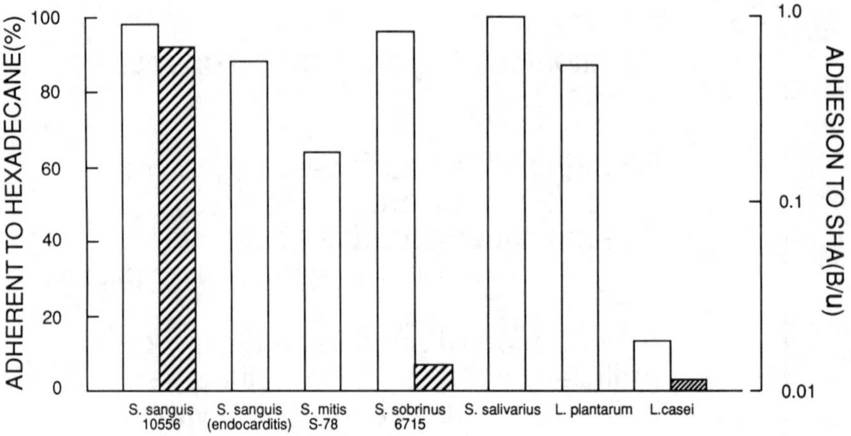

Figure 2–2. Adhesion of oral bacteria to hexadecane or to saliva-coated hydroxylapatite (SHA), The details for the adhesion experiments have been described by Nesbitt et al. (1982). The figure shows that there is a correlation between the adhesion to hexadecane and to the saliva-coated hydroxylapatite. Only one cell density was used in the hydroxylapatite experiments. *B* represents the bound cells, whereas *U* represents unbound cells. The text further describes *B–U* relationships. The parallel bars represent adhesion to hexadecane. The open box represents adhesion to the hydroxylapatite.

receptors and adhesins or stabilize their interactions. Another factor to consider in the partitioning of cells by the hydrocarbon system is that there may be varying degrees of hydrophobicity of cells in the population.

Other systems are available that can be used to assay the functional expression of adhesins or receptors. For bacterial adhesins, the binding to surfaces coated with macromolecules derived from the environment is a simple and useful technique, especially when the receptor specificities of the adhesins are not yet characterized (Table 2–1). For bacteria suspected of possessing surface lectins, aggregation of glycoconjugate-coated beads or cells with well-defined carbohydrate structures on their surfaces is a model system of choice (Ofek and Beachey, 1978; de Man et al., 1987). Although hemagglutination reactions are frequently used as monitoring systems for certain enterobacterial species, they are more related to the natural systems because the erythrocyte surface resembles that of tissue cells normally colonized by the organisms. Bacteria-induced hemagglutination is usually easy to perform and readily inhibited by simple molecules (see Chapter 1). Moreover, because the surfaces of erythrocytes are rich in glycoconjugates, the red cells constitute a valuable tool to monitor lectin activities on bacteria.

The other model systems illustrated in Figure 2–1 are designed primarily to develop means to characterize the adhesion to natural environmental surfaces or to living cells. In some cases the same target surface serves as a monitoring system for particular species and as a natural target surface for other bacterial species. An example is the case of partitioning in hydrocarbon- and oil-degrading bacteria (Rosenberg et al., 1980) and *S. streptococcus pyogenes* (Rosenberg et al., 1981), respectively. In general, unlike the monitoring systems, the adhesion to natural target surfaces at the molecular level is more complex, especially in the case of adhesion to animal cells or other surfaces coated with macromolecules from the environment. Advantages of the tissue culture system include cellular uniformity, reproducibility between assays, and ease and simplicity of experimentation. Disadvantages include the fact that the culture cells may have surfaces distinct from their origin, especially in terms of distribution and accessibility of receptors. The tissue culture system may be very useful on occasions where specific issues are addressed (Elbein et al., 1981; Firon et al., 1985; Stanislawski et al., 1985). Stanislawski et al. (1985) employed tissue culture cells possessing various amounts of fibronectin on their surfaces as substrata for *spyogenes*. The results showed that adhesion was proportional to the amount of surface fibronectin, a finding suggesting fibronectin to be a receptor for streptococcal ligands.

Firon et al. (1985) employed glycosylation-defective mutants to study the adhesion of *Enterobacteriaceae* to tissue culture cells. The role of carbohydrates

Table 2–1. Inert surfaces as targets for the study of microbial adhesion

Surface	Comments
Hydroxylapatite (buffer- or saliva-coated)	Related to bacterial colonization of teeth. Specificity may be established by using inhibitors (i.e., antibodies, polysaccharide, chaotropes, cosmotropes).
Implants or prosthetic devices (contact lenses, peritoneal dialysis tubing, catheters, etc.)	Surfaces coated with natural fluids may have selectivity for certain microorganisms.
Beads or plastics	Much information can be obtained with respect to the binding characteristics of microbial surfaces with constitutents resembling those of surfaces normally colonized by test organisms. For example, when coated with defined glycoconjugates they can be used to monitor carbohydrate-binding properties of microbial surfaces. When coated with buffers or hydrophobic molecules they can be used in assessing changes in the physiocochemical characteristics of microbial surfaces.
Nitrocellulose sheets	Macromolecules derived from surfaces commonly colonized by test organisms may interact with microbial surface constitutents after they have been separated by gel electrophoresis and blotted onto nitrocellulose. The method is useful to define the properties of macromolecules that specifically interact with microbial surfaces.
Silica gel sheets	Glycolipids are readily separated on silica gels (thin-layer chromatography). Microbial adhesins can then bind to the complementary glycolopid. This method is especially useful in determining the specificities of certain microbial lectins.

in adhesion can also be studied by treating the tissue culture cells with swainsonine, an inhibitor of protein glycosylation (Elbein et al., 1981). Others have used drugs, antibodies, and enzymes to modify the surfaces of tissue culture cells prior to adhesion experiments. All of these variables have had the goal of defining the nature of the receptors for bacterial adhesins.

Exfoliated and scraped cells are most useful when the property of adhesion to tissues normally colonized by the test bacteria is the aim of the study. This is why the system became the most widely used since the early work of Duguid and Gillies (1957) and during the first decade starting with the work of Ellen and Gibbons (1973), who systematically studied the adhesion of *S. pyogenes* to scraped buccal epithelial cells. A disadvantage is that it is difficult to maintain viability of these cells. Furthermore, these cells may also harbor normal microbiota, which may occupy sites under investigation. It is likely that excised tissue approximates the most natural target surface model without resorting to in vivo inoculation. This model, however, suffers from limited sources (biopsies, etc.) and has generally been confined to segments of intestinal tract tissue (Baselsky and Parker, 1978; Zilberberg et al., 1983). The in vivo system employing the inoculation of experimental animals or human volunteers may seem to approach the optimal natural system, but also suffers from many disadvantages. After inoculation of animals it is frequently necessary shortly thereafter to sacrifice the animals, then remove tissue and determine the extent of adhesion. It is also necessary to have a population large enough to provide statistical validation of the data. In contrast, the in vivo system is excellent to confirm data obtained from other models, which utilize artificial targets (Aronson et al., 1979; Svanborg-Eden et al., 1982). Under certain conditions when a set of specific questions are asked the system may provide an adequate but limited answer (Duguid et al., 1976; Goldhar et al., 1986). The same kinds of considerations can be applied to clays, soils, plants, teeth, and any other surfaces normally colonized by bacteria. Model systems have been valuable in achieving the present understanding of bacterial adhesion. It is certain that models will continue to be developed that come closer and closer to the natural substrata.

Methods for the Separation of Adherent from Nonadherent Bacteria

The study of bacterial adhesion to surfaces usually is dependent on the reliability of the techniques used to separate test target surfaces with attached bacteria from nonadherent bacteria. Some of these separation procedures are outlined in Table 2–2. Whereas most of the procedures are rapid and effective, there are some inherent problems to be considered. For example, when unbound cells are removed from a substratum by differential centrifugation or filtration and washing, some weakly bound cells may be removed by shear forces. The use of density gradients to separate animal from bacterial cells offers one means that largely

Table 2–2. Methods commonly employed to separate adherent from nonadherent microorganisms

Filtration through filter paper with a pore size that selectively allows passage of microbial cells, but not animal cells
Differential centrifugation
Density gradient centrifugation
Immobilization of target cells or macromolecules and washing with buffer solution
Partitioning between aqueous and hydrophobic probes

circumvents shear forces (Valentin-Weigand et al., 1987), but the methods are restricted to cells in suspensions and any potential effects of the gradient medium on adhesion must be ruled out.

The determination of adhesion of bacteria to hydroxylapatite crystals (or beads) is greatly simplified by the fact that the hydroxylapatite is dense and does not have to be centrifuged to separate it from bacteria in suspension. The settled particles can be washed by gentle inversion and decantation, thereby minimizing shear forces of high magnitude. Other problems must be considered, however, in the washing of crystals, beads, or any surface. Surfaces have imperfections and bacteria may become physically trapped in pits, fissures, or serrated areas. This is one case in which it may be advisable to introduce shear as a means to dislodge nonadherent bacteria. Shear, however, may remove specifically bound, albeit weakly, bacteria that may be important in colonization. As yet, there is no acceptable method to distinguish between weakly bound, shear-sensitive, and firmly bound, shear-resistant bacteria. It is also possible that in a shear-sensitive population, a gradient of binding affinities may exist. The following experiment illustrates some of the difficulties encountered in establishing the extent of adhesion of a population of bacteria with an animal cell. A mixed population of 55% nonfimbriated and 45% type 1 fimbriated *E. coli* (5×10^8 cells total) were added to scraped human buccal epithelial cells (5×10^6 cells). After incubation of the cellular mixture for 30 min at 37°C, the nonadherent bacteria were separated from the epithelial cells containing adherent *E. coli* by differential centrifugation. The supernatant containing the nonadherent bacteria was aspirated, counted under the microscope using a Petroff–Hausser chamber, and the percentage of fimbriated organisms determined by electron microscopy. The pellet, which contained the epithelial cells with associated bacteria, was washed three times by differential centrifugation to remove residual nonadherent bacteria or adherent bacteria loosely bound to the epithelial cells. The washed epithelial cell suspension was counted and a sample was dried, fixed, and stained onto a glass to determine by light microscopy the average number of *E. coli* adhering per epithelial cell. Using this procedure, it was found that after the first washing there were no detectable fimbriated bacteria in the supernatant. After three washings the pellet contained an average of 35 *E. coli* per epithelial cell. Assuming that all adherent *E. coli* (times 5×10^6 epithelial cells) and nonadherent *E. coli* from the supernatant can account for 2.8×10^8 bacteria, it then follows that about $40–50 \times 10^7$ fimbriated *E. coli* were lost during the washing procedures. The lost bacteria were presumably loosely bound to the epithelial cells and most likely represent a shear-sensitive population.

Shear forces may be avoided provided that the adhesive event produces a measurable phenomenon or biological response. Phagocytes, for example, may release antimicrobial substances on interaction with bacteria. A phagocyte undergoes a burst of oxidative metabolism on interacting with bacteria (see Chapter 7). Aggregation of cells or beads by bacteria also does not need a separation system because the interacting bacteria are readily visible to the naked eye as macroscopic aggregates.

Filtration methods are rapid, but susceptible to high shear and high backgrounds of nonadherent bacteria. Many exponential phase bacteria form chains because of incomplete separation and insufficient time for autolysins to separate completed crosswalls, and therefore may be retained in high numbers on the filters. In addition, bacteria may autoagglutinate in many growth media, forming clumps incapable of penetrating filters. To overcome clumping or chain formation problems it is advisable to prefilter the test bacteria before the adhesion assay. Once the problem of clumping or chain formation is resolved one or the other method gives an excellent separation of adherent from nonadherent bacteria. Since its first introduction by Ellen and Gibbons (1972) in the study of streptococcal adhesion many investigators have adopted it, especially in cases where target cells in suspension are employed.

The differential centrifugation technique suffers from the same types of disadvantages as those discussed for the filtration method. Precentrifugation steps may be necessary to remove aggregates. Shear forces depend on the number of differential washes and centrifugations. The differential centrifugation technique has been useful in studying the adhesion of both Gram-positive and Gram-negative organisms to animal cells in suspension (Beachey and Ofek, 1976; Ofek et al., 1977).

When immobilized target cells or macromolecules are employed as substrata for adhesion, there must be a means to reduce adventitious (random, nonspecific) adhesion to the surfaces that are not occupied by the immobilized elements. To achieve this, it is mandatory to use blocking agents. Examples of blocking agents that have been successfully used to prevent such nonspecific adhesion of streptococci to plastics are shown in Table 2–3. It is emphasized that the choice of a particular blocking agent must be guided by its nonreactivity with the test bacteria and its high affinity for the immobilization matrix. Once the degree of nonspecific binding has been determined, the separation of nonadherent bacteria from adherent bacteria on a specific target is easy to achieve. The washings may be gentle to minimize shear forces. Macromolecules are relatively easy to immobilize by dissolving them in the alkaline buffers usually recommended for enzyme-linked immunosorbent assay (ELISA) tests. Target cells are typically tissue culture cells attached to the surface matrix. Because separation is relatively

Table 2–3. Use of selected agents to block the adventitious adhesion of Streptococcus pyogenes *to activated plastic surfaces*

Blocking agent	Inhibition of adhesion (%)[a]
Amino acids mixture (1 mg/ml)[b]	10
Bovine serum albumin (50 mg/ml)	84
Casein (20 mg/ml)	43
Gelatin (2 mg/ml)	99
Hemoglobin (20 mg/ml)	94
Horse serum (100 mg/ml)	39
Tween 20 (0.01%, v/v)	95

[a]Comparison with unblocked plate.

[b]Amino acid mix contained 20 different amino acids in equal weight proportions. The amino acid mix and the other blocking agents were prepared in phosphate-buffered saline. (Modified from Ofek et al., 1986.)

easy to achieve, methods to immobilize target cells in suspensions prior to mixing them with potentially adherent bacteria have been developed (Ofek et al., 1986). In these studies it was found that the sequential coating of polystyrene plates with polylysine or with lysine and glutaraldehyde resulted in a surface that irreversibly bound buccal epithelial cells and enterocytes. The cells were then used as substrata for the adhesion of bacteria. The polylysine or lysine served as a support for the bifunctional glutaraldehyde, which in turn could covalently tether the cells. The irreversible coupling of target cells to insoluble supports has the advantage of providing an exceptionally stable substratum for bacteria. One possible disadvantage is that the glutaraldehyde may induce surface modifications and result in a loss of viability. Following the binding of target cells onto a glutaraldehyde-activated surface, it may be useful to wash the surfaces in ethanolamine- or albumin-containing buffers to sequester any unreacted glutaraldehyde. Removal of nonadherent bacteria is somewhat simplified in the polystyrene–polylysine or lysine–glutaraldehyde system because target cells are unlikely to be dislodged during washing procedures.

The adhesion of bacteria to a hydrophobic probe as a measure of their surface hydrophobicity can be assessed by either partitioning into two biphasic polymers (Albertsson, 1958) or into hydrocarbon droplets (Rosenberg et al., 1980) or to octyl-Sepharose beads (Hjerten et al., 1974; Smyth et al., 1978) or simply to plain plastics (Rosenberg, 1981; Schadow et al., 1988). Good separation of test bacterial suspensions into two populations of cells is usually achieved. These populations possess certain hydrophobic properties that may be related to their ability to adhere to surfaces normally colonized by the test bacteria.

Quantitation of Bacterial Adhesion (or Determination of Extent of Bacterial Adhesion)

Determination of the extent of bacterial adhesion depends on the model system employed (Figure 2–1). Some experimental designs will yield only a qualitative estimate, whereas others yield quantitative values for the extent of adhesion (Table 2–4). Although it is difficult to formulate general rules governing the measurement of bacterial adhesion to surfaces, it appears that the most quantitative measurements can be made only on model systems far removed from the natural environment. For example, reasonably good estimates of adhesion can be obtained from viable cell counts of organisms adherent to macerated tissues or other substrata, whereas the direct microscopic examination of the tissues would yield poor estimates of adhesion. Electron microscopic techniques are valuable in establishing the qualitative extent of adhesion, but are not readily amenable to defining quantitative relationships of bacteria–substratum interactions (Costerton, 1980). Electron microscopic observations are of great value in showing adhesion of bacteria to excised tissues after in vivo inoculation. When quantitative estimates are required, the method for enumeration must provide results that are proportional to adhesion (Rosenstein et al., 1985). Viable counts may not reflect the true extent of adhesion, as dead cell count determinations are almost always achieved by use of exponential phase cells. Adhesin expression may require the attainment of the stationary phase, however, in which case a mixture of viable and dead cells may coexist. In many Gram-negative rods, type 1 fimbrial adhesins are best expressed in late stationary phase (Duguid and Old, 1980). It should be noted that viable counts are usually based on a logarithmic scale and only gross differences can be appreciated by this method of enumeration. For certain model systems, such as adhesion of bacteria to tissues excised after in vivo inoculation, the enumeration of viable counts is mandatory (Goldhar et al., 1986). The high standard deviation normally encountered in viability determinations detracts from the precision in quantitation of adhesion.

Microscopic estimates of adhesion reveal the presence of both dead and viable bacteria. Furthermore, the distribution of bacteria on target cells can be measured by microscopic techniques. In this regard, no other technique is as valuable. Microscopic counts are arithmetic and increments of less than twofold increase or decrease are readily detected. Another advantage is that it is possible to distinguish between truly adherent bacteria and the background of nonadherent

Table 2–4. Quantitation of adhesive capacity of microorganisms

Aggregometry
Blotting
Electron microscopy
Enzyme-linked biotin–avidin assay (ELBA)
Eznyme-linked immunosorbent assay (ELISA)
Enzyme-linked lectinosorbent assay (ELLA)
Fluorescence labeling
Measurement of metabolites, such as CO_2, free radical production
Microscopic counts
Radiolabeling
Viable counts

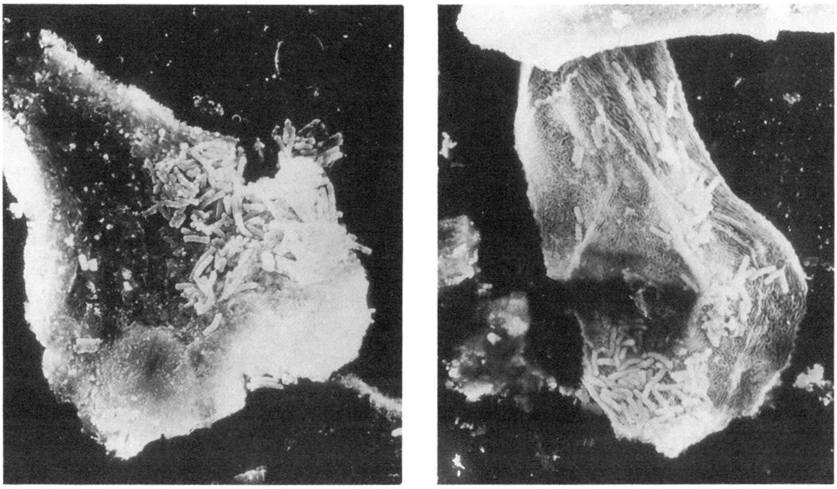

Figure 2–3. Adhesion of lactobacilli to human epithelial cells. A, adhesion occurring in a microcolony formation (localized adhesion); B, adhesion in a diffuse monolayer. Courtesy Gregor Reid (Univ. of Western Ontario) and the University of Chicago Press [*Rev. Infect. Dis.* 9:470–487 (1987)].

bacteria. A disadvantage is that the method is tedious, somewhat subjective, and amenable to the introduction of many experimental variables (see also Mackowiak and Marling-Cason, 1984; Allison and Sutherland, 1984). In spite of its limitations, one of the powers of the microscopic determination of adhesion is that it is possible to distinguish between local and diffuse adhesion (Reid, 1989). Patterns of adhesion, showing the distribution of adherent bacteria on the substratum, can be established only by microscopic observations (Scaletsky et al., 1984) (Figure 2–3). Furthermore, only by electron microscopic techniques can the structures involved in adhesion be observed (Chapter 4).

Aggregometry (and other related spectrophotometric methods) has been widely used in bacterial adhesion. The method is based on light scattering changes when turbid suspensions of bacteria interact with particulate (or soluble) receptors or receptor analogues. The method is useful for monitoring the expression of bacterial adhesins. For example, aggregometry can be employed to study the expression of type 1 fimbriae of the *Enterobacteriaceae* by employing yeast cells as substrate (Ofek and Beachey, 1978). Once the adhesin has been identified, aggregometry can be used to study the fine specificity of the adhesin and the factors that govern its activity. Because turbidity changes in bacterial suspensions may be proportional to the concentrations of adhesins and receptors, it is possible to construct first-order rate plots to characterize the extent of adhesin–receptor interaction (Drake et al., 1988). There are different kinds of particles that can be aggregated by bacteria. These include other microorganisms, red cells, red cells sensitized with receptor analogues, platelets, tissue culture cells, and Latex beads coated with receptors or receptor analogues. The one criterion for successful aggregometry experiments is that the bacteria bearing the adhesins and the corresponding particle-bearing receptors must form even suspensions.

All other methods involved in enumeration of bacterial adhesion require some type of bacterial labeling procedure. Labeling of cells with radioactive isotopes offers a considerable advantage because there is a linear relationship between numbers of bacteria and radioactive counts. Although the method is attractive and widely used, there are some factors that must be taken into account. First of all, label is sometimes lost from bacteria due to metabolic turnover or lysis. This is why tritiated thymidine (a DNA label) has become a label of choice (Zilberberg et al., 1984). The use of this label is, however, limited because it is difficult to obtain cells with high specific activity (Wyatt et al., 1990). Metabolic activities of the bacteria have also been used to study adhesion to substrates (Ludwicka et al., 1985). Bioluminescent assays, using luciferin–luciferase reactions, have been widely used.

Recently, the ELISA technique has been introduced as a convenient and sensitive means to enumerate adherent bacteria. There are several advantages to the method, including the fact that all manipulations follow the actual adhesion process. The bacteria therefore are not modified in any manner during the adhesion events. The method is highly sensitive because the bound enzyme can produce substrate product continuously over a long period of time. In addition, the method is reproducible, easy to perform and allows the handling of multiple test samples in a single experiment. The method has been successfully used by several investigators employing scraped cells (McEachran and Irwin, 1986; Ofek et al., 1986; Filler et al., 1987) as well as tissue culture cells (Stanislawski et al., 1985). For the method to be reliable, it is essential to standardize several steps. Color development can be equated with bacterial numbers. This can be done by immobilizing a known quantity of bacteria (predetermined by counting chambers) onto surfaces of plastic wells followed by the addition of the ELISA reagents. Optical densities from ELISA readings can then be plotted against bacterial numbers to obtain a standard curve. The standard curve must be generated simultaneously with the adhesion experiment (Athamna and Ofek, 1988). Other enzyme-linked assay procedures, such as enzyme-linked lectinosorbent assay (ELLA) and enzyme-linked biotin–avidin (ELBA) (Table 2–4), are similar to the ELISA method and the same principles are involved in obtaining quantitative adhesion data.

Some adhesion experiments make use of metabolic responses of cells bearing the receptors. For example, phagocytic cells and lymphocytes undergo bursts of oxidative metabolism, or proliferation, respectively, when bacteria interact with their surfaces. The metabolic bursts are detectable in liquid scintillation counters, whereas cellular proliferation can be measured by thymidine uptake (Pooniah et al., 1989). One of the attractive features of measuring oxidative bursts is that it is not required to remove nonadherent bacteria from the suspensions. Care must be

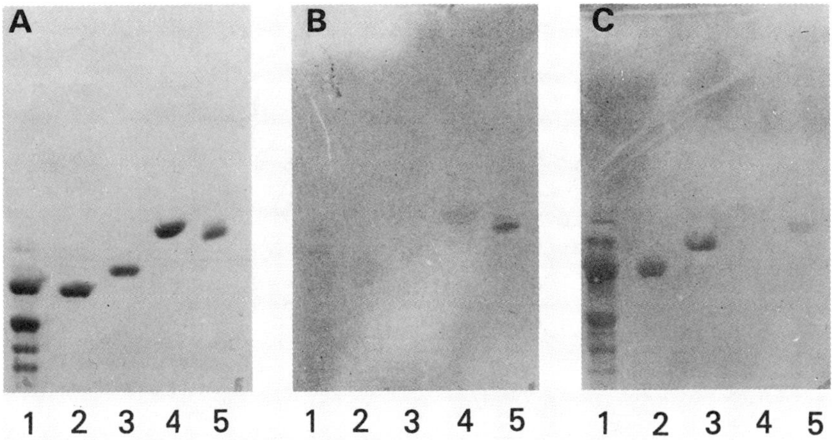

Figure 2–4. Binding of B. *pertussis* 460 to gangliosides separated by thin-layer chromatography. Approximately 2 μg of purified gangliosides or 5 μg of bovine brain gangliosides were applied to each lane as specified, and chromatograms were stained with orcinol (A), or incubated with radiolabeled bacteria (B and C). Chromatogram C was treated with sialidase before the addition of bacteria as described in the text. Lanes contain: 1, bovine brain gangliosides; 2, GM1; 3, GM2; 4, GM3; and 5, asialo GM1. (See Table 3–1 for a description of the carbohydrate structures associated with the gangliosides.) Adhesion in B and C was determined by autoradiography. Figure courtesy of M.J. Brennan, NIH, Bethesda, MD and the American Society of Biochemistry and Molecular Biology (see Brennan et al., 1991).

taken, however, to avoid metabolic bursts that may be associated with endogenous metabolism or temperature changes in the target cells.

In blotting experiments, receptors are separated from other proteins (or glycolipids) by electrophoresis or thin-layer chromatography. The bacteria (or adhesins) are then layered onto the mixtures, incubated, and washed with buffers. Adherent cells can then be visualized by use of staining techniques or by autoradiography. The method is useful in the identification of receptors (Karlsson, 1986) (Figures 2–4 and 2-5).

Treatment of Adhesion Data

Careful analyses of adhesion data can yield important insights into adhesion mechanisms. It is frequently useful to determine the numbers of combining sites (receptors) on a substratum, the affinity constant(s) describing the complex between an adhesin and its receptor, and the amount of an inhibitor required to cause a 50% reduction in adhesion. It is further important to determine the rate of an adhesion reaction and its corresponding desorption or dissociation. These

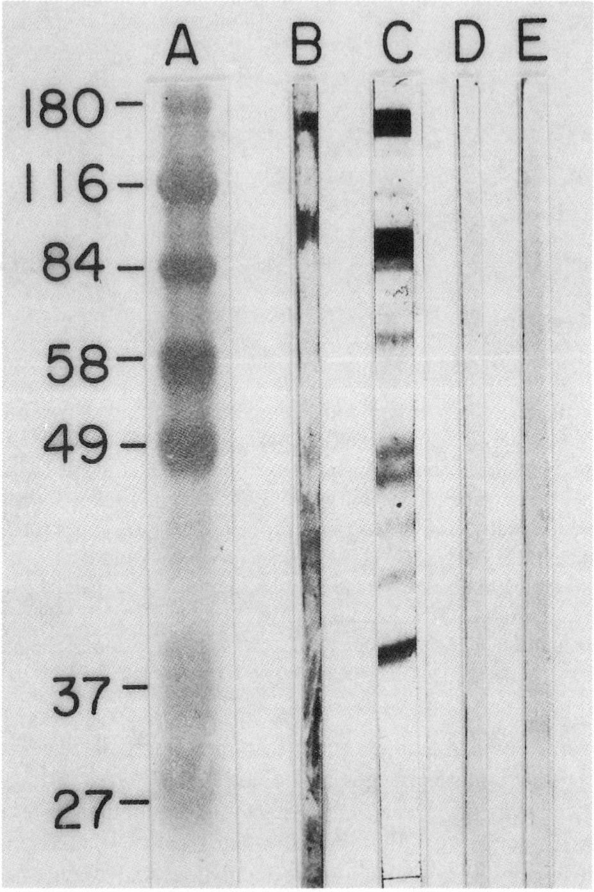

Figure 2–5. Identification of a proteinacous receptor of human granulocytes for the type 1 fimbrial adhesin of *E. coli*. Glycoproteins in granulocyte lysates were separated on SDS-PAGE and transferred to nitrocellulose. The blots were incubated with type 1 fimbriated *E. coli* without (lane C) or with mannosides (lane D), and concanavalin A without (lane B) or with mannosides (lane E). After washing the unbound bacteria, the bacteria bound to the gels were visualized by ELISA using anti *E. coli* serum and anti-rabbit horseradish peroxidase labeled antisera. Note that although the mannose-specific lectin concanavalin A detected as many as 10 glycoproteins in granulocytes, the mannose-specific type 1 fimbriated *E. coli* bound to only two mannosylated glycoproteins, corresponding to molecular weights in the range of 90,000 to 100,000 and 160,000 to 170,000. From Gbarah et al. (1991) by permission. Lane A, mol wt. standards.

quantitative values can then be employed in experimental designs to modulate adhesion reactions (Doyle, 1991).

The simple recording of numbers of adherent bacteria on a defined surface area using one single density of bacteria in triplicate is common in adhesion experiments. The process of adhesion, however, generally obeys the rules of equilibria and should be treated as such. Several empirical expressions have been derived that can be used to illustrate adhesion data graphically. The equations are simple, yet of great value in providing and understanding of some microbe–substratum interactions. In a given system the total (T) number of cells is equivalent to the combined bound (B) and unbound (U) or free cells. Knowledge of T and B makes it possible to calculate U, or conversely, knowledge of any two of the three entities provides a means of computing the third. In most adhesion experiments, T is the known and B is determined. The units for T, B, and U can be in counts per minute (c/m) colony-forming units (CFU), cells per surface area as determined microscopically, etc.

The general strategy in most properly designed adhesion experiments is to determine the extent of adhesion using several different bacterial cell densities but employing a constant amount of receptor material. If adhesion is directly proportional to numbers of cells added, then Henry's Law:

$$B = K \cdot U$$

is obeyed. Thus, a plot of B vs U will yield a straight line with the slope equal to K and the intercept at zero (Figure 2–6). This type of plot is rarely found in results from adhesin experiments (see Klotz, 1982). This is because most substrata have limiting numbers of sites onto which microbes may bind. Furthermore, many substrata possess heterogeneous sites and there may be an influence of one site on another (cooperativity). Figure 2–6d shows a system where adhesion is proportional to cell density when the number of cells is relatively low, but when cell numbers become large, saturation of sites occurs. In actual practice, it is preferable to plot B vs log U or B vs log T. Saturation is achieved when B is insensitive to changes in log T (or U). The importance of achieving saturating cell densities will be obvious when other types of plots are discussed below. It must also be kept in mind that at some very low cell densities adhesion may not be proportional to numbers of cells added because of cooperative effects or site heterogeneity.

One common derivative of the Law of Mass Action, the Langmuir adsorption isotherm, has found wide usage in treating adhesion data (Clark et al., 1978; Doyle et al., 1985; Kluepfel and Pueppke, 1985; Rosan et al., 1985). The Langmuir equation:

$$U/B = \frac{1}{KN} + \frac{U}{N}$$

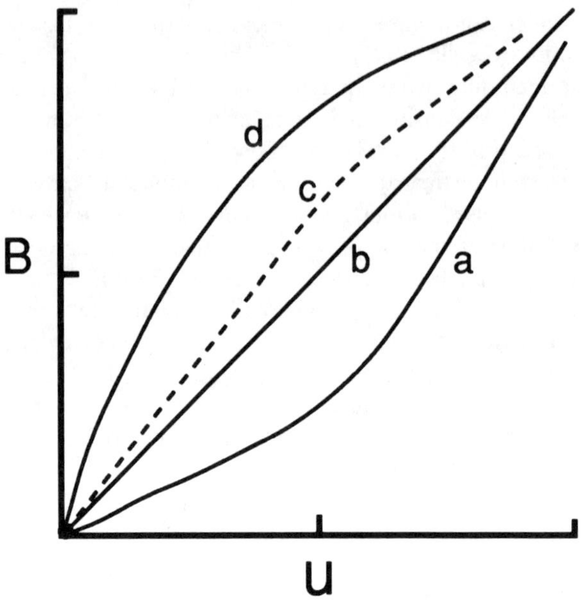

Figure 2–6. Binding isotherms of adhesion data. In regular binding isotherms, bound (*B*) cells are plotted against unbound (*U*) cells. In curve a, adhesion is not proportional to cell density. This kind of curve suggests that adhesion may depend on cooperative interactions. In curve b, Henry's Law is obeyed because adhesion is directly proportional to cell density. Curve c represents adhesion that does not approach saturation, whereas curve d suggests that a saturation of available binding sites can be achieved.

is in the form of the familiar $Y = mX + b$ for a straight line. Plots for U/B vs U (Figure 2–7) are convenient to compute N(1/slope) and $-1/K$ (the x-intercept). The N value represents the number of receptor sites, whereas K is the average association constant. The equation is valid providing certain assumptions are made. All sites must be identical and the occupancy of one site must not affect another. Langmuir, in the early 1900s, observed that when gases were adsorbed onto presumably homogeneous surfaces deviations from linearity were frequently observed. This is evidently because no surface is truly homogeneous. In the adhesion of microorganisms to surfaces, linearity may be observed only when reasonably narrow cell densities are employed.

The Scatchard equation:

$$B/U = KN - KB$$

is probably of more value in microbial adhesion. A plot of B/U vs B yields a straight line with a slope of $-K$ and an ordinate intercept yields $N \cdot K$. As with the Langmuir equation, linearity is achieved only with a system that possesses site–site homogeneity and is free of cooperative phenomena. The greatest value of the Scatchard equation in adhesion studies is its ability to detect nonlinear changes that may occur at various cell densities. A continuously changing negative slope (Figure 2–8) suggests that the population of sites is heterogeneous or that when one site is filled it inhibits the filling of additional sites (this is one definition of negative cooperativity). Negative cooperativity does not reflect competition between adherent cells; rather, it exists when the binding of a site by one cell results in a dimunition in the affinity of another site for another cell. It is a prerequisite that the sites change before any kind of cooperative process may occur. Positive cooperativity occurs when the filling of one site enhances the filling of additional sites. This type of cooperativity has been observed for the adhesion of oral streptococci and *Haemophilus* to saliva-coated surfaces (Nesbitt et al., 1982; Liljemark et al., 1985). A predictable feature of the Scatchard plot describing positive cooperativity in adhesion is that there must be a positive slope (see Figure 2–8). The *B*/*U* ratio increases with the binding of more cells, resulting in an upward slope. One meaning of the positive cooperativity

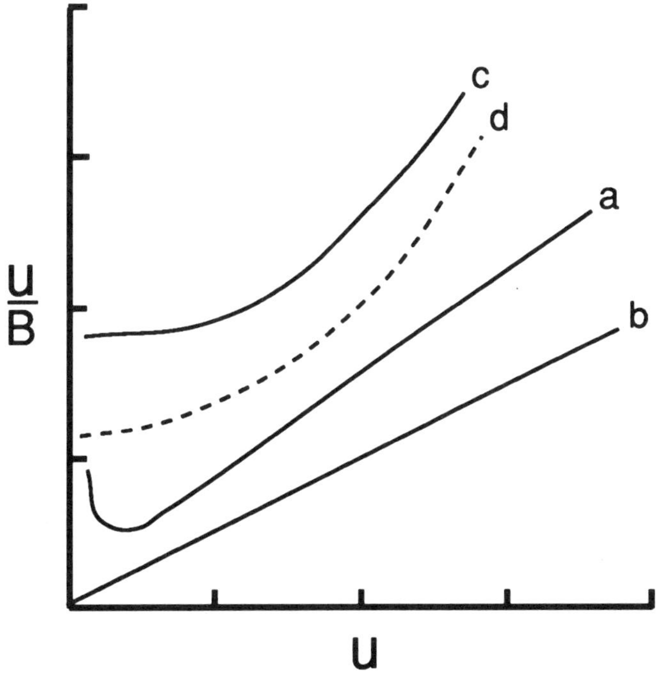

Figure 2–7. Langmuir isotherm describing the adhesion of bacteria to a substratum. The Langmuir isotherm conveniently linearizes adhesion data, making it possible to compute an association constant and the numbers of available receptor sites. The plot is generally insensitive to changes in binding and is not recommended when cooperative effects are suspected. Curves a–d are analogous to those in Figure 2–6.

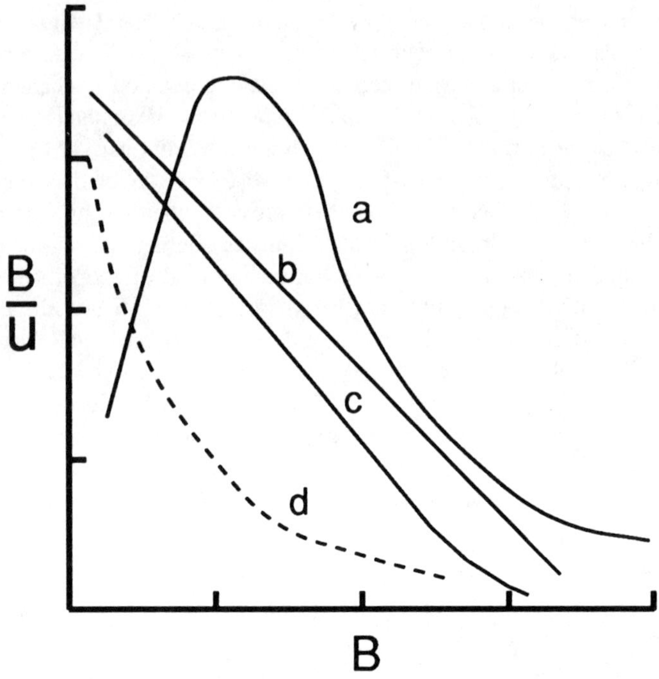

Figure 2–8. Scatchard plots of adhesion data. These idealized curves suggest positive cooperativity (a) and negative cooperativity (d), respectively. Curve b suggests that cooperative effects are absent over the range of cell densities used. Curve c may represent one set of binding sites, as described by the constant slope, followed by another set of adhesion sites or by negative cooperativity. In order to obtain the numbers of binding sites on the substratum, linear portions of the curves may be extrapolated to the abscissa ($B/U = 0$). The Scatchard plot is especially sensitive to site changes or to site heterogeneity. Curves a–d are analogous to those in Figure 2–6.

is that all sites are different. It therefore becomes meaningless to obtain composite K or N values. Every site has its own K. In a practical sense, there are infinite numbers of K values, as the slope of the curve at any point represents a K value. The extent of cooperativity (positive or negative) may be assessed by use of the Hill equation:

$$B/N - B = KU^{n_H}$$

where n_H is the Hill coefficient or index of cooperativity (many immunologists will recognize the Hill equation as identical with the Sips equation used to determine binding site heterogeneity in antihapten antibodies). When $\log(B/N - B)$ is plotted against $\log U$, the coefficient n_H may be obtained from the slope of the resulting curve. When $n_H > 1$, there is positive cooperativity. Similarly, negative cooperativity is indicated when $n_H < 1$.

It is suggested that the proper plotting of adhesion data can be of great value in understanding microorganism–substratum interactions. For example, if competitive adhesion experiments between various organisms are planned, it is best to use nonsaturating densities of cells where true competitive effects can be assessed. At very high or saturating cell densities, competition by other organisms would not be readily detected, especially at low or moderate densities of the competing organisms. Saturating cell densities in experiments designed to measure the effects of inhibitors of adhesion are not recommended. A plot of *B* vs log *U* to determine where saturation occurs would permit the selection of a cell density not approaching saturation. By using various cell densities and various means to plot adhesion results, conclusions may be reached that would have remained obscure otherwise. For example, Firon et al. (1985) provided a binding isotherm for the adhesion of *E. coli* to BHK tissue culture cells. When the data are replotted according to Scatchard, a curve is obtained suggestive of positive cooperativity (Figure 2–9). This, of course, raises a question about the mechanism of positive cooperativity. When *E. coli* binds to an epithelial cell at a particular site, there may be some type of local rearrangement, so that other *E. coli* are more easily accommodated.

This local rearrangement may or may not require metabolic energy, but at least this can be tested. Oral streptococci have been observed to adhere to clusters on saliva-coated glass, and as randomly spaced cells on buffer-coated glass

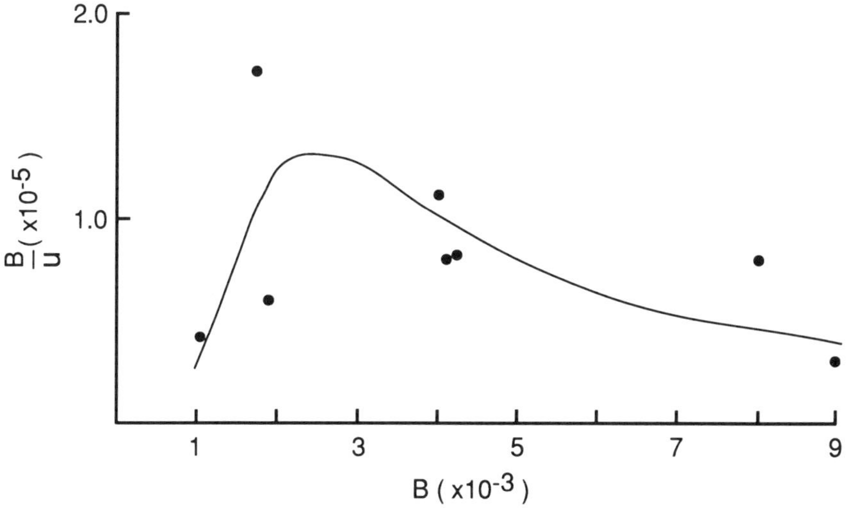

Figure 2–9. Scatchard plot for the binding of *E. coli* to BHK tissue culture cells. These data suggest that adhesion is promoted by positive cooperativity. The data were taken from a binding isotherm of Firon et al. (1985) and replotted.

Table 2–5. Information gained by plotting results of adhesion experiments according to several mass action equations

Homogeneity or heterogeneity of sites mediating adhesion
Association (or dissociation) constants
Numbers of receptor sites on a substratum
Detection of cooperative interactions
Amenable to computer simulations (computer analyses)
Determination of the mode of action of inhibitors (competitive, non-competitive)
Predictions about pattern of adhesion of microorganism onto substratum (random or clustered)
Selection of proper cell densities for performing inhibition or kinetics experiments
Determination of cell density at saturation
Determination of change in affinity with extent of binding

(Cowan et al., 1987b). It seems unlikely that an adherent cell could create a change on a substratum site far removed from the site of initial adhesion. Positively cooperative events probably would result in the presence of adherent cells adjacent to each other. In this regard, the binding of *E. coli* to uroepithelial cells revealed a skewed or non-normal distribution (Reid, 1989). Some uroepithelial cells contained many *E. coli,* whereas others contained only a few or none at all. This kind of distribution may reflect positive cooperativity of adhesion or it may reflect a nonrandom distribution of receptor sites. If the adhesion had been measured by use of a single density of radioactive *E. coli* instead of microscopy, it would have been impossible to show that the adhesion was non-Poisson (normal distribution).

Questions may be raised about the validity of employing equations of equilibria to describe the adhesion of microbes to surfaces. In most cases, both the microbe and the substratum possess highly heterogeneous surface characteristics. The equation that analyzes the adsorption of a gas onto a relatively simple surface may not be amenable to such complex systems as those involving microorganisms. It must be kept in mind, however, that the firm adhesion of a microorganism to a surface represents a composite of interactions at the molecular level. It is emphasized that certain kinds of information can be gained by various plots from adhesion experiments (Table 2–5). The analysis of the equilibria data can be useful in helping to define the characteristics of the cell–substratum system. Comparative results of adhesion of various bacterium to various substrata become clear if the data are properly plotted. The comparative results are usually of greater value than absolute values of affinity constants or numbers of combining sites. Finally, adhesion results should be subjected to statistical analyses, such as the parametric (unpaired *I* test) or unpaired (Mann–Whitney *U* test) especially when comparing bacteria and/or substrata (Woolfson et al., 1987).

It may be useful to study the rates of adhesion and desorption (Myerthall and Thomas, 1983; Cowan et al., 1986) of bacteria to substrata. Kinetic measurements frequently lead to the same conclusions as equilibria measurements. One major usefulness of kinetic measurements is that it may be possible to determine if a potential inhibitor affects adhesion or desorption. Studies of kinetics or equilibria at various temperatures can lead to an understanding of the thermodynamics of adhesion (Cowan et al., 1987a,b). To date, however, there have only been a few reports on the kinetics and thermodynamics of bacterial adhesion.

Objectives and Variables in Experimental Design

Some of the major objectives in the study of bacteria–surface interactions are listed in Table 2–6. The variables in the experimental design employed to achieve the objectives are also listed. The functional expression of most bacterial adhesins is not an essential activity for the survival of the bacteria. In fact, as discussed later, most organisms display differences in adhesive qualities during growth. These qualities can be modulated by pH, composition and physical state of growth medium, and presence of sublethal concentrations of antibiotics. The variables may either affect the functional expression of the adhesin or alter the bacterial surface to express an alternative mechanism of adhesion. These variables can be identified and manipulated in vitro, but it can only be assumed that similar kinds of changes in adhesive properties are elicited under natural conditions. The natural target surface also is probably dynamic, constantly changing with each new condition it faces. A typical adhesion experiment is limited to the study of only one variable at a time. Failure to observe adhesion does not mean that a bacterium is incapable of binding to a surface. It means that the conditions for observing adhesion have not been found.

Once a particular set of conditions has been established for promoting adhesion, then variables such as mutants, chemical or enzymatic modification of bacterial

Table 2–6. Summary of variables and objectives in the experimental design of microbe–substrata interactions

Variables	Objectives
Growth conditions of test bacteria (pH, type of medium, antibiotics, etc . . .)	Functional expression of adhesins
Density of cells	Cooperativity, numbers of binding sites, affinity of binding, kinetics
Solvents, pH, and ions of reaction mixtures	Effect of bulk phase
Mutants (or phenotypes) of bacteria or target cells; chemical or enzymatic modification of the bacterial or substrata surfaces; inhibitors; types of inert surfaces and coating molecules	Identification and role of surface components in adhesion
Temperature	Thermodynamic quantities involved in adhesion

Table 2–7. Identification, isolation, and characterization of cell surface receptors for bacteria

Methods	Results and Interpretations
Prevention of adhesion by use of low molecular weight inhibitors (haptens). The inhibitors must be of a known structure, soluble, and have no untoward effects on bacteria or target cells. The inhibition can be determined by direct binding assays or by measuring biological activity.	If inhibition is achieved, it is likely that the structure of the receptor is closely related to that of the inhibitor. This can be confirmed by showing that the inhibitor binds directly to adhesin.
Inhibition of bacterial adhesion by macromolecules. The inhibitors may be lectins or antibodies directed against the target surface receptors. Monoclonal antibodies are preferred because of their restricted (known epitopes) specificities.	Inhibition may indicate that the lectin or antibody receptor is also the bacterial receptor.
Adhesion following chemical or enzymatic modification of the target surface. Proteases, lipases, glycosidases, or reagents that chemically modify covalent linkages may be used to alter receptor surfaces.	Proteolytic digestion may suggest that the receptor is a protein; glycosidase digestion may indicate a carbohydrate-containing structure. Modification with agents known to derivatize or alter known residues is preferable.
Adhesion of bacteria to surfaces possessing known structural modification. Mutants (selected for resistance to lectins) or tunicamycin-treated tissue culture cells are examples of cells with altered glycoproteins that may serve as receptor donors.	When the bacterium loses its affinity for the altered target, it may be possible to empirically identify the target as glycoprotein, etc.
Immobilization of receptors by affinity chromatography. Bacterial adhesins covalently attached to solid matrices may complex with solubilized receptor molecules. Elution may be by chaotropes, low pH, or hapten-like inhibitors.	Procedure may provide receptors with high activities for adhesins or whole bacteria.
Gel overlay methods. Following separation by electrophoresis or by thin-layer chromatography, bacteria or bacterial adhesins may bind to the separated receptor(s). Detection may be by use of viable cell counts, by autoradiography, or by selected enzymatic reactions.	Convenient to separate and identify receptor molecules in a complex mixture.
Incorporation of receptor in artificial membranes or in membranes of cells devoid of receptor, followed by adhesion studies.	This method is frequently difficult, but the results may provide compelling evidence that the constitutent is a receptor for an adhesin or bacterium.

or target surfaces, solvents, salts, inhibitor agents, and temperature changes can be introduced into the experimental scheme to achieve the outlined objectives.

An ultimate goal in adhesion studies is to identify and characterize adhesins and receptors. Some of the methodology developed during the past few decades to identify biological receptors of living cells can be successfully applied to the identification of receptors for bacterial adhesins. These methods are summarized in Table 2–7.

References

Albertsson, P.A. 1958. Particle fractionation in liquid two-phase system. The composition of some phase systems and the behavior of some model particles in them. Application to the isolation of cell walls from microorganisms. *Biochim. Biophys. Acta* **27**:378–395.

Allison, D.G. and I.W. Sutherland. 1984. A staining technique for attached bacteria and its correlation to extracellular carbohydrate production. *J. Microbiol. Meth.* **2**:93–99.

Aronson, M., O. Medalia, L. Schori, D. Mirelman, N. Sharon, and I Ofek. 1979. Prevention of colonization of the urinary tract of mice with *Escherichia coli* by blocking of bacterial adherence with methyl alpha-D-mannopyranoside. *J. Infect. Dis.* **139**:329–332.

Athamna, A. and I Ofek. 1988. Enzyme-linked immunosorbent assay for quantitation of attachment and ingestion stages of bacterial phagocytosis. *J. Clin. Microbiol.* **26**:62–66.

Baselsky, V.S. and C.D. Parker. 1978. Intestinal distribution of *Vibrio cholerae* in orally infected infant mice: kinetics of recovery of radiolabel and viable cells. *Infect. Immun.* **21**:518–525.

Beachey, E.H. and I. Ofek. 1976. Epithelial cell binding of group A streptococci by lipoteichoic acid on fimbriae denuded of M proteins. *J. Exp. Med.* **143**:759–771.

Brennan, M.J., J.H. Hannah and E. Leininger. 1991. Adhesion of *Bordetella pertussis* to sulfatides and to the GalNacβ4Gal sequence found in glycosphingolipids. *J. Biol. Chem.* **266**:18827–18831.

Clark, W.B., L.L. Bammann, and R.J. Gibbons. 1978. Comparative estimates of bacterial affinities and adsorption sites on hydroxyapatite surfaces. *Infect. Immun.* **19**:846–853.

Costerton, J.W. 1980. Some techniques involved in study of adsorption of microorganisms to surfaces. In: Bitton, G. and K.C. Marshall (eds.), *Adsorption of Microorganisms to Surfaces*. John Wiley & Sons, New York, pp. 403–423.

Cowan, M.M., K.G. Taylor, and R.J. Doyle. 1986. Kinetic analysis of *Streptococcus sanguis* adhesion to artificial pellicle. *J. Dent. Res.* **65**:1278–1283.

Cowan, M.M., K.G. Taylor, and R.J. Doyle. 1987a. Role of sialic acid in the kinetics of *Streptococcus sanguis* adhesion to artificial pellicle. *Infect. Immun.* **55**:1552–1557.

Cowan, M.M., K.G. Taylor, and R.J. Doyle. 1987b. Energetics of the initial phase of adhesion of *Streptococcus sanguis* to hydroxylapatite. *J. Bacteriol.* **169**:2995–3000.

de Man, P., B. Cedergren, S. Enerback, A.C. Larsson, H. Leffler, A.L. Lundell, B. Nilsson, and C. Svanborg-Eden. 1987. Receptor-specific agglutination tests for detection of bacteria that bind globoseries glycolipids. *J. Clin. Microbiol.* **25**:401–406.

Doyle, R.J. 1991. Strategies in experimental microbial adhesion research. In: Mozes, N, P.S. Handley, H.J. Busscher, and P.G. Rouxhet (eds.), *Microbial Cell Surface analysis—Structural and Physiocochemical Methods*. VCH Publishers, New York, pp. 293–316.

Doyle, R.J., J.D. Oakley, K.R. Murphy, D. McAlister, and K.G. Taylor. 1985. Graphical analyses of adherence data. In: Mergenhagen, S.E. and B. Rosan (eds.), *Molecular Basis of Oral Microbial Adhesion*. American Society for Microbiology, Washington, pp. 109–113.

Drake, D., K.G. Taylor, A.S. Bleiweis, and R.J. Doyle. 1988. Specificity of the glucan-binding lectin of *Streptococcus cricetus*. *Infect. Immun.* **56**:1864–1872.

Duguid, J.P. and R.R. Gillies. 1957. Fimbriae and adhesive properties of dysentery bacilli. *J. Pathol. Bacteriol.* **74**:397–411.

Duguid, J.P. and D.C. Old. 1980. Adhesive properties of *Enterobacteriaceae*. In: Beachey, E.H. (ed.), *Bacterial Adherence* (*Receptors and Recognition, Vol. 6*). London, Chapman and Hall, pp. 184–217.

Duguid, J.P., M.R. Darekar, and D.W.F. Wheater. 1976. Fimbriae and infectivity in *Salmonella typhimurium*. *J. Med. Microbiol.* **9**:459–473.

Elbein, A.D., B.A. Sanford, M.A. Ramsey, and Y.T. Pan. 1981. Effect of inhibitors of glycoprotein biosynthesis and bacterial adhesion. In: Elliot, K., M. O'Connor, and J. Whelan (eds.), *Adhesion and Microorganism Pathogenicity* (*Ciba Foundation Symposium 80*). Pitman Medical, London, pp. 270–282.

Ellen, R.P. and R.J. Gibbons. 1972. M protein-associated adherence of *Streptococcus pyogenes* to epithelial surfaces: prerequisite for virulence. *Infect. Immun.* **5**:826–830.

Ellen, R.P. and R.J. Gibbons. 1973. Parameters affecting the adherence and tissue tropisms of *Streptococcus pyogenes*. *Infect. Immun.* **9**:85–91.

Filler, S.G., L.G. Der, C.L. Mayer, P.D. Christenson, and J.E. Edwards, Jr. 1987. An enzyme-linked immunosorbent assay for quantifying adherence of *Candida* to human vascular endothelium. *J. Infect. Dis.* **156**:561–566.

Firon, N., D. Duksin, and N. Sharon. 1985. Mannose-specific adherence of *Escherichia coli* to BHK cells that differ in their glycosylation patterns. *FEMS Microbiol. Lett.* **27**:161–165.

Goldhar, J., A. Zilberberg, and I. Ofek. 1986. Infant mouse model of adherence and colonization of intestinal tissues by enterotoxigenic strains of *Escherichia coli* isolated from humans. *Infect. Immun.* **52:**205–208.

Hjerten, S., J. Rosengren, and S. Pahlman. 1974. Hydrophobic interaction chromatography. The synthesis and the use of some alkyl and aryl derivatives of agarose. *J. Chromatogr.* **101**:281–288.

Karlsson, J. 1986. Animal glycolipids as attachment sites for microbes. *Chem. Phys. Lipids* **42**:153–172.

Klotz, I.M. 1982. Numbers of receptor sites from Scatchard graphs: facts and fantasies. *Science* **217**:1247–1249.

Kluepfel, D.A. and S.G. Pueppke. 1985. Isotherm for adsorption of *Agrobacterium tumefaciens* to susceptible potato (*Solanum tuberosum* L.) tissues. *Appl. Environ. Microbiol.* **49**:1351–1355.

Liljemark, W.F., C.G. Bloomquist, and L.J. Fenver. 1985. Characteristics of the adherence of oral *Hemophilius* species to an experimental salivary pellicle and to other oral bacteria, In: Mergenhagen, S. and B. Rosan (eds.), *Molecular Basis of Oral Microbial Adhesion*. American Society for Microbiology, Washington, pp. 94–102.

Ludwicka, A., L.M. Switalski, A. Lundin, G. Pulverer, and T. Wadstrom. 1985. Bioluminescent assay for measurement of bacterial attachment to polyethylene. *J. Microbiol. Methods* **4**:169–177.

Mackowiak, P.A. and M. Marling-Cason. 1984. A comparative analysis of *in vitro* assays of bacterial adherence. *J. Microbiol. Methods* **2**:147–158.

McEachran, D.W. and R.T. Irvin. 1986. A new method for the irreversible attachment of cells or proteins to polystyrene tissue culture plates for use in the study of bacterial adhesion. *J. Microbiol. Methods* **5**:99–111.

Myerthall, D.L. and T.H. Thomas. 1983. Kinetics of adherence of *Actinomyces viscosus* to saliva-coated silica and hydroxyapatite beads. *J. Gen. Microbiol.* **129**:1387–1395.

Nesbitt, W.E., R.J. Doyle, K.G. Taylor, R.H. Staat, and R.R. Arnold. 1982. Positive cooperativity in the binding of *Streptococcus sanguis* to hydroxylapatite. *Infect. Immun.* **35**:157–165.

Ofek, I. and E.H. Beachey. 1978. Mannose binding and epithelial cell adherence of *Escherichia coli*. *Infect. Immun.* **22**:247–254.

Ofek, I., D. Mirelman, and N. Sharon. 1977. Adherence of *Escherichia coli* to human mucosal cells mediated by mannose receptors. *Nature* **265**:923–625.

Ofek, I., H.S. Courtney, D.M. Schifferli, and E.H. Beachey. 1986. Enzyme-linked immunosorbent assay for adherence of bacteria to animal cells. *J. Clin. Microbiol.* **24**:512–516.

Pooniah, S., S.N. Abraham, M.E. Dokter, C.D. Wall, and R.D. Endres. 1989. Mitogenic stimulation of human lymphocytes by the mannose-specific adhesin on *Escherichia coli* type 1 fimbriae. *J. Immunol.* **142**:992–998.

Reid, G. 1989. Local and diffuse bacterial adherence on uropithelial cells. *Curr. Microbiol.* **18**:93–97.

Rosan, B., R. Eifert, and E. Golub. 1985. Bacterial surfaces, salivary pellicles and plaque formation. In: Mergenhagen, S.E. and B. Rosan (eds.), *Molecular Basis of Oral Microbial Adhesion*. American Society for Microbiology, Washington, pp. 69–76.

Rosenberg, M. 1981. Bacterial adherence to polystyrene: a replica method of screening for bacterial hydrophobicity. *Appl. Environ. Microbiol.* **42**:375–377.

Rosenberg, M., D. Gutnick, and E. Rosenberg. 1980. Adherence of bacteria to hydrocarbons: a simple method for measuring cell surface hydrophobicity. *FEMS Microbiol. Lett.* **9**:29–33.

Rosenberg, M.A., E.A. Perry, D.L. Gutnick, E. Rosenberg, and I. Ofek. 1981. Adherence of *Acinetobacter calcoacetius* RAG-1 to human epithelial cells and to hexadecane. *Infect. Immun.* **33**:29–33.

Rosenstein, I.J., D. Grady, J.M.T. Hamilton-Miller, and W. Brumfitt. 1985. Relationship between adhesion of *Escherichia coli* to uro-epithelial cells and the pathogenesis of urinary infection: problems in methodology and analysis. *J. Med. Microbiol.* **20**:335–344.

Scaletsky, C.A., M. Silva, and R. Trabulsi, 1984. Distinctive patterns of adherence of enteropathogenic *Escherichia coli* to HeLa cells. *Infect. Immun.* **45**:534–536.

Schadow, K.H., W.A. Simpson, and S.D. Christensen. 1988. Characteristics of adherence to plastic tissue culture plates of coagulase negative staphylococci exposed to subinhibitory concentrations of antibiotics. *J. Infect. Dis.* **157**:71–77.

Smyth, C.J., P. Jonsson, E. Olsson, O. Soderlind, R. Rosengren, S. Hjerten, and T. Wadstrom. 1978. Differences in hydrophobic surface characteristics of porcine enteropathogenic *Escherichia coli* K88 antigen as revealed by hydrophobic interaction chromatography. *Infect. Immun.* **22**:462–472.

Stanislawski, L., W.A.Simpson, D.L. Hasty, N. Sharon, E.H. Beachey, and I. Ofek. 1985. Role of fibronectin in attachment of *Streptococcus pyogenes* and *Escherichia coli* to human cell lines and isolated oral epithelial cells. *Infect. Immun.* **48**:257–259.

Svanborg-Eden, C., R. Freter, L. Hagberg, R. Hull, S. Hull, H. Leffler, and G. Schoolnik. 1982. Inhibition of experimental ascending urinary tract infection by an epithelial cell-surface receptor analogue. *Nature* **298**:560–562.

Valentin-Weigand, P., G.S. Chhatwal, and H. Blobel. 1987. A simple method for quantitative determination of bacterial adherence to human and animal epithelial cells. *Microbiol. Immunol.* **31**:1017–1023.

Woolfson, A.D., S.P. Gorman, D.F. McCafferty, and D.S. Jones. 1987. On the statistical evaluation of adherence assays. *J. Appl. Bacteriol.* **63**:147–151.

Wyatt, J.E., S.M. Poston, and W.C. Noble. 1990. Adherence of *Staphylococcus aureus* to cell monolayers. *J. Appl. Bacteriol.* **69**:834–844.

Zilberberg, A., J. Goldhar, and I. Ofek. 1983. Adherence of enterotoxigenic *Escherichia coli* (ETEC) strains to mouse intestine segments analyzed by Langmuir adherence isotherms. *FEMS Microbiol. Lett.* **16**:225–228.

Zilberberg, A., I. Ofek, and J. Goldhar. 1984. Affinity of adherence *in vitro* and colonization of mice intestine by enterotoxigenic *Escherichia coli* (ETEC). *FEMS Microbiol. Lett.* **23**:103–106.

3

Animal Cell Membranes as Substrata for Bacterial Adhesion

The purpose of this chapter is to review briefly the composition and organization of animal cell surface structures that may be potential receptors for adhesins of bacteria. The understanding of the specificity of animal cell–bacteria interactions requires a basic knowledge of the molecular structure of the animal cell surface, especially of those molecules that serve as receptors for ligands in general and for bacterial adhesins in particular. All animal cell membranes share common compositional and organizational features (Figure 3–1): (1) The major membrane lipids are arranged in a planar bilayer configuration that is predominantly in a "fluid" state under physiological conditions. The membrane lipids are commonly composed of glycerolphospholipids, sphingolipids, and sterols. (2) The bilayer membrane contains integral membrane constituents composed of both glycolipids and glycoproteins that are inserted or "intercalated" into the bilayer structure. (3) Other glycoproteins and proteins are bound to the surface of the plasma membrane by weak ionic interactions, hydrogen bonding, or the hydrophobic effect. These surface-associated glycoproteins and proteins bound to integral membrane structures are referred to as peripheral or extrinsic components. (4) In many animal cells there is a substantial layer of carbohydrate-containing materials of variable thicknesses outside the plasma membrane but in close or intimate association with the membrane. This layer is known as the cell coat or extracellular matrix. The distinction between membrane constituents as being integral, peripheral, or belonging to the cell coat is based on the method required to dissociate the constituent in question from the cell membrane. The integral constituents may be released only after disruption or perturbation of the phospholipid bilayer, usually by detergents (Lichtenberg et al., 1983). Nonintegral surface constituents are commonly released by washing the cells with buffers of different pH or ionic strength, or by using chelating agents, such as ethylenediaminetetraacetic acid (EDTA). There is no general method, however, to release selectively either peripheral or extracellular matrix constituents. As a result, the distinction

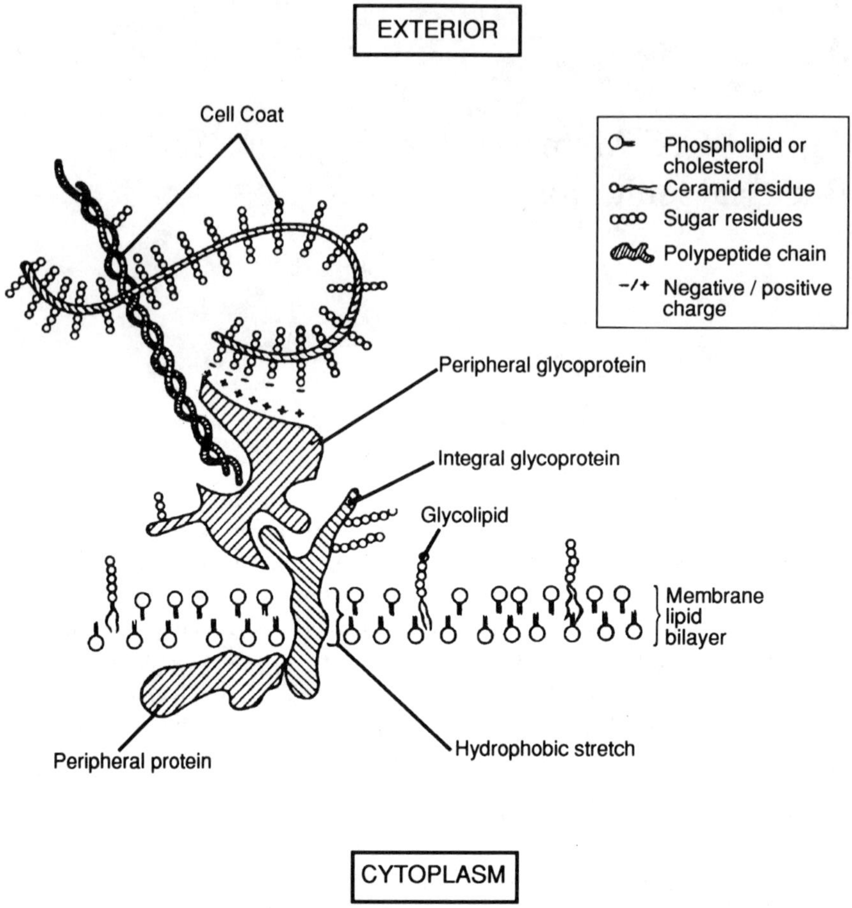

Figure 3–1. Idealized representation of the surface of an animal cell membrane. Bacteria may bind to the cell coat, peripheral glycoproteins, integral glycoproteins, or glycolipids. Adapted from Ofek et al. (1985).

between the two classes of membrane constituents is sometimes difficult to resolve and very often they are referred to as nonintegral membrane constituents. One of the key features of the membrane is its asymmetry. For nonglycosylated lipids the asymmetry is only partial, in that every phospholipid is present on both sides of the bilayer but in different amounts. In human erythrocytes, for example, lipids with positively charged head groups (e.g., phosphatidylethanolamine and phosphatidylserine) are predominant in the internal leaflet facing the cytoplasm (Marinetti and Crain, 1978). The asymmetry with respect to proteins, glycoproteins, and glycolipids is absolute: every molecule of a given membrane constituent has the same orientation across the lipid bilayer, with the carbohydrate moieties of the glycosylated compounds always exposed on the outer surface. For further information on the organization of the animal cell membrane, the reader is referred to reviews (Lodish et al., 1981; Lotan and Nicolson, 1981; Singer, 1981; Aplin and Hughes, 1982) and a book (Sim, 1982).

Integral Membrane Constituents

As mentioned in the foregoing, both glycolipids and glycoproteins belong to this class of membrane constituents. With respect to their organization in the cell membrane, these components share several properties. Of special interest is their mobility within the plane of the membrane, made possible by the fluidity of the lipids of the membrane bilayer (Edidin, 1974; Cherry, 1979). Not all integral constituents possess the same degree of mobility. Their movement may be restricted by interactions with extrinsic proteins on the inner surface or by cell–cell junctions (Nicolson, 1976; Geiger, 1983). As a result, the topography of integral glycoproteins is nonrandom and some of these constituents are confined to specific regions of the membrane. Nevertheless, the concept that the membrane lipids behave as a continuous fluid with regions of low fluidity, in which integral constituents are free to move (Singer and Nicolson, 1972; Singer, 1974), has been very useful in describing many phenomena associated with the cell surface. For example, "patching" and "capping" of integral membrane macromolecules follow their interaction with multivalent extracellular agents (Karnovsky and Unanu, 1973; Raff and dePetris, 1973).

All glycolipids of the cell membrane are constituents of the outer half of the bilayer, where they comprise 30%–60% of the total lipid content. The ceramide group of the glycolipids is responsible for anchoring the molecules to the membrane by intercalating within the lipid bilayer, whereas the oligosaccharide chains extend to the outer surface. Chemically, the glycolipids are either neutral or acidic (Table 3–1). The acidic glycolipids contain sialic acids (i.e., gangliosides), which contribute significantly to the net negative charge of the animal cell surface. When isolated in pure form from the cell membrane, glycolipids tend to form micelles. When mixed with animal cells the glycolipids can "coat" most cells and assume an orientation comparable to the original one, i.e., with the oligosaccharide portion exposed to the outer surface (Sedlacek et al., 1976; Callies et al, 1977). This behavior is similar to that of bacterial glycolipids, such as lipopolysaccharides of Gram-negative bacteria and lipoteichoic acids of Gram-positive bacteria (Wicken and Knox, 1981). Many investigators take advantage of this property for experiments in which the type of oligosaccharide and the number of molecules to be inserted in the animal cell can be controlled (Wiegandt et al., 1981).

Virtually all proteins of animal membranes are glycosylated (Sharon and Lis, 1981, 1982). In many respects, membrane glycoproteins are similar to soluble

Table 3–1. Structures of major neutral and acidic glycolopids in animal cells[a]

Neutral	
Glucosylceramide	Glc-Cer
Lactosylceramide	Galβ1,4Glc-Cer
Trihexosylceramide	Galα1,4Galβ1,4Glc-Cer
Globoside	GalNAcβ1,3Galα1,4Galβ1,4Glc-Cer
Forssman hapten	GalNAcα1,3GalNAcβ1,3Galα1,4Galβ1,4Glc-Cer
Acidic (gangliosides)	
GM3	Galβ1,4Glc-Cer ↑ α2,3 NeuNAc
GM2	Galβ1,3GalNAcβ1,4Galβ1,4Glc-Cer ↑ α2,3 (on Galβ1,4) NeuNAc
GM1	Galβ1,3GalNAcβ1,4Galβ1,4Glc-Cer ↑ α2,3 (on Galβ1,4) NeuNAc
GD1a	Galβ1,3GalNAcβ1,4Galβ1,4Glc-Cer ↑ α2,3 ↑ α2,3 NeuNAc NeuNAc
GT1	Galβ1,3GalNAcβ1,4Galβ1,4Glc-Cer ↑ α2,3 ↑ α2,3 NeuNAc NeuNAcα 2,8NeuNAc

[a]All sugars are of the D configuration. (Modified from Ofek et al., 1985.)

glycoproteins. For example, they do not differ significantly in the overall amino acid composition and carbohydrate content and they contain the same monosaccharide constituents: the hexoses D-galactose and D-mannose and the methylpentose, L-fucose; *N*-acetylhexosamines; and the sialic acids. D-Glucose, found in the collagens, does not appear to occur in membrane glycoproteins. The uronic acids, D-glucuronic acid and L-iduronic acid, are confined to proteoglycans, such as heparan sulfate, when these are associated with membranes. Several types of carbohydrate–peptide linkages are common to both groups of glycoproteins: the *N*-glycosyl bond between *N*-acetyl-D-glucosamine and asparagine (*N*-acetyl-D-glucosaminyl-asparagine) and the *O*-glycosidic bond between *N*-acetyl-D-galactosamine and a serine or threonine of the protein polypeptide backbone. A distinctive feature of the primary amino acid sequence of membrane glycoproteins is the presence of region(s) rich in hydrophobic amino acid(s). The hydrophobic region(s) is (are) responsible for anchoring the molecules in the membrane. Most, if not all, integral glycoproteins extend across the entire thickness of the bilayer so that they are exposed to both the integral and the external environment. Some of the glycoproteins, such as glycophorin in the human erythrocyte (Marchesi et al., 1976; Tanner, 1978) and the E_2 glycoprotein of Semliki Forest virus (Garoff, 1979), span the membrane only once, whereas others traverse it more than once (Steck, 1978). The oligosaccharide moieties are always exposed on the cell surface. The internal segment of the membrane-spanning intrinsic glycoprotein may be in close contact with some of the proteins located on the cytoplasmic face of the membrane. The intrinsic glycoproteins may therefore play an essential role in processes that require communication between the outside of the cell and its interior, such as transport phenomena and the responses of cells to external stimuli, such as hormones, toxins, and other cells (Nicolson, 1976, 1979; Finean et al., 1984). Depending on the sizes and numbers of their hydrophobic regions, the membrane glycoproteins may display low to moderate solubility in detergent-free aqueous solutions. Unlike glycolipids, the integral glycoproteins in isolated form do not reassociate or "coat" cell membranes, but can be reassembled into liposomes (Eytan, 1982). The orientation of the reconstituted glycoproteins is, however, symmetrical in that half of the molecules insert with the glycosylated part facing outwardly and half facing inwardly.

One class of integral membrane proteins is referred to as integrins. The integrins are glycosylated, membrane-spanning heterodimers. One of the heterodimers is a β subunit, whereas the other is an α subunit (Ruoslahti, 1991). The β subunit is classified further as β_1, β_2, or β_3. A single β subunit may be paired with one of several α subunits. The β subfamily contains the major fibronectin, laminin, and collagen receptors. The most common fibronectin receptor is $\alpha_5\beta_1$, which binds to the RGD sequence of fibronectin. A few α subunits may combine with any one of the three subunits. The $\alpha_m\beta_2$, previously called CR3, integrin is found on phagocytic cells and is capable of binding directly with certain bacteria, such as *Escherichia coli* (rough strains), *Bordetella pertussis*, and *Yersinia pseudotuberculosis*. The adhesion of these bacteria to integrins is discussed in later chapters.

Peripheral Membrane Components

The peripheral proteins and glycoproteins are anchored to the surface of the membrane by weak ionic interactions or by hydrogen bonding with integral constituents of the cell membranes, e.g., glycoproteins, glycolipids, or polar head groups of phospholipids. Some proteins are associated mainly with the inner face. Some of them interact with elements of the cytoskeleton of the cell, thereby providing a link between integral membrane constituents and the interior of the cells. Because of the difficulties mentioned earlier in distinguishing between peripheral glycoproteins of the cell surface and glycoproteins of the extracellular matrix, very few compounds have been unequivocally identified as peripheral glycoproteins. The best example of a membrane constituent of the peripheral class is fibronectin, widely distributed on the surfaces of normal cells. The structure of fibronectin is characterized by the presence of several distinct domains with binding sites for specific ligands, such as heparan sulfate, collagen, and hyaluronic acid (Hynes and Yamada, 1982: Ruoslahti, 1988, 1991; Ruoslahti et al., 1982). Fibronectin also contains an RGD (Arg-Gly-Asp) sequence recognized by members of a class of integral membrane glycoproteins called integrins. By virtue of this structure, and the outermost location of fibronectin on the cell surface, this glycoprotein is eminently suitable to mediate contacts between cells and their environment. Many different bacteria are able to bind to membrane-associated fibronectin.

Cell Surface Coat Constituents

Most cells are coated with a layer of glycoproteins that are bound to the cell surface by noncovalent bonds with either a peripheral or an integral constituent. The coat can be of variable thickness and chemical complexity. It may take the form of a matrix, as in mature cartilage where it is composed of a collagen–proteoglycan-hyaluronic acid complex that is associated with the cell membrane via fibronectin (Aplin and Hughes, 1982). Other cells, such as mucosal cells of the respiratory and urinary tracts, are coated with a layer rich in highly sialylated, high molecular weight glycoproteins known as mucins.

The different types of coats, produced by specialized cells at specific sites in the host, are well adapted to their specific physiological functions. Thus, the matrix of the cartilage is able to withstand stress and to bear considerable loads, whereas the mucins form viscous solutions that perform protective, lubricant, and transport roles in the appropriate tissues. Regardless of the thickness or composition of the cell coat, for a bacterium to colonize tissues, it must either bind to or penetrate through the cell coat.

Cell Surface Receptors

The surface constituents of animal cells are important for controlling many cellular activities, such as development, differentiation, intracellular communication, and recognition or clearance of foreign particles, as well as the activation of cells and the triggering of defined biological responses. When a particular cell membrane constituent mediates one or more of such activities by specifically combining with an extracellular biological agent (or ligand), it becomes known as a receptor for that agent. Two criteria characterize an animal cell membrane receptor: (1) it should be able to interact specifically with a ligand; and (2) this interaction should lead to a biologically relevant response (Cuatrecasas, 1974; Schulster and Levitski, 1980). The interaction step follows general rules that are common to all receptor molecules and their ligands, and is analogous to the interaction between antibody and antigen, lectin and specific sugar or saccharide, or enzyme and substrate. In contrast, the biological consequences are numerous, depending on the type of cell, the nature of the ligand, and the receptor.

The active chemical group of the receptor to which the ligand binds is referred to as the attachment or recognition site or the determinant group. It is a structure consisting, as a rule, of either carbohydrate residues or amino acid residues. Occasionally, both the peptide backbone and the carbohydrate moiety of a glycoprotein may serve as a determinant attachment site (Duk et al., 1982; Tollefsen and Kornfeld, 1983). The carbohydrate attachment sites are always glycoprotein and/or glycolipid molecules. Table 3–2 give examples of receptor molecules and attachment sites or different ligands. As can be seen, the same cell membrane constituents may possess recognition sites for numerous types of agents, and therefore serve as a receptor for more than one ligand. Conversely, identical attachment sites for a particular ligand may be found on different constituents. Some distinct receptors on different types of animal cells for bacterial adhesins share common determinants that bind the adhesin. These receptors are referred to as isoreceptors (Chapter 1). This is particularly common in Gram-negative bacteria which bind to animal tissues and cause hemagglutination.

Of the two types of attachment sites on receptor molecules, i.e., saccharides and amino acids, the former are by far the more versatile. This is because of the enormous numbers of specific structures that can be formed from relatively few monosaccharide units, compared with the numbers of peptides that can be formed from the same number of amino acids (Table 3–3). Several reviews have appeared during the past few years on cell surface saccharides as recognition sites in biological interactions (Yamakawa and Nagai, 1978; Gahmberg, 1981; Hakomori, 1981; Sharon, 1981; Sharon and Lis, 1989; Fishman, 1982). It is not surprising, therefore, that glycoconjugates commonly serve as receptors for numerous bacterial adhesins (Mirelman and Ofek, 1986).

Table 3–2. Chemical nature of representative receptors isolated from animal cell membranes

Biological agent (ligand)	Source	Receptor: Chemical Nature	Receptor: Recognition Site	Reference
Acetylcholine	Postsynaptic cells	Glycoprotein (mol wt 200,000)	A 40,000-mol wt subunit	Prives (1980)
Cholera toxin	Several types of cells	Glycoprotein (GM1)	NeuNAcGalGalNAc	Critchley et al. (1982); Yamakawa and Nagai (1978)
Concanavalin A	Human erythrocytes	Glycoprotein (band 3)	Mannose residues	Findlay (1974)
Blood group A antibody	Human erythrocytes	Glycoprotein (band 3) Macroglycolipids	*N*-Acetyl-D-galactosamine	Finne (1980) Koscielak et al. (1976)
Asialoglycoprotein (e.g., asialofetuin)	Hepatic cells	Glycoprotein (mol wt 45,000)	Galactose specific binding site	Ashwell and Harford (1982)

Table 3–3. Comparison of the number of possible isomeric peptides and oligosaccharides (pyranose ring only)

		Number of Isomers	
Monomer composition	Product	Peptides	Saccharide
X_2	Dimer	1	11
X_3	Trimer	1	176
XYZ	Trimer	6	1056

From Sharon (1975); calculations by Clamp (1974).

Expression of Receptors on Cell Surfaces

On living tissues, there is a dynamic change in the production and expression of any particular cell membrane constituent as a function of age and the physiological state of the cell. Changes are also bound to occur in cells exposed to the actions of drugs, some of which may affect the biosynthesis and expression of cell membrane constituents. Such changes have been best documented in carbohydrate residues of glycoproteins and glycolipids (Gahmberg, 1977; Fukuda and Fukuda, 1978; Critchley, 1979; Hokomori, 1981; Gahmberg and Andersson, 1982), largely due to the availability of specific lectin and glycosidase probes (Flowers and Sharon, 1979; Roth, 1980; Sharon and Lis, 1989). Moreover, the proximity of other cell surface constituents may interfere with the ability of receptors to bind the ligand. When comparing the binding of a given ligand to cells under varying physiological conditions, it is therefore not always possible to determine whether any difference observed is due to a change in the numbers of receptor molecules, alterations in the recognition site, or interference by other surface constituents. Receptors for bacterial adhesins are probably no exception. Unfortunately, factors that govern the presentation and distribution of receptors that bind bacteria have not been studied in depth.

Characterization of Cell Surface Receptors

Identification and characterization of a cell surface constituent as a receptor for a biological agent requires (1) characterization of the nature of the attachment site(s); (2) isolation of the receptor molecule that contains the attachment site(s) from the cell membrane and its subsequent physicochemical characterization; (3) demonstration that it is available to the agent on the cell surface; and (4) proof that it mediates the biological activity of the agent. The procedures available for approaching some of the above problems are given in Chapter 2.

Table 3–4. Receptor molecules and attachment sites for bacteria on animal cells

Bacteria	Receptor molecule	Location in membrane	Attachment site	Reference
E. coli (pyelonephritogenic)	Glycolipid	Integral	Galα1,4-Gal	Kallenius et al. (1981); Leffler and Svanborg-Eden (1981)
Mycoplasma	Glycophorin	Integral	Sialic acid	Feldner et al. (1979)
Streptococci	Fibronectin	Peripheral	Fatty acid binding region	Beachey et al. (1983)
Shigellae	Colonic mucus	Cell coat	L-Fuc/D-Glc-specific binding site	Izhar et al. (1982)

Animal Cell Surface Receptors in Bacterial Adhesion

Animal cell receptors of bacterial adhesins fulfill the criteria of a biological receptor. They exhibit specific binding followed by physiologically relevant responses, in some cases manifested as infections. Receptors for bacterial adhesins are found in all three classes of membrane constituents, namely, integral, peripheral, and cell coat components (Table 3–4). Chemically, they may be either proteins, glycoproteins, or glycolipids. The chapters that follow review the knowledge accumulated in identifying the bacterial attachment sites and the receptor molecules.

References

Aplin, J.D. and R.C. Hughes. 1982. Complex carbohydrates of the intracellular matrix: structures, interactions and biological roles. *Biochim. Biophys. Acta* **694**:375–418.

Ashwell, G. and J. Harford. 1982. Carbohydrate-specific receptors of the liver. *Annu. Rev. Biochem.* **51**:531–554.

Beachey, E.H., W.A. Simpson, I. Ofek, D.K. Hasty, J.B. Dale, and E. Whitnack. 1983. Attachment of *Streptococcus pyogenes* to mammalian cells. *Rev. Infect. Dis.* **5**:5670–5677.

Callies, R., G. Schwarzmann, K. Radsak, R. Siegert, and H. Wiegandt. 1977. Characterization of the cellular binding of exogenous gangliosides. *Eur. J. Biochem.* **80**:425–423.

Cherry, R.J. 1979. Rotational and lateral diffusion of membrane proteins. *Biochim. Biophys. Acta* **559**:289–327.

Clamp, J. 1974. Analysis of glycoproteins. *Biochem. Soc. Symp.* **40**:3–16.

Critchley, D.R. 1979. Glycolipids as membrane receptors important in growth regulation.

In: Hynes, R.O. (ed.), *Surfaces of Normal and Malignant Cells*. John Wiley & Sons, New York, pp. 63–101.

Critchley, D.R., C.H. Streuli, S. Kellie, S. Ansell, and B. Patel. 1982. Characterization of the cholera toxin receptor on Balb/C3T3 cells as a ganglioside similar to, or identical with, ganglioside GM_1: no evidence for galactoproteins with receptor activity. *Biochem. J.* **204**:209–219.

Cuatrecasas. P. 1974. Membrane receptors. *Annu. Rev. Biochem.* **43**:169–214.

Duk, M., E. Lisowska, M. Kordowicz, and K. Wasniowska. 1982. Studies on the specificity of the binding site of *Vicia graminea* anti-N lectin. *Eur. J. Biochem.* **123**:105–112.

Edidin, M. 1974. Rotational and translational diffusion in membranes. *Annu. Rev. Biophys. Bioeng.* **8**:165–193.

Eytan, G.D. 1982. Use of liposomes for reconstruction of biological function. *Biochim. Biophys. Acta* **694**:185–202.

Feldner, J., W. Bredt, and I. Kahane. 1979. Adherence of erythrocytes to *Mycoplasma pneumoniae*. *Infect. Immun.* **30**:554–561.

Findlay, J.B.C. 1974. The receptor proteins for concanavalin A and *Lens culinaris* phytohemagglutinin in the membrane of the human erythrocyte. *J. Biol. Chem.* **249**:4398–4403.

Finean, J.B., R. Coleman, and R.H. Mitchell. 1984. *Membranes and Their Cellular Function,* 3rd ed. Blackwell, Oxford.

Finne, J. 1980. Identification of the blood-group ABO-active glycoprotein components of human erythrocyte membrane. *Eur. J. Biochem.* **104**:181–189.

Fishman, P.H. 1982. Role of membrane gangliosides in the binding and action of bacterial toxins. *J. Membr. Biol.* **69**:85–97.

Flowers, H.M. and N. Sharon. 1979. Glycosidases-properties and application to the study of complex carbohydrates and cell surfaces. *Adv. Enzymol.* **48**:29–95.

Fukuda, M. and M.N. Fukuda. 1978. Changes in cell surface glycoproteins and carbohydrate structures during the development and differentiation of human erythroid cells. *J. Supramol. Struct.* **8**:313–324.

Gahmberg, C.G. 1977. Cell surface proteins: changes during cell growth and malignant transformation. In: Poste, G. and G.L. Nicholson (eds.), *Cell Surface Reviews*. North-Holland, Amsterdam, pp. 371–421.

Gahmberg, C.G. 1981. Membrane glycoproteins and glycolipids: structure, localization and function of carbohydrates. In: Finean, J.B. and R.H. Mitchell (eds.), *Membrane Structure*. Elsevier/North-Holland, Amsterdam, pp. 127–160.

Gahmberg, C.G. and L.C. Anderson. 1982. Surface glycoproteins of malignant cells. *Biochim. Biophys. Acta* **651**:65–83.

Garoff, H. 1979. Structure and assembly of the Semliki Forest virus membrane. *Biochem. Soc. Trans.* **7**:301–306.

Geiger, B. 1983. Membrane cytoskeleton interactions. *Biochim. Biophys. Acta* **737**:305–341.

Hakomori, S. 1981. Glycosphingolipids in cellular interaction, differentiation and oncogenesis. *Annu. Rev. Biochem.* **50**:733–764.

Hynes, R.O. and K.M. Yamada. 1982. Fibronectins: multifunctional molecular glycoproteins. *J. Cell Biol.* **95**:369–377.

Izhar, M., Y. Nuchamowitz, and D. Mirelman. 1982. Adherence of *Shigella flexneri* to guinea pig intestinal cells is mediated by a mucosal adhesin. *Infect. Immun.* **35**:1110–1118.

Kallenius, G., S.B. Svensson, R. Mollby, B. Cedergren, H. Hultberg, and J. Winberg. 1981. Structure of carbohydrate part of receptor on human uroepithelial cells for pyelonephritogenic *Escherichia coli. Lancet* **2**:604–606.

Koscielak, J., H. Miller-Podraza, R. Krauze, and A. Piasek. 1976. Isolation and characterization of poly(glycosyl) ceramides (megaloglycolipids) with A,H and I blood-group activities. *Eur. J. Biochem.* **71**:9–18.

Leffler, H. and C. Svanborg-Eden. 1981. Glycolipid receptors for uropathogenic *Escherichia coli* on human erythrocytes and uroepithelial cells. *Infect. Immun.* **34**:920–929.

Lichtenberg, D., R.J. Robson, and E.A. Dennis. 1983. Solubilization of phospholipids by detergents: structural and kinetic aspects. *Biochim. Biophys. Acta* **737**:285–304.

Lodish, H.F., W.A. Braell, A.L. Schwartz, G.J.A.M. Strous, and A. Zilberstein. 1981. Synthesis and assembly of membrane and organelle proteins. *Int. Rev. Cytol. Suppl.* **12**:247–307.

Lotan, R. and G.L. Nicolson. 1981. Plasma membrane of eukaryotes, In: Schwartz, L.M. and M.M. Azar (eds.), *Advanced Cell Biology*. Van Nostrand-Reinhold, Princeton, NJ, pp. 129–154.

Marchesi, V.T., H. Furthmayr, and M. Tomita. 1976. The red cell membrane. *Annu. Rev. Biochem.* **45**:667–698.

Marinetti, G.V. and R.C. Crain. 1978. Topology of amino-phospholipids in the red-cell membrane. *J. Supramol. Struct.* **8**:191–213.

Mirelman, D. and I. Ofek. 1986. Introduction to microbial lectins and agglutinins. In: Mirelman, D. (ed.), *Microbial Lectins and Agglutinins*. John Wiley & Sons, New York, pp. 1–19.

Nicolson, G.L. 1976. Trans-membrane control of the receptors on normal and tumor cells. I. Cytoplasmic influence of cell surface components. *Biochim. Biophys. Acta* **457**:57–108.

Nicolson, G.L. 1979. Topographic display of cell surface components and their role in trans-membrane signaling. *Curr. Top. Dev. Biol.* **3**:305–338.

Ofek, I., H. Lis, and N. Sharon. 1985. Animal cell surface membranes. In: Savage, D.C. and M. Fletcher (eds.), *Bacterial Adhesion: Mechanisms and Physiological Significance,* Plenum Press, New York, pp. 71–88.

Prives, J.M. 1980. Nicotinic acetylcholine receptors. In: Schulster, D. and A. Levitski (eds.), *Cellular Receptors for Hormones and Neurotransmitters*. John Wiley & Sons, New York, pp. 331–351.

Raff, M.C. and S. dePetris. 1973. Movement of lymphocyte surface antigens and receptors: the fluid nature of the lymphocyte plasma membrane and its immunological significance. *Fed. Proc.* **32**:48–54.

Roth, J. 1980. The use of lectins as probes for carbohydrates-cytochemical techniques and their application in studies on cell surface dynamics. *Acta Histochem. Suppl.* **22**:113–121.

Ruoslahti, E. 1988. Fibronectin and its receptors. *Annu. Rev. Biochem.* **57**:375–413.

Ruoslahti, E. 1991. Integrins. *J. Clin. Invest.* **87**:1–5.

Ruoslahti, E., M. Pierschbacher, E.G. Hayman, and E. Engvall. 1982. Fibronectin: a molecule with remarkable structural and functional diversity. *Trends Biochem. Sci.* **7**:188–190.

Schulster, D. and A. Levitski (eds.). 1980. *Cellular Receptors for Hormones and Neurotransmitters*. John Wiley & Sons, New York.

Sedlacek, H.H., J. Stark, F.R. Seiler, W. Ziegler, and H. Wiegandt. 1976. Cholera toxin induces redistribution of sialoglycolipid receptor at the lymphocyte membrane. *FEBS Lett.* **61**:272–276.

Sharon, N. 1975. *Complex Carbohydrates: Their Chemistry, Biosynthesis, and Functions*. Addison-Wesley, Reading, MA.

Sharon, N. 1981. Glycoproteins in membranes. In: Balian, R., M. Chabre, and P.F. Devaux (eds.), *Membranes and Intercellular Communications*. North-Holland, Amsterdam, pp. 117–182.

Sharon, N. and H. Lis 1981. Glycoproteins: research booming on long-ignored, ubiquitous compounds. *Chem. Engr. News* **59**:21–24.

Sharon, N. and H. Lis. 1982. Glycoproteins. In: Neurath, H., and R.L. Hill (eds.), *The Proteins, Vol. V,* 3rd ed. Academic Press, New York, pp. 1–144.

Sharon, N. and H. Lis. 1989. Lectins as cell recognition molecules. *Science* **246**:227–234.

Sim, E. 1982. *Membrane Biochemistry*. Chapman and Hall, London.

Singer, S.J. 1974. The molecular organization of membranes. *Annu. Rev. Biochem.* **43**:805–833.

Singer, S.J. 1981. The cell membrane. In: Balian, R., M. Chabre, and P.F. Devaux (eds.), *Membranes and Intercellular Communication*. North-Holland, Amsterdam, pp. 1–16.

Singer, S.J. and G.L. Nicolson. 1972. The fluid mosaic model of cell membranes. *Science* **175**:710–731.

Steck, T.L. 1978. Band 3 protein of the human red cell membrane: a review. *J. Supramol. Struct.* **8**:311–324.

Tanner, M.J.A. 1978. Erythrocyte glycoproteins. *Curr. Top. Membr. Transp.* **11**:279–325.

Tollefsen, S.E. and R. Kornfeld. 1983. The B_4 lectin from *Vicia villosa* interacts with *N*-acetylgalactosamine residues linked to serine or threonine residues in cell surface glycoproteins. *J. Biol. Chem.* **258**:5172–5176.

Wicken, A.J. and K.W. Knox. 1981. Composition and properties of amphiphiles. In: Shockman, G.D. and A.J. Wicken (eds.), *Chemistry and Biological Activities of Bacterial Surface Amphiphiles*. Academic Press, New York, pp. 1–7.

Wiegandt, H., S. Kanda, K. Inoue, K. Utsumi, and S. Nojima. 1981. Studies on the cell association of exogenous glycolipids. *Adv. Exp. Med. Biol.* **152**:343–352.

Yamakawa, T. and Y. Nagai. 1978. Glycolipids at the cell surface and their biological functions. *Trends Biochem. Sci.* **3**:128–131.

4

Relationship Between Bacterial Cell Surfaces and Adhesins

The central dogma of bacterial adhesion requires that the adhesin(s) function from the bacterial surface. In most cases, the adhesins are assembled on the surface, but in a few cases, the adhesins are initially secreted in the soluble form and then associate with the bacterial surface (Tuomanen, 1986; Baker et al., 1991; Wentworth et al., 1991). In either case, the adhesin must dock or anchor on the bacterial surface before it can participate in adhesive processes. Because adhesion is a property of most bacteria, especially of tissue-colonizing bacteria, it follows that evolution has selected specific structures that function as adhesins or onto which adhesins can assemble. In this chapter, a concise review of bacterial surface structures, with special emphasis on macromolecules involved in adhesion, is given. More comprehensive discussions of the bacterial cell surface are provided elsewhere (Rogers et al., 1980; Nikaido and Vaara, 1985; Krell and Beveridge, 1987; Doyle and Sonnenfeld, 1989; Handley, 1990; Hancock, 1991; Irvin, 1990; Gilbert et al., 1991).

Figures 4–1 and 4-2 depict idealized surface structures of Gram-positive and Gram-negative bacteria. There are a number of common structures and surface components in both cell types. However, each cell type has some distinguishing characteristics. Although certain features of bacterial cell surfaces remain constant, it must be kept in mind that the surface composition may vary with age and medium composition, and the presence of antimicrobial agents, all of which may influence directly or indirectly adhesin function. Both Gram-positive and Gram-negative bacteria possess peptidoglycan, the stress-bearing and shape-determining structure. In Gram-positive bacteria, the peptidoglycan is relatively thick, of the order 20–40 nm, whereas in the Gram-negative bacterium, the peptidoglycan is relatively thin, of the order 5–10 nm. The peptidoglycan of the Gram-negative bacterium is surrounded by a structure called the outer membrane, whereas peptidoglycan of the Gram-positive cell is generally exposed to the environment. The cell wall of the Gram-positive bacterium is composed of

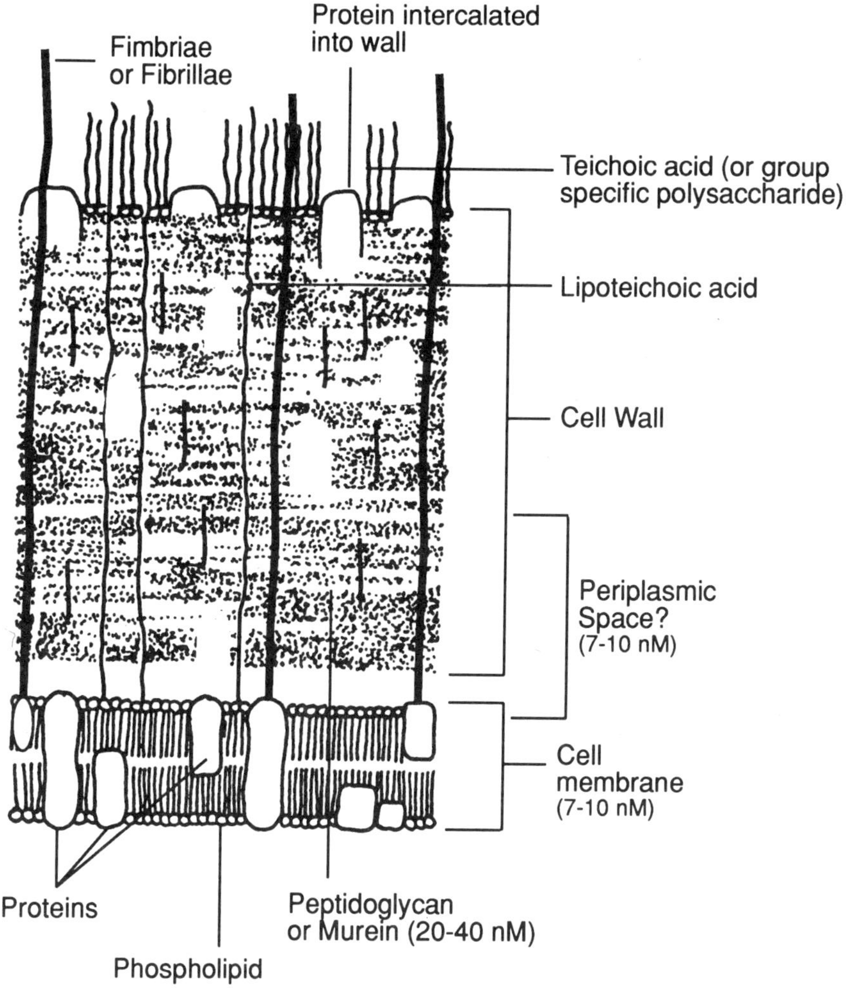

Figure 4–1. Idealized representation of the cell surface of a Gram-positive bacterium. Adhesion function may be ascribed to virtually any surface structure or macromolecule. In Gram-positive cells, the most prominent adhesins appear to be fibrils, fimbriae and wall-associated, amorphous-appearing (amorphin) proteins. (Modified from DiRienzo *et al.*, 1978.)

peptidoglycan and its auxiliary polymers, teichoic acid or teichuronic acid. The cell wall of the Gram-negative bacterium is classically regarded as peptidoglycan plus outer membrane, although some maintain that the outer membrane should not be classified as a wall component. Peptidoglycan is sometimes referred to as "murein." The entire insoluble network surrounding the cytoplasmic membrane of bacteria is called the sacculus. In Gram-positive bacteria, the osmotic pressure inside the cell is equivalent to approximately 20–25 atmospheres, roughly equiva-

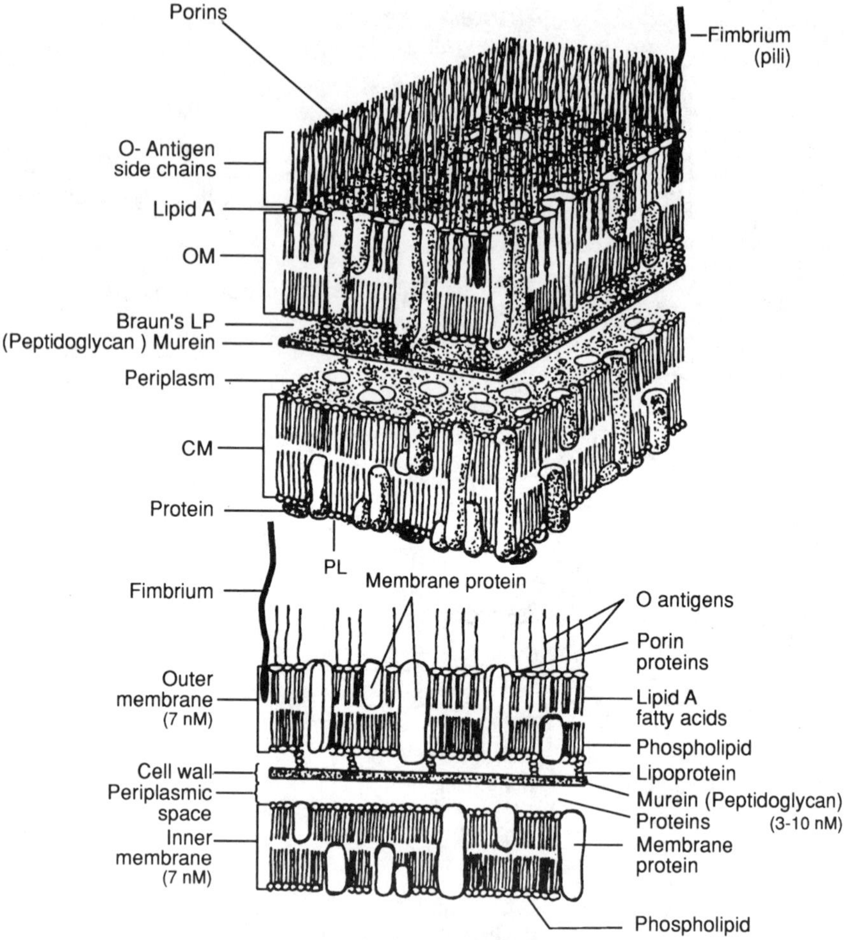

Figure 4–2. Cross-sectional representation of the surface of an idealized Gram-negative bacterium. The Gram-negative bacterial adhesins are most often found to be on fimbriae, although fibrils, capsule, slimectin, outer membrane and flagella have been reported to possess adhesin characteristics. Fimbriae may be flexible, rigid or curled (curli). The fimbrial adhesin may be located at the very tip of the structure or may be found all along the filamentous appendage. Outer membrane adhesins seem to be integral components of the OM structure. (Modified from DiRienzo *et al.*, 1978.)

lent to a mole per liter of colligative molecules. The high turgor of Gram-positive cells in most environments requires that the cells withstand the internal pressure without bursting. The thick cell wall is able to accommodate the turgor pressure of the bacteria (Koch et al., 1982). Gram-negative bacteria have much lower internal osmolarities and, therefore, do not require the relatively thick peptidoglycan.

There are four major cell "compartments" that serve to anchor adhesins onto the bacterial surface. Three compartments, cytoplasmic membrane, peptidoglycan, and the S layer, are found in both Gram-positive and Gram-negative bacteria, and one compartment, the outer membrane, is found only in Gram-negative bacteria. Table 4–1 lists examples of adhesins associated with various cell-surface compartments. A common strategy evolved by bacteria is to scaffold an adhesin onto some other structures on the bacterial surface. The adhesin then assumes a "bridging" position between the bacterium and its substratum. Strategies for adhesion of pathogenic bacteria to surfaces are similar for both Gram-positive and Gram-negative cells. Both cell types produce filamentous appendages called fimbriae and fibrillae, which serve as adhesins or act to position adhesins. The fimbriae or fibrillae seem to be anchored in the sacculus and/or cytoplasmic membrane in Gram-positive, whereas in the Gram-negative cells, the fimbriae are anchored on the outer leaflet of the outer membrane. In some cases of Gram-

Table 4–1. Selected examples of adhesive polymers and structures associated with cell-surface compartments

Cell-surface compartment	Bacteria	Adhesin	References
Outer membrane	*Neisseria gonorrhoeae*	P11, PIa Fimbriae	Layh-Schmitt et al., 1989; Swanson et al., 1975; Swanson et al., 1988
	Yersinia enterocolitica	YP01 protein	Heesemann and Gruter 1987
	Escherichia coli	LPS	Perry and Ofek, 1984; Wright et al., 1989
Cytoplasmic membrane	*Escherichia coli*	Flagella-MS	Eshdat et al., 1981b
	Vibrio cholerae	Flagella	
	Streptococcus pyogenes	LTA-M protein complex	Ofek et al., 1982; Fischetti, 1991
Peptidoglycan	*Staphylococcus aureus*	Cell-wall TA	Aly and Levit, 1987; Aly et al., 1980
	Streptococcus cricetus and *S. sobrinus*	Glucan binding protein	Drake et al., 1988 a, b
S-Layer	*Aeromonas* spp.	Protein?	Janda and Duffy, 1988
Capsule	*Escherichia coli*	NFA I protein	Jann and Hoschutzky, 1990
	Klebsiella pneumoniae	Polysaccharide	Athamna et al., 1991
Slimectin	*Staphylococcus epidermidis*	Polysaccharide	Christensen et al., 1982; Hogt et al., 1983
	Pseudomonas aeruginosa	Polysaccharide	Ramphal et al., 1987

MS, mannose specific adhesin; TA, teichoic acid, LPS, lipopolysaccharide, LTA, lipoteichoic acid.

positive cells there may be nonfimbrial or fibrillar protein molecules intercalated into the peptidoglycan matrix.

Duguid et al. (1955) observed thread-like appendages in *Escherichia coli* and coined the word "fimbriae," from the Latin *fiber* (meaning thread). In 1959, Brinton (1959) suggested the name "pili" (from the Latin to describe hair-like structures) for nonflagellar surface appendages in *E. coli*. Both words remain in common use and frequently mean the same thing. The word fimbriae seems to be more widely accepted at the present time. The word pilus (singular of pili) is now used most commonly to describe the conjugative filaments of many Gram-negative bacteria. In this book the word fimbriae is used to describe the thread-like surface appendages required for adhesion of some bacteria. The word pili remains firmly fixed in the literature to describe adhesins of neisseriae. Fibrillae are also surface extensions on bacteria, but they have some morphological features distinguishing them from fimbriae (Handley, 1990).

Some of the adhesive structures and polymers are dispensable for the bacteria under normal growth conditions in the laboratory (e.g., fimbriae). Others are dispensable only under complex growth conditions (e.g., outer membrane), whereas other may not be dispensable under any growth conditions (e.g., cytoplasmic membrane lipoteichoic acid). Structures and molecules belonging to the first category are subject to modification under various conditions that affect cell wall dynamics and can usually be manipulated by genetic techniques, resulting in mutants with altered adhesive capacities. Each of the compartments listed in Table 4–1 contains, in addition to the adhesive polymers or structures, other molecules or structures that may affect the function of the adhesins. Certain structures, such as capsules, may be adhesive and antiadhesive. Some molecules associated with membranes and/or the sacculus also seem to possess multiple functions. For example, lipoteichoic acid, a known adhesin for many Gram-positive pathogens, also may affect cell wall turnover (Doyle et al., 1988). A brief description of the various cell wall compartments and their association with adhesins and adhesin-associated structures of representative bacteria is given in the following paragraphs.

Outer Membrane

Outer membranes (OM) of Gram-negative bacteria cover the peptidoglycan layer. The OM is a phospholipid–lipopolysaccharide–(LPS)protein bilayer with typical fluid mosaic characteristics (Nikaido and Vaara, 1985). The phospholipid forms most of the inner leaflet, whereas the polysaccharide or LPS forms much of the outer bilayer leaflets. The polysaccharide moiety of the LPS extends away from the outer membrane and is a prominent antigen. In addition to LPS, the OM contains intercalated proteins, all of which contain one or more hydrophobic regions responsible for anchoring the molecules to the cell membrane. Most of the OM proteins extend across the entire thickness of the bilayer so that they are exposed to both the periplasmic space and external environment. Many of the classic OM proteins, known as OMPs, are porins that serve as channels to provide a selective permeability to the membrane (Nikaido, 1985). Some of the other functions of the outer membrane include acting as a selective permeability barrier and binding certain bacteriophages (the binding of bacteriophages is one means leading to exchange of genetic information). The outer membrane is an excellent compartment for anchoring and positioning adhesins to the outer surface. Three types of outer membrane adhesins may be distinguished: LPS, fimbrial structures, and outer membrane proteins. The anchoring of the outer membrane adhesin proteins is via a hydrophobic stretch and that of LPS via the lipid A portion.

Most lipopolysaccharides from wild-type bacteria have very long polysaccharide structures consisting of repeating units of tri-or tetrasaccharides. Mutations in polysaccharide synthesis may lead to truncated polysaccharide molecules giving rise to so-called rough strains. LPS molecules have been suggested to function as adhesins that mediate the binding of *Campylobacter jejuni* to epithelial cells (McSweegan and Walker, 1986). Preparations of LPS were able to prevent the adhesion of *C. jejuni* to epithelial cells or to mucus. The inhibitory effect may be due to hydrophobic interaction between the lipid A moiety of the soluble LPS and receptor molecules. The LPS has also been shown to mediate interaction of bacteria with phagocytic cells (Perry and Ofek, 1984; Wright et al., 1989).

By far the most common way of positioning adhesins to bacterial surfaces of Gram-negative bacteria is via surface appendages termed fimbriae or pili, structures that are invariably anchored in the outer membrane (Ottow, 1975; Oudega and DeGraaf, 1988; DeGraaf, 1990; Hacker, 1990; Jann and Hoschützky, 1990; Meyer, 1990; Paranchych, 1990). There have been numerous attempts to devise rational classification schema for bacterial fimbriae. All have met with limited success. Duguid et al. (1966) suggested that the fimbriae of Gram-negative rods can be classified according to types. A particular type reflected a particular morphology and hemagglutination pattern. Later, this "type" classification was revised to include newly discovered fimbriae that did not easily fit into the previous scheme (Ottow, 1975; Duguid and Old, 1980; Pearce and Buchanan, 1980). Type 1 refers to fimbriae that are rigid and exhibit mannose-sensitive hemagglutination; type 2 is serologically related to type 1 but does not cause hemagglutination; type 3 refers to mannose-resistant, flexible fimbriae common in *Enterobacteriaceae,* save *E. coli* and *Shigella* spp.; type 4 are fimbriae possessing *N*-methylphenylalanine in the amino terminus region of the major subunit; and type 5 are mannose sensitive, thinner than type 1, and usually few in numbers per cell. Recently, Orskov and Orskov (1990) used morphological and serological markers to devise a new scheme of fimbrial classification. Fimbrial Group I are those with rigid (approximately 7 nm in diameter) structures in which the major fimbrial unit is distinct from the subunit mediating the adhesive function (see

Chapter 9). Fimbrial Group II comprise thin (<7 nm in diameter) and flexible structures that frequently appear capsular in the electron microscope, and usually the major subunit also functions as an adhesin. Serotyping within Groups I and II further classifies individual fimbriae. Although this classification is satisfactory for a number of bacteria, many fimbriae do not conform to the scheme. Some of these are the fimbriae that belong to the type 4 category discussed earlier and all of the Gram-positive fimbrial adhesins. At present, it does not appear that an ideal classification scheme is available. In Table 4–2, rather than classifying the fimbriae, some important physical and chemical characteristics of selected Gram-positive and Gram-negative fimbriae are given, and wherever possible, the receptor specificities are shown as well. Figures 4–3–4–6 show fimbriae of several types from Gram-negative bacteria. As far as is known, the fimbriae of all Gram-negative bacteria are assembled on the outer leaflet of the outer membrane.

The fimbrial adhesins are heteropolymeric structures (Table 4–2). The bulk of the fimbriae is composed of about 1000 identical repeating monomers known as the major subunits and 1–10 of different minor subunits. The major subunit or one of the minor monomeric subunits may serve as the adhesin, whereas other subunits, usually the largest, are located at the base of the fimbrial structures, intercalated into the membrane and serve to assemble the anchor for the fimbrial structure onto the outer membrane (see Chapter 9). Electron microscopy has revealed that fimbrial adhesive structures can vary in diameter and length. The length of fimbriae can be up to 10 μm, although most fimbrial preparations have a distribution of length ranging from 0.5 to 10 μm. Fimbriae can be detached and purified by shearing in a Waring blender and recovering the insoluble structures from the supernatant following centrifugation to remove cells (Jann and Hoschützky, 1990). The crude fimbrial preparation can then be centrifuged in sucrose or cesium chloride density gradients to yield the pure forms. They can then be solubilized by concentrated chaotropic agents. Removal of the chaotropes by dialysis frequently results in reassembly of the fimbrial structures. Purified fimbriae usually retain their ability to bind to their regular substrata (Korhonen et al., 1990). Because most fimbriae are thin and fragile, they are fragmented during preparation procedures. The diameters of fimbriae are much more uniform or constant for a given bacterial clone regardless of preparative procedures. Based on diameter there are two major categories of fimbriae. One category represents a diameter of 7–11 nm, whereas the other category is represented by fimbriae with diameters of 1–4 nm. The fimbriae are classified as either rigid or flexible, depending on the ultrastructural appearance (Table 4–2 and Figures 4–3 and 4-4). Because some fimbrial adhesins are very thin, the distinction between fimbrial and nonfimbrial adhesin may be due to the limitation of the electron microscopic resolution.

A number of outer membrane proteins have been shown to serve as adhesins (Table 4–1). In *Yersinia enterocolitica* and *Y. pseudotuberculosis* insertional mutagenesis led to a loss of plasmid protein WaA from their outer membranes and the concomitant loss in the ability to adhere to rabbit intestinal tissues and to polystyrene (Paerregaard et al., 1991). The results clearly demonstrate a role for outer membrane protein in adhesion to animal cells. Outer membrane proteins, unrelated to fimbriae, have been implicated in the adhesion reactions of *N. gonorrhoeae* (Dekker et al., 1990), *Pseudomonas aeruginosa* (Ramphal et al., 1991), cytotoxin-producing *E. coli* (Sherman et al., 1991), *Campylobacter* spp. (Fauchere et al., 1989), and other bacteria. There are enough reports to suggest that adhesion of a variety of Gram-negative bacteria does not depend solely on fimbriae or fimbrial lectins (see Chapter 10). It seems likely that in some of these cases the outer membrane-associated proteins can extend their combining sites to such a distance to enable successful contact with their animal cell receptors. Sometimes an adhesin of the same receptor specificity may be associated with the outer membrane in the fimbrial form as well as in a nonfimbrial form. A typical example is the mannose-specific adhesin of *E. coli*. This adhesin was first identified as type 1 fimbriae (Duguid and Old, 1980). Eshdat et al. (1981a) showed that an outer membrane preparation of a nonfimbriated *E. coli* strain possessed mannose binding activity. It seems likely that the mannose-specific adhesin of the strain employed originated as a result of a faulty assembly process of the type 1 fimbriae. In support of this is the finding that a genetically manipulated gene cluster of the type 1 fimbrial adhesin results in a nonfimbriated strain causing mannose-specific hemagglutination (Hultgren et al., 1990).

Peptidoglycan

Peptidoglycan typically contains nonreducing GlcNAc residues. Many peptidoglycans also contain unsubstituted glucosamine and muramic acid (Rogers et al., 1980). These nonreducing residues could conceivably serve as ligands for animal cell surface proteins. There is at present no evidence for this, although soluble peptides and murein sacculus interact with lymphocyte subpopulations, possibly via a lectin-like substance on the lymphocyte surfaces (Dziarski, 1991). Otherwise, peptidoglycan is not known as such to act as an adhesin that mediates the binding of bacteria to animal cells. Its main function in adhesion is to support or form a matrix for adhesins. There is a great deal of heterogeneity in composition of peptidoglycan from various bacteria. Most of the heterogeneity arises in the peptide portion of the peptidoglycans. The peptides are very short sequences of three or four amino acids in both the D- and L-configurations. Crosslinking of the amino acids by a single amino acid or a short peptide also contributes to peptidoglycan heterogeneity (Rogers et al., 1980). In the Gram-negative bacterium, protein molecules, covalently attached to peptidoglycan, may extend into and partially stabilize the outer membrane. These proteins, frequently referred to as Braun lipoproteins (Braun, 1975), have no apparent role in adhesion. In

Table 4–2. Some characteristics of fimbriae and fibrils of selected bacterial pathogens

Bacterium	Type of surface structure	Rigid (R) or Flexible (F)	Diameter (nM)	Mol wt major subunit (KDa)	Receptor
Actinomyces naeslundii	Type 2 fimbriae	F	5	54	Galactosides
Actinomyces viscosus	Type 1 fimbriae	F	5	65	SHA
	Type 2 fimbriae	F	5	54	Galactosides
Bordetella pertussis	Fimbriae (FHA)	F	2	220	Galactose
Escherichia coli	Type 1 fimbriae	R	7	17–18	Oligomannosides
	Type P fimbriae	R	7	17–22	Galabiose
	Type 1C fimbriae	R	7	16	Unknown
	Type K88 fimbriae	F	2–3	23–28	Galactosides
	Type K99 fimbriae	F	4–5	18–19	Sialic acids
	Type 987P fimbriae	R	7	20	Galactosides
	Type F-41 fimbriae	F	3.2	29.5	GalNAc, mucins
	Type CFA/I fimbriae	F	3.2	15.1	Sialic acids
	Type CFA/III fimbriae	F		18	Complex
	Type S fimbriae	R	5–7	17	Sialic acidα2,3 Galα
	Curli fimbriae	F	2.4	17	Fibronectin
	Prs Fimbriae	R	5–7	18–19	GalNAcα1,3 αGalNAc
	075(Dr)fimbriae	F	2–4	21	Dr blood group
	M fimbriae	Ring-like	N/A	21	Glycophorin
	G fimbriae	R		19–20	GlcNAc
	CFA/II (CS1) fimbriae	R	5–7	16–17	Sialyl galactosides
	(CS2) fimbriae	R	3.2	15–16	Complex

	(CS3) fimbriae	F	3.2	14–16	Complex
	CFA/IV(CS4) fimbriae	R	7	22	Complex
	(CS5) fimbriae	R	7	15–16	Complex
	(CS6) fimbriae	F	7	24	Complex
	(AF/R1) fimbriae	F	5	19	Protein
	F17(FY) fimbriae	F	2–3	18–21	Complex
	F1845 fimbriae	F	2–3	14–15	Dr blood group
Klebsiella pneumoniae	Type 3 fimbriae	F	4–5	18–22	Complex
Moraxella bovis	Type 4 fimbriae	F	4–5	19–23	Complex
Neisseria gonorrhoeae	Type 4 fimbriae	F	6–7	18–24	Complex
Porphyromonas gingivalis	Fimbriae	F	4–6	43	Complex
Pseudomonas aeruginosa	Type 4 fimbriae	F	4–5	17–21	L-Fucose
Streptococcus anginosus	Fimbriae	F	3–4	U	Complex
Streptococcus salivarius	Fibrils	NA		320	Unknown
Streptococcus sanguis	Fimbriae	2			SHA
	Fibrils	NA		>300	SHA
Streptococcus pyogenes	Amorphin	NA		Variable	Fibronectin
Vibrio cholerae	Type 4 fimbriae	F	3–5	20–21	L-Fucose

Adapted from Baddour et al., 1989; Handley, 1990; Paranchych and Frost, 1988; Paranchych, 1990; Hacker, 1990; Meyer, 1990; Oudega and DeGraaf, 1988; Irvin, 1990; Olsen et al., 1989; Moon, 1990; Krogfelt, 1991; Krogfelt et al., 1990; Fulks et al., 1990; DeGraaf, 1990; DeGraaf and Mooi, 1986.

The "M" adhesin is classified as fimbriae by some, but not all, authors. "Amorphin" refers to amorphous appearing surface structures that cannot be classified as fibrils or fimbriae. Further details on the receptor specificities of fimbrial adhesins are found in Chapter 5.

Abbreviations: SHA saliva-coated hydroxylapatite; BEC, buccal epithelial cells; NA, not applicable; FHA, filamentous hemagglutinin; U, unknown. Complex means complex carbohydrates of unknown composition and sequence.

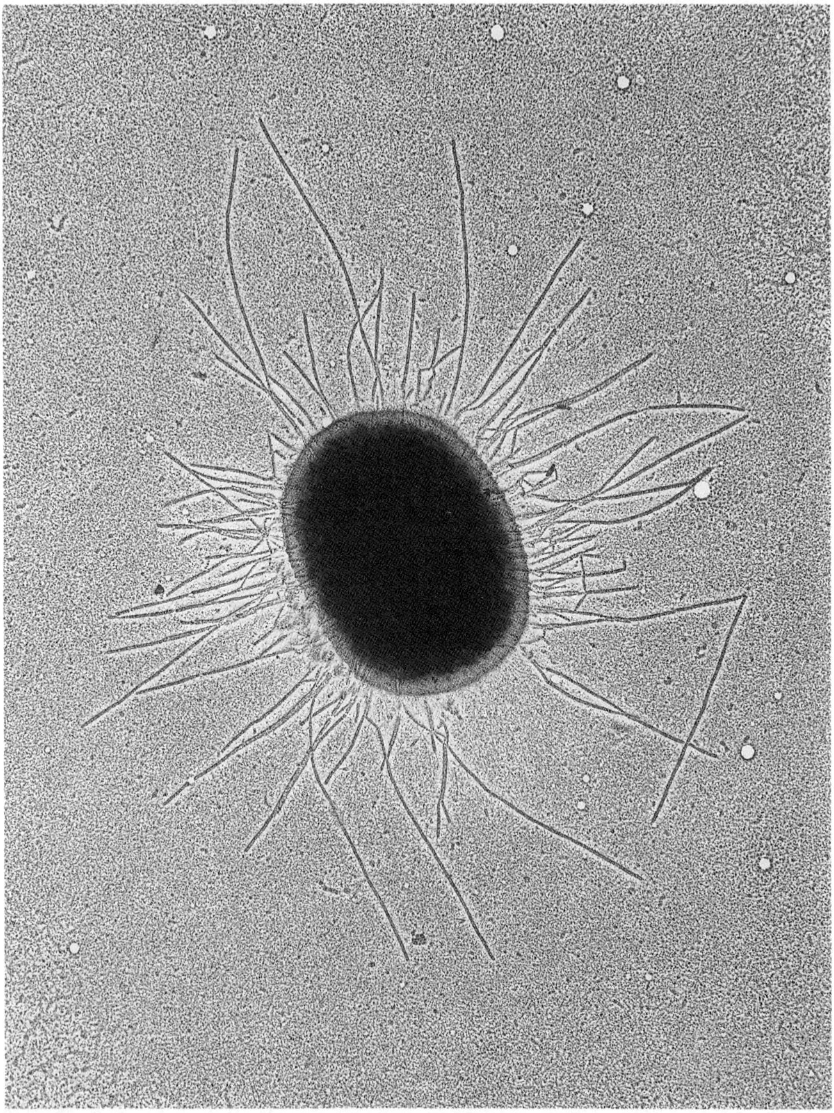

Figure 4–3. Type 1 fimbriae of *Escherichia coli*. In type 1 fimbriae, the mannose-binding site is at the tip of the assembled structure. When a fimbrium is broken by shearing, frequently a new mannose-binding site is exposed, suggesting that the lectin is located on subunits of protein at various sites along the length of the filament. Type 1 fimbriae are characterized by rigid structures. Micrograph courtesy Sam To (Univ. Pittsburgh) and H.W. Moon (USDA, Ames, IA). Reprinted courtesy of the authors and Springer-Verlag.

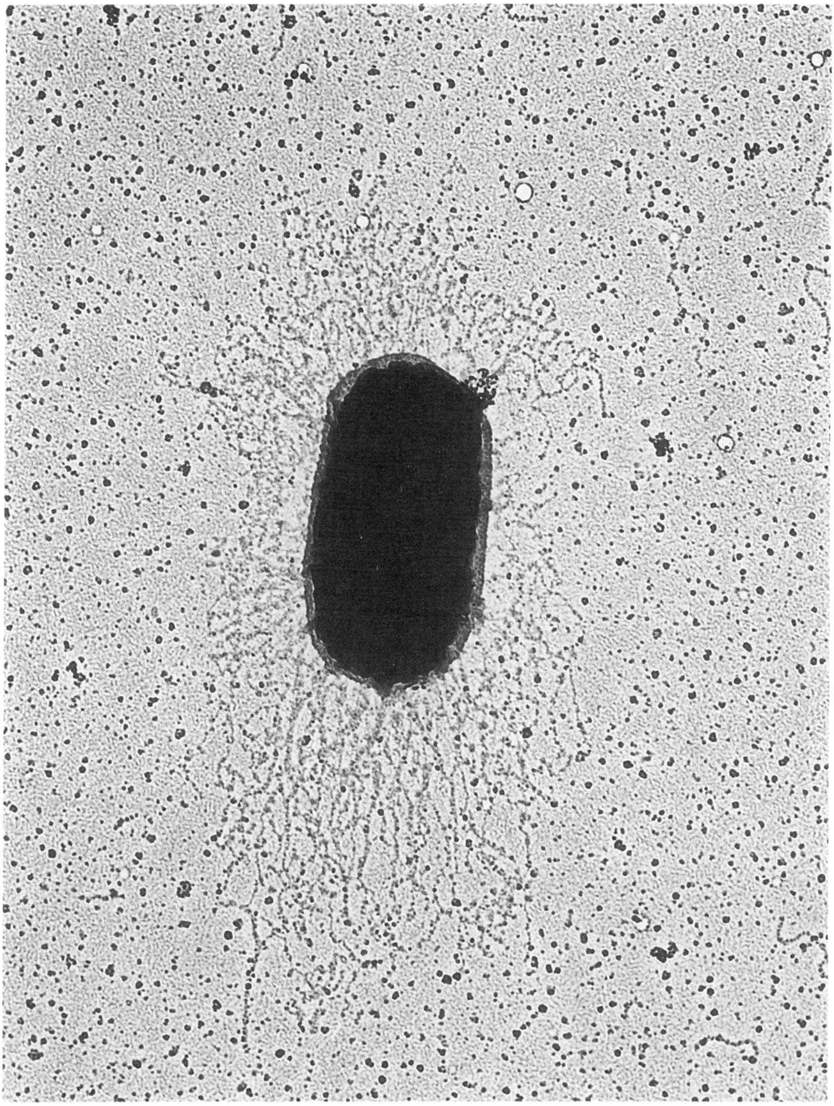

Figure 4–4. Type F–41 fimbriae of *Escherichia coli*. The F–41 fimbriae of *E. coli* are situated peritrichously around the cell. The fimbriae are thin and flexible. Micrograph courtesy of Sam To (Univ. Pittsburgh) and H.W. Moon (USDA, Ames, IA) and Springer-Verlag Publishers.

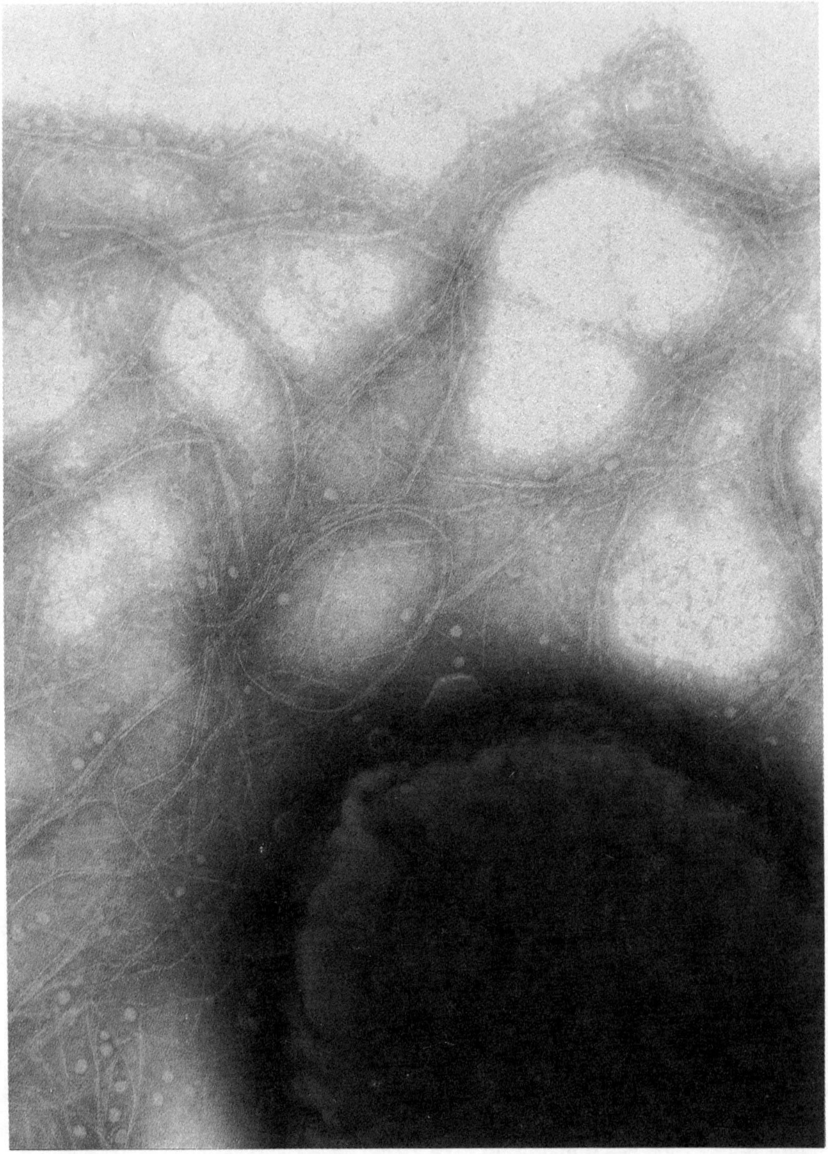

Figure 4–5. Curli fimbriae of *Escherichia coli*. The curli fimbriae are very thin and highly flexible. These fimbriae have an affinity for a domain on fibronectin. Micrograph courtesy of S. Normark (Olson et al., 1989) and *Nature*.

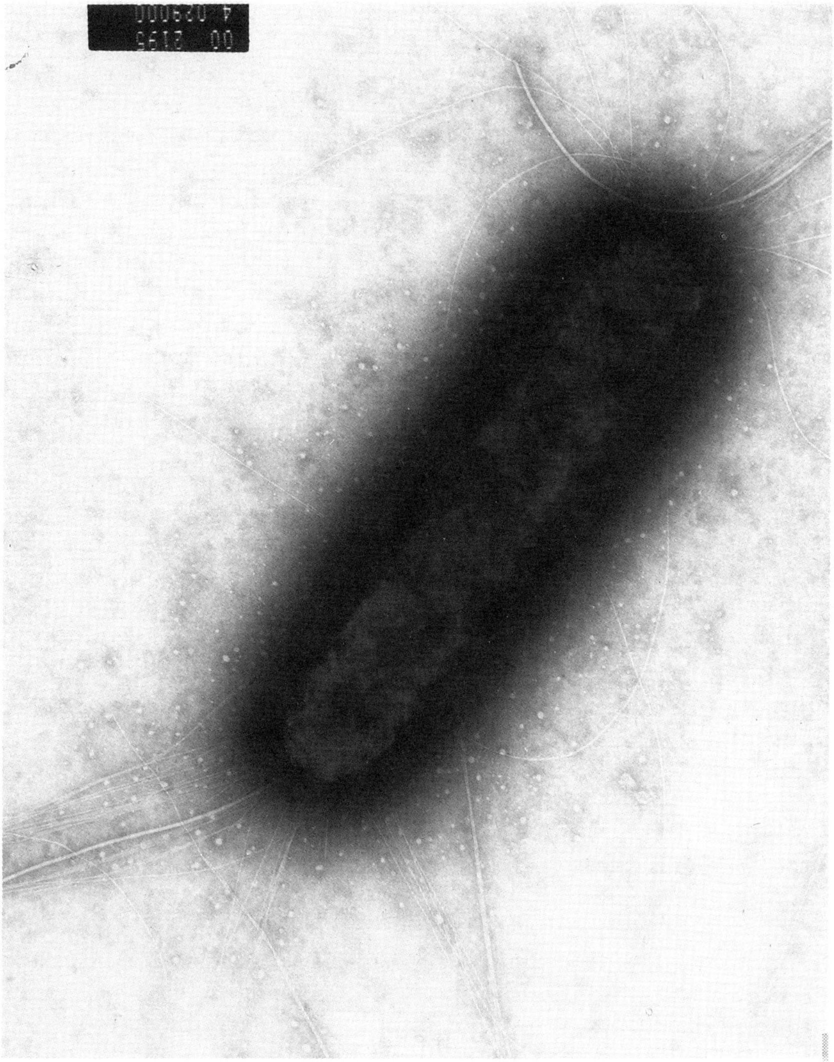

Figure 4–6. Type 4 pili from *Pseudomonas aeruginosa*. In *P. aeruginosa*, the fimbriae are very sparse and are usually located in polar regions. Micrographs courtesy of R.T. Irvin and R. Sherburne, University of Alberta. Note that the pili are thinner and less flexible than flagella. The numbers of pili in *P. aeruginosa* are very much lower than the numbers of fimbriae in the *Enterobacteriaceae*.

Gram-positive bacteria, proteins covalently linked to peptidoglycan are unusual, although it appears that wall preparations from many bacteria possess nonfimbrial or fibril proteins bound to or intercalated into the peptidoglycan mat (Nesbitt et al, 1980; Anderson et al., 1983; Mobley et al., 1983; Gunnarsson et al., 1984; Liang et al., 1989). In dealing with peptidoglycan-associated adhesins, several experimental problems arise. First, the adhesin may be sheared from the peptidoglycan during preparation of the wall material. Most wall materials are obtained by bursting cells by sonaration or by breakage in a pressure cell. Both of these methods are harsh and may result in loss of a labile adhesin. The wall may adventitiously bind cytoplasmic components on cellular breakage. The binding of proteins released from the inside of the bacterium results in added complexity to isolated walls. The removal of the proteins from the wall requires high concentrations of chaotropes, frequently resulting in loss of adhesin activity. Efforts are now being made in several laboratories to obtain cell wall adhesins from bacteria suspended in high osmotic strength buffers and dissolving away the peptidoglycan by lysozyme or other glycosaminidases (Fischetti, 1991). At present, details on the insertion of adhesins into the wall matrix of bacteria are lacking. In addition, there is no general knowledge regarding how the adhesins are stabilized in the wall matrix. In general, a proteinaceous adhesin that is not organized in a readily identifiable surface structure (Handley, 1990) or that lacks characteristic cytoplasmic anchoring sequences (Fischetti, 1991), but is associated with cell wall material, should be considered as peptidoglycan-associated in Gram-positive bacteria. The glucan-binding proteins of *S. cricetus* and *S. sobrinus* may be examples of such peptidoglycan-associated adhesins. The glucan binding proteins are bound to their surfaces via noncovalent interactions, possibly by association with the peptidoglycan matrix. The proteins can be partially solubilized by chaotropes or salts. Normal growth processes release small amounts of glucan-binding proteins into the medium. These streptococci do not have detectable fimbrial or fibril structures. The glucan-binding protein is uniformly distributed over the surface of the bacteria as revealed by fluorescein α-1,6 glucan binding.

The wall-bound glucan-binding protein may have a domain for interaction with peptidoglycan in addition to a domain responsible for binding the organisms to glucan matrices. Interestingly, growth of *S. sobrinus* in subinhibitory concentrations of fluoride results in loss of the lectin from the cell surfaces, possibly a tangent to fluoride-induced cell-wall turnover (Lesher et al., 1977). It is possible that the proteinaceous adhesins of *S. pneumoniae* (Anderson et al., 1983), *S. saprophyticus* (Gunnarsson et al., 1984), *S. pyogenes* (Tylewska and Hryniewicz, 1987), and *Lactobacillus fermentum* (Conway and Kjelleberg, 1989) are also peptidoglycan associated.

Nonproteinaceous adhesins may covalently bind to peptidoglycan in Gram-positive bacteria. These include polysaccharide and teichoic and teichuronic acids. Both the latter polymers are highly negatively charged and are synthesized coordinately with peptidoglycan. Because of their polyelectrolyte characteristics, teichoic and teichuronic acids may serve to cause expansion of sacculi. Teichoic acids are surface exposed and serve as antigens, but as far as is known, they are not protective antigens. Teichuronic acid residues contain no phosphate other than linkage sites to peptidoglycan. Although covalently linked teichoic acid to peptidoglycan is a common feature of many Gram-positive bacteria its role in adhesion has been implicated only in *S. aureus*. In most pathogenic *S. aureus* strains, the wall teichoic acid is of the ribitol type, substituted with α and β *N*-acetylglucosaminyl residues (Rogers et al., 1980). Aly et al. (1980) found that when nasal epithelial cells were pretreated with teichoic acid, the treated cells bound poorly to *S. aureus* as compared to untreated cells. Even though the teichoic adhesin is covalently bound to peptidoglycan, it is turned over into the medium along with peptidoglycan (Doyle et al., 1988). Unlike lipoteichoic acids, the wall teichoic acids are not able to bind directly to animal cells.

Cytoplasmic Membrane

The cytoplasmic membrane as a matrix that anchors adhesins to the outer surface is seen only in Gram-positive bacteria and in cell-wall-deficient bacteria. The Gram-positive bacteria must position their adhesins at sites removed from the cytoplasmic membrane. It is now clear that many surface proteins produced by staphylococci and streptococci are anchored to the cytoplasmic membrane via a hydrophobic stretch and extend through the peptidoglycan layer to the outer surface at their amino termini regions (Fischetti, 1991) (Figure 4–1). The proteins may either function as adhesins or may complex other molecules that function as adhesins. Electron microscopy has revealed that in a number of cases the surface exposure of the adhesin is either in the form of fimbriae, fibrillae, or amorphous projections of undefined morphology. Table 4–1 lists some of the surface appendages involved in adhesion. Gram-positive fimbrial adhesins have been described in a number of species (Handley, 1990). Compared to Gram-negative fimbriae, the Gram-positive fimbrial adhesins are thin (about 1–3 nm in diameter), with variable length ranging from 100 to 400 nm long. Gram-positive fimbrial adhesins were first described for *Actinomyces viscosus* (McIntire et al., 1978). Electron microscopic surveys of a number of Gram-positive bacteria showed that almost all *Actinomyces* species and many streptococci express fimbrial structures of the thin type (Handley, 1990). Not all the fimbrial structures described have been proven to function as adhesins. Furthermore, in most cases, the cell surface compartment that anchors the fimbrial structure has not been defined. Based on studies with Gram-negative bacteria, it can be predicted that the Gram-positive fimbriae possess cytoplasmic membrane anchoring subunits (see Chapter 9).

Fibrils are bristle-like appendages emanating from the cytoplasmic membrane

(Handley, 1990) (Figure 4–7). They are much shorter in length than fimbriae, usually no more than 2 nm long, but because they tend to aggregate, it is impossible to obtain accurate diameters (Handley, 1990). Fibrils may be either protein or glycoprotein, depending on the strain. In most cases, the fibrils are arranged petritrichously, although in a few strains of oral streptococci, the fibrils are polar. Adhesin function for fibrils has been documented for a limited number of Gram-positive species, including streptococci and actinomycetes, but has been suggested for other species as well (Handley, 1990). Fibrils appear to have evolved for the sole purpose of adhesion. Adhesion mediated by fibrils has been demonstrated in aggregating pairs of bacteria, in bacterial binding to various animal cells, and in bacterial binding to protein-coated surfaces (see Chapter 8). Unlike the amount of data on Gram-negative fimbriae, there is a paucity of information on the regulation or genetics of fibril formation. The fibrils seem to be anchored by one end to the cytoplasmic membrane, penetrate the peptidoglycan, and extend to the exterior environment. Purified fibrils tend to have globular ends and a fibrillar rod shape (Handley, 1990). Staining of whole Gram-positive cells with methylamine tungstate results in the pulling away of the cytoplasmic membrane from the cell wall revealing that fibrils are membrane associated at one end.

The amorphous projections of undefined morphology that appear to function as adhesins have been described for *S. pyogenes*. These undefined structures, termed amorphin, are probably composed of M protein–lipoteichoic acid (LTA) complexes (Figure 4–8). The LTA adhesins are associated with the cytoplasmic compartment, but function as adhesins only when they are associated with surface-anchored protein. The bulk of the lipoteichoic acids are bound to the cytoplasmic membrane of the bacteria via their glycolipid moiety, whereas the polyglycerol phosphate backbone may penetrate into and through the wall matrix (Fischer, 1988; Knox and Wicken, 1973). Under certain conditions, LTA is released from its anchoring site in the cytoplasmic membrane and associates with LTA binding proteins (such as M-protein of *S. pyogenes*). The LTA bound to the cytoplasmic membrane does not appear to participate directly as an adhesin but on its release from the membrane it becomes positioned on the outer surface by complexing with cytoplasmic proteins and assumes a capacity to function as an adhesin. Because many Gram-positive streptococci and staphylococci produce LTA (Knox and Wicken, 1973) and M protein is not the only protein capable of complexing LTA (Doyle et al., 1975; Ofek et al., 1982), it is likely that amorphins are more common on the surfaces of Gram-positive bacteria than has been generally recognized.

The cell-wall-deficient pathogens, particularly members of the genus *Mycoplasma,* do not produce long fimbriae or fibrils capable of serving as adhesins. They bind to host cells via direct interaction between they cytoplasmic membrane constituents and host cell membranes. *Mycoplasma* spp., including *M. pneumon-*

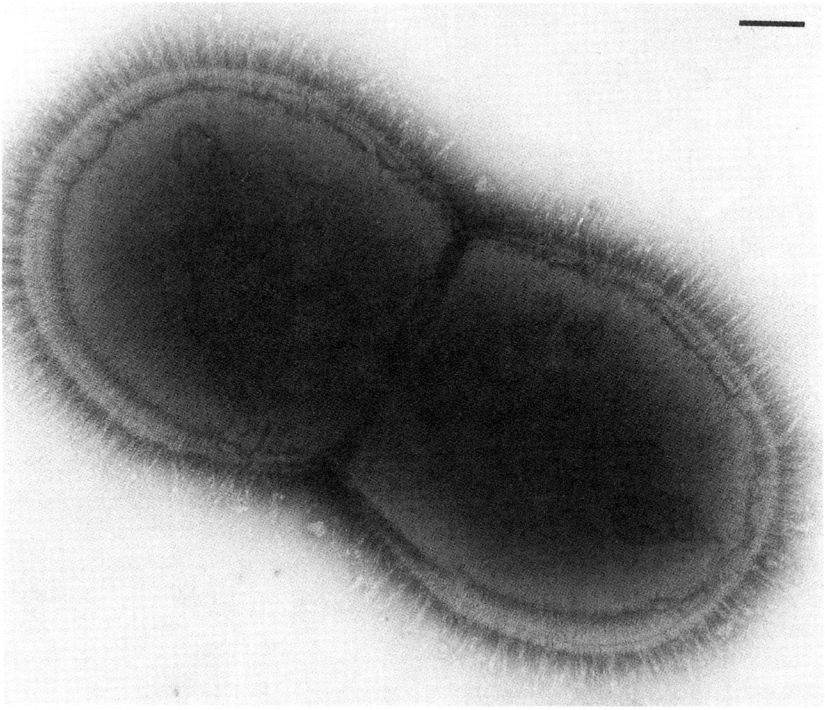

Figure 4–7. Fibrils associated with the cell wall of *Streptococcus sanguis*. The cells were stained with methylamine tungstate. Fine, sparsely distributed fibrils can be observed as being bound to the cytoplasmic membrane and penetrating through the wall matrix. Micrograph courtesy of Pauline Handley (Univ. Manchester).

iae, M. gallisepticium, and *M. genitalium,* adhere to the epithelial linings of respiratory and urogenital tracts (Razin, 1986). The bacteria are small and pleomorphic and do not have cell walls. The adhesins of most of the *Mycoplasma* are located in elongated or protruding areas of their cytoplasmic membranes. *M. pneumoniae* is known to possess several adhesins (Razin, 1986) of which a protein called P1 is the best studied. Antibodies directed against P1 show that the adhesin is located in high concentrations at the tips of the membrane protrusions. Furthermore, anti-P1 prevents adhesion of *M. pneumoniae* to tracheal epithelia.

Flagella

Bacterial flagella are thread-like appendages originating in the cytoplasmic membrane and extend through the peptidoglycan and outer membrane into the extracellular environment (Macnab, 1989). They are composed solely of proteins, possess characteristic antigenic determinants, and are found in both Gram-positive and Gram-negative bacteria. Their function is to propel the bacterium toward a source of nutrients (positive chemotaxis) or away from a source of potentially deleterious agents (negative chemotaxis). Because they are much longer than fimbriae, they may be ideal structures to position adhesins, but such functions have been reported in only a few Gram-negative species. Eshdat et al. (1978, 1981b) observed that strains of *E. coli* and *Serratia marcescens* possess mannose-specific adhesins, associated with flagella. The adhesin preparation of *E. coli* gave an amino acid composition consistent with that of flagella. McSweegan and Walker (1986) have suggested that flagellar proteins participate in the adhesion of *Campylobacter jejuni* to epithelial cells. They observed that purified flagella were able to bind to cultured epithelial cells but not to mucus. The binding was reduced when the flagella were sheared from the cells (it must be kept in mind that shearing may remove structures other than flagella from bacterial surfaces). Interestingly, and possibly importantly, when the cells were treated with cyanide, adhesion was increased. This raises the possibility that energized membrane may have a role in adhesion of some bacterial pathogens. *Vibrio cholerae* flagella have also been implicated in adhesion of the organisms (Attridge and Rowley, 1983). Flagellar function usually requires that the flagella bundle up during chemotaxis. Adhesive function associated with flagella would probably impede cellular motility. It is not surprising, therefore, that evolutionary pressure selected separate structures for motion and adhesion.

Surface Array

Many bacteria, including members of the archae and prokarya, possess proteinaceous layers outside the boundary of outer membrane or rigid cell walls (pseudomurein for archaebacteria and peptidoglycan for eubacteria). This layer is usually arrayed in geometric forms (Koval, 1988) (Figure 4–9). In many cases, the surface arrays are composed of hexagonal subunits. The abbreviations S or RS refer to regular or repeating surface arrays. Many human and animal colonizers, including members of the genera *Actinobacillus, Aeromonas, Bacillus, Bacteroides, Campylobacter,* and *Pseudomonas* exhibit or express R or RS layers. The S layers appear to serve many functions, depending on the bacterium. They have been suggested as shape determinants, selective permeability barriers, or organelles to protect against osmotic shock and protection against parasitism by viruses or *Bdellovibrio*. Ishiguro et al. (1981) found a correlation between virulence of the fish pathogen *Aeromonas salmonicida* and the presence of surface array. McCoy et al. (1975) have suggested that the S layer of *Campylobacter fetus* is antiphagocytic. Species and strains of *Aeromonas* observed to have surface array layers tended to be more pathogenic in mice (Janda et al., 1987; Altwegg and Geiss, 1989). A role of adhesion in all these activities remains to be better described.

Capsules and Other Secreted Polymers

Capsules are usually acidic polysaccharides secreted by bacteria, although a few organisms secrete polypeptide capsules. The capsule is best defined as the polymeric structure that remains cell bound following its secretion (Figure 4–10). This differs from the so-called slime, acidic polymers secreted into the growth medium, but instead of remaining cell-bound, become loosely associated with the cell surface. Because the chemical structures of capsular and slime materials are frequently similar, the distinction between them is based solely on cytological observations. Chemically, the polysaccharide capsule is a polymer of repeating units of oligosaccharide sequences. They are highly hydrated, a property that may prevent bacterial desiccation. Furthermore, capsules have the ability to bind to metal ions and positively charged amino acids. These properties may aid in maintaining nutrients near the cells (Sutherland, 1977). Slime refers to secreted polysaccharides. The word glycocalyx has also been used to describe secreted anionic polysaccharides which could include slime and capsules (Costerton et al., 1981). Capsules and slime material are probably the only cell-surface "compartments" that have been ascribed dual functions: one adhesive and the other antiadhesive. It has long been recognized that capsules are antiphagocytic because of their uniformly high density of negative charges (Gotschlich, 1983). They may also interfere with the presentation or accessibility of the adhesin to interact with the receptor. St. Geme and Falkow (1991) recently found that when *Haemophilus influenzae* type B lost its ability to express capsule, the organism adhered

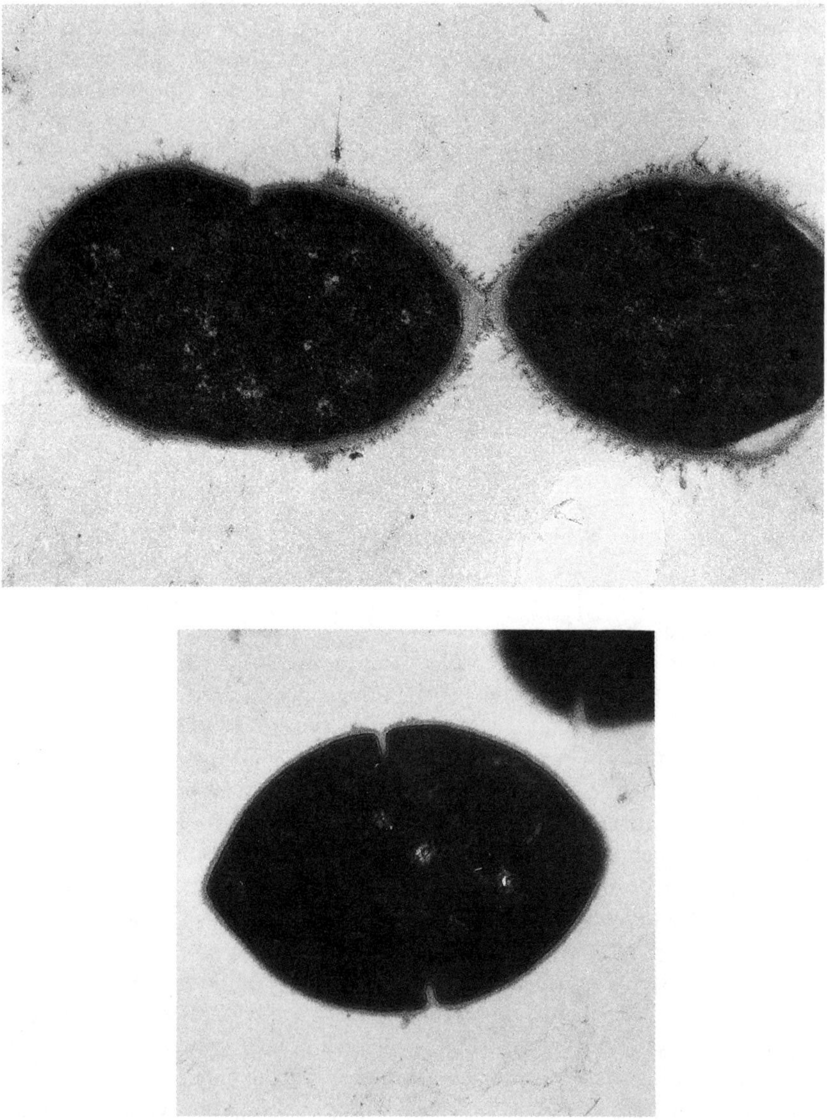

Figure 4–8. Surface structures of *Streptococcus pyogenes*. Non-adherent variants are nearly bald, whereas the adherent phenotype shows a diffuse staining pattern extending away from the cell wall. A, wild-type *S. pyogenes*; B, bald mutant *S. pyogenes*. Magnification: 80,000. Courtesy D. Hasty, UT, Memphis.

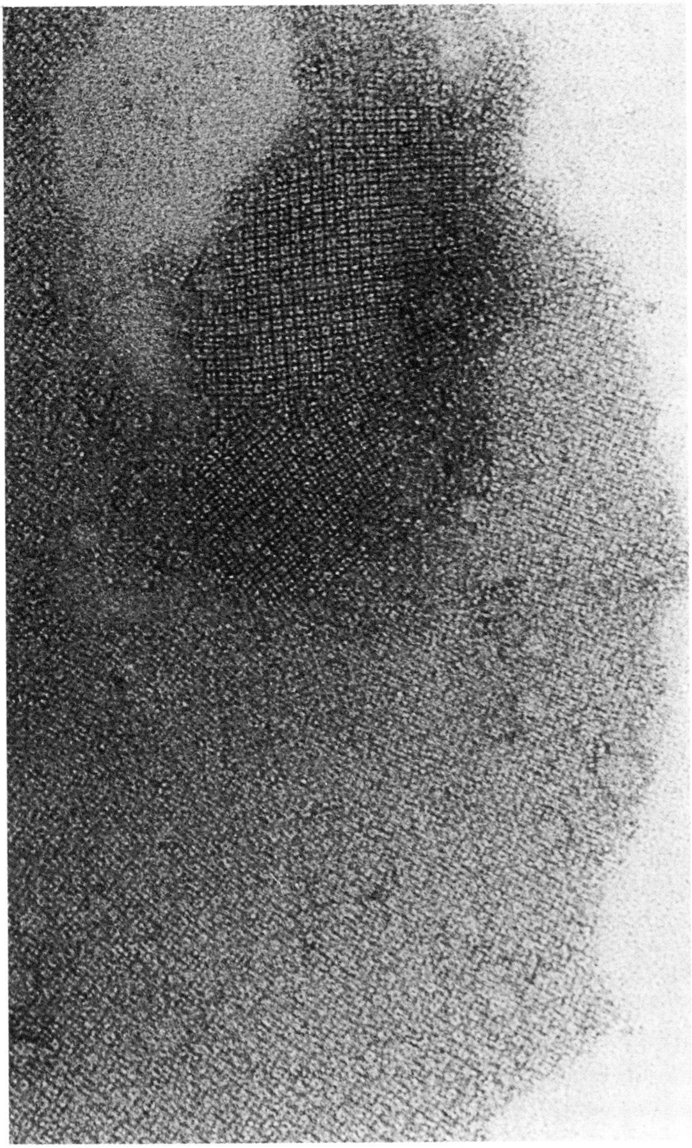

Figure 4–9. Surface array structures of *Aeromonas salmonicida*. Two morphological surface array patterns are observed. Magnification, 296,180. Courtesy of T.J. Trust and R. Garduno, University of Victoria.

better to cultured human epithelial cells. It is possible that surface adhesins of some encapsulated bacteria are cryptic, becoming exposed only on removal of capsule. The presence of capsule may obscure the hydrophobicity of certain bacteria. For example, Hogt et al. (1983) found that encapsulated strains of coagulase-negative staphylococci were more hydrophilic than nonencapsulated strains. The presence of capsule was independent of slime production. Capsular materials are frequently not synthesized during exponential growth, so growing bacteria without capsules may have a greater opportunity to adhere to complementary substrata (Ofek et al., 1983). Once adherent, the bacteria can then secrete capsules that may aid the nutritional status of the bacteria. Brook and Myhal (1991), in contrast, found that adhesion of *Bacteroides fragilis* to intestinal cells is dependent on both fimbriae and capsules. Oyston and Handley (1991) observed that when the capsular surface carbohydrate was chemically modified, *B. fragilis* lost its ability to hemagglutinate rabbit red cells. When fimbriae were removed by proteases, the cells retained their hemagglutinating activity, suggesting that capsules are a critical component of adhesion. Perhaps the most convincing evidence that capsular polysaccharide can act as an adhesin is the study of the interaction of *Klebsiella pneumoniae* with macrophages (Athamna et al., 1991). *Klebsiella pneumoniae* undergoes phagocytosis mediated by capsular polysaccharides recognized by the mannose/*N*-acetylglucosamine-specific lectin of macrophages (Figure 4–10). Strains of enteropathogenic *E. coli* capable of producing capsule are less adherent than unencapsulated mutant strains (Runnels and Moon, 1984). Fab fragments directed against the *E. coli* capsule fail to render the cells more adherent to intestinal epithelial cells of pigs. It is likely that the capsular materials are masking potential adhesins. Composition of capsule may not be limited to secreted polysaccharides. Protein-containing adhesins that are secreted to the outer surface and demonstrate the appearance of capsule when stabilized with antibody (Figure 4–11) have been shown in *E. coli* producing non-fimbrial adhesins (NFA) (Orskov et al., 1985; Jann and Hoschutzky, 1990; Kroncke et al., 1990).

Slime, such as alginate of *Pseudomonas,* may also function as an antiphagocytic to prevent attachment of the organism to phagocytic cells. Onderdonk et al. (1978) found that secreted polysaccharide would inhibit the adhesion of *B. fragilis* to rat mesenthelial tissue. The polysaccharide material was shown to bind only loosely to cells and, therefore, can be classified as slime. It is possible that the secreted polysaccharide masks the presentation of the adhesin. In contrast to the capsular material, however, the ability of slime to promote adhesion of bacteria to various substrata is now well established. We define the word "slimectin" to represent slime-like materials involved in microbial adhesive events. The name slimectin assumes its role in adhesion, whereas the name "slime" refers to secreted polysaccharides that do not participate as adhesins. Polymers such as α-1,3 and α-1,6 glucans produced by oral streptococci qualify as slimectin,

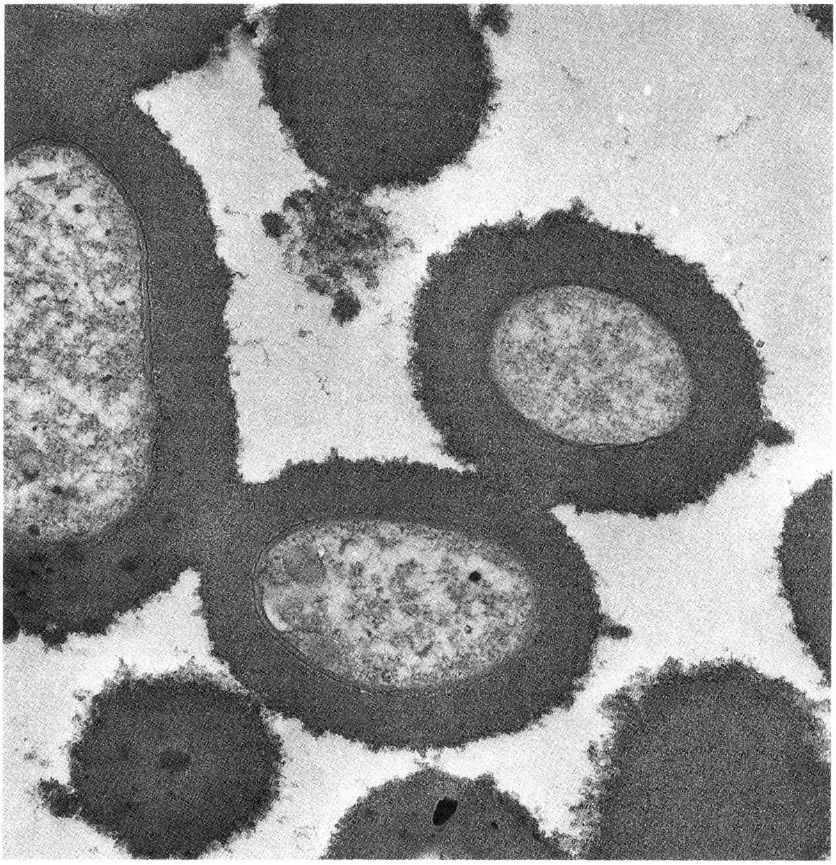

Figure 4–10. Cross section of a *Klebsiella pneumoniae* showing a thick capsular layer. In many bacteria, the capsule may serve as a receptor for phagocyte lectins, whereas in others the capsule may not be recognized by phagocytes. In only a few cases have capsules been implicated as adhesins.

although they are not charged nor are they secreted. These glucans participate as receptors for glucan binding lectins of some oral streptococci. Slimectin production is under metabolic regulation. Many bacteria produced extracellular anionic polysaccharides in copious quantities in the stationary phase, whereas in the exponential phase, production of the polymers is relatively low (Sutherland, 1977). Furthermore, the composition of the growth medium affects production of extracellular polymers (Christensen et al., 1982). Many researchers have isolated bacteria from biofilms or from infected tissues and found the colonies

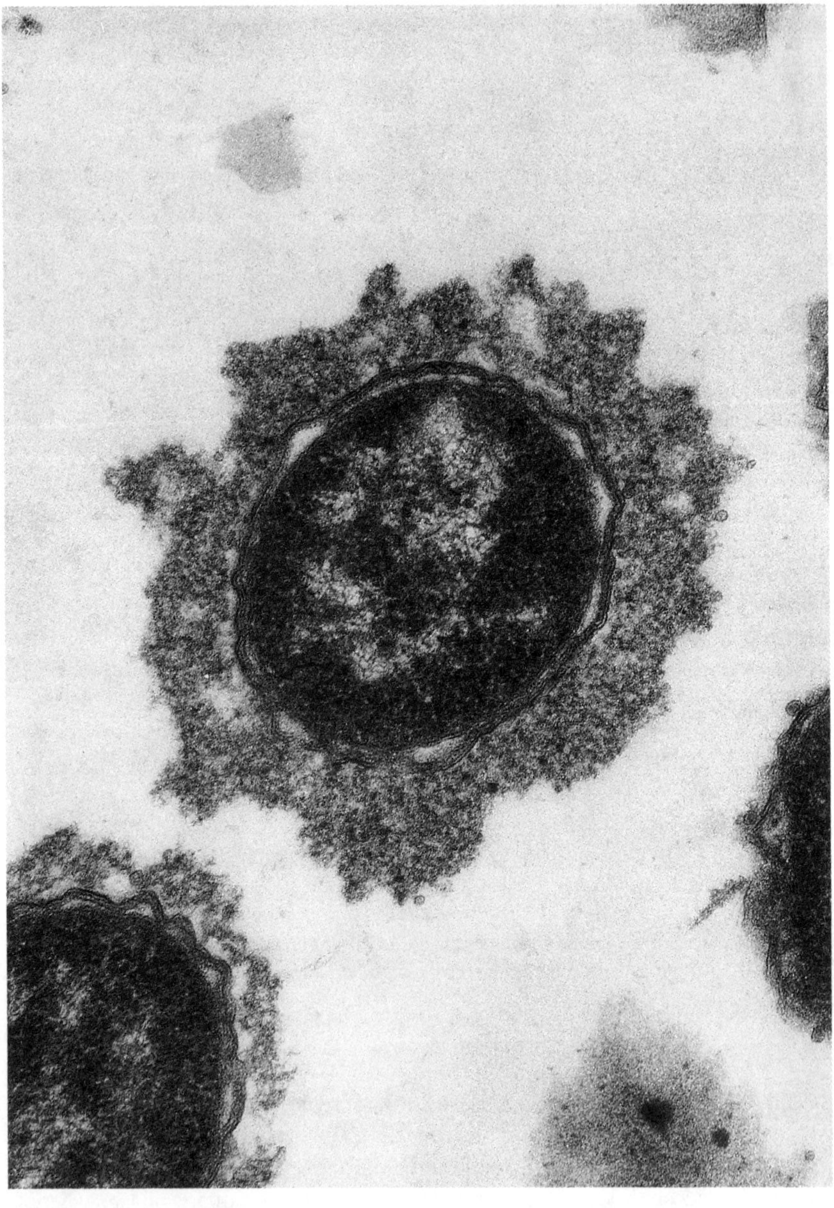

Figure 4–11. Capsular protein adhesin of *Escherichia coli*. A few strains of *E. coli* appear to require a secreted proteinaceous adhesin in order to bind to mucosa. Micrograph courtesy of K. Jann, Max Planck Institute, Freiburg. Magnification 92,500. Taken from Goldhar et al., 1987.

to be mucoid. They have attributed adhesion of the bacteria to the various surfaces to the extracellular polymers associated with mucoid colony characteristics. However, some bacteria that produce mucoid colonies, produce very small amounts of anionic polysaccharide in rapidly growing populations. Bacteria free of capsule or polysaccharide can frequently adhere to a surface and then produce the capsules or polysaccharide. This results in the formation of "films" and "biofilms" or adherent masses of bacteria and extracellular polymers. In a strict sense, the exopolysaccharides do not contribute to initial adhesion, but they stabilize the adherent bacteria by reducing shear effects and/or excluding toxic materials.

The "Soluble" Compartment

In a few cases, bacteria may secrete soluble adhesins, which in turn associate with the intact bacteria, giving rise to cells capable of binding to appropriate substrata via the absorbed adhesin. Tuomanen (1986) showed that a secreted form of the filamentous hemagglutinin of *Bordetella pertussis* can become bound not only by the secreting bacteria but also to other bacteria, such as *Haemophilus* and *Staphylococcus* spp., to yield adherent bacteria. It was suggested that the other species "pirated" an adhesin from *B. pertussis* by presumably expressing receptors for the secreted *Bordetella* adhesin. Wentworth et al. (1991) showed that soluble fucose- and galactose-specific lectins from *Pseudomonas aeruginosa* conferred on the bacteria the ability to adhere to cultured rabbit corneal epithelial cells. The adhesion was specific because it was inhibited by the carbohydrates complementary to the absorbed lectin. Baker et al. (1991) noted that exotoxin A could bind to *P. aeruginosa* to render the cells adherent. It appears that structures on the bacterial surface are capable of binding to a soluble adhesin with a relatively high affinity. The bound adhesin is then capable of interacting with its receptor on animal cells.

Characterization of Bacterial Cell Surfaces

Adhesion of bacteria to surface substrata is dictated by the physicochemical properties of the bacterial cell surfaces. So far as is known, the only way a bacterium can adhere to a surface is to use its own surface structures, whether they be appendages such as fimbriae or fibrillae, or capsules or slimectin, cytoplasmic or outer membrane, or peptidoglycan and its associated molecules. In order to adhere, bacteria commonly employ "bridges" that connect some bacterial surface component to some substratum molecule. In the past few years, methods have been developed to characterize bacterial cell surfaces to a higher level of understanding than in the past. In early studies on bacterial surfaces, the most common approaches were directed at fractionating the cells and then determining the composition of the fractions by chemical assays. These approaches, coupled with mutant selection and immunochemical procedures, were important in developing an appreciation of the properties of bacterial cell surfaces.

Figure 4–12 provides an outline of some of the prominent methods involved in bacterial surface characterization. The methods range from electron microscopic techniques to X-ray assays for elementary composition to adhesion to hydrocarbon (as an index of hydrophobicity) to migration in an electric field (for zeta potentials). None of the assays identify specific adhesins, but all of the assays reveal important information about cell surfaces.

Electron microscopic techniques are versatile in that a variety of stains and probes can be employed to detect morphological features of the locations of specific components. One of the most common methods is negative staining in which an electron-dense (high atomic number) compound is mixed with the specimen. The specimen "stands out" in the electron-dense milieu and can be observed in the electron microscope. Structures such as fibrillae, fimbriae, flagella, capsules, S-layers, membranes, and peptidoglycan (or pseudomurein) can be readily identified by viewing suitably prepared samples. It is also possible to determine which of the surface structures form direct contacts with substrata. Handley (1991a) has recently reviewed the staining techniques employed in bacterial cell surface studies. Negative staining techniques are simple, the specimen maintains stability for lengthy times, and only small volumes of cells are required. Furthermore, negative staining techniques can be used on isolated structures, such as membranes or fimbriae. The resolution available with negative staining methods is about 2 nm. In adhesion research, negative staining has been used most extensively to study surface appendages and to identify adhesin-supporting structures. The technique has been essential in defining the sizes of various types of fimbriae and fibrillae. Another type of stain, ruthenium red, has become an important probe for surface polysaccharides. Ruthenium red does not seem to have an affinity for teichoic acids, but binds strongly to charged polysaccharides and also to neutral polysaccharides, such as dextrans (Handley, 1991b.) Fibrillar glycoproteins, identified as adhesins, can be detected in ruthenium red stained preparations of *Streptococcus salivarius* (Handley et al., 1988). Conventional osmium staining did not reveal the fibrils. Ruthenium red is widely employed in studying slimectin, capsules, and other secreted materials surrounding bacterial cell surfaces.

In recent years, it has become increasingly clear that hydrophobicity is involved in many bacterial adhesion phenomena (reviewed in Doyle and Rosenberg, 1990). Hydrophobicity may not be readily defined, but in a broad sense means that a substance has a greater attraction to a hydrocarbon than to water. Recently, Rosenberg and Doyle (1990) reviewed some of the major methods in establishing the hydrophobicity of a bacterium. The most common method is to measure the

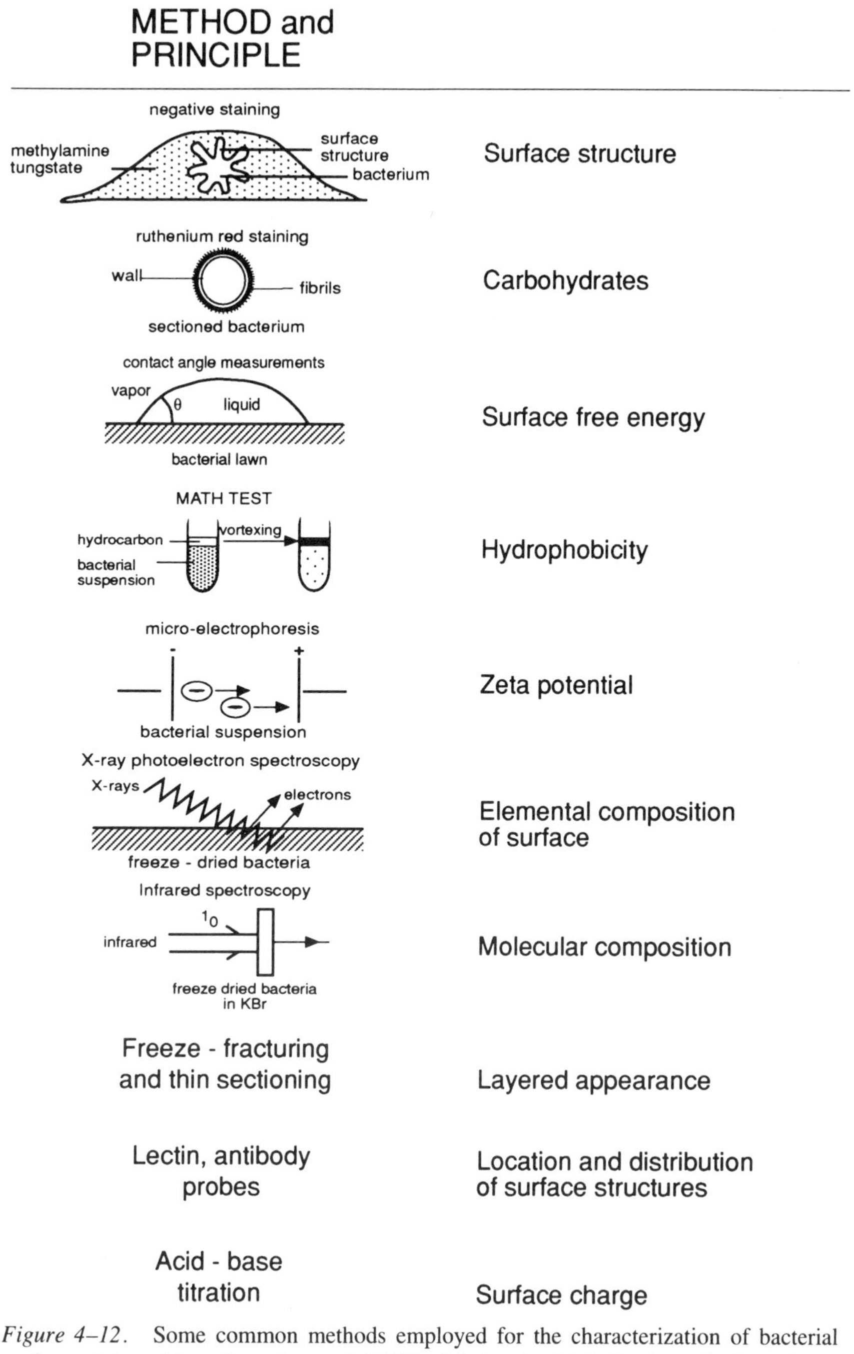

Figure 4–12. Some common methods employed for the characterization of bacterial surfaces. Adapted from Busscher et al. (1991). (Math: microbial adhesion to hydrocarbon.)

adhesion of an aqueous suspension to a hydrocarbon, such as hexadecane. This procedure requires that the aqueous layer be mixed with the hydrocarbon layer to form droplets of hydrocarbon at the interface between the two liquids. Bacteria tend to adhere to the droplets, resulting in a reduced opacity of the bacterial suspension. This method is popular because of its simplicity, reproducibility, and nondestructiveness to the bacteria. Another widely employed method involves contact angle measurements. Just as a water droplet does not spread on parafilm, water may not spread on lawns of hydrophobic bacteria. The contact angle, α, between the water droplet and the bacterial lawn is a measure of cell surface hydrophobicity (Figure 4–12). Another technique used by many adhesion researchers is to determine the extent of binding of bacteria to octyl-Sepharose or some other hydrocarbon-substituted bead. Almost all of these methods give rise to cellular hydrophobicity values rather than values that can be ascribed to individual structural components. However, similar technologies, employing isolated structures on molecules, can be used in assessing the relative hydrophobicities of individual surface components. Figure 4–13 provides an arbitrary outline of the relative hydrophobicities of surface molecules and surface structures found in bacteria. In all cases, except for teichuronic acids, the structures or molecules

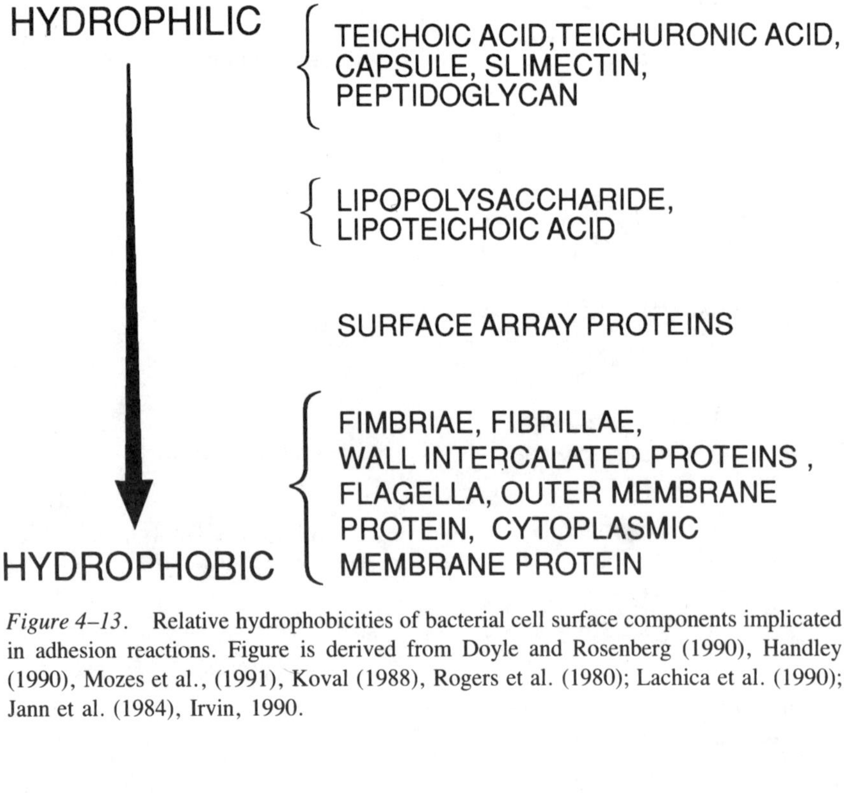

Figure 4–13. Relative hydrophobicities of bacterial cell surface components implicated in adhesion reactions. Figure is derived from Doyle and Rosenberg (1990), Handley (1990), Mozes et al., (1991), Koval (1988), Rogers et al. (1980); Lachica et al. (1990); Jann et al. (1984), Irvin, 1990.

have been shown to participate in adhesion reactions. Statistically, the most common adhesins tend to be associated with the more hydrophobic entities.

Surface charges of a bacterium may govern its ability to adhere to a substratum. Most bacterial surfaces are negatively charged at physiological pH values. Teichoic, lipoteichoic, and teichuronic acids and peptidoglycan contribute to the negative charges of Gram-positive bacteria, whereas in Gram-negative bacteria, proteins, peptidoglycan, and lipopolysaccharide combine to create net negative surface charges. In some Gram-positive bacteria, negatively charged molecules, probably teichoic acids, are more concentrated on the outer face of the wall than on the smooth inner face (Birdsell et al., 1975; Sonnenfeld et al., 1985). Cationized ferritin could be seen to concentrate on the external portion of the wall of *Bacillus subtilis* (Sonnenfeld et al., 1985).

The zeta potential is a measure of overall surface charge of a bacterium and is affected by solvent pH and composition, temperature, surfactants, and cell surface composition. James (1991) reviewed the significance and measurement of bacterial zeta potentials. Low zeta potentials reflect high acidities of the bacterial surface. A surface with a high content of teichoic acid would be expected to have a lower zeta potential than cells with no teichoic acid. Van der Mei et al. (1988) observed that nonfibrillar and low adherent mutants of *Streptococcus salivarius* had lower zeta potentials than the parent strain. Loss of fibrils was accompanied by exposure of teichoic acids on the surfaces of the mutants.

Aside from fractionation and wet chemical analyses, spectroscopic techniques can be employed for assay of bacterial cell surface composition (Figure 4–12). X-ray photoelectron spectroscopy (XPS) has now become widely accepted as a means of measuring the relative amounts of O, N, C, and P on bacterial surfaces (Rouxhet and Genet, 1991). When a beam of X-rays irradiate a dried bacterial specimen, electrons are emitted which can be detected in a suitable spectrometer. The spectrum of the emitted electrons is a characteristic for a particular element. The intensity of the spectrum reflects the relative amount of the element sample. Only electrons emitted on or near the cell surface can be assayed, due to inelastic scattering in the irradiated samples (Rouxhet and Genet, 1991). The technique yields information on the relative amounts of teichoic acid (high P/N ratios), carbohydrate (high C/N ratios), and protein (high amounts of N compared to the other elements). Cells with relatively high N/C ratios tend to be hydrophobic, probably reflecting the role of protein in bacterial hydrophobicity.

Infrared spectroscopy (IR) can also be a useful method to assess the presence of components on bacterial surfaces. The method, however, like XPS, requires a dried bacterial specimen. Infrared radiation causes stretching and bending of chemical bonds. The absorption spectra reflect the kinds of chemical bonds present in the specimen. The method is most useful in determining the presence of amide (protein), —C—OH (polysaccharide), and —P═O bonds. Application of special techniques, such as attenuated reflectance spectroscopy, ensures that surface characteristics are measured, instead of bulk bacterial components. In terms of employing XPS and IR to study bacterial adhesins of pathogens, most results have come from characterization of cells capable of adhering to biomaterials.

Dynamic Characteristics of Bacterial Cell Surfaces

It is no longer tenable to view the cell surface of the bacterium as a static, immortal entity. The cell wall of some bacteria is metabolically active. During cell division the wall is cleaved by autolysins so as to accommodate new growth sites. Up to 50% of the wall may be lost into the growth medium per generation. Loss of wall into the growth medium has been termed "turnover" (reviewed by Doyle et al., 1988) (Figure 4–14). Turnover products in Gram-positive bacteria are not reuti-

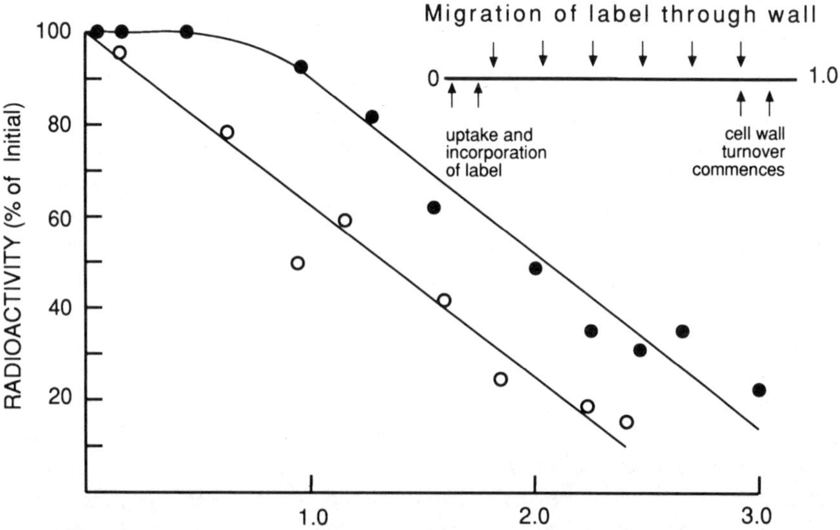

Figure 4–14. Cell wall turnover in bacteria. Some pathogens, including *Bacillus anthracis, Staphylococcus aureus* and other staphylococci, members of the genus *Listeria* and *Neisseria,* turn over or shed their wall materials into the growth medium. Turnover also occurs in Gram-negative rods, but the turnover products are trapped by the outer membrane and are re-utilized during growth. Turnover seems to be a requirement for surface expansion by some bacteria (Doyle et al., 1988). It is not known if wall associated adhesins are turned over during growth. In this figure, a culture of *B. subtilis* was pulsed for 0.1 generation with [^{3}H]GlcNAc, a wall precursor, then chased with 100 K_m of unlabeled GlcNAc. Wall begins to turn over after about one generation of growth, post-incorporation of the label (solid circles). Another part of the culture was pulsed 1.0 gen with the label, then chased (open circles). The inset shows the disposition of side wall during the cell cycle of *B. subtilis*.

lized for subsequent growth, although in Gram-negative rods turnover products are trapped by the outer membrane and are returned to the cell for growth.

Expression of cell surface adhesins also is not always a stringent requirement. Growth of a population of bacteria under one condition may prevent the expression of the adhesin. All of the conditions that lead to presentation of an adhesin on a bacterial surface have not been defined, although now many conditions are known that affect production of surface adhesins. For example, the K99 and K88 fimbrial adhesins of *E. coli* are produced in relatively higher quantities at 36°C compared to temperatures below 20°C (Isaacson, 1980, 1983; Gaastra and DeGraaf, 1982). Exponential growth in shake cultures enhances the K99 fimbrial expression, but expression of type 1 fimbriae in *E. coli* or *Salmonella typhimurium* seems to be much lower under the same conditions (Old and Duguid, 1970).

Table 4–3. Environmental factors influencing the expression of adhesins on bacterial cell surfaces

Bacterium	Adhesin	Adhesin expression	Reference(s)
Escherichia coli	K99 fimbriae	Adhesin is produced poorly at >20°C.	Isaacson, 1980
Escherichia coli	K99 fimbriae	Exponential growth growth in shake cultures enhances expression.	Isaacson, 1983
Escherichia coli	K88 fimbriae	Temperatures lower than 20°C result in low expression of adhesin.	Gaastra and DeGraaf, 1982
Escherichia coli and *Salmonella typhimurium*	Type 1 fimbriae	Stationary phase, static cultures are best for expression of MS fimbriae.	Old and Duguid, 1970
Streptococcus cricetus	Glucan-binding lectin	Mn^{2+} aquo ion is required for expression.	Drake et al., 1988b
Streptococcus mutans	Hydrophobins	Repeated subculture results in loss of hydrophobins.	Westergren and Olsson, 1983
Streptococcus pyogenes	Lipoteichoic acid	Sublethal amounts of penicillin result in excretion of LTA.	Alkan and Beachey, 1978
Pseudomonas aeruginosa	PA-I and PA-II lectins	The lectins are synthesized only in late stationary phase.	Gilboa-Garber, 1982
Streptococcus pyogenes	Hyaluronic acid capsule	Capsule is expressed poorly in exponential phase.	Plummer et al., 1962
Various bacteria	Fimbrial and nonfimbrial adhesins	Sublethal concentrations of protein synthesis inhibitors or others antibiotics may cause an increase or decrease in the expression of adhesins.	Schifferli and Beachey, 1988a, b

When antibiotics are included in growth media, adhesin expression may be enhanced or inhibited (Alkan and Beachey, 1978), depending on the bacterium and antibiotic (reviewed by Schifferli and Beachey, 1988a,b). Trace quantities of Mn^{2+} are required for the expression of the glucan-binding lectin of *Streptococcus cricetus* (Drake et al., 1988b). The cytoplasmic lectins of *Pseudomonas aeruginosa,* known to function as adhesins (Wentworth et al., 1991), are produced only in late stationary phase cultures (Gilboa-Garber, 1982). Adhesins may be masked in bacteria obtained from certain culture conditions. The hyaluronic acid capsule of group A streptococci is better expressed in stationary phase than in exponential phase (Plummer et al., 1962). Repeated subculture is also known to alter the hydrophobicities of streptococci (Westergren and Olsson, 1983). Adhesin expression is under genetic regulation (Chapter 9 provides a comprehensive discussion of how some adhesins may be regulated at the genetic level) as well as environmental regulation (Table 4–3). Because a particular isolate from a diseased site fails to yield progeny capable of adhering to a substratum, it does not mean that the adhesin is always absent. Conditions need to be varied in order to find the proper environment for expression of adhesins.

Although peptidoglycan and its covalently bound auxiliary polymers, teichoic and teichuronic acids, turn over during growth, there is no evidence that cell surface proteins turn over. Protein turnover in bacteria is very low, except in sporulating bacteria (Doyle et al., 1988). If protein turnover accompanied peptidoglycan turnover it would be expected that adhesins, such as fimbriae, would be found in culture supernatants. Such soluble adhesins might then compete with intact bacteria for substrata.

References

Alkan, M.L. and E.H. Beachey, 1978. Excretion of lipoteichoic acid by group A streptococci: influence of penicillin on excretion and loss of ability to adhere to human oral mucosal cells. *J. Clin. Invest.* **61:**671–677.

Altwegg, M. and H.K. Geiss. 1989. *Aeromonas* as a human pathogen. *CRC Crit. Rev. Microbiol.* **16:**253–286.

Aly, R. and S. Levit. 1987. Adherence of *Staphylococcus aureus* to squamous epithelium: role of fibronectin and teichoic acid. *Rev. Infect. Dis.* **9:**S341–S350.

Aly, R., H.R. Shinefield, C. Litz, and H.I. Maibach. 1980. Role of teichoic acid in the binding of *Staphylococcus aureus* to nasal epithelial cells. *J. Infect. Dis.* **141:**463–465.

Anderson, B., J. Dahmen, T. Frejd, H. Leffler, G. Magnusson, G. Noori, and C. Svanborg-Eden. 1983. Identification of an active disaccharide unit of a glycoconjugate receptor for pneumococci attaching to human pharyngeal epithelial cells. *J. Exp. Med.* **158:**559–570.

Athamna, A., I. Ofek, Y. Keisari, S. Markowitz, G.D. S. Dutton, and N. Sharon. 1991. Lectinophagocytosis of encapsulated *Klebsiella pneumoniae* mediated by surface lectins of guinea pig alveolar macrophages and human monocyte-derived macrophages. *Infect. Immun.* **59:**1673–1682.

Attridge, S.R. and D. Rowley. 1983. The role of the flagellum in the adherence of *Vibrio cholerae*. *J. Infect. Dis.* **147:**864–872.

Baddour, L.M., G.D. Christensen, W.A. Simpson, and E.H. Beachey. 1989. Microbial adherence. In: Mandell, G.L., R.G. Douglas, Jr. and J.E. Bennett (eds.), *Principles and Practices of Infectious Disease*. Churchill Livingstone, New York, pp. 9–25.

Baker, N.R., V. Minor, C. Deal, M.S. Shahrabadi, D.A. Simpson, and D.E. Woods. 1991. *Pseudomonas aeruginosa* exoenzyme S is an adhesin. *Infect. Immun.* **59:**2859–2863.

Birdsell, D.C., R.J. Doyle, and M. Morgenstern. 1975. Organization of teichoic acid in the cell wall of *Bacillus subtilis*. *J. Bacteriol.* **121:**726–734.

Braun, V. 1975. Covalent lipoprotein from the outer membrane of *Escherichia coli*. *Biochim. Biophys. Acta* **415:**335–377.

Brinton, C.C., Jr. 1959. Non-flagellar appendages of bacteria. *Nature* **183:**782–786.

Brook, I. and M.L. Myhal. 1991. Adherence of *Bacteroides fragilis* group species. *Infect. Immun.* **59:**742–744.

Busscher, H.J., P.S. Handley, P.G. Rouxhet, L.M. Hesketh, and H.C. van der Mei. 1991. The relationship between structural and physicochemical surface properties of tufted *Streptococcus sanguis* strains. In: Mozes, N., P.S. Handley, H.J. Busscher, and P.G. Rouxhet (eds.), *Microbial Cell Surface Analysis: Structural and Physicochemical Methods*. VCH publishers, New York, pp. 317–338.

Christensen, G.D., W.A. Simpson, A.L. Bisno, and E.H. Beachey. 1982. Adherence of slime-producing strains of *Staphylococcus epidermidis* to smooth surfaces. *Infect. Immun.* **37:**318–326.

Conway, P.L. and S. Kjelleberg. 1989. Protein-mediated adhesion of *Lactobacillus fermentum* strain 737 to mouse stomach squamous epithelium. *J. Gen. Microbiol.* **135:**1175–1186.

Costerton, J.W., R.T. Irvin, and K.J. Cheng. 1981. The bacterial glycocalyx in nature and disease. *Annu. Rev. Microbiol.* **35:**299–324.

DeGraaf, F.K. 1990. Genetics of adhesive fimbriae of intestinal *Escherichia coli*. *Curr. Top. Microbiol. Immunol.* **151:**29–53.

DeGraaf, F.K. and F.R. Mooi. 1986. The fimbrial adhesins of *Escherichia coli*. *Adv. Microbial Physiol.* **28:**65–143.

Dekker, N.P., C.J. Lammel, R.E. Mandrell, and G.F. Brooks. 1990. Opa (protein II) influences gonococcal organization in colonies, surface appearance, size and attachment to human fallopian tube tissues. *Microb. Pathogen.* **9:**19–31.

DiRienzo, J.M., K. Nakamura, and M. Inouye. 1978. The outer membrane proteins of Gram-negative bacteria: biosynthesis, assembly, and functions. *Annu. Rev. Biochem.* **47:** 481–532.

Doig, P., P.A. Sastry, R.S. Hodges, K.K. Lee, W. Paranchych, and R.T. Irvin. 1990. Inhibition of pilus-mediated adhesion of *Pseudomonas aeruginosa* to human buccal epithelial cells by monoclonal antibodies directed against pili. *Infect. Immun.* **58:**124–130.

Doyle, R.J. and M. Rosenberg (eds.). 1990. *Microbial Cell Surface Hydrophobicity.* American Society for Microbiology, Washington.

Doyle, R.J. and E.M. Sonnenfeld. 1989. Properties of the cell surfaces of pathogenic bacteria. *Int. Rev. Cytol.* **18**:33–92.

Doyle, R.J., M.L. McDannel, J.R. Helman, and U.N. Streips, 1975. Distribution of teichoic acid in the cell wall of *Bacillus subtilis. J. Bacteriol.* **122**:152–158.

Doyle, R.J., J. Chaloupka, and V. Vinter. 1988. Turnover of cell walls in microorganisms. *Microbiol. Rev.* **52**:554–567.

Drake, D., K.G. Taylor, A.S. Bleiweis, and R. J. Doyle. 1988a. Specificity of the glucan binding lectin of *Streptococcus cricetus. Infect. Immun.* **56**:1864–1872.

Drake, D., D.G. Taylor, and R.J. Doyle. 1988b. Expression of glucan-binding lectin of *Streptococcus cricetus* requires manganous ion. *Infect. Immun.* **56**:2205–2207.

Duguid, J.P. and D.C. Old. 1980. Adhesive properties of *Enterobacteriaceae*. In: Beachey, E.H. (ed.), *Bacterial Adherence (Receptors and Recognition, Vol. 6).* Chapman and Hall, London, pp. 187–217.

Duguid, J.P., I.W. Smith, G. Dempster, and P.N. Edmunds. 1955. Non-flagellar filamentous appendages ("fimbriae") and hemagglutinating activity in *Bacterium coli. J. Pathol. Bacteriol.* **70:**335–358.

Duguid, J.P., E.S. Anderson, and I. Campbell. 1966. Fimbriae and adhesive properties in salmonellae. *J. Pathol. Bacteriol.* **92**:107–138.

Dziarski, R. 1991. Peptidoglycan and lipopolysaccharide bind to the same binding site on lymphocytes. *J. Biol. Chem.* **266:**4719–4725.

Eshdat, Y., N. Sharon, I. Ofek and D. Mirelman. 1978. Isolation of a mannose-specific lectin from *Escherichia coli* and its role in the adherence of the bacteria to epithelial cells. *Biochem. Biophys. Res. Commun.* **85:**1551–1559.

Eshdat, Y., V. Speth, and K. Jann. 1981a. Participation of pili and cell wall adhesin in the yeast agglutination activity of *Escherichia coli. Infect. Immun.* **34:**980–986.

Eshdat, Y., M. Izhar, N. Sharon, and D. Mirelman. 1981b. Structural association of the outer surface mannose-specific lectin of *Escherichia coli* with bacterial flagella. *Isr. J. Med. Sci.* **16:**479.

Fauchere, J.L., M. Kervella, A. Rosenau, K. Mohanna, and M. Vernon. 1989. Adhesion to HeLa cells of *Campylobacter jejuni* and *C. coli* outer membrane components. *Res. Microbiol.* **140:**379–392.

Fischer, W. 1988. Physiology of lipoteichoic acids in bacteria. *Adv. Microbial Physiol.* **29:**233–302.

Fischetti, V. 1991. Streptococcal M protein. *Sci. Am.,* June pp. 58–65.

Fulks, K.A., C.F. Marrs, S.P. Stevens, and M.R. Green. 1990. Sequence analysis of the inversion region containing the pilin genes of *Moraxella bovis. J. Bacteriol.* **172:**310–316.

Gaastra, W. and F.K. DeGraaf. 1982. Host-specific fimbrial adhesins and noninvasive enterotoxigenic *Escherichia coli* strains. *Microbiol. Rev.* **46:**129–161.

Gilbert, P., D.J. Evans, I.G. Duguid, E. Evans, and M.R. W. Brown. 1991. Cell surface properties of *Escherichia coli* and *Staphylococcus epidermidis*. In: Mozes, N., P.S. Handley, H.J. Busscher, and P.G. Rouxhet (eds.). *Microbial Cell Surface Analysis: Structural and Physicochemical Methods*. VCH Publishers, New York, pp. 339–356.

Gilboa-Garber, N. 1982. *Pseudomonas aeruginosa* lectins. *Meth. Enzymol.* **83:**378–385.

Goldhar, J., R. Perry, J.R. Golecki, H. Hoschutzky, B. Jann and K. Jann. 1987. Nonfimbrial, mannose-resistant adhesins from uropathogenic *Escherichia coli* 083:K1:H4 and 014:K?:H11. *Infect. Immun.* **55:**1837–1842.

Gotschlich, E.C. 1983. Thoughts on the evolution of strategies used by bacteria for evasion of host defenses. *Rev. Infect. Dis.* **5**(*Suppl.* 4):S778–S783.

Gunnarsson, A., P.A. Mardh, A. Lundblad, and S. Svensson. 1984. Oligosaccharide structures mediating agglutination of sheep erythrocytes by *Staphylococcus saprophyticus. Infect. Immun.* **45:**41–46.

Hacker, J. 1990. Genetic determinants coding for fimbrial adhesins of extra-intestinal *Escherichia coli. Curr. Top. Microbiol. Immunol.* **151:**1–27.

Hancock, I.C. 1991. Microbial cell surface architecture. In: Mozes, N., P.S. Handley, H.J. Busscher, and P.G. Rouxhet (eds.), *Microbial Cell Surface Analysis: Structural and Physicochemical Methods*. VCH publishers, New York, pp. 21–59.

Handley, P.S. 1990. Structure, composition and functions of surface structures of oral bacteria. *Biofouling* **2:**239–264.

Handley, P.S. 1991a. Negative staining. In: Mozes, N., P.S. Handley, H.J. Busscher, and P.G. Rouxhet (eds.), *Microbial Cell Surface Analysis: Structural and Physicochemical Methods*. VCH Publishers, New York, pp. 63–86.

Handley, P.S. 1991b. Detection of cell surface carbohydrate components. In: N. Mozes, P.S. Handley, H.J. Busscher, and P.G. Rouxhet (eds.), *Microbial Cell Surface Analysis: Structural and Physicochemical Methods*. VCH Publishers, New York, pp. 87–107.

Handley, P.S. J. Hargreaves, and D.W. Harty. 1988. Ruthenium red staining reveals surface fibrils and a layer external to the cell wall in *Streptococcus salivarius* HB and adhesion deficient mutants. *J. Gen. Microbiol.* **134:**3165–3172.

Heesemann, J. and L. Gruter. 1987. Genetic evidence that the outer membrane protein YOP1 of *Yersinia enterocolitica* mediates adherence and phagocytosis resistance to human epithelial cells. *FEMS Microbiol. Let.* . **40:**37–41.

Hogt, A.H., J. Dankert, and J. Feijen. 1983. Encapsulation, slime production and surface hydrophobicity of coagulase-negative staphylococci. *FEMS Microbiol. Lett.* **18:**211–215.

Hultgren, S.J., J.L. Duncan, A.J. Schaeffer, and S.K. Amundsen. 1990. Mannose-sensitive haemagglutination in the absence of piliation in *Escherichia coli. Mol. Microbiol.* **4:**1311–1318.

Irvin, R.T. 1990. Hydrophobicity of proteins and bacterial fimbriae. In: Doyle, R.J. and M. Rosenberg (eds.), *Microbial Cell Surface Hydrophobicity*. American Society for Microbiology, Washington, pp. 137–177.

Isaacson, R.E. 1980. Factors affecting expression of the *Escherichia coli* pilus K99. *Infect. Immun.* **28:**190–194.

Isaacson, R.E. 1983. Regulation of expression of *Escherichia coli* pilus K99. *Infect. Immun.* **40:**633–639.

Ishiguro, E.E., W.W. Kay, T. Ainsworth, J.B. Chamberlain, R.A. Austen, J.T. Buckley, and T.J. Trust. 1981. Loss of virulence during culture of *Aeromonas salmonicida* at high temperature. *J. Bacteriol.* **148:**333–340.

James, A.M. 1991. Charge properties of microbial cell surfaces. p. 221–262. In: Mozes, N., P.S. Handley, H.J. Busscher, and P.G. Rouxhet (eds.), *Microbial Cell Surface Analysis: Structure and Physicochemical Methods*. VCH Publishers, New York, pp. 339–356.

Janda, J.M. and P.S. Duffy. 1988. Mesophilic aeromonads in human disease: current taxonomy, laboratory identification and infection disease spectrum. *Rev. Infect. Dis.* **10:**980–997.

Janda, J.M., L.S. Oshiro, S.K. Abbott, and P.S. Duffey. 1987. Virulence markers of the mesophilic aeromonads: association of the autoagglutination phenomenon with mouse pathogenicity and the presence of the peripheral cell-associated layer. *Infect. Immun.* **55:**3070–3077.

Jann, K. and H. Hoschützky. 1990. Nature and organization of adhesins. *Curr. Top. Microbiol. Immunol.* **151:**55–70.

Jann, K., G. Schmidt, E. Blumenstock, and K. Vosbeck. 1981. *Escherichia coli* adhesion to *Saccharomyces cerevisiae* and mammalian cells: role of piliation and surface hydrophobicity. *Infect. Immun.* **32:**484–489.

Knox, K.W. and A.J. Wicken. 1973. Immunological properties of teichoic acids. *Bacteriol. Rev.* **37:**215–257.

Koch, A.L., M.L. Higgins, and R.J. Doyle. 1982. The role of surface stress in the morphology of microorganisms. *J. Gen. Microbiol.* **128:**927–945.

Korhonen, T.K., R. Virkola, B. Westerlund, H. Holthofer, and J. Parkkinen. 1990. Tissue tropism of *Escherichia coli* adhesins in human extraintestinal infection. *Curr. Top. Microbiol. Immunol.* **151:**115–127.

Koval, S.F. 1988. Paracrystalline protein surface arrays on bacteria. *Can. J. Microbiol.* **34:**407–414.

Krell, P.J. and T.J. Beveridge. 1987. The structure of bacteria and molecular biology of viruses. *Int. Rev. Cytol.* **175:**15–88.

Krogfelt, K.A. 1991. Bacterial adhesion: genetics, biogenesis and role in pathogenesis of fimbrial adhesins of *Escherichia coli*. *Rev. Infect. Dis.* **13:**721–735.

Kroncke, K.D., I. Orskov, F. Orskov, B. Jann, and K. Jann. 1990. Electron microscopic study of coexpression of adhesive protein capsules and polysaccharide capsules in *Escherichia coli*. *Infect. Immun.* **58:**2710–2714.

Lachica, R.V. 1990. Significance of hydrophobicity in the adhesion of pathogenic Gram-negative bacteria: In: Doyle, R.J. and M. Rosenberg (eds.), *Microbial Cell Surface Hydrophobicity:* American Society for Microbiology, Washington, pp. 297–313.

Layh-Schmitt, G., S. Schmitt, and T.M. Buchanan. 1989. Interaction of non-pilated *Neisseria gonorrhoeae* strain 7122 and protein 1A with an epithelial cell monolayer. *Int. J. Med. Microbiol.* **271:**158–170.

Lesher, R.J., G.R. Bender, and R.E. Marquis. 1977. Bacteriolytic action of fluoride ions. *Antimicrob. Agents Chemother.* **12:**339–345.

Liang, L., D. Drake and R.J. Doyle. 1989. Stability of the glucan-binding lectin of oral streptococci. *J. Dent. Res.* **68**(Spec. Issue):1677.

Macnab, R.M. 1989. Motility and chemotaxis. In: Neidhardt, F.C., J.L. Ingraham, K.B. Low, B. Magasanik, M. Schaechter, and H.E. Umbarger (eds.), *Escherichia coli and Salmonella typhimurium: Cellular and Molecular Biology*. American Society for Microbiology, Washington, pp. 732–759.

McCoy, E.C., D. Doyle, K. Burda, L.B. Corbeil, and A.J. Winter. 1975. Superficial antigens of *Campylobacter (Vibrio) fetus:* characterization of the antiphagocytic component. *Infect. Immun.* **11:**517–525.

McIntire, F.C., A.E. Vatter, J. Baros, and J. Arnold. 1978. Mechanism of coaggregation between *Actinomyces viscosus* T14V and *Streptococcus sanguis* 34. *Infect. Immun.* **21:**978–988.

McSweegan, E. and R.I. Walker. 1986. Identification and characterization of two *Campylobacter jejuni* adhesins for cellular and mucous substrates. *Infect. Immun.* **53:**141–148.

Meyer, T.F. 1990. Variation of pilin and opacity-associated protein in pathogenic *Neisseria* species. In: Iglewski, B. and V.L. Clark (eds.), *Molecular Basis of Bacterial Pathogenesis*. Academic Press, San Diego, pp. 137–153.

Mobley, H.L. T., L.K. Jolliffe, and R.J. Doyle. 1983. Cell wall-polypeptide complexes in *Bacillus subtilis*. *Carbohydr. Res.* **116:**113–125.

Moon, H. 1990. Colonization factor antigens of enterotoxigenic *Escherichia coli* in animals. *Curr. Topics Microbiol. Immunol.* **151:**147–165.

Mozes, N., P.S. Handley, H.J. Busscher, and P.G. Rouxhet (eds.). 1991. *Microbial Cell Surface Analysis: Structure and Physicochemical Methods*. VCH Publishers, New York.

Nesbitt, W.E., R.H. Staat, B. Rosan, K.G. Taylor, and R.J. Doyle. 1980. Association of proteins with the cell wall of *Streptococcus mutans*. *Infect. Immun.* **28:**118–126.

Nikaido, H. and M. Vaara. 1985. Molecular basis of bacterial outer membrane permeability. *Microbiol. Rev.* **49:**1–32.

Ofek, I., W.A. Simpson, and E.H. Beachey. 1982. Formation of molecular complexes between a structurally defined M protein and acylated or deacylated lipoteichoic acid of *Streptococcus pyogenes*. *J. Bacteriol.* **149:**426–433.

Ofek, I., E. Whitnack, and E.H. Beachey. 1983. Hydrophobic interactions of group A streptococci with hexadecane droplets. *J. Bacteriol.* **154:**139–145.

Old, D.C. and J.P. Duguid. 1970. Selective outgrowth of fimbriate bacteria in static liquid medium. *J. Bacteriol.* **103:**447–456.

Olsen, A., A. Jonsson, and S. Normark. 1989. Fibronectin binding mediated by a novel class of surface organelles on *Escherichia coli*. *Nature* **338:**652–655.

Onderdonk, A.B., N.E. Moon, D.L. Kasper, and J.G. Bartlett. 1978. Adherence of *Bacteroides fragilis in vivo*. *Infect. Immun.* **19:**1083–1087.

Orskov, I. and F. Orskov. 1990. Serologic classification of fimbriae. *Curr. Top. Microbiol. Immunol.* **151:**71–90.

Orskov, I., A. Birch-Anderson, J.P. Duguid, J. Stenderuys, and F. Orskov. 1985. An adhesive protein capsule of *Escherichia coli*. *Infect. Immun.* **47:**191–200.

Ottow, J.C. G. 1975. Ecology, physiology and genetics of fimbriae and pili. *Annu. Rev. Microbiol.* **29:**79–108.

Oudega, B. and F.K. DeGraaf. 1988. Genetic organization and biogenesis of adhesive fimbriae of *Escherichia coli*. *Antonie van Leeuwenhoek* **54:**285–299.

Oyston, P.C. and P.S. Handley. 1991. Surface components of *Bacteroides fragilis* involved in adhesion and hemagglutination. *J. Med. Microbiol.* **34:**51–55.

Paerregaard, A., F. Espersen, and M. Skurnik. 1991. Role of *Yersinia* outer membrane protein YadA in adhesion to rabbit intestinal tissue and rabbit intestinal brush border membrane vesicles. *APMIS* **99:**226–232.

Paranchych, W. 1990. Molecular studies on *N*-methylphenylalanine pili. In: Iglewski, B. and V.L. Clark (eds.), *Molecular Basis of Bacterial Pathogenesis*. Academic Press, San Diego, pp. 61–78.

Paranchych, W. and L.S. Frost. 1988. The physiology and biochemistry of pili. *Adv. Microbial Physiol.* **29:**53–114.

Pearce, W.A. and T.M. Buchanan. 1980. Structure and cell membrane-binding properties of bacterial fimbriae. In: Beachey, E.H. (ed.), *Bacterial Adherence (Receptors and Recognition, Vol. 6)*. Chapman and Hall, London, pp. 288–344.

Perry, A. and I. Ofek. 1984. Inhibition of blood clearance and hepatic tissue binding of *Escherichia coli* by liver lectin-specific sugars and glycoproteins. *Infect. Immun.* **43:**257–262.

Plummer, D.T., A.M. James, and W.R. Maxted. 1962. Some physical investigations of the behavior of bacterial surfaces. V. The variation of the surface structure of streptococci during growth. *Biochim. Biophys. Acta* **60:**595–603.

Ramphal, R., C. Gray, and G. Pier. 1987. *Pseudomonas aeruginosa* adhesins for tracheobronchial mucin. *Infect. Immun.* **55:**600–603.

Ramphal, R., L. Koo, K.S. Ishimoto, P.A. Totten, J.C. Lara, and S. Lory. 1991. Adhesion of *Pseudomonas aeruginosa* pilin-deficient mutants to mucin. *Infect. Immun.* **59:**1307–1311.

Razin, S. 1986. Mycoplasmal adhesins and lectins. In: D. Mirelman (ed.). *Microbial Lectins and Agglutinins: Properties and Biological Activity*. John Wiley & Sons, New York, pp. 217–235.

Rogers, H.J., H.R. Perkins, and J.B. Ward. 1980. *Microbial Cell Walls and Membranes*. Chapman and Hall, London.

Rosenberg, M. and R.J. Doyle. 1990. Microbial cell surface hydrophobicity: history, measurement and significance. In: Doyle, R.J. and M. Rosenberg (eds.), *Microbial Cell Surface Hydrophobicity*. American Society for Microbiology, Washington, pp. 1–37.

Rouxhet, P.G. and M.J. Genet. 1991. Chemical composition of the microbial cell surface by x-ray photoelectron spectroscopy. In: Mozes, N., P.S. Handley, H.J. Busscher, and P.G. Rouxhet (eds.), *Microbial Cell Surface Analysis: Structural and Physicochemical Methods*. VCH Publishers, New York, pp. 173–220.

Runnels, P.L. and H.W. Moon. 1984. Capsule reduces adherence of enterotopathogenic *Escherichia coli* to isolated intestinal epithelial cells of pigs. *Infect. Immun.* **45:**737–740.

Schifferli, D.M. and E.H. Beachey. 1988a. Bacterial adhesion: modulation by antibiotics which perturb protein synthesis. *Antimicrob. Agents Chemother.* **32:**1603–1608.

Schifferli, D.M. and E.H. Beachey. 1988b. Bacterial adhesion: modulation by antibiotics with primary targets other than protein synthesis. *Antimicrob. Agents Chemother.* **32:**1609–1613.

Sherman, P., F. Cockerill III, R. Soni, and J. Brunton. 1991. Outer membranes are competitive inhibitors of *Escherichia coli* 0157:H7 adherence to epithelial cells. *Infect. Immun.* **59:**890–999.

Sonnenfeld, E.M., T.J. Beveridge, A.L. Koch, and R.J. Doyle. 1985. Asymmetric distribution of charge on the cell wall of *Bacillus subtilis*. *J. Bacteriol.* **163:**1167–1171.

St. Geme, J.W. III and S. Falkow. 1991. Loss of capsule expression by *Hemophilus influenzae* type b results in enhanced adherence to and invasion of human cells. *Infect. Immun.* **59:**1325–1333.

Sutherland, I.W. 1977. Bacterial exopolysaccharides, their nature and production. In: Sutherland, I.W. (ed.), *Surface Carbohydrates of the Prokaryotic Cell*. Academic Press, London, pp. 27–96.

Swanson, J. 1988. Genetic mechanisms responsible for changes in pilus expression by gonococci. *UCLA Symp. Mol. Cell Biol. New Ser.* **20:**347–363.

Swanson, J., E. Sparks, D. Young, and G. King. 1975. Studies on gonococcus infection. X. Pili and leukocyte association factor as mediators of interactions between gonococci and eukaryotic cells *in vitro*. *Infect. Immun.* **11:**1352–1361.

Tuomanen, E. 1986. Piracy of adhesins: attachment of superinfecting pathogens to respiratory cilia by secreted adhesins of *Bordetella pertussis*. *Infect. Immun.* **54:**905–908.

Tylewska, S. and W. Hryniewicz. 1987. *Streptococcus pyogenes* cell wall protein responsible for binding to pharyngeal epithelial cells. *Zbt. Bakt. Microb. Hyg.* **A265:**146–150.

Van der Mei, H.C., A.J. Leonard, A.H. Weerkamp, P.G. Rouxhet, and H.J. Busscher. 1988. Surface properties of *Streptococcus salivarius* HB and nonfibrillar mutants. Measurement of zeta potential and elemental composition with x-ray photoelectron spectroscopy. *J. Bacteriol.* **170:**2462–2466.

Wentworth, J., F.E. Austin, N. Garber, N. Gilboa-Garber, C. Paterson, and R.J. Doyle. 1991. Cytoplasmic lectins contribute to the adhesion of *Pseudomonas aeruginosa*. *Biofouling* **4:**99–104.

Westergren, G. and J. Olsson. 1983. Hydrophobicity and adherence of oral streptococci after repeated subculture *in vitro*. *Infect. Immun.* **40:**432–435.

Wright, S.D., S.M. Levine, M.C. T. Jong, Z. Chad, and L.G. Kabbash. 1989. CR3 (CD11b/CD18) expresses one binding site for Arg-Gly-Asp-containing peptides and a second site for bacterial lipopolysaccharide. *J. Exp. med.* **169:**175–183.

5

Bacterial Lectins as Adhesins

In 1977, Ofek et al. suggested that proteins with lectin-like properties on bacterial surfaces could serve as adhesins that bind the organisms to animal cells. It was found that *E. coli,* bearing type 1 fimbriae specific for mannose, could agglutinate red cells. The adhesins of many pathogenic bacteria are now thought to be carbohydrate-binding proteins, possibly lectins (Table 5–1). Although some members of the genera *Staphylococcus* and *Streptococcus* appear to express adhesins that lack lectin activity, other members of the same genera are known to possess surface lectins with adhesin functions. Lectins that serve as adhesins may be associated with the cell wall, the outer membrane, or with fimbrial structures. More detailed information on the molecular biology of these lectins is provided in Chapter 9. In this chapter information is presented concerning the occurrence and specificities of some bacterial surface lectins and their role in infection. Many bacterial lectins have not yet been defined as adhesins. A comprehensive review of bacterial lectins is found in the book edited by Mirelman (1986).

Specificities of Bacterial Lectins

Carbohydrate-containing lectin receptors on animal cell surfaces fulfill one of the criteria for biological receptors: they mediate specific binding which is followed by a physiologically relevant response (Chapter 3). The approaches used to determine the carbohydrate specificities of the bacterial lectins are based on procedures commonly used in the studies on ligand–animal cell receptor interactions (Sharon et al., 1981; Ofek et al., 1985). The initial step involves the testing of the effects of a panel of carbohydrates (monosaccharides, simple glycosides, and oligosaccharides) on bacterial adhesion to, or agglutination of, erythrocytes or other cells. This step is based on the well-known hapten inhibition techniques used in the study of the specificities of antibodies and plant lectins (Kabat, 1978).

It should be considered that carbohydrate specificities of bacterial lectins may depend not only on the primary structure of the lectin but also on the relationship of the lectin to other constituents of the bacterial surface. For example, in some fimbriae the lectins are thought to exist on the very tips of the organelles (see Chapter 9). The structural proteins of the fimbriae may affect the combining site of the lectin. In solution, purified lectins may or may not be able to bind carbohydrates. Most of the studies related to the carbohydrate specificities (Table 5–1), as well as in the following discussions, employed intact bacteria. For a few, the complex form of the lectin, such as purified fimbrial preparations, has been employed. This may be relevant to the adhesin function of the bacterial lectins which are presented on the bacterial surface in complex form.

The specificity of the lectin is ultimately defined by the carbohydrate that will occupy all or part of the lectin combining sites. Usually, it is determined by the carbohydrates that is the best inhibitor of the adhesion or agglutination reaction (Table 5–1). Alternatively, agglutination of cells, such as yeasts, or particles, such as Latex beads (Ofek and Beachey, 1978; de Man et al., 1987) carrying defined carbohydrates, may serve as substrata. Important information about the structures of the receptors on the animal cells may be obtained by examining the effects on the agglutination or adhesion following the treatment of the cells with enzymes such as glycosidases or galactose oxidase or with chemical reagents (e.g., periodate) that modify carbohydrates (reviewed in Ofek et al., 1985).

For further identification of the carbohydrates that serve as attachment sites on cell surfaces the receptor must be isolated and characterized. In general, the receptor molecules that carry the carbohydrate structures complementary to the bacterial lectins may be glycoproteins and/or glycolipids (Rauvala and Finne, 1979). For receptors that are glycoproteins, techniques using affinity chromatography on immobilized bacterial lectins (Rodriguez-Ortega et al., 1987; Giampapa et al., 1988), or overlay of blots of electrophoretograms of cell membrane glycoproteins, are used (Prakobphol et al., 1987). For receptors that are glycolipids, membrane glycolipids are separated on thin-layer silica gel plates and the chromatograms are probed with the lectin-carrying bacteria (Hansson et al., 1985). Radioactive bacteria bind to the sites on the gel plates that contain molecules complementary to the surface lectins. Autoradiography will reveal the areas of bacterial adhesion to the plate. Although bacterial lectins may bind to glycoconjugates on silica gels, the absolute carbohydrate specificity may not easily be determined. For example, Stromberg and Karlsson (1990b) found that the adhesion of *Propionibacterium granulosum* to lactosylceramide is affected by the ceramide structure. They observed that the bacteria require the presence of a 2-hydroxy fatty acid and/or a trihydroxy base on the lipid structure. When a nonhydroxy fatty acid or a dihydroxy base is substituted, the bacteria fail to adhere. Neither free lactose nor lactose conjugated to bovine serum albumin

Table 5–1. Carbohydrate specificities of common bacterial surface lectins

Carbohydrate	Bacteria	Fimbriae	Reference(s)
Mannose	*Citrobacter freundii, Enterobacter* spp., *Erwinia carotorora, Escherichia coli, Klebsiella aerogenes, Klebsiella pneumoniae, Salmonella* spp., *Serratia marcescens, Shigella flexneri, Pseudomonas aeruginosa*	Type 1	Duguid and Old 1980; Nilsson et al., 1983
Galactose	*Escherichia coli, Fusobacterium nucleatum, Pseudomonas aeruginosa, Bordetella pertussis*		Nilsson et al., 1983; Murray et al., 1988; Tuomanen et al., 1988
L-Fucose	*Vibrio cholerae, Pseudomonas aeruginosa*		Jones and Freter, 1976; Gilboa-Garber, 1986
L-Rhamnose, D-Fucose	*Capnocytophaga ochracea*		Weiss et al., 1987
N-Acetylgalactosamine	*Eikenella corrodens, Escherichia coli*		Yamazaki et al., 1981; Faris et al., 1980
N-Acetylglucosamine	*Escherichia coli*	Type G	Vaisanen-Rhen et al., 1983
Galα1,4Galβ	*Escherichia coli*	Type P	Leffler and Svanborg-Eden, 1986; Vaisanen-Rhen et al., 1984
Galβ1,3GalNAc	*Actinomyces naeslundii, Actinomyces viscosus, Bacteroides* spp., *Clostridium* spp., *Lactobacillus* spp.	Type 2	Cisar, 1986; Brennan et al., 1987; Hansson et al., 1983
Gal(SO_3)	*Helicobacter pylori*		Saitoh et al., 1991; Lingwood et al., 1989; Slomiany et al., 1989
Galβ1,4GlcNAc	*Erwinia rhapontici, Staphylococcus saprophyticus, Pseudomonas aeruginosa*		Korhonen et al., 1988; Ramphal et al., 1991; Gunnarsson et al., 1984
GalNAcβ1,4Gal	*Haemophilus influenzae, Klebsiella pneumoniae, Neisseria gonorrhoeae, Pseudomonas aeruginosa, Pseudomonas cepacia*		Krivan et al., 1988a,b; Stromberg et al., 1988
GlcNAcβ1,3Gal	*Streptococcus pneumoniae*		Andersson et al., 1983; Pulverer et al., 1987; Smit et al., 1984
NeuGcα2,3Galβ1,4Glc	*Escherichia coli*	K99	Smit et al., 1984

continued

Table 5–1. Continued

Carbohydrate	Bacteria	Fimbriae	Reference(s)
NeuNAcα2,3Galβ	*Escherichia coli, Bordetella bronchiseptica, Campylobacter (Helicobacter) pyelori, Mycoplasma gallisepticum, Mycoplasma pneumoniae, Steptococcus sanguis*	Type S	Parkinnen et al., 1986; Korhonen et al., 1984; Ishikawa and Isayama, 1987; Evans et al., 1988; Glasgow and Hill, 1980; Loomes et al., 1984; Murray et al., 1982
$(Glc\alpha 1,6)_{6-9}$	*Streptococcus cricetus, Streptococcus sobrinus*		Drake et al., 1988; Landale and McCabe, 1987
Galβ1,3GlcNAc	*Pseudomonas aeruginosa*		Ramphal et al., 1991
Galβ1,4Glcβ	*Escherichia coli Propionibacterium* spp.	Prs	Senior et al., 1988; Karlsson, 1989; Stromberg and Karlsson, 1990b
GalNAcα1,3GalNAcβ	*Escherichia coli*	Prs	Lindstedt et al., 1989, 1991; Stromberg et al., 1990
Thiogalactosides	*Pseudomonas aeruginosa*		Garber et al., 1992

prevent the adhesin from binding lactosylceramide. Hydrophobicity and substitutions on ceramides may affect the affinity of a bacterial lectin for glycoconjugates commonly found in membranes. Although the number of known bacterial lectin receptors is still very small, it is already clear that they differ among the various cell types that bind the bacterial lectin (Karlsson, 1986). In this case the term "isoreceptor" was suggested to describe such a family of molecules. This kind of heterogeneity has been shown for the P-fimbrial lectin which has an absolute specificity for Galα1,4Gal (galabiose). The same may also be true for the mannose-specific type 1 fimbrial lectin, which binds to glycoproteins from guinea pig erythrocytes (Giampapa et al., 1988), and to three types of glycoproteins from polymorphonuclear leukocytes (Rodriguez-Ortega et al., 1987), all of which contain mannose residues. The accessibility of such isoreceptors may vary with the types of tissue. As a consequence, the affinity of the different bacterial strains carrying the same lectin for the isoreceptors of various tissues may not be the same (Goldhar et al., 1980; Sherman et al., 1985).

Bacteria that carry the genes coding for certain types of lectins may not always express the lectins on their surfaces (see Chapter 9). Due to the phenomenon of phase variation the bacteria switch back and forth from one phenotype that expresses the lectin to one that does not. Outgrowth of one phenotype over the other may occur in a bacterial population growing under specific conditions. In addition, bacterial cultures from the same clone frequently contain two or more types of lectins. This may be the result of either coexpression on the cell surface of multiple types of lectins or the presence of heterogeneous populations of cells, each bearing a different lectin. It may be helpful to establish the percentage of the total test population of bacteria that express a particular lectin (Maayan et al., 1985; Goochee et al., 1987). Bacterial surface lectins occur on both Gram-positive and Gram-negative cells. The lectins may be found on cocci or bacilli. The detailed carbohydrate specificities of all bacterial lectins are not known. Of the bacterial lectins noted in Table 5–1, a few are reviewed in more detail this chapter. These are the lectins for which the specificities have been reported most extensively.

Fimbrial Lectins of *Enterobacteriaceae*

Much of the knowledge on the surface lectins of *Enterobacteriaceae* is based on the pioneering studies performed in the 1950s and 1960s by Duguid and Old (1980) and Brinton (1965). Duguid and Old were the first to systematically study the hemagglutination properties of enteric bacteria. They observed that many of the bacteria agglutinated with erythrocytes of different animal species and classified the agglutinating bacteria into two groups: (1) One group agglutinated guinea pig erythrocytes very strongly, the agglutination of which was inhibited by low concentrations of mannose, methyl-α-mannoside, or yeast mannan. These bacteria were designated as mannose-sensitive (MS in brief); (2) The second group agglutinated human erythrocytes strongly, and were not inhibited by mannose, were designated as mannose-resistant (MR). Bacteria that caused MS or mannose-specific hemagglutination were shown to carry on their surfaces multiple filamentous protein-containing appendages that have been designed as fimbriae (or pili). Chapter 4 reviews the surface structures of bacteria. Several other types of fimbriae were identified at that time, most of which were associated with MR bacteria. The mannose-specific fimbriae are designated as "type 1 fimbriae." It is now well established that many of the MR strains exhibit distinct carbohydrate specificities (Table 5–1). Most bacterial strains are genotypically capable of producing more than one type of lectin (see also Chapter 9). This is particularly pronounced in the case of pathogenic enterobacteria which possess the ability to produce type 1 fimbriae as well as other fimbrial lectins or hemagglutinins (Lund et al., 1988). The hemagglutinins may or may not be carbohydrate-specific. Some of the hemagglutinins with a defined carbohydrate specificity may not always function as an adhesin. Some lectins may be involved in binding of carbohydrates for nutritional purposes.

Type 1 Fimbriae

Most of the information on the carbohydrate specificities of type 1 fimbriae is based on studies of the inhibitory effects of a variety of glycosides and oligosaccharides of mannose on the agglutination of yeasts by type 1 fimbriated *E. coli, K. pneumoniae,* and *Salmonellae* spp. (Firon et al., 1983, 1984). The best inhibitors of type 1 fimbriae of *E. coli* are the branched oligosaccharides Manα1,6 (Manα1,3) Manα1,6 (Manα1,3) ManαMeManα1,6 (Manα1,3) (Manα1, 6Manα1,2(Manα1,3)ManαOMe, and the trisaccharide Manα1,3Manβ1,4GlcNAc (Table 5–2). A similar pattern of specificity has recently been found by Neeser et al. (1986), who used guinea pig erythrocytes instead of yeasts as indicator cells. The disaccharide Manα1,3Man, as well as the tetrasaccharide Manα1,2Manα1,3Manβ1,4GlcNAc and the pentasaccharide Manα1,2Manα1, 2Manα1,3Manβ1,4GlcNAc, are poor inhibitors of yeast agglutination by *E. coli* (Firon et al., 1983). This has led to the suggestion that the combining site of the type 1 fimbrial lectin corresponds to the size of a trisaccharide (Figure 5–1) and that it is in the form of a depression or pocket on the surface of the lectin. Extended carbohydrate binding sites have been described for enzymes (e.g., lysosyme) and several plant lectins and antibodies (Sharon and Lis, 1989). In the case of the *E. coli* lectin, there are probably three adjacent subsites, each of which accommodates a monosaccharide residue.

The presence of a hydrophobic binding region adjacent to the binding site of type 1 fimbriae is indicated by the finding that aromatic α-mannosides are powerful inhibitors of the agglutination of yeasts by *E. coli* and of the adhesion of the bacteria to guinea pig ileal epithelial cells (Firon et al., 1984, 1987) (Table 5–3). In both systems, the best inhibitors are 4-methylumbelliferyl α-mannoside and *p*-nitro-*o*-chlorophenyl-α-mannoside (500–1000 times more inhibitory than methyl-α-mannoside). 4-Methylumbelliferyl-α-mannoside was also more effective than methyl-α-mannoside in removing adherent *E. coli* from ileal epithelial cells. The interaction with the hydrophobic glycosides depends primarily on the presence of α-1inked mannose because (1) 4-methylumbelliferone is inactive even at a concentration that is well over 100 times higher than the concentration of 4-methylumbelliferyl-α-mannoside required for 50% inhibition of the *E. coli* lectin; and (2) both methylumbelliferyl-α-glucoside and *p*-nitrophenyl-β-mannoside are not inhibitory.

Although all strains of *E. coli,* as well as *K. pneumoniae,* exhibit essentially the same patterns of specificity, this is not the case with other enterobacteria (Firon et al., 1984). For example, with several *Salmonella* species, aromatic α-mannosides, as well as the trisaccharide Manα1,3Manβ1,4GlcNAc, are weaker inhibitors than methyl-α-mannoside. The combining site of different salmonellae species is probably smaller than that of *E. coli* or *K. pneumoniae,* and it is devoid of a hydrophobic region. Different combining sites appear to be expressed by several other mannose-specific bacterial lectins. Therefore, although classified under the general term mannose-specific (or mannose-sensitive), the fimbrial lectins of different genera and species are distinct with respect to their carbohydrate specificities. Within a given genus, however, all strains exhibit the same specificity.

Table 5–2. Relative inhibition by mannose derivatives of yeast agglutination by Escherichia coli *346 and by its type 1 fimbriae*

Inhibitor	*E. coli* 346	Isolated fimbriae
MeαMan	1.00	1.00
Mannose	0.8	—
Manα1,6Man	0.5	—
Manα1,2Man	1.3	1.8
Manα1,2Man	1.2	—
Manα1,2Manα1,2Man	1.4	—
Manβ1,4GlcNAc	21	30
Manα1,6Manβ1,4GlcNAc	0.7	—
Manα1,2Manα1,3Manβ1,4GlcNAc	0.7	5.5
Manα1,2Manα1,2Manα1,3Manβ1,4GlcNAc	0.7	—
Manα1,6 \\ Manα1,6 ManαOMe \\ / Manα1,3 / Manα1,3	3.5	—
Manα1,2Manα1,6 \\ Manα1,6 ManαOMe \\ / Manα1,3 / Manα1,3	4.7	4.8
Manα1,6 \\ ManαOMe / Manα1,3	10.5	—
Manα1,6 \\ Manα1,6 / \\ Manα1,3 ManαOMe / Manα1,3	30	—
Manα1,6 \\ Manα1,6 / \\ Manα1,3 ManαOMe / Manα1,2Manα1,3	30	36
Galβ1,4GlcNAcβ1,2Manα1,6Manβ1,4GlcNAc	0.25	—
Galβ1,4GlcNAcβ1,2Manα1,6 Manβ1,4GlcNAc Galβ1,4GlcNAcβ1,2Manα1,3	0.6	—

Adapted from Firon et al., 1984.

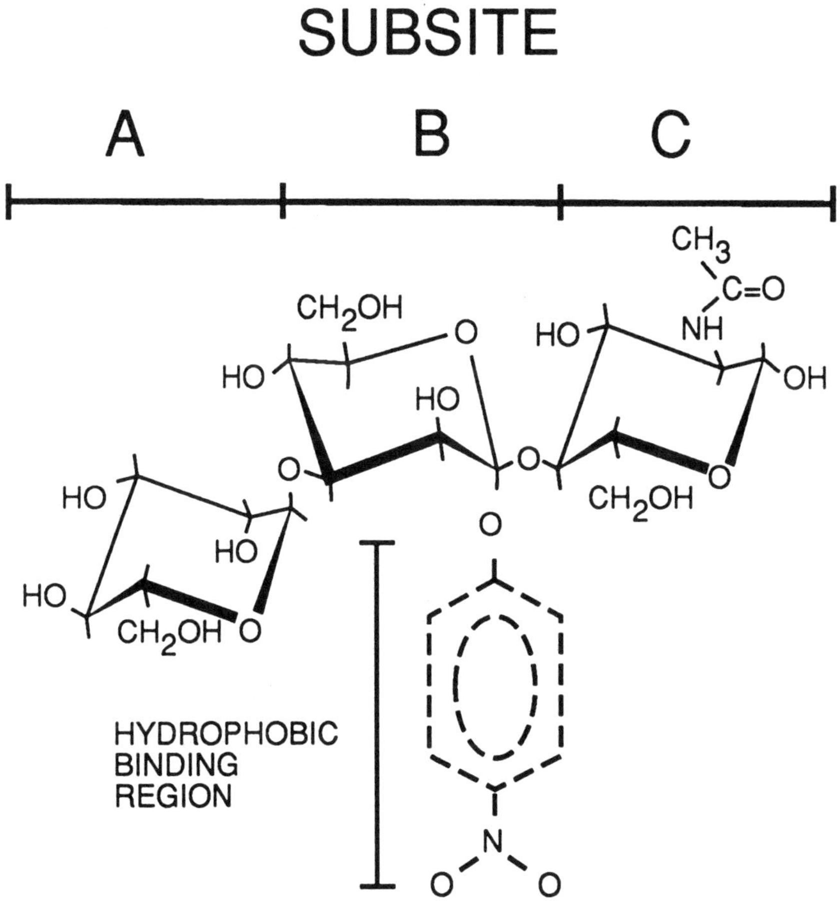

Figure 5–1. Postulated combining site stereochemistry of the mannose-specific fimbriae of *Escherichia coli*. (Adapted from Ofek and Sharon, 1990.)

In general, it appears that mannose-specific bacteria preferentially bind structures found in short oligomannose chains (and in hybrid units, see below) of *N*-linked glycoproteins. Such structures are common constituents of many eukaryotic cell surfaces (Sharon and Lis, 1982), which accounts for the fact that mannose-specific bacteria bind to a wide variety of cells (Chapter 11). Animal membrane glycolipids are unlikely to serve as receptors for mannose-specific bacteria, because they are devoid of mannose residues. These conclusions are supported by studies of the binding of mannose-specific *E. coli* to mammalian cells that differ in the level of oligomannose units on their surfaces. Animal cells treated with swainsonine, an inhibitor of the processing of asparagine-linked oligosaccharide units of glycoproteins, express increased levels of oligomannose or hybrid

Table 5–3. Inhibition by aromatic α-mannosides of the interaction of Escherichia coli *with yeasts and intestinal epithelial cells*

	Relative inhibitory activity	
	Agglutination of yeasts	Adhesion to epithelial cells
Aglycon	*E. coli* 025	*E. coli* 0128
Methyl	1	1
Phenyl	40	ND
p-Nitrophenyl	69	70
p-Bromophenyl	72	150
p-Ethylphenyl	77	250
p-Methoxyphenyl	140	70
p-Ethoxyphenyl	154	240
p-Nitro-*o*-chlorophenyl	717	470
p-Methylumbelliferyl	600	1015

Adapted from Firon et al., 1987; ND, not determined.

type oligosaccharides on their surfaces, and decreased levels of complex oligosaccharides. Such cells bind increased numbers (1.5 to 2-fold) of *E. coli,* but binding to the cells of *E. coli* expressing a mannose-resistant lectin is unaffected (Elbein et al., 1981, 1983).

Mutants of baby hamster kidney (BHK) cells with increased levels of *N*-linked oligomannose or hybrid units in their glycoproteins bind large numbers of mannose-specific *E. coli* (Firon et al., 1985) (Figure 5–2). These cells are also more sensitive to agglutination of these bacteria than the parental wild-type cells. The best example is a mutant (RicR14) that lacks the enzyme *N*-acetylglucosaminyltransferase 1, which catalyzes the first step in the conversion of oligomannose units into complex ones. This mutant binds four times more type 1 fimbriated *E. coli* and is agglutinated at a rate more than 10 times faster than the parental BHK cells. Preference for binding to hybrid units is indicated from results of experiments with mutants that express high levels of such units on their surface. The *N*-linked oligomannoside receptors specific for type 1s fimbrial lectins probably reside on more than one type of molecule on the animal cell surface. For example, type 1 fimbriae bind to three types of glycoproteins derived from the cell membrane of human polymorphonuclear leukocytes (Rodriguez-Ortega et al., 1987).

P Fimbriae

Many pyelonephritogenic isolates of *E. coli* can express a fimbrial lectin specific for Galα1,4Gal residues. The lectin may serve as an adhesin that binds the bacteria to cells of the urinary tract. The fimbrium specific for Galα1,4Gal

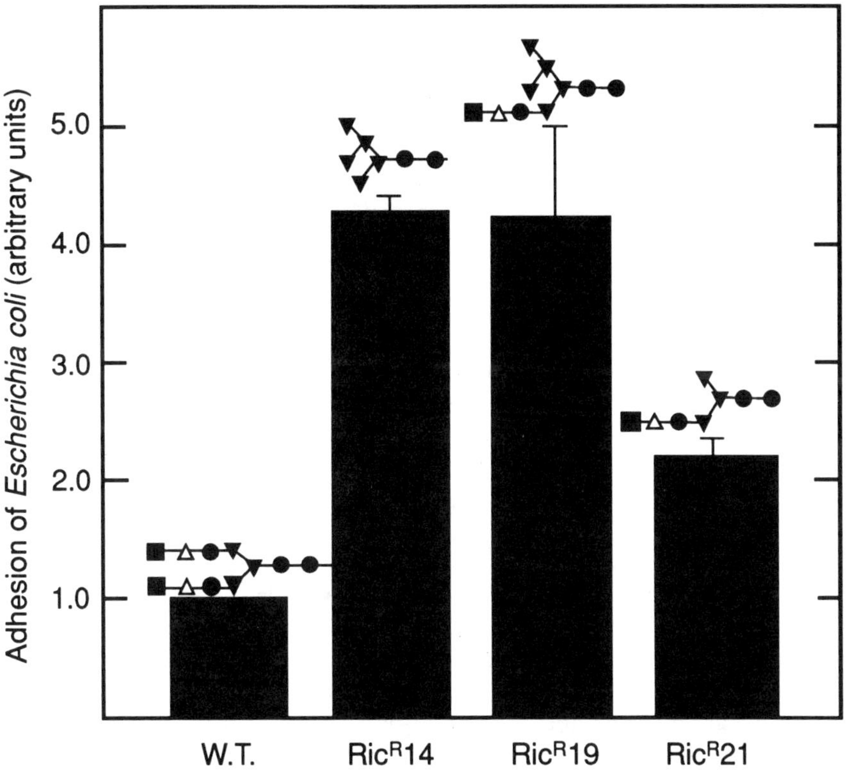

Figure 5–2. Binding of mannose-specific fimbriated *Escherichia coli* to BHK cells. The symbols above represent oligosaccharide expressed on the surface of BHK cells. ▲, mannose; ●, *N*-acetyl-D-glucosamine; △, galactose; ■, sialic acid. WT, wild-type cells; Ric^R, ricin-resistant mutants. The type 1 fimbriae bind optimally to exposed mannosyl residues. (Taken from Firon et al., 1985.)

residues is designated as type P because this disaccharide is common to the P blood group antigen. Only erythrocytes that carry the P blood group antigen are agglutinated by *E. coli* expressing P fimbriae. The type P fimbrial lectin is capable of binding the Galα1,4Gal sequence, regardless of whether the disaccharide is linked internally or by its reducing end (Leffler and Svanborg-Eden, 1986). It should be pointed out that most plant lectins recognize carbohydrates by their nonreducing termini, whereas bacterial lectins may bind internal as well as terminal linkages. Compounds containing the Galα1,4Gal disaccharide are the best inhibitors of agglutination of human erythrocytes by type P fimbriated *E. coli* and of adhesion of these bacteria to epithelial cells (Table 5–4). Monosaccharides do not inhibit the attachment of the bacteria to the cells. Globotetraose (but not methyl-α-mannoside) blocks the agglutination of human type P erythrocytes by the isolated fimbriae. The higher activity of Galα1,4GalβOEt, in comparison

Table 5–4. Inhibition of adhesion of type P fimbriated Escherichia coli *to human urinary epithelial cells*

Inhibitor	Inhibitory Concentration[a]
GalNAcβ1,3Galα1,4Galβ1,4Glcβ1-Cer (globotetraosylceramide)	0.1
Galα1,4Galβ1,4Glcβ1,O$-(CH_2)_2-$S$-(CH_2)_{17}-CH_3$	0.2
GalNAcβ1,3Galα1,4Galβ1,4Glc	0.2
Galα1,4Galβ1,4Glcβ1,OEt	2
Galα1,4GalβOEt	2
Galα1,4Gal	11
GalNAcβ1,3GalαOMe	>50
Galβ1,4GlcβOEt	>50

Data adapted from Leffler and Svanborg-Eden, 1986.

[a]Concentration (m*M*) required for complete inhibition of the adhesion of 50 bacteria/cell.

with Galα1,4Gal, indicates the importance of groups linked at the reducing end of the disaccharide (in this case the ethyl group) or the configuration of the nonterminal galactose. The considerably higher activity of globotetraose demonstrates the importance of the terminal *N*-acetylgalactosamine for interaction with the type P fimbrial lectin.

The receptor for P-specific *E. coli* on epithelial cells appears to be globotetraosylceramide. Such bacteria do not bind to epithelial cells (e.g., urinary tract cells of p individuals) lacking this substance. Also, erythrocytes not agglutinated by the bacteria or by the isolated type P fimbriae become susceptible to agglutination after they have been coated with globotetraosylceramide (or trihexosylceramide) (Leffler and Svanborg-Eden, 1980; Korhonen et al., 1982). Latex beads coated with Galα1,4Gal, or suitable derivatives of this disaccharide, have been used to detect the presence of P fimbriae in clinical isolates of *E. coli* (de Man et al., 1987). Additional information on the specificity of type P fimbriae has been obtained in experiments in which the binding of the bacteria to glycolipids separated by thin-layer chromatography has been examined (Bock et al., 1985; Karlsson, 1986). In these studies, glycolipids obtained from various sources were separated on thin-layer gels and overlayed with radiolabeled bacteria carrying the P fimbrial lectin. The specificity of the bacterial binding to the chromatogram appears to be absolute for the Galα1,4Gal, irrespective of the source of the glycolipids. This disaccharide is carried by several types of glycolipid molecules, all of which bind the fimbrial lectin whether the disaccharide is placed terminally or internally (Karlsson, 1986). It has been suggested that cells contain on their surfaces a series of "isoreceptors," namely glycolipids having the active part in different saccharide environments (Karlsson, 1986). Computer-based calculations and molecular models of the preferred conformations of the Galα1,4Gal-containing glycolipids reveal that the disaccharide forms a bend, or knee, in the chain on which there is a continuous nonpolar surface surrounded by polar oxygens. The carbohydrate-combining site of the P-fimbrial lectin seems to bind to this side of the disaccharide.

Carbohydrate Specificity of Prs Fimbriae

The Prs fimbriae, closely related to P fimbriae, exhibit a carbohydrate specificity slightly different from that of P fimbriae. Whereas internal or terminal Galα1,4Gal sequence is sufficient for binding P fimbriae, the carbohydrate specificity of the Prs fimbriae is enhanced by the disaccharide GalNAcβ1,3Gal (Table 5–5). Thus, Prs carrying *E. coli* bind avidly to sequences containing GalNAc β1,3Galα1,4Gal. Sequences lacking the GalNAcβ1,3 Gal bind the organisms poorly and Galα1,4Gal sequences do not suffice to form a receptor for the Prs adhesin.

The *pap* gene cluster codes for adhesins specific for Galα1,3Galβ, but some Pap^{+} cells, called *prs* or *pap*-2, encode for a Pap-like adhesin. Two major studies have been performed with Prs-specific adhesins. Both studies have taken advantage of the fact that bacteria will bind their complementary receptors on silica gel plates or will aggregate Latex beads coated with glycoconjugates. Lindstedt et al. (1989) found that the minimal requirement for interaction with the Prs adhesin was the GalNAcβ1,3Galα1,4Galα structure (Table 5–5). Only weak interaction was found for Galα1,4Gal structures. The Forssman and the globo-A glycolipids both possess GalNAcβ1,3Galα1,4Galβ sequences. Somewhat similar results, obtained by Stromberg et al. (1990), revealed that even though Prs-bearing strains possess a relatively high affinity for Forssman and globo-A structures, there is considerable variation in affinities when clinical

Table 5–5. Receptors for the Prs adhesin of uropathogenic Escherichia coli

Glycolipid substratum	Relative adhesion
GalNAcα1,3GalNAcβ1,3Galα1,4Galβ1,4GlcCer (Forssman)	H
Galβ1,3GalNAcβ1,3Galα1,4Galβ1,4GlcCer	H
Fucα1,2Galβ1,3GalNAcβ1,3Galα1,4Galβ1,4GlcCer	H
GalNAcα1,3(Fucα1,2)Galβ1,3GalNAcβ1,3Galα1,4Galβ1,4GlcCer	H
GlcCer	N
GalCer	N
Galα1,4Galβ1,4GlcCer	N
Galα1,3Galβ1,4GlcCer	N
Galα1,4Galβ1,4GlcNAcβ1,3Galβ1,4GlcCer	L
GalNAcβ1,3Galα1,4Galβ1,4GlcCer	L
Galα1,4GalLatex	L

Results adapted from Stromberg et al. (1990), Lindstedt et al. (1989, 1991). Adhesion was based on binding of intact bacteria to glycolipids on thin-layer plates or aggregation of latex coated beads. H, high; L, low; N, no adhesion.

isolates are compared with cloned Prs adhesins. Stromberg et al. (1990) suggested that the variations in affinity for isoreceptors is related to the source of the isolate.

Sialic Acid-binding Lectins

Some strains of *E. coli* isolated from human and farm animals have been found to express fimbrial hemagglutinins specific for glycoconjugates containing sialic acids. This conclusion is based primarily on the observation that hemagglutination caused by fimbriated bacteria is decreased or completely abolished after sialidase treatment of the erythrocytes. Several fimbrial lectins with this type of specificity are known. They include K99 and F41, the adhesins expressed by enterotoxigenic strains isolated from piglets, calves, and lambs suffering from diarrhea; CFA/I and CFA/II, expressed by human enterotoxigenic isolates; and type S expressed by strains frequently isolated from newborn infants suffering from sepsis and meningitis. Sialic acid-specific bacterial lectins seem to have a common motif in amino acid sequences. Morschhauser et al. (1990) analyzed the sequences of several sialic acid binding adhesins and found that the basic amino acids lysine and arginine were common in putative active sites of the K99 adhesin, CFA1 adhesin, cholera B subunit, the *E. coli* labile toxin B subunit, and the fimbrial Sfa S adhesin (Figure 5–3).

The detailed carbohydrate specificity of the K99 fimbrial lectin was studied by Lindahl et al. (1987, 1988), who examined the effects of various derivatives of sialic acid-containing oligosaccharides on hemagglutination. The affinity of K99 fimbriae for *N*-glycolylneuraminic acid (NeuGc) was found to be twice that of *N*-acetylneuraminic acid (NeuNAc), and the monosaccharides were found to be less than one-tenth effective as inhibitors (Table 5–6). 2-Benzyl-*N*-acetylneur-

SfaS:	116	Lys	Ala	Arg	Ala	Val	Ser	Lys
K99:	132	Lys	-	Lys	Asp	-	Asp	Lys
CFAI:	56	Lys	-	Lys	Val	Ile	Val	Lys
CT-B:	62	Lys	-	Lys	Ala	Ile	Glu	Arg
LTI-B:	62	Lys	-	Lys	Ala	Ile	Glu	Arg

Figure 5–3. Comparison of the amino acid sequence segment between Lys-116 and Lys-122 of Sfa S to amino acid sequences of the sialic acid-binding proteins K99 adhesin, CFA/1 adhesin, cholera B subunit, and *E. coli* LT1 B subunit. Identical or functionally identical amino acids are in boxes. Dashes represent gaps introduced for optimal alignment. (Taken from Hacker, 1990.)

Table 5–6. Inhibition of hemagglutination caused by K99-carrying Escherichia coli *by monosaccharides and oligosaccharides*

Inhibitor	Inhibitory concentrations[a]
D-GlcNAc	300
D-GalNAc	230
L-Fucose	160
D-Galactose	120
NeuNAc	12
NeuGc	6.8
1-α-Benzylnonulosamine	5.2
2-α-Methyl-NeuNAc	3.7
2-β-Methyl-NeuNAc	14
2-α-Benzyl-NeuNAc	0.4
2-α-Benzyl-NeuNAc-methylester	20
2-α-Benzyl-NeuNAc-amide	115
3-Hydroxy-NeuNAc	6.9
4-*O*-Acetyl-NeuNAc	4.1
4-Epi-NeuNAc	3.5
2-α-Benzyl-5-noracetyl-NeuNAc	>6.0
4,7-di-*o*-Acetyl-NeuNAc	>2.5
2-α-Benzyl-8,9-isopropylidene-NeuNAc	18
2-α-Benzyl-9-*O*-acetyl-NeuNAc	3.2
9-*o*-Acetyl-NeuNAc	>3.0
NeuNAcα2,3Galβ1,4Glc	3.5
NeuNAcα2,6Galβ1,4Glc	3.1
NeuNAcα2,8NeuNAcα2,3Galβ1,4Glc	0.8

Data adapted from Lindahl et al., 1987.

[a]Concentration (m*M*) required for 50% inhibition of hemagglutination.

aminic acid was a better inhibitor than 2-methyl-*N*-acetylneuraminic acid, suggesting that a hydrophobic site is near the carbohydrate-binding site of the fimbrial lectin, similar to *E. coli* type 1 fimbriae and many plant lectins. Axially oriented hydroxyl groups, as in 4-epi-*N*-acetylneuraminic acid and 3-hydroxy-*N*-acetylneuraminic acid, seem to optimally fit the carbohydrate combining site. The fimbrial lectin shows a higher affinity for 4-*O*-acetylneuraminic acid than to *N*-acetylneuraminic acid with *O*-substituents at C7–C9.

Smit et al. (1984) have isolated glycolipids from horse erythrocytes that contain the attachment sites for binding the K99 fimbrial lectin. The receptor was identified as being the glycolipid NeuNAcα2,3Galβ1,4Glcβ1-ceramide. The purified glycolipid inhibits hemagglutination caused by the whole bacteria to intestinal cells. To characterize further the cellular receptor for K99 fimbrial adhesin, the ability of whole radiolabeled *E. coli* to bind glycolipids from intestinal epithelial cells of pigs was examined. No binding was observed for acid or nonacid glycolipids of adult pigs, whereas the organisms bind avidly to glycolipids obtained from piglets. Physicochemical methods suggested that the receptor structure of the piglet glycolipids is a NeuGcα1,3Galβ1,4Glcβ1-Cer (Teneberg et al., 1990). In other studies, it was shown that the best inhibitors of horse eythrocytes caused by whole bacteria as well as by isolated fimbriae were NeuGcα2,3Galβ1,4Glcβ1-Cer and NeuGcα2,3Galβ1,4GlcNAcβ1, 3Galβ1,4Glcβ1-Cer (Ono et al., 1989).

The inhibitory effects of various naturally occurring glycopeptides and oligosaccharides on hemagglutination caused by bacteria carrying CFA/I and CFA/II fimbriae were examined by Neeser et al. (1988) and Pieroni et al. (1988). Although defined carbohydrates were not employed to determine precisely the specificities of the lectins, several interesting findings emerged. First, only complex-type *N*-linked or human milk oligosaccharides were inhibitory. The inhibitory activity was dependent on the presence of sialic acids in the compounds. Second, there was a close analogy between the carbohydrate specificities of the CFA/I and CFA/II fimbrial lectins. Third, of the various sialylated glycoproteins of human erythrocyte membranes, CFA/I carrying *E. coli* bind a sialoglycoprotein with an apparent molecular weight of 26,000. Finally, the combining site of the lectins seems to best fit oligosaccharide structures of glycopeptides derived from human red blood cells, in spite of the fact that CFA/II expressing bacteria do not agglutinate human erythrocytes. The CS3 fimbriae of the CFA/II complex appear to have a high affinity for sialylgalactosides (Sjoberg et al., 1988). NeuAcLac is a potent inhibitor of hemagglutination of bovine erythrocytes caused by whole bacteria, whereas NeuGc and NeuNAc are marginal inhibitors. Neutral monosaccharides do not inhibit hemagglutination. The results emphasize the notion that even though the target cells contain glycoconjugates specific for the bacterial lectin, the cells may not be agglutinated by whole bacteria expressing the lectin, probably due to inaccessibility of the receptors or lectins on the surfaces of both types of cells (Ofek et al., 1981).

The specificity of the S-fimbrial lectin was investigated by using the overlay method in which erythrocyte cell membrane glycoproteins are electrophoresed, blotted, and incubated with radiolabeled *E. coli* bacteria expressing these fimbriae (Parkkinen et al., 1983). The only band that bound the bacteria was glycophorin A, suggesting that the latter contains an attachment site for the bacterial lectin. Binding of bacteria to glycophorin A-coated plates was inhibited by several sialylated oligosaccharides, the best inhibitor of which was the oligosaccharide NeuNAcα2,3GalNAc, which is also found in glycophorin A (Table 5–7).

The F41 Adhesin of Enterotoxigenic *E. coli*

Only a limited number of studies has been performed on *E. coli* strains bearing the F41 adhesin. The adhesin is fimbria-associated and is found as a causative agent of diarrhea in both humans and animals. The F-41 strains typically produce shiga-like toxins. The F41-specific *E. coli* do not agglutinate sialidase-treated red cells, suggesting the involvement of a sialic acid contribution to the receptor

Table 5–7. Inhibition of type S fimbriated Escherichia coli *binding to glycophorin A by oligosaccharides*

Saccharide	Concentration (m*M*) causing 50% inhibition
NeuNAcα2,3Galβ1,4Glc	4.9
NeuNAcα2,3Galβ1,4Glcol	5.0
NeuNAcα2,3Galβ1,4GlcNAc	4.8
NeuNAcα2,3Galβ1,3(NeuNAc2,6)GalNAc	2.1
NeuNAcα2,6Galβ1,3GalNAc	2.3
NeuNAcα2,6Galβ1,4Glc	>10
NeuNAcα2,6Galβ1,4GlcNAc	>10
NeuNAcα2,8NeuNAcα2,3Galβ1,4Glc	>10
NeuNAc	>50
Galβ1,4Glc	>50

Data from Parkkinen et al., 1986.

for the bacteria (Faris et al., 1980, 1981). In hemagglutination–inhibition assays, Wadstrom and Baloda (1986) showed that the receptor for the F41 adhesin probably contains both sialic acid and GalNAc (Table 5–8). Only a few carbohydrates have been tested for hemagglutination inhibition, so virtually nothing is known about linkages, size of combining site, or contribution of hydrophobicity.

Vibrio Lectins

Cholera vibrios produce a variety of hemagglutinins (Hanne and Finkelstein, 1982, Finkelstein and Hanne, 1982). Agglutination of human group O erythrocytes by *V. cholerae* and adhesion of the organism to brush borders is specifically inhibited by L-fucose and various glycosides of L-fucose, and to a lesser extent by mannose (Jones and Freter, 1976). The bacteria adhere specifically to agarose beads that carry covalently linked L-fucose on their surfaces. It has been suggested that structures containing L-fucose on eukaryotic cell surfaces may function as receptors for the vibrio hemagglutinins and may therefore be an important determinant of host susceptibility to these bacteria. The vibrio El-Tor biotype

Table 5–8. Specificity of the F41 adhesin of enteropathogenic Escherichia coli

GalNAc	0.16
GalN	0.35
GlcN	1.1
NeuNAc	1.1
Poly-α2,8NeuNAc	>100

Adapted from Wadstrom and Baloda (1986). Results are based on hemagglutination inhibition tests. Values shown are concentrations required to give 50% inhibition. Sialidase-treated red cells are not aggregated by F41 bearing *Escherichia coli*.

produces a cell-associated mannose-specific hemagglutinin that is also inhibited by fructose. The lectin is active on all human (A, B, O) and all chicken erythrocytes. A lectin specific for L-fucose can be detected in early exponential-phase growth in several strains of *V. cholerae,* including a number of mutants that are not inhibited by mannose.

Mycoplasma pneumoniae Lectin

Mycoplasma pneumoniae expresses a lectin specific for *N*-acetylneuraminic acid attached by a NeuNAcα2,3Galβ linkage to the terminal galactose residue of the poly-*N*-acetyllactosamine sequence of blood type I/i antigen (Loomes et al., 1984, 1985). This conclusion is based in part on measurements of the adhesion of sialidase-treated human erythrocytes that have been resialylated by specific sialytransferases to sheet cultures of the organism. Highest levels of binding are observed with erythrocytes having the sequence NeuNAcα2,3Galβ1,4GlcNAc on their surfaces. The binding was nearly 70-fold higher than with the desialylated erythrocytes, and sevenfold higher than with the untreated erythrocytes which have higher levels of sialic acid than any of the modified cells tested. Because a major sialylated product obtained with the Galβ1,4GlcNAc sialyltransferase corresponds to band 3 glycoprotein, the very high binding of the erythrocytes derivatized with this enzyme may be accounted for by assuming extra sialylation of band 3 oligosaccharides.

The specificity of the *M. pneumoniae*–erythrocyte interaction has also been investigated using oligosaccharides and glycoproteins as inhibitors of binding (Table 5–9). The preference for sialic acid α2,3-linked, rather than α2,6-linked, to galactose was confirmed by the finding that the two α2,3-linked isomers were more inhibitory than the corresponding α2,6-linked isomers. Sialic acid α2,6-linked to *N*-acetylgalactosamine was ruled out as a receptor sequence for *M. pneumoniae,* as none of the three submaxillary mucins known to contain this sequence gave inhibition. In its strong preference for sialic acid α2,3-linked to galactose residues, *M. pneumoniae* differs from *M. gallisepticum*. Because glycophorin is the major sialoglycoprotein of human erythrocyte membranes, it is not surprising that it binds readily to *M. gallisepticum,* and that asialoglycophorin binds poorly (Banai et al., 1978; Glasgow and Hill, 1980). Other sialoglycoproteins (and fetuin) bind poorly to *M. gallisepticum* although they inhibit to varying extents the binding of glycophorin to the organisms (Glasgow and Hill, 1980). Sialic acid, linked either α2,3 to galactose or α2,6 to *N*-acetylgalactosamine, binds equally well to the organism. Monosaccharides, including *N*-acetylneuraminic acid, are not inhibitory in any of the system. Clustering of sialic acid residues (as in glycophorin) therefore appears to be required for effective binding to *M. gallisepticum* (Glasgow and Hill, 1980).

The data summarized in Table 5–9 shows the strong inhibitory activity of

Table 5–9. Inhibition of binding of ^{51}*Cr-labeled erythrocytes to sheet cultures of* Mycoplasma pneumoniae *by oligosaccharides and glycoproteins*

Compound	Inhibitory Concentration[a]
Oligosaccharides	
NeuNAcα2,3Galβ1,4GlcNAcβ1,3Galβ1,4GlcNAc-Asn	16
NeuNAcα2,3Galβ1,4GlcNAc-Asn	88
NeuNAcα2,6Galβ1,4GlcNAc-Asn	640
NeuNAcα2,3Galβ1,4Glc	245
NeuNAcα2,6Galβ1,4Glc	2136
Glycoproteins and glycopeptides	
Polyglycosyl peptides	0.9
Bovine erythrocyte sialoglycoprotein (GP-2)	2.5
Glycophorin A	12.6
α_1-Acid glycoprotein	47.7

Data derived from Loomes et al., 1984.

[a]Concentration is expressed as nanomoles of sialic acid per milliliter giving 50% inhibition. No inhibition was observed with the following compounds (the highest micromolar concentrations tested are given in parentheses): NeuNAc (3236); bovine fetuin (292); submaxillary mucins—bovine (1070), ovine (1050), and porcine (470). The relative inhibitory activities of the glycoproteins may reflect the cooperative effects of multivalence of their carbohydrates.

oligosaccharides and glycoproteins containing the repeating *N*-acetyllactosamine sequence. Among the glycoproteins and glycolipids examined, the most potent inhibitors are the polyglycosyl peptides of human erythrocytes (largely derived from bands 3 and 4.5) and the bovine erythrocyte glycoprotein GP-2, both of which are rich in branched poly-*N*-acetyllactosamine sequences of type I antigen. Human glycophorin A and α_1-acid glycoprotein are substantially less active and fetuin is inactive, in accordance with their lack of poly-*N*-acetyllactosamine sequences. Inhibition experiments with glycolipids have also indicated that sialylated poly-*N*-acetyllactosamine sequences are the preferred sequences for *M. pneumoniae* binding. The availability of the variant erythrocytes of i-blood groups provided the strongest evidence for the importance of sialylated poly-*N*-acetyllactosamine sequences as receptors for *M. pneumoniae*. Cells of this blood group have a high content of linear poly-*N*-acetyllactosamine sequences which are more susceptible to digestion with endo-β-galactosidase than the corresponding branched sequences on erythrocytes of Group I. The binding of these cells to *M. pneumoniae* is decreased by 85% following treatment with endo-β-galactosidase, although only 5% of the total erythrocyte sialic acid was released by the glycosidase. Therefore, minor sialylated oligosaccharides of the poly-*N*-acetyllactosamine series (such as those carried by glycoprotein band 3 and 4.5) and glycolipids are the main receptors for *M. pneumoniae,* rather than the carbohydrate chains of the major sialoglycoproteins, which are not susceptible to digestion by endo-β-galactosidase.

Attempts to identify the cellular receptor that carries the attachment site for the *Mycoplasma* adhesin have employed immobilized glycoproteins overlayed with radiolabeled *Mycoplasma*. It was observed that adhesion occurs to proteins containing α2,3-linked sialic acid residues, confirming the sialic acid specificity of the *Mycoplasma* adhesin (Roberts et al., 1989). It was shown that *M. pneumoniae,* as well as other *Mycoplasma* spp., could bind to Gal (3SO_4)β1 residues (Table 5–10). Removal of sulfate from the glycolipids results in virtual elimination of bacterial adhesion. The data provide evidence that *M. pneumoniae* contains at

Table 5–10. Binding of Mycoplasma pneumoniae *to glycolipids on thin-layer chromatograms*

Name[a]	Structure	Binding[b]
Sulfatide	Gal(3SO_4)β1,1Cer	+++
Sulfatide	Gal(6SO_4)β1,1Cer	+++
Lactosylsulfatide	Gal(3SO_4)β1,4Glcβ1,1Cer	+++
Glucosylcer (CHM)	Glcβ1,1Cer	+
Lactosylcer (CDH)	Galβ,1,4Glcβ1,1Cer	++
Lacto-*N*-triaosylcer	GlcNAcβ1,3Galβ,1,4Glcβ1,1Cer	+
Paragloboside	Galβ1,4GlcNAcβ1,3Galβ1,4Glcβ1,1Cer	+
Galactosylcer (CMH)	Galβ1,1Cer	−
Trihexosylcer (CTH)	Galα1,4Galβ1,4Glcβ1,1Cer	−
Asialo GM2	GalNAcβ1,4Galβ1,4Glcβ1,1Cer	−
Globoside (GLA)	GalNAcβ1,3Galα1,4Galβ1,4Glcβ1,1Cer	−
Asialo GM1	Galβ1,3GalNAcβ1,4Galβ1,4Glcβ1,1Cer	−
GM3	NeuNAcα2,3Galβ1,4Glcβ1,1Cer	−
GM3 (NeuGc)	NeuGcα2,2Galβ1,4Glcβ1,1Cer	−
GM2	GalNAcβ1,4[NeuNAcα2,3]Galβ1,4Glcβ1,1Cer	−
GM1	Galβ1,3GalNAcβ1,4[NeuNAcα2,3]Galβ1,4Glcβ1,1Cer	−
Sialylparagloboside	NeuNAcα2,3Galβ1,4GlcNAcβ1,3Galβ1,4Glcβ1,1Cer	−
Sialylparagloboside (NeuGc)	NeuGcα2,3Galβ1,4GlcNAcβ1,3Galβ1,4Glcβ1,1Cer	−
GD1A	NeuNAcα2,3Galβ1,3GalNAcβ1,4[NeuNAcα2,3]Galβ1,4Glcβ1,1Cer	−
GD1b	Galβ1,3GalNAcβ1,4[NeuNAcα2,8NeuNAcα2,3]Galβ1,4Glcβ1,1Cer	−
GT1b	NeuNAcα2,3Galβ1,3GalNAcβ1,4[NeuNAcα2,8NeuNAcα2,3] Galβ1,4Glcβ1,1Cer	−
acI-active siactoisooctaosylcer	Galα1,3Galβ1,4GlcNAcβ1,6 \ >Galβ1,4GlcNAcβ1,4Glcβ1,1Cer NeuNAcα2,3Galβ1,4GlcNAcβ1,3 /	

[a]Cer, ceramide; CHM, ceramide monohexoside; CDH, ceramide dihexoside; CTH, ceramide trihexoside; GLA, globoside.

[b]Negative binding (−) indicates no binding to 4μg of lipid and positive binding to <0.5 μg(+++), 2μg(++), 0.5–2μg(+). Derived from Krivan et al. (1989).

Table 5–11. Sialyl glycoconjugates inhibit the adhesion and hemagglutination of Haemophilus influenzae[a]

Glycoconjugate	Adhesion[b,c]	Hemagglutination
NeuNAcα2,3Gal-protein (α-2-macroglobulin)	>5.5	>5.5
NeuNAcα2,3Galβ1,3GalNAc-protein (glycophorin)	>100 mg L^{-1}	ND
NeuNAcα2,3Galβ1,4Glcβ (sialyllactose)	1200	2000
Galβ1,4Glc (lactose)	>1000	1000
Galβ1,4Glcβ1,4Cer (lactosylceramide)	NI	NI
Galβ1,3GalNAcβ1,4(NeuNAcα2,3)Galβ1,4Glcβ1,1Cer(GM1)	15	<0.36
Galβ1,3GalNAcβ1,4Galβ1,1Cer (asialo GM1)	NI	NI
GalNAcβ1,4(NeuNAcα2,3)Galβ1,4Glcβ1,1Cer(GM2)	17	<0.41
NeuNAcβ2,3Galβ1,4Glcβ1,1Cer(GM3)	200	40
NeuNAcα2,3Galβ1,3GalNAcβ1,4(Neuα2,3)Galβ1,4Glcβ1,4Ce (GDla or diasialoganglioside)	26	<0.62

[a]Unless otherwise stated, values are in n*M* required for 10-fold reductions in adhesion or in complete inhibition of hemagglutination.

[b]Adhesion was to human oropharyngeal cells, whereas hemagglutination was assayed employing human AnWj-positive red cells. ND, not determined; NI, no inhibition.

[c]Results adapted from van Alphen et al. (1991).

least two separate lectin-adhesins, one for sialylglycoproteins and the other for sulfated glycolipids.

Binding of Sialyl Gangliosides by *Haemophilis influenzae*

H. influenzae appears to adhere to animal cells by fimbrial lectins specific for sialyl gangliosides (van Alphen et al., 1991). The organism adheres to oropharyngeal epithelial cells and erythrocytes, the binding on which can be inhibited by glycolipids and glycopeptides, but not by simple sugars or saccharides. Inhibition can be attained by GMI, GM2, GM3, and GDla gangliosides in nanomolar concentrations (Table 5–11). In contrast, the asialo-GMI, sialyllactose, and sialoglycoproteins are poor inhibitors. The results suggest that the sialyllactosylceramide (GM3) is the receptor for the *H. influenzae* fimbrial adhesin.

Glycolipid Receptors for *Neisseria gonorrhoeae*

Neisseria gonorrhoeae belongs to a series of organisms that bind to the lacto- and ganglio-series of glycolipids (Karlsson, 1989). The specificity of the *N. gonorrhoeae* binding has been examined by thin-layer chromatography techniques and it was found that the organisms specifically bind to terminal and internal GlcNAcβ1,3Galβ1,4Glc, and GalNAcβ1,4Galβ1,4Glc, sequences in the lacto- and ganglio-series of glycolipids (Deal and Krivan, 1990; Table 5–

12). The gonococci do not bind to Galβ1,4Glcβ1,1-Cer or to neutral glycolipids. Neither fimbriae nor Opa outer membrane proteins are involved in the binding of the gonococci to the lacto-ganglioside series.

Lectins of *Pseudomonas aeruginosa*

P. aeruginosa is an opportunistic pathogen capable of causing infections of the eye, lung, skin, and other parts of the body. The organism is often isolated from peritoneal dialysis membranes. There are numerous articles reporting various mechanisms of adhesion for *P. aeruginosa*. These include the hydrophobic effect (Garber et al., 1985), adhesion to and by alginates, and lectin-dependent adhesion. Some reports maintain that the adhesion of *P. aeruginosa* is fimbriae-dependent (Doig et al., 1989). Lectins specific for sialic acids (Hazlett et al., 1986), fucose, galactose, mannose, and *N*-acetylmannosamine have been reported by one or more authors as participants in adhesion. Krivan et al. (1988a,b) and Baker et al. (1990) have shown that *P. aeruginosa* can bind to glycosphingolipids on thin-layer plates. Baker et al. (1990) suggest that adhesion is strain-dependent and dependent on the blocking agents. In the case of blocking agents, bovine serum albumin prevented interaction with some glycosphingolipids, whereas gelatin afforded maximal interaction. Some of the results of Baker et al. are summarized in Table 5–13. One strain, 244NM, bound sialyl glycosphingolipids and lactosylceramide and in separate experiments was shown to adhere to human buccal epithelial cells and weakly to tracheal epithelial cells. Another strain, 0705M, bound to gangliotria- and gangliotetraosylceramide and adhered to tracheal epithelial cells more strongly than to buccal cells. At present, it is unknown if the *P. aeruginosa* expresses at least two adhesins capable of binding glycosphingolipids, or if a single adhesion can bind the various glycolipids. It is also unknown where the adhesin(s) are located on the cell. Because of its interesting evolutionary history *P. aeruginosa* has adapted to survive in animal tissues, plant tissues, and streams. The organism may be able to undergo phase

Table 5–12. Adhesion of Neisseria gonorrhoeae *to glycolipids*

Receptor	Adhesion
Galβ1,4Glcβ1,-1ceramide (with nonhydroxy fatty acid and dihydroxy long-chain base)	–
Galβ1,4Glcβ1,1ceramide (with 2-hydroxy fatty acid and/or trihydroxy long-chain base	+
GalNAcβ1,4Galβ1,4Glcβ1,1cer (asialo GM2)	++
Galβ1,3GalNAcβ1,4Galβ1,4Glcβ1,1cer (asialo GM1)	++
GlcNAcβ1,3Galβ1,4Glcβ1,1cer (lactotriaosylcer)	+
Galα1,4Galβ1,4Glcβ1,1cer	–

Modified from Karlsson (1989), Deal and Krivan (1990). Adhesion was assayed by use of thin-layer overlays of radioactive *N. gonorrhoeae*.

Table 5–13. Glycosphingolipid receptors for Pseudomonas aeruginosa *strains on TLC plates*

Glycosphingolipid	Binding to *P. aeruginosa*	
	Strain 244NM	Strain 0705M
Galβ1,4Glcβ1,1Cer(h)	+	−
GalNAcβ1,4Galβ1,4Glcβ1,1Cer	+	+
Galβ1,3GalNAcβ1,4Galβ1,4Glcβ1,1Cer	+	+
Galα1,4Galβ1,4Glcβ1,1Cer	−	−
Galα1,3Galβ1,4Glcβ1,1Cer	−	−
NeuNAcα2,3Galβ1,4Glcβ1,1Cer	+	−
Galβ1,3GalNAcβ1,4Galβ1,4Glcβ1,1Cer	+	−
3		
NeuNAc2		
NeuNAc2,3Galβ1,4GlcNAcβ1,4Glcβ1,1Cer	+	−
Galβ1,3GlcNAcβ1,3Galβ1,1Cer	−	−
Galβ1,3GlcNAcβ1,3Galβ1,4Glcβ1,1Cer	−	−
4		
Fucα1		
Galβ1,3GlcNAcβ1,4Galβ1,4Glcβ1,1Cer (Forssman)	−	−
GalNAcβ1,3Galα1,4Galβ1,4Glcβ1,1Cer (globoside)	−	−
GalNAcα1,3Galβ1,4GlcNAcβ1,3Galβ1,4Glcβ1,1Cer	−	−
2		
Fucα1		
Galβ1,4GlcNAcβ1,3Galβ1,4Glcβ1,1Cer	−	−
3		
Fucα1		
Fucα1-2Galβ1,4GlcNAcβ1,3Galβ1,4Glcβ1,1Cer	−	−
3		
Fucα1		

Adapted from Baker et al., 1990.

Glycosphingolipids were separated on thin-layer plates, then radioactive *P. aeruginosa* was overlaid. The plates were washed and autoradiographed.

[b]Hydroxylated ceramide.

variations (Chapter 9), leading to the expression of adhesins that are complementary with receptors in its environment.

Wentworth et al. (1991) have recently proposed an entirely new mechanism to account for lectin-dependent adhesion of *P. aeruginosa*. The bacterium produces two lectins (Gilboa-Garber, 1982, 1986), but the lectins are found mainly in the cytoplasm. One lectin, PA-I, is specific for D-galactose, but binds thio- and hydrophobic galactosides better than unsubstituted D-Gal (Table 5–14). The other lectin, PA-II, binds L-fucose and D-mannose and their hydrophobic derivatives (Table 5–15). Wentworth et al. (1991) observed that protonmotive force dissipating agents resulted in the partial lysis of the bacteria and the concomitant release of the lectins. The released lectlins, in turn, could bind with intact *P. aeruginosa,* giving the bacteria the ability to adhere to cultured rabbit corneal

Table 5–14. Inhibition of Pseudomonas aeruginosa *PA-I lectin activity by sugars and saccharides*

Saccharides	Inhibitory concentration (m*M*)
L-Fucose	>200
D-Galactose	0.5
D-Galactosamine	>100
N-Acetyl-D-galactosamine	2.0
Methyl-α-D-galactoside	0.25
Methyl-β-D-galactoside	1.0
L-Galactose	>100
Isopropyl-β-D-thiogalactoside	0.2
Methyl-β-D-thiogalactoside	0.25
p-Nitophenyl-α-D-galactoside	0.1
o-Nitophenyl-β-D-galactoside	0.2
Phenyl-β-D-thiogalactoside	0.03
D-Glucose	>300
D-Glucosamine	>100
N-Acetyl-D-glucosamine	>300
p-Nitrophenyl-α-D-glucopyranoside	>20
D-Mannose	>100
p-Nitrophenyl-α-D-mannopyranoside	>20
L-Rhamnose	50

From Garber et al., 1992.

Assays were based on hemagglutination inhibition.

epithelial cells. Figure 5–4 provides a description of the mechanism proposed for the adhesion of *P. aeruginosa* to glycoconjugates. The mechanism requires the participation of cytoplasmic lectins, but requires that the lectins become exteriorized and bound to the bacterial surface before they act as adhesin(s). The mechanism of lysis is presumably from a reduced concentration of protons surrounding the bacteria when the protonmotive force is reduced (Jolliffe et al., 1981). It is unknown if lectins PA-1 and PA-II can bind to glyco-sphingolipids. The synthesis of PA-1 and PA-II depend on cell age as well as culture medium (Gilboa-Garber, 1982). Differential rates of expression of PA-I and PA-II could explain the selectivity for glycoconjugates exhibited by the organism.

Lectins of Gram-positive Bacteria

There seems to be an increasing awareness that in Gram-positive bacteria surface lectins play a role in adhesion. Members of the genus *Streptococcus,* such as *S. pneumoniae* and *S. sobrinus*, possess carbohydrate-specific proteins on their surfaces. In addition, *Staphylococcus saprophyticus* and members of the genus *Actinomyces* express surface lectins.

Table 5–15. Inhibition of L-[6−³H]*fucose binding to lectin PA-II of* Pseudomonas aeruginosa *by various saccharides*

Sugar or saccharide	Inhibition(%)
N-Acetyl-D-galactosamine	1.5
N-Acetyl-D-glucosamine	0.9
N-Acetyl-D-mannosamine	0.9
D-Fucose	0.6
L-Fucose	66.5
L-Fucosylamine	74.5
D-Galactosamine	0.0
D-Galactose	46.5
D-Glucose	0.0
D-Mannose	0.0
6-*O*-Methyl-α-D-galactose	9.8
p-Nitrophenyl-α-D-fucose	6.4
p-Nitrophenyl-α-L-fucose	84.4
p-Nitrophenyl-α-D-galactose	1.1
p-Nitrophenyl-α-D-glucose	1.6
p-Nitrophenyl-α-D-mannose	0.9
L-Rhamnose	2.2

Adapted from Garber et al., 1987.

Inhibition of L-[6−³H]fucose (2 μ*M*) binding to PA-II (37 μg protein/ml, equivalent to approximately 3 μ*M*) by various carbohydrates (each at 25 μ*M*) was determined by equilibrium dialysis performed at 3°C in phosphate-buffered saline, pH 7.3 for 44–46h.

Although not isolated yet, convincing evidence for adhesive function was obtained for the surface lectin of *Streptococcus pneumoniae*. *N*-Acetylglucosamine inhibited binding of the organisms to tissue cells only at relative high concentrations (25 m*M*) (Beuth et al., 1987). The most active inhibitory compound was found to be GlcNAcβ1,3Gal (Andersson et al., 1983). The receptor is likely to belong to the neolacto- and lacto-series of glycolipids containing this disaccharide unit (Table 5–16). Whole bacteria bind to glycoconjugates bearing structures with the disaccharide. For example, coating of guinea pig erythrocytes, normally not agglutinated by the organisms, with glycolipids containing GlcNAcβ1,3Gal makes them susceptible to agglutination by the streptococci. The pneumococcal lectin is inhibited strongly by the tetrasaccharide Galβ1,4GlcNAcβ1,3Galβ1,4Glc. This tetrasaccharide is also present in human milk and it has been suggested that it may protect breast-feeding infants from pneumococcal infections (Andersson et al., 1986) (Chapter 11).

The discovery by McIntire et al. (1978, 1983) that lactose, galactose, methyl-β-galactoside, and *N*-acetylgalactosamine (but not several other sugars and saccharides) completely inhibit the interaction of *Actinomyces viscosus* T14V with a carbohydrate on *Streptococcus sanguis* 34 provided the first evidence for lectin–carbohydrate binding as a mechanism of coaggregation between oral actinomyces and streptococci (Table 5–17). Subsequent studies (reviewed by Cisar, 1986) have shown that most, although not all, human strains of *A. viscosus* and *A.*

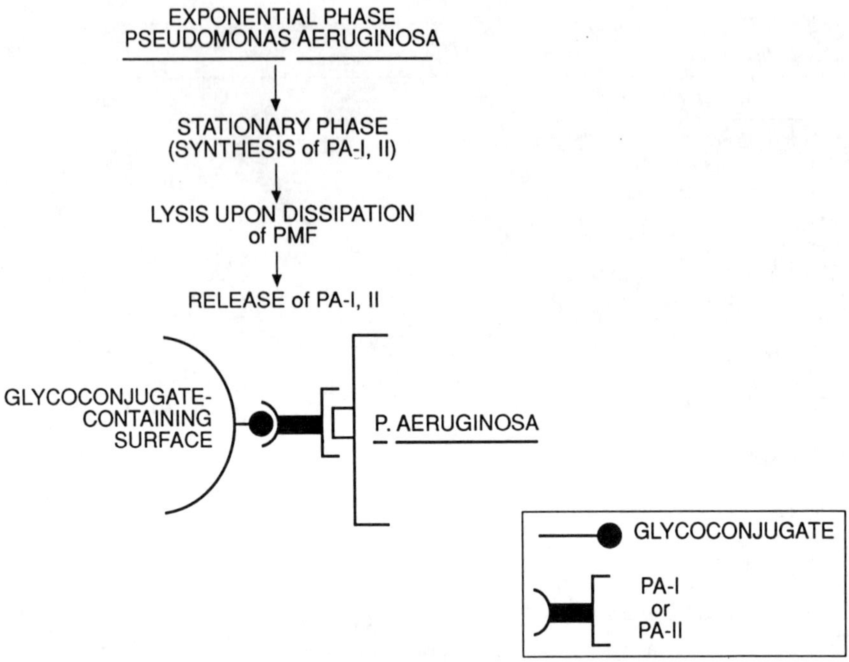

Figure 5–4. A proposed lectin-dependent mechanism for the adhesion of *Pseudomonas aeruginosa* to animal cells. Stationary phase *P. aeruginosa* synthesize PA-I and PA-II lectins. The lectins may be released when the cells become anaerobic or encounter metabolic poisons. Very aged cells may spontaneously externalize the lectins. The released lectins can bind to surviving intact bacteria, creating an organism now capable of complexing with glycoconjugates containing D-galactose or D-mannose (or L-fucose). (Reprinted from Wentworth et al., 1991 with permission.)

naeslundii exhibit a lectin activity that accounts for the coaggregation of these bacteria with certain streptococci (Cisar et al., 1979, 1984; Kolenbrander and Williams, 1981) and for the bacteria-mediated hemagglutination reactions with sialidase-treated erythrocytes (Costello et al., 1979; Ellen et al., 1980). Moreover, a large number of other oral bacteria exhibit galactose- or lactose-inhibitable adhesion to various other bacterial and mammalian cells, and this adhesion is often enhanced by treatment of the latter cells with sialidase. The enhancement of *Actinomyces* adhesion by sialidase treatment is of interest because these organisms contain a cell-associated form of the enzyme and secrete soluble sialidase. In contrast to the lactose-sensitive coaggregation, specific bacterial pairs, including *A. viscosus* T14V and *S. sanguis* H1, coaggregate by a mechanism that is lactose resistant (Cisar et al., 1979).

The ability of lactose to inhibit binding of the actinomyces lectin to receptors on streptococci and mammalian cells depends primarily on the β-1inked galactose moiety of this disaccharide. When compared on a molar basis, methyl-α-galacto-

Table 5–16. Inhibition of adhesion of Streptococcus pneumoniae *to pharyngeal cells by oligosaccharides*

Oligosaccharide	Source	Concentration (mg/ml) causing 50% inhibition
Galβ1,4GlcNAcβ1,4Glc	Human milk	0.2
Galβ1,4GlcNAβ1,3Galβ1,4Glc	Synthetic	0.6
Galβ1,4GlcNAcβ1,3Galβ1,4Glc	Human milk	1.3
Galβ1,4GlcNAcβ1,3Galβ1,4βGlc	Synthetic	2.0
Galβ1,4GlcNAcβ1,3GalβOMe	Synthetic	2.5
GlcNAcβ1,3GalβOMe	Synthetic	3.5
GlcNAcβ1,4GalβOMe	Synthetic	<10.0
Galβ1,4Glcβ1-OEt	Synthetic	3.0
Galβ1,3GlcNAc	Synthetic	1.0
Galβ1,3GlcNAc	Synthetic	5.0
Galβ1,4GlcNAcβ1-OMe	Synthetic	4.5

Data derived from Andersson et al., 1983.

side is at least fourfold less active than galactose as an inhibitor of the *A. viscosus* T14V lectin, whereas methyl-β-galactoside and lactose are from two to five times more active than galactose (Heeb et al., 1982; McIntire et al., 1983). The specificities of the lectins of *A. viscosus* T14V and *A. naeslundii* WVU45 have been further defined using galactose-containing oligosaccharides as inhibitors of coaggregation with *S. sanguis* 34 (Table 5–17). The most effective disaccharide inhibitor is Galβ1,3GalNAc, which is more than 10 times as active as lactose on a molar basis and also more active than other galactose disaccharides. Examination of the adhesion of *Actinomyces viscosus* and *Actinomyces naeslundii* to glycolipids in thin-layer plates shows that both bacterial species avidly bind to Galβ1,4Glcβ1-Cer, whereas several glycolipids with the internal Galβ1,4Glc sequence bind the organisms with a lower affinity (Stromberg and Karlsson, 1990a).

Information on the specificity of the *Actinomyces* lectins has been obtained in studies of bacterial adhesion to glycoprotein-coated Latex beads (Heeb et al., 1982). *A. viscosus* specifically binds to beads coated with asialofetuin, which has Galβ1,4GlcNAc termini and with asialo-bovine submaxillary mucin in which *N*-acetylgalactosamine is β-linked to serine residues of the protein. Adhesion is not observed when beads are coated with glycoproteins having other terminal carbohydrates (ovalbumin) or with proteins containing no carbohydrate (bovine serum albumin). The direct involvement of the galactose termini in lectin recognition of surface-associated asialofetuin has been demonstrated by the loss of receptor activity on treatment of soluble asialofetuin with galactose oxidase prior to its adsorption onto beads and by the subsequent recovery of receptor activity on reduction of the adsorbed glycoprotein with sodium borohydride. However, similar attempts to modify cell surface receptors on *S. sanguis* 34 or sialidase-treated erythrocytes with galactose oxidase or to remove them with β-galactosidase have not been successful.

Table 5–17. Inhibition of coaggregation of Streptococcus sanguis *34 with* Actinomyces viscosus *T14V and* A. naeslundii *WvU45*

	Relative Potency	
Galactoside	*A. viscosus*	*A. naeslundii*
Galactose	1.0	1.0
N-Acetylgalactosamine	0.7	1.7
Galβ1,4GalNAc	<1.5	1.2
Galβ1,4GlcNAc	<1.5	1.2
Galβ1,3GlcNAcβ1,3Galβ1,4Glc	4.6	7.3
Galβ1,4Glc	5.5	3.8
Galβ1,6GalβOpNp	9.1	5.8
Galβ1,3GalβOpNp	11	17
Galβ1,3GalNAc	64	60
Galβ1,3GalNAcβOpNp	91	172
Galβ1,3GalNAcβOMEUmb	195	165

Data from McIntire et al., 1983. OpNP denotes *p*-nitrophenyl; OMeUmb, methylumbelliferyl. Activity of galactose arbitrarily set as 1.0.

Adhesion of radiolabeled *A. naeslundii* WVU45 to sialidase-treated monolayers of an epithelial cell line was inhibited effectively by the binding of *Bauhinia purpurea* lectin and peanut agglutinin to the cell monolayers (Brennan et al., 1984). Whereas these plant lectins differ in specificity, both react well with Galβ1,3GalNAc and, similar to the *Actinomyces* lectin, their binding to epithelial cells is enhanced by the action of sialidase. In contrast, the *Erythrina cristagalli agglutinin,* specific for Galα1,4GlcNAc, failed to inhibit the adhesion although the lectin binds to the sialidase-treated epithelial cells. These and other experiments with lectins led to the conclusion that the receptor for actinomyces is most likely the *O*-linked Galβ1,3GalNAc termini of mammalian cell-surface glycoproteins. A receptor that binds the fimbriae of *A. naeslundii* was identified using bacterial binding assays to glycosphingolipids chromatographed on thin-layer silica gel plates (Brennan et al., 1987). The organisms bound GMl, GDlb, and globoside. The attachment site recognized by the lectin was suggested to be Galβ1,3GalNAc for the ganglioside and GalNAcβ1,3Gal termini for the globoside.

Several studies have suggested the presence of specific glucan-binding proteins, akin to lectins, on the surfaces of *Streptococcus mutans* (McCabe et al., 1977; Russell, 1979; Russell et al., 1983). It was suggested that α-1, 6-linked glucans of 7–10 glucose residues seemed to fit the active site of the lectin (Drake et al., 1988).

Gunnarsson et al. (1984) observed that the agglutination of sheep erythrocytes by *S. saprophyticus* could be inhibited by certain carbohydrates. The best inhibitors were GlcNAc-containing saccharides (Table 5–18). Members of the genus *Propionibacterium* also seem to be able to bind lactosylceramides on thin-layer plates (Karlsson, 1989; Stromberg and Karlsson, 1990b). The presentation of

Table 5–18. Specificity of the surface lectin of Staphylococcus saprophyticus

Inhibitor	MIC (m*M*)
Galβ1,4GlcNAc	0.72
Galβ1,4GlcNAcβ-1-*O*-methyl	0.58
Galβ1,4GlcNAcβ1,3Galβ1,4Glc	0.32
Galβ1,4GlcNAcα1,2Manβ1,6Manβ1,4GlcNAc	0.25
Galα1,4Galβ1,4Glc	4.6
Galβ1,4Gal	7.0
Galβ1,4Glc	7.0

Data from Gunnarsson et al., 1984.

MIC, minimum inhibitory concentration.

the disaccharide by the ceramide group appears to be important for adhesion of the bacteria to the glycolipids. *Propionibacterium freudenreichii* binds to Galβ,1-4Glcβ1-Cer when the ceramide residues contain nonhydroxy fatty acids and a dihydroxy base (Table 5–19). *P. granulosum,* however, has no affinity for this ceramide structure even though it does adhere to the galactoside linked to ceramide containing two hydroxy fatty acids. Conversely, *P. freudenreichii* has no affinity for the glycolipid. The data were interpreted to suggest that the ceramide group orients the requisite disaccharide receptor for interaction with the bacteria.

Significance of Surface Lectins in Infection

In three separate systems, each employing different type 1 fimbriated enterobacterial species, it was demonstrated that D-mannose and methyl-α mannoside inhib-

Table 5–19. Ceramide structure affects the binding of two variants of Propionibacterium *to lactosylceramide*

Isoreceptor	Ceramide components	Localization		*Propionibacterium freudenreichii*	*Propionibacterium granulosum*
		Epithelial	Subepithelial		
Galβ1,4GlcβCer	Nonhydroxy fatty acid Dihydroxy base	−	+	B	N
	2-Hydroxy fatty acid Dihydroxy base	+	−	N	B
	Nonhydroxy fatty acid Trihydroxy base	+	−	N	B
	2-Hydroxy fatty acid Trihydroxy base	+	−	N	B
Galβ1,3GalNAcβ1, 4Galβ1,4GlcβCer	Nonhydroxy fatty acid Dihydroxy base	−	+	N	B
	2-Hydroxy fatty acid Trihydroxy base	+	−	N	B

Adapted from Karlsson, 1989.

B, binds to the ceramide on thin layers; N, no binding.

Table 5–20. Inhibitors of carbohydrate-specific adhesion prevent infection

Organism	Animal, site of infection	Inhibitor	Reference
Type 1 fimbriated			
Escherichia coli	Mice, UT	MeαMan	Aronson et al., 1979
	Mice, GI	Mannose	Goldhar et al., 1986
	Mice, UT	Anti-mannose antibody	Abraham et al., 1985
Klebsiella pneumonia	Rats, UT	MeαMan	Fader and Davis, 1980
Shigella flexneri	Guinea pigs, eye	Mannose	Andrade, 1980
Type P fimbriated			
Escherichia coli	Mice, UT	Globotetraose	Svanborg-Eden et al., 1982a,b
	Monkeys, UT	Galα1,4GalβOMe	Roberts et al., 1984

UT, urinary tract; GI, gastrointestinal tract; MeαMan, methyl-α-D-mannopyranoside.

ited experimental infection by the organisms (Table 5–20). In each of these, glucose (or methyl-α-glucoside), which is not an inhibitor of type 1 fimbriae, did not affect the infectivity of injected bacteria. Similarly, Galα1,4Gal-containing saccharides were shown to prevent urinary tract infection in both mice and monkeys by P-fimbriated *E. coli*. In addition, antibodies against the mannooligosaccharide attachment site of the receptor on epithelial cells also prevented urinary tract infection in mice by type 1 fimbriated *E. coli*. These findings provide some of the most convincing evidence for the central role of bacterial lectins in infection, and particularly in mucosal colonization. They also illustrate the great potential of simple carbohydrates in the prevention of infections caused by bacteria that express surface lectins. For example, the studies on the carbohydrate specificity of type 1 fimbrial lectin (Table 5–2) suggest aromatic glycosides such as 4-methylumbelliferyl-α-mannoside and *p*-nitro-*o*-chlorophenyl-α-mannoside can provide a basis for the design of therapeutic agents that may prevent adhesion in vivo and infection by *E. coli* (Firon et al., 1987).

Another experimental approach to assess the role of enterobacterial lectins in colonization of mucosal surfaces includes injection of organisms bearing the fimbrial lectin into animals that are passively or actively immunized against the fimbrial antigen. This approach has been employed in the case of *E. coli* suspensions bearing types P and 1 fimbrial lectins. In all cases, infection is prevented in the immunized animals challenged with strains expressing the fimbrial lectin (Table 5–21). It should be pointed out that, although the antifimbrial antibodies inhibit adhesion in vitro (Silverblatt and Cohen, 1979; Svanborg-Eden et al., 1982a); there is no compelling evidence these antibodies are directed against the carbohydrate-combining sites of the fimbrial lectins.

Comparison between the fimbriated phenotype and nonfimbriated ones show in all cases that the infectivity of the former is significantly higher. Type 1 fimbriated strains of *S. typhimurium, K. pneumoniae,* and *E. coli* are found to be more infective than their nonfimbriated isogens in mice and rats (Duguid et al., 1976; Fader and Davis, 1982; Iwahi et al., 1983; Maayan et al., 1985). Similar findings have been obtained from mice for types P and S fimbriated strains (Hagberg et al., 1983a,b; Marré and Hacker, 1987). In one study (Keith et al., 1986), mice were infected intravesicularly with isogenic mutants, both of which were type 1 fimbriated, but only one of which possessed mannose-binding activity. Only the latter isogen induced urinary tract infections. In another study (Hagberg et al., 1983a,b), mice were injected intraurethrally with mixtures of fimbriated and nonfimbriated, as well as with mixtures of types 1 and type P fimbriated isogens. Examination of the bladders one day after challenge showed a marked preponderance of type 1 fimbriated bacteria, whereas in the kidney there was a preponderance of type P fimbriated bacteria. It was concluded that the type 1 fimbrial lectin confers an advantage to organisms during growth in the bladder. In contrast, in deep tissue (e.g., kidney) the type P fimbrial lectin seems advantageous for infectivity (further discussion can be found in Chapter 9).

Although experiments in laboratory animals have provided evidence for the role of bacterial lectins in the infectious process, evidence for such a role in humans is circumstantial. Epidemiological studies show that a specific symptom is associated with the isolation of an enteropathogen bearing a particular fimbrial lectin. Whereas most *E. coli* isolates are genotypically capable of expressing type 1 fimbriae, only those from patients presenting pyelonephritis in the absence of predisposing factors, such as vesicoureteric reflux, express in addition the type P fimbriae (reviewed in Leffler and Svanborg-Eden, 1986). The fact that there is no such association in strains isolated from pyelonephritic patients with defects in the urine flow has been interpreted to mean type P fimbriae confer an advantage in the ability of the organisms to induce pyelonephritis in otherwise healthy persons.

Table 5–21. Prevention of infection caused by enterobacteria carrying fimbrial lectin by antifimbrial immunity

Type of infection	Fimbrial antigen	Animal, site of action	Reference
Passive	Type 1	Mice, UT	Abraham et al., 1985
		Porcine, GI	Jayappa et al., 1985
Active	Type 1	Rats, UT	Silverblatt and Cohen, 1979
		Rats, GI	Guerina et al., 1983
		Porcine	Jayappa et al., 1985
	Type P	Monkeys, UT	Kaack et al., 1988
			Roberts et al., 1984
		Mice, UT	O'Hanley et al., 1985
			Schmidt et al., 1988
	Type K99	Pigs, GI	Isaacson et al., 1980

Somewhat compelling evidence for a role of the P fimbrial lectin in human infection has been provided by Lindstedt et al. (1991). They studied 1473 children with urinary tract infections and found those who were infected with strains expressing globo-A-specific adhesin lectins were of the blood group A. This compares with 45% of the population at large. Globo-A is a determinant of the A blood group. Some *E. coli* urinary isolates, after having been kept on agar, shift to a population rich in type 1 fimbriae within one 48-hour passage in static broth, whereas others do so only after multiple passages (Ofek et al., 1982). It has been shown that most isolates from cases of long-term catheterized urinary tract patients suffering multiple episodes of infections require only one broth passage to shift into a population rich in the type 1 fimbrial lectin, as compared to isolates from a single episode of an urinary tract infection (Mobley et al., 1987). It is concluded that expression of type 1 fimbriae in vivo is required to produce multiple episodes of urinary tract infections in catheterized patients.

A significant rise in antifimbrial antibodies following the natural course of infection with bacteria was demonstrated for types 1 and type P fimbriae in patients suffering urinary tract infections caused by *E. coli* (Rene and Silverblatt, 1982; de Ree and van den Bosch, 1987). These findings demonstrate that the fimbrial lectins are expressed in the course of natural infection. Studies on human tissue biopsies have shown the presence of fimbrial lectin receptors to which the bacteria presumably bind during the infection. The distribution and relative density of fimbrial receptor sites in a frozen section of human kidney was determined by using purified fimbriae or fluorescein-labeled bacteria expressing the fimbriae (Virkola, 1987; Korhonen, 1986a,b). The distribution of type 1 fimbriae binding to sections of human kidney was found to be different from that of the P and S fimbriae of *E. coli*. Type 1 fimbriae bind strongly to the luminal and cytoplasmic aspects of the proximal tubules and to the connective tissue layers of the veins and arteries. In contrast, the P and S fimbriae bind strongly to proximal and distal tubular cells, as well as to the apical and cytoplasmic sites of collecting ducts. The S fimbriae also bind to the visceral epithelium of the glomeruli. Interestingly, O'Hanley et al. (1985) did not find any differences in receptor distribution between types 1 and P fimbriae in mouse kidney, but found receptors for both fimbriae on cells of vagina and bladder. The finding of receptors for the P-fimbrial lectin in the upper urinary tract of human tissue may explain their association with pyelonephritis. The lack of receptors for type 1 fimbriae in this part of the kidney and their presence in the bladder tissue of the lower urinary tract, as well as in the vaginal epithelium, emphasizes the importance of this bacterial lectin in bladder and vaginal colonization. This differential distribution of receptors for two different fimbrial lectins produced by the same clone are in line with the results in experimental infections discussed earlier where phenotypes expressing type 1 fimbriae dominate in the bladder, whereas those expressing type P fimbriae survive best in the kidney. Some new results also support a role for bacterial lectins in natural infections. Glycoprotein sialoglycans, administrated orally, protect colostrum-deprived, newborn calves against lethal doses of enterotoxigenic *E. coli* K99 (Mouricout et al., 1990).

The studies on bacterial lectins expressed on the surfaces of enterobacteria and, in particular, those associated with types 1 and type P fimbriae, clearly demonstrate that they affect infectious processes in more than one way. This is by virtue of their ability to serve as adhesins that tether the organisms to host cells and to glycoconjugates in body fluids. It should be possible to intervene in the infectious process by specific carbohydrates that inhibit the lectin activity.

Expression of the bacterial lectins during the infection process is not always beneficial to the survival of the bacteria. For example, any adverse effect of type 1 fimbriae on the survival of the organisms due to its interaction with phagocytes or mannosylated soluble glycoproteins (e.g., Tamm–Horsfall glycoprotein) will not result necessarily in the termination of the infection, because these bacteria may undergo phase variations in vivo (Maayan et al., 1985) (phase variations are considered in detail in Chapter 9). This would ensure that a nonfimbriated phenotype or phenotype expressing another surface lectin of the same infecting strain be present to proceed with the infection in an environment hostile to the type 1 fimbriated phenotype. We expect that similar mechanisms of survival may be found in other bacteria that express surface lectins capable of interacting with phagocytes and/or with soluble glycoconjugates in body fluids.

There must be additional roles for the bacterial lectins in vivo besides those discussed above. For example, why among all *E. coli* strains expressing the mannose-specific lectin associated with type 1 fimbriae, is it only those that, in addition, express the Gal-Gal specific lectin associated with P fimbriae can induce pyelonephritis in healthy individuals? Clearly, the P-fimbriae lectins possess some role that is unique for the survival of the organisms during an infection that leads to pyelonephritis. Another example is the type 1 fimbrial lectin. The genes coding for the surface expression of this lectin are conserved among many enteric bacteria, suggesting that the lectin confers on the bacterial cell advantages not shared by other bacterial lectins. It is possible that such advantages are related to the distribution of receptors in the host tissues. In contrast to receptors for mannose-resistant lectins, receptors containing *N*-linked mannooligosaccharide side chains to which type 1 fimbriated bacteria bind are ubiquitous and abundant on many types of cells (see Chapters 3, 11). It is possible that each bacterial lectin may have several functions, some of which are shared by other types of lectins and one or more are characteristic to that lectin. Further studies are therefore expected to reveal new and important functions for these proteins in host–pathogen interactions.

References

Abraham, S.N., J.P. Babu, C.S. Giampapa, D.L. Hasty, W.A. Simpson, and E.H. Beachey. 1985. Protection against *Escherichia coli*-induced urinary tract infections with hybridoma antibodies directed against type 1 fimbriae or complementary D-mannose receptors. *Infect. Immun.* **48**:625–628.

Andersson, B., J. Dahmen, T. Frejd, H. Leffler, G. Magnusson, G. Noori, and C. Svanborg-Eden. 1983. Identification of an active disaccharide unit of a glycoconjugate receptor for pneumococci attaching to human pharyngeal epithelial cells. *J. Exp. Med.* **158**:559–570.

Andersson, B., O. Porras, L.A. Hanson, T. Lagergard, and C. Svanborg-Eden. 1986. Inhibition of attachment of *Streptococcus pneumoniae* and *Haemophilus influenzae* by human milk and receptor oligosaccharides. *J. Infect. Dis.* **153**:232–237.

Andrade, J.R.C. 1980. Role of fimbrial adhesiveness in guinea pig keratoconjuctivitis by *Shigella flexneri*. *Rev. Microbiol. (S. Paulo)* **11**:117–125.

Aronson, M., O. Medalia, L. Schori, D. Mirelman, N. Sharon, and I. Ofek. 1979. Prevention of colonization of the urinary tract of mice with *Escherichia coli* by blocking of bacterial adherence with methyl α-D-mannopyranoside. *J. Infect. Dis.* **139**:329–332.

Baker, N., G.C. Hannson, H. Leffler, G. Riise, and C. Svanborg-Eden, 1990. Glycosphingolipid receptors for *Pseudomonas aeruginosa*. *Infect. Immun.* **58**:2361–2366.

Banai, M., I. Kahane, S. Razin, and W. Bredt. 1978. Adherence of *Mycoplasma gallisepticum* to human erythrocytes. *Infect. Immun.* **21**:365–372.

Beuth, J., H.L. Ko, G. Uhlenbruck, and G. Pulverer. 1987. Lectin-mediated bacterial adhesion to human tissue. *Eur. J. Clin. Microbiol. Infect. Dis.* **6**:591–593.

Bock, K., M.E. Breimer, A. Brignole, G.C. Hansson, K.A. Karlsson, G. Larson, H. Leffler, B.E. Samuelsson, N. Stromberg, and C. Svanborg-Eden. 1985. Specificity of binding of a strain of uropathogenic *Escherichia coli* to Galα1,4Gal-containing glycosphingolipids. *J. Biol. Chem.* **260**:8545–8551.

Brennan, M.J., J.O. Cisar, A.E. Vatter, and A.L. Sandberg. 1984. Lectin-dependent attachment of *Actinomyces naeslundii* to receptors on epithelial cells. *Infect. Immun.* **46**:459–464.

Brennan, M.J., R.A. Joralmon, J.O. Cisar, and A.L. Sandberg. 1987. Binding of *Actinomyces naeslundii* to glycosphingolipids. *Infect. Immun.* **55**:487–489.

Brinton, C.C., Jr. 1965. The structure, function, synthesis and genetic control of bacterial pili and a molecular model for DNA and RNA transport in gram-negative bacteria. *Trans. NY Acad. Sci.* **27**:1003–1054.

Cisar, J.O. 1986. Fimbrial lectins of the oral actinomyces. In: Mirelman, D. (ed.), *Microbial Lectins and Agglutinins: Properties and Biological Activity*. John Wiley & Sons, New York, pp. 183–196.

Cisar, J.O., P.E. Kolenbrander, and F.C. McIntire. 1979. Specificity of coaggregation reactions between human oral streptococci and strains of *Actinomyces viscosus,* or *Actinomyces naeslundii*. *Infect. Immun.* **24**:742–752.

Cisar, J.O., A.L. Sandberg, and S.E. Mergenhagen. 1984. The function and distribution of different fimbriae on strains of *Actinomyces viscosus* and *Actinomyces naeslundii*. *J. Dent. Res.* **63**:393–396.

Costello, A.H., J.O. Cisar, P.E. Kolenbrander, and O. Gabriel. 1979. Neuraminidase-dependent hemagglutination of human erythrocytes by human strains of *Actinomyces viscosus* and *Actinomyces naeslundii*. *Infect. Immun.* **26**:563–572.

Deal, C.D. and H.C. Krivan. 1990. Lacto- and ganglio-series glycolipids are adhesin receptors for *Neisseria gonorrhoeae*. *J. Biol. Chem.* **265**:12774–12777.

de Man, P., B. Cedergren, S. Enerback, A.C. Larsson, H. Leffler, A.L. Lundell, B. Nilsson, and C. Svanborg-Eden. 1987. Receptor specific agglutination tests for detection of bacteria that bind globoseries glycolipids. *J. Clin. Microbiol.* **25**:401–406.

de Ree, J.M. and J.F. van den Bosch. 1987. Serological response to the P fimbriae of uropathogenic *Escherichia coli* in pyelonephritis. *Infect. Immun.* **55**:2204–2207.

Doig, P., W. Paranchych, P.A. Sastry, and R.T. Irvin. 1989. Human buccal epithelial cell receptors of *Pseudomonas aeruginosa:* identification of glycoproteins with pilus binding activity. *Can. J. Microbiol.* **35**:1141–1145.

Drake, D., K.G. Taylor, A.S. Bleiweis, and R.J. Doyle. 1988. Specificity of the glucan-binding lectin of *Streptococcus cricetus*. *Infect. Immun.* **56**:1864–1872.

Duguid, J.P. and D.C. Old. 1980. Adhesive properties of *Enterobacteriaceae*. In: Beachey, E.H. (ed.), *Bacterial Adherence (Receptors and Recognition, Vol. 6)*. Chapman and Hall, London, pp. 185–217.

Duguid, J.P., M.R. Darekar, and D.W.F. Wheather. 1976. Fimbriae and infectivity in *Salmonella typhimurium*. *J. Med. Microbiol.* **9**:459–473.

Elbein, A.D., R. Solf, P.R. Dorling, and K. Vosbeck. 1981. Swainsonine: an inhibitor of glycoprotein processing. *Proc. Natl. Acad. Sci. USA* **78**:7393–7397.

Elbein, A.D., Y.T. Pahn, R. Solf, and K. Vosbeck. 1983. Effect of swainsonine, an inhibitor of glycoprotein processing, on cultured mammalian cells. *J. Cell. Physiol.* **115**:265–275.

Ellen, R.P., E.D. Fillery, K.H. Chan, and D.A. Grove. 1980. Sialidase-enhanced lectin-like mechanisms for *Actinomyces viscosus* and *Actinomyces naeslundii* hemagglutination. *Infect. Immun.* **27**:335–343.

Evans, D.G., D.J. Evans, Jr., J.J. Moulds, and D.Y. Graham. 1988. *N*-Acetylneuraminyllactose-binding fibrillar hemagglutinin of *Campylobacter pylori:* a putative colonization factor antigen. *Infect. Immun.* **56**:2896–2906.

Fader, R.C. and C.P. Davis. 1980. Effect of piliation on *Klebsiella pneumoniae* infection in rat bladders. *Infect. Immun.* **30**:554–561.

Fader, R.C. and C.P. Davis. 1982. *Klebsiella pneumonia*-induced experimental pyelitis: the effect of piliation on infectivity. *J. Urol.* **128**:197–201.

Faris, A., M. Lindahl, and T. Wadstrom. 1980. GM_2-like glycoconjugate as possible receptor for the CFA/I and K99 hemagglutinins of entero-toxigenic *Escherichia coli*. *FEMS Microbiol. Lett.* **7**:265–269.

Faris, A., T. Wadstrom, and J.H. Freer. 1981. Hydrophobic absorptive and haemagglutinating properties of *Escherichia coli* possessing colonization factor antigens (CFA/I or CFA/II), type 1 or other pili. *Curr. Microbiol.* **5**:67–72.

Finkelstein, R.A. and L.F. Hanne. 1982. Purification and characterization of the soluble hemagglutinin (cholera lectin) produced by *Vibrio cholerae*. *Infect. Immun.* **36**:1199–1208.

Firon, N., I. Ofek, and N. Sharon. 1983. Carbohydrate specificity of the surface lectins of *Escherichia coli, Klebsiella pneumoniae,* and *Salmonella typimurium*. *Carbohydr. Res.* **120**:235–249.

Firon, N., I. Ofek, and N. Sharon. 1984. Carbohydrate-binding sites of the mannose-specific fimbrial lectins of enterobacteria. *Infect. Immun.* **43**:1088–1090.

Firon, N., D. Duskin, and N. Sharon. 1985. Mannose-specific adherence of *Escherichia coli* to BHK cells that differ in their glycosylation patterns. *FEMS Microbiol. Lett.* **27**:161–165.

Firon, N., S. Ashkenazi, D. Mirelman, I. Ofek, and N. Sharon. 1987. Aromatic alpha-glycosides of mannose are powerful inhibitors of the adherence of type 1 fimbriated *Escherichia coli* to yeast and intestinal epithelial cells. *Infect. Immun.* **55**:472–476.

Garber, N., N. Sharon, D. Shohet, J.S. Lam, and R.J. Doyle. 1985. Contribution of hydrophobicity to hemagglutination reactions of *Pseudomonas aeruginosa*. *Infect. Immun.* **50**:336–337.

Garber, H., U. Guempel, N. Gilboa-Garber, and R.J. Doyle. 1987. Specificity of the fucose-binding lectin of *Pseudomonas aeruginosa*. *FEMS Microbiol. Lett.* **48**:331–334.

Garber, N., U. Guempel, A. Belz, N. Gilboa-Garber, and R.J. Doyle. 1992. On the specificity of the D-galactose-binding lectin (PA-I) of *Pseudomonas aeruginosa* and its strong binding to hydrophobic derivatives of D-galactose and thiogalactose. *Biochim. Biophys. Acta* **1116**:331–333.

Giampapa, C.S., S.N. Abraham, T.M. Chiang, and E.H. Beachey. 1988. Isolation and characterization of a receptor for type 1 fimbriae of *Escherichia coli* from guinea pig erythrocytes. *J. Biol. Chem.* **263**:5362–5367.

Gilboa-Garber, N. 1982. *Pseudomonas aeruginosa* lectins. *Methods Enzymol.* **83**:378–385.

Gilboa-Garber, H. 1986. Lectins of *Pseudomonas aeruginosa*. Properties, effects and applications. In: Mirelman, D. (ed.), *Microbial Lectins and Agglutinins. Properties and Biological Activity*. John Wiley & Sons, New York, pp. 255–269.

Glasgow, L.R. and R.L. Hill. 1980. Interaction of *Mycoplasma gallisepticum* with sialyl glycoproteins. *Infect. Immun.* **30**:353–361.

Goldhar, J., R. Peri, R. Zilberberg, and M. Lahau. 1980. Enterotoxigenic *Escherichia coli* isolated in the Tel-Aviv area. *Med. Microbiol.* **169**:53–61.

Goldhar, J., A. Zilberberg, and I. Ofek. 1986. Infant mouse model of adherence and colonization of intestinal tissues by enterotoxigenic strains of *Escherichia coli* isolated from humans. *Infect. Immun.* **52**:205–208.

Goochee, C.F., R.T. Hatch, and T.W. Cadman. 1987. Some observations on the role of type 1 fimbriae in *Escherichia coli* autoflocculation. *Biotech. Bioeng.* **29**:1024–1034.

Guerina, N.G., T.W. Kessler, V.J. Guerina, M.R. Neutra, H.W. Clegg, S. Lagermann,

F.A. Scannapieco, and D.A. Goldmann. 1983. The role of pili and capsule in the pathogenesis of neonatal infection with *Escherichia coli* K1. *J. Infect. Dis.* **148**:395–405.

Gunnarsson, A., P.A. Mardh, A. Lundblad, and S. Svensson. 1984. Oligosaccharide structures mediating agglutination of sheep erythrocytes by *Staphylococcus saprophyticus*. *Infect. Immun.* **45**:41–46.

Hacker, J. 1990. Genetic determinants coding for fimbriae and adhesins of extra-intestinal *Escherichia coli*. *Curr. Top. Microbiol. Immunol.* **151**:1–27.

Hagberg, L., I. Engberg, R. Freter, J. Lam, S. Olling, and C. Svanborg-Eden. 1983a. Ascending, unobstructed urinary tract infection in mice caused by pyelonephritogenic *Escherichia coli* of human origin. *Infect. Immun.* **40**:273–283.

Hagberg, L., R. Hull, S. Hull, S. Falkow, R. Freter, and C. Svanberg-Eden. 1983b. Contribution of adhesion to bacterial persistence in the mouse urinary tract. *Infect. Immun.* **40**:265–272.

Hanne, L.F. and R.A. Finkelstein. 1982. Characterization and distribution of the hemagglutinins produced by *Vibrio cholerae*. *Infect. Immun.* **36**:209–214.

Hansson, G.C., K.A. Karlsson, G. Larson, A. Lindberg, N. Stromberg, and J. Thurin. 1983. Lactosylceramide is the probable adhesion site for major indigenous bacteria of the gastrointestinal tract. In: Chester, M.A., D. Heingegard, A. Lundblad, and S. Svensson (eds.), *Glycoconjugates. 7th International Symposium on Glycoconjugates*. Lund, Sweden, July 17–23, p. 631.

Hansson, G.C., K. -A Karlsson, G. Larson, N. Stromberg, and J. Thurin. 1985. Carbohydrate-specific adhesion of bacteria to thin-layer chromatograms: a rationalized approach to the study of host cell glycolipid receptors. *Anal. Biochem.* **146**:158–163.

Hazlett, L.D., M. Moon, and R.S. Berk 1986. *In vivo* identification of sialic acid as the ocular receptor for *Psuedomonas aeruginosa*. *Infect. Immun.* **51**:687–689.

Heeb, M.J., A.H. Costello, and O. Gabriel. 1982. Characterization of a galactose-specific lectin from *Actinomyces viscosus* by a model aggregation system. *Infect. Immun.* **38**:993–1002.

Isaacson, R.E., E.A. Dean, R.L. Morgan, and H.W. Moon. 1980. Immunization of suckling pigs against enterotoxigenic *Escherichia coli*-induced diarrheal disease by vaccinating dams with purified K99 and 987P pili: Antibody production in response to vaccination. *Infect. Immun.* **29**:824–826.

Ishikawa, H. and Y. Isayama. 1987. Evidence for sialyl glycoconjugates as receptors for *Bordetella bronchiseptica* on swine nasal mucosa. *Infect. Immun.* **55**:1607–1609.

Iwahi, T., Y. Abe, M. Nakao, A. Imada, and K. Tsuchiya. 1983. Role of type 1 fimbriae and the pathogenesis of ascending urinary tract infection induced by *Escherichia coli* in mice. *Infect. Immun.* **39**:1307–1315.

Jayappa, H.G., R.A. Goodnow, and S.J. Geary. 1985. Role of *Escherichia coli* type 1 pilus colonization of porcine ileum and its protective nature as a vaccine antigen in controlling colibacillosis. *Infect. Immun.* **48**:350–354.

Jolliffe, L.K., R.J. Doyle, and U.N. Streips. 1981. The energized membrane and cellular autolysis in *Bacillus subtilis*. *Cell* **25**:753–763.

Jones, G.W. and R. Freter. 1976. Adhesive properties of *Vibrio cholerae:* nature of the interaction with isolated rabbit brush border membranes and human erythrocytes. *Infect. Immun.* **14**:240–245.

Kaack, M.B., J.A. Roberts, G. Baskin, and G.M. Patterson. 1988. Maternal immunization with P-fimbriae for the prevention of neonatal pyelonephritis. *Infect. Immun.* **56**:1–6.

Kabat, E.A. 1978. Dimensions and specificities of recognition sites on lectins and antibodies. *J. Supramol. Struct.* **8**:79–88.

Karlsson, K.A. 1986. Animal glycolipids as attachment sites for microbes. *Chem. Phys. Lipids* **42**:153–172.

Karlsson, K.A. 1989. Animal glycosphingolipids as membrane attachment sites for bacteria. *Annu. Rev. Biochem.* **58**:309–350.

Keith, B.R., L. Maurer, P.A. Spears, and P.E. Orndorff. 1986. Receptor-binding function of type 1 pili effects bladder colonization by a clinical isolate of *Escherichia coli. Infect. Immun.* **53**:693–696.

Kolenbrander, P.E. and B.L. Williams. 1981. Lactose-reversible coaggregation between oral actinomycetes and *Streptococcus sanguis. Infect. Immun.* **33**:95–102.

Korhonen, T.K., V. Vaisanen, H. Saxen, H. Hultberg, and S.B. Svenson. 1982. P-Antigen-recognizing fimbriae from human uropathogenic *Escherichia coli* strains. *Infect. Immun.* **37**:286–291.

Korhonen, T.K., V. Vaisanen-Rhen, M. Rhen, A. Pere, J. Parkkinen, and J. Finne. 1984. *Escherichia coli* recognizing sialyl galactosides. *J. Bacteriol.* **159**:762–766.

Korhonen, T.K., R. Virkola, and H. Holthofer. 1986a. Localization of binding sites for purified *Escherichia coli* P fimbriae in the human kidney. *Infect. Immun.* **54**:328–332.

Korhonen, T.K., J. Parkkinen, J. Hacker, J. Finne, A. Pere, M. Rhen, and H. Holthofer. 1986b. Binding of *Escherichia coli* S fimbriae to human kidney epithelium. *Infect. Immun.* **54**:322–327.

Korhonen, T.K., K. Haahtela, A. Pirkola, and J. Parkkinen. 1988. A *N*-acetyllactosamine-specific cell-binding activity in a plant pathogen. *Erwinia rhapontici. FEBS Lett.* **236**:163–166.

Krivan, H.C., D.D. Roberts, and V. Ginsburg. 1988a. Many pulmonary pathogenic bacteria bind specifically to the carbohydrate sequence GAlNAcβ1,4Gal found in some glycolipids. *Proc. Natl. Acad. Sci. USA* **85**:6157–6161.

Krivan, H.C., V. Ginsburg, and D.D. Roberts. 1988b. *Pseudomonas aeruginosa* and *Pseudomonas cepacia* isolated from cystic fibrosis patients bind specifically to gangliotetraosylceramide (asialoGM1) and gangliotetraosylceramide (asialo GM2). *Arch. Biochem. Biophys.* **260**:493–496.

Krivan, H.C., L.D. Olson, M.F. Barile, V. Ginsburg, and D.D. Roberts. 1989. Adhesion of *Mycoplasma pneumoniae* to sulfated glycolipids and inhibition by dextran sulfate. *J. Biol. Chem.* **264**:9283–9288.

Landale, E.C. and M.M. McCabe. 1987. Characterization by affinity electrophoresis of an α-1, 6-glucan-binding protein from *Streptococcus sobrinus. Infect. Immun.* **55**:3011–3016.

Leffler, H. and C. Svanborg-Eden. 1980. Chemical-identification of a glycosphingolipid receptor for *Escherichia coli* attaching to human urinary-tract epithelial cells and agglutinating human-erythrocytes. *FEMS Microbiol. Lett.* **8**:127–134.

Leffler, H. and C. Svanborg-Eden. 1986. Glycolipids as receptors for *Escherichi coli* lectins or adhesins. In: Mirelman, D. (ed.), *Microbial Lectins and Agglutinins: Properties and Biological Activity*. John Wiley & Sons, New York, pp. 83–111.

Lindahl, M., R. Brossmer, and T. Wadstrom. 1987. Carbohydrate receptor specificity of K99 fimbriae of enterotoxigenic *Escherichia coli*. *Glyconjugate J.* **4**:51–58.

Lindahl, M., R. Brossmer, and T. Wadstrom. 1988. Sialic acid and *N*-acetylgalactosamine specific bacterial lectins of enterotoxigenic *Escherichia coli* (ETEC). *Adv. Exp. Med. Biol.* **228**:123–152.

Lindstedt, R., N. Baker, P. Falk, R. Hull, S. Hull, J. Karr, K. Leffler, C. Svanborg-Eden, and G. Larson. 1989. Binding specificities of wild-type and cloned *Escherichia coli* strains that recognize globo-A. *Infect. Immun.* **57**:3389–3394.

Lindstedt, R., G. Larson, P. Falk, U. Jodal, H. Leffler, and C. Svanborg. 1991. The receptor repertoire defines the host range for attaching *Escherichia coli* strains that recognize globo-A. *Infect. Immun.* **59**:1086–1092.

Lingwood, C.A., A. Pellizzari, H. Law, P. Sherman, and B. Drumm. 1989. Gastric glycerolipid as a receptor for *Campylobacter pylori*. *Lancet* **ii**:238–241.

Loomes, L.M., K. Uemura, R.A. Childs, J.C. Paulson, G.N. Rogers, P.R. Scudder, J.-C. Michalski, E.F. Hounsell, D. Taylor-Robinson, and T. Feizi. 1984. Erythrocyte receptors for *Mycoplasma pneumoniae* are sialylated oligosaccharides of Ii antigen type. *Nature* **307**:560–563.

Loomes, L.M., K. Uemura, and T. Feizi. 1985. Interaction of *Mycoplasma pneumoniae* with erythrocyte glycolipids and I and i antigen types. *Infect. Immun.* **47**:15–20.

Lund, G., B. -I. Marklund, N. Stromberg, F. Lindberg, K.-A. Karlsson, and S. Normark. 1988. Uropathogenic *Escherichia coli* can express serologically identical pili of different receptor binding specificities. *Mol. Microbiol.* **2**:255–263.

Maayan, M.C., I. Ofek, O. Medalia, and M. Aronson. 1985. Population shift in mannose specific fimbriated phase of *Klebsiella pneumoniae* during experimental urinary tract infection in mice. *Infect. Immun.* **49**:785–789.

Marré, R. and J. Hacker. 1987. Role of S-type and common type 1 fimbriae of *Escherichia coli* in experimental upper and lower urinary tract infection. *Microb. Pathog.* **2**:233–226.

McCabe, M.M., R.M. Hamelik, and E.E. Smith. 1977. Purification of dextran-binding protein from cariogenic *Streptococcus mutans*. *Biochem. Biophys. Res. Commun.* **78**:273–278.

McIntire, F.C., A.E. Vatter, J. Baros, and J. Arnold. 1978. Mechanism of coaggregation between *Acinomyces viscosus* T14V and *Streptococcus sanguis* **34**. *Infect. Immun.* **21**:978–988.

McIntire, F.C., L.K. Crosby, J.J. Barlow, and K.L. Matta. 1983. Structural preferences of β-galactoside-reactive lectins on *Actinomyces viscosus* T14V and *Actinomyces naeslundii* WVU45. *Infect. Immun.* **41**:848–850.

Mirelman, D. (ed). 1986. *Microbial Lectins and Agglutinins: Properties and Biological Activity*. John Wiley & Sons, New York.

Mobley, H.L.T., G.R. Chippendale, J.H. Tenney, R.A. Hull, and J.W. Warren. 1987. Expression of type 1 fimbriae may be required for persistence of *Escherichia coli* in the catherized urinary-tract. *J. Clin. Microbiol.* **25**:2253–2257.

Morschhauser, J., H. Hoschutzky, K. Jann, and J. Hacker. 1990. Functional analysis of the sialic acid-binding adhesin Sfa S of pathogenic *Escherichia coli* by site-specific mutagenesis. *Infect. Immun.* **58**:2133–2138.

Mouricout, M., J.M. Petit, J.R. Carias, and R. Julien. 1990. Glycoprotein glycans that inhibit adhesion of *Escherichia coli* mediated by K99 fimbriae: Treatment of experimental colibacillosis. *Infect. Immun.* **58**:98–106.

Murray, P.A., M.J. Levine, L.A. Tobak, and M.S. Reddy. 1982. Specificity of salivary–bacterial interactions: II. Evidence for a lectin on *Streptococcus sanguis* with specificity for a NeuAcα2,3,Galβ1,3GalNAc sequence. *Biochem. Biophys. Res. Commun.* **106**:390–396.

Murray, P.A., D.G. Kern, and J.R. Winkler. 1988. Identification of a galactose-binding lectin on *Fusobacterium nucleatum* FN-2. *Infect. Immun.* **56**:1314–1319.

Nesser, J. -R., B. Koellreutter, and P. Wuersch. 1986. Oligomannoside-type glycopeptides inhibiting adhesion of *Escherichia coli* strains mediated by type 1 pili: preparation of potent inhibitors from plant glycoproteins. *Infect. Immun.* **52**:428–436.

Nesser, J. -R., A. Chambaz, K.Y. Hoang, and H. Link-Amster. 1988. Screening for complex carbohydrates inhibiting hemagglutinations by CFA/I-and CFA/II-expressing enterotoxigenic *Escherichia coli* strains. *FEMS Microbiol. Lett.* **49**:301–307.

Nilsson, G., S. Svensson, and A. A. Lindberg. 1983. The role of the carbohydrate portion of glycolipids for the adherence of *Escherichia coli* K88$^+$ to pig intestine. In: Chester, M. A., D. Heinegärd, A. Lundbead, and S. Svensson (eds.), *Glycoconjugates: 7th International Symposium on Glycoconjugates,* Lund, pp. 637–638.

Ofek, I. and E.H. Beachey. 1978. Mannose binding and epithelial cell adherence of *Escherichia coli*. *Infect. Immun.* **22**:247–254.

Ofek, I. and N. Sharon. 1990. Adhesins as lectins: specificity and role in infection. *Curr. Top. Microbiol. Immunol.* **151**:91–113.

Ofek, I., D. Mirelman, and N. Sharon. 1977. Adherence of *Escherichia coli* to human mucosal cells mediated by mannose receptors. *Nature* **265**:625.

Ofek, I., A. Mosek, and N. Sharon. 1981. Mannose-specific adherence of *Escherichia coli* freshly excreted in the urine of patients with urinary-tract infections, and of isolates subcultured from the infected urine. *Infect. Immun.* **34**:708–711.

Ofek, I., J. Goldhar, Y. Eshdat, and N. Sharon. 1982. The importance of mannose specific adhesins (lectins) in infections caused by *Escherichia coli*. *Scand. J. Infect. Dis. Suppl.* **33**:61–77.

Ofek, I., H. Lis, and N. Sharon. 1985. Animal cell surface membranes. In: Savage, D.C. and M. Fletcher (eds.), *Bacterial Adhesion*. Plenum Publishing Corp., New York, pp. 71–88.

O'Hanley, P., D. Lark, S. Falkow, and G. Schoolnik. 1985. Molecular basis of *Escherichia coli* colonization of the upper urinary tract in BALB/c mice. Gal-Gal pili immunization prevents *Escherichia coli* pyelonephritis in the BALB/C mouse model of human pyelonephritis. *J. Clin. Invest.* **75**:347–360.

Ono, E., K. Abe, M. Nakazawa, and M. Naiki. 1989. Ganglioside epitope recognized by K99 fimbriae from enterotoxigenic *Escherichia coli. Infect. Immun.* **57**:907–911.

Parkkinen, J., J. Finne, M. Achtman, V. Vaisanen, and T.K. Korhonen. 1983. *Escherichia coli* strains binding neuraminyl α2-3 galactosides. *Biochem. Biophys. Res. Commun.* **111**:456–461.

Parkkinen, J., G.N. Rogers, T. Korhonen, W. Dahr, and J. Finne. 1986. Identification of *O*-linked sialyloligosaccharides of glycophorin A as the erythrocyte receptors for S-fimbriated *Escherichia coli. Infect. Immun.* **54**:37–42.

Pieroni, P., E.A. Worobec, W. Paranchych, and G.D. Armstrong. 1988. Identification of a human erythrocyte receptor for colonization factor antigen I pili expressed by H10407 enterotoxigenic *Escherichia coli. Infect. Immun.* **56**:1334–1340.

Prakobphol, A., P.A. Murray, and S.J. Fisher. 1987. Bacterial adherence on replicas on sodium dodecyl sulfate-polyacrylamide gels. *Anal. Biochem.* **164**:5–11.

Pulverer, G., J. Beuth, H.L. Ko, J. Solter, and G. Uhlenbruck. 1987. Modification of glycosylation by tunicamycin treatment inhibits lectin-mediated adhesion of *Streptococcus pneumoniae* to various tissues. *Zbl. Bakt. A* **266**:137–144.

Ramphal, R., C. Carnoy, S. Fievre, J.-C. Michalski, N. Houdret, G. Lamblin, G. Strecker, and P. Roussel. 1991. *Pseudomonas aeruginosa* recognizes carbohydrate chains containing type 1 (Galβ1-3GlcNAc) or type 2 (Galβ1-4GlcNAc) disaccharide units. *Infect. Immun.* **59**:700–704.

Rauvala, H. and J. Fine. 1979. Structural similarity of the terminal carbohydrate sequence of glycoproteins and glycolipids. *FEBS Lett.* **97**:1–8.

Rene, P. and F.J. Silverblatt. 1982. Serological response to *Escherichia coli* pili in pyelonephritis. *Infect. Immun.* **37**:749–754.

Roberts, J.A., B. Kaack, G. Kallenius, R. Mollby, J. Winberg, and S.B. Svenson. 1984. Receptors for pyelonephritogenic *Escherichia coli* in primates. *J. Urol.* **131**:163–168.

Roberts, D.D., L.D. Olson, M.F. Barile, V. Ginsburg, and H.C. Krivan. 1989. Sialic acid-dependent adhesion of *Mycoplasma pneumoniae* to purified glycoproteins. *J. Biol. Chem.* **264**:9289–9293.

Rodriguez-Ortega, M., I. Ofek, and N. Sharon. 1987. Membrane glycoproteins of human polymorphonuclear leukocytes that act as receptors for mannose-specific *Escherichia coli. Infect. Immun.* **55**:968–973.

Russell, R.R.B. 1979. Glucan-binding proteins of *Streptococcus mutans* serotype c. *J. Gen. Microbiol.* **112**:197–201.

Russell, R.R.B., A.C. Donald, and C.W.I. Douglas. 1983. Fructosyltransferase activity of a glucan-binding protein from *Streptococcus mutans. J. Gen. Microbiol.* **129**:3243–3250.

Saitoh, T., H. Natomi, W. Zhao, K. Okuzumi, K. Sugano, M. Iwamori, and Y. Nagai. 1991. Identification of glycolipid receptors for *Helicobacter pylori* by TLC-immunostaining. *FEBS Lett.* **282**:385–387.

Schmidt, M.A., P. O'Hanley, D. Lark, and G.K. Schoolnik. 1988. Synthetic peptides corresponding to protective epitopes of *Escherichia coli* digalactoside binding pilin prevent infection in a murine pyelonephritis model. *Proc. Natl. Acad. Sci. USA* **85**:1247–1251.

Senior, D., N. Baker, B. Cedergren, P. Falk, G. Larson, R. Lindstedt, and C. Svanborg-Eden. 1988. Globo-A: a new receptor specificity for attaching *Escherichia coli*. *FEBS Lett.* **237**:123–127.

Sharon, N. and H. Lis. 1982. Glycoproteins-research booming on long-ignored ubiquitous compounds. *Mol. Cell. Biochem.* **42**:167–187.

Sharon, N. and H. Lis. 1989. *Lectins*. Chapman and Hall, London, pp. 1–127.

Sharon, N., Y. Eshdat, F.J. Silverblatt, and I. Ofek, 1981. Bacterial adherence to cell surface sugars. In: Elliott, K.H., M. O'Connor, and J. Whelan (eds.), *Adhesion and Microorganisms*. Pitman Medical Publishers, Belmont, CA, pp. 119–135.

Sherman, P.M., W.L. Houston, and E.C. Boedeker, 1985. Functional heterogeneity of intestinal *Escherichia coli* strains expressing type 1 somatic pili (fimbriae): assessment of bacterial adherence to intestinal membranes and surface hydrophobicity. *Infect. Immun.* **49**:797–804.

Silverblatt, F.J. and L.S. Cohen. 1979. Antipili antibody affords protection against experimental ascending pyelonephritis. *J. Clin. Invest.* **64**:333–336.

Sjoberg, P.O., M. Lindahl, J. Porath, and T. Wadstrom. 1988. Purification and characterization of CS2, a sialic acid-specific haemagglutinin of enterotoxigenic *Escherichia coli*. *Biochem. J.* **255**:105–111.

Slomiany, J. Piotrowski, A. Samanta, K. VanHorn, V.L.N. Murty, and A. Slomiany. 1989. *Campylobacter pylori* colonization factor shows specificity for lactosylceramide sulfate and GM3 ganglioside *Biochem. Int.* **19**:929–936.

Smit, H., W. Gaastra, J.P. Kamerling, J.F.G. Vliegenthart, and F.K. DeGraaf. 1984. Isolation and structural characterization of the equine erythrocyte receptor for enterotoxigenic *Escherichia coli* K99 fimbrial adhesin. *Infect. Immun.* **46**:578–584.

Stromberg, N. and K.A. Karlsson. 1990a. Characterization of the binding of *Actinomyces naeslundii* (ATCC 12104) and *Actinomyces viscosus* (ATCC 19246) to glycosphingolipids, using a solid-phase overlay approach. *J. Biol. Chem.* **265**:11251–11258.

Stromberg, N. and K. -A. Karlsson. 1990b. Characterization of the binding of *Propionibacterium granulosum* to glycosphingolipids adsorbed on surfaces. An apparent recognition of lactose which is dependent on the ceramide structure. *J. Biol. Chem.* **265**:11244–11250.

Stromberg, N., C. Deal, G. Nyberg, S. Normark, M. So, and K. -A. Karlsson. 1988. Identification of carbohydrate structures that are possible receptors for *Neisseria gonorrhoeae*. *Proc. Natl. Acad. Sci. USA* **85**:4902–4906.

Stromberg, N., B.-I. Marklund, B. Lund, D. Ilver, A. Hamers, W. Gaastra, K.-A Karlsson, and S. Normark. 1990. Host-specificity of uropathogenic *Escherichia coli* depends on differences in binding specificity to Galα1-4Gal-containing isoreceptors. *EMBO J.*9:2001–2010.

Svanborg-Eden, C., S. Marild, and T.K. Korhonen. 1982a. Adhesion inhibition by antibodies. *Scand. J. Infect. Dis. (Suppl.)* **33**:72–78.

Svanborg-Eden, C., R. Freter, L. Hagberg, R. Hull, S. Hull, H. Leffler, and G. Schoolnik. 1982b. Inhibition of experimental ascending urinary tract infection by an epithelial cell-surface receptor analogue. *Nature* **298**:560–562.

Teneberg, S., P. Willemsen, F.K. de Graaf, and K.A. Karlsson. 1990. Receptor-active glycolipids of epithelial cells of the small intestine of young and adult pigs in relation to susceptibility to infection with *Escherichia coli* K99. Dept. Med. Biochem. Univ. Goteborg. *FEBS Lett.* **263**:10–14.

Tuomanen, E., H. Towbin, G. Rosenfelder, D. Braun, G. Larson, G.C. Hansson, and R. Hill. 1988. Receptor analogues and monoclonal antibodies that inhibit adherence of *Bordetella pertussis* to human ciliated respiratory epithelial cells. *J. Exp. Med.* **168**:267–277.

Vaisanen-Rhen, V., T.K. Korhonen, and J. Finne. 1983. Novel cell-binding activity specific for *N*-acetyl-D-glucosamine in an *Escherichia coli* strain. *FEBS Lett.* **159**:233–236.

Vaisanen-Rhen, V., J. Elo, E. Vaisanen, A. Siitonen, I. Orskov, F. Orskov, S.B. Svenson, H. Makela, and T.K. Korhonen. 1984. P-fimbriated clones among uropathogenic *Escherichia coli* strains. *Infect. Immun.* **43**:149–155.

van Alphen, L., L.G. van den Broek, L. Blaas, M. van Horn, and J. Dankert, 1991. Blocking of fimbria-mediated adherence of *Haemophilus influenzae* by sialyl gangliosides. *Infect. Immun.* **59**:4473–4477.

Virkola, R. 1987. Binding characteristics of *Escherichia coli* type 1 fimbriae in the human kidney. *FEMS Microbiol. Lett.* **40**:257–262.

Wadstrom, T. and S.B. Baloda. 1986. Molecular aspects of small bowel colonization by enterotoxigenic *Escherichia coli*. *Mikrookol. Ther.* **16**:243–255.

Weiss, E.I., J. London, P. Kolenbrander, A.S. Kagemeier, and R. N. Andersen. 1987. Characterization of lectin like surface components on *Capnocytophaga ochracea* 33596 that mediate coaggregation with gram-positive oral bacteria. *Infect. Immun.* **55**:1198–1202.

Wentworth, J.S., F.E. Austin, N. Garber, N. Gilboa-Garber, C.A. Patterson, and R.J. Doyle. 1991. Cytoplasmic lectins contribute to the adhesion of *Pseudomonas aeruginosa*. *Biofouling* **4**:99–104.

Yamazaki, Y., S. Ebisu, and H. Okada. 1981. *Eikenella corrodens* adherence to human buccal epithelial cells. *Infect. Immun.* **31**:21–27.

6

Gram-Positive Pyogenic Cocci

The pyogenic Gram-positive cocci include members of the genera *Streptococcus* and *Staphylococcus*. It appears that there are no general rules regarding the adhesion of the pyogenic cocci. A few, including strains of *S. pyogenes, S. agalactiae, S. pneumoniae*, and *S. aureus*, express cell surface lectins, whereas others depend on lipoteichoic acids, cell-wall-bound proteins, secreted polysaccharides, and hydrophobins. There is a growing body of evidence to suggest that all of the pyogenic Gram-positive cocci rely on multiple adhesins in order to adhere avidly to substrata (Hasty et al., 1992). Most adhesion studies have been carried out on Gram-positive cocci, the normal habitat of which is the animal host. The successful pyogenic cocci have the ability to adhere to and grow on mucosal tissues without causing symptoms. This is commonly known as the carrier state.

Streptococcus pyogenes

S. pyogenes has a complex cell surface containing many different structures. The bacterium possesses a typical Gram-positive cell wall in terms of thickness and peptidoglycan content (Wagner and Ryc, 1978) (Chapter 4). Covalently linked to the peptidoglycan is the group-specific C-polysaccharide. In contrast to many other Gram-positive bacteria there is no cell wall poly(glycerol phosphate) or poly(ribitol phosphate) teichoic acid. A lipoteichoic acid (LTA), composed primarily of a repeating poly(glycerol phosphate) and covalently attached fatty acids, is anchored in the plasma membrane by hydrophobic interactions. The organisms also secrete proteins, some of which seem to be at least partially embedded in the peptidoglycan matrix (Pancholi and Fischetti, 1988). The best known of these proteins are the M proteins and T proteins. Most strains secrete a hyaluronic acid capsule. The surface components, including LTA, M-protein, and C-polysaccharide, form a matrix for the secreted capsular material. All the above-mentioned polymers, except for hyaluronic acid, are antigens accessible to specific antibodies. During normal growth, LTA is released from the plasma membrane in fully acylated and deacylated forms. The poly(glycerol phosphate) portion of the LTA is capable of complexing a number of cell-wall-associated proteins, including M-protein molecules. Thus the streptococcus may also have exposed fatty acids on its surface, a trait that confers hydrophobic properties to whole bacteria (Rosenberg et al., 1981; Ofek et al., 1983; Courtney et al., 1990). It is this surface that must form a stable union with an epithelial (or endothelial) cell in order for the bacterium to colonize complementary mucosal (or endothelial) surfaces.

Strategies for studying adhesion were reviewed in Chapter 2. Adhesion experiments with pyogenic streptococci are similar. The first step is to find conditions under which the bacteria and the target cells interact to form stable complexes, resulting in firm adhesion of the bacteria. It is then possible to introduce inhibitors or to modify either the surface of the bacterium or that of the target cells. For streptococcal bacterial adhesion experiments, the target surface has frequently been washed buccal epithelial cells or tissue culture cells. Occasionally, scraped pharyngeal cells or phagocytic cells have also been employed. The chemical identity of the streptococcal adhesins and their complementary receptors has been deduced from experiments that employ various inhibitors and potentiators of adhesion of *S. pyogenes* to buccal epithelial cells (Table 6–1). Of the bacterial surface components studied, only LTA is an effective inhibitor. Similarly, anti-LTA, but not normal serum, is also a good inhibitor. LTA, however, does not inhibit adhesion of Gram-negative bacteria to animal cells (Beachey and Ofek, 1976; Botta, 1981). Interestingly, antifibronectin is an inhibitor but soluble fibronectin is a potent promoter of adhesion when used in a coating solution for buccal cells. In contrast, soluble fibronectin at very low concentrations is an effective inhibitor when used in direct competition experiments. One of the best competitive agents is serum albumin, which binds to the fatty acid moieties of LTA. An outline of the results showing that LTA is an adhesin is provided in Table 6–2. Similarly, a summary of results describing the role of fibronectin as a receptor is found in Table 6–3.

Various cell lines have been tested for their ability to express surface fibronectin and to bind to streptococci. In general, the extent of adhesion could be correlated with fibronectin on the tissue culture cells. Fibronectin is a complex protein that contains domains capable of binding several kinds of ligands (Yamada, 1983). The protein contains carbohydrates and disulfide linkages and has a molecular weight of about 440,000. The protein exists in a soluble form in plasma and is insoluble when bound to cell surfaces. Fibronectin is known to bind collagen and gelatins, hyaluronic acid, heparin, various proteoglycans, and the hexose, glucose. It is not surprising that many other LTA-producing streptococci and staphylococci bind fibronectin, the binding ligands of which are apparently differ-

Table 6–1. Factors affecting the adhesion of Streptococcus pyogenes *to buccal epithelial cells*

	Inhibition (% of control)	Reference
Coating solution for buccal cells		
C-Polysaccharide	0	Ofek et al., 1975
M-Protein	0	Ofek et al., 1975
Peptidoglycan (sonicated suspension)	0	Ofek et al., 1975
Lipoteichoic acid	78	Ofek et al., 1975
Rabbit serum	6	Courtney et al., 1986
Rabbit antifibronectin	75	Courtney et al., 1986
Fibronectin	About 200% increase over control	Simpson and Beachey, 1983
β-Mercaptoethanol	46	Simpson and Beachey, 1983
Inhibitors		
Albumin	97	Beachey et al., 1980
Fibronectin	57	Simpson and Beachey, 1983
Gelatin	60	Simpson and Beachey, 1983
Collagen 1(I) chain	57	Simpson and Beachey, 1983
Collagen 1(I) chain	8	Simpson and Beachey, 1983
Coating solutions for the streptococci		
Anti-LTA serum	73	Ofek et al., 1975
Normal rabbit serum	0	Ofek et al., 1975
Trypsin (followed by washing away the enzyme)	100	Ofek et al., 1975

ent from LTA. The multiple types of surface ligands that are capable of binding fibronectin are illustrated by intraspecies aggregation by a monomeric derivative of fibronectin (Simpson et al., 1987). Fibronectin can be reduced and alkylated under mild conditions to yield a monomeric form of the protein. This form has a molecular weight of about 220,000 and retains the ability to bind various bacteria. If the bacteria possess unique combining sites on the monomeric fibronectin, then it should be possible to coat one bacterium with the protein and elicit an agglutination reaction with another uncoated bacterium. Conversely, if all the sites on the fibronectin are identical for the various kinds of bacteria, then agglutination would not be expected to occur. Fibronectin-coated *S. pyogenes* agglutinates uncoated *S. aureus, S. sanguis, S. mitis,* or *S. pneumoniae*. This experiment can be taken as evidence for distinct combining sites on the fibronectin molecule for *S. pyogenes* and the other cocci. The coating of *S. aureus* with fibronectin also results in the coaggregation of *S. pyogenes*. When LTA is added to the reaction mix of fibronectin-coated *S. aureus* and uncoated *S. pyogenes*, there is an inhibition of coaggregation. This is likely because the fibronectin is

Table 6–2. Summary of results supporting LTA as an adhesin of Streptococcus pyogenes

Soluble LTA, but not deacylated LTA, binds spontaneously to a wide variety of animal cells and inhibits the adhesion of *S. pyogenes* to phargyngeal (Botta, 1981; Tylewska et al., 1988); buccal (Ofek et al., 1975; Beachey and Ofek, 1976; Carruthers and Kabat, 1983), skin (Alkan et al., 1977), phagocytic (Ofek and Beachey, 1979; Courtney et al., 1981), and tissue culture (DeVuono and Panos, 1978; Grabovskaya et al., 1980) cells.
Antipoly (glycerol phosphate) inhibits the adhesion of *S. pyogenes* to buccal cells (Beachey and Ofek, 1976).
Surface components other than LTA or antibodies to such components do not inhibit streptococcal adhesion to buccal cells (Ofek et al., 1975; Beachey and Ofek, 1976) nor does LTA inhibit adhesion of Gram-negative bacteria to pharyngeal cells (Botta, 1981) or to phagocytic cells (Ofek and Beachey, 1979; Courtney et al., 1981).
Loss of LTA induced by penicillin or berberine sulfate results in reduced adhesion of *S. pyogenes* to buccal cells (Alkan and Beachey, 1978; Nealon et al., 1986; Sun et al., 1988).
Serum albumin complexes with the acylated portion of LTA via its fatty acid binding sites and inhibits streptococcal adhesion (Beachey et al., 1980; Simpson et al., 1980).
There is a direct relationship between surface exposure of LTA and the ability of the bacteria to adhere to epithelial cells (Miorner et al., 1983; Ofek et al., 1983; Courtney et al., 1990).
Development of LTA receptors on buccal epithelial cells during the neonatal period coincides with the ability of the cells to bind streptococci (Ofek et al., 1977).
Although most LTA is anchored in the cytoplasmic membrane, some is exposed on the outer surface (Beachey and Ofek, 1976; Mattingly and Johnston, 1987; Ryc et al., 1988). Surface proteins form electrostatic complexes with the poly(glycerolphosphate) backbone of LTA and orient the molecule for its adhesin-receptor function (Ofek et al., 1982). Quantitative and qualitative changes in the surface constitutents may cause a disarray in the supramolecular structure of the envelope. This prevents the correct cell-surface presentation of the LTA and results in altered adhesion.

able to complex with the poly(ribitol phosphate) backbone on the *S. aureus* and the soluble LTA via its hydrophobic sites (Aly and Levit, 1987). The fatty acid ends of the LTA molecules on the *S. pyogenes* cells would therefore be in competition with soluble LTA for the fibronectin. Even though there are compelling results showing that LTA and fibronectin can interact and contribute to the adhesion process there may be other surface components on the streptococci that also complex with fibronectin. Some of these components are presented below in a discussion on multiple adhesins in streptococci.

In spite of the solid evidence that LTA and fibronectin are centrally involved in binding of *S. pyogenes* to target tissue cells, it is likely the organisms are capable of adhering to human cells by additional mechanisms. At least six different mechanisms of adhesion have been proposed for *S. pyogenes* (Table 6–4). It would appear that detection of the different mechanisms has depended primarily on the type of target substratum used in the assay. For example, fibrinogen-mediated adhesion was detected in tissue culture cells infected with

Table 6–3. Summary of results favoring a role for fibronectin as a receptor for Streptococcus pyogenes

The adhesion of *S. pyogenes* to tissue culture cell lines and phagocytic cells is proportional to the amounts of fibronectin on the cell surface (Stanislawski et al., 1985; Savoia and Landolfo, 1987).
Removal of fibronectin from buccal (Simpson and Beachey, 1983) or tissue culture cells (Bartelt and Duncan, 1978) by trypsin reduces streptococcal adhesion.
Antibody directed against fibronectin reduces adhesion of streptococci to buccal (Simpson and Beachey, 1983; Courtney et al., 1986) and to phagocytic cells (Savoia and Landolfo, 1987).
Fibronectin-coated buccal (Simpson and Beachey, 1983) and phagocytic (Savoia and Landolfo, 1987) cells bind more streptococci than uncoated cells.
Soluble and immobilized fibronectin bind whole streptococci and LTA inhibit the binding (Simpson and Beachey, 1983; Courtney et al., 1986; Valentin-Weigand et al., 1988; Wagner et al., 1988).
Fibronectin contains a fatty acid binding domain in its amino-terminal region that binds the lipid moiety of LTA (Courtney et al., 1983; Stanislawski et al., 1987; Valentin-Weigand et al., 1988).
Group A streptococci bind selectively to fibronectin-containing buccal epithelial cells and specifically to the fibronectin deposits on the cells (Abraham et al., 1983).

Table 6–4. Multiple mechanisms of adhesion of Streptococcus pyogenes

Adhesin/Receptor	Target Substrata	References
LTA/fibronectin or ?	Human buccal cells	Beachey and Ofek, 1976; Simpson and Beachey, 1983
	Human pharyngeal cells	Botta, 1981
	Mouse pharyngeal cells	Baird et al., 1990
	HEp-2 cells	Grabovskaya et al., 1980
	Human and mouse fibroblasts	Leon and Panos, 1990
	Human leukocytes	Courtney et al., 1981
FBP[a]/fibronectin	Human fibronectin	Speziale et al., 1984; Talay et al., 1991
M-protein/galactose and fucose	Human pharyngeal cells	Ellen and Gibbons, 1972; Tylewska et al., 1987, 1988; Wadström and Tylewska, 1982
M-protein/?	HEp-2 cells	Grabovskaya et al., 1980
	Buccal cells	Sjernquist et al., 1986
M-protein/fibrinogen	Virus-infected MDCK cells	Sanford et al., 1982
VBP?[b]/vitronectin	Human endothelial cells	Valentin-Weigand et al., 1988
C-carbohydrate/?	Human pharyngeal cells	Botta, 1981

[a]FBP, fibronectin-binding protein.

[b]VBP, vitronectin-binding protein.

the influenza virus, whereas M protein-mediated adhesion was detected in assays employing pharyngeal epithelial cells and HEp-2 cells. The possibility of LTA/M-protein-independent adhesion was recently suggested in experiments studying adhesion of organisms purported to be isogenic M-positive and M-negative strains to buccal and tonsillar epithelial cells (Caparon et al., 1991). Although the M-protein gene is not expressed in these organisms, it is not known whether other surface structures might be affected by this deletion. Endothelial cells, on the other hand, reveal the presence of a vitronectin-binding structure on the streptococcal surface. An exception is adhesion mediated by LTA, which appears to be important for bacterial attachment to many different types of substrata. If the adhesion of group A streptococci obeys a two-step kinetic model as in the case of *S. sangius* (Chapter 8) (Figure 6–1), then a plausible explanation for these findings can be suggested. The first step would be mediated by a hydrophobic moiety, such as LTA, capable of recognizing a relatively wide range of molecules, whereas the second step would be mediated by another, possibly more specific adhesin for which receptors are accessible on a particular substratum (Figure 6–

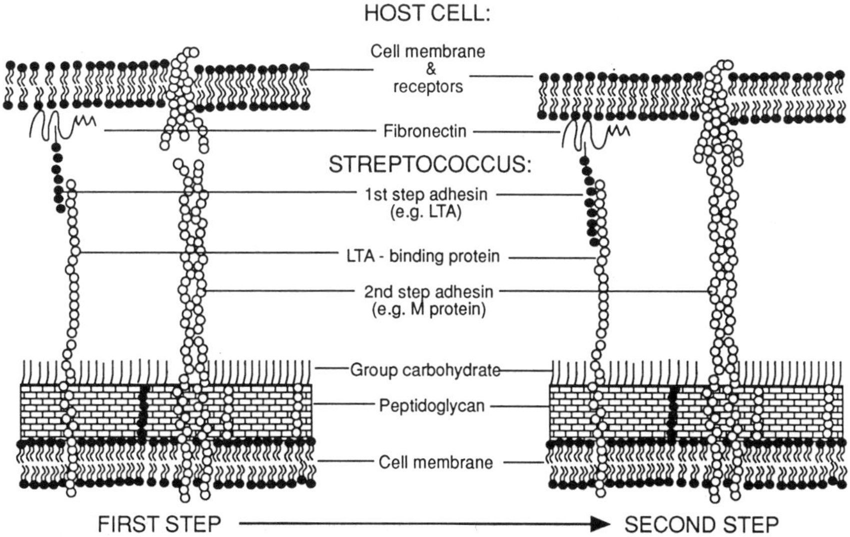

Figure 6–1. Proposed two-step model for the adhesion of group A streptococci to epithelial cells. (Adapted from Simpson et al., 1987; Hasty *et al.*, 1992 and Ofek *et al.*, 1982). Details of the model are provided in this text.

1). The first step, therefore, would involve a reversible, weak hydrophobic interaction between LTA, oriented with its fatty acid moiety toward the streptococcal surface by LTA-binding proteins (Ofek et al., 1982) and the fatty acid binding domains that apparently are present on the membranes of many different cell types and different molecules (Ofek et al., 1975; Beachey et al., 1977; Alkan and Beachey, 1978; Beachey et al., 1979; Simpson et al., 1980a,b,c; Courtney et al., 1981, 1983; Hasty et al., 1989). The correlation between hydrophobicity of the streptococci mainly contributed by LTA (Miorner et al., 1983) and the ability of the bacteria to adhere to epithelial cells (Courtney et al., 1990), and also the ability of LTA to inhibit adhesion of streptococci to many types of cells (Table 6–4), support this premise. Variants of group A streptococci produce different amounts of LTA, yet retain their hydrophobic characteristics, suggesting that hydrophobicity is governed by factors other than LTA (Leon and Panos, 1990).

To examine further the validity of a two-step adhesion model in group A streptococci, the M-protein-mediated adhesion to human epithelial cells will be examined. Grabovskaya et al. (1980) postulated that group A streptococci adhere to HEp-2 cells by two different mechanisms, one involving LTA and the other involving M-protein. According to an interpretation of the two-step hypothesis, the first step of adhesion would be inhibited by LTA and the second step would be inhibited by M-protein. On the other hand, the second step of adhesion to an M-protein receptor would not occur at all if an M-negative strain were used (LTA-negative strains have not been reported and it is thought that an LTA-deletion may be a lethal mutation). Thus, the overall effect in any of these instances where the first step is prevented would be reduction or elimination of adhesion. Any adhesion that occurred as a result of successful completion of the first step should be easily reversible by multiple washing. The findings of Grabovskaya et al. support this notion: the M-negative strain adhered to the cells, but in a weak and reversible fashion, as evidenced by the rapid decrease in the number of adherent streptococci with sequential washings; the M-positive strain adhered more avidly to the cells, being resistant to multiple washings. Utilizing an assay in which adherent bacteria are enumerated only after numerous washings, it was recently determined that both LTA and a fragment of M-protein inhibit adhesion of group A streptococci to HEp-2 cells (Courtney et al., 1990). These data are consistent with a two-step model of adhesion. Attempts to show binding of M-protein to pharyngeal and buccal epithelial cells have not been successful, although other cell-wall-associated proteins have been shown to bind to the cells (Tylewska and Hryniewicz, 1987).

The adhesin(s) participating in the first step of adhesion of *S. pyogenes* (e.g., LTA) may be generally reactive with a number of cell types, whereas the adhesins participating in the second step (e.g., M-protein, FBP, VBP, etc.) may be specific for a particular tissue. For example, skin isolates of *S. pyogenes* adhere in higher numbers to skin cells, whereas throat isolates adhere better to buccal cells (Alkan et al., 1977). But the adhesion of both the skin and the throat isolates to cells is inhibited by LTA. Although at face value the results suggest that LTA-mediated adhesion is the only mechanism involved in both skin and pharyngeal strains, the results are difficult to interpret based on a single adhesion mechanism. Because the adhesion assays were subject to multiple washings before enumeration of adherent bacteria, the results may be interpreted according to the two-step adhesion model. It is postulated that LTA participates in the first step of adhesion with both types of isolates, whereas distinct adhesins expressed by skin or throat isolates participate in the second step of adhesion to skin or throat cells, respectively. Thus, LTA would be expected to inhibit adhesion in both cases because it is involved in the first step, whereas it is the second-step adhesin that determines tissue tropism. It is also possible that the model applies to other LTA-producing streptococci and staphylococci, as LTA was found to inhibit adhesion of several members of these species to epithelial or tissue culture cells (Botta, 1981; Bramley and Hogben, 1982; Carruthers and Kabat, 1983; Nealon and Mattingly, 1984; Teti et al., 1987a,b; von Hunolstein et al., 1987; Chugh et al., 1989b, 1990). Further, each of these bacterial species has been found to produce adhesins other than LTA.

Adhesion results from *S. pyogenes*–epithelia interactions have largely been derived from bacteria that are unencapsulated. Bartelt and Duncan (1978) first noticed that hyaluronic acid capsule interfered with the ability of the streptococci to adhere to tissue culture cells. It is thought that capsule is generated only after the bacteria have adhered or colonized. Exponential-phase *S. pyogenes* produces large quantities of capsule but do not adhere as well as the stationary phase bacteria (Figure 6–2). Exponential phase *S. pyogenes* is much less hydrophobic than stationary phase cells, probably because hyaluronic acid masks hydrophobic residues in the growing bacteria. The assay for hydrophobicity may be a useful indicator of the extent of exposure of LTA on streptococcal surfaces, and by inference, on the percent of cells that are encapsulated with poor adhering ability. The presence of absence of hyaluronic acid must be regulated, but at present the mechanism of its regulation is unknown. Hyaluronidase is a good candidate for a regulatory enzyme but this needs a critical examination.

The importance of streptococcal adhesion to tonsillar cells during the infectious process is evident from direct observations of tonsillar cells from patients with group A streptococcal infections. Cells from these patients were heavily laden with adherent streptococci (Stenfors and Raisanen, 1991). Following a 24-hour treatment with penicillin, cells from the same patients were largely free of adherent group A streptococci. Healthy carriers of group A streptococci also harbor adherent streptococci onto their tonsillar cells but in significantly fewer numbers than the symptomatic patients (Stenfors and Raisanen, 1991; Stenfors et al., 1991). In another study, it was shown that adhesion of group A streptococci to HEp-2 cells was greater when the streptococci were isolated from recurrent infections (Galioto et al., 1988). These results suggest that phenotypic variation

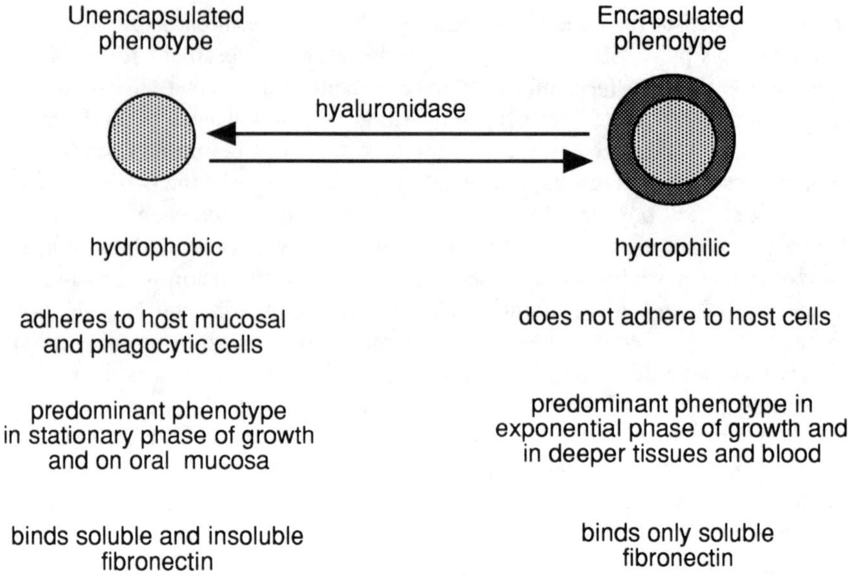

Figure 6–2. Hypothetical role of hyaluronic acid capsule in adhesion of streptococci to host cells in different environments. Encapsulated streptococci are not hydrophobic and do not adhere well to immobilized fibronectin or to host cells. Therefore, it would be expected that in deeper tissues encapsulated streptococci would be the predominant phenotype because they would be able to avoid attachment to phagocytes more effectively. Conversely, in the oral cavity the streptococci must attach to the host epithelium to avoid being swept away by the cleansing mechanisms of the host. Thus, in the oral cavity the prevailing bacteria would be of the nonencapsulated phenotype because they would be able to attach to the host epithelium. (From Courtney et al., 1990).

in adhesin may be important in infectivity of the streptococci. Two separate experimental approaches support the view that particular streptococcal adhesins contribute to the colonization of the upper respiratory tract of mice. In one protocol, mice were immunized intranasally with M-protein peptide linked to a carrier and then challenged intranasally with virulent streptococci (Bessen and Fischetti, 1988; Fischetti et al., 1989). It was observed that the immunized mice exhibited a lower level of streptococcal pharyngeal colonization. In another protocol, the nasopharyngeal area was coated with streptococcal LTA as a potential receptor blocking agent (Baird et al., 1990). When challenged with virulent streptococci, the LTA-treated mice exhibited a lower level of streptococcal colonization than untreated mice. The results are consistent with the model requiring multiple adhesins (Figure 6–1). The importance of streptococcal adhesion in triggering symptomatic infection may be due to targeting potent cytolytic toxins, such as streptolysin S. The degree of cytolytic activity of streptolysin S-bearing streptococci toward tissue culture cells was proportional to the ability of the streptococci to adhere to the cells (Ofek et al., 1990). Furthermore, the adhesion-dependent cytotoxicity of the streptococci could not be neutralized by streptolysin S inhibitors, probably because streptolysin S toxin, proximal to the eucaryotic cell surface, was inaccessible (see Figure 11–4). Group A streptococci are capable of binding many host-derived proteins including albumin, fibrinogen, α_2-macroglobulin, immunoglobulin G, fibronectin, laminin, plasmin, collagen type IV, salivary glycoproteins, salivary mucins, basement membrane proteins, and haptoglobin (Switalski et al., 1984; Chhatwal et al., 1985, 1987a,b, 1990; Schalen et al., 1986; Yarnall and Boyle, 1986; Chhatwal and Blobel, 1987; Lottenberg et al., 1987; Nardella et al., 1987; Schmidt et al., 1987; Speziale et al., 1987; Broeseker et al., 1988; Yamada and Matsumoto, 1988; Bergey and Stinson, 1988; Broeder et al., 1989; Kostrzynska et al., 1989; Lammler et al., 1988; Bessen and Fischetti, 1990; Courtney and Hasty, 1991; Raeder et al., 1991). Only a few (albumin, fibronectin, and salivary glycoprotein) of these host-derived proteins have been shown to have an effect on streptococcal adhesion and except for perhaps fibronectin their effect on the adhesion process in vivo remains to be defined.

Streptococcus agalactiae (Group B Streptococci)

S. agalactiae is a common bacterial pathogen for neonates and infants and is a colonizer of the human vagina. The group B streptococcus possesses surface appendages morphologically similar to those of group A streptococci (Bramley and Hogben, 1982). The factors governing the adhesion of group B streptococci to animal cells have many features resembling those of group A streptococci. These are inhibition of adhesion by LTA, anti-poly(glycerol phosphate), bovine serum albumin, and proteolytic enzymes but not with deacylated LTA (Bramley and Hogben, 1982; Goldschmidt and Panos, 1984; Nealon and Mattingly, 1984; Teti et al., 1987a). Because group B streptococci possess surface-associated proteins (Doran and Mattingly, 1982), it is possible that LTA may be intercalated in the wall periphery analogous to the LTA of group A streptococci. Work by Mattingly and Johnston (1987) has provided evidence to suggest that unlike *S. pyogenes*, the LTA remains largely membrane-associated during growth. This precludes a contribution of LTA to the hydrophobicity of the bacteria. Isolates from infants have been reported to contain relatively high amounts of LTA (Nealon and Mattingly, 1983). Moreover, deacylated LTA was an inhibitor of adhesion to fetal lung tissue culture cells (Nealon and Mattingly, 1985). In contrast, adhesion to neonatal buccal epithelia was inhibited by coating the cells with LTA. The deacylated form of LTA was a poor inhibitor of adhesion (Teti et al., 1987a). A conclusion is that LTA plays some role in the adhesion of *S. agalactiae* to neonatal vaginal and epithelial cells. The studies reported by Teti et al. (1987a) employed exfoliated epithelial cells, whereas those from Mattingly's laboratory used embryonic and fetal tissue culture cells as well as exfoliated buccal epithelial cells from adults. A thesis developed in this book is that a bacterium has more than one means to adhere to a surface. Changes in substrata may permit different modes of adhesion, even when employing bacterial cells from the same culture. It may be that *S. agalactiae* depends on polyanionic binding using poly(glycerol phosphate) as an adhesin for one cell type (e.g., embryonic tissue culture cells) and the fatty acid end of LTA as an adhesin for another cell type (e.g., exfoliated buccal epithelial cells). LTA may be important for adhesion of the bacteria to polyethylene intrauterine contraceptive devices because adhesion of the bacteria to biomaterials was inhibited by fibronectin and albumin (Jacques and Costerton, 1987). It is also possible that binding of group B streptococci to vaginal cells involves a proteinaceous adhesin that recognizes structures masked by terminal sialic acid because vaginal cells treated with sialidase bind more bacteria (Bulgakova et al., 1986). Vaginal and pharyngeal cells bind more *S. agalactiae* than HEp-2 cells, whereas HEp-2 cells bind more *S. pyogenes* than vaginal or pharyngeal cells, suggesting that *S. agalactiae* binds via a mechanism distinct from that of *S. pyogenes* (Totolian et al., 1985). The state of the animal cell cultures may affect their ability to bind *S. agalactiae* (Kubin and Ryc, 1988). In one study it was found that adhesion of group B streptococci to synchronously growing human cell monolayers involves a proteinaceous adhesin rather than surface-exposed LTA (Miyazaki et al., 1988). *S. agalactiae* also binds fibrinogen and strains that bind fibrinogen induce platelet aggregation in the presence of plasma (Usui et al., 1987). *S. agalactiae* isolated from preterm infants with sepsis bind in higher numbers to chorionic epithelial cells as compared to isolates from uncomplicated pregnancies (Helmig et al., 1990).

Jelinkova et al., (1986) found that differences in adhesion of group B streptococci could not be attributed to the type-specific proteins. Electron microscopic observations have indicated that the peripheral coat associated with the capsule of the bacteria is involved in adhesion to vaginal epithelial cells (Bulgakova et al., 1986). Although group B streptococci are known to cause infection in farm animals, little is known defining the molecular bases of adhesion to their epithelial cells (Valentine et al., 1988).

The availability of poly(glycerol phosphate) receptors for binding group B streptococci at a particular site in human tissue remains to be determined. Cox (1982) obtained evidence, unique of its kind, indicating that LTA-mediated adhesion plays a role in neonatal group B streptococcal infection in mice. He treated 3-day-old mice with topical applications of LTA in the oral cavity, perineum, and nape. The mother was painted with a slurry of group B streptococci over her oral cavity, vagina, and nipples. After 3 days, 47% of the control pups were culture-positive for group B streptococci whereas none of the treated pups were culture-positive. Cox (1982) further noted that the oral application of LTA prevented colonization of the oropharynx with *S. agalactiae* in a dose-dependent manner. Topical LTA probably interferes with colonization of neonatal mice by flooding their tissue with excess bacterial adhesin and thereby covering receptor sites and inhibiting group B streptococcal adhesion.

Streptococcus pneumoniae

S. pneumoniae is a normal inhabitant of the upper respiratory tract of humans and frequently causes respiratory tract infections. There is now convincing evidence to suggest that among the surface lectins expressed by the few Gram-positive bacterial species (Mirelman and Ofek, 1986), the adhesion of *Streptococcus pneumoniae* to buccal and pharyngeal cells is mediated by saccharide–protein interactions (see also Chapter 5).

Andersson et al. (1983) found that various oligosaccharides containing glucose, galactose, and *N*-acetylglucosamine were inhibitors of adhesion of the bacteria to pharyngeal cells. The best inhibitor was Galβ1,4GlcNAcβ1,3Galβ1,4Glc. Coating of guinea pig erythrocytes with glycolipids rendered the red cells susceptible to agglutination by *S. pneumoniae* (Andersson et al., 1983; Beuth et al., 1987). Structures containing GlcNAcβ1,3Gal were required to sensitize the cells to agglutination by bacteria. Glycolipids containing the GalNAcβ1,4Galβ1,4Glc sequence were also able to bind *S. pneumoniae* on thin-layer chromatographic plates (Krivan et al., 1988). The supernatants of heat-treated *S. pneumoniae* caused aggregation of Latex particles covalently coupled with oligosaccharides, suggesting that organisms possess a peripheral loosely bound lectin capable of functioning as an adhesin by bridging between the bacteria and the epithelial cells (Andersson et al., 1988a,b). There may be additional pneumococcal adhesins, as shown in experiments employing frog palate mucosa as a substratum for *S. pneumoniae* adhesion. The pneumococci bound in higher numbers to the mucus-coated cells than to mucus-depleted cells of the palate mucosa (Plotkowski et al., 1988, 1989). Of interest is the observation that human milk contains glycolipids, which inhibit adhesion of the organisms to pharyngeal cells (Andersson et al., 1986; Aniansson et al., 1990). This finding provides evidence that natural protection may be achieved by breast milk not involving antibodies.

In attempts to understand the importance of adhesion in vivo a number of studies have compared the adhesive capacity of various pneumococcal isolates and of epithelia from patients. Strains isolated from healthy carriers of *S. pneumoniae* adhered to nasopharyngeal cells equally well as strains isolated from acute otitis media patients (Andersson et al., 1988b). Also, the magnitude of adhesion was independent of pneumococcal capsular serotype. In other studies, it has been observed that pneumococci bind significantly higher to buccal epithelial cells from smokers as compared to nonsmokers (Mahajan and Panhotra, 1989). Also, nasopharyngeal epithelial cells from children bind more pneumococci than similar cells from adults (Shimamura et al., 1990). Viral infections may also predispose mucosa to bind increased numbers of pneumococci (Plotkowski et al., 1986).

The results raise the possibility that viral infection may induce pneumococcal receptors in tracheal cells.

Staphylococci

Humans serve as a common reservoir for staphylococci. The bacteria rapidly colonize newborns within hours to a few days postpartum. Eyes, skin, intestines, and nasopharyngeal areas provide sites of attachment for staphylococci. Because the staphylococci can survive on so many different tissues, it seems plausible that the bacteria have evolved multiple molecular mechanisms for adhesion. Most isolates can be classified as either coagulase-positive or coagulase-negative. *S. aureus* is the only coagulase-positive member of the genus and is frequently isolated from clinical specimens. Coagulase-negative organisms are typical opportunistic pathogens of which more than 20 species have been identified (Christensen et al., 1983). Among the coagulase-negative species, *S. saprophyticus* is a major cause of urinary tract infections in young females.

Staphylococcus aureus

S. aureus does not appear to express a surface lectin. The cell surface of *S. aureus* is characterized by a peptidoglycan that contains a pentaglycine bridge as a crosslinking peptide. The muramic acid residue is substituted with a ribitol teichoic acid. The ribitol moieties may possess esterified D-alanine and α-or β-*N*-acetylglucosaminyl residues (Oeding and Grov, 1972).

Although antigenic variations in *S. aureus* are not based on surface M-like proteins, the organisms carry on their surfaces cell-associated proteins that may share common structural and functional traits with those of *S. pyogenes*. Most notably is protein A, an immunoglobulin-binding protein. The C-terminal region of protein A was found to possess a high sequence homology with M-proteins of *S. pyogenes* (Pancholi and Fischetti, 1988; Fischetti, 1991). This conserved region, also known as the "anchor region," contains a hydrophobic stretch that spans the cytoplasmic membrane. Immunoelectron microscopy has revealed that LTA molecules are surface-exposed in *S. aureus* (Aasjord and Grov, 1980). It is also known that the organisms exhibit hydrophobic surfaces (Colleen et al., 1979; Jonsson and Wadström, 1983, 1984), and similar to *S. pyogenes*, hydrophobicity is abolished by treating the organisms with proteolytic enzymes (Ljungh et al., 1985). The hydrophobicity may be due to either LTA or to protein(s) or to both (Wadström, 1990). The origin of the surface protein hydrophobin may be the result of surface exposure of some of the anchor region proteins, but at present there is no clear evidence for that.

In 1978, Kuusela (1978) showed that *S. aureus* could bind to fibronectin. Since that discovery, most investigators have concluded that fibronectin serves as a receptor that binds the organisms to various cell types as well as to biomaterials that become coated with fibronectin on insertion into host tissues. Kuusela (1978) showed that intact *S. aureus*, as well as heat-treated, formalin-fixed, or trichloroacetic acid-aggregated *S. aureus* could form stable complexes with human plasma fibronectin. It is likely that the staphylococcal surface presents different components to fibronectin after each treatment. The molecular bases for *S. aureus*–fibronectin interactions have not been completely worked out because the organisms are capable of expressing multiple ligands that complement the various domains of the fibronectin molecule.

The fibronectin–*S. aureus* complex formation could be inhibited by amino sugars, urea, and lysine, but not by protein A (Kuusela, 1978). Furthermore, gelatin was only a weak inhibitor, suggesting that the gelatin-binding domain of fibronectin is not involved. The dissociation constant for both heat-killed or living *S. aureus* has been reported to be of the order of 5×10^{-9} *M* fibronectin, suggesting a very strong association. Froman et al. (1987) loaded a lysostaphin digest of heat-killed *S. aureus* cells over a column of 29,000 mol wt fibronectin fragment–Sepharose and eluted a 210,000 mol wt protein with 2 *M* guanidine hydrochloride. The protein was purified to homogeneity and found to bind fibronectin. When 210,000 mol wt protein was hydrolyzed by staphylococcal V8 protease, several peptides were isolated that contained fibronectin binding activities, suggesting that the protein may contain functionally repeating sequences. Previously, Espersen and Clemmensen (1982) isolated a protein of approximately 200 kDa from sonic extracts of *S. aureus* and found that the protein could precipitate with fibronectin in crossed immunoelectrophoresis. Antibody against the 210-kDa protein would bind to *S. aureus* cells, providing evidence for a surface location of the protein (Froman et al., 1987). A fibronectin-binding protein of *S. aureus* has been cloned into *E. coli* (Flock et al., 1987). One clone secreted proteins of 87,000 and 165,000 mol wt into the periplasm. Both gene products were capable of binding fibronectin and both possessed amino acid compositions similar to that of the 210,000 mol wt protein of Froman et al. (1987). There are several lines of evidence that the 210-kDa protein possesses attachment sites for binding soluble and insoluble fibronectin. First, protein from *S. aureus* lysates recovered from fibronectin–Sepharose columns inhibited the adhesion of the bacteria to fibronectin-coated surfaces. Second, the density of fibronectin binding sites on clinical isolates of *S. aureus* was roughly proportional to invasiveness of the organisms (Wadström, 1987). Third, when *S. aureus* was treated with sublethal concentrations of penicillin, the density of fibronectin binding sites increased, the increase of which paralleled adhesion to fibronectin-coated surfaces (Proctor et al., 1983). Fibronectin seems to promote the adhesin of *S. aureus* to blood clots (Raja et al., 1990). The adhesion was observed only in fibronectin-containing plasma and could be inhibited by a fibronectin peptide of the amino-terminal fragment. Antibody directed against fibronectin also seems to effectively inhibit the adhesion of *S. aureus* to biomaterial implants, suggesting that fibronectin is a receptor for the bacteria (Vaudaux et al., 1984). Although it now appears certain that fibronectin mediates the adhesion of *S. aureus* to various surfaces, the molecular aspects of the adhesive events remain to be defined. The exact role of the 210,000 mol wt protein of *S. aureus* is uncertain. How does this protein recognize fibronectin? Lectin-like interactions probably are not involved. Protein–protein interactions mediated by complementary hydrophobic amino acid side chains are a possibility. Collectively, the findings suggest that heat-killed *S. aureus* expresses a 210,000 mol wt protein that possesses repeating subunits and is capable of complexing with the fibronectin molecule. Because heating of *S. aureus* results in altered surface hydrophobicity (Ljungh et al., 1985), it is not known whether the heat treatment may expose cytoplasmic proteins so as to induce reactivity with the anti-210,000 mol wt protein or with fibronectin. Nevertheless, the amino terminus 27,000 mol wt domain of fibronectin appears to be a major receptor for the 210,000 mol wt protein of heat-killed or viable *S. aureus* (Maxe et al., 1986). The gelatin-binding domain (carboxy terminal region) of fibronectin displays only a weak affinity for *S. aureus* (Ryden et al., 1983). Although the same amino terminal fragment of fibronectin binds *S. aureus* and *S. pyogenes*, Simpson et al., (1987) provided evidence that the two organisms bind to distinct regions of the protein. A series of experiments conducted by Kuypers and Proctor (1989) provide the most convincing results that *S. aureus* binding to fibronectin is important in infectivity. They created by transposon mutagenesis, several mutants of *S. aureus* that exhibited a weak ability to bind fibronectin. They then infected the heart valves of rats with the isogenic strain and the mutants. Of the organisms recovered from the heart valves most were of the isogenic parent origin. The low fibronectin-binding mutants were recovered in poor numbers.

S. aureus also binds collagen (Holderbaum et al., 1985, 1986), gelatin (Carret et al., 1985), laminin (Lopes et al., 1985; Mota et al., 1988), a bone sialoprotein (Ryden et al., 1989), fibrinogen (Wadstrom, 1991), thrombospondin (Herrmann et al., 1991), vitronectin (Chhatwal et al., 1987b), serum-spreading factor (Fuquay et al., 1986), mucin-secretory IgA complexes (Biesbrock et al., 1991), and lactoferrin (Naidu et al., 1991). *S. aureus* bound to soluble collagen with a *K*d of 9.7×10^{-9} *M* collagen and was not inhibited by fibronectin, fibrinogen, Clq component of complement, or IgG, suggesting that the collagen binding site is distinct. It was suggested that collagen may be a receptor for *S. aureus* in connective tissues (Carret et al., 1985). It was found that gelatin would clump 2 of 98 strains of *S. aureus* tested. It may be possible that these two strains interacted with renatured or collagen-like domains on the gelatin molecules. The role of laminin as a receptor for *S. aureus* in the basement membrane is not well resolved, but it underscores the thesis that the bacterium can express multiple mechanisms for interacting with tissue components. Thrombospondin has been suggested to be important in mediating staphylococcal adhesion to activated platelets, to blood clots, and to extracellular matrices. In contrast, interaction of staphylococci with salivary mucin–secretory IgA complexes may prevent adhesion of the bacteria to oral mucosa. Numerous soluble host-derived proteins, including fibronectin, fibrinogen, and collagen bind to biomaterials and form substratum receptors for *S. aureus* (Wadström, 1991; Miedzobrodzki et al., 1989). Fibrinogen (clumping factor) aggregates *S. aureus* and fraction D presumably represents the staphylococcal recognition site (Kloczewiak et al., 1987). The fibrinogen-binding protein has been identified as a 59,000 mol wt protein (Usui et al., 1986). The fibrinogen–staphylococcal interaction may be one of the factors leading to colonization of wounds (Wadström, 1991). Catheter devices removed from patients are superior substrata to uninserted catheters for adhesion of *S. aureus* (and *S. epidermidis*) (Vaudaux et al., 1989). Adhesion of *S. aureus* to skin wounds in pigs was shown to be a complex phenomenon involving many factors that either promote or inhibit the staphylococcal adhesion (Mertz et al., 1987). The adhesion to wounds appears to be not affected by fibronectin. A laminin-binding protein of 57,000 mol wt from the surface of *S. aureus* was identified and found to antigenically crossreact with a 67,000 mol wt protein from mouse melanoma cells (Mota et al., 1988).

Teichoic acids may be ligands for receptors in nasal and vulvar epithelial cells (Aly and Levit, 1987; Bibel et al., 1987). It was found that washed epithelial cells, when incubated with streptococcal LTA or with *S. aureus* wall teichoic acid, exhibited a reduced affinity for intact *S. aureus* cells. In this regard, it has been found (Carruthers and Kabat, 1983) that staphylococcal LTA, but not deacylated LTA preparations, inhibited adhesion of *S. aureus* to buccal cells. A human mesothelial cell line served as a substratum for *S. aureus*, the adhesion of which could be inhibited by LTA, but not by wall teichoic acid, and protein A (Wyatt *et al.*, 1990). The composite interpretation of the data suggests that LTA, wall teichoic acid, and protein A serve as adhesins for epithelial cells.

S. aureus harvested during the exponential phase are less hydrophobic and adhere poorly to human buccal epithelial cells, compared to the more hydrophobic and adherent stationary phase cells (Beck et al., 1989). Taking advantage of a sensitive nitrate reductase colorimetric assay, it was observed that *S. aureus* adhered avidly to fibrin clots but poorly to uroepithelial and HeLa cells (Shoeb et al., 1991)

Endovascular infections are frequently associated with *S. aureus*. It is thought that initiation of these infections requires the adhesion of *S. aureus* to endovascular cell surfaces. Electron microscopic studies reveal that staphylococcal infection of human endothelial cells requires three separable steps: adhesion, endocytosis, and intracellular replication, suggesting that the organism produces adhesin/invasin on its surface (Hamill et al., 1986; Vann and Proctor, 1987; Lowy et al., 1988). In vitro studies have shown that fibronectin increased the adhesion of *S. aureus* to cultured bovine endothelial cells, but fibronectin had no effect on endocytosis of the bound bacteria (Hamill et al., 1986). In contrast, fibronectin has no effect on adhesion of *S. aureus* to cultured porcine endothelial cells

(Johnson et al., 1988). Soluble extracts of the staphylococci, however, inhibited the adhesion to the porcine cells. Adhesion of *S. aureus* to human endothelial cells is modulated by a number of interleukins that are likely to be released during the infectious process. IL-1 and acidic fibroblast growth factor were found to enhance the adhesion of *S. aureus* human umbilical vein endothelial cells, but α-interferon had no effect on the adhesion (Thomas et al., 1988). In contrast, Blumberg et al. (1988) found that IL-1 had no effect on the adhesion of *S. aureus* to cultured human endothelial cells. Tumor necrosis factor and other plasma cofactors were shown to significantly increase the adhesion of *S. aureus* to cultured human umbilical vein cells (Cheung et al., 1991b). Although both the stimulated and unstimulated endothelial cells bind *S. aureus* (albeit the latter in higher numbers), it is likely that the receptor specificities of the staphylococcal adhesins involved in adhesion to unstimulated cells are distinct from those participating in the adhesion to stimulated cells. A 50,000 mol wt protein from the membrane of unstimulated endothelial cells was found to bind to *S. aureus* in the absence of plasma factors (Tompkins et al., 1990). Unstimulated cardiac valve subendothelial cells contain a protein of 220,000 mol wt, possibly fibronectin, which acts as a staphylococcal receptor (Campbell and Johnson, 1990). Fibronectin is specifically implicated as a receptor that binds *S. aureus* to cultured human endothelial cells. Cardiac valve endothelial cells expressed a protein of 120,000 mol wt capable of interacting with *S. aureus*. In another study, it was found that fibrinogen promotes adhesion of *S. aureus* to unstimulated cultured human endothelial cells (Cheung et al., 1991a). In summary, the molecular mechanisms underlying the adhesion of *S. aureus* to endothelial cells probably involve multiple adhesins and receptors, some of which are expressed only by particular types of endothelial cells and others may require activation of the endothelial cells to be expressed.

Other substrata encountered by *S. aureus* are the corneal epithelium, peritoneal cavity and mucus. *S. aureus* adheres to corneas, but neither fibronectin nor tear lysozyme has an effect on this adhesion (Miller and Inglis, 1987; Matoba et al., 1991). Adhesion of *S. aureus* to human corneocytes (scraped skin cells) has been suggested to depend on protein A (Cole and Silverberg, 1986). *S. aureus* also adheres to mesothelial cell monolayers and tissue culture cells (Haagen et al, 1990). A plasmid-encoded adhesin was found to bind *S. aureus* to HeLa cells (Dunkle et al., 1986). When cells are infected with viruses, new substrata may be induced. For example, it is known that influenza predisposes patients to bacterial superinfections. It has been observed that *S. aureus* binds to multiple distinct proteins on uninfected and influenza A virus-infected cells (Sanford et al., 1986; Sanford and Ramsey, 1986). During experimental intranasal infection of ferrets, it was observed that staphylococci were mainly associated with mucus gels, rather than the underlying epithelia (Sanford et al., 1989). Interestingly, bovine submaxillary mucins, but not Tamm–Horsfall glycoprotein, was a strong inhibitor of the adhesion to the mucus. *S. aureus* has a high affinity for bones and biomaterials, including steels, plastics, glass, and fibers (Gristina et al., 1985; Guevara et al., 1987; Cheung and Fischetti, 1989; Keller et al., 1988; Wadström, 1991). In general, the molecular mechanisms underlying the adhesion of staphylococci to these solid surfaces have been poorly characterized. Cheung and Fischetti (1989) showed that cell-wall-associated proteins (120,000–220,000 mol wt) function as adhesin to cotton fibers, via a hydrophobic effect (Rosenberg and Doyle, 1990) mechanism. There is substantial evidence that the hydrophobic effect is involved in adhesion of staphylococci to many solid surfaces (Wadström, 1991). Bovine mastitis isolates of *S. aureus* have adhesins that appear to be specific for bovine tissues. The organisms are highly hydrophobic (Mamo et al., 1988), bind to bovine mammary epithelial cells, and cause hemagglutination of erythrocytes from human, bovine, equine, and sheep, characteristics not commonly ascribed to human isolates (Lindahl et al., 1989). Two staphylococcal proteins of mol wt 116,000 and 145,000 were found to serve as adhesins for bovine epithelial cells (Lindahl et al., 1990). The 145,000 mol wt protein was identified as the hemagglutinin, whereas the 116,000 mol wt protein could bind fibronectin. The bovine isolates also bind to components of the extracellular matrix, such as collagen type II (Speziale et al., 1984) and to lactoferrin (Naidu et al., 1991). Staphylococcal proteins of mol wt 67,000 and 92,000 have been identified as adhesins for bovine lactoferrin.

The general conclusion made in this book relative to the adhesion of *S. aureus* to animal cells and biomaterials is that the bacterium has the ability to bind strongly to various proteins, such as fibronectin, fibrinogen, collagen, and laminin. The binding may be a result of bacterial surface proteins, LTA, hydrophobins (Rosenberg and Doyle, 1990), and possibly wall teichoic acid. The adaptability of the organisms to reside on various surfaces may be a reflection of numerous adhesins. How the *S. aureus* may regulate its multiple adhesins is unknown.

Staphylococcus epidermidis

S. epidermidis and other coagulase-negative staphylococci have a remarkable ability to bind to various surfaces. *S. epidermidis* has several characteristics in common with *S. aureus* (Archer, 1984). Both bind to fibronectin and to collagen (Wadström, 1987). Both may secrete extracellular polysaccharide materials. Both express surface hydrophobins (Rozgonyi et al., 1985). Unlike *S. aureus, S. epidermidis* is not known to secrete a variety of enzymes or toxins. *S. epidermidis* is also incapable of producing protein A and its cell wall teichoic acid is composed of poly(glycerol phosphate) chains. The most common habitat of *S. epidermidis* is the skin tissue. Symptomatic infections frequently occur in compromised hosts, involving internal organs such as the heart, peritoneum, eye, meninges, and bones (Graevenitz et al., 1985; Pulverer et al., 1987). In many instances, these infections are associated with biomaterial implants. It is not surprising, therefore, that most efforts have been directed to understanding how *S. epidermidis* colonizes endothelial cells and uncoated or body fluids coated biomaterials.

Most clinical isolates of *S. epidermidis* do not synthesize extracellular polysaccharides, but these polymers appear to be associated with strains from in-dwelling catheters. Strain-to-strain variations in the ability of *S. epidermidis* to adhere to polypropylene have been noted (Dunne et al., 1987). Other studies show that adhesion to glass of clinical isolates is not related to slime production or to the clinical outcome of the infection (Kotilainen, 1990). In addition, strain-to-strain variations in the surface hydrophobicity among various clinical isolates have been noted (Pascual et al., 1988). Christensen et al. (1982) found that 63% of *S. epidermidis* isolates from intravascular catheters contained characteristic polysaccharide materials. It was observed (Christensen et al., 1987) that from a single isolate, at least three stable phenotypes could be generated. The phenotypes were characterized by their ability to synthesize extracellular polysaccharide and to adhere to plastic. A strongly adherent phenotype gave rise to a weakly adherent phenotype, which in turn gave rise to an adherent phenotype. The results show very nicely that *S. epidermidis* can undergo phase variations resulting in phenotypes with altered adhesion characteristics. Catheters and other biomedical devices may select for traits that produce surface characteristics favoring adhesion to noncellular substrata. Hogt et al. (1986) studied the adhesion of *S. epidermidis* to fluorinated poly(ethylenepropylene) and found that none of the encapsulated strains tended to bind to the polymer better than encapsulated strains. When the organism's surface was modified by antibiotics, there was a good correlation between hydrophobicity as determined by a standard assay and adhesion to noncoated plastic (Ludwicka et al., 1984; Hogt et al., 1986; Schadow et al., 1988). This is not surprising because the two assays probably detect hydrophobic residues on the bacterial surface. An acidic polysaccharide material may be found associated with microcolonies of *S. epidermidis* colonizing catheters (Peters et al., 1982; Sheth et al., 1983; Franson et al., 1984). Molecular evidence for the role of polysaccharide in mediating adhesion to smooth surfaces is lacking, although a polysaccharide isolated by Tojo et al. (1988) inhibited the attachment of *S. epidermidis* to catheter materials.

The so-called "foreign body" infections caused by *S. epidermidis* involve dissemination of the organisms leading to colonization of traumatized tissue (Dankert et al., 1986 Becker *et al.*, 1988). It has been suggested that the ability of these organisms to bind fibronectin and type II collagen may play a role in colonizing these tissues (Wadström et al., 1987). Chugh et al. (1990) observed that LTA could inhibit the adhesion of *S. epidermidis* to fibrin-platelet clots in vitro, suggesting that LTA may be a major adhesin for some substrata. It has been suggested that fibronectin, fibrinogen, and laminin may also mediate adhesion of staphylococci (including *S. epidermidis*) to foreign materials (Herrmann et al., 1987). The source of biomedical device-associated infections is probably the skin. The device may select for phenotypic variants that are hydrophobic and best suited to cause deep tissue infections.

The adhesiveness of *S. epidermidis* to glass surfaces is directly proportional to surface electronegativity and hydrophobicity of the bacteria (Gilbert et al., 1991). Hydrophobicity may not be the only factor that governs adhesion to biomaterials. Adhesion of coagulase-negative staphylococci to silicon rubber and polystyrene is increased when the cells are grown in the presence of subinihibitory concentrations of antibiotics (Wilcox et al., 1991). The antibiotic-induced changes did not correlate with surface hydrophobicity. Fibrinogen enhanced the binding of the staphylococci to various biomaterial surfaces but whole plasma and serum albumin inhibited adhesion (Carballo et al., 1991). Another study describing adhesion of coagulase-negative staphylococci to human blood protein coated biomaterials noted that the coating blood proteins had no effect on the ability of the bacteria to adhere to the solid surfaces (Muller et al., 1991). A proteinaceous adhesin of approximately 200,000 mol wt from the cell wall of *S. epidermidis* has been identified as the adhesin for polystyrene (Timmerman et al., 1991). *S. epidermidis* also binds to pharyngeal cells (Chugh et al., 1989a,b), mucus (Sanford et al., 1989), and extended wear contact lenses (Slusher et al., 1987). Like the bovine isolates of *S. aureus*, bovine isolates of coagulase-negative staphylococci also bind to type II collagen (Speziale et al., 1987), fibronectin (Mamo et al., 1988), and lactoferrin (Naidu et al., 1990). It seems reasonable to interfere in the ability of the bacteria to adhere to keratinized epithelial cells in order to reduce the population of *S. epidermidis* on the skin and thereby the chance of attracting foreign body infection. Unfortunately, very little is known about the mechanism of *S. epidermidis*–keratinized tissue interactions. It may also be possible to design biomaterials that exhibit a surface with low affinity for hydrophobic bacteria.

Staphylococcus saprophyticus

Except for some bovine isolates of *S. aureus, S. saprophyticus* is the only species found to consistently cause hemagglutination of erythrocytes. Gunnarson et al. (1984) found hemagglutination of sheep erythrocytes could be inhibited by GlcNAc-containing saccharides. Among the various oligosaccharides employed as inhibitors of hemagglutination, only structures containing disaccharide Galβ-1,4GlcNAc were inhibitors (Table 5–18, Chapter 5). Beuth et al. (1988a,b) found GlcNAc-inhibitable hemagglutination of sheep erythrocytes in a series of 31 urinary isolates of *S. saprophyticus*. They found that about one-half of the strains exhibited GalNAc-inhibitable hemagglutination of horse erythrocytes and the other half exhibited sialic acid inhibitable hemagglutination of rabbit erythrocytes. It is possible therefore that each *S. saprophyticus* isolate is capable of expressing two (or more) lectins. One of these lectins shared by all isolates is GlcNAc-specific and detectable by hemagglutination of sheep erythrocytes. Hemagglutination by *S. saprophyticus* has been proposed to be related to the tendency of the bacteria to colonize kidney tissue (Gatermann et al., 1988). The presentation and orientation of the carbohydrate-containing receptor, rather than lack of receptors, may determine the type of erythrocyte that can react with a particular lectin on the bacterial surface. Adhesion of *S. saprophyticus* isolates expressing either GalNAc- or sialic acid-specific hemagglutinins to human tubular and uroepithelial cells was inhibited by the corresponding glycoconjugates, suggesting that the putative receptors for these two lectins are accessible on urinary cells (Beuth et al., 1988a,b; Gaterman et al., 1991). The receptors are also accessible on human phagocytic cells and they may mediate lectinophagocytosis of the bacteria (Beuth et al., 1987, 1988a,b).

LTA derived from *S. saprophyticus*, at a concentration of 100–250 μg/ml, was reported to inhibit the adhesion of the organisms to uroepithelial cells, suggesting that LTA on the surface of the organisms mediates adhesion to urinary cells (Teti et al., 1987; Beuth et al., 1988a,b). There are a number of similarities between this mechanism of adhesion and those of streptococci. Deacylated LTA did not inhibit adhesion of the organisms, whereas albumin did, suggesting that the acyl groups mediate the binding of the organisms to urinary cells (Teti et al., 1987b). Albumin also inhibited adhesion of the bacteria to hexadecane droplets, providing evidence for the premise that LTA is surface-exposed on the bacteria (Teti et al., 1987b). LTA, however, did not inhibit *S. saprophyticus*-induced hemagglutination of sheep erythrocytes (Teti et al., 1987b). The results, taken together, indicate that the organisms are capable of exhibiting at least two mechanisms of adhesion: LTA- and lectin-mediated adhesion (Beuth et al., 1988a,b). It is assumed that the expression of each adhesin is subject to regulatory mechanisms. As with other bacteria (see Chapters 9 and 11) regulation of the expression of adhesins may prove to be essential for the ability of *S. saprophyticus* to colonize various tissues during a natural course of urinary tract infection.

References

Aasjord, P. and A. Grov. 1980. Immunoperoxidase and electron microscopy studies of staphylococcal lipoteichoic acid. *Acta Pathol. Microbiol. Scand. B* **88**:47–52.

Abraham, S.N., E.H. Beachey, and W.A. Simpson. 1983. Adherence of *Streptococcus pyogenes, Escherichia coli* and *Pseudomonas aeruginosa* to fibronectin-coated and uncoated epithelial cells. *Infect. Immun.* **41**:1261–1268.

Alkan, M. and E.H. Beachey. 1978a. Excretion of lipoteichoic acid by group A streptococci: influence of penicillin on excretion and loss of ability to adhere to human oral mucosal cells. *J. Clin. Invest.* **61**:671–677.

Alkan, M. and E.H. Beachey. 1978b. Excretion of lipoteichoic acid by group A streptococci: influence of penicillin on excretion and loss of ability to adhere to human oral epithelial cells. *J. Clin. Invest.* **61**:671–677.

Alkan, M., I. Ofek, and E.H. Beachey. 1977. Adherence of pharyngeal and skin strains of group A streptococci to human skin and oral epithelial cells. *Infect. Immun.* **18**:555–557.

Aly, R. and S. Levit. 1987. Adherence of *Staphylococcus aureus* to squamous epithelium: role of fibronectin and teichoic acid. *Rev. Infect. Dis.* **9**:S342–S350.

Andersson, B., J. Dahmen, T. Freijd, H. Leffler, G. Magnusson, G. Noori, and C. Svanborg-Eden. 1983. Identification of an active disaccharide unit of a glycoconjugate receptor for pneumococci attaching to human pharyngeal epithelial cells. *J. Exp. Med.* **158**:559–570.

Andersson, B., O. Porras, L.A. Hanson, T. Lagerberg, and C. Svanborg-Eden. 1986. Inhibition of attachment of *Streptococcus pneumoniae* and *Haemophilus influenzae* by human milk receptor oligosaccharides. *J. Infect. Dis.* **153**:232–237.

Andersson, B., E.H. Beachey, A. Tomasz, E. Tuomanen, and C. Svanborg-Eden. 1988a. A sandwich adhesion on *Streptococcus pneumoniae* attaching to human oropharyngeal epithelial cells in vitro. *Microbial Path.* **4**:267–278.

Andersson, B., B.M. Gray, C.H. Dillon Jr., A. Bahrmand, and C. Svanborg-Eden. 1988b. Role of adherence of *Streptococcus pneumoniae* in acute otitis media. *Pediatr. Infect. Dis. J.* **7**:476–480.

Aniansson, G., B. Andersson, R. Lindstedt, and C. Svanborg. 1990. Anti-adhesive activity of human casein against *Streptococcus pneumoniae* and *Haemophilus influenzae*. *Microb. Pathol.* **8**:315–323.

Archer, G.L. 1984. *Staphylococcus epidermidis*: the organism, its diseases, and treatment. In: Remington, J.S. and M.N. Swartz (eds.), *Current Topics in Infectious Diseases, Vol. 5*. McGraw-Hill, New York, pp. 25–48.

Baird, R.W., H.S. Courtney, M.S. Bronze, and J.B. Dale. 1990. Passive protection against group A streptococcal infection in mice by lipoteichoic acid. *Clin. Res.* **38**:353A.

Bartelt, M.A. and J.L. Duncan. 1978. Adherence of group A streptococci to human epithelial cells. *Infect. Immun.* **20**:200–208.

Beachey, E.H. and I. Ofek. 1976. Epithelial cell binding of group A streptococci by lipoteichoic acid on fimbriae denuded of M protein. *J. Exp. Med.* **143**:759–771.

Beachey, E.H., T.M. Chiang, I. Ofek, and A.H. Kang. 1977. Interaction of lipoteichoic acid of group A streptococci with human platelets. *Infect. Immun.* **16**:649–654.

Beachey, E.H., J. Dale, S. Grebe, A. Ahmed, W.A. Simpson, and I. Ofek. 1979a. Lymphocyte binding and T-cell mitogenic properties of group A streptococcal lipoteichoic acid. *J. Immunol.* **122**:189–195.

Beachey, E.H., J.B. Dale, W.A. Simpson, J.D. Evans, K.W. Knox, I. Ofek, and A.J. Wicken. 1979b. Erythrocyte binding properties of streptococcal lipoteichoic acids. *Infect. Immun.* **23**:618–625.

Beachey, E.H., W.A. Simpson, and I. Ofek. 1980. Interaction of surface polymers of *Streptococcus pyogenes* with animal cells. In: Berkeley, R.C.W., J.M. Lynch, J.

Melling, P.R. Rutter, and B. Vincent, (eds.), *Microbial Adhesion to Surfaces*. Horwood, Chichester, pp. 389–405.

Beck, G., E. Puchelle, C. Plotkowski, and R. Peslin. 1989. *Streptococcus pneumoniae* and *Staphylococcus aureus* surface properties in relation to their adherence to human buccal epithelial cells. *Res. Microbiol.* **140**:563–567.

Becker, R.C., P.M. DiBello, and F.V. Lucas. 1988. Bacterial tissue tropism: an in vitro model for infective endocarditis. *Cardiovasc. Res.* **21**:813–820.

Bergey, E.J. and M.W. Stinson. 1988. Heparin-inhibitable basement membrane-binding protein of *Streptococcus pyogenes*. *Infect. Immun.* **56**:1715–1721.

Bessen, D. and V.A. Fischetti. 1988. Influence of intranasal immunization with synthetic peptides corresponding to conserved epitopes of M protein on mucosal colonization by group A streptococci. *Infect. Immun.* **56**:2666–2672.

Bessen, D. and V.A. Fischetti. 1990. Human IgG receptor of group A streptococci is associated with tissue site infection and streptococcal class. *J. Infect. Dis.* **161**:747–754.

Beuth, J., H.L. Ko, H. Schroten, J. Solter, G. Uhlenbruck, and G. Pulverer. 1987. Lectin mediated adhesion of *Streptococcus pneumoniae* and its specific inhibition in vitro and in vivo. *Zbl. Bakt. Hyg. A* **265**:160–168.

Beuth, J., H.L. Ko, F. Schumacher-Perdreau, G. Peters, P. Heczko, and G. Pulverer. 1988a. Hemagglutination by *Staphylococcus saprophyticus* and other coagulase-negative staphylococci. *Microbial Pathogen* **4**:379–383.

Beuth, J., H.L. Ko, and G. Pulverer. 1988b. The role of staphylococcal lectins in human granulocyte stimulation. *Infect.* **16**:46–48.

Bibel, D.J., R. Aly, L. Lahti, H.R. Shinefield, and H.I. Maibach. 1987. Microbial adherence to vulvar epithelial cells. *J. Med. Microbiol.* **23**:75–82.

Biesbrock, A.R., M.S. Reddy, and M.J. Levine. 1991. Interaction of a salivary mucin-secretory immunoglobulin A complex with mucosal pathogens. *Infect. Immun.* **59**:3492–3497.

Blumberg, E.A., V.B. Hatcher, and F.D. Lowy. 1988. Acidic fibroblast growth factor modulates *Staphylococcus aureus* adherence to human endothelial cells. *Infect. Immun.* **56**:1470–1474.

Botta, G.A. 1981. Surface components in adhesion of group A streptococci to pharyngeal epithelial cells. *Curr. Microbiol.* **6**:101–104.

Bramley, A.J. and E.M. Hogben. 1982. The adhesion of human and bovine isolates of *Streptococcus agalactiae* (group B) to bovine mammary gland epithelial cells. *J. Comp. Pathol.* **92**:131–137.

Broder, C.C., R. Lottenberg, and M.D. P. Boyle. 1989. Mapping of the human plasmin domain recognized by the unique plasmin receptor of group A streptococci. *Infect. Immun.* **57**:2597–2605.

Broeseker, T.A., M.D.P. Boyle, and R. Lottenberg. 1988. Characterization of the interaction of human plasmin with its specific receptor on a group A streptococcus. *Microbial Pathol.* **5**:19–27.

Bulgakova, T.N., K.B. Grabovskaya, M. Ryc, and J. Jelinkova. 1986. The adhesin structures involved in the adherence of group B streptococci to human vaginal cells. *Folia Microbiol.* **31**:394–401.

Campbell, K.M. and C.M. Johnson. 1990. Identification of *Staphylococcus aureus* binding proteins on isolated porcine cardiac valve cells. *J. Lab. Clin. Med.* **115**:217–223.

Caparon, M.G., D.S. Stephens, A. Olsen, and J.R. Scott. 1991. Role of M protein in adherence of group A streptococci. *Infect. Immun.* **59**:1811–1817.

Carballo, J., C.M. Ferreiros, and M.T. Criado. 1991. Importance of experimental design in the evaluation of the influence of proteins in bacterial adherence to polymers. *Med. Microbiol. Immunol.* **180**:149–155.

Carret, G., H. Emonard, G. Fardel, M. Druguet, D. Herbage, and J.P. Flandrois. 1985. Gelatin and collagen binding to *Staphylococcus aureus* strains. *Ann. Inst. Pasteur/ Microbiol.* **136A**:241–245.

Carruthers, M.M. and W.J. Kabat. 1983. Mediation of staphylococcal adherence to mucosal cells by lipoteichoic acid. *Infect. Immun.* **40**:444–446.

Cheung, A.L. and V.A. Fischetti. 1989. Role of surface proteins in staphylococcal adherence to fibers in vitro. *J. Clin. Invest.* **83**:2041–2049.

Cheung, A.L., M. Krishnan, E.A. Jaffe, and V.A. Fischetti. 1991a. Fibrinogen acts as a bridging molecule in the adherence of *Staphylococcus aureus* to cultured human endothelial cells. *J. Clin. Invest.* **87**:2236–2245.

Cheung, A.L., J.M. Koomey, S. Lee, E.A. Jaffe, and V.A. Fischetti. 1991b. Recombinant human tumor necrosis factor alpha promotes adherence of *Staphylococcus aureus* to cultured human endothelial cells. *Infect. Immun.* **59**:3827–3831.

Chhatwal, G. and H. Blobel. 1987. Heterogeneity of fibronectin reactivity among streptococci as revealed by binding of fibronectin fragments. *Comp. Immun. Microbiol. Infect. Dis.* **10**:99–108.

Chhatwal, G., C. Lammler, and H. Blobel. 1985. Interactions of plasma proteins with group A, B, C and G streptococci. *Zbl. Bakt. Hyg. A* **259**:219–227.

Chhatwal, G., G. Albohn, and H. Blobel. 1987a. Novel complex formed between a nonproteolytic cell wall protein of group A streptococci and α_2-macroglobulin. *J. Bacteriol.* **169**:3691–3695.

Chhatwal, G., K.T. Preissner, G. Muller-Berghaus, and H. Blobel. 1987b. Specific binding of the human S protein (vitronectin) to streptococci, *Staphylococcus aureus* and *Escherichia coli*. *Infect. Immun.* **55**:1878–1883.

Chhatwal, G., P. Valentin-Weigand, and K.N. Timmis. 1990. Bacterial infection of wounds: fibronectin-mediated adherence of group A and C streptococci to fibrin thrombi in vitro. *Infect. Immun.* **58**:3015–3019.

Christensen, G.D., W.A. Simpson, A.L. Bisno, and E.H. Beachey. 1982. Adherence of slime-producing strains of *Staphylococcus epidermidis* to smooth surfaces. *Infect. Immun.* **37**:318–326.

Christensen, G.D., J.T. Parisi, A.L. Bisno, W.A. Simpson, and E.H. Beachey. 1983. Characterization of clinically significant strains of coagulase-negative staphylococci. *J. Clin. Microbiol.* **18**:258–269.

Christensen, G.D., L.M. Baddour, and W.A. Simpson. 1987. Phenotypic variation of *Staphylococcus epidermidis* slime production *in vitro* and *in vivo*. *Infect. Immun.* **55**:2870–2877.

Chugh, T.D., E. Babaa, G. Burns, and H. Shuhaiber. 1989a. Effect of sublethal concentration of antibiotics on the adherence of *Staphylococcus epidermidis* to eukaryotic cells. *Chemotherapy* **35**:113–118.

Chugh, T.D., G.J. Burns, H.J. Shuhaiber, and E.A. Bishbishi. 1989b. Adherence of *Staphylococcus epidermidis* to human pharyngeal epithelial cells: evidence for lipase-sensitive adhesin and glycoprotein receptor. *Curr. Microbiol.* **18**:109–112.

Chugh, T.D., G.J. Burns, H.J. Shuhaiber, and G.M. Bahr. 1990. Adherence of *Staphylococcus epidermidis* to fibrin-platelet clots *in vitro* mediated by lipoteichoic acid. *Infect. Immun.* **58**:315–319.

Cole, G.W. and N.L. Silverberg. 1986. The adherence of *Staphylococcus aureus* to human corneocytes. *Arch. Dermatol.* **122**:166–169.

Colleen, S., B. Hovelius, A. Wieslander, and P.A. Mardh. 1979. Surface properties of *Staphylococcus saprophyticus* and *Staphylococcus epidermidis* as studied by adherence tests and two-polymer, aqueous phase systems. *Acta Pathol. Microbiol.* Scand. B **87**:321–328.

Courtney, H.S. and D.L. Hasty. 1991. Aggregation of group A streptococci by human saliva and effect of saliva on streptococcal adherence to host cells. *Infect. Immun.* **59**:1661–1666.

Courtney, H.S., I. Ofek, W.A. Simpson, and E.H. Beachey. 1981. Characterization of lipoteichoic acid binding to polymorphonuclear leukocytes of human blood. *Infect. Immun.* **32**:625–631.

Courtney, H.S., W.A. Simpson, and E.H. Beachey. 1983. Binding of streptococcal lipoteichoic acid to fatty acid-binding sites on human plasma fibronectin. *J. Bacteriol.* **153**:763–770.

Courtney, H.S., I. Ofek, W.A. Simpson, D.L. Hasty, and E.H. Beachey. 1986. Binding of *Streptococcus pyogenes* to soluble and insoluble fibronectin. *Infect. Immun.* **53**:454–459.

Courtney, H.S., D.L. Hasty, and I. Ofek. 1990. Hydrophobicity of group A streptococci and its relationship to adhesion of streptococci to host cells. In: Doyle, R.J. and M. Rosenberg (eds.), *Microbial Cell Surface Hydrophobicity*. American Society for Microbiology, Washington, pp. 361–386.

Cox, F. 1982. Prevention of group B streptococcal colonization with topically applied lipoteichoic acid in a maternal-newborn mouse model. *Pediatr. Res.* **16**:816–819.

Dankert, J., A. Hogt, and J. Feijen. 1986. Biomedical polymers, bacterial adhesion, colonization and infection. *CRC Crit. Rev. Biocompat.* **2**:219–300.

DeVuono, J. and C. Panos. 1978. Effect of L-form *Streptococcus pyogenes* and of lipoteichoic acid on human cells in tissue culture. *Infect. Immun.* **22**:255–265.

Doran, T.I. and S.J. Mattingly. 1982. Association of type- and group-specific antigens with the cell wall of serotype III group B streptococci. *Infect. Immun.* **36**:1115–1122.

Dunkle, L.M., L.L. Blair, and K.P. Fortune. 1986. Transformation of a plasmid encoding and adhesin of *Staphylococcus aureus* into a nonadherent staphylococcal strain. *J. Infect. Dis.* **153**:670–675.

Dunne, W.M., Jr., N.K. Sheth, and T.R. Franson. 1987. Quantitative epifluorescence assay of adherence of coagulase-negative staphylococci. *J. Clin. Microbiol.* **25**:741–743.

Ellen, R.P. and R.J. Gibbons. 1972. "M" protein-associated adherence of *Streptococcus pyogenes* to epithelial surfaces: prerequisite for virulence. *Infect. Immun.* **5**:826–830.

Espersen, F. and I. Clemmensen. 1982. Isolation of a fibronectin-binding protein from *Staphylococcus aureus. Infect. Immun.* **37**:526–531.

Fischetti, A. 1991. Streptococcal M protein. *Sci. Am.* June, pp. 58–65.

Fischetti, V.A., W.M. Hodges, and D.E. Hruby. 1989. Protection against streptococcal pharyngeal colonization with a vaccinia: M protein recombinant. *Science* **244**:1487–1490.

Flock, J.I., G. Froman, B. Nilsson, C. Signas, G. Raucci, B. Guss, K. Jonsson, M. Hook, T. Wadström, and M. Lindbert. 1987. Cloning and expression of the gene for fibronectin-binding protein from *Staphylococcus aureus. EMBO J.* **6**:2351–2357.

Franson, T.R., N.K. Sheth, H.D. Rose, and P.G. Sohnle. 1984. Scanning electron microscopy of bacteria adherent to intravascular catheters. *J. Clin. Microbiol.* **20**:500–505.

Froman, G., L.M. Switalski, P. Speziale, and M. Hook. 1987. Isolation and characterization of a fibronectin receptor from *Staphylococcus aureus. J. Biol. Chem.* **262**:6564–6571.

Fuquay, J.I., D.T. Loo, and D.W. Barnes. 1986. Binding of *Staphylococcus aureus* by human serum spreading factor in an in vitro assay. *Infect. Immun.* **52**:714–717.

Galioto, G.B., E. Mevio, R. Maserati, P. Galioto, S. Galioto, C. Dos Santos, and I. Pedrotti. 1988. Bacterial adherence and upper respiratory tract disease: a correlation between *S. pyogenes* attachment and recurrent throat infections. *Acta Otolaryngol.* **454**:167–174. (*Suppl.*).

Gatermann, S., R. Marre, J. Heesemann, and W. Henkel. 1988. Hemagglutinating and adherence properties of *Staphylococcus saprophyticus*: epidemiology and virulence in experimental urinary tract infection of rats. *FEMS Microbiol. Immunol.* **47**:179–186.

Gatermann, S., M. Kretschmar, B. Kreft, E. Straube, H. Schmidt, and R. Marre. 1991. Adhesion of *Staphylococcus saprophyticus* to renal tubular epithelial cells is mediated by an *N*-acetylgalactosamine-specific structure. *Int. J. Med. Microbiol.* **275**:358–363.

Gilbert, P., D.J. Evans, E. Evans, I.G. Duguid, and M.R. Brown. 1991. Surface characteristics and adhesion of *Escherichia coli* and *Staphylococcus epidermidis. J. Appl. Bacteriol.* **71**:72–77.

Goldschmidt, J.C. and C. Panos. 1984. Teichoic acids of *Streptococcus agalactiae*: chemistry, cytotoxicity, and effect on bacterial adherence to human cells in tissue culture. *Infect. Immun.* **43**:670–677.

Grabovskaya, K.B., A.A. Totolian, M. Ryc, J. Havlicek, L.A. Burova, and R. Bicova.

1980. Adherence of group A streptococci to epithelial cells in tissue culture. *Zbl. Bakt. Mikrobiol. Hyg. A* **247**:303–314.

Graevenitz, A. von. 1985. Coagulase-negative staphylococci in wounds: pathogens or contaminants? *Infection* **13**:2–3.

Gristina, A.G., M. Oga, L.X. Webb, and C.D. Hobgood. 1985. Adherent bacterial colonization in the pathogenesis of osteomyelitis. *Science* **228**:990–993.

Guevara, J.A., G. Zuccaro, A. Trevisan, and C.D. Denoya. 1987. Bacterial adhesion to cerebrospinal fluid shunts. *J. Neurosurg.* **67**:438–445.

Gunnarson, A., P.A. Mardh, A. Lundblad, and S. Svensson. 1984. Oligosaccharide structures mediating agglutination of sheep erythrocytes by *Staphylococcus saprophyticus*. *Infect. Immun.* **45**:41–46.

Haagen, I.A., H.C. Heezius, R.P. Verkooyen, J. Verhoef, and H.A. Verbrugh. 1990. Adherence of peritonitis-causing staphylococci to human peritoneal mesothelial cell monolayers. *J. Infect. Dis.* **161**:266–273.

Hamill, R.J., J.M. Vann, and R.A. Proctor. 1986. Phagocytosis of *Staphylococcus aureus* by cultured bovine aortic endothelial cells: model for postadherence events in endovascular infections. *Infect. Immun.* **54**:833–836.

Hasty, D.L., E.H. Beachey, H.S. Courtney, and W.A. Simpson. 1989. Interactions between fibronectin and bacteria. In: Carsons S.E. (ed.), *Fibronectin in Health and Disease*. CRC Press, Boca Raton, FL, pp. 89–112.

Hasty, D.L., I. Ofek, H.S. Courtney, and R.J. Doyle. 1992. Multiple adhesins of streptococci. *Infect. Immun.* **60**:2147–2152.

Helmig, R., J.T. Halaburt, N. Uldbjert, A.C. Thomsen, and A. Stenderup. 1990. Increased cell adherence of group B streptococci from preterm infants with neonatal sepsis. *Obstet. Gynecol.* **76**:825–827.

Herrmann, M., P.E. Vaudaux, D. Pittet, R. Auckenthaler, P.D. Lew, F. Schumacher-Perdreau, G. Peters, and F.A. Waldvogel. 1987. Fibronectin, fibrinogen, and laminin act as mediators of adherence of clinical staphylococcal isolates to foreign material. *J. Infect. Dis.* **158**:693–701.

Herrmann, M., S.J. Suchard, L.A. Boxer, F.A. Waldvogel, and P.D. Lew. 1991. Thrombospondin binds to *Staphylococcus aureus* and promotes staphylococcal adherence to surfaces. *Infect. Immun.* **59**:279–288.

Hogt, A.H., J. Dankert, C.E. Hulstaert, and J. Feijen. 1986. Cell surface characteristics of coagulase-negative staphylococci and their adherence to fluorinated polyethylenepropylene. *Infect. Immun.* **51**:294–301.

Holderbaum, D., R.A. Spech, and L.A. Ehrhart. 1985. Specific binding of collagen to *Staphylococcus aureus*. *Collagen Rel. Res.* **5**:261–271.

Holderbaum, D., G.S. Hall, and L.A. Ehrhart. 1986. Collagen binding to *Staphylococcus aureus*. *Infect. Immun.* **54**:359–364.

Jacques, M. and J.W. Costerton. 1987. Adhesion of group B *Streptococcus* to a polyethylene intrauterine contraceptive device. *FEMS Microbiol. Lett.* **41**:23–28.

Jelinkova, J., K.B. Grabovskaya, M. Ryc, T.N. Bulgakova, and A.A. Totolian. 1986.

Adherence of vaginal and pharyngeal strains of group B streptococci to human vaginal and pharyngeal epithelial cells. *Zbl. Bakt. Hyg. A* **62**:492–499.

Johnson, C.M., G.A. Hancock, and G.D. Goulin. 1988. Specific binding of *Staphylococcus aureus* to cultured porcine cardiac valvular endothelial cells. *J. Lab. Clin. Med.* **112**:16–22.

Jonsson, P. and T. Wadström. 1983. High surface hydrophobicity of *Staphylococcus aureus* as revealed by hydrophobic interaction chromatography. *Curr. Microb.* **8**:347–353.

Jonsson, P. and T. Wadström. 1984. Cell surface hydrophobicity of *Staphylococcus aureus* measured by the salt aggregation test (SAT). *Curr. Microbiol.* **10**:203–209.

Keller, J.D., J. Falk, H.S. Bjornson, E.B. Silberstein, and R.F. Kempczinski. 1988. Bacterial infectibility of chronically implanted endothelial cell-seeded expanded polytertrafluoroethylene vascular grafts. *J. Vasc. Surg.* **524**:524–530.

Kloczewiak, M., S. Timmons, and J. Hawiger. 1987. Reactivity of chemically cross-linked fibrinogen and its fragments D toward the staphylococcal clumping receptor. *Biochemistry* **26**:6152–6156.

Kostrzynska, M., C. Schalen, and T. Wadström. 1989. Specific binding of collagen type IV to *Streptococcus pyogenes*. *FEMS Microbiol. Lett.* **59**:229–234.

Kotilainen, P. 1990. Association of coagulase-negative staphylococcal slime production and adherence with the development and outcome of adult septicemias. *J. Clin. Microbiol.* **28**:2779–2785.

Krivan, H.C., D.D. Roberts, and V. Ginsburg. 1988. Many pulmonary pathogenic bacteria bind specifically to the carbohydrate sequence GalNAcβGal found in some glycolipids. *Proc. Natl. Acad. Sci. USA* **85**:6157–6161.

Kubin, V. and M. Ryc. 1988. Adherence of group B streptococci to human buccal epithelial cells, its dependence on the biological state of the culture. *Folia Microbiol.* **33**:224–229.

Kuusela, P. 1978. Fibronectin binds to *Staphylococcus aureus*. *Nature* **276**:718–720.

Kuypers, J.M. and R.A. Proctor. 1989. Reduced adherence to traumatized rat heart valves by a low fibronectin-binding mutant of *Staphylococcus aureus*. *Infect. Immun.* **57**:2306–2312.

Lämmler, C., T. Guszczynski, and W. Dobryszycka. 1988. Further characterization of haptoglobin binding to streptococci of serological group A. *Zbl. Bakt. Hyg.* **269**:454–459.

Leon, O. and C. Panos. 1990. *Streptococcus pyogenes* clinical isolates and lipoteichoic acid. *Infect. Immun.* **58**:3779–3787.

Lindahl, M., P. Jonsson, and P.A. Mardh. 1989. Hemagglutination by *Staphylococcus aureus*. Studies on strains isolated from bovine mastitis. *Acta Path. Microbiol. Immunol. Scand.* **97**:175–180.

Lindahl, M., O. Holmberg, and P. Jonsson. 1990. Adhesive proteins of hemagglutinating *Staphylococcus aureus* isolated from bovine mastitis. *J. Gen. Microbiol.* **136**:935–939.

Ljungh, A., S. Hjerten, and T. Wadström. 1985. High surface hydrophobicity of autoaggregating *Staphylococcus aureus* strain isolated from human infections studied with salt aggregation test. *Infect. Immun.* **47**:522–526.

Lopes, J.D., M. dos Reis, and R.R. Brentani. 1985. Presence of laminin receptors in *Staphylococcus aureus. Science* **229**:275–277.

Lottenberg, R., C.C. Broder, and M.D.P. Boyle. 1987. Identification of a specific receptor for plasmin on a group A streptococcus. *Infect. Immun.* **55**:1914–1928.

Lowy, F.D., J. Fant, L.L. Higgins, S.K. Ogawa, and V.B. Hatcher. 1988. *Staphylococcus aureus*–human endothelial cell interactions. *J. Ultrastruct. Mol. Struct. Res.* **98**:137–146.

Ludwicka, A., B. Jansen, T. Wadström, L.M. Switalski, G. Peters, and G. Pulverer. 1984. Attachment of staphylococci to various synthetic polymers. In: Shalaby, S.W. and A.S. Hoffman (eds.), *Polymers as Biomaterials*. Plenum Publishing Corp., New York, pp. 241–256.

Mahajan, B. and B.R. Panhotra. 1989. Adherence of *Streptococcus pneumoniae* to buccal epithelial cells of smokers and nonsmokers. *Indian J. Med. Res.* **89**:381–383.

Mamo, W., G. Fröman, and T. Wadström. 1988. Interaction of sub-epithelial connective tissue and components with *Staphylococcus aureus* and coagulase-negative staphylococci from bovine mastitis. *Vet. Microbiol.* **18**:163–176.

Matoba, A.Y., R.J. Hamill, and M.S. Osato. 1991. The effects of fibronectin on the adherence of bacteria to corneal epithelium. *Cornea* **10**:387–389.

Mattingly, S.J. and B.P. Johnston. 1987. Comparative analysis of the localization of lipoteichoic acid in *Streptococcus agalactiae* and *Streptococcus pyogenes*. *Infect. Immun.* **55**:2383–2386.

Maxe, I., D. Ryden, T. Wadström, and K. Rubin. 1986. Specific attachment of *Staphylococcus aureus* to immobilized fibronectin. *Infect. Immun.* **54**:695–704.

Mertz, P.M., J.M. Patti, J.J. Marcin, and D.A. Marshall. 1987. Model for studying bacterial adherence to skin wounds. *J. Clin. Microbiol.* **25**:1601–1604.

Miedzobrodzki, J., A.S. Naidu, J.L. Watts, P. Ciborowski, K. Palm, and T. Wadström. 1989. Effect of milk on fibronectin and collagen type I binding to *Staphylococcus aureus* and coagulase-negative staphylococci isolated from bovine mastitis. *J. Clin. Microbiol.* **27**:540–544.

Miller, M.R. and T. Inglis. 1987. Influence of lysozyme on aggregation of *Staphylococcus aureus. J. Clin. Microbiol.* **25**: 1587–1590.

Miorner, H., G. Johansson, and G. Kronvall. 1983. Lipoteichoic acid is the major cell wall component responsible for surface hydrophobicity of group A streptococci. *Infect. Immun.* **39**:336–343.

Mirelman, D. and I. Ofek. 1986. Introduction to microbial lectins and agglutinins. In: Mirelman, D. (ed.), *Microbial Lectins and Agglutinins*. John Wiley & Sons, New York, pp. 1–20.

Miyazaki, S., O. Leon, and C. Panos. 1988. Adherence of *Streptococcus agalactiae* to synchronously growing human cell monolayers without lipoteichoic acid involvement. *Infect. Immun.* **56**:505–512.

Mota, G.F.A., C.R.W. Carneiro, L. Gomes, and J.D. Lopes. 1988. Monoclonal antibodies to *Staphylococcus aureus* laminin-binding proteins cross-react with mammalian cells. *Infect. Immun.* **56**:1580–1584.

Mouricout, M., J.M. Petit, J.R. Carias, and R. Julien. 1990. Glycoprotein glycans that inhibit adhesion of *Escherichia coli* mediated by K99 fimbriae: treatment of experimental colibacillosis. *Infect. Immun.* **58**:98–106.

Muller, E., S. Takeda, D.A. Goldmann, and G. Pier. 1991. Blood proteins do not promote adherence of coagulase-negative staphylococci to biomaterials. *Infect. Immun.* **59**:3323–3326.

Naidu, A.S., J. Miedzobrodzki, M. Andersson, L-E. Nilsson, A. Forsgren, and J.L. Watts. 1990. Bovine lactoferrin binding to six species of coagulase-negative staphylococci isolated from bovine intramammary infections. *J. Clin. Microbiol.* **28**:2312–2319.

Naidu, A.S., M. Andersson, J. Miedzobrodzki, A. Forsgren, and J.L. Watts. 1991a. Bovine lactoferrin receptors in *Staphylococcus aureus* isolated from bovine mastitis. *J. Dairy Sci.* **74**:1218–1226.

Naidu, A.S., J. Miedzobrodzki, J.M. Musser, V.T. Rosdahl, S.A. Hedstrom, and A. Forsgren. 1991b. Human lactoferrin binding in clinical isolates of *Staphylococcus aureus*. *J. Med. Microbiol.* **34**:323–328.

Nardella, F.A., A.K. Schroder, M.L. Svensson, J. Sjoquist, C. Barber, and P. Christensen. 1987. T15 group A streptococcal Fc receptor binds to the same location on IgC as staphylococcal protein A and IgG rheumatoid factors. *J. Immunol.* **138**:922–926.

Nealon, T.J. and S.J. Mattingly. 1983. Association of elevated levels of cellular lipoteichoic acids of group B streptococci with human neonatal disease. *Infect. Immun.* **39**:1243–1251.

Nealon, T.J. and S.J. Mattingly. 1984. Role of cellular lipo-teichoic acids in mediating adherence of serotype III strains of group B streptococci to human embryonic, fetal, and adult epithelial cells. *Infect. Immun.* **43**:523–530.

Nealon, T.J. and S.J. Mattingly. 1985. Kinetic and chemical analyses of the biologic significance of lipoteichoic acids in mediating adherence of serotype III group B streptococci. *Infect. Immun.* **50**:107–115.

Nealon, T.J., E.H. Beachey, H.S. Courtney, and W.A. Simpson. 1986. Release of fibronectin-lipoteichoic acid complexes from group A streptococci with penicillin. *Infect. Immun.* **51**:529–535.

Oeding, P. and A. Grov. 1972. Cellular antigens. In: Cohen, J.O. (ed.), *The Staphylococci*. Wiley Inter-Science, New York, pp. 333–356.

Ofek, I. and E.H. Beachey. 1979. Lipoteichoic acid-sensitive attachment of group A streptococci to phagocytes. In: Parlser, M.T. (ed.), *Pathogenic Streptococci*. Reed-Books Ltd., Chertsey, Surrey, England, pp. 44–46.

Ofek, I., E.H. Beachey, W. Jefferson, and G.L. Campbell. 1975. Cell membrane-binding properties of group A streptococcal lipoteichoic acid. *J. Exp. Med.* **141**:990–1003.

Ofek, I., E.H. Beachey, F. Eyal, and J.C. Morrison. 1977. Postnatal development of binding of streptococci and lipoteichoic acid by oral mucosal cells of humans. *J. Infect. Dis.* **135**:267–274.

Ofek, I., W.A. Simpson, and E.H. Beachey. 1982. Formation of molecular complexes between a structurally defined M-protein and acylated or deacylated lipoteichoic acid of *Streptococcus pyogenes*. *J. Bacteriol.* **149**:426–433.

Ofek, I., E. Whitnack, and E.H. Beachey. 1983. Hydrophobic interactions of group A streptococci with hexadecane droplets. *J. Bacteriol.* **154**:139–145.

Ofek, I., D. Zafriri, J. Goldhar, and B.I. Eisenstein. 1990. Inability of toxin inhibitors to neutralize enhanced toxicity caused by bacteria adherent to tissue culture cells. *Infect. Immun.* **58**:3737–3742.

Pancholi, V. and V.A. Fischetti. 1988. Isolation and characterization of the cell wall associated region of group A streptococcal M6 protein. *J. Bacteriol.* **170**:2618–2624.

Pascual, A., A. Fleer, N.A.C. Westerdaal, M. Berghuis, and J. Verhoef. 1988. Surface hydrophobicity and opsonic requirements of coagulase-negative staphylococci in suspension and adhering to a polymer substratum. *Eur. J. Clin. Microbiol. Infect. Dis.* **7**:161–166.

Peters, G., R. Locci, and G. Pulverer. 1982. Adherence and growth of coagulase-negative staphylococci on surfaces of intravenous catheters. *J. Infect. Dis.* **146**:479–482.

Plotkowski, M.C., E. Puchelle, G. Beck, J. Jacquot, and C. Hannoun. 1986. Adherence of type I *Streptococcus pneumoniae* to tracheal epithelium of mice infected with influenza A/PR8 virus. *Am. Rev. Respir. Dis.* **134**:1040–1044.

Plotkowski, M.C., G. Beck, and E. Puchelle. 1988. A new model for studying bacterial adherence to the respiratory epithelium. *Braz. J. Med. Biol. Res.* **21**:285–288.

Plotkowski, M.C., G. Beck, J. Jacquot, and E. Puchelle. 1989. The frog palate mucosa as a model for studying bacterial adhesion to mucus-coated respiratory epithelium. *J. Comp. Pathol.* **100**:37–46.

Proctor, R.A., Christman, and D.F. Mosher. 1983. Fibronectin-induced agglutination of *Staphylococcus aureus* correlates with invasiveness. *J. Lab. Clin. Med.* **104**:455–469.

Pulverer, G., G. Peters, and F. Schumacher-Perdreau. 1987. Coagulase-negative staphylococci. *Zbl. Bakt. Hyg. A* **264**:1–28.

Raeder, R., R. A. Otten, and M.D.P. Boyle. 1991. Comparison of albumin receptors expressed on bovine and human group G streptococci. *Infect. Immun.* **59**:609–616.

Raja, R.H., G. Raucci, and M. Hook. 1990. Peptide analogs to a fibronectin receptor inhibit attachment of *Staphylococcus aureus* to fibronectin-containing substrates. *Infect. Immun.* **58**:2593–2598.

Rosenberg, M. and R.J. Doyle. 1990. Microbial cell surface hydrophobicity: history, measurement, and significance. In: R.J. Doyle and M. Rosenberg (ed.), *Microbial Cell Surface Hydrophobicity*. American Society for Microbiology, Washington, pp. 1–37.

Rosenberg, M., A. Perry, E.A. Bayer, D.L. Gutnick, E. Rosenberg, and I. Ofek. 1981. Adherence of *Acinetobacter calcoaceticus* RAG–1 to human epithelial cells and to hexadecane. *Infect. Immun.* **33**:29–33.

Rozgonyi, F., K.R. Szitha, S. Hjerten, and T. Wadström. 1985. Standardization of salt aggregation test for reproducible determination of cell-surface hydrophobicity with special reference to *Staphylococcus* species. *J. Appl. Bacteriol.* **59**:451–457.

Ryc, M., B. Wagner, M. Wagner, and R. Bicova. 1988. Electron microscopic localization of lipoteichoic acid on group A streptococci. *Zentralbl. Bakteriol. Mikrobiol. Hyg. [A]* **269**:168–178.

Ryden, C., K. Rubin, P. Speziale, M. Hook, M. Lindberg, and T. Wadström. 1983. Fibronectin receptors from *Staphylococcus aureus*. *J. Biol. Chem.* **258**:3396–3401.

Ryden, C., A.I. Yacoub, I. Maxe, D. Heinegard, A. Oldberg, A. Franzen, A. Ljungh, and K. Rubin. 1989. Specific binding of bone sialoprotein to *Staphylococcus aureus* isolated from patients with osteomyelitis. *Eur. J. Biochem.* **184**:331–336.

Sanford, B.A. and M.A. Ramsay. 1986. Detection of staphylococcal membrane receptors on virus-infected cells by direct adhesin overlay. *Infect. Immun.* **52**:671–675.

Sanford, B.A., V.E. Davison, and M.A. Ramsay. 1982. Fibrinogen-mediated adherence of group A streptococcus to influenza virus-infected cell cultures. *Infect. Immun.* **38**:513–520.

Sanford, B.A., V.E. Davison, and M.A. Ramsay. 1986. *Staphylococcus aureus* adherence to influenza A virus-infected and control cell cultures: evidence for multiple adhesins. *Proc. Soc. Exp. Biol. Med.* **181**:104–111.

Sanford, B.A., V.L. Thomas, and M.A. Ramsay. 1989. Binding of staphylococci to mucus *in vivo* and *in vitro*. *Infect. Immun.* **57**:3735–3742.

Savoia, D. and S. Landolfo. 1987. Modulation of the adherence of group A streptococci to murine cells. *Microbiologia.* **10**:281–290.

Schadow, K.H., W.A. Simpson, and G.D. Christensen. 1988. Characteristics of adherence to plastic tissue culture plates of coagulase-negative staphylococci exposed to subinhibitory concentrations of antimicrobial agents. *J. Infect. Dis.* **157**:71–77.

Schalen, C., D. Kurl, and P. Christensen. 1986. Independent binding of native and aggregated IgG in group A streptococci. *Acta Path. Microbiol. Immunol. Scand. Sect. B.* **94**:333–338.

Schmidt, K-H., O. Kuhnemund, T. Wadström, and W. Kohler. 1987. Binding of fibrinogen fragment D to group A streptococci causes strain dependent decrease in cell surface hydrophobicity as measured by the salt aggregation test (SAT) and cell clumping in polyethylene glycol. *Zbl. Bakt. Hyg.* **A264**:185–195.

Sheth, N.K., H.D. Rose, T.R. Franson, F.L. Buckmire, and P.G. Sohnle. 1983. *In vitro* quantitative adherence of bacteria to intravascular catheters. *J. Surg. Res.* **34**:213–218.

Shimamura. K., H. Shigemi, Y. Kurono, and G. Mogi. 1990. The role of bacterial adherence in otitis media with effusion. *Arch. Otolaryngol. Head Neck Surg.* **116**:1143–1146.

Shoeb, H.A., A.F. Tawfik, and A.M. Shibl. 1991. A nitrate reductase-based colorimetric assay for the study of bacterial adherence. *J. Appl. Bacteriol.* **71**:270–276.

Simpson, W.A. and E.H. Beachey. 1983. Adherence of group A streptococci to fibronectin on oral epithelial cells. Infect. Immun. *39*:275–279.

Simpson, W.A., I. Ofek, C. Sarasohn, J.C. Morrison, and E.H. Beachey. 1980. Characteristics of the binding of streptococcal lipoteichoic acid to human oral epithelial cells. *J. Infect. Dis.* **141**:457–462.

Simpson, W.A., I. Ofek, and E.H. Beachey. 1980a. Binding of streptococcal lipoteichoic acid to the fatty acid binding sites on serum albumin. *J. Biol. Chem.* **255**:6092–6097.

Simpson, W.A., I. Ofek and E.H. Beachey. 1980b. Fatty acid binding sites of serum albumin as membrane receptor analogs for streptococcal lipoteichoic acid. *Infect. Immun.* **29**:119–122.

Simpson, W.A., I. Ofek, C. Sarasohn, J.C. Morrison and E.H. Beachey. 1980c. Characteristics of the binding of streptococcal lipoteichoic acid to human oral epithelial cells. *J. Infect. Dis.* **141**:457–462.

Simpson, W.A., H.S. Courtney, and I. Ofek. 1987. Interactions of fibronectin with streptococci: the role of fibronectin as a receptor for *Streptococcus pyogenes. Rev. Infect. Dis.* **9**:S351–S359.

Slusher, M.M., Q.N. Myrvik, J.C. Lewis, and A.G. Gristina. 1987. Extended-wear lenses, biofilm, and bacterial adhesion. *Arch. Ophthalmol.* **105**:110–115.

Speziale, P., M. Hook, L. Switalski, and T. Wadström. 1984. Fibronectin binding to a *Streptococcus pyogenes* strain. *J. Bacteriol.* **157**:420–427.

Speziale, P., G. Raucci, S. Meloni, M.L. Meloni, and T. Wadström. 1987. Binding of collagen to group A,B,C,D and G streptococci. *FEMS Microbiol. Lett.* **48**:47–51.

Stanislawski, L., W.A. Simpson, D.L. Hasty, N. Sharon, E.H. Beachey, and I. Ofek. 1985. Role of fibronectin in attachment of *Streptococcus pyogenes* and Escherichia coli to human cell lines and isolated oral epithelial cells. *Infect. Immun.* **48**:259–259.

Stanislawski, L., H.S. Courtney, W.A. Simpson, D.L. Hasty, E.H. Beachey, L. Robert, and I. Ofek. 1987. Hybridoma antibodies to the lipid-binding site(s) in the amino-terminal region of fibronectin inhibits binding of streptococcal lipoteichoic acid. *J. Infect. Dis.* **156**:344–349.

Stenfors, L.E. and S. Raisanen. 1991. *In vivo* attachment of beta-haemolytic streptococci to tonsillar epithelial cells in health and disease. *Acta Otolaryngol.* **111**:562–568.

Stenfors, L.E., S. Raisanen, and I. Rantala. 1991. *In vivo* attachment of group A streptococci to tonsillar epithelium during acute tonsillitis. *Scand. J. Infect. Dis.* **23**:309–313.

Stjernquist, D.A., D.N. Kurl, and P. Christensen. 1986. Repeated passage of freshly isolated group A streptococci on blood agar. II. Effect on adherence capacity. *Acta Pathol. Microbiol. Immunol. Scand. [B]* **96**:405–408.

Sun, D., H.S. Courtney, and E.H. Beachey. 1988. Berberine sulfate blocks adherence of *Streptococcus pyogenes* to epithelial cells, fibronectin and hexadecane. *Antimicrobiol. Agents Chemother.* **32**:1370–1374.

Switalski, L.M., P. Speziale, M. Hook, T. Wadström, and R. Timpl. 1984. Binding of *Streptococcus pyogenes* to laminin. *J. Biol. Chem.* **259**:3734–3738.

Talay, S.R., E. Ehrenfeld, G.S. Chhatwal, and K.N. Timmis. 1991. Expression of fibronectin binding components of *Streptococcus pyogenes* in *Escherichia coli* demonstrates that they are proteins. *Mol. Microbiol.* **5**:1727–1734.

Teti, G., F. Tomasello, M.S. Chiofalo, G. Orefici, and P. Mastroeni. 1987a. Adherence of group B streptococci to adult and neonatal epithelial cells mediated by lipoteichoic acid. *Infect. Immun.* **55**:3057–3064.

Teti, G., M.S. Chiofalo, F. Tomasello, C. Fava, and F. Mastroeni. 1987b. Mediation of *Staphylococcus saprophyticus* adherence to uroepithelial cells by lipoteichoic acid. *Infect. Immun.* **55**:839–842.

Timmerman, C.P., A. Fleer, J.M. Besnier, L. DeGraaf, F. Cremers, and J. Verhoaef. 1991. Characterization of a proteinaceous adhesin of *Staphylococcus epidermidis* which mediates attachment to polystyrene. *Infect. Immun.* **59**:4187–4192.

Thomas, P.D., F.W.Hampson, and G.W. Hunninghake. 1988. Bacterial adherence to human endothelial cells. *J. Appl. Physiol.* **65**:1372–1376.

Tojo, M., N. Yamashita, D.A. Goldmann, and G.B. Pier. 1988. Isolation and characterization of a capsular polysaccharide adhesin from *Staphylococcus epidermidis*. *J. Infect. Dis.* **157**:713–722.

Tompkins, D.C., V.B. Hatcher, D. Patel, A. Orr, L.L. Higgins, and F.D. Lowy. 1990. A human endothelial cell membrane protein that binds *Staphylococcus aureus in vitro*. *J. Clin. Invest.* **85**:1248–1254.

Totolian, A., T. Bulgakova, K. Grabovskaya, M. Ryc, J. Jelinkova, and R. Bicova. 1985. Factors influencing adherence of group-B streptococci (GBS) to human epithelial cells. In: Kimura, Y., S. Kotani, and Y. Shiokawa (eds.), *Recent Advances in Streptococci and Streptococcal Diseases*. Reedbooks Ltd., Bracknell, Berkshire, pp. 135–139.

Tylewska, S. and W. Hryniewicz. 1987. *Streptococcus pyogenes* cell wall protein responsible for binding to pharyngeal epithelial cells. *Zentralbl. Bakteriol. Mikrobiol. Hyg.* [*A*] **265**:146–150.

Tylewska, S.K., V.A. Fischetti, and R.J. Gibbons. 1987. Application of Percoll density gradients in studies of the adhesion of *Streptococcus pyogenes* to human epithelial cells. *Curr. Microbiol.* **16**:129–135.

Tylewska, S.K., V. Fischetti, and R.J. Gibbons. 1988. Binding selectivity of *Streptococcus pyogenes* and M-protein to epithelial cells differs from that of lipoteichoic acid. *Curr. Microbiol.* **16**:209–216.

Usui, Y. 1986. Biochemical properties of fibrinogen binding protein (clumping factor) of the staphylococcal cell surface. *Zbl. Bakt. Hyg. A* **263**:287–297.

Usui, Y., Y. Ohshima, and K. Yoshida. 1987. Platelet aggregation by group B streptococci. *J. Gen. Microbiol.* **133**:1593–1600.

Valentin-Weigand, P., J. Grulich-Henn, G.S. Chhatwal, G. Muller-Berghaus, H. Blobel, and K.T. Preissner. 1988. Mediation of adherence of streptococci to human endothelial cells by complement S protein (vitronectin). *Infect. Immun.* **56**:2851–2855.

Valentine, W.P., G.S. Chhatwal, and H. Blobel. 1988. Adherence of streptococcal isolates from cattle and horses to their respective host epithelial cells. *Am. J. Vet. Res.* **49**:1485–1488.

Vann, J.M. and R.A. Proctor. 1987. Ingestion of *Staphylococcus aureus* by bovine endothelial cells results in time-and inoculum-dependent damage to endothelial cell monolayers. *Infect. Immun.* **55**:2155–2163.

Vaudaux, P., R. Suzuki, F.A. Waldvogel, J.J. Morgenthaler, and U.E. Nydegger. 1984. Foreign body infection: role of fibronectin as a ligand for the adherence of *Staphylococcus aureus*. *J. Infect. Dis.* **150**:546–553.

Vaudaux, P., D. Pittet, A. Haeberli, U.E. Nydegger, D.P. Lew, and F.A. Waldvogel. 1989. Host factors selectively increase staphylococcal adherence on inserted catheters: a role for fibronectin and fibrinogen or fibrin. *J. Infect. Dis.* **160**:865–875.

von Hunolstein, C., M.L. Ricci, R. Scenati, and G. Orefici. 1987. Adherence of *Streptococcus bovis* to adult buccal epithelial cells. *Microbiologica* **10**:385–392.

Wadström, T. 1987. Molecular aspects on pathogenesis of wound and foreign body infections due to staphylococci. *Zbl. Bakt. Hyg. A* **266**:191–211.

Wadström, T. 1990. Hydrophobic characteristics of staphylococci: role of surface structures and role in adhesion and host colonization. In: Doyle, R.J. and M. Rosenberg (eds.), *Microbial Cell Surface Hydrophobicity* American Society for Microbiology, Washington, pp. 315–333.

Wadström, T. 1991. Molecular aspects on pathogenesis of staphylococcal wound and foreign body infections: Bacterial cell surface hydrophobicity, fibronectin, fibrinogen, collagen, binding surface proteins, determine ability of staphylococci to colonize in damaged tissues and on prosthesis materials. *Zbl. Bakt. Suppl.* **21**:37–52.

Wadström, T. and S.K. Tylewska. 1982. Glycoconjugates as possible receptors for *Streptococcus pyogenes*. *Curr. Microbiol.* **7**:343–346.

Wadström, T., P. Speziale, F. Rozgonyi, A. Ljungh, I. Maxe, and C. Ryden. 1987. Interactions of coagulase-negative staphylococci with fibronectin and collagen as possible first step of tissue colonization in wounds and other tissue trauma. *Zbl. Bakt. Suppl.* **A16**:83–91.

Wagner, B. and M. Ryc. 1978. Electron microscopic study of location of peptidoglycans in group A and C streptococcal cell wall. *J. Gen. Microbiol.* **108**:283–294.

Wagner, B., K-H. Schmidt, M. Wagner, and T. Wadström. 1988. Localization and characterization of fibronectin-binding to group A streptococci: an electron microscopic study using protein-gold complexes. *Zbl. Bakt. Hyg. A* **269**:479–491.

Wilcox, M.H., R.G. Finch, D.G. Smith, P. Williams, and S.P. Denyer. 1991. Effects of carbon dioxide and sub-lethal levels of antibiotics on adherence of coagulase-negative staphylococci to polystyrene and silicone rubber. *J. Antimicrob. Chemother.* **27**:577–587.

Wyatt, J.E., S.M. Poston, and W.C. Noble. 1990. Adherence of *Staphylococcus aureus* to cell monolayers. *J. Appl. Bacteriol.* **69**:834–844.

Yamada, K.M. 1983. Cell surface interactions with extracellular materials. *Annu. Rev. Biochem.* **52**:761–799.

Yamada, S. and A. Matsumoto. 1988. Hemagglutination activity and localization of Fc receptor of group A and G streptococci. *Microbiol. Immunol.* **32**:15–23.

Yarnall, M. and M.D.P. Boyle. 1986. Isolation and characterization of type IIa and type IIb Fc receptors from a group A streptococcus. *Scand. J. Immunol.* **24**:549–557.

7

Interaction of Bacteria with Phagocytic Cells

Phagocytic cells, unlike other cells of soft and hard tissue, are preordained to engulf organisms. The interaction of bacteria with phagocytic cells may be either beneficial or harmful to the bacteria. In some cases, the bacteria survive in phagocytes, thereby escaping from environmental challenge, whereas in other cases the outcome is lethal. In order for the phagocyte to recognize and ingest a bacterium it must possess receptors complementary to the bacterial surface. Several bacterial species have been found to express adhesins for which receptors are accessible on the phagocytic membrane. Three major nonopsonic mechanisms of interaction of bacteria with phagocytic cells in a serum-free system have been described (Table 7–1). One of these, termed lectinophagocytosis, is based on recognition between surface lectins on one cell and carbohydrates on the opposing cell. The second mechanism involves protein–protein interactions via the Arg-Gly-Asp (RGD) sequence. The final mechanism involves hydrophobic interactions between the two cell types.

Lectinophagocytosis

The lectin-dependent uptake of bacteria has been defined as lectinophagocytosis (Ofek and Sharon, 1988). The general features of lectinophagocytosis are summarized in Figure 7–1. In one case, bacteria possessing surface lectins interact with complementary carbohydrates on the membranes of phagocytic cells. In the other case, membrane-associated lectins of the phagocytic cells on macrophages interact with complementary carbohydrates on the bacterial surfaces. The methodology for the study of lectinophagocytosis is similar to that discussed in Chapter 2, particularly as regards lectin-mediated adhesion.

Unique features of lectinophagocytosis require that several controls be studied in order to confirm a role of bacterial and/or macrophage lectin in the adhesion

Table 7–1. Nonopsonic modes of recognition in phagocytosis of bacteria

Type of interaction	Bacterial ligand	Phagocytic receptor	Reference(s)
Lectin–carbohydrate	Lectin (e.g., type 1 fimbriae)	Glycoprotein (e.g., integrins)	Ofek and Sharon, 1988
	Polysaccharide (e.g., capsule)	Lectin (e.g., Man,GlcNAc receptors)	Athamna et al., 1991
Protein–protein	RGD-containing proteins (e.g., filamentous hemagglutinin)	RGD[a] receptor (e.g., integrin)	Saukkonen et al., 1991
Hydrophobin–protein	Hydrophobin (e.g., lipoteichoic acids)	Unknown (e.g., fatty acid site of fibronectin)	Ofek and Beachey, 1979; Courtney et al., 1981

[a]RGD, arginine-glycine-aspartic acid sequence.

process. These controls include: (1) use of antibodies directed against the bacteria to show that hapten inhibitors (e.g., carbohydrates) have no effect on opsonophagocytosis; (2) bacterial mutants devoid of lectin should be employed whenever possible; (3) because phagocytosis is a process associated with a measurable biological response initiated immediately following the interaction of the bacteria with the phagocytes, any inhibitor of the recognition step should also inhibit any subsequent biological response. The inhibitor should have no direct effect on the response, however.

The presence of lectins on bacterial surfaces was discussed in a preceding chapter. There are several lines of evidence to support a premise that phagocytic cells also express surface lectins. A critical question is whether phagocytic cells contain surface-associated glycoconjugates that act as receptors for the bacterial lectins, or conversely, do surface lectins on phagocytic cells act as ligands for carbohydrates on bacterial surfaces?

Because lectinophagocytosis may be mediated either via bacterial surface lectins or via macrophage surface lectins, inhibition of adhesion by carbohydrates or glycoconjugates does not allow one to assign the lectin to any one of the cell types. Multivalent ligands, such as mannans or glycoproteins, may first be added to the bacteria or to the phagocytes and the cells then washed to remove unreacted ligands (Bar-Shavit et al., 1977). Usually, polymeric ligands bind lectins with tight avidity and may result in the blocking of active sites of the lectins.

Lectinophagocytosis Mediated by Bacterial Surface Lectins

A number of bacteria interact via their lectins with phagocytic cells (Table 7–2). The most thoroughly investigated lectinophagocytosis is that mediated by the

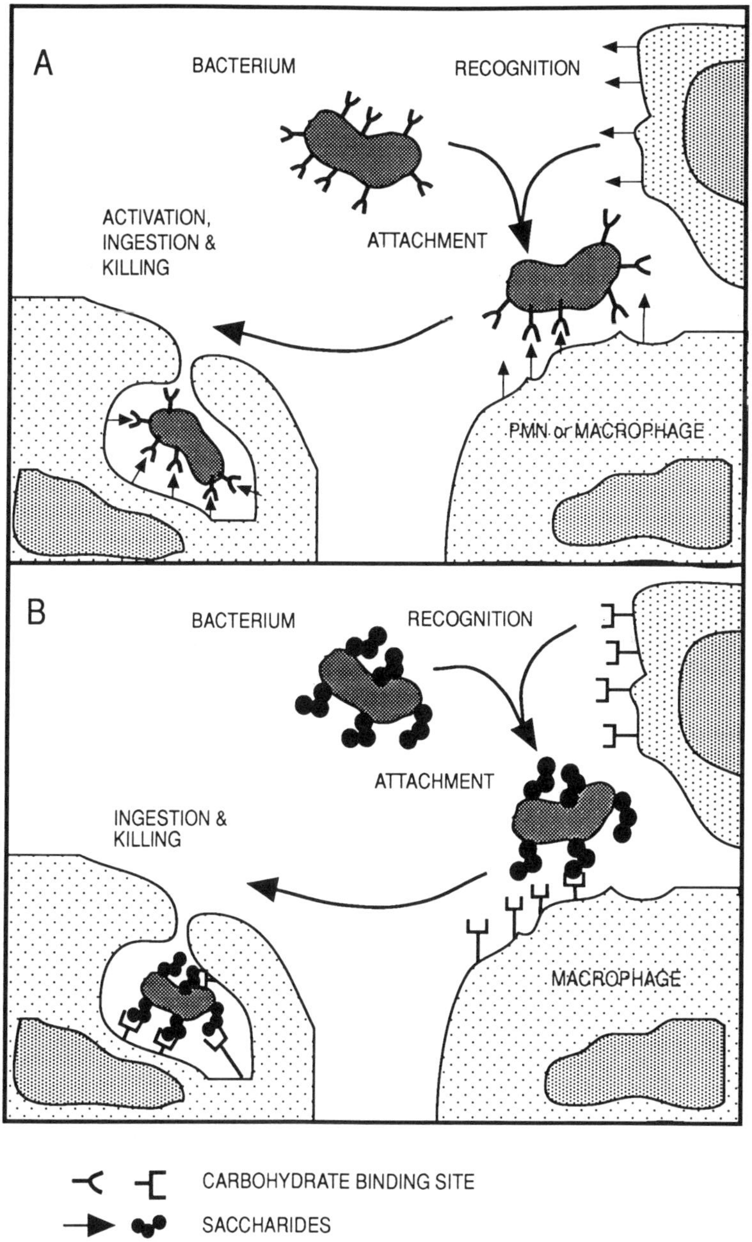

Figure 7–1. General features of lectinophagocytosis. Lectinophagocytosis may be mediated by bacterial surface lectins that recognize complementary carbohydrate residues on phagocytes (A) or by phagocyte surface lectins that complex with carbohydrates on bacterial surfaces (B). PMN, polymorphonuclear leukocyte. (Adapted from Ofek and Sharon, 1988.)

Table 7–2. Examples of lectinophagocytosis mediated by bacterial cell surface lectins

Type and location of lectin	Carbohydrate specificity	Bacterium	Phogocytic activity assayed	Types of cells tested
Type 1 fimbriae	Mannose	*Escherichia coli*	Attachment	Human G and PΦ
			Ingestion	Mouse PΦ
			Stimulation	Rat PΦ
			Killing	Human G & PΦ
		Salmonella typhimurium	Association	Mouse PΦ
		Klebsiella pneumoniae	Attachment Ingestion	Mouse PΦ
Type 2 fimbriae	Lactose	*Actinomyces spp.*	Killing Stimulation	Human G
Outer membrane	GalNAc	*Eikenella corrodens*	Attachment	Guinea pig Φ
Outer membrane	Galactose	*Fusobacterium nucleatum*	Association Stimulation	Human G
Type P fimbriae	Gal-Gal	*Escherichia coli*	Stimulation	Human G, Gal-Gal coated
Outer membrane protein (Opa)	$GlcNH_2$	*Neisseria gonorrhoeae*	Attachment Ingestion Stimulation	Human G
Cell surface	GalNAc	*Staphylococcus saprophyticus*	Stimulation	Human G
	Sialic acid	*Staphylococcus saprophyticus*	Stimulation	Human G
Pertussis toxin	Sialic acid	*Bordetella pertussis*	Attachment	Human AΦ Rabbit AΦ
Filamentous hemagglutinin	Galactose	*Bordetella pertussis*	Attachment	Human AΦ Rabbit AΦ

GalNAc, N-acetylglucosamine; $GlcNH_2$, glucosamine; Gal-Gal, galactose α-1,4 galactose; Φ, macrophage; AΦ, alveolar macrophage; PΦ, peritoneal macrophage; G, granulocytes. Derived from Bar-Shavit et al., 1980; Athamna and Ofek, 1988; Sandberg et al., 1988; Kurashima et al., 1991; Svanborg-Eden et al., 1984; Rest et al., 1985; Beuth et al., 1988; Saukkonen et al., 1991; Miki et al., 1986; Passo et al., 1982; Ofek, 1989; Ofek and Sharon, 1988.

mannose-specific (MS) lectins, associated with type 1 fimbriae. The evidence that the recognition of type 1 fimbriated *E. coli* by phagocytes is mediated by interaction of the fimbrial lectin with mannose-containing glycoproteins on the surface of the phagocyte is based on several lines of evidence: (1) the specificity pattern of inhibition of bacteria–host cell interaction observed is the same for phagocytic and nonphagocytic target cells (Firon et al., 1983, 1985); (2) when carbohydrates other than mannose have been examined, they inhibit poorly, if at all, the interaction of type 1 fimbriated *E. coli* with mouse peritoneal macrophages or human polymorphonuclear leukocytes (PMNs) (Bar-Shavit et al., 1977; Silverblatt et al., 1979); (3) a very good correlation has been found between the mannose binding activity of the bacteria and the extent of their attachment to mouse peritoneal macrophages (Bar-Shavit et al., 1980); (4) the finding that pretreatment of type 1 fimbriated bacteria with yeast mannan inhibits their attachment to mouse and human phagocytes, whereas pretreatment of the phagocytes does not have such an effect (Bar-Shavit et al., 1977), shows that the receptor for the bacterial lectin is on the surface of the phagocytes; and (5) Latex particles coated with purified type 1 fimbriae stimulate human PMNs, the activity of which can be inhibited by D-mannose (Goetz and Silverblatt, 1987). Because the only class of mannose-containing compounds in animal membranes are glycoproteins (Rauvala and Finne, 1979), the receptor for mannose-specific bacteria must belong to this class.

The mannose-sensitive (MS) lectin-mediated binding to phagocytic cells leads to ingestion, stimulation of antimicrobial systems (e.g., oxygen bursts, degranulation), and killing of the bacteria (Table 7–3). All of the activities listed in Table 7–3 are initiated by specific interaction of the MS fimbrial lectin with mannose-containing receptors on the phagocytic cell (Perry et al., 1983; Soderstrom and

Table 7–3. Stages of lectinophagocytosis of Escherichia coli *mediated by type 1 fimbriae*

Stage	Temperature	Cell	Assay system	Reference(s)
Attachment	4°C	Macrophage, mouse	CFU	Ofek and Sharon, 1988
Ingestion	37°C	PMN, human	EM	Silverblatt et al., 1979; Rottini et al., 1979
			FITC-*E. coli*	Ohman et al., 1985
		Macrophage, mouse	FITC-anti-*E. coli*	Bar-Shavit et al., 1980
Stimulation	37°C	PMN, human	0_2 consumption	Goetz and Silverblatt 1987; Rottini et al., 1979
			Chemiluminescence	Mangan and Snyder, 1979; Bjorksten and Wadström, 1982; Soderstrom and Ohman, 1984
			Lysozyme release	Svanborg-Eden et al., 1984; Mangan and Snyder, 1979
			Protein iodination	Perry et al., 1983
		Macrophage, rat	Chemiluminescence	Blumenstock and Jann, 1982
Killing	37°C	PMN, human	CFU	Silverblatt et al., 1979; Ohman et al., 1982
		Macrophage, mouse	CFU	Ofek and Sharon, 1988
		Macrophage, human	CFU	Boner et al., 1989

Abbreviations: CFU, colony-forming units; EM, electron microscopy; FITC, fluorescein isothiocyanate labeled; PMN, polymorphonuclear leukocyte.

Ohman, 1984). This is confirmed by experiments showing that (1) methyl-α-mannoside inhibits activation of phagocytes by MS-fimbriated *E. coli;* (2) the saccharide does not inhibit activation of phagocytes induced by opsonized bacteria; (3) nonfimbriated bacteria do not induce activation of phagocytic cells; and (4) Latex particles coated with MS fimbriae stimulate high phagocytic activities that are inhibited by mannosides.

Maximum stimulation of antimicrobial systems in phagocytes appears to require cross-linking of the MS fimbriae on the bacterial surfaces (Figure 7–2). Such cross-linking may cause aggregation of the receptors on the phagocytic cells. Aggregation of receptors is important for many other membrane-initiated events (Figure 7–2). Stimulation of human blood neutrophils and peritoneal macrophages involves protein kinase C activation (Gbarah et al., 1989). A unique pattern of degranulation of human neutrophils by type 1 fimbriae has also been observed (Steadman et al., 1988).

In quantitative terms, lectinophagocytosis and opsonophagocytosis are comparable (Figure 7–3). The rates of ingestion of type 1 fimbriated *E. coli* by the phagocyte vary markedly with growth conditions of the bacteria and the presence of capsules on the bacteria. Electron microscopy has revealed that the interaction of type 1 fimbriated *E. coli* with PMN, leads to internalization of the organisms in intracellular vacuoles (Silverblatt et al., 1979). Ingestion can be distinguished from adhesion by the use of dyes or antibodies attached to the bacteria, as has been shown for both *E. coli* and *Klebsiella pneumoniae*. Recently, an enzyme-linked immunosorbent assay (ELISA) test was developed to study the binding, ingestion, and killing of type 1 fimbriated *K. pneumoniae* by mouse peritoneal leukocytes (Athamna and Ofek, 1988). The method is easy to perform and allows for the enumeration of bound and ingested bacteria by phagocytic cells. Using this method, it has been found that phagocytosis of *K. pneumoniae* via type 1 fimbriae is essentially the same as that of *E. coli*. When methyl-α-D-mannoside is added to bacteria–phagocyte suspensions, both attachment and ingestion are inhibited. The rate of ingestion of *E. coli* mediated by type 1 fimbriae is reduced when the organisms contain K antigen (a highly negatively charged surface polysaccharide), even though the extent of adhesion of the organisms is not affected as compared to *E. coli* without this polysaccharide (Soderstrom and Ohman, 1984). Sublethal concentrations of β-lactam antibiotics may lead to *E. coli* populations with a reduced expression of type 1 fimbriae. The antibiotic-grown cells are ingested by phagocytes much more slowly than fimbriated bacteria grown in antibiotic-free medium (Bar-Shavit et al., 1980). Iwahi and Imada (1988) showed that the bacterial surface components involved in interference with internalization of *E. coli* attached to phagocytic cells can be removed by pretreating the organisms with ethylenediaminetetraacetic acid. Whether the mannose-specific attachment of phagocytes will lead to ingestion and killing depends on the overall surface characteristics of the bacteria. A rough hydrophobic surface generally facilitates phagocytosis, whereas a smooth hydrophobic

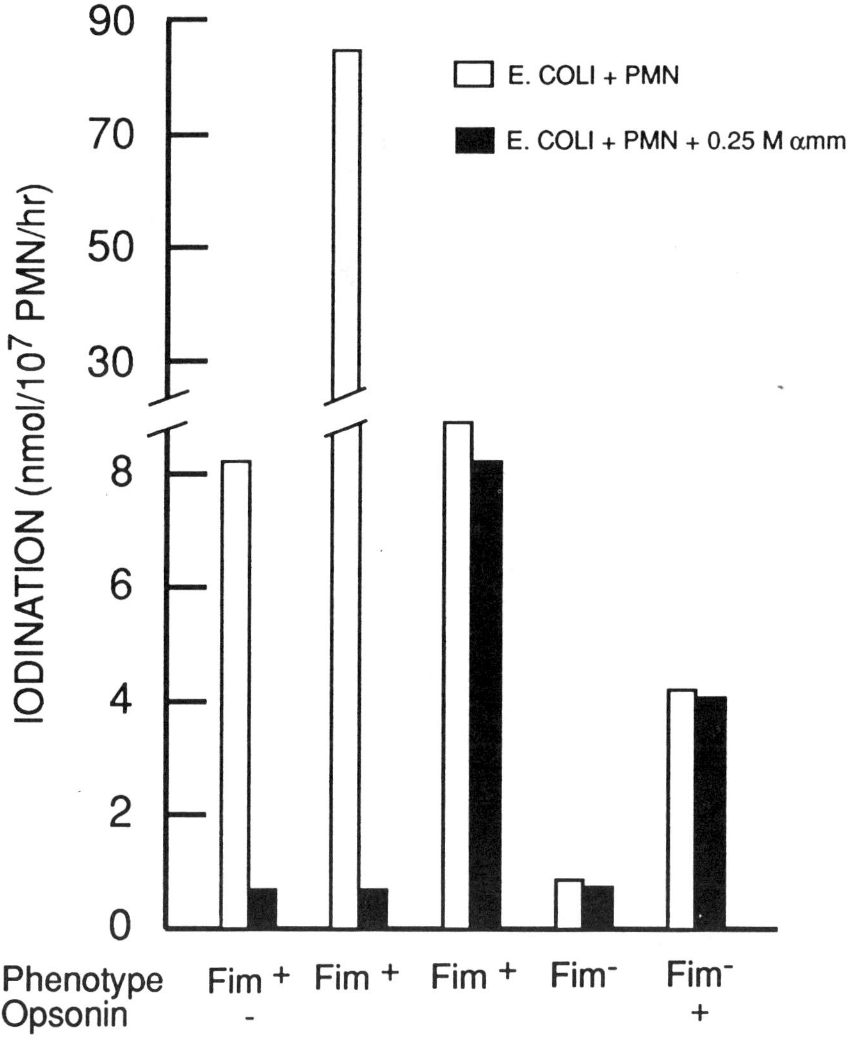

Figure 7–2. Stimulation of protein iodination by type 1 (Fim^+) and nonfimbriated (Fim^-) *Escherichia coli* strains in human polymorphonuclear leukocytes (PMNs). Fimbriae were cross-linked with antifimbrial $F(ab_2')$ antibodies or by glutaraldehyde. Opsonization was initiated by coating the bacteria with antisomatic antibodies. Broken bar, results for cross-linked fimbriae. Methyl-α-mannoside, αMM. Results were adapted from Perry et al., 1983.

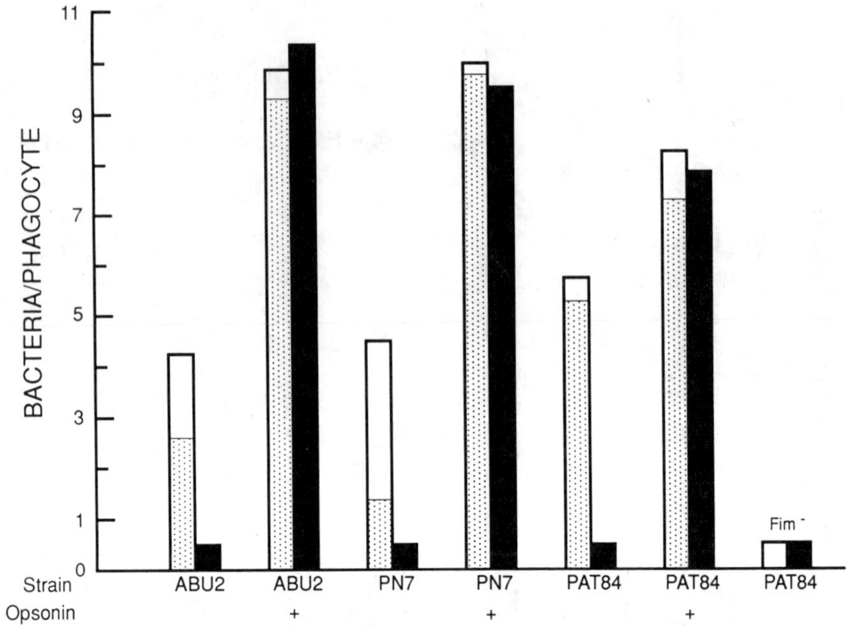

Figure 7–3. *Escherichia coli* Fim^+ and Fim^- strains undergoing lectinophagocytosis and opsonophagocytosis. Fim^+ strains were type 1, mannose-sensitive. Open bars represent association of antibody-coated (opsonized) or nonopsonized bacteria. Hatched bars represent ingested bacteria. Solid bars represent controls run in the presence of methyl-α-mannoside. Human PMNs were used with *E. coli* strains ABU2 and PH7, whereas mouse peritoneal macrophages were used with *E. coli* strain PAT84. Data were derived from Bar-Shavit et al., 1977, 1980 and Ohman et al., 1982, 1988.

surface will impede ingestion. For example, hydrophobic variants are avidly ingested with limited release of oxidative metabolites. Hydrophobic variants, in contrast, that may resist phagocytosis will frequently stimulate PMNs metabolically and cause release of oxidative metabolites. In both cases, type 1 fimbriae are essential for mediating the bacterium–phagocyte interactions. A number of studies have noted diminished killing of type 1 fimbriated *E. coli* by phagocytic cells. In one study, it was concluded that a lower bactericidal activity against type 1-carrying bacteria may be due to a decrease in intraphagolysosomal myeloperoxidase rather than to inadequate ingestion of the organisms by human neutrophils (Goetz et al., 1987). In another study, it was found that certain strains of *E. coli* attached to human neutrophils via type 1 fimbriae were readily killed, whereas other strains were killed only intracellularly (Lock et al., 1990). Recombinant derivatives of *E. coli* K-12 expressing type 1 fimbriae attached to mouse peritoneal phagocytes are killed to a much lower degree than their nonfimbriated derivatives, which presumably are phagocytized via a nondefined adhesin (Keith et al., 1990). Collectively, the studies suggest that the fate of *E. coli* bound to phagocytic cells via type 1 fimbrial lectin depends on the density of the fimbriae on the bacterial surface as well as on the relative hydrophobicities of the bacterial and phagocytic surfaces. Other surface constituents may interfere with the ability of the organisms to initiate internalization and induction of antimicrobial systems in the phagocytic cells.

Attempts to identify the receptor(s) on PMNs for type 1 fimbriae have used affinity chromatographic techniques. Type 1 fimbriae of *E. coli* were immobilized on agarose and employed as the primary affinity trapping agent. Three glycoproteins have been isolated from human granulocytes, two of which migrate on sodium dodecyl sulfate-polyacrylamide gel electrophoresis (SDS-PAGE) gels as the adhesion molecules or antigens CD11/CD18 (Rodriguez-Ortega et al., 1987). More recently, Gbarah et al. (1991) obtained results to suggest that the leukocyte integrins (CD11/CD18) serve as the major receptors for the type 1 fimbrial lectin of *E. coli*. The bacteria bind in a dose-dependent, saturable manner to the isolated integrin molecules. Nonfimbriated *E. coli* fail to interact with the integrin preparation. In addition, binding is inhibited by mannosides and by monoclonal antibodies to CD11A, CD11B, or CD11/8, but not by galactose. Furthermore, periodate-oxidized integrin is a poor substratum for *E. coli*. Because CD116/CD18 and CD11C/CD118 serve as receptors for complement fragment C3bi, the findings indicate the receptor molecule engaged in opsonophagocytosis may also function in lectinophagocytosis.

In other studies it was found that a glycoprotein called NCA-50, which belongs to a family of nonspecific cross-reacting antigens known to be associated with granulocyte membranes, specifically binds type 1 fimbriated bacteria (Sauter et al., 1991). On the basis of its molecular weight (ca. 50,000), NCA-50 does not belong to the integrin family. It remains to be seen whether this glycoprotein participates in the process of phagocytosis of type 1 fimbriated *E. coli*. Salmon et al. (1987) suggested that Fc receptors of human granulocytes are involved in the process of ingestion, but not attachment, of type 1 fimbriated *E. coli* by the phagocytes. They also suggested that the process is mediated by specific interaction between the oligosaccharide side chain of the Fc receptor and the type 1 fimbriae. It has been postulated that following attachment of the fimbriated bacteria to granulocytes via the CD11/CD18 glycoproteins, the fimbrial lectin becomes concentrated in proximity of the Fc receptor, which interacts with the lectin to trigger ingestion of the bacteria (Gbarah et al., 1991).

Extensive studies have been carried out on the involvement of Opa or PII outer membrane proteins, of *Neisseria gonorrhoeae*, particularly Opa B, in mediating the interaction of the gonococci with human blood neutrophils in a serum-free system (Swanson et al., 1974, 1975; King et al., 1978; King and Swanson, 1978; Virji and Hechels, 1986; Shafer and Rest, 1989; Elkins and Rest, 1990; Farrel and Rest, 1990; Naids and Rest, 1991; Naids et al., 1991). The evidence that recognition of gonococci by human neutrophils is mediated by Opa proteins is based on the following findings. First, Opa-carrying gonococci bind to and stimulate vigorous chemiluminescence responses in human neutrophils, whereas the Opa-deficient gonococci do not bind to the phagocytes nor do they stimulate a detectable response. Second, liposomes carrying purified Opa B or peptides comprising the second hypervariable region of Opa B bind to neutrophils and inhibit binding of gonococci expressing Opa B, whereas liposomes carrying other outer membrane proteins, such as Opa A, lack these activities. Third, monoclonal antibodies against Opa B inhibit gonococci from binding to and stimulating human granulocytes. Stimulation of PMN by Opa (or PII)-bearing gonococci is inhibited by several monosaccharides, such as glucosamine, mannose, methyl-α-mannoside, and *N*-acetylneuraminic acid, as well as by pretreating the granulocytes with concanavalin A or mannosidase, whereas other carbohydrates or sialidase have no effects. Although the data, taken together, suggest that lectin-like components, possibly Opa B proteins, on the surface of the gonococci mediate the binding of the bacteria to the neutrophils, a lectin activity of Opa B or other surface components has not been proven. Other studies have shown that gonococci bind galactose-containing glycoconjugates (see Chapter 7), but there is no evidence that such carbohydrates can serve as attachment sites for binding Opa-carrying gonococci to human neutrophils. The findings that the receptors on phagocytic cells can be upregulated by agents such as formyl peptides or by phorbol myristate acetate are intriguing because PMN sites of bacterial invasion are probably activated to express the putative receptor(s) capable of binding gonococci (Shafer and Rest, 1989). In summary, although the nature of the bacterial lectin(s) and its carbohydrate specificity involved in binding of gonococci to human neutrophils has not been defined, the gonococci–phagocyte interaction mediated by Opa B, and possibly other outer membrane proteins, is nonopsonic and may be considered as another example of lectinophagocytosis.

The case of *Bordetella pertussis* is unusual in that it employs two nonfimbrial lectins in order to bind and be internalized by human and rabbit alveolar macrophages (Relman et al., 1990; Van 'T Wout et al., 1991; Saukkonen et al., 1991). One lectin is the filamentous hemagglutinin (FHA) and the other is the subunit S3 of the pertussis toxin. FHA, a 220,000 mol wt protein, mediates the binding of the organisms to galactose residues on macrophages. The toxin is a 105,000 mol wt hexameric protein composed of five different types of subunits, S1–S5. The toxin is secreted by *Bordetella* and binds to the surface of the organisms by an as yet unknown mechanism in a manner that allows its subunit S3 to mediate the binding of the organisms to macrophage gangliosides. *B. pertussis* deficient in filamentous hemagglutinin (FHA) and toxin do not bind to macrophages but binding can be restored by adding toxin to reaction mixtures. Whereas FHA or toxin alone is sufficient for binding the organisms to macrophages, subsequent ingestion of the bacteria requires a cooperative (or coordinate) process in which the macrophages appear to undergo "priming" by preligation to the toxin. The process of internalization also appears to require the participation of protein–protein interactions between FHA and macrophage integrins (see below). Internalization and survival of *B. pertussis* within alveolar macrophages is markedly inhibited by lactose, which binds specifically to FHA. Thus, although not sufficient by themselves, the lectin–carbohydrate interactions are essential steps in the internalization by (and subsequent survival of) *B. pertussis* by alveolar macrophages. In that sense the process can also be considered as a variant of lectinophagocytosis.

Whereas the occurrence of lectinophagocytosis mediated by bacterial lectins has been established unequivocally in vitro, little is known about its occurrence in vivo. Indirect evidence that mannose-specific lectinophagocytosis may take place in vivo has been obtained in experimental infections with mixed bacterial phenotypes (or isogens), one of which expresses a mannose-specific lectin associated with type 1 fimbriae (MS^+) and the other does not (MS fim^-). Such experiments reveal that whenever the organisms reach phagocyte-rich sites, the MS fim^- phenotype (or isogen) survived, whereas at phagocyte-poor sites, the MS fim^+ phenotype (or isogen) survived, irrespective of the bacterial species or experimental animal employed or of the route or site of the infection (Table 7–4). The results are interpreted as being due to selective survival of MS fim^- phenotype in deep tissues (e.g., kidney or peritoneal cavity) from which the MS fim^+ phenotype is eliminated by macrophages and/or polymorphonuclear leukocytes which usually are present in deep tissues. It was suggested that the

Table 7–4. Survival of mannose-specific fimbriated (MS Fim^+) and nonfimbriated (MS Fim^-) phenotypes or isogens in phagocyte-rich and -poor sites following infection with mixed strains.

Organism	Animal	Route of infection	Site of infection	Level of phagocytes at infection site	Phenotype	Reference
Escherichia coli	Mouse	Intravesicular	Kidney	Rich	Fim^-	Hagberg, 1983; Schaeffer et al., 1987
			Bladder	Poor	Fim^+	
		Oral	Blood	Rich	Fim^-	Guerina et al., 1983
			Oral cavity	Poor	Fim^+	
		Peritoneal	Peritoneum	Rich	Fim^-	Alkan et al., 1986
	Rat	Intravesicular	Kidney	Rich	Fim^-	Marre and Hacker, 1987
		Peritoneal	Blood	Rich	Fim^-	Saukkonen et al., 1988
Klebsiella pneumoniae	Mouse	Intraviscular	Kidney	Rich	Fim^+	Maayan et al., 1985
			Bladder	Poor	Fim^+	

phase variation, a random on–off switching process that allows the cells to alternate between fimbriated and nonfimbriated states, is an important virulence trait of type 1 fimbriated *E. coli* (Ofek and Silverblatt, 1982) as well as for other fimbriated bacterial species (Silverblatt and Ofek, 1983). Soluble glycoconjugates may interfere with lectinophagocytosis. For example, the lectinophagocytosis of mannose-specific *E. coli,* coated with Tamm–Horsfall glycoprotein, abundant in the urine, is considerably reduced, suggesting that soluble glycoconjugates containing saccharides to which bacterial lectins bind can potentially interfere with the process of lectinophagocytosis (Kuriyama and Silverblatt, 1986).

It is now known that the galactose-specific lectin of certain Gram-positive bacteria, such as members of the genus *Actinomyces,* may mediate phagocytosis of the organism. Adhesion, ingestion, and killing of *A. viscosus* TIV14 and *A. naeslundii* WVU 45 by human PMNs are inhibited by methyl-β-D-galactosides or lactose (Sandberg et al., 1988). The treatment of the PMNs with sialidase preparations results in an enhanced adhesion and ingestion of the bacteria, suggesting that some of the receptors are not accessible on nontreated PMNs. Presumably, the sialidase exposes penultimate β-D-galactoside residues, which then serve as receptors for the *Actinomyces* lectin. Other bacterial species that have been found to interact with macrophages via their surface lectins are shown in Table 7–2.

Lectinophagocytosis Mediated by Macrophage Surface Lectins

The second mechanism of lectinophagocytosis involves recognition of carbohydrate residues on the bacterial surfaces by lectins that are integral components of the phagocytic cell membrane. Evidence for the presence of lectins on phagocytic cells began to emerge in the late 1960s with the work of Ashwell and Morell, who studied blood clearance of asialoglycoproteins, mediated by receptors on the surfaces of liver cells (hepatocyte and Kupffer cells). At least three types of lectins expressed on the surface of liver cells and on tissue macrophages have been isolated and characterized (Ashwell and Morell, 1974; Ashwell and Hartford, 1982). All three are glycoproteins. The Gal type lectin (or asialoglycoprotein receptor) is found on Kupffer cells, hepatocytes, and subpopulations of peritoneal macrophages. The human hepatic receptor appears to be a single polypeptide with a mol wt of 41,000 (Baenziger and Maynard, 1980) or 46,000 (Schwartz and Rup, 1983) as determined by SDS-PAGE. The Gal type lectin is specific for *N*-acetylgalactosaminne and galactose. The Man type lectin (or GlcNAc/Man lectin) is found on tissue macrophages (e.g., Kupffer cells and alveolar macrophages). It has a mol wt of 175,000 (Shepherd et al., 1982; Wileman et al., 1986) and is specific for mannose, *N*-acetylglucosamine, glucose, and fucose. The Fuc type lectin is found only on Kupffer cells. It contains two subunits of mol wts of 88,000 and 77,000 (Lehrman and Hill, 1986) and is specific for L-fucose. Animal lectins are now classified as "C-type" and "S-type" (Drickamer, 1988). The C-type lectins require Ca^{2+} for activity, possess several disulfide bonds, and exhibit various specificities. In contrast, the S-type lectins do not require Ca^{2+} ions and they are generally rich in thiol groups. Lectins of the S-type usually are specific for β-galactosides. The macrophage lectins are of the C-type. Taylor et al. (1990) and Ezekowitz et al. (1990) have recently sequenced a human macrophage C-type lectin. The lectin, specific for Man/GlcNAc/Fuc, is unusual in the sense that it is thought to contain at least eight carbohydrate recognition domains. The large number of carbohydrate combining sites would be expected to increase the avidity of the lectin for glycoproteins or other glycoconjugates. These lectins have been found to serve as receptors for serum asialoglycoproteins, various other glycoproteins, and perhaps asialoerythrocytes. They may therefore possess a physiological role in clearance of these elements from the blood.

The idea that bacterial surface carbohydrates are important for nonopsonic binding to macrophages was first suggested by Ogmundsdottir and Weir (1976) and by Sutherland et al. (1978). However, the exact carbohydrate specificities of these interactions were not defined and the authors did not refer to any of the macrophage lectins as being involved in phagocytosis of a particular strain of bacteria. Warr (1980) was the first to implicate the involvement of the Man type lectin of alveolar macrophages in the binding of mannan-containing yeast cells. It now appears that only two of the macrophage lectins can participate in lectinophagocytosis; these are the Gal- and the GlcNAc/Man-specific lectins.

The involvement of macrophage lectins in blood clearance in mice, of Gram-negative (nonfimbriated *E. coli*) and Gram-positive (group B streptococci) has been studied in some detail. The bacteria resist killing by whole blood of mice in vitro and are sequestered in the liver after intravenous injections (Perry and Ofek, 1984; Perry et al., 1985). The *E. coli* strain employed agglutinated with concanavalin A but not with wheat germ, peanut, or *Ricinus communis* lectins, indicating that glucose or mannose residues, or both, were exposed on the bacterial surface. The blood clearance of the bacteria in mice was strongly inhibited by derivatives of D-mannose, D-glucose, or L-fucose, but not of D-galactose or L-rhamnose, suggesting that the blood clearance of the *E. coli* was mediated by the Man-specific receptor of liver phagocytes (e.g., Kupffer cells). In contrast, blood clearance of antibody-coated *E. coli* was not inhibited by D-mannose (Table 7–5). The involvement of the Gal-specific type lectin in the blood clearance of bacteria was tested by employing a strain of type IB *Streptococcus agalactiae*. The structure of the type IB capsular polysaccharide is composed of repeating units in which all the side chain β-D-galactose residues are masked by terminal sialic acid residues (Kasper et al., 1983). Removal of the sialic acid with sialidase results in exposure of the galactose residues. Indeed, *Ricinus communis* agglutinin, a galactose-specific lectin, reacted only with sialidase-treated bacteria. The blood clearance of the desialylated polysaccharide, which

Table 7–5. Inhibition by saccharides and neoglycoproteins of blood clearance and phagocytic attachment of Escherichia coli *and* Streptococcus agalactiae.

	Blood Clearance in Mice (% Inhibition)		Phagocytic Attachment (% Inhibition of)	
Inhibitor	*E. coli* 025	Type Ib	*S. agalactiae* Type II	*S. agalactiae* Type II
Me-Fuc or Fuc-BSA	70	20	20	ND
Me-man or Man-BSA	70	20	20	44
Me-Gal or Gal-BSA	20	70	20	23
Gal-BSA or Glc-BSA	ND	ND	70	71
Me-Glc or Glc-BSA	70	ND	ND	40
Macrophage Lectin Specificity	Man	Gal	Gal and Man	Gal and Man

Results shown in this table were derived from Perry and Ofek (1984) and Perry et al. (1985). Inhibition of blood clearance of *E. coli* was determined using methylglycosides and that for the streptococci using the neoglycoprotein indicated. The results were obtained by use of thioglycollate-elicited mouse peritoneal macrophages. For *S. agalactiae* type II, the degree of attachment of the opsonized bacteria was several-fold higher than that for the nonopsonized cells. Type Ib *S. agalactiae* is free of terminal sialic acid residues on its group-specific polysaccharide. Abbreviations are: Me-Fuc, methyl-α-fucoside; Me-Gal, methyl-α-galactoside; Me-Man, methyl-α-mannoside; BSA, bovine serum albumin. *E. coli* 025 and *S. agalactiae* type II could be agglutinated by concanavalin A, whereas *Ricinus communis* lectin agglutinated only *S. agalactiae* type I.

had terminal galactose residues exposed on its surface, was strongly inhibited by galactosylated but not mannosylated or fucosylated bovine serum albumin (BSA) derivatives (Table 7–5), suggesting the involvement of the Gal type lectin of liver cells in the blood clearance of the streptococci. The blood clearance of type II group B streptococci, the surfaces of which contain both galactose and glucose residues (Kasper et al., 1983), is strongly inhibited only by a combination of galactosyl-bovine serum albumin and mannosyl-BSA but not with the individual neoglycoproteins (Table 7–5), indicating that neither one of the liver macrophage lectins (i.e., the Gal type or the Man type) can mediate the blood clearance of the bacteria. More direct results that lectins on the surface of macrophages mediate binding of bacteria was obtained from studies employing attachment of type II group B streptococci, expressing both galactose and mannose residues on their surfaces, to a monolayer of mouse thioglycollate-elicited peritoneal macrophages. As with blood clearance of these organisms, only the combination of Gal-BSA and Man-BSA and Glc-BSA strongly inhibited the attachment of the bacteria to the macrophage monolayer (Table 7–5). The specificity of the inhibition of attachment of the mixture of neoglycoproteins is shown further by the inability of these compounds to inhibit attachment to phagocytes of bacteria coated with antibodies (Table 7–5). The involvement of liver lectins in binding of bacterial surface carbohydrates can be demonstrated further by coaggregation experiments in which mouse liver homogenates induce agglutination of *E. coli*.

The agglutination is inhibited by the same carbohydrates that inhibit the blood clearance of the bacteria, as well as by lipopolysaccharides extracted from *E. coli* (Perry and Ofek, 1984).

In summary, several lines of evidence implicate interactions of macrophage lectins with bacterial surface carbohydrates as mechanisms for blood clearance in mice. First, the patterns of carbohydrate derivatives and neoglycoproteins inhibiting blood clearance and attachment of the bacteria to macrophages correspond to the carbohydrate residues exposed on the surfaces of the organisms and to the carbohydrate specificities of the types of liver or macrophage lectins involved (Perry and Ofek, 1984; Perry et al., 1985). Second, the inhibitory carbohydrates or glycoproteins do not inhibit the blood clearance or attachment to macrophages of bacteria precoated with antibodies (Perry and Ofek, 1984; Perry et al., 1985). Third, mouse liver preparations induce agglutination of *E. coli,* which is inhibited by the same carbohydrates that inhibit the blood clearance of the bacteria, as well as by lipopolysaccharides extracted from the bacterial strain employed (Perry and Ofek, 1984).

The Man/GlcNAc-specific lectin is expressed on alveolar peritoneal and hepatic macrophages. Several studies have shown that the lectin mediates the adhesion and phagocytosis of a number of bacteria, including *K. pneumoniae* (Athamna et al., 1991), *S. agalactiae,* and *E. coli* (as discussed above), *Pseudomonas aeruginosa* (Speert et al., 1988), and *Mycobacterium avium* (Bermúdez et al., 1991). A possible role in the human for lectinophagocytosis of microorganisms dependent on the Man-specific macrophage lectin has been suggested in pneumonia caused by *K. pneumoniae*.

Athamna et al. (1991) studied the binding of guinea pig alveolar macrophages in a serum-free medium with 16 *K. pneumoniae serogroups* and with seven serogroups of polysaccharides from the bacteria. Only five polysaccharides containing the repeating sequence Manα2,3Man or L-Rhaα2,1Rha bind to the macrophages. Various glycoconjugates containing reactive disaccharides, including mannans and neoglycoproteins, are inhibitors of the binding. The binding is also dependent on the presence of Ca^{2+} in the medium. Most of the bacteria bound to the macrophages were ingested and killed. It is generally accepted that polysaccharide capsules are antiphagocytic (Gotschlich, 1983). However, as shown for *K. pneumoniae,* certain polysaccharide sequences may actually facilitate phagocytosis, especially when they contain carbohydrates complementary to macrophage lectins. That macrophage lectin–capsular polysaccharide interactions play a role in vivo is supported by epidemiological observations on the distribution of the various serogroups of isolates of *K. pneumoniae*. The majority of the serogroups of *K. pneumoniae* isolated from the blood do not contain polysaccharides capable of binding with C-type macrophage lectins (Athamna et al., 1991). It was also observed that human monocyte-derived macrophages are more efficient than monocytes in binding and killing the bacteria. It seems possible that the C-type lectin-dependent attachment of bacteria to macrophages may be a normal defense mechanism against bacterial infections in the lung. The ingestion of *Pneumocystis carinii* by human and rabbit alveolar macrophages in a serum-free medium presents an interesting observation in the lectinophagocytosis process (Ezekowitz et al., 1990, 1991). Transfection of the Cos cell line with cDNA coding for the Man-specific lectin of the human macrophage results in a macrophage phenotype possessing the ability to bind and ingest *P. carinii*, a heavily surface mannosylated organism. The transformed receptor may be regarded as a "professional" receptor for phagocytosis, in that it can now specifically bind oligomannoside-containing organisms and subsequently lead to their ingestion.

RGD-dependent Phagocytosis

The interaction of phagocytic cells with *Bordetella pertussis* is thought to involve at least two adhesion mechanisms (Relman et al., 1990; Saukkonen et al., 1991). One mechanism is dependent on a lectin–carbohydrate complex (as discussed above), whereas the other requires an RGD sequence on the filamentous hemagglutinin that is recognized by macrophage integrin CR3. The latter adhesive mechanism does not lead to oxidative bursts, a factor favoring survival of the bacteria. It is interesting to note that the binding of pertussis toxin to macrophage surface carbohydrates upregulates CR3. In turn, the "activated" CR3 can bind the FHA of *B. pertussis* via the RGD sequence. "Activation" of the macrophages by the toxin results in increased efficiency in the binding of both the toxin and the FHA. *Leishmania donovani*, a protozoan that expresses a surface glycoprotein containing an RGD sequence, can also bind to RGD receptors on macrophages (Wilson and Pearson, 1988).

Bacteria–macrophage Interactions Mediated by the Hydrophobic Effect

Van Oss (1978) has observed that hydrophobic bacteria (as determined by contact angle measurements) tend to be readily engulfed by phagocytic cells. The importance of bacterial hydrophobicity in the interaction with phagocytic cells in a serum-free system is similarly emphasized when various partition methods are employed (Absolom, 1988). It is possible that surface hydrophobic interactions facilitate the binding of bacteria with phagocytic cells to permit more specific secondary complexes to occur. The implication of a specific hydrophobin to mediate the attachment of a bacterium to phagocytic cells has been suggested for only a few bacteria. Lipoteichoic acid (LTA) receptors on the surfaces of PMNs have been suggested to mediate the adhesion of group A streptococci (Ofek and Beachey, 1979; Courtney et al., 1981). Escape from phagocytosis may be possible when the streptococci secrete capsular materials to mask accessibility of LTA (Chapter 6).

Speert et al. (1986) found that the binding of *Pseudomonas aeruginosa* to phagocytic cells was inhibited by hydrophobic compounds, such as *p*-nitrophenol. They were unable to find carbohydrates that would inhibit the bacterium–phagocyte interactions and concluded that the interaction was not dependent on a *Pseudomonas* surface lectin. Strains that were hydrophobic and piliated were taken up more readily than hydrophilic, nonpiliated strains. Irvin (1990) has discussed the role of fimbriae in bacterial hydrophobicity. Earlier, Garber et al. (1985) found that hemagglutination reactions of *P. aeruginosa* could be inhibited by low molecular weight hydrophobic compounds, but not by carbohydrates. Some species of *P. aeruginosa* may express lectins on their fimbriae, but as of yet, there are no clear results describing fimbrial lectins similar to those of *E. coli*.

Other Nonopsonic Mechanisms of Phagocytosis Requiring Adhesion Reactions

The literature suggests that even more mechanisms may exist for serum-free interactions between phagocytes and microorganisms. For example, the nonfimbrial hemagglutinins (NFAs) of some *E. coli* isolates seem to mediate the attachment of the bacteria to human granulocytes (Goldhar et al., 1991). Because NFA^+ *E. coli* can undergo phase variation to yield the type 1 fimbriated phenotype, this organism is susceptible to phagocytosis via at least two mechanisms. *Pseudomonas* pili have been found to serve as ligands for nonopsonic phagocytosis of fibronectin-stimulated mouse macrophage cell line $P388^{01}$ (Kelly et al., 1989). *E. coli* expressing various types of hemagglutinins stimulate phagocytosis by human neutrophils (Ventur et al., 1990).

It has been argued (Farries and Atkinson, 1991) that phagocytosis mediated by the binding of C3bi, generated by activation of the alternative pathway of complement, as a host defense against a wide range of pathogens, developed early in evolution, before the appearance of antibodies and Fc receptors. The findings that CR3, through a distinct molecular region, also participates in nonopsonic phagocytosis mediated by type 1 fimbriae or by RGD-containing proteins support the view that these modes of nonopsonic phagocytosis may have developed even earlier, because they do not require complement. The necessity for antibodies to function as opsonins may have developed later in evolution, to cope with the emergence of mutations in microorganisms that either render them unable to activate complement or result in constituents not recognized by any of the receptors for nonopsonic phagocytosis.

The fact that deficiencies in complement components or CR3 adhesion molecules result in recurrent infections by only a limited number of microbial species suggests that the host relies on different types of nonopsonic and opsonic recognition of the invading microorganisms to cope with such infections.

References

Absolom, D.R. 1988. The role of bacterial hydrophobicity in infection: bacterial adhesion and phagocytic ingestion. *Can. J. Microbiol.* **34**:287–298.

Alkan, M.L., L. Wong, and F.J. Silverblatt. 1986. Change in degree of type 1 piliation of *Escherichia coli* during experimental peritonitis in the mouse. *Infect. Immun.* **52**:549–554.

Ashwell, C. and H. Hartford. 1982. Carbohydrate-specific receptors of the liver. *Annu. Rev. Biochem.* **51**:531–554.

Ashwell, G. and A.G. Morell. 1974. The role of surface carbohydrates in the hepatic recognition and transport of circulating glycoproteins. *Adv. Enzymol.* **42**:99–128.

Athamna, A. and I. Ofek. 1988. Enzyme-linked immunosorbent assay for quantitation of attachment and ingestion stages of bacterial phagocytosis. *J. Clin. Microbiol.* **26**:62–66.

Athamna, A., I. Ofek, Y. Keisari, S. Markowitz, G.G.S. Dutton, and N. Sharon. 1991. Lectinophagocytosis of encapsulated *Klebsiella pneumoniae* mediated by surface lectins of guinea pig alveolar macrophages and human monocyte-derived macrophages. *Infect. Immun.* **59**:1673–1682.

Baenzinger, J.N. and Y. Maynard. 1980. Human hepatic lectin. *J. Biol. Chem.* **255**:4607–4613.

Bar-Shavit, Z., I. Ofek, R. Goldman, D. Mirelman, and N. Sharon. 1977. Mannose residues on phagocytes as receptors for the attachment of *Escherichia coli* and *Salmonella typhi*. *Biochem. Biophys. Res. Commun.* **78**:455–460.

Bar-Shavit, Z., R. Goldman, I. Ofek, N. Sharon, and D. Mirelman. 1980. Mannose-binding activity of *Escherichia coli:* a determinant of attachment and ingestion of the bacteria by macrophages. *Infect. Immun.* **29**:417–424.

Bermudez, L.E., L.S. Young, and H. Enkel. 1991. Interaction of *Mycobacterium avium* complex with human macrophages: roles of membrane receptors and serum proteins. *Infect. Immun.* **59**:1697–1702.

Beuth, I., H.L. Ko, and G. Pulverer. 1988. The role of staphylococcal lectins in human granulocyte stimulation. *Infection* **16**:46–48.

Bjorksten, B. and T. Wadström. 1982. Interaction of *Escherichia coli* with different fimbriae and polymorphonuclear leukocytes. *Infect. Immun.* **38**:298–305.

Blumenstock, E. and K. Jann. 1982. Adhesion of piliated *Escherichia coli* strains to phagocytes. Differences between bacteria with mannose-sensitive pili and those with mannose-resistant pili. *Infect. Immun.* **35**:264–269.

Boner, G., A.M. Mhashilkar, M. Rodriguez-Ortega, and N. Sharon. 1989. Lectin-mediated, non-opsonic phagocytosis of type 1 *Escherichia coli* by human peritoneal macrophages of uremic patients treated by peritoneal dialysis. *J. Leuk. Biol.* **46**:239–245.

Courtney, H.S., I. Ofek, W.A. Simpson, and E.H. Beachey. 1981. Characterization of the binding of lipoteichoic acid to polymorphonuclear leucocytes of human blood. *Infect. Immun.* **32**:625–631.

Drickamer, K. 1988. Two distinct classes of carbohydrate-recognition domains in animal lectins. *J. Biol. Chem.* **263**:9557–9560.

Elkins, C. and R.F. Rest. 1990. Monoclonal antibodies to outer membrane protein PII block interactions of *Neisseria gonorrhoeae* with human neutrophils. *Infect. Immun.* **58**:1078–1084.

Ezekowitz, R.A.B., K. Sastry, P. Bailly, and A. Warner. 1990. Molecular characterization of the human macrophage mannose receptor: demonstration of multiple carbohydrate-recognition domains and phagocytosis of yeasts in Cos-I cells. *J. Exp. Med.* **172**:1785–1794.

Ezekowitz, R.A.B., D.J. Williams, H. Koziel, M.Y.K. Armstrong, A. Warner, F.F. Richards, and R.M. Rose, 1991. Uptake of *Pneumocystis carinii* mediated by the macrophage mannose receptor. *Nature* **351**:155–158.

Farrell, C.F. and R.F. Rest. 1990. Up-regulation of human neutrophil receptors for *Neisseria gonorrhoeae* expressing PII outer membrane proteins. *Infect. Immun.* **58**:2777–2784.

Farries, T.C. and J.P. Atkinson. 1991. Evolution of the complement system. *Immun. Today* p. 295–306.

Firon, N., I. Ofek, and N. Sharon. 1983. Carbohydrate specificity of the surface lectins of *Escherichia coli, Klebsiella pneumoniae* and *Salmonella typhimurium*. *Carbohydr. Res.* **120**:235–249.

Firon, N., D. Duksin, and N. Sharon. 1985. Mannose specific adherence of *Escherichia coli* to BHK cells that differ in their glycosylation patterns. *FEMS Microbiol. Lett.* **27**:161–165.

Garber, H., N. Sharon, D. Shohet, J.S. Lam, and R.J. Doyle. 1985. Contribution of hydrophobicity to hemagglutination reactions of *Pseudomonas aeruginosa*. *Infect. Immun.* **50**:336–337.

Gbarah, A., A.M. Mhashilkar, G. Boner, and N. Sharon. 1989. Involvement of protein kinase C in activation of human granulocytes and peritoneal macrophages by type 1 fimbriated (mannose specific) *Escherichia coli*. *Biochem. Biophys. Res. Commun.* **165**:1243–1249.

Gbarah, A., C.G. Gahmberg, I. Ofek, U. Jacobi, and N. Sharon. 1991. Identification of the leukocyte adhesion molecules CD11/CD18 as receptors for type 1 fimbriated (mannose specific) *Escherichia coli*. *Infect. Immun.* **59**:4524–4530.

Goetz, M.B. and F.J. Silverblatt. 1987. Stimulation of human polymorphonuclear leukocyte oxidate metabolism by type 1 pili from *Escherichia coli*. *Infect. Immun.* **55**:534–540.

Goetz, M.B., S.M. Kuriyama, and F.J. Silverblatt. 1987. Phagolysosome formation by polymorphonuclear neutrophilic leukocytes after ingestion of *Escherichia coli* that express type 1 pili. *J. Infect. Dis.* **156**:229–233.

Goldhar, J., M. Yavzori, Y. Keisari, and I. Ofek. 1991. Phagocytosis of *Escherichia coli* mediated by mannose resistant non-fimbrial haemagglutinin (NFA-1). *Microb. Pathogen.* **11**:171–178.

Gotschlich, E. 1983. Thoughts on the evolution of strategies used by bacteria for evasion of host defenses. *Rev. Infect. Dis.* **5**(*suppl 4*):S778–S783.

Guerina, N.G., T.W. Kessler, V.J. Guerina, M.R. Neutra, H.W. Clegg, S. Langermann, F.A. Scannapieco, and D.A. Goldmann. 1983. The role of pili and capsule in the pathogenesis of neonatal infection with *Escherichia coli* K1 *J. Infect. Dis.* **148**:395–405.

Hagberg, L., R. Hull, S. Hull, S. Falkow, R. Freter, and C. Svanborg-Eden. 1983. Contribution of adhesion to bacterial persistence in the mouse urinary tract. *Infect. Immun.* **40**:265–272.

Irvin, R.T. 1990. Hydrophobicity of proteins and bacterial fimbriae. In: Doyle, R.J. and M. Rosenberg (eds.), *Microbial Cell Surface Hydrophobicity*. American Society for Microbiology, Washington, pp. 137–177.

Iwahi, T. and A. Imada. 1988. Interaction of *Escherichia coli* with polymorphonuclear leukocytes in pathogenesis of urinary tract infection in mice. *Infect. Immun.* **57**:947–953.

Kasper, D.L., C.J. Baker, B. Galdes, and E. Katzenellenbogen. 1983. Immunochemical analysis and immunogenicity of the type II group B streptococcal capsular polysaccharide. *J. Clin. Invest.* **72**:260–269.

Keith, B.R., S.L. Harris, P.W. Russell, and P.E. Orndorff. 1990. Effect of type 1 piliation on *in vitro* killing of *Escherichia coli* by mouse peritoneal macrophages. *Infect. Immun.* **58**:3448–3454.

Kelly, N.M., J.L. Kluftinger, B.L. Pasloske, W. Paranchych, and R.E.W. Hancock. 1989. *Pseudomonas aeruginosa* pili as ligands for nonopsonic phagocytosis by fibronectin-stimulated macrophages. *Infect. Immun.* **57**:3841–3845.

King, G.J. and J. Swanson. 1978. Studies on gonococcus infection. XV. Identification of surface proteins of *Neisseria gonorrhoeae* correlated with leukocyte association. *Infect. Immun.* **21**:575–583.

King, G.J., J.F. James, and J. Swanson. 1978. Studies of gonococcus infection. II. Comparison of *in vivo* and *in vitro* association of *Neisseria gonorrhoeae* with human neutrophils. *J. Infect. Dis.* **137**:38–43.

Kurashima, C., A.L. Sandberg, J.O. Cisar, and L.L. Mudrick, 1991. Cooperative complement and bacterial lectin initiated bactericidal activity of polymorphonuclear leuocytes. *Infect. Immun.* **59**:216–221.

Kuriyama, S.M. and F.J. Silverblatt. 1986. Effect of Tamm–Horsfall urinary glycoprotein on phagocytosis and killing of type 1-fimbriated *Escherichia coli*. *Infect. Immun.* **51**:193–198.

Naids, F.L. and R.F. Rest. 1991. Stimulation of human neutrophil oxidative metabolism by nonopsonized *Neisseria gonorrhoeae*. *Infect. Immun.* **59**:4383–4390.

Naids, F.L., B. Belisle, N. Lee, and R.F. Rest. 1991. Interactions of *Neisseria gonorrhoeae* with human neutrophils: studies with purified PII (opa) outer membrane proteins and synthetic opa peptides. *Infect. Immun.* **59**:4628–4635.

Ofek, I. 1989. Lectinophagocytosis mediated by bacterial surface lectins. *Zbl. Bakt. Hgy. A* **270**:449–455.

Ofek, I. and E.H. Beachey. 1979. Lipoteichoic acid-sensitive attachment of group A streptococci to phagocytes. In: Parker, M.T. (ed.), *Pathogenic Streptococci,* Redbooks, Ltd., Chertsey, Surrey, England, pp. 44–46.

Ofek, I. and N. Sharon. 1988. Lectinophagocytosis: a molecular mechanism of recognition between cell surface sugars and lectins in the phagocytosis of bacteria. *Infect. Immun.* **56**:539–547.

Ofek, I. and F.J. Silverblatt. 1982. Bacterial surface structures involved in adhesion to phagocytic and epithelial cells. In: D. Schlessinger (ed.), *Microbiology-1982*. American Society for Microbiology, Washington, pp. 296–300.

Lehrman, A.M. and R.L. Hill. 1986. The binding of fucose-containing glycoproteins by hepatic lectins. Purification of a fucose-binding lectin from rat liver. *J. Biol. Chem.* **16**:7419–7425.

Lock, R., C. Dahlgren, M. Linden, O. Stendahl, A. Svensbergh, and L. Ohman. 1990. Neutrophil killing of two type 1 fimbria-bearing *Escherichia coli* strains: dependence on respiratory burst activation. *Infect. Immun.* **58**:37–42.

Maayan, M.L., I. Ofek, O. Medalia, and M. Aronson. 1985. Population shift in mannose-specific fimbriated phase of *Klebsiella pneumoniae* during experimental urinary tract infection in mice. *Infect. Immun.* **79**:785–689.

Mangan, D.F. and J.S. Snyder. 1979. Mannose-sensitive interactions of *Escherichia coli* with human peripheral leukocytes in vitro. *Infect. Immun.* **26**:520–527.

Marre, R. and J. Hacker. 1987. Role of S- and common type 1 fimbriae of *Escherichia coli* in experimental upper and lower urinary tract infection. *Microb. pathogen.* **2**:223–226.

Miki, Y., S. Ebisu, and H. Okada. 1987. The adherence of *Eikenella corrodens* to guinea pig macrophages in the absence and presence of anti-bacterial antibodies. *J. Periodont. Res.* **22**:359–365.

Ogmundsdottir, H.M. and D.M. Weir. 1976. The characteristics of binding of *Corynebacterium parvum* to glass-adherent mouse peritoneal exudate cells. *Clin. Exp. Immunol.* **26**:334–339.

Ohman, L., J. Hed, and O. Stendahl. 1982. Interaction between human polymorphonuclear leukocytes and two different strains of type 1 fimbriae-bearing *Escherichia coli. J. Infect. Dis.* **146**:751–757.

Ohman, L., K.E. Magnusson, and O. Stendahl. 1985. Mannose-specific and hydrophobic interaction between *Escherichia coli* and polymorphonuclear leukocytes-influence of bacterial culture period. *Acta Pathol. Microbiol. Immunol. Scand. Sect. B* **93**:125–131.

Ohman, L., G. Maluszynska, K.-E. Magnusson, and O. Stendal. 1988. Surface interactions between bacteria and phagocytic cells. *Prog. Drug Res.* **32**:131–147.

Passo, S., S.A. Syed, and J. Silva. 1982. Neutrophil chemiluminescence in response to *Fusobacterium nucleatum. J. Periodont. Res.* **17**:604–613.

Perry, A. and I. Ofek. 1984. Inhibition of blood clearance and hepatic tissue binding of *Escherichia coli* by liver lectin-specific sugars and glycoproteins. *Infect. Immun.* **43**:257–262.

Perry, A., Y. Keisari, and I. Ofek. 1985. Liver cell and macrophage surface lectins as determinants of recognition in blood clearance and cellular attachment of bacteria. *FEMS Microbiol. Lett.* **27**:345–350.

Perry, A., I. Ofek, and F.J. Silverblatt. 1983. Enhancement of mannose-mediated stimulation of human granulocytes by type 1 fimbriae aggregated with antibodies on *Escherichia coli* surfaces. *Infect. Immun.* **39**:1334–1345.

Rauvala, H. and J. Finne. 1979. Structural similarity of the terminal carbohydrate sequence of glycoproteins and glycolipids. *FEBS Lett.* **97**:1–8.

Relman, D., E. Tuomanen, S. Falkow, D.T. Golenbock, K. Saukkonen, and S.D. Wright. 1990. Recognition of a bacterial adhesin by an integrin: macrophage CR3 ($\alpha_m\beta_2$, CD11b/CD18) binds filamentous hemagglutinin of *Bordetella pertussis*. *Cell* **61**:1375–1382.

Rest, R.F., N. Lee, and C. Bowden. 1985. Stimulation of human leukocytes by protein II^+ gonococci is mediated by lectin-like gonococcal components. *Infect. Immun.* **50**:116–122.

Rodriguez-Ortega, M., I. Ofek, and N. Sharon. 1987. Membrane glycoproteins of human polymorphonuclear leukocytes that act as receptors for mannose-specific *Escherichia coli*. *Infect. Immun.* **55**:968–973.

Rottini, G., F. Cian, M.R. Soranzo, R. Albrigo,and P. Patriarc. 1979. Evidence for the involvement of human polymorphonuclear leukocyte mannose-like receptors in the phagocytosis of *Escherichia coli*. *FEBS Lett.* **105**:307–312.

Salmon, J.E., S. Kapur, and R.P. Kimberly. 1987. Opsonin-independent ligation of Fc_2 receptors: the 3G8-bearing receptors on neutrophils mediate the phagocytosis of concanavalin A-treated erythrocytes and nonopsonized *Escherichia coli*. *J. Exp. Med.* **166**:1783–1813.

Sandberg, A.L., L.L. Mudrick, J.O. Cisar, J.A. Metcalf, and H.L. Malech. 1988. Stimulation of superoxide and lactoferrin release from polymorphonuclear leucocytes by the type 2 fimbrial lectin of *Actinomyces viscosus* T14V. *Infect. Immun.* **56**:267–269.

Saukkonen, K.M.J., B. Nowicki, and M. Leinonen. 1988. Role of type 1 and S fimbriae in the pathogenesis of *Escherichia coli:* 018:K1 bacteremia and 018:K1 meningitis in the infant rat. *Infect. Immun.* **56**:892–897.

Saukkonen, K., C. Cabellos, M. Burroughs, S. Prasad, and E. Tuomanen. 1991. Integrin-mediated localization of *Bordetella pertussis* within macrophages: role in pulmonary colonization. *J. Exp. Med.* **173**:1143–1149.

Sauter, S.L., S.M. Rutherfurd, C. Wagener, J.E. Shively, and S.A. Hefta. 1991. Binding of nonspecific cross-reacting antigen, a granulocyte membrane glycoprotein, to *Escherichia coli* expressing type 1 fimbriae. *Infect. Immun.* **59**:2485–2493.

Schaeffer, A.J., W.R. Schwan, S.J. Hultgren, and J.L. Duncan. 1987. Relationship of type 1 pilus expression in *Escherichia coli* to ascending urinary tract infection. *Infect. Immun.* **55**:373–380.

Schwartz, A.L. and D. Rup. 1983. Biosynthesis of the human asialoglycoprotein receptor. *J. Biol. Chem.* **258**:11249–11255.

Shafer, W.M. and R.F. Rest. 1989. Interactions of gonococci with phagocytic cells. *Annu. Rev. Microbiol.* **43**:121–145.

Shepherd, V.L., E.J. Campbell, R.M. Senior, and P.D. Stahl. 1982. Characterization of the mannose/fucose receptor on human mononuclear phagocytes. *J. Reticuloendothelial Soc.* **32**:423–431.

Silverblatt, F.J. and I. Ofek. 1983. Interaction of bacterial pili and leukocytes. *Infection* **11**:235–238.

Silverblatt, F.J., J.S. Dreyer, and S. Schauer. 1979. Effect of pili on susceptibility of *Escherichia coli* type 1 pili and capsular polysaccharides on the interaction between bacteria and human granulocytes. *Scand. J. Immunol.* **20**:299–305.

Soderstrom, T., and L. Ohman. 1984. The effect of monoclonal antibodies against *Escherichia coli* type 1 pili and capsular polysaccharides on the interaction between bacteria and human granulocytes. *Scand. J. Immunol.* **20**:299–305.

Speert, D.P., B.A. Loh, D.A. Cabral, and I.E. Salit. 1986. Nonopsonic phagocytosis of nonmucoid *Pseudomonas aeruginosa* by human neutrophils and monocyte-derived macrophages is correlated with bacterial piliation and hydrophobicity. *Infect. Immun.* **53**:207–212.

Speert, D.P., D.W. Samuel, S.C. Silverstein, and M. Bernadette. 1988. Functional characterization of unopsonized *Pseudomonas aeruginosa. J. Clin. Invest.* **82**:872–879.

Steadman, R., N. Topley, D.E. Jenner, M. Davies, and J.D. Williams. 1988. Type 1 fimbriate *Escherichia coli* stimulates a unique pattern of degranulation by human polymorphonuclear leukocytes. *Infect. Immun.* **56**:815–822.

Sutherland, I.W., L. Graham, and D.M. Weir. 1978. The role of cell wall carbohydrates in binding microorganisms to mouse peritoneal exudate macrophages. *Acta. Pathol. Microbiol. Scand. Sect. B* **86**:53–57.

Svanborg-Eden, C.F., L.J. Bjursten, R. Hull, K.E. Magnusson, Z. Meldoveno, and H. Leffler. 1984. Influence of adhesins on the interaction of *Escherichia coli* with human phagocytes. *Infect. Immun.* **44**:672–680.

Swanson, J., E. Sparks, B. Zeligs, M.A. Siam, and C. Parrott. 1974. Studies on gonococcus infection. V. Observations on *in vitro* interactions of gonococci and human neutrophils. *Infect. Immun.* **10**:633–644.

Swanson, J., E. Sparks, D. Young, and G. King. 1975. Studies on gonococcus infection. X. Pili and leukocyte association factor as mediators of interactions between gonococci and eukaryotic cells *in vitro*. *Infect. Immun.* **11**:1352–1361.

Taylor, M.E., J.T. Conary, M.R. Lennartz, P.D. Stahl, and K. Drickamer, 1990. Primary structure of the mannose receptor contains multiple motifs resembling carbohydrate-recognition domains. *J. Biol. Chem.* **265**:12156–12162.

Van Oss, C.J. 1978. Phagocytosis as a surface phenomenon. *Annu. Rev. Microbiol.* **32**:19–39.

Van'T Wout, J., W.N. Burnette, V. Mar, P. Gordon, S.D. Wright, and E. Tuomanen. 1991. The role of pertussis toxin subunits in adherence of *Bordetella pertussis* to human macrophages. *Abstr., General meeting,* American Society for Microbiology, 113.

Ventur, Y., J. Scheffer, J. Hacker, W. Goebel, and W. Konig. 1990. Effects of adhesins from mannose-resistant *Escherichia coli* on mediator release from human lymphocytes, monocytes and basophils and from polymorphonuclear granulocytes. *Infect. Immun.* **58**:1500–1508.

Virji, M. and J.E. Heckels. 1986. The effect of protein II and pili on the interaction of *Neisseria gonorrhoeae* with human polymorphonuclear leukocytes. *J. Gen. Microbiol.* **132**:503–512.

Warr, G.A. 1980. A macrophage receptor for (mannose/glucosamine) glycoprotein of potential importance in phagocytic activity. *Biochem. Biophys. Res. Commun.* **93**:737–945.

Wileman, T.E., M.R. Lennartz, and P.D. Stahl. 1986. Identification of the macrophage mannose receptor as a 175-kDa membrane protein. *Proc. Natl. Acad. Sci. USA* **83**:2501–2505.

Wilson, M.E. and R.D. Pearson. 1988. Roes of CR3 and mannose receptor in the attachment and ingestion of *Leishmania donovani* by human mononuclear phagocytes. *Infect. Immun.* **56**:363–369.

8

Adhesion of Bacteria to Oral Tissues

The oral environment contains many kinds of bacteria, including both Gram-negative and Gram-positive cocci, bacilli, and spirochetes. More than 300 distinct species of bacteria may exist in the oral cavity. Some of these can be readily cultured and identified, whereas others can be cultured only with difficulty. Some may not be cultured at all. In ecological terms, the oral environment is a perfect niche for some bacteria. Frequently, the bacteria encountered in the oral environment are not found elsewhere in the body. In the mouth, bacteria are challenged by the turbulent effects of saliva; antibacterial proteins in saliva, such as lysozyme and lactoferrin; immunoglobulins; products from other microorganisms; and dietary constituents. Some of these factors influence adherent reactions, details of which will be discussed below.

The most common infectious diseases known are those that affect teeth and periodontal (gums, gingival) tissues. Dental caries is a disease resulting in the local demineralization of teeth. Periodontitis is an inflammation of the gingival tissue and periodontal disease is the loss of integrity of attachment sites between the tooth and gingiva. In all cases, before pathology can be observed, bacteria must adhere and form dental plaques. Dental plaque is usually composed of salivary constituents, bacteria (characteristically a mixed microbiota), bacterial metabolites, and foodstuffs.

Dental plaque forms when salivary constituents bathe the surfaces of teeth and gingival tissues to form a conditioned surface, or pellicle (Leach, 1970; Abbott and Hayes, 1984). The pellicle forms a substratum for the adhesion of bacteria. Once adherent, the bacteria may divide, trap foodstuffs and salivary components, and thereby colonize the tissues. There are different kinds of dental plaque. Coronal plaque refers to deposits on the smooth surfaces of teeth. Figure 8–1 shows an idealized representation of a tooth and its association with gingival tissue. Dental plaque resides only on the hard tissues, and surfaces such as enamel, dentures, fillings, and artificial crowns. The mucosal tissues of the oral

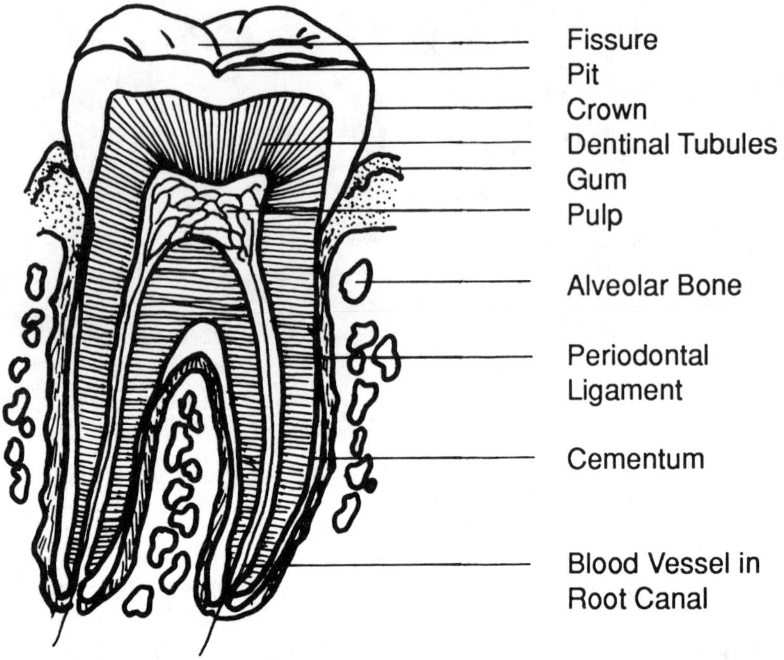

Figure 8–1. Idealized cross-section of a mandibular molar tooth. Adhesion of bacteria on the hard surfaces may lead to dental decay. Periodontal disease may occur as a result of bacterial adhesion near the gingival (gum)–tooth interface. (Adapted from the NIH-NIDR Long-Range Research Plan, 1990.)

cavity constantly shed or desquamate, whereas the hard tissues are reasonably stable. Sites particularly prone for plaque development include fissures, approximal surfaces (surfaces between teeth), and gingival crevices (Figure 8–1).

Normal microbiota (sometimes called commensal, autochthonous, or resident microbiota) is a term used to describe the spectrum of microorganisms found in normal healthy sites (Table 8–1). It is now clear that certain generalizations can be made concerning the normal microbiota of the oral cavity. Streptococci predominate on hard tissues, as well as on mucosa. The mutans streptococci, which include *S. mutans, S. sobrinus, S. cricetus,* and *S. rattus,* are frequently found in carious lesions, but other streptococci are also commonly isolated from plaques and areas of tooth decay. *S. salivarius* is found in high numbers in saliva and on oral mucosa. The oral streptococci are usually aerotolerant. Members of the genus *Actinomyces* are also common inhabitants of hard tissues, but are frequently found in high numbers in diseased gingival and subgingival sites. The gingival crevice contains the largest numbers of bacteria, at least in terms of genera and species.

Table 8–2 shows the major early and secondary colonizers of the tooth surface.

Table 8–1. Bacteria frequently found in the oral environment and their association with oral diseases

Gram-positive bacteria		Gram-negative bacteria	
Aerobes, facultative anaerobes, microaerophils	Anaerobes	Aerobes, facultative anaerobes, microaerophils	Anaerobes
		Cocci	
Enterococcus			
E. faecium		*Branhamella*	
Staphylococcus (TB)		*B. catarrhalis*	
S. epidermidis		*Moraxella* spp.	
Streptococcus (CC)	*Peptostreptococcus*	*Neisseria*	*Veillonellae (PD)*
S. anginosus	*P. anaerobius*	*N. flavescens*	*V. atypica*
S. cricetus	*P. micros*	*N. mucosa*	*V. dispar*
S. gordonii		*N. sicca*	*V. parvula*
S. mitis		*N. subflava*	
S. morbillorum			
S. mutans			
S. oralis			
S. parasanguis			
S. rattus			
S. salivarius			
S. sanguis			
S. sobrinus			
S. vestibularis			
Stomatococcus			
S. mucilaginosus			
		Rods	
Actinomyces (RSC)	*Actinomyces*	*Haemophilus*	*Bacteroides (PD)*
A. naeslundii	*A. israelii*	*H. aphrophilus*	*B. buccae*
A. odontolyticus	*A. meyeri*	*H. parahemolyticus*	*B. buccalis*
A. viscosus	*Bifidobacterium*	*H. parainfluenzae*	*B. capillosus*
Arachnia	*B. dentium*	*H. paraphrophilus*	*B. denticola*
A. propionica		*H. segnis*	*B. endodontalis*
			B. forsythus
Bacillus (TB)			*B. gracilis*
B. cereus			*B. heparinolyticus*
Corynebacterium		*Actinobacillus (PD)*	
C. matruchotii	*Eubacterium*	*A. actinomycetemcomitans*	
Bacterionema	*E. alactolyticum*	*Eikenella (PD)*	
Rothia	*E. brachy*	*E. corrodens*	
R. dentocariosa	*E. nodatum*		
	E. saburreum		
	E. timidum		
	Propionibacterium		
	P. acnes		
Lactobacillus		*Campylobacter*	
L. brevis		*C. concisus*	
L. buchneri		*C. sputorum*	

continued

Table 8–1. Continued

Gram-positive bacteria		Gram-negative bacteria	
Aerobes, facultative anaerobes, microaerophils	Anaerobes	Aerobes, facultative anaerobes, microaerophils	Anaerobes
Lactobacillus			
L. casei		*Capnocytophaga (PD)*	*B. oralis*
L. fermentum		*C. gingivalis*	*B. oris*
L. plantarum		*C. ochracea*	*B. oulorum*
L. salivarius		*C. sputigena*	*B. pneumosintes*
		Pseudomonas (TB)	*B. ureolyticus*
		P. aeruginosa	*B. veroralis*
		P. cepacia	*B. zoogleoformans*
			Centipeda
			C. periodontii
		Escherichia (TB)	*Fusobacterium*
		E. coli	*F. naviforme*
		Simonsiella (TB)	*F. nucleatum*
			F. periodonticum
			Leptotrichia
			L. buccalis
			Porphyromonas (Bacteroides) gingivalis
			Prevotella (PD)
			P. loescheii (formerly Bacteroides loescheii)
			Selenomonas (PD)
			S. flueggli
			S. infelix
			S. noxia
			S. sputigena
			Wolinella
			W. curva
			W. recta
			W. succinogenes
		Wall-less bacteria	Spirochetes
		Mycoplasma	*Treponema*
		M. orale	*T. denticola*
		M. pneumoniae	*T. macrodentium*
		M. salivarium	*T. oralis*
			T. pectinovorum
			T. scoliodontium
			T. socranskii
			T. vincentii

Abbreviations: CC, coronal caries; RSC, root surface caries; PD, probable periodontopathogen; TB, probable transient bacterium.

Adapted from Theilade, 1990.

Table 8–2. Primary and secondary colonizers of the tooth surface

Bacterium	Fimbriae	Fibrils	Surface Lectin	Glucan (G) or Levan (L) from sucrose
Primary Bacterial Colonizers				
Streptococcus sanguis	F	M	F	G,L
S. oralis	F	M	F	G,L
S. gordonii	F	M	F	G,L
S. mitis	F	M	F	G,L
S. mutans	N	F	F	G,L
S. sobrinus	N	N	M	G,L
S. cricetus	N	N	M	G,L
S. milleri (S. anginosus)	M	N	M	G,L(F)
S. salivarius	M	N	N	G,L
Actinomyces israelii	M	F	F	L(F)
Actinomyces naeslundii	M	N	F	L(F)
Actinomyces viscosus	M	N	F	L(F)
Haemophilus spp.	F	N	F	—
Veillonella spp.	F	N	F	—
Secondary Bacterial Colonizers				
Prevotella loescheii[a]	M	M	F	N
Porphyromonas gingivalis[b]	M	M	F	N
Actinobacillus actinomycetemcomitans	M	N	N	N
Fusobacterium nucleatum	M	N	M	N
Eubacterium nodatum	F	F	N	N
Capnocytophaga gingivalis (or C. ochracea)	F	F	F	N
Selenomonas spp.			F	N
Wolinella spp.	F	N	F	N
Propionibacterium acnes				N
Corynebacterium (Bacterionema)				N
matruchotii	M	N	M	N
Eikenella corrodens				

Adapted from Socransky et al., 1977; Slots et al., 1978; Syed and Loesche, 1978; Gibbons, 1984; Cisar et al., 1985; Loesche et al., 1985; Kolenbrander and Andersen, 1986, 1988; Nyvad and Kilian, 1987; Nyvad and Fejerskov, 1987a,b; Kolenbrander, 1988, 1989, 1991; and Handley 1990.

Abbreviations: M, Most strains; F, few strains; N, none of the strains. See also Figure 8–2.

[a]Formerly *Bacteroides loescheii.*

[b]Formerly *B. melaninogenicus* subsp. *asaccharolyticus.*

Early colonization refers to those adherent bacteria found on the tooth surface following professional cleaning. *S. sanguis* and *S. sanguis*-like (*S. gordonii, S. mitis, S. oralis, S. parasanguis*) streptococcci seem to have the greatest ability to adhere to saliva-coated teeth. The mutans streptococci adhere in great numbers only when sucrose is present. Secondary colonizers tend to be anaerobic to aerotolerant and include both Gram-positive and Gram-negative bacteria. As described later, the secondary colonizers may adhere to primary colonizing bacteria, creating complex ecological interrelationships between numerous bacterial types. When teeth are extracted, only the organisms that normally adhere to mucosa remain in large numbers.

Organisms such as *S. oralis, S. salivarius, S. mitis, Veillonella* spp, *Peptostreptococcus* spp, and *Bacteroides* spp, do not require teeth or subgingival or supragingival plaques in order to maintain their presence in the mouth. Table 8–3 outlines some of the major characteristics of the better studied oral bacteria. It is clear that sucrose promotes the adhesion of the mutans streptococci to teeth. The reasons for the sucrose-dependent adhesion are discussed in later sections, but the adhesion seems to depend on the synthesis of glucans (dextrans, predominantly α-1,6-rich glucans, although α-1,3-rich glucans are synthesized on some streptococci).

There are different kinds of plaque, the microbiota of which depends on the site of the plaque deposit. Coronal or smooth surface plaque consists of the early colonizing bacteria, but as the plaque ages, numerous genera and species may be found (Figure 8–2). Approximal surface plaque (between teeth) consists of streptococci, *Actinomyces* spp., *Veillonella, Bacteroides (Porphyromonas), Fusobacterium* spp, and others. Supragingival plaque, situated at and just above the gingival margin, contains mainly members of the genera *Actinomyces, Streptococcus, Bacteroides, Prevotella, Veillonella,* and many others. Subgingival plaque, situated below the margin or in the gingival crevice and periodontal pocket, is also abundant in anaerobic actinomycotic forms and streptococci, *Actinobacillus actinomycetemcomitans,* treponemes, fusiforms, *Eubacterium,* and other anaerobes. Gingivitis may arise from plaques associated with the gingival tissues, although the exact mechanisms eliciting the inflammation are not well-defined. Periodontal disease (Figure 8–3) may result if the plaque depos-

Table 8–3. Characteristics of selected oral bacteria

	Adhesion to Pellicle						
Organisms	With Sucrose	Absence of Sucrose	Adhesion to Cheeks, Tongue	GTF(s)	FTF(s)	GBL	AP
Mutans steptococci	H	L	L	+	+	S	H
Sanguis group	H	H	I	+	+	–	H
Streptococcus salivarius	I	L	H	+(L)	+	–	I
Lactobacilli	L	L	L	–	–	–	H
Actinomyces spp.	H	H	L	–	+	–	L
Neisseria (Moraxella) spp.	L	L	L	–	–	–	–
Veillonella spp.	L	L	I	–	–	–	–
Haemophilus spp.	L	L	I	–	–	–	–

Arbitrary rankings: H = high; L = low; I = intermediate; S = some species and strains. Phenotypic characteristics are highly strain dependent. GTF, glucosyltransferase; FTF, fructosyltransferase; GBL, glucan-binding lectin; AP, acid production.

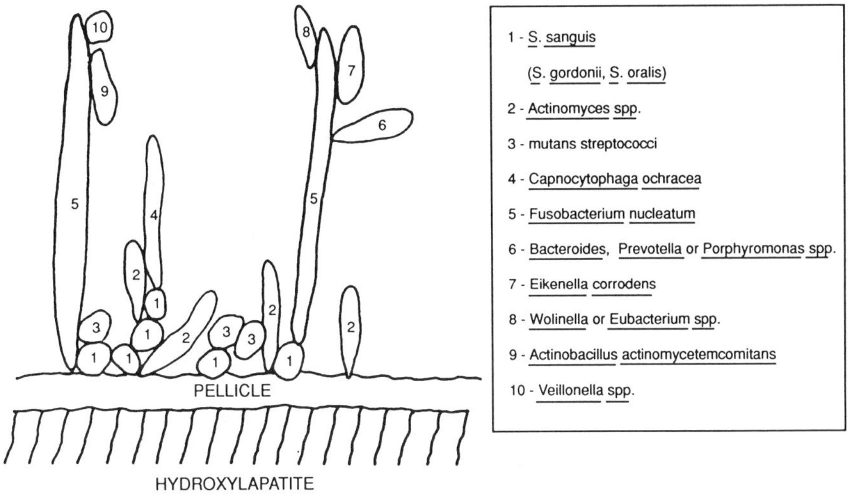

Figure 8–2. Ecology of microbiota on the surface of a tooth. Streptococci and actinomycetes tend to be the initial colonizers of freshly cleaned areas of the tooth. Subsequent adhesion of bacteria to bacteria contributes to the formation of dental plaque. Plaque is required for the development of dental caries or periodontal disease (Figure 8–3.)

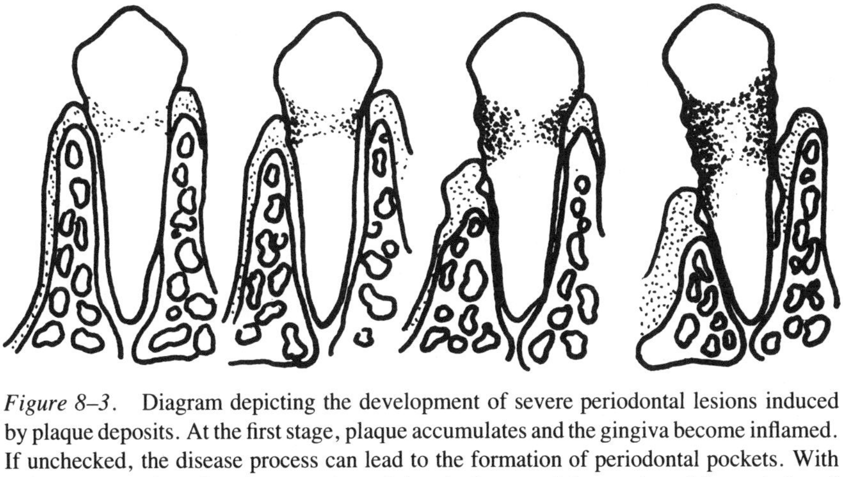

Figure 8–3. Diagram depicting the development of severe periodontal lesions induced by plaque deposits. At the first stage, plaque accumulates and the gingiva become inflamed. If unchecked, the disease process can lead to the formation of periodontal pockets. With further progression, there is resorption of alveolar bone and destruction of the periodontal ligament. As loss of attachment proceeds, the tooth loosens and eventually is lost. (Adapted from the NIH-NIDR Long-Range Research Plan, 1990.)

it(s) is (are) remain intact. Adherent reactions, regardless of the type of plaque, are important in steady-state maintenance of plaque (Hudson and Curtiss, 1990).

Some Mechanisms of Oral Streptococcal Adhesion to Hard Tissues

Oral bacteria have a variety of substrata with which to interact. The bacteria can bind to saliva-coated enamel, cementum surfaces, periodontal tissues, root surfaces of teeth, the dorsum of the tongue, the cheeks, to other bacteria, and to fungi and protozoa (Gibbons and Qureshi, 1978; Cisar et al., 1979, 1989; Doyle et al., 1982; Cisar, 1986; Lantz et al., 1986; Seow et al., 1987, 1989; Winkler et al., 1987; Babu et al., 1991). Research on microbial adhesion to various oral tissues is now entering the fourth decade. Yet, there are very few adhesion mechanisms that have been well-characterized. In most of the studies on oral adhesion, two model systems have been employed. The most widely used model has been saliva-coated hydroxylapatite (SHA) or hydroxylapatite coated with buffers, proteins, or other substances. This system is amenable to a variety of manipulations, including the addition of inhibitors, kinetic constant (on and off constants) determinations, and competition. The other model consists of intergeneric coaggregation between various oral bacteria. This model is also amenable to the introduction of experimental variables. The model permits the determination of the specificity of coaggregation, genetic manipulations, and the introduction of inhibitory molecules. Importantly, it may represent a model that reflects the accretion of microorganisms during the maturation of plaques.

In dental caries, there is a compelling need to characterize the adhesion of the mutans streptococci, *S. sanguis* and *S. sanguis*-related bacteria. These bacteria, more than any other(s), seem to be responsible for coronal caries. Clark et al. (1978), employing SHA beads, found that several strains of the mutans streptococci adhered at very low levels. Nesbitt et al. (1982a) found that the adhesion of *S. sanguis* was not a function of the density of cells employed in the assay. It is generally accepted that *S. sanguis* adheres much better to SHA than the mutans streptococci (Hamada and Slade, 1980). Several mechanisms have been reported to account for the adhesion of oral streptococci (Table 8–4). These include the participation of surface lectins, glucosyltransferases (GTFs) and fructosyltransferases (FTFs), lipoteichoic acids, and hydrophobins. No one mechanism satisfactorily explains streptococcal adhesion. The most prominent mechanisms will be discussed in the following sections.

Lectin-dependent Streptococcal Adhesion

Oral streptococci are confronted with numerous carbohydrate-containing molecules. They may interact with host salivary and cellular glycoproteins, other microorganisms, polymers synthesized by other microorganisms, or with food-

Table 8–4. Some mechanisms reported to be involved in the adhesion of streptococci to the surfaces of teeth

Mechanism	Comment(s)
Lectin-dependent streptococcal adhesion	Lectins have been reported to be involved in adhesion of various oral streptococci to pellicle and coaggregation with *Actinomyces* spp.
GTF-dependent adhesion	GTF on pellicle synthesizes glucan in presence of sucrose, to which some mutans streptococci may bind.
Glucan-dependent composite adhesion	Adherent *S. sanguis* may produce α-1,6 glucan in presence of sucrose. The glucan may serve as substratrum for adhesion by streptococci expressing the glucan-binding lectin.
Lipteichoic acid-dependent adhesion	Lipoteichoic acid may mediate between pellicle and bacterium. There may be a role for fibronectin in oral streptococcal adhesion.
Hydrophobic effect-dependent multiple site mechanism	Adhesion of *S. sanguis* to pellicle is thought to involve multiple sites, possibly initiated and/or stabilized by the hydrophobic effect.

stuffs. There are now several reports confirming the lectin activities of oral streptococci. Levine et al. (1978) found that strains of *S. sanguis* and *S. mutans* were aggregated by a salivary protein of 150,000 mol wt. When the salivary protein was treated with sialidase, aggregation reactions with *S. sanguis* were abolished. It was suggested that *S. sanguis* possessed a surface lectin capable of complexing with terminal sialic acid residues of certain salivary glycoproteins. Later, Murray et al. (1982) used hemagglutination inhibition to determine that *S. sanguis* possessed a lectin specific for short-chain oligosaccharides. The established affinities were NeuNAcα2,3Galβ1, 3GalNAcol $>>$ NeuNAcα2,3Galβ1, 4Glc $>>$ NeuNAc $>$ Gal. Later, it was found (Morris and McBride, 1983) that serum and crevicular fluids were capable of aggregating *S. sanguis*. The streptococci failed to aggregate when the proteins were treated with sialidase or when gangliosides were present in the mixtures. Further studies (Cowan et al., 1987a) revealed that sialic acid contributed very little to the initial interaction between pellicle protein and the bacteria, but sialic acid did confer stability to *S. sanguis* adhesion to SHA.

Demuth et al. (1990) have been able to provide a deduced amino acid sequence of a sialic acid-binding protein of *S. sanguis*. The lectin (160,000 mol wt) consists of three unique structural domains, two of which possess repetitive amino acid sequences. The N-terminal domain is comprised of four tandem copies of 82 residues and is similar to a domain of M-protein of group A streptococci. On the basis of sequence, this region is predicted to be highly α-helical. A second domain of 39 residues is repeated three times and is rich in proline. The third domain, near the C-terminus, is nearly one-half proline. The *S. sanguis* lectin sequence is highly homologous to a surface antigen of some of the mutans streptococci. Aggregation assays confirmed that the lectin was specific for sialoglycopeptides. The lectin may be an attractive candidate for a vaccine, particularly because of its structural similarly with a surface antigen of the mutans streptococci.

A saliva-binding protein of an *S. sanguis* strain has recently been sequenced (Ganeshkumar et al., 1991). This protein, which exhibits lectin-like properties, but has not yet been proven to complex with carbohydrates, has a mol wt of about 36,000. The protein is highly hydrophilic and does not appear to have a hydrophobic membrane anchor sequence. Earlier, Fenno et al. (1989) cloned a fimbrial protein of *S. sanguis* and determined it to be 87% homologous to the saliva-binding protein described by Ganeshkumar et al. Antiserum directed against the saliva-binding protein also reacted with an *S. sanguis* protein capable of binding to *Actinomyces naeslundii*. These results suggest that the saliva-binding protein of *S. sanguis* is an adhesin for other oral bacteria.

It has been reported (Gibbons and Qureshi, 1979) that the mutans streptococci could be prevented from adhering to SHA by galactose and melibiose and by amines, such as spermine. These and other studies (Gibbons et al., 1983b) suggest that at least two mechanisms could be involved in adhesion of the mutans streptococci to SHA. One mechanism is lectin-dependent and is specific for galactose residues. The other is probably dependent on ionic interactions between basic and acidic groups on the cell surface and pellicle. Nesbitt et al. (1982b) would add hydrophobicity (see below) to the charge–charge interactions to cause stabilization of oppositely charged pairs. As far as is known, other workers have not followed up the possibility that the mutans streptococci express galactophilic lectins. Desialyated salivary proteins would possess many terminal galactose residues with which lectins could combine.

Many of the mutans streptococci are known to express cell surface α-glucan-binding proteins or lectins (Gibbons and Fitzgerald, 1969; Kelstrup and Funder-Nielsen, 1974; McCabe and Smith, 1975; Russell, 1979, 1990; Russell et al., 1983, 1985, 1986; Drake et al., 1988a,b; Lü-Lü *et al.*, 1992). The glucan-binding lectins (GBL) serve as adhesins for the bacteria. The GBLs may be assayed by aggregation of intact bacteria following addition of a high mol wt α-1,6 glucan. Changes in absorbance of the cellular suspensions can be plotted as first-order rate plots, the slopes of which can be used as an index of lectin activity (Drake et al., 1988a). Figure 8–4 shows the aggregation of *S. cricetus* by various concentrations of a high molecular weight linear glucan. The assays are amenable to inhibition studies employing the appropriate oligosaccharides (Figure 8–5). Drake et al. (1988a) found that the GBLs of *S. sobrinus* 6715 and *S. cricetus* AHT could accommodate isomaltosaccharides of 6–10 hexose residues. In independent experiments, Landale and McCabe (1987), employing electrophoretic mobility measurements of a purified GBL, concluded that only α-1,6 glucans of 8–11 hexose residues would optimally fit the combining site of *S. sobrinus*. The large size of the lectin combining site seems logical, as foodstuffs would be unlikely

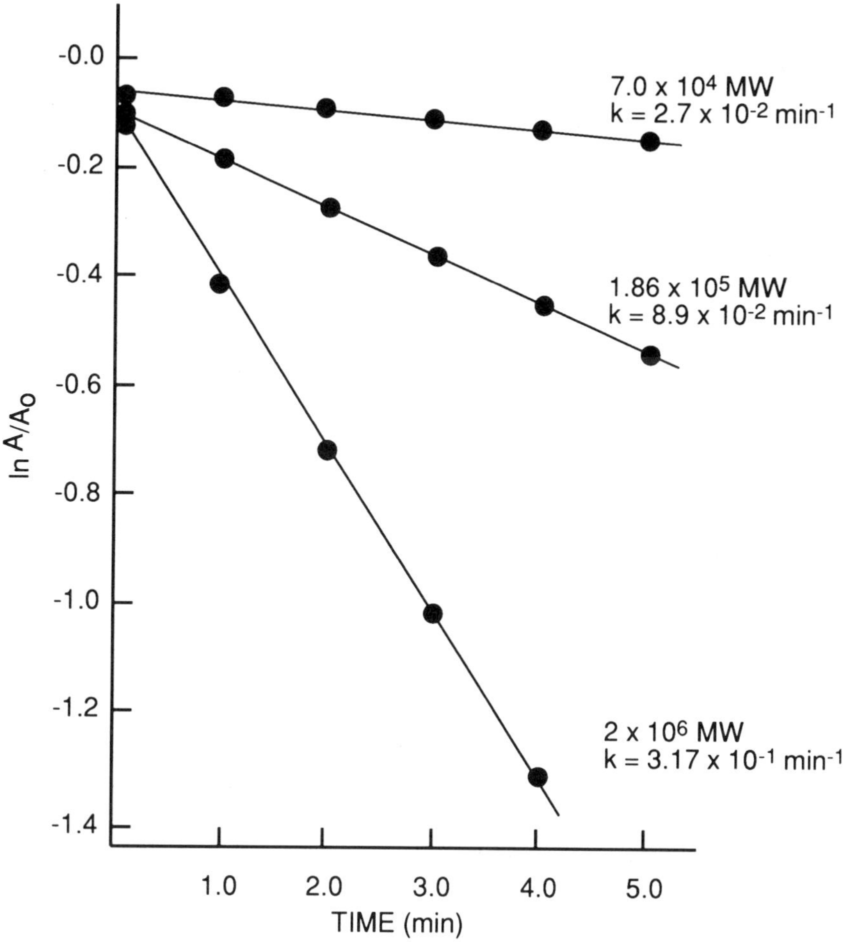

Figure 8–4. Agglutination of *S. cricetus* AHT mediated by high molecular weight glucans. *S. cricetus* AHT was cultured in a sucrose-free medium in a 5% CO_2 atmosphere at 37°C. Cell suspensions in buffer were assayed by the standard rate assay. Rate constants (k) are slopes calculated from the linear regression statistical program. A, absorbance at each time point, A_0, absorbance at time zero; MW, molecular weight. (Taken from Drake et al., 1988a.)

to compete effectively. Starches and amyloses rich in α-1,4 glucose linkages, and common dietary constituents, were unable to complex with the streptococcal GBLs (Table 8–5).

The GBLs of oral streptococci not only are inhibited by low molecular weight α-1,6 glucans, but also their activities are abolished or inhibited by proteases, heat, chaotropes, and chelating agents (Drake et al., 1988b; Liang et al., 1989).

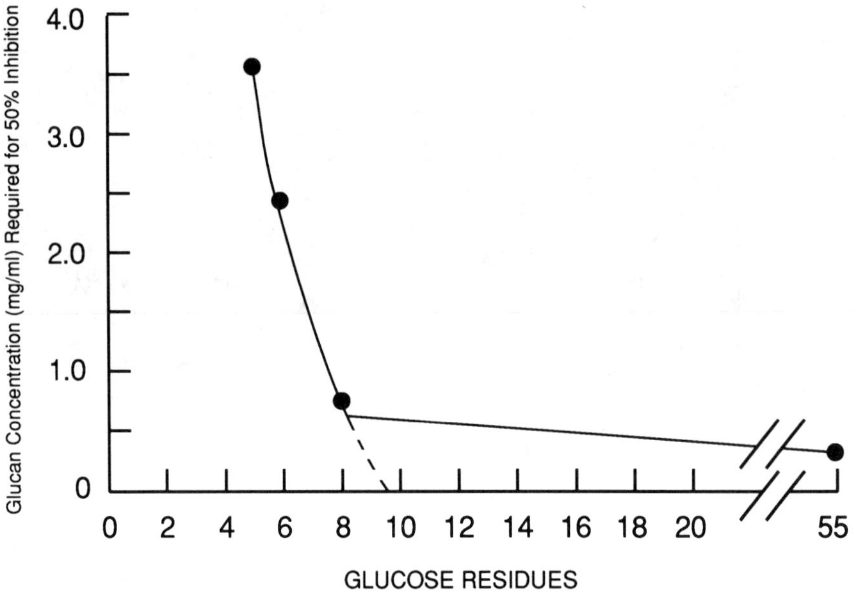

Figure 8–5. Determination of the minimal isomaltooligosaccharide chain length to inhibit glucan T2000-mediated aggregation of *S. cricetus*. Suspensions of *S. cricetus* AHT were incubated with the prospective inhibitors and assayed for glucan T2000-induced aggregation. IM-8, isomaltooctaose; IM-6, isomaltohexaose; IM-5, isomaltopentaose; MW, molecular weight. Inhibition data obtained with glucan T10 and the isomaltosaccharides were plotted in terms of the concentration of inhibitor needed to achieve a 50% level of inhibition versus the number of glucose residues of each inhibitor. Note that isomaltotriose (data not plotted) essentially represents infinity as no inhibition was observed even at concentrations up to 33 mg/ml. (From Drake et al., 1988a. Reprinted with permission of the authors and the American Society for Microbiology.)

Glucan partially protects against inactivation by these agents. Interestingly, the inhibition of the GBLs by chelating agents is reversible. Agents such as EDTA, citrate, lactate, and acetylacetone are capable of reversibly inhibiting the GBLs. The manganous ion is also essential for expression of GBL activity (Drake et al., 1988b). At present, it is not known if the GBLs of the mutans streptococci are manganoproteins, although many plant lectins require Mn^{2+} for activity. The literature supports a role for Mn^{2+} in promoting dental decay (Beighton, 1982, 1983; Curzon, 1983). It seems possible that there is a correlation between dental decay and the requirement for Mn^{2+} ion in GBL activity.

Preparations of GBL can be obtained by affinity chromatographic procedures employing insolubilized dextran (Sephadex) (McCabe et al., 1976, 1977; McCabe and Hamelik, 1978; Russell, 1979; Russell et al., 1983, 1985). Usually, culture supernatants are poured over the dextran column which is then eluted with buffers. The GBL may then be dissociated from the column by chaotropes. In

Table 8–5. Specificity of the glucan-binding lectin of Streptococcus sobrinus *6715*

Polysaccharide	Linkage	Aggregates with *S. sobrinus*
Amylose	Glucan (α-1,4)	–
Arabinan	Branched (α-1,5; α-1,3)	–
Cellulose	Glucan (β-1,4)	–
Dextran (insoluble) from *S. sobrinus*	Glucan (>90% α-1,3)	–
Dextran T-2000	Glucan (high molecular weight α-1,6 glucan)	+
Dextran B-1208	Glucan containing 95% α-1,6 and 5% other linkages	+
Dextran B-1355(s)	Soluble glucan containing 45% α-1,6 and 55% other linkages	–
Fucan (sulfated)	Fucose (α-1,4 linkages)	–
Galactan	Galactose (β-1,3)	–
Laminaran	Glucan (β-1,3,β-1,6)	–
Levan (fructan)	Fructose (2,6 linkages)	–
Lichenan	Glucan [$(\alpha\text{-}1,4)_2\alpha\text{-}1,3$]	–
Mannan	Branched (Man α-1,2; Man α-1,3; Man α-1,6)	–
Nigeran	Glucan (α-1,3,α-1,4)	–
Pullulan	Glucan (α-1,4,α-1,6)	–

Taken partially from Drake et al., 1988a.

When the polysaccharides were insoluble, they were sonicated until even suspensions were obtained. Aggregation was measured by both turbidimetric techniques and microscopic methods.

some cases, GFT and/or FTF enzymes contaminate the eluted GBL fraction. Elution with low molecular weight α-1,6 glucan is the superior method for preparing GBL. A GBL preparation from a mutans streptococcus has been reported to contain FTF activity (Russell et al., 1983). More recently, Aduse-Opoku et al. (1989) cloned a GBL from *S. mutans* and found that the protein could not synthesize levan from sucrose. They also showed that the gene for the GBL was distinct from an FTF gene in the same bacterium. Drake et al. (1988a) observed that high molecular weight glucan would not aggregate *S. mutans* Ingbritt, the strain used by Russell et al. (1983). Banas et al. (1990) sequenced the GBL of *S. mutans* Ingbritt using molecular biology techniques. The protein had a calculated molecular weight of 59,039. The carboxyl-terminal region had two series of repeated sequences which were homologous to sequences found in a gene from *S. downei* that is responsible for soluble glucan synthesis (Gilmore et al., 1990).

A common feature of the GTFs is that they possess glucan-binding domains (Mooser and Wong, 1988; Russell, 1990). It is unknown if these glucan-binding domains of GTF (and FTF) enzymes can participate in adhesion reactions. Frequently, in the presence of sucrose, extracellular sucrose-metabolizing enzymes become cell-bound (Townsend-Lawman and Bleiweis, 1991). The cell-bound

GTFs (or FTF) could then function as lectins providing the glucan-binding domains were exposed. The participation of GBL in streptococcal adhesion depends on the conversion of sucrose to glucans by glucosyltransferases of the streptococci. As such, the GTFs provide a substratum for the GBLs.

Coaggregation between members of the genus *Actinomyces* and oral streptococci has been known for several years (a later section reviews coaggregation reactions of oral bacteria). The *Actinomyces* are early colonizers of freshly cleaned tooth surfaces (Table 8–2) and may provide substrata for other bacteria. Bourgeau and McBride (1976) found that *A. viscosus* could bind to *S. mutans* in the presence of α-1,6 glucan. This binding reaction was determined to be glucan-dependent and is probably not related to direct co-aggregation reactions. In the absence of glucan, the two bacteria would not coaggregate. This kind of reaction mechanism could enhance streptococcal adhesion in the presence of sucrose, but has not been studied in detail.

Glucosyltransferase (GTF)-dependent Adhesion

Most oral streptococci secrete GTFs and/or FTFs. These enzymes convert sucrose into glucans and fructans, respectively. It is now clear that FTFs can form a part of normal pellicle (Rolla et al., 1980, 1984, 1985; Schilling and Bowen, 1988). In the presence of sucrose, glucans may be formed on the pellicle surface and create new receptors for bacteria bearing glucan-binding proteins. *S. mutans* GS-5 adhered in much higher numbers to SHA in the presence of sucrose than in the absence of the disaccharide (Schilling et al., 1989). The sucrose-dependent adhesion was increased at all starting cell densities. Interestingly, low molecular weight α-1,6 glucan tended to inhibit the sucrose-induced adhesion to SHA and to glass (Figure 8–6). This is consistent with the view that low molecular weight glucans inhibit the glucan-binding lectins of oral streptococci (Drake et al., 1988b).

S. mutans GS-5 is not readily aggregated by high molecular weight glucans (Drake et al., 1988a), although the bacterium and other mutans streptococci may secrete a GBL into the growth medium (Banas et al., 1990). It is possible that *S. mutans* can adhere to glass, and synthesize glucans in the presence of sucrose which then can interact with secreted GBLs. The end result would be more analogous to entrapment of organisms rather than true adhesion to a substratum. Although speculative, it does explain the massive build-up of cells and polysaccharides onto surfaces such as glass or steel. Figure 8–7 outlines a possible means for GTF-dependent adhesion of streptococci to SHA. The key feature is that adhesion depends on glucan and glucan-binding proteins. *S. sobrinus* and *S. cricetus* possess the most active GBLs, but some isolates of *S. mutans* may also express GBL activities (Drake et al., 1988a).

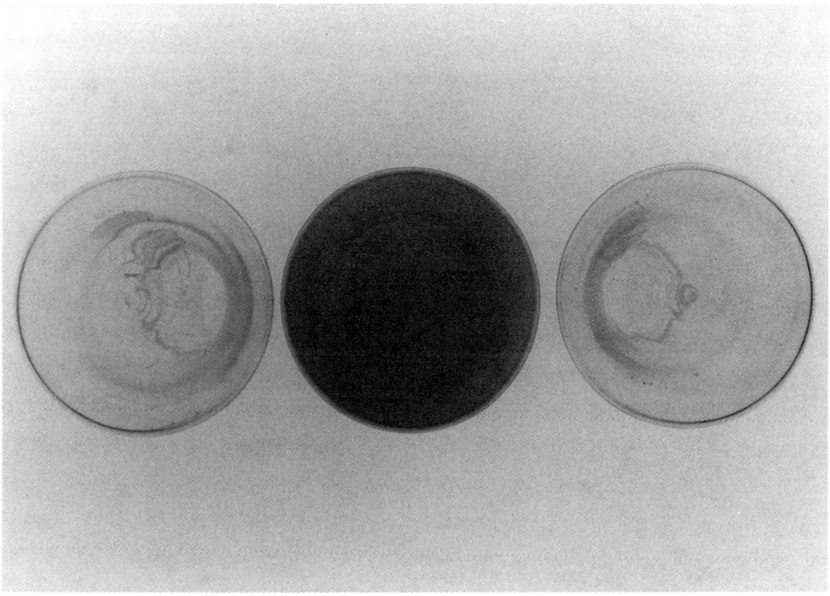

Figure 8–6. Low molecular weight α-1,6 glucan prevents sucrose-dependent adhesion of *Streptococcus sobrinus*. Left, adhesion to glass in absence of sucrose. Center, bacterial adhesion to glass vials following incubation in a sucrose-supplemented growth medium. Right, sucrose-containing growth medium was supplemented with 10 mg/ml glucan T-10.

Glucan-dependent Composite Adhesion Model

S. sanguis (including *S. oralis* and *S. gordonii*) adheres to pellicle much better than any of the mutans streptococci (Hamada and Slade, 1980; Loesche, 1986). Members of the *S. sanguis* group of oral streptococci possess GTFs which yield α-1,6 glucans. Because *S. sanguis* is an early colonizing bacterium (Figure 8–2 and Table 8–2), it may occupy sites on the pellicle that are capable of interacting with other bacteria. If *S. sanguis* or related bacteria occupy sites potentially available for other streptococci, such as members of mutans streptococci, there must be some means for mutans streptococci to adhere to the tooth if the bacteria are to colonize the tooth surface. Recent studies (Doyle et al., unpublished, 1992) have revealed a possible new mechanism for adhesion of the mutans streptococci. Hydroxylapatite beads were mixed with a suspension of *S. sanguis* and the nonadherent bacteria removed by washing the beads in a dilute phosphate buffer. The *S. sanguis*–bead mixture was then incubated with sucrose. Finally, the sucrose-treated beads were again washed with the buffer. Radioactive cultures of *S. sobrinus* or *S. cricetus* were then added to the washed beads. The results revealed that the mutans streptococci were more than one hundred times as

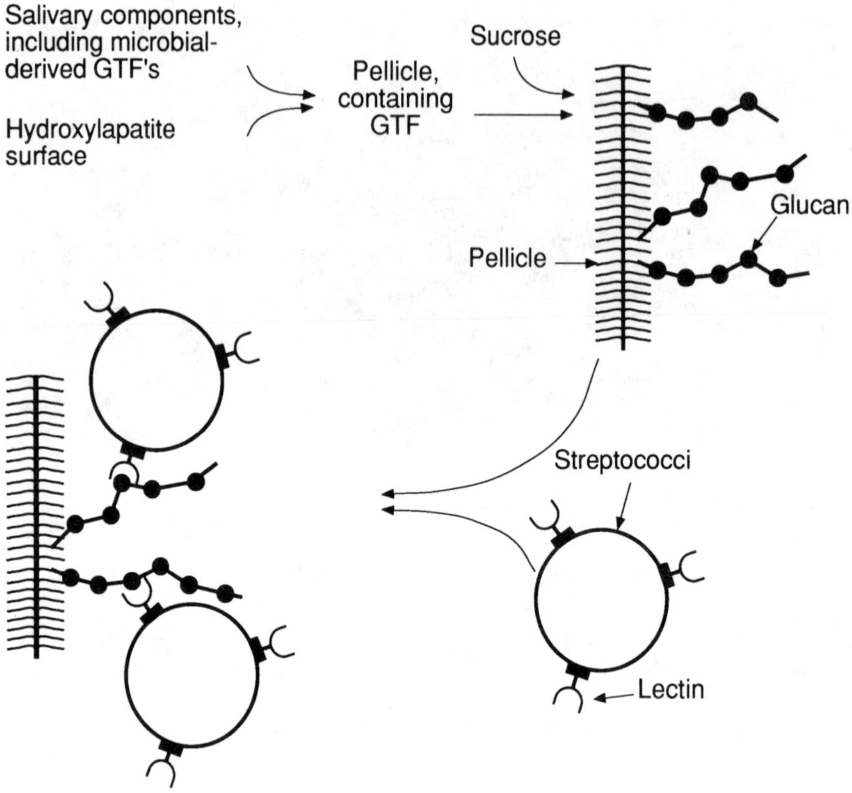

Figure 8–7. Glucosyltransferase (GTF) in saliva may promote streptococcal adhesion. GTF bound to the hydroxylapatite (tooth) surface, when in the presence of sucrose, gives rise to α-1,6 glucan, a substratum for the streptococci. Parotid saliva, which contains no GTF, would not support adhesion. Only the mutans streptococci are capable of adhering via glucan-binding lectins.

adherent to the sucrose-treated, *S. sanguis*–bead mixture than to untreated beads or beads treated with sucrose, but in the absence of *S. sanguis*. A likely explanation (Figure 8–8) is that the adherent *S. sanguis* synthesized α-1,6 glucans from sucrose which then formed substrata for the *S. sobrinus* bacteria. Control experiments showed that GBL^- mutans streptococci failed to adhere to the extent of the GBL^+ mutans streptococci. This model proposed for the interspecies adhesion of streptococci has some attractive features: (1) adhesion of some mutans streptococci requires (or takes advantage of) the adhesion of *S. sanguis,* a known early colonizing bacterium; and (2) adhesion of mutans streptococci is dependent on sucrose, a finding consistent with much of the literature (Hamada and Slade, 1980; Loesche, 1986); and (3) there is no requirement for a particular pellicle protein(s) as the mutans streptococci utilize glucan produced by *S. sanguis* as

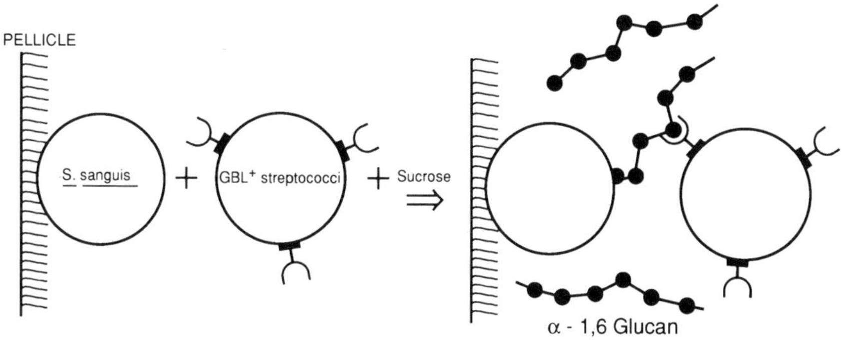

Figure 8–8. How *Streptococcus sanguis* may promote adhesion of GBL$^+$ mutans streptococci. *S. sanguis,* or closely related bacteria firmly adherent to pellicle, synthesize α-1,6 glucan in the presence of sucrose. The glucan then forms a substratum for GBL$^+$ streptococci.

the substratum. The model described above does not adequately explain why some GBL$^-$ mutans streptococci are frequently found in plaque deposits or in carious lesions. Nor does the model explain why plaque materials are so heterogeneous in microbial populations. The model assumes that most adhesion of the mutans streptococci is indirectly dependent on sucrose. Ultimately, the model is a form of the GTF-dependent adhesion discussed above, but with the added complication that a composite of streptococcal species is required for development of plaques.

Lipoteichoic Acid-dependent Adhesion of Oral Streptococci

Some bacteria, including oral members of the genus *Streptococcus,* may secrete lipoteichoic acids (LTAs) during balanced or perturbed growth. The LTAs, in turn, may become a part of pellicle and provide receptor sites for bacterial adhesion. There is convincing evidence for the presence of LTAs in normal saliva (Ciardi et al., 1977; Rolla et al., 1980). It is believed that pellicle LTAs can bridge directly to various streptococci, thereby causing adhesion of the bacteria. In this regard, Rolla et al. (1980) found that sucrose-induced plaque contained much more LTAs than non-sucrose-induced plaque in human volunteers. Because the binding of LTA to hydroxylapatite can be inhibited by fluoride or phosphate anions, it has been assumed that the poly(glycerol phosphate) backbone of the LTA complexes with the calcium of hydroxylapatite (Ciardi et al., 1977). It follows that the glycolipid end of the LTA would be available for interaction with streptococci, probably via hydrophobic interactions. Fibronectin (Fn) is capable of interacting with LTA via the hydrophobic end of the LTA (see Chapter 6), leaving the hydrophilic polyglycerol phosphate) extending into the solvent environment. Hogg and Manning (1988) coated hydroxylapatite beads with fibronectin and observed that most oral streptococci adhered better to Fn-hydroxylapatite than to hydroxylapatite alone. They further showed that exogeneously added LTA would inhibit the adhesion of most of the streptococci to fibronectin-hydroxylapatite. Many oral members of the genus *Streptococcus* bind fibronectin (Babu et al., 1983; Babu and Dabbous, 1986), but the role of LTA-fibronectin interactions in mediating oral streptococcal adhesion remains speculative.

In an evolutionary sense, it seems that multiple sites on pellicle may be favored by oral streptococci. The bacteria are confronted with many different pellicle proteins (Leach and Saxton, 1966; Leach et al., 1967; Wilkes and Leach, 1979), some of which may be receptors for bacterial adhesins. The bacterium possessing a single adhesin for one or even a few pellicle receptors would probably be less successful in colonizing teeth (or gingiva) than bacteria with multiple adhesins. Furthermore, the bacterium with multiple adhesins would probably be less affected by dietary constituents than a bacterium expressing a single adhesin. Adhesins of oral streptococci are under genetic control (Tardiff et al., 1989) but it is not known if more than one adhesin at a time can be genetically regulated.

Adhesion of Oral Bacteria to Soft Tissues

Gibbons (1984) has emphasized an apparent tropism in oral microbial adhesion. Whereas *S. gordonii, S. oralis, S. sanguis,* and the mutans streptococci are frequently found on the tooth pellicle, organisms such as *S. salivarius* and *S. mitis* are more common on soft tissues. Furthermore, many other oral bacteria adhere well to soft tissues. These include members of the genera *Prevotella, Porphyromonas, Bacteroides, Eikenella, Fusobacterium, Leptotrichia, Streptococcus, Actinomyces, Veillonella,* and most of the bacteria thought to be involved in periodontal disease (Marsh and Martin, 1984). Some of the oral bacteria capable of adhering to animal cells possess galactose-specific lectins. As regards the lectin-expressing oral bacteria, most will bind to any galactosylated substratum, assuming of course that the anomeric requirements have been met. A theme in this book is that tropism occurs because of a composite of factors, not because a particular bacterium possesses a specific lectin or some other specific adhesin.

Fusobacterium nucleatum adheres avidly to human polymorphonuclear neutrophils (PMNs) (Mangan et al., 1989). The adhesion is readily inhibited by β-galactosides, Gal, and GalNAc. Adhesion of *F. nucleatum* to the neutrophils results in PMN aggregation and loss of membrane protonic potential, leading to an increase in the PMN free Ca^{2+} concentration, an increase in superoxide anion, and the release of lysozyme. Assays revealed that most of the adherent bacteria are killed within 1 hour. When the bacteria are pretreated with high temperatures (80°) or proteases, adhesion is abolished. The authors speculated that the nonopsonic attachment of the bacteria by PMNs may be important in the killing of *F.*

nucleatum in the gingival sulcus and the subsequent release of PMN factors associated with tissue inflammation. The attachment of bacteria to PMNs may be one of the initial steps leading to periodontal disease. Studies on the specificity of the *F. nucleatum* surface lectin were provided by Mongiello and Falkler (1979), who observed that Gal and GalNAc inhibited hemagglutination of human and sheep red cells by the bacteria. Interestingly, they also found that lysed suspensions of the bacteria were capable of causing hemagglutination. In addition, when the red cells were initially incubated with plant lecins, the bacteria could no longer hemagglutinate the erythrocytes. Human buccal epithelial cells were also able to complex with *F. nucleatum* via the Gal-specific lectin. The lectin may function in adhesion of the bacteria not only to oral mucosa, but to other bacteria as well (the section on coaggregation phenomena of oral bacteria lists some of the organisms with which *F. nucleatum* can bind).

Eikenella corrodens can also adhere to human buccal epithelial cells via a (α or β) galactose (or GalNAc)-dependent mechanism (Yamazaki et al., 1981). Treatment of the bacteria at 100° or with trypsin abolished the ability of the cells to bind the buccal epithelia. Furthermore, the adhesion could be inhibited by EDTA, an inhibitor of many bacterial and plant lectins. It seems likely that the buccal cells possess exposed galactose residues for interaction with the bacterial lectin(s). *E. corrodens* is often found in deep periodontal pockets and is thought to be a major periodontopathogen.

Earlier, Saunders and Miller (1980) found that the attachment of *A. naeslundii* to human buccal epithelial cells could be inhibited by Gal or GalNAc. These results are in accord with other findings that showed that *A. naeslundii* and *A. viscosus* were potent hemagglutinating bacteria only when the red cells had been pretreated with a sialidase (Costello et al., 1979). The sialidase exposed penultimate Gal residues which could then serve as receptors for the *Actinomyces* species. The hemagglutination reactions were reversed by β-galactosides and completely abolished by treating the bacteria at 80° for 30 minutes. Heeb et al. (1982) employed a model system to study the surface lectin of *A. viscosus*. They coated Latex beads with various glycoproteins and observed aggregation following the addition of the bacteria. The strength of this system is the ease with which the various glycoproteins can be manipulated. It was possible to determine the minimum amount of protein on the beads to elicit aggregation. It was also possible to show that galactose oxidase destroyed aggregation with the bacteria, but when the oxidized substrata were reduced with sodium borohydride, aggregation was achieved. The model system was also readily amenable to inhibition by various sugars and saccharides.

In an early article on the hemagglutinating activity of oral bacteria, Kondo et al. (1976) found that there may be a lectin involvement in the interaction between *Leptotrichia buccalis* and red cells. They observed that hemagglutination could be inhibited by β-galactosides and GalNAc. Electron microscopy of the bacteria revealed the presence of fimbriae on the cell surface. These fimbriae presumably carry the lectin activity although it is not known if the bacteria have their lectins at the very tips of fimbriae, in a manner analogous to that of *E. coli* (Chapter 9). Kondo et al. (1976) also observed that the bacteria could bind to saliva-coated hydroxylapatite. The adhesion was readily inhibited by lactose, suggesting that the same adhesin was responsible for hemagglutination and binding to saliva-coated surfaces.

A type of lectin-independent reaction between *Poryphyromonas gingivalis* and *Actinomyces* strains has been observed (Bourgeau and Mayrand, 1990). Extracellular vesicles of *P. gingivalis* were able to aggregate various *Actinomyces* strains. The aggregation was inhibited by arginine, but not by carbohydrates. Antibody directed against *P. gingivalis* surface hemagglutinin also inhibited the interaction with the actinomycete. Vesicles form as a result of natural growth processes and seem to contain virulence determinants such as proteases and hemagglutinins. The vesicles may aid in colonization of the bacteria by serving to trap coaggregating partners or by adhering to mucosal surfaces.

The periodontopathogen, *Prevotella* (*Bacteroides*) *loescheii,* has been reported to express two distinct surface adhesins simultaneously. London et al. (1989) provided evidence to suggest that *P. loescheii* can express adhesins capable of interacting with *Actinomyces israelii* and *S. sanguis*. A monoclonal antibody preparation specific for the *Actinomyces* adhesin was labeled with 10-nm gold particles, whereas another monoclonal antibody specific for the streptococcus was labeled with 5-nm gold particles. When the antibody preparations were allowed to mix with the intact *P. loescheii,* evidence was seen for dual labeling. Even though the experiments employed bacteria as substrata for the *P. loescheii* adhesins, it is likely that the adhesins could bind with other substrata. In fact, the interaction of *P. loescheii* with *S. sanguis* is inhibited by galactosides, whereas the adhesin specific for the actinomycete is not inhibited by simple carbohydrates. The lectin involved in coaggregation with *S. sanguis* is also involved in hemagglutination (Weiss et al., 1989).

Treponema denticola, a suspected periodontopathogen, has been demonstrated to bind to fibronectin (Thomas et al., 1986; Dawson and Ellen, 1990). The bacterium produces a number of virulence determinants, including endotoxins, proteases, chemotactic factors, ammonia, and putrescine, the latter two of which may be cytotoxic for gingival tissues. Interestingly, the bacteria seem to adhere to fibronectin-coated surfaces by their tips. When laminin was employed as a substratum, adhesion was at a level similar to that observed for fibronectin. Collagen and non-RGD-containing polypeptides were incapable of supporting *T. denticola* adhesion.

Oral streptococci are known to cause bacterial endocarditis. Recent studies have been directed to determining how the bacteria adhere to endothelial tissues and to blood elements likely to be involved in vascular diseases. Certain strains of *S. sanguis* cause the aggregation of platelets in vitro. Two different classes of *S. sanguis* cell surface proteins have been shown to mediate adhesion with platelets. One class of surface protein (class I antigen) served as an adhesin that binds the bacteria to platelets (Erickson and Herzberg, 1987, 1990; Herzberg et al., 1990). The second surface protein antigen (class II) interacts with platelets in the presence of plasma and triggers the aggregation induced by *S. sanguis*. A sequential series of reactions seems to be required for the platelet aggregation: the first involves the adhesion and the second involves the triggering response. A fragment of the class II antigen in solution inhibits platelet aggregation induced by *S. sanguis*. Receptor proteins are found to be functionally and immunologically cross-reactive with type I collagen. The platelet receptors for the bacterial proteins and the collagen are likely identical (Erickson and Herzberg, 1987). The implication of these findings is that the platelet–*S. sanguis* interaction contributes to bacterial endocarditis and disseminated intravascular coagulation.

Another approach implicating the role of adhesion of *S. sanguis* to soft tissue in endocarditis employed mutants incapable of binding fibronectin. *S. sanguis* was found to adhere to fibronectin immobilized on gelatin-coated plastic (Lowrance et al., 1988). The adhesion was conformation-dependent as the bacteria were shown to adhere only to immobilized fibronectin fragments not inhibited by fibronectin in solution. A mutant derived by transposon insertional mutagenesis was obtained and shown to have reduced binding to immobilized fibronectin. The mutant also has a diminished ability to induce infective endocarditis in a rat model (Lowrance et al., 1990). The results suggest that fibronectin on the surface of heart tissues may act as a receptor for *S. sanguis*. Soluble plasma fibronectin would be unable to interfere in this kind of adhesion.

Invasion of oral bacteria into mucosal cells has not been examined in detail. The best studies are preliminary and deal with the invasion of *A. actinomycetemcomitans* into a human oral cell line (Meyer et al., 1991). Invasion was detected by treating KB cell line–*A. actinomycetemcomitans* mixtures with gentamicin. The antibiotic killed the extracellular bacteria, whereas the internalized bacteria could be released by lysing washed KB cells. A smooth variant was more invasive than a rough strain, although the basis of the smooth–rough transition is unknown. *A. actinomycetemcomitans* is a bacterium implicated in juvenile periodontitis. The studies of Meyer et al. (1991) did not address how *A. actinomycetemcomitans* interacted with the cell line prior to internalization of the bacteria. The prospect that oral bacteria can invade epithelial cells is a new lead into the mechanism of tissue destruction associated with periodontal disease.

In terms of literature citations the most oral bacteria that adhere to soft tissues do so primarily by galactose-binding lectins. These lectins may also be involved in coaggregation reactions between various bacteria. It is likely that the lectins serve both functions to the advantage of the bacteria. As discussed below, coaggregation between different bacterial species provides selective advantages to the species involved. A few bacteria, such as *T. denticola,* may possess nonlectin adhesins capable of binding to fibronectin or laminin. One bacterium, *P. gingivalis,* has an adhesin capable of interacting with arginine side chains of proteins. Oral epithelial cells not only contain their own surface glycoconjugates, but also are in contact with salivary proteins, foodstuffs, and other microorganisms. It is not surprising that the various oral bacteria have evolved many adhesion mechanisms. The advantages conferred on bacteria capable of participating in multiple mechanisms of adhesion are discussed in Chapter 11.

Coaggregation Phenomena in Oral Bacteria

Coaggregation is a result of two or more bacteria interacting to form a stable composite aggregate. It now seems clear that most human oral bacteria can participate in coaggregation reactions. In many cases, coaggregation results from the interactions between a lectin on the surface of one bacterium and a complementary carbohydrate structure on the surface of the second. In a few cases, one bacterium may express a cell surface lectin and a carbohydrate specific for a lectin expressed by another bacterium. Some coaggregations are not inhibited by carbohydrates and are presumably not mediated by lectins. As might be expected, coaggregation is independent of cell viability. Coaggregating bacteria can be readily observed by the light or electron microscopic examination of dental plaque samples. Furthermore, coaggregation can be observed visually when coaggregating pairs are mixed in a test tube. In Figure 8–9, a suspension of *Streptococcus sanguis* is mixed with a suspension of *Actinomyces viscosus*. Within seconds, large macroscopic aggregates appear. The aggregates can be dispersed by lactose or other β-galactosides. It is convenient to monitor coaggregation by light scattering, employing a spectrophotometer or a platelet aggregometer.

Some bacteria, such as streptococci or bacteroides, may coaggregate with large rod-shaped bacteria, giving rise to corn-cob-like configurations (Figure 8–10). Some authors refer to these peculiar geometrical formations as "corn-cobs" (Jones, 1972). Although this term is well-established in the literature, a more descriptive name for the aggregates may be proposed. It is suggested to use the word "microbiadh" not only for "corn-cob" but also for all coaggregating microorganisms.

Many factors combine to affect coaggregation. They include adhesin and receptor density and distribution; hydrophobic character of receptor, adhesin, or receptor nearest neighbors; medium pH and composition; and chelating agents. Surface lectins can usually be inactivated by heat, proteases, or concentrated protein denaturants. In some cases, the lectin is inhibited by EDTA or other chelating agents, suggesting that the lectin is a metalloprotein. When one member of an interacting pair is present at a much higher cell density then the corresponding partner, the cell of the higher density seems to adhere circumferentially to its partner. This type of adhesion is usually referred to as rosette formation.

Only for a few bacteria have the molecular bases of the coaggregation reactions

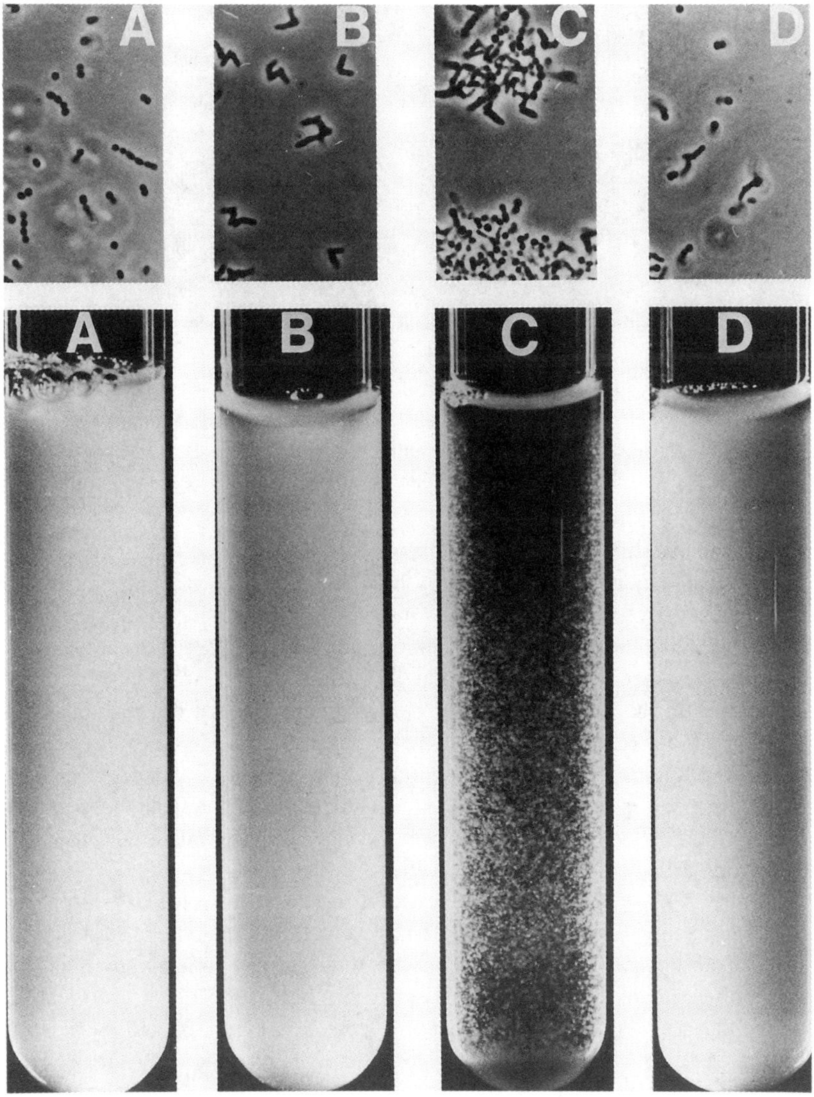

Figure 8–9. Coaggregation between *S. sanguis* and *A. viscosus*. Homogeneous suspensions containing about 1×10^0 cells/ml of (A) streptococci, or (B) actinomyces are vortex-mixed, and mixed-cell type coaggregates are formed (C, upper panel), which immediately settle to the bottom of the tube (C, lower panel). Addition of lactose (100 m*M* final concentration) reverses coaggregation to individual cell types side-by-side (D, upper panel), giving the appearance of a homogeneous cell suspension (D, lower panel) which contains both cell types. (Magnification, upper panel × 1000). (Reprinted courtesy of P. Kolenbrander, 1989 and CRC Press.)

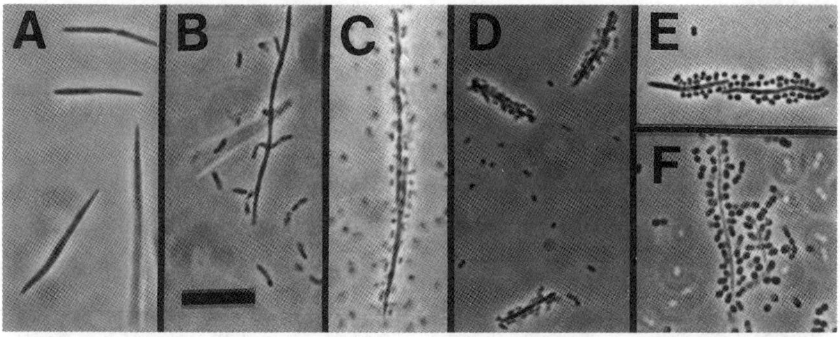

Figure 8–10. Intergeneric coaggregations among oral bacteria. *F. nucleatum* PK1594 (A) forms microbioadh configurations with its partners *S. flueggei* PK1958 (B), *P. gingivalis,* PK1924 (C), *A. actinomycetemcomitans* Y4(D), V. atypica PK1910 (E), and *S. sanguis* C104 (F) when the partners are present in a 10-fold excess. Bar, 10 μm. (Courtesy of P. Kolenbrander (Kolenbrander, 1991) and the American Society for Microbiology.)

been studied in detail. They can be divided into two categories. The coaggregation of one category is readily inhibited by low concentrations of specific carbohydrates and is therefore mediated by an interaction between lectin-like molecules on the surface of one cell and a complementary carbohydrate structure on the surface of the other type of cell (Tables 8–6 and 8–7). The molecular structures participating in the other category of coaggregation are not defined, but based on the sensitivity to heat and/or proteases, may be subdivided into three groups (Tables 8–6 and 8–7). The coaggregation of one group is dependent on heat- and/or protease-sensitive structures on only one type of cell in the pair. The coaggregation of the third group of pairs involves as yet unidentified molecules on the cell surfaces of both pairs of cells.

McIntire et al. (1978, 1982, 1983; McIntire, 1985) were the first to propose that coaggregation between some bacteria was lectin-dependent. They found that lactose or other β-galactosides could prevent the *A. viscosus–S. sanguis* adhesion. They further showed that when *A. viscosus* was heated (85°C), it lost its ability to bind the streptococci. In contrast, *S. sanguis* was unaffected by heating. McIntire et al. concluded that the *Actinomyces* must possess a cell surface lectin capable of complexing with a carbohydrate receptor of *S. sanguis*. The structure of the carbohydrate receptor has now been determined. It consists of a GalNAcRhaGlcGalGalNAcGal repeating sequence (McIntire et al., 1987) (Figure 8–11). The *Actinomyces* lectin presumably recognizes the penultimate and terminal residues. In this regard, Cassels et al. (1990) have recently sequenced a polysaccharide from another *S. sanguis* strain capable of aggregation with *Capnocytophaga ochracea*. The hexasaccharide repeating unit was found to consist of a α-Rhaα-1,2Rhaα-1,3Galα-1,3Galβ-1,4Glcβ-1,3Gal(α/β). Interestingly, the best inhibitor of coaggregation between *C. ochracea* and the *S. sanguis* is the monosac-

Table 8–6. Oral bacteria participating in intergeneric coaggregation of undefined specificity[a]

Coaggregation between bacteria mediated by proteinaceous (P, heat and/or protease-sensitive) or (U) undefined ligands expressed on:

Bacterial species No. 1		Bacterial species No. 2	
Actinobacillus actinomycetemcomitans	(U)	*Fusobacterium nucleatum*	(P)
Actinomyces israelii	(U)	*Bacteroides buccalis*	(U)
		Bacteroides denticola	(P)
		Bacteroides intermidius	(P)
		B. intermidius (8944 group)	(U)
		Bacteroides oralis	(P)
		Bacteroides veroralis	(U)
		Prevotella loescheii	(P)
		P. loescheii (D1C-20) group	(U)
		Veillonella atypica	(U)
		Veillonella dispar	(U)
		Veillonella parvula	(U)
Actinomyces naeslundii	(U)	*B. buccalis*	(P)
		B. denticola	(P)
		B. intermedius	(P)
		B. intermedius (group 8944)	(P)
		B. veroralis	(P)
		F. nucleatum	(P)
		Pseudomonas aeruginosa	(P)
		V. atypica	(P)
		V. dispar	(P)
		V. parvula	(P)
		Streptococcus anginosus	(P)
Actinomyces odontolyticus	(U)	*B. oralis*	(U)
		F. nucleatum	(P)
		Prevotella loescheii	(U)
		P. loescheii	(U)
		V. atypica	(U)
		V. dispar	(P)
Actinomyces viscosus		*B. buccalis*	(P)
		Bacteroides gingivalis	(U)
		B. intermedius	(P)
		B. intermedus	(P)
		F. nucleatum	(P)
		P. loescheii	(P)
		P. aeruginosa	(P)
		S. anginosus	(P)
		V. atypica	(P)
		V. dispar	(P)
		V. parvula	(P)

continued

Table 8–6. Continued

Coaggregation between bacteria mediated by proteinaceous (P, heat and/or protease-sensitive) or (U) undefined ligands expressed on:

Bacterial species No. 1		Bacterial species No. 2	
Actinomyces WVa963	(U)	*F. nucleatum*	(P)
Bacterionema matruchotti	(U)	*Streptococcus sanguis*	(P)
B. gingivalis	(U)	*Streptococcus salivarius*	(P)
B. intermedius	(P)	*F. nucleatum*	(P)
Bacteroides melaninogenicus	(U)	*F. nucleatum*	(P)
B. veroralis	(U)	*S. sanguis*	(P)
B. denticola	(U)	*F. nucleatum*	(P)
Capnocytophaga gingivalis	(U)	*F. nucleatum*	(P)
Capnocytophaga ochracea	(U)	*F. nucleatum*	(P)
Capnocytophaga sputigena	(U)	*F. nucleatum*	(P)
Eikenella corrodens	(U)	*B. gingivalis*	(U)
F. nucleatum	(U)	*S. salivarius*	(P)
P. loescheii	(U)	*F. nucleatum*	(P)
Propionibacterium acnes	(U)	*F. nucleatum*	(P)
Rothia dentocariosa	(U)	*B. buccalis*	(U)
		B. denticola	(U)
		B. intermedius	(U)
		B. intermedius	(U)
		B. veroralis	(U)
		F. nucleatum	(P)
		P. loescheii	(U)
		V. atypica	(U)
		V. parvula	(P)
Streptococcus agalactiae	(U)	*A. viscosus*	(U)
Streptococcus cricetus	(P)	*Actinomyces serovar* WVa963	(P)
Streptococcus mitior	(P)	*F. nucleatum*	(P)
Streptococcus mitis	(U)	*Haemophilus parainfluenzae*	(U)
		P. aeruginosa	(P)
Streptococcus morbillorum	(U)	*P. loescheii*	(U)
Streptococcus mutans	(U)	*V. parvula*	(U)
Streptococcus pyogenes	(U)	*A. viscosus*	(U)
Streptococcus salivarius	(U)	*V. atypica*	(P)
		V. dispar	(P)
		V. parvula	(U)
Streptococcus sanguis	(U)	*A. naeslundii*	(P)
		A. odontolyticus	(U)
		H. parainfluenzae	(U)
		P. aeruginosa	(P)
		V. atypica	(U)
		V. dispar	(U)
		V. parvula	(U)
S. sanguis	(P)	*F. nucleatum*	(P)
V. atypica	(U)	*F. nucleatum*	(P)
V. dispar	(U)	*F. nucleatum*	(P)
V. parvula	(U)	*S. salivarius*	(P)

[a]Adapted from Kolenbrander (1988, 1989, 1990).

Table 8–7. Intragenic carbohydrate specific coaggregation (microbioad formation) of oral bacteria[a]

Coaggregation between bacteria expressing		
Polysaccharide ligand	Proteinaceous ligand (lectin-like)	Inhibitors
Actinomyces naeslundii	*S. sanguis*	β-Galactosides
Actinomyces odontolyticus	*Veillonella parvula*	β-Galactosides
Bacteroides gingivalis	*Fusobacterium nucleatum*	β-Galactosides
Eubacterium nodatum	*F. nucleatum*	β-Galactosides
Haemophilus aphrophilus	*Streptococcus mitis*	β-Galactosides
Haemophilus seguis	*S. mitis*	β-Galactosides
Peptostreptococcus anaerobius	*F. nucleatum*	β-Galactosides
Propionibacterium acnes	*Veillonella atypica*	β-Galactosides
	Veillonella dispar	β-Galactosides
	V. parvula	β-Galactosides
S. mitis	*A. naeslundii*	β-Galactosides
	Actinomyces viscosus	β-Galactosides
S. morbillorum	*V. parvula*	β-Galactosides
	V. dispar	β-Galactosides
	V. atypica	β-Galactosides
	A. naeslundii	β-Galactosides
Streptococcus sanguis	*A. naeslundii*	β-Galactosides
	A. viscosus	β-Galactosides
	Actinomyces serovar WVa 963	β-Galactosides
	A. viscosus T14V	β-Galactosides
	Bacteroides loescheii PK1295	β-Galactosides
	B. loescheii	β-Galactosides
	Capnocytophaga ochracea	β-Galactosides
	Capnocytophaga sputigena	β-Galactosides
	Haemophilus aphrophilus	β-Galactosides
	Haemophilus segnis	β-Galactosides
	Streptococcus morbillorum	β-Galactosides
A. israelii	*Capnocytophaga gingivalis*	GalNAc, sialic acids
	C. sputigena	Sialic acids, rhamnose
A. odontolyticus	*Streptococcus salivarius*	Sialic acids
A. viscosus	*Eikenella corrodens*	GalNAc
Propionibacterium acnes	*S. sanguis*	
S. sanguis	*E. corrodens*	
A. viscosus	*C. sputigena*	Sialic acids, rhamnose
Rothia dentocariosa	*C. sputigena*	Fucose, rhamnose
A. israelii	*C. ochracea*	
A. naeslundii	*C. ochracea*	
A. viscosus	*C. ochracea*	
R. dentocariosa	*C. ochracea*	
S. sanguis	*C. ochracea*	

[a]Adapted from Kolenbrander (1988, 1989, 1990).

Receptor Structure	Major Microbiological Partner
Streptococcus sanguis 10557 [→ 6)Gal α-1,3Rhaβ-1,4Glcβ-1,3Gal$_f$β–1,6Galβ–1,3– 2 ↑ $(OAc)_{0.33}$ (on Gal$_f$) -Gal NAcα-1-PO_4^-]$_n^-$ 6 ↑ $(OAc)_{0.33}$ (on Gal NAc)	Actinomyces spp, others
Streptococcus sanguis 34 [→PO_4^- →6)GalNAcβ-1,3Rhaβ-1,4Glcβ-1,6- Gal$_f$β-1,6GalNAcα-1,3Galα] $_n$	Actinomyces spp, Prevotella, others
Streptococcus mitis J22 [-PO_4→6)GalNAcβ-1,3Rhaβ-1,4Glcβ-1,6- Rhaα-1,2 (on Glc) -Gal$_f$β-1,6Galβ-1,3GalNAcα-1 →] $_n$	Actinomyces spp, Prevotella, Haemophilus, Veillonella, others
Streptococcus sanguis (oralis) HI Rhaα-1,2Rhaα-1,3Galα-1,3Galβ-1,4- -Glcβ-1,3Galα / β	Capnocytophaga ochracea, others

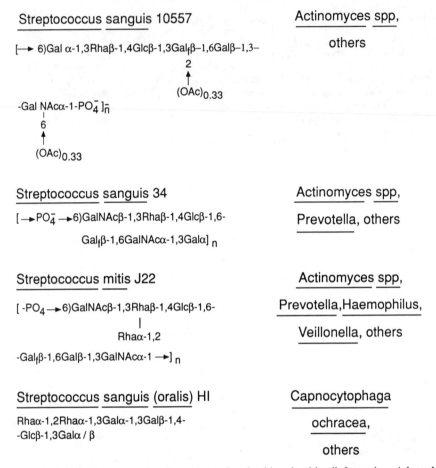

Figure 8–11. Structures of lectin receptors involved in microbioadh formation. Adapted from Kolenbrander, 1991.

charide rhamnose (Figure 8–11 shows the known streptococcal carbohydrate receptors for adhesins of oral bacteria).

It is now known that numerous oral bacteria express cell surface lectin adhesins (reviewed by Kolenbrander, 1988, 1989). Tables 8–6 and 8–7 list some of these species and strains. Lactose and other *β*-galactosides are the most common inhibitors, suggesting that galactoside-specific lectins mediate microbiadh formation. Rhamnose, fucose, and sialic acid-specific lectins seem to be on the surfaces of some bacteria.

The most widely studied coaggregating pairs belong to the genera *Actinomyces* and *Streptococcus*. Members of these genera are early colonizers of cleaned teeth (Figure 8–2; Table 8–2). Once an *Actinomyces* or a *Streptococcus* becomes adherent to the pellicle, additional bacteria can then bind, forming microbiadhs. As more and more bacteria become adherent, plaque is developed. The mutans streptococci do not form microbiadhs with *Actinomyces* spp. as often as with *S. sanguis*. Similarly, *S. salivarius* only infrequently coaggregates with an *Actinomyces*. In most cases, the lectin appears to be on the *Actinomyces* surface, as evidenced by its inactivation by protease and heat. In a few cases, the lectin seems to be expressed by the streptococci. Oral *Actinomyces* express two types of fimbriae, defined by function and not by electron microscopic appearance. *A. viscosus* expresses both types, called type 1 and type 2. *A. naeslundii* carries only type 2. Type 1 fimbriae of *A. viscosus* exhibit an affinity for pellicle receptors, particularly proline-rich salivary proteins (Clark et al., 1989) and are not lactose-sensitive. Type 2 fimbriae bind β-galactosides and are probably responsible for most microbiadh formations, as well as adhesion to animal cells. Types 1 and 2 fimbriae are distinct, as they do not share antigenic cross-reactivity (Cisar *et al.*, 1981).

Gibbons and Nygaard (1970) noted that members of the genus *Veillonella* were capable of coaggregating with various oral bacteria. In 1981, McBride and van der Hoeven reported that mutans streptococci and *V. parvula* can interact in vivo. A strain of *S. mutans* was implanted into gnotobiotic rats and then superinfected with *V. parvula*. Control experiments established that *V. parvula* would not mono-infect the rats. However, the organism could be shown adherent to the streptococci on the tooth surfaces when both bacteria were inoculated. Furthermore, a strain of *S. mutans,* incapable of in vitro coaggregation with *V. parvula,* did not support adhesion of the latter organism in vivo. The results support a view that coaggregation reactions observed in vitro are important in plaque development. *Veillonella* spp, regarded as early colonizing bacteria to cleaned tooth surfaces, may utilize already adherent streptococci as substrata for adhesion. For the streptococci and veillonellae there are ecological advantages to microbiadh formation *in vivo*. For example, streptococci frequently ferment carbohydrates to lactic acid. The lactic acid, in turn, can be utilized by *Veillonella* as a carbon source (reviewed by Kolenbrander 1988, 1989).

Another group of organisms that rapidly colonize cleaned teeth include members of the genus *Haemophilus*. Furthermore, *Haemophilus* spp. are common in human supragingival plaque, but do not appear to form microbiadhs with *Actinomyces* spp. (Liljemark et al., 1985). *H. parainfluenzae* usually coaggregates with early-colonizing streptococci, such as *S. sanguis*. Furthermore, the coaggregations are usually inhibited by lactose and other β-galactosides. It is thought that the streptococci bear a lectin capable of interacting with the *Haemophilus* spp., because the latter is heat resistant and the former is heat sensitive (Liljemark et al., 1985). The role of *Haemophilus* spp. in oral diseases is poorly understood.

The bacteria are not thought to contribute to coronal dental caries, but may have a role in root surface caries.

Most of the "secondary colonizers" of the tooth seem to participate in microbiadh formation, either by providing adhesin or receptors. *Bacteroides* spp. (and the similar *Prevotella* and *Porphyromonas* spp.) are among the most prominent secondary colonizers (Tables 8–2 and 8–3). *Bacteroides* coaggregate not only with streptococci, but also with *Actinomyces*. Furthermore, *Bacteroides* frequently coaggregate with other secondary colonizers, such as *Fusobacterium nucleatum* (Table 8–7). Recently, it was reported that *B. (Prevotella) loescheii* expressed two adhesins simultaneously (London et al., 1989). One adhesin mediates aggregation with *S. sanguis,* whereas the other is responsible for microbiadh formation with *A. israelii*. The *Bacteroides* adhesin for the streptococcus was responsible for hemagglutination of sialidase-treated red cells. The *Bacteroides* adhesin specific for the streptococci could be inhibited by β-galactoside, but the interaction of *B. loescheii* with the *Actinomyces* was not inhibited by carbohydrates. *B. loescheii* seems to be unique, in that it expresses two adhesins simultaneously (London et al., 1989). Other bacteria have a genetic switching mechanism that usually provides for the expression of only a single lectin at a time (Chapter 9). The adhesins of *B. loescheii* make it possible for the organism to coaggregate with two different genera at the same time, forming multigeneric coaggregates (Kolenbrander, 1989). Two adhesins on a single bacterium may provide a means for the bacterium to compete successfully in a microorganism-rich environment, such as the oral cavity. It seems likely that other oral bacteria will be found to express multiple adhesins simultaneously. One late colonizing organism, *Eubacterium nodatum,* coaggregates with only a few oral bacteria. However, its common coaggregation partner seems to be *Porphyromonas gingivalis,* a probable periodontopathogen. Thus, the colonization of *E. nodatum* may be dependent on the prior colonization of *Porphyromonas gingivalis* in periodontally affected sites (Kolenbrander, 1989). Kolenbrander (1989) has brought out the interesting possibility that a coaggregation partner may assist the metabolism of the other partner. *P. gingivalis* can release amino acids (via proteolysis) of host tissues. The coaggregation partner can use the amino acids to aid in the transport of sugars, such as glucose, galactose, or fructose (Robrish et al., 1987).

In some cases, a lectin on the cell surface of a bacterium needed for microbiadh formation may prove to be deleterious to the bacterium. *F. nucleatum* expresses a β-galactoside-dependent adhesin on its surface. The lectin, however, causes the bacteria to bind to PMNs, resulting in ingestion and killing of the bacteria (Chapter 7). The lectinophagocytosis of *F. nucleatum* may be important in the gingival sulcus, where opsonins are readily degraded by proteases. The lectinophagocytosis of *F. nucleatum* (and other bacteria) may not only result in bacterial killing, but tissue destruction may also accompany the ingestion of the organism(s). The extent of lectinophagocytosis in promoting gingival inflammation, leading to periodontal disease, is unknown.

The Hydrophobic Effect and Adhesion of Streptococci and Other Oral Bacteria

A number of approaches have been used to study the involvement of the hydrophobic effect in adhesion of oral bacteria to hard tissues (Table 8–8). In only a few studies, however, have attempts been made, using defined mutants, to identify the bacterial hydrophobin(s) responsible for promoting or facilitating adhesion. Although many oral isolates have been found to exhibit hydrophobic surfaces

Table 8–8. Studies showing the involvement of the hydrophobic effect on adhesion of oral bacteria to hard surfaces

Type of Study	Type of surface/ Infection	Organisms	References
Inhibition of adhesion by hydrophobic agents	SHA/HA	*S. sanguis* *S. mutans*	Nesbitt et al., 1982a,b; Zhang et al., 1990; Liljemark et al., 1978
Positive correlation between hydrophobicity and adhesion/or infectivity	SHA/HA	*S. sanguis* *S. salivarius* *S. mutans*	Cowan et al., 1987a,b; Fachon-Kalweit et al., 1985; van der Mei et al., 1987; Rosenberg et al., 1991; Svanberg et al., 1984; Olsson and Westergren, 1982; Westergren and Olsson, 1983
	Biomaterials	*S. mutans* *S. sanguis* *B. catarrhalis*	Cowan et al., 1987; Satou et al., 1988; Gotoh et al., 1989
	Cariogenicity Virulence	*S. mutans* *F. necrophorum*	Olsson and Emilson, 1988; Shinjo et al., 1987
Positive correlation between hydrophobicity of target surface and its ability to bind bacteria	SHA/HA Polymers Teeth	*S. mutans* *S. sanguis* Plaque microbiota	Busscher et al., 1990; Reynolds and Wong, 1983; van Pelt et al., 1985; Leach and Agalamanyi, 1984
Identification of bacterial hydrophobins using defined mutants	SHA/HA Pellicle	*S. sanguis* *S. mutans* *S. sanguis*	Fives-Taylor and Thompson, 1985; Morris et al., 1985; Jenkinson and Carter, 1988; Lee et al., 1989; Gibbons et al., 1983
Identification of bacterial hydrophobins that inhibit adhesion	SHA	*S. sanguis*	Hogg and Manning, 1988; Morris and McBride, 1984
Treatment of bacteria with agents that affect hydrophobicity and adhesion	SHA Pellicle	*S. sanguis* *S. salivarius* *A. viscosus* *P. gingivalis*	Oakley et al., 1985; Weerkamp et al., 1987; Peros et al., 1985; Peros and Gibbons, 1982

(Doyle et al., 1990), the hydrophobicity of the bacterial surface is subject to changes either due to natural turnover of cell wall during growth or to the presence of proteases that abolish or inhibit hydrophobicity (Table 8–8). In general, the more hydrophobic the bacterial surface, the greater the ability of organism to bind to saliva-coated hydroxylapatite.

In 1982, Nesbitt et al. (1982a,b) observed that *S. sanguis* could adhere to hydrocarbons. They also showed that inhibitors of the hydrophobic effect were effective inhibitors of adhesion of the bacteria to saliva-coated hydroxylapatite (SHA). It followed that adhesion was dependent on hydrophobic interactions between surface adhesins and pellicle protein. Since that time, numerous studies have appeared describing hydrophobic properties of various oral bacteria.

For *S. sanguis,* adhesion occurs in two kinetically distinct steps. The first step is reversible and described by pseudo-first-order rate constants for adhesion and desorption (Busscher et al., 1986; Cowan et al., 1986). Cells bound in the first phase may become more strongly adherent. The proposed mechanism of adhesion of *S. sanguis* to SHA is (Cowan et al., 1986, 1987a,b):

$$\text{Cell (C)} + \text{pellicle (P)} \rightleftharpoons \text{CP}^* \rightarrow \text{CP}$$

where CP* represents the weakly tethered initial adhesion. When kinetic constants for the formation of the CP complex were obtained as a function of temperature, it was found that adhesion was enthalpy and entropy dependent. Low concentrations of salts, such as ammonium sulfate, decreased the rate of CP formation, but low concentrations of nonpolar *p*-dioxane increased binding. The composite interpretation is that the initial adhesion of *S. sanguis* to SHA is dependent on both the hydrophobic effect and electrostatic interactions. These results provide a firm basis for considering the importance of the hydrophobic effect in oral microbial adhesion. If the hydrophobic effect is essential in streptococcal adhesion to oral surfaces, then prevention of hydrophobic interactions should lead to agents that prevent plaque formation.

Adhesion of streptococci to biomaterials also seems to be influenced by the hydrophobic effect. *S. mutans* adhered better to various surfaces when the surfaces had low free energies (Fujioka et al., 1987; Busscher et al., 1990). Strains of *S. sanguis* also adhered better to surfaces when the bacteria possessed high hydrophobicities (Satou et al., 1988). Furthermore, van der Mei et al. (1987) found that strains of *S. salivarius* bound better to a hydrophobic plastic when the bacteria were hydrophobic.

When hydrophilic mutants of *S. sanguis* were studied, it was found that hydrophobic parent strains adhered better to SHA (Fives-Taylor and Thompson, 1985). In this regard, fresh isolates of streptococci tend to be hydrophobic and adhere better than passaged hydrophilic variants (Westergren and Olsson, 1983). Furthermore, most bacterial isolates from teeth are hydrophobic (Weiss et al.,

1982; Rosenberg et al., 1983). It has been found that isolates from intraoral sites in *Macaca fascicularis* were hydrophobic and the hydrophilic isolates could be rendered hydrophobic by saliva (Beighton, 1984).

A 160,000 mol wt protein on the surface of *S. sanguis* is known to contribute to adhesion of the bacterium to hexadecane or to SHA (Morris et al., 1985). Thus, the adhesin of *S. sanguis* for a hydrocarbon appears to also be responsible for adhesion to artificial pellicle. Proteolysis of *S. sanguis* reduces its ability to adhere to octyl-Sepharose and SHA (Oakley et al., 1985).

Recently, Olsson and Emilson (1988) observed that hydrophobic strains of *S. mutans* tended to be more cariogenic than hydrophilic strains in experimental animals. McBride et al. (1984) showed that high molecular weight proteins were present in wall preparations of hydrophobic *S. mutans* strains. These proteins were absent in hydrophilic strains. In *S. mutans,* it has been found that a hydrophobic protein serves as an adhesin (Lee et al., 1989). Recombinant techniques were employed to insert the protein into adhesion-deficient recipients which then assumed the ability to adhere.

Other studies support a view that adhesion to various surfaces by oral streptococci is dependent on hydrophobic interactions. A possible pellicle receptor for *S. sanguis* has been identified (Babu et al., 1986). It was observed that two proteins in saliva were capable of preventing adhesion of *S. sanguis* to hexadecane. Proteolysis of *S. salivarius* resulted in the loss of the ability of the organism to bind hydrocarbons and SHA (Weerkamp et al., 1987). Mutants of *S. sanguis* concomitantly lose their ability to adhere to hexadecane and SHA (Gibbons et al., 1983). When hydroxylapatite (HA) disks are treated with proteins, the proteins with higher amounts of nonpolar residues create superior substrata for adhesion of mutans streptococci (Reynolds and Wong, 1983). Salivary components also adhere to hexadecane (Leach and Agalamanyi, 1984). It was suggested that the stabilizing force of plaque integrity was the hydrophobic effect.

When adhesion to SHA or hydrocarbons has been studied, hydrophobic effect inhibitors, such as thiocyanate, urea, fatty acids, and hydrophobic proteins, tend to inhibit streptococci from binding (Liljemark et al., 1978; Nesbitt et al., 1982a,b). In any studies on inhibition of adhesion by chaotropic agents or by hydrophobic effect inhibitors, it must be kept in mind that the inhibitor may be denaturing the adhesin (or associated molecules) or the substratum. This is an argument frequently forgotten when inhibition occurs at much lower concentrations of inhibitor than are needed to unfold proteins, such as lysozyme or ribonuclease. Furthermore, adhesion is more sensitive to changes in temperature than are thermal transitions of most proteins (Cowan et al., 1987b). Finally, the inhibition of *S. sanguis* adhesion to SHA can be brought about by sulfolane, a reagent that "dilutes" the hydrophobic effect at concentrations not capable of unfolding proteins (Zhang et al., 1990).

Adhesion of streptococci to SHA, hydrocarbons, and other surfaces depends on several factors (Busscher and Weerkamp, 1987). Growth of bacteria in certain media results in populations with distinct surface characteristics (Rogers et al., 1984; Knox et al., 1985; Weerkamp and Handley, 1986; Drake et al., 1988b). Proteases may reduce hydrophobicity of bacteria (Oakley et al., 1985). Antibiotics may also result in the transient or permanent loss of cell surface hydrophobicity (Peros and Gibbons, 1982; Peros et al., 1985). Mutations (Fives-Taylor and Thompson, 1985; Lee et al., 1989; Willcox and Drucker, 1989; Willcox et al., 1989; Harty et al., 1990) frequently result in loss (rarely an acquisition, although some deep-rough mutants of *Enterobacteriaceae* are more hydrophobic than parent strains) of hydrophobicity of bacteria. Other factors, such as buffer, temperature, exogenously added proteins, or saliva may modify the ability of bacteria to adhere to hydrophobic substrata (Nesbitt et al., 1982a,b; Beighton, 1984; Babu et al., 1986.

Streptococci are not the only hydrophobic oral bacteria. Members of the genera *Actinomyces, Bacteroides* (and related species), *Branhamella, Actinobacillus, Hemophilus, Fusobacterium,* and *Veillonella* have been reported to bind to hydrophobic substrata (reviewed by Gibbons and Etherden, 1983; Clark et al., 1985; Doyle et al., 1990). Hydrophobic strains tended to bind to SHA better than hydrophilic variants. *Bacteroides* spp. generally are capable of adhering to hydrocarbons, but wide variations in adhesion have been reported (van Steenbergen et al., 1985; Okuda et al., 1986). Virulent stains of *Fusobacterium necrophorum* tend to be more hydrophobic than avirulent strains (Shinjo et al., 1987).

The foregoing encompasses only a portion of the literature supporting a role for hydrophobicity in oral bacterial adhesion. The substrata range from saliva-coated teeth, saliva-coated hydroxylapatite, hydrocarbons, biomaterials, such as plastics, to mucosa and gingival tissues. It would be expected, therefore, that appropriate inhibitors of the hydrophobic effect would be useful in the treatment or prevention of diseases dependent on oral microbial adhesion. In this regard, Goldberg and Rosenberg (1991) have shown that two-phase oil–water amphiphile mixtures are superior to commercial mouthwashes in removing adherent bacteria from solid surfaces. The use of such desorbing agents may have promise in removing adherent bacteria from various oral sites.

References

Abbott, A., and M.L. Hayes. 1984. The conditioning role of saliva in streptotoccal attachment to hydroxyapatite surfaces. *J. Gen. Microbiol.* **130**:809–816.

Aduse-Opoku, J., M.L. Gilpin, and R.R.B. Russell. 1989. Genetic and antigenic comparison of *Streptococcus mutans* fructosyltransferase and glucan-binding protein. *FEMS Microbiol. Lett.* **59**:279–282.

Babu, J.P. and M.K. Dabbous. 1986. Interaction of salivary fibronectin with oral streptococci *J. Dent. Res.* **65**:1094–1098.

Babu, J.P., W.A. Simpson, H.S. Courtney, and E.H. Beachey. 1983. Interaction of human plasma fibronectin with cariogenic and noncariogenic oral streptococci. *Infect. Immun.* **41**:162–168.

Babu, J.P., E.H. Beachey, and W.A. Simpson. 1986. Inhibition of the interaction of *Streptococcus sanguis* with hexadecane droplets by 55-and 66-kilodalton hydrophobic proteins of human saliva. *Infect. Immun.* **53**:278–284.

Babu, J.P., M.K. Dubbous, and S.N. Abraham. 1991. Isolation and characterization of a 180-kilodalton salivary glycoprotein which mediates the attachment of *Actinomyces naeslundii* to human buccal epithelial cells. *J. Periodont. Res.* **26**:97–106.

Banas, J., R.R.B. Russell, and J.J. Ferretti. 1990. Sequence analysis of the gene for the glucan-binding protein of *Streptococcus mutans Ingbritt. Infect. Immun.* **58**:667–673.

Beighton, D. 1982. The influence of manganese on carbohydrate metabolism and caries induction by *Streptococcus mutans* strain *Ingbritt. Caries Res.* **16**:189–192.

Beighton, D. 1983. Manganese, trace elements and dental disease. In: Curzon, M.E.J., T.W. Curtress, and A.F. Gardner (eds.), *Postgraduate Dental Handbook Series, vol. 9*. John Wright PSG, Inc. Littleton, MA, pp. 237–244.

Beighton, D. 1984. The influence of saliva on the hydrophobic surface properties of bacteria isolated from oral sites of macaque monkeys. FEMS *Microbiol. Lett.* **21**:239–242.

Bourgeau, G. and D. Mayrand. 1990. Aggregation of *Actinomyces* strains by extracellular vesicles produced by *Bacteroides gingivalis. Can J. Microbiol.* **36**:362–365.

Bourgeau, G. and B.C. McBride. 1976. Dextran-mediated interbacterial aggregation between dextran-synthesizing streptococci and *Actinomyces viscosus. Infect. Immun.* **13**:1228.

Busscher, H.J. and A.H. Weerkamp. 1987. Specific and non-specific interactions in bacterial adhesion to solid substrata. *FEMS Microbiol. Rev.* **46**:165–173.

Busscher, H.J., H.M. Uyen, A.W.J. van Pelt, A.H. Weerkamp, and J. Arends. 1986. Kinetics of adhesion of the oral bacterium *Streptococcus sanguis* CH3 to polymers with different surface free energies. *Appl. Environ. Microbiol.* **51**:910–914.

Busscher, H.J., J. Sjollema, and H.C. van der Mei. 1990. Relative importance of surface free energy as a measure of hydrophobicity in bacterial adhesion to solid surfaces. In: Doyle, R.J. and M. Rosenberg (eds.), *Microbial Cell Surface Hydrophobicity*. American Society for Microbiology, Washington, pp. 335–359.

Cassels, F.J., H.M. Fales, J. London, R.W. Carlson, and H. van Halbeek. 1990. Structure of a streptococcal adhesin carbohydrate receptor. *J. Biol. Chem.* **265**:14127–14135.

Ciardi, J.E., G. Rolla, W.H. Bowen, and J.A. Reilly. 1977. Adsorption of *Streptococcus mutans* lipoteichoic acid to hydroxyapatite. *Scand. J. Dent. Res.* **85**:387–391.

Cisar, J.O. 1986. Fimbrial lectins of the oral *Actinomyces*. In: Mirelman, D. (ed.), *Microbial Lectins and Agglutinins; Properties and Biological Activity*. John Wiley & Sons, New York, pp. 183–196.

Cisar, J.O., P.E. Kolenbrander, and F.C. McIntire. 1979. Specificity of coaggregation reactions between human oral streptococci and strains of *Actinomyces viscosus* and *Actinomyces naeslundii. Infect. Immun.* **24**:742–752.

Cisar, J.O., E.L. Barsumian, S.H. Curl, A.E. Vatter, A.L. Sandberg, and R.P. Siraganian. 1981. Detection and localization of a lectin of *Actinomyces viscosus* T14V by monoclonal antibodies. *J. Immunol.* **127**:1318–1324.

Cisar, J.O., M.J. Brennan, and A.L. Sandberg. 1985. Lectin-specific interaction of *Actinomyces* fimbriae with oral streptococci. In: Mergenhagen, S.E., and B. Rosan (eds.), *Molecular Basis of Oral Microbial Adhesion*. American Society for microbiology, Washington, pp. 159–163.

Cisar, J.O., A.L.Sandberg, and W.B. Clark. 1989. Molecular aspects of adherence of *Actinomyces viscosus* and *Actinomyces naeslundii* to oral surfaces. *J. Dent. Res.* **68**:1558–1559.

Clark, W.B., L.L. Bammann, and R.J. Gibbons. 1978. Comparative estimates of bacterial affinities and adsorption sites on hydroxyapatite surfaces. *Infect. Immun.* **19**:846–853.

Clark, W.B., M.D. Lane, J.E. Beem, S.L. Bragg, and T.T. Wheeler. 1985. Relative hydrophobicities of *Actinomyces viscosus* and *Actinomyces naeslundii* strains and their adsorption to saliva-treated hydroxyapatite. *Infect. Immun.* **47**:730–736.

Clark, W.B., J.E. Beem, W.E. Nesbitt, J.O. Cisar, C.C. Tseng, and M.J. Levine. 1989. Pellicle receptors for *Actinomyces viscosus* type 1 fimbriae *in vitro*. *Infect. Immun.* **57**:3003–3008.

Costello, A.H., J.O. Cisar, P.E. Kolenbrander, and O. Gabriel. 1979. Neuraminidase-dependent hemagglutination of human erythrocytes by human strains of *Actinomyces viscosus* and *Actinomyces naeslundii*. *Infect. Immun.* **26**:563–572.

Cowan, M.M., K.G. Taylor, and R.J. Doyle. 1986. Kinetics analysis of *Streptococcus sanguis* adhesion to artificial pellicle. *J. Dent. Res.* **65**:1278–1280.

Cowan, M.M., K.G. Taylor, and R.J. Doyle. 1987a. Role of sialic acid in the kinetics of *Streptococcus sanguis* adhesion to artificial pellicle. *Infect. Immun.* **55**:1552–1557.

Cowan, M.M., K.G. Taylor, and R.J. Doyle. 1987b. Energetics of the initial phase of adhesion of *Streptococcus sanguis* to hydroxylapatite. *J. Bacteriol.* **169**:2995–3000.

Curzon, M.E.J. 1983. Epidemiology of trace elements and dental caries. In: M.E. Curzon, T.W. Curtress, and A.F. Gardner (eds.), *Trace Elements and Dental Disease. Postgraduate Dental Handbook Series, vol. 9*. John Wright PSG, Inc., Littleton, MA, pp. 11–30.

Dawson, J.R. and R.P. Ellen. 1990. Tip-oriented adherence of *Treponema denticola* to fibronectin. *Infect. Immun.* **58**:3924–3928.

Demuth, D.R., E.E. Golub, and D. Malamud. 1990. Streptococcal-host interactions: structural and functional analysis of a *Streptococcus sanguis* receptor for a human salivary glycoprotein. *J. Biol. Chem.* **265**:7120–7126.

Doyle, R.J., W.E. Nesbitt, and K.G. Taylor. 1982. On the mechanism of adherence of *Streptococcus sanguis* to hydroxylapatite. FEMS *Microbiol. Lett.* **15**:1–5.

Doyle, R.J., M. Rosenberg, and D. Drake. 1990. Hydrophobicity of oral bacteria. In: Doyle, R.J. and M. Rosenberg (eds.), *Microbiol Cell Surface Hydrophobicity*. American Society for Microbiology, Washington, pp. 387–419.

Drake, D., K.G. Taylor, A.S. Bleiweis, and R.J. Doyle. 1988a. Specificity of the glucan-binding lectin of *Streptococcus cricetus*. *Infect. Immun.* **56**:1864–1872.

Drake, D., K.G. Taylor, and R.J. Doyle. 1988b. Expression of glucan-binding lectin of *Streptococcus cricetus* requires manganous ion. *Infect. Immun.* **56**:2205–2207.

Erickson, P.R. and M.C. Herzberg. 1987. A collagen-like immunodeterminant on the surface of *Streptococcus sanguis* induces platelet aggregation. *J. Immunol.* **138**:3360–3366.

Erickson, P.R. and M.C. Herzberg. 1990. Purification and partial characterization of a 65 kDa platelet aggregation-associated protein antigen from *Streptococcus sanguis*. *J. Biol. Chem.* **265**:14080–14087.

Fachon-Kalweit, S., B.L. Elder, and P. Fives-Taylor. 1985. Antibodies that bind to fimbriae block adhesion of *Streptococcus sanguis* to saliva-coated hydroxyapatite. *Infect. Immun.* **48**:617–624.

Fenno, J.C., D.J. LeBlanc, and P.M. Fives-Taylor. 1989. Nucleotide sequence of a type 1 fimbrial gene of *Streptococcus sanguis* FW213. *Infect. Immun.* **57**:3527–3533.

Fives-Taylor, P.M. and Thompson, D.W. 1985. Surface properties of *Streptococcus sanguis* FW 213 mutants non-adherent to saliva-coated hydroxyapatite. *Infect. Immun.* **47**:752–759.

Fujioka, Y., Y. Akagawa, S. Minagi, H. Tsuru, Y. Miyake, and H. Suginaka. 1987. Adherence of *Streptococcus mutans* to implant materials. *J. Biomed. Mater. Res.* **21**:913–920.

Ganeshkumar, N., P.M. Hannam, P.E. Kolenbrander, and B.C. McBride. 1991. Nucleotide sequence of a gene coding for a saliva-binding protein (Ssa B) from *Streptococcus sanguis* 12 and possible role of the protein in coaggregation with *Actinomyces*. *Infect. Immun.* **59**:1093–1099.

Gibbons, R.J. 1984. Adherent interactions which may affect microbial ecology in the mouth. *J. Dent. Res.* **63**:378–385.

Gibbons, R.J. and I. Etherden. 1983. Comparative hydrophobicities of oral bacteria and their adherence to salivary pellicles. *Infect. Immun.* **41**:1190–1196.

Gibbons, R.J. and R.J. Fitzgerald. 1969. Dextran-induced agglutination of *Streptococcus mutans* and its potential role in the formation of microbial dental plaques. *J. Bacteriol.* **98**:341–346.

Gibbons, R.J. and M. Nygaard. 1970. Interbacterial aggregation of plaque bacteria. *Arch. Oral Biol.* **15**:1397–1400.

Gibbons, R.J. and J.B. Qureshi. 1978. Selective binding of blood-group reactive salivary mucins by *Streptococcus mutans* and other oral organisms. *Infect. Immun.* **22**:665–671.

Gibbons, R.J. and J.V. Qureshi. 1979. Inhibition of adsorption of *Streptococcus mutans* strains to saliva-treated hydroxyapatite by galactose and certain amines. *Infect. Immun.* **26**:1214–1217.

Gibbons, R.J., I. Etherden, and Z. Skobe. 1983a. Association of fimbriae with the hydrophobicity of *Streptococcus sanguis* FC-1 and adherence to salivary pellicles. *Infect. Immun.* **41**:414–417.

Gibbons, R.J., E.C. Moreno, and I. Etherden. 1983b. Concentration-dependent multiple binding on saliva-treated hydroxyapatite for *Streptococcus sanguis*. *Infect. Immun.* **39**:280–289.

Gilmore, K.S., R.R.B. Russell, and J.J. Ferretti. 1990. Analysis of the *Streptococcus downei gtfS* gene, which specifies a glucosyltransferase that synthesizes soluble glucans. *Infect. Immun.* **58**:2452–2458.

Goldberg, S. and M. Rosenberg. 1991. Bacterial desorption by commercial mouthwashes vs two-phase oil:water formulations. *Biofouling* **3**:193–198.

Gotoh, N., S. Tanaka, and T. Nishino. 1989. Supersusceptibility to hydrophobic antimicrobial agents and cell surface hydrophobicity in *Branhamella catarrhalis*. FEMS *Microbiol. Lett.* **59**:211–214.

Hamada, S. and H. Slade. 1980. Biology, immunology, and cariogenicity of *Streptococcus mutans*. *Microbiol. Rev.* **44**:331–384.

Handley, P.S. 1990. Structure, composition and functions of surface structures on oral bacteria. *Biofouling* **2**:239–264.

Harty, D.W.S., M.D.P. Wilcox, J.E. Wyatt, P.C.F. Oyston, and P.S. Handley. 1990. The surface ultrastructure and adhesive properties of a fimbriate *Streptococcus sanguis* strain and six non-fimbriate mutants. *Biofouling* **2**:75–86.

Heeb, M.J., A.H. Costello, and O. Gabriel. 1982. Characterization of a galactose-specific lectin from *Actinomyces viscosus* by a model aggregation system. *Infect. Immun.* **38**:993–1002.

Herzberg, M.C., P.R. Erickson, P.K. Kane, D.J. Clawson, C.C. Clawson, and F.A. Hoff. 1990. Platelet-interactive products of *Streptococcus sanguis* protoplasts. *Infect. Immun.* **58**:4117–4125.

Hogg, S.D. and J.E. Manning. 1988. Inhibition of adhesion of viridans streptococci to fibronectin-coated hydroxyapatite beads by lipoteichoic acid. *J. Appl. Bacteriol.* **65**:483–489.

Hudson, M. and R. Curtiss, III. 1990. Regulation of expression of *Streptococcus mutans* genes important to virulence. *Infect. Immun.* **58**:464–470.

Jenkinson, H.F. and D.A. Carter. 1988. Cell surface mutants of *Streptococcus sanguis* with altered adherence properties. *Oral Microbiol. Immunol.* **3**:53–57.

Jones, S.J. 1972. A special relationship between spherical and filamentous microorganisms in mature human dental plaque. *Arch. Oral Biol.* **17**:613–616.

Kelstrup, J. and T.D. Funder-Nielsen. 1974. Aggregation of oral streptococci with *Fusobacterium* and *Actinomyces*. *J. Biol. Buccale* **2**:347–362.

Knox, K.W., L.N. Hardy, L.J. Markevics, J.D. Evans, and A.J. Wicken. 1985. Comparative studies on the effect of growth conditions on adhesion, hydrophobicity, and extracellular protein profile of *Streptococcus sanguis* G9B. *Infect. Immun.* **50**:545–554.

Kolenbrander, P.E. 1988. Intergeneric coaggregation among human oral bacteria and ecology of dental plaque. *Annu. Rev. Microbiol.* **42**:627–656.

Kolenbrander, P.E. 1989. Surface recognition among oral bacteria: multigeneric coaggregations and their mediators. *CRC Crit. Rev. Microbiol.* **17**:137–159.

Kolenbrander, P.E. 1991. Coaggregation: adherence in the human oral microbial ecosystem. In: M. Dworkin (ed.), *Microbial Cell–Cell Interactions. American Society for Microbiology,* Washington, pp. 303–329.

Kolenbrander, P.E. and R.N. Andersen. 1986. Multigeneric aggregations among oral bacteria: a network of independent cell-to-cell interactions. *J. Bacteriol.* **168**:851–859.

Kolenbrander, P.E. and R.N. Andersen. 1988. Intergeneric rosettes: sequestered surface recognition among human periodontal bacteria. *Appl. Environ. Microbiol.* **54**:1046–1050.

Kondo, W., M. Sato and H. Ozawa. 1976. Haemagglutinating activity of *Leptotrichia buccalis* cells and their adherence to saliva-coated enamel powder. *Arch. Oral Biol.* **21**:363–369.

Landale, E.C. and M.M. McCabe. 1987. Characterization by affinity electrophoresis of an α1,6 glucan-binding protein from *Streptococcus sobrinus. Infect. Immun.* **55**:3011–3016.

Lantz, M.S., R.W. Rowland, L.M. Switalski, and M. Hook. 1986. Interactions of *Bacteroides gingivalis* with fibrinogen. *Infect. Immun.* **54**:654–658.

Leach, S.A. 1970. A review of the biochemistry of dental plaque. In: McHugh, W.D. (ed.), *Dental Plaque.* Churchill Livingstone, Edinburgh, pp. 143–156.

Leach, S.A. and E.A. Agalamanyi. 1984. Hydrophobic interactions that may be involved in the formation of dental plaque. In: ten Cate, J.M., S.A. Leach, and J. Arends (eds.), *Bacterial Adhesion and Preventive Dentistry.* IRL Press, Oxford, pp. 43–50.

Leach, S.A. and C.A. Saxton. 1966. An electron microscopic study of the acquired pellicle and plaque formed on the enamel of human incisors. *Arch. Oral Biol.* **11**:1081–1094.

Leach, S.A., P. Critchley, A.B. Kolendo, and C.A. Saxton. 1967. Salivary glycoproteins as components of the enamel integuments. *Caries Res.* **1**:104–111.

Lee, S.F., A. Progulske-Fox, G.W. Erdos, D.A. Piacentini, G.Y. Aya-kawa, P.J. Crowley, and A.S. Bleiweis. 1989. Construction and characterization of isogenic mutants of *Streptococcus mutans* deficient in major surface protein antigen P1 (I/II). *Infect. Immun.* **57**:3306–3313.

Levine, M.J., M.C. Herzberg, M.S. Levine, S.A. Ellison, M.W. Stinson, and T. Van Dyke. 1978. Specificity of salivary-bacterial interaction: role of terminal sialic acid residues in the interaction of salivary glycoproteins with *Streptococcus sanguis* and *Streptococcus mutans. Infect. Immun.* **19**:107–115.

Liang, L., D. Drake, and R.J. Doyle. 1989. Stability of the glucan-binding lectin of oral streptococci. *J. Dent. Res.* **68**:1677.

Liljemark, W.F., S.V. Schauer, and C.G. Bloomquist. 1978. Compounds which affect the adherence of *Streptococcus sanguis* and *Streptococcus mutans* to hydroxyapatite. *J. Dent. Res.* **57**:373–379.

Liljemark, W.F., C.G. Bloomquist, and L.J. Fenner. 1985. Characteristics of the adherence of oral *Haemophilis* species to an experimental salivary pellicle and to other oral bacteria. In: Mergenhagen, S.E. and B. Rosan (eds.), *Molecular Basis of Oral Microbial Adhesion.* American Society for Microbiology, Washington, pp. 94–102.

Loesche, W.J. 1986. Role of *Streptococcus mutans* in human dental decay. *Microbiol. Rev.* **50**:353–380.

Loesche, W.J., S.A. Syed, E. Schmidt, and E.C. Morrison. 1985. Bacterial profiles of subgingival plaque in periodontitis. *J. Periodontol.* **56**:447–456.

London, J., A.R. Hand, E.I. Weiss, and J. Allen. 1989. *Bacteriodes loeschei* PK1295 cells express two distinct adhesins simultaneously. *Infect. Immun.* **57**:3940–3944.

Lowrance, J.H., D.L. Hasty, and W.A. Simpson. 1988. Adherence of *Streptococcus sanguis* to conformationally specific determinants in fibronectin. *Infect. Immun.* **56**:2279–2285.

Lowrance, J.H., L.M. Baddour, and W.A. Simpson. 1990. The role of fibronectin binding in the rat model of experimental endocarditis caused by *Streptococcus sanguis*. *J. Clin. Invest.* **86**:7–13.

Lü-Lü, J.S. Singh, M.Y. Galperin, D. Drake, K.G. Taylor, and R.J. Doyle. 1992. Chelating agents inhibit activity and prevent expression of streptococcal glucan-binding lectins. *Infect. Immun.* **60**:3807–3813.

Mangan, D.F., M.J. Novak, S.A. Vora, J. Mourad, and P.S. Kriger. 1989. Lectin-like interactions of *Fusobacterium nucleatum* with human neutrophils. *Infect. Immun.* **57**:3601–3611.

Marsh, P. and M. Martin. 1984. *Oral Microbiology,* 2nd ed. American Society for Microbiology, Washington.

McBride, B.C. and J.S. van der Hoeven. 1981. Role of interbacterial adherence in the colonization of the oral cavities of gnotobiotic rats infected with *Streptococcus mutans* and *Veillonella alcalescens*. *Infect. Immun.* **33**:467–474.

McBride, B.C., M. Song, B. Krasse, and J. Olsson. 1984. Biochemical and immunological differences between hydrophobic and hydrophilic strains of *Streptococcus mutans*. *Infect. Immun.* **44**:68–75.

McCabe, M.M. and R. Hamelik. 1978. Multiple forms of dextran-binding proteins from *Streptococcus mutans*. *Adv. Exp. Biol. Med.* **107**:749–759.

McCabe, M.M. and E.E. Smith. 1975. Relationship between cell-bound dextransucrase and the agglutination of *Streptococcus mutans*. *Infect. Immun.* **12**:512–520.

McCabe, M.M., A.U. Haynes, and R.M. Hamelik. 1976. Cell adherence of *Streptococcus mutans*. In: Stiles, H.M., W.J. Loesche, and T.C. O'Brien (eds.), *Proceedings Microbial Aspects of Dental Caries* (a special supplement to Microbiology Abstracts Vol. 3. Information Retrieval, Inc., Washington, pp. 413–424.

McCabe, M.M., R. Hamelik, and E. Smith. 1977. Purification of dextran-binding protein from cariogenic *Streptococcus mutans*. *Biochem. Biophys. Res. Commun.* **78**:273–278.

McIntire, F.C. 1985. Specific surface components and microbial coaggregation. In: Mergenhagen, S.E. and B. Rosan (eds.), *Molecular Basis of Oral Microbial Adhesion*. American Society for Microbiology, Washington, pp. 153–158.

McIntire, F.C., A.E. Vatter, J. Baros, and J. Arnold. 1978. Mechanism of coaggregation between *Actinomyces viscosus* T14V and *Streptococcus sanguis* 34. *Infect. Immun.* **21**:978–988.

McIntire, F.C., L.K. Crosby, and A.E. Vatter. 1982. Inhibitors of coaggregation between *Actinomyces viscosus* T14V and *Streptococcus sanguis* 34: β-galactosides, related sugars and anionic amphipathic compounds. *Infect Immun.* **36**:371–78.

McIntire, F.C., L.K. Crosby, J.J. Barlow, and K.L. Matta. 1983. Structural preferences of β-galactoside-reactive lectins on *Actinomyces viscosus* T14V and *Actinomyces naeslundii* WVU45. *Infect. Immun.* **41**:848–850.

McIntire, F.C., C.A. Bush, S.S. Wu, S.C. Li, Y.T. Li, M. McNeil, S.S. Tjoa, and P.V. Fennessey. 1987. Structure of a new hexasaccharide from the coaggregation polysaccharide of *Streptococcus sanguis* 34. *Carbohydr. Res.* **166**:133.

Meyer, D.H., P.K. Sreenivasan, and P.M. Fives-Taylor. 1991. Evidence for invasion of a human oral cell line by *Actinobacillus actinomycetemcomitans*. *Infect. Immun.* **59**:2719–2726.

Mongiello, J.R. and W.A. Falkler, Jr. 1979. Sugar inhibition of oral *Fusobacterium nucleatum* haemagglutination and cell binding. *Arch Oral Biol.* **24**:539–545.

Mooser, G. and C. Wong. 1988. Isolation of a glucan-binding domain of glucosyltransferase (1,6-α-glucan synthase) from *Streptococcus sobrinus*. *Infect. Immun.* **56**:880–884.

Morris, E.J. and B.C. McBride. 1983. Aggregation of *Streptococcus sanguis* by a neuraminidase-sensitive component of serum and crevicular fluid. *Infect. Immun.* **42**:1073–1080.

Morris, E.J. and B.C. McBride. 1984. Adherence of *Streptococcus sanguis* to saliva-coated hydroxyapatite: evidence for two binding sites. *Infect. Immun.* **43**:656–663.

Morris, E.J., N. Ganeshkumar, and B.C. McBride. 1985. Cell surface components of *Streptococcus sanguis:* relationship to aggregation, adherence, and hydrophobicity. *J. Bacteriol.* **164**:255–262.

Murray, P.A., M.J. Levine, L.A. Tabak, and M.S. Reddy. 1982. Specificity of salivary-bacterial interactions: II. Evidence for a lectin on *Streptococcus sanguis* with specificity for a NeuAcα2,3Gβ1,3GalNAc sequence. *Biochem. Biophys. Res. Commun.* **106**:390–396.

Nesbitt, W.E., R.J. Doyle, K.G. Taylor, R.H. Staat, and R.R. Arnold. 1982a. Positive cooperativity in the binding of *Streptococcus sanguis* to hydroxylapatite. *Infect. Immun.* **35**:157–165.

Nesbitt, W.E., R.J. Doyle, and K.G. Taylor. 1982b. Hydrophobic interactions and the adherence of *Streptococcus sanguis* to hydroxylapatite. *Infect. Immun.* **38**:637–644.

Nyvad, B. and O. Fejerskov. 1987a. Scanning electron microscopy of early microbial colonization of human enamel and root surfaces *in vivo*. *Scand. J. Dent. Res.* **95**:287–296.

Nyvad, B. and O. Fejerskov. 1987b. Transmission electron microscopy of early microbial colonization of human enamel and root surfaces *in vivo*. *Scand. J. Dent. Res.* **95**:297–307.

Nyvad, B. and M. Kilian. 1987. Microbiology of the early colonization of human enamel and root surfaces *in vivo*. *Scand. J. Dent. Res.* **95**:369–380.

Oakley, J.D., K.G. Taylor, and R.J. Doyle. 1985. Trypsin-susceptible cell surface characteristics of *Streptococcus sanguis*. *Can. J. Microbiol.* **31**:1103–1107.

Okuda, K., A. Yamamoto, Y. Naito, I. Takazoe, J. Slots, and R.J. Genco. 1986. Purification and properties of a hemagglutinin from culture supernatants of *Bacteroides gingivalis*. *Infect. Immun.* **54**:659–665.

Olsson, J. and C.G. Emilson. 1988. Implantation and cariogenicity in hamsters of *Streptococcus mutans* with different hydrophobicity. *Scand. J. Dent. Res.* **96**:85–90.

Olsson, J. and G. Westergren. 1982. Hydrophobic surface properties of oral streptococci. *FEMS Microbiol. Lett.* **15**:319–323.

Peros, W.J. and R.J. Gibbons. 1982. Influence of sublethal antibiotic concentrations on bacterial adherence to saliva-treated hydroxyapatite. *Infect. Immun.* **35**:326–334.

Peros, W.J., I. Etherden, R.J. Gibbons, and Z. Skobe. 1985. Alteration of fimbriation and cell hydrophobicity by sublethal concentrations of tetracycline. *J. Periodont. Res.* **20**:24–30.

Reynolds, E.C. and A. Wong. 1983. Effect of absorbed protein on hydroxyapatite zeta potential and *Streptococcus mutans* adherence. *Infect. Immun.* **39**:1285–1290.

Robrish, S.A., C. Oliver, and J. Thompson. 1987. Amino acid-dependent transport of sugars by *Fusobacterium nucleatum* ATCC 10953. *J. Bacteriol.* **169**:3891–3897.

Rogers, A.H., K. Pilowsky, and P.S. Zilm. 1984. The effect of growth rate on the adhesion of the oral bacteria *Streptococcus mutans* and *Streptococcus milleri*. *Arch. Oral Biol.* **29**:147–150.

Rolla, G., R.V. Oppermann, W.H. Bowen, J.E. Ciardi, and K.W. Knox. 1980. High amounts of lipoteichoic acid in sucrose-induced plaque *in vivo*. *Caries Res.* **14**:235–238.

Rolla, G., J.E. Ciardi, M. Deas, A. Lau, and W.H. Bowen. 1984. Adherence of active glucosyltransferase from *Streptococcus mutans* to ionic, hydrophobic and dextran surfaces. In: ten Cate, J.M., S.A. Leach, and J. Arends (eds.), *Bacterial Adhesion and Preventive Dentistry*. IRL Press, Oxford, pp. 133–142.

Rolla, G., A.A.Scheie, and J.E. Ciardi. 1985. Role of sucrose in plaque formation. *Scand J. Dent Res.* **93**:105–111.

Rosenberg, M., H. Judes, and E. Weiss. 1983. Cell surface hydrophobicity of dental plaque microorganisms *in situ*. *Infect. Immun.* **42**:831–834.

Rosenberg, M., A. Buivids, and P. Ellen. 1991. Adhesion of *Actinomyces viscosus* to *Porphyromonas (Bacteroides) gingivalis*-coated hexadecane droplets. *J. Bacteriol.* **173**:2581–2589.

Russell, R.R.B. 1979. Glucan-binding proteins of *Streptococcus mutans* serotype c. *J. Gen. Microbiol.* **112**:197–201.

Russell, R.R.B. 1990. Molecular genetics of glucan metabolism in oral streptococci. *Arch. Oral Biol.* **35**:53S–58S.

Russell, R.R.B., A.C. Donald, and C.W.I. Douglas. 1983. Fructosyltransferase activity of the glucan-binding protein from *Streptococcus mutans*. *J. Gen. Microbiol.* **129**:3243–3250.

Russell, R.R.B., D. Coleman, and G. Dougan. 1985. Expression of a gene for glucan-binding protein from *Streptococcus mutans* in *Escherichia coli*. *J. Gen. Microbiol.* **131**:295–299.

Russell, R.R.B., E. Abdulla, M.L. Gilpin, and K. Smith. 1986. Characterization of *Streptococcus mutans* surface antigens. In: Hamada, S., S.M. Michalek, H. Kiyono, L. Menaker, and J.R. McGhee (eds.), *Molecular Microbiology and Immunobiology of Streptococcus mutans*. Elsevier, Amsterdam, pp. 61–70.

Satou, J., A. Fukunaga, N. Satou, H. Shintani, and K. Okuda. 1988. Streptococcal adherence on various restorative materials. *J. Dent. Res.* **67**:588–591.

Saunders, J.M. and C.H. Miller. 1980. Attachment of *Actinomyces naeslundii* to human buccal epithelial cells. *Infect. Immun.* **29**:981–989.

Schilling, K.M. and W.H. Bowen. 1988. The activity of glucosyltransferase absorbed onto saliva-coated hydroxyapatite. *J. Dent. Res.* **67**:2–8.

Schilling, K.M., M.H. Blitzer, and W.H. Browen. 1989. Adherence of *Streptococcus mutans* to glucans formed *in situ* in salivary pellicle. *J. Dent. Res.* **68**:1678–1680.

Seow, W.K., G.J. Seymour, and Y.H. Thong. 1987. Direct modulation of human neutrophil adherence of coaggregating periodontopathic bacteria. *Int. Arch. Allergy Appl. Immun.* **83**:121–128.

Seow, W.K., P.S. Bird, G.J. Seymour, and Y.H. Thong. 1989. Modulation of human neutrophil adherence by periodontopathic bacteria: reversal by specific monoclonal antibodies. *Int. Arch Allergy Appl. Immun.* **9**:24–30.

Shinjo, T., H. Hazu, and H. Kiyoyama. 1987. Hydrophobicity of *Fusobacterium necrophorum* biovars A and B. *FEMS Microbiol. Lett.* **48**:243–247.

Slots, J., D. Moenbo, J. Langebaek, and A. Frandsen. 1978. Microbiota of gingivitis in man. *Scand J. Dent. Res.* **86**:174–181.

Socransky, S.S., A.D. Manganiello, D. Propas, V. Oram, and J. van Houte. 1977. Bacteriological studies of developing supragingival plaque. *J. Periodontal Res.* **12**:90–106.

Svanberg, M., G. Westergren, and J. Olsson. 1984. Oral implantation in humans of *Streptococcus mutans* strains with different degrees of hydrophobicity. *Infect. Immun.* **43**:817–821.

Syed, S.A. and W.J. Loesche. 1978. Bacteriology of human experimental gingivitis: effect of plaque age. *Infect. Immun.* **21**:821–829.

Tardiff, G., M.C. Sulavik, G.W. Jones, and D.B. Clewell. 1989. Spontaneous switching of the sucrose-promoted colony phenotype in *Streptococcus sanguis*. *Infect. Immun.* **57**:3945–3948.

Theilade, E. 1990. Factors controlling the microflora of the healthy mouth. In: Hill, M.J. and P.D. Marsh (eds.), *Human Microbial Ecology*. CRC Press, Boca Raton, FL, pp. 1–56.

Thomas, D.D., B. Baseman, and F. Alderete. 1986. Enhanced levels of attachment of fibronectin-primed *Treponema pallidum* to extracellular matrix. *Infect. Immun.* **52**:736–741.

Townsend-Lawman, P. and A.S. Bleiweis. 1991. Multilevel control of extracellular sucrose metabolism in *Streptococcus salivarius* by sucrose. *J. Gen. Microbiol.* **137**:5–13.

van Pelt, A.W.J., A.H. Weerkamp, M.H.W.J.C. Uyen, H.J. Busscher, H.P. deJong, and J. Arends. 1985. Adhesion of *Streptococcus sanguis* CH3 to polymers with different surface free energies. *Appl. Environ. Microbiol.* **49**:1270–1275.

van Steenbergen, T.J.M., F. Namavar, and J. de Graaff. 1985. Chemiluminescence of human leukocytes by black-pigmented *Bacteroides* strains from dental plaque and other sites. *J. Periodont. Res.* **20**:58–71.

van der Mei, H.C., A.H. Weerkamp, and H.J. Busscher. 1987. Physico-chemical surface characteristics and adhesive properties of *Streptococcus salivarius* strains with defined cell surface structures. *FEMS Microbiol. Lett.* **40**:15–19.

Weerkamp, A.H. and P.S. Handley. 1986. The growth rate regulates the composition and density of the fibrillar coat on the surface of *Streptococcus salivarius* K^+ cells. *FEMS Microbiol. Lett.* **33**:179–183.

Weerkamp, A.H., H.C. van der Mei, and J.W. Slot. 1987. Relationship of cell surface morphology and composition of *Streptococcus salivarius* K^+ to adherence and hydrophobicity. *Infect. Immun.* **55**:438–445.

Weiss, E., M. Rosenberg, H. Judes, and E. Rosenberg. 1982. Cell-surface hydrophobicity of adherent oral bacteria. *Curr. Microbiol.* **7**:125–128.

Weiss, E.I., J. London, P.E. Kolenbrander, and R.N. Andersen. 1989. Fimbria-associated adhesin of *Bacteroides loeschei* that recognizes receptors on procaryotic and eucaryotic cells. *Infect. Immun.* **57**:2912–2913.

Westergren, G. and J. Olsson. 1983. Hydrophobicity and adherence of oral streptococci after repeated subculture in vitro. *Infect. Immun.* **40**:432–435.

Wilkes, P.D. and Leach, S.A. 1979. The factors involved in the adsorption of glycoproteins from saliva onto hydroxyapatite surfaces. *J. Dent.* **7**:213–220.

Willcox, M.D.P. and D.B. Drucker. 1989. Surface structures, coaggregation and adherence phenomena of *Streptococcus oralis* and related species. *Microbios* **59**:19–29.

Willcox, M.D.P., J.E. Wyattt, and P.S. Handley. 1989. A comparison of the adhesive properties and surface ultrastructure of the fibrillar *Streptococcus sanguis* 12 and an adhesion deficient nonfibrillar mutant 12 na. *J. Appl. Bacteriol.* **66**:291–299.

Winkler, J.R., S.R. John, R.H. Kramer, C.I. Hoover, and P.A. Murray. 1987. Attachment of oral bacteria to a basement-membrane-like matrix and to purified matrix proteins. *Infect. Immun.* **55**:2721–2726.

Yamazaki, Y., S. Ebisu, and H. Okada. 1981. *Eikenella corrodens* adherence to human buccal epithelial cells. *Infect. Immun.* **31**:21–27.

Zhang, X.-hua, M. Rosenberg, and R.J. Doyle. 1990. Inhibition of the cooperative adhesion of *Streptococcus sanguis* to hydroxylapatite *FEMS Microbiol. Lett.* **71**:315–318.

9

Regulation and Expression of Bacterial Adhesins

There are various levels of complexity in bacterial cell surfaces. In Chapter 4, the main features of the Gram-negative and Gram-positive surfaces were reviewed. It was seen for both cell types that their surfaces were composed of components that associate noncovalently to form a supramolecular structure. These noncovalently associated molecules may be essential for survival of the bacteria in a particular environment. In general, it seems that adhesins are of the "associated" molecules. For example, in the case of *S. pyogenes,* M-proteins possess a domain that enables the proteins to be anchored to the wall matrix, whereas another domain is exposed and capable of interacting with lipoteichoic acid (LTA). In this sense then, the adhesin (LTA) is "carried" by M-protein or other surface proteins in order for the bacterium to adhere to an appropriate substratum (see also Chapter 6 for details). Similarly, in the Gram-negative *Enterobacteriaceae,* the adhesin(s) molecules are associated with fimbrial subunits forming a complex in which the fimbrial subunit is the major protein (Figure 9–1) (see Chapter 5). This is especially true in *E. coli* and may be true for other *Enterobacteriaceae* as well. It may, therefore, be a common strategy for bacteria to assemble the adhesin onto some other surface structure that serves as a carrier thereby forming an adhesive complex. This strategy would ensure that the adhesin is one of the most exposed molecules on the bacterial surface. The adhesin complex not only must be expressed in functional form on the bacterial surfaces but the organisms must also be able to regulate the expression or presentation of the adhesins in a functional form during the infectious process. Indeed, in all systems studied, one can always find adhesive and nonadhesive phenotypes of the same bacterial clone, and in many cases growth conditions that favor growth of one phenotype over the other have been defined.

The aim of the study of the molecular biology of adhesion is to understand the structure–function relationships of genetic elements encoding for products that enable the organisms to assemble the adhesin complexes on their surfaces.

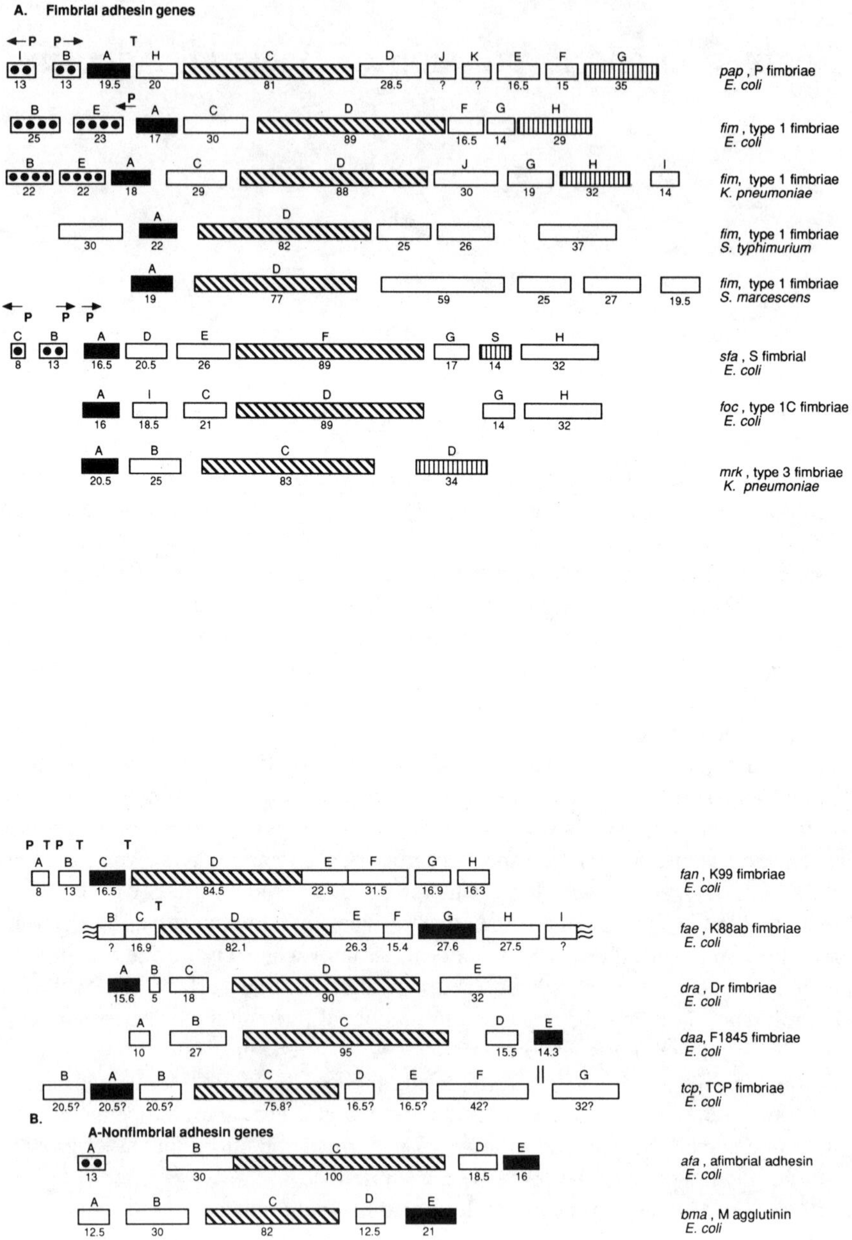

Figure 9–1 Legend on opposite page

This is best achieved by cloning the genetic elements coding for each of the components in the complex. Analysis of gene expression in minicells or maxicells, the gene nucleotide sequences, and the behavior of bacterial constructs harboring the genetic elements would reveal the chemical nature, molecular organization, biogenesis, regulation, and expression of the adhesin complexes. The study of the molecular biology of a particular system of adhesion is also best achieved when the receptor specificity of the adhesin has been defined. It is therefore not surprising that extensive studies on the molecular biology of adhesins have begun only during the last decade, focussing mainly on pathogenic bacteria. In particular, the proteinaceous adhesins expressed by several pathogens possessing lectin or agglutinin activities have attracted the most interest. This is probably because these adhesins and their carrier molecules are the products of a single gene cluster. Furthermore, their functional activities (e.g., ability to bind cell surface glycoconjugages) can be monitored by simple procedures with relatively high degrees of accuracy (see Chapter 2). The assay procedures also allow for selection of better defined mutants.

It has become apparent, as expected, that besides the structural genes of the

Figure 9–1. Comparison of the physical map of the gene cluster encoding for fimbrial and nonfimbrial adhesin complexes. The gene clusters for *pap* (*E. coli*), type 1 (*E. coli, K. pneumoniae, S. typhimurium,* and *S. marcescens*), type S (*E. coli*), type 1C (*E. coli*), type 3, MR/K (K. *pneumoniae*), K99 (*E. coli*), K88ab (*E. coli*), and TCP (V. cholerae) fimbrial adhesins and for the nonfimbrial adhesin complexes AFA and M agglutinin, respectively, are shown. The locations of the genes are indicated by boxes and the approximate molecular weights of the proteins encoded by the genes are given in kilodaltons below the boxes. The established or postulated functions of the gene products are indicated as follows: black boxes represent genes encoding for the major subunit of the adhesin complexes; the diagonal shaded boxes represent genes encoding for the large outer membrane protein anchor of the adhesin complexes; the vertical shaded boxes represent genes encoding for minor structural proteins that mediate adhesion; the boxes with closed circles represent genes encoding for proteins involved in regulation; and the open boxes represent genes encoding for either minor structural proteins or periplasmic proteins involved in transport of the structural proteins (chaperones) and initiation and termination of the assembly of the adhesin complexes. The major subunits of the nonfimbrial adhesin complexes (AFA and M agglutinin) and of the K88ab, K99, Dr, and F1845 fimbrial adhesin complexes also function as adhesins that determine the receptor specificity. The main direction of transcription is from left to right and whenever available the position and transcriptional orientation of promoter and terminator DNA sequences within the gene cluster have been shown. Adopted from Tennent et al. (1990) for *pap;* Beachey et al. (1988) for *fim* of *E. coli;* Gerlach et al. (1989, 1988) for *fim* of K. *pneumoniae, S. typhimurium, and S. marcescens and* for *mrk of* K. *pneumoniae;* Hacker (1990) and Ott et al. (1987) for *Sfa;* Riegman et al. (1990) for *foc;* DeGraaf (1990) for *fae* and *fan;* Nowicki et al. (1989) for *dra;* Bilge et al. (1989) for *daa;* Taylor et al. (1988) for *tcp;* Labigne-Roussel and Falkow (1988) for *afa;* and Rhen et al. (1986b) for *bma*.

adhesins, the biosynthesis and expression of the adhesins on bacterial surfaces in a functional form is complex. This complexity requires regulatory genes and genes coding for the carrier molecules and products essential for adhesin transport, assembly, and anchoring on cell surfaces. Table 9–1 summarizes the cloned genetic elements coding for the adhesin complexes and their biogenesis in various bacteria. In some cases only the genes coding for the carrier molecules have been cloned (e.g., M-protein of *S. pyogenes*), whereas in others (e.g., fimbrial adhesins of *E. coli*) most of the genes in the gene clusters required for the biosynthesis and expression of the adhesin complexes have been cloned. The most salient examples that have been studied in more detail are presented below. These include adhesins of *E. coli,* one of the species that has been studied extensively and *N. gonorrhoeae,* which contains a unique mechanism for regulating the expression of its adhesin complex. General principles for the formation and function of adhesin complexes will emerge from the following detailed discussion of the molecular biology of the adhesins.

E. coli Adhesins

E. coli adhesins can be classified according to a number of properties, the most important of which is their receptor specificity. Structural and serological properties of *E. coli* clones that colonize a particular site and a particular animal species are also important. According to their receptor specificities, most *E. coli* isolates studied to date are genotypically capable of expressing two types or classes of adhesins, one of which is always the MS adhesin, whereas the other is referred to as an MR adhesin (Duguid and Old, 1980; Sharon and Ofek, 1986; Jann and Hoschützky, 1990). The MR adhesins are classified further by their receptor specificities. Although only a few MR adhesin receptors have been identified, it is assumed that each *E. coli* clone associated with a definite site of colonization and type of infection in a particular host usually produces an MR adhesin with a distinct receptor specificity (Mirelman and Ofek, 1986). In most cases the receptor that binds adhesins is present on erythrocytes of one or more species, and hence, genetically manipulated phenotypes can be monitored for their ability to express functional adhesins by hemagglutination reactions.

The adhesins produced by *E. coli* may be associated with fimbrial or nonfimbrial structures. The fimbrial adhesins may be of two morphological types. One type is represented by rigid structures with a diameter of approximately 7 nm and an apparent axial hole (e.g., types 1 and P fimbriae). The other type is represented by flexible and very thin filamentous appendages of a diameter of 2–5 nm. These latter appendages have been termed "fibrillar" fimbriae (Jones and Isaacson, 1983). The length of the fimbrial adhesins varies between 0.5 and 10 μm (Chapter 4). The nonfimbrial or afimbrial adhesins are also in complex form with other surface structures. Immunoelectron microscopy has shown that these adhesins surround the bacterial surface, reminiscent of capsules, and have been termed "adhesive protein capsules" (Kröncke et al., 1990) (Chapter 4).

Immunologically, the adhesin complex may be identified by epitopes contained either in the adhesin molecule or in the carrier molecules. Today, the serological classification is based on the antigenic relatedness of the major fimbrial subunit. Fimbrial surface antigens are known as "F" serotypes (Orskov and Orskov, 1990). In some cases, one single serotype is given to a fimbrial adhesin in one receptor specificity (e.g., F1 for MS type 1 fimbrial adhesin) although, as shall be seen later, the major subunit of type 1 fimbriae of *E. coli* may exhibit a number of immunological variants. In other cases a number of fimbrial serotypes are designated to a group of fimbrial adhesins of one receptor specificity (e.g., F7 to F16 for the P-fimbrial adhesins specific for Galα1,4Gal-containing receptors). It is also possible to identify one fimbrial serotype that carries adhesins of a different receptor specificity (e.g., F13 fimbrial serotype may be associated with either the Galα1,4Gal-specific adhesin or with an adhesin specific for Forssman antigen (Table 9–1). Many *E. coli* adhesins have not been serologically classified. Serotyping of fimbrial adhesins, therefore, is of little help in understanding the complexity of the molecular biology of *E. coli* adhesion, but serotyping may serve to illustrate some major principles involved in the characterization of adhesin complex structures.

The nature and organization of the *E. coli* adhesin complex was not understood until sophisticated genetic studies on a number of fimbrial (and later nonfimbrial) adhesins revealed that the adhesiveness of a phenotype is encoded by a gene cluster. In the following, the molecular nature and organization of specific fimbrial adhesins that have been studied the most extensively will be discussed. Later, the emphasis will be on the common features of the gene cluster encoding the fimbrial adhesin. Some information on nonfimbrial or afimbrial adhesins will be discussed for comparison purposes.

Suprastructures of adhesin–fimbriae complexes and their expression on the surfaces of an *E. coli* are under the control of a single set of genes, or a gene cluster. The gene cluster is composed of regulatory genes and genes encoding for major fibrillin and minor fibrillin components, including those that function as adhesins and proteins essential for expression and assembly. Fig. 9-1 depicts the gene clusters coding for the adhesin–fimbriae complexes of different *E. coli* clones that have been studied in the most detail. The clones have been examined as regards the cloning of the respective genetic determinants in a suitable vector, construction of deletion and transposon–insertion mutants, analysis of gene expression in minicells/maxicells, and nucleotide sequence analyses.

Mannose-specific Type 1 Fimbrial Adhesins

Although most members of *Enterobacteriaceae* (Duguid and Old, 1980; Mirelman and Ofek, 1986) express the MS type 1 fimbrial adhesin, many of the

Table 9–1. Cloned determinants coding for adhesin complexes of bacteria

Bacteria	Source of isolation	Adhesin type (serotype)	Gene abbreviation and location (C=chromosome, P=plasmid)		References
Actinomyces viscosus	Oral, human	Type 1 fimbriae	*fim*	(C)	Yeung et al., 1987; Yeung and Cisar, 1988, 1990
		Type 2 fimbriae	*fim*A	(C)	Donkersloot et al., 1990
Actinomyces naeslundi	Oral, human	Type 2 fimbriae	*fim*A	(C)	Yeung and Cisar, 1990
Bordetella pertusis	Respiratory tract, human	Filamentous hemagglutinin	*fha*B	(C)	Mattei et al., 1986; Brown and Parker, 1987; Jacob et al., 1988; Stibitz et al., 1988; Relman et al., 1989; Delisse-Gathoye et al., 1990
Enterobacter cloacae	Wound, human	Type 1 fimbriae	*fim*	(C)	Clegg et al., 1985a,b,
Escherichia coli	All isolates	Type 1 fimbriae	*pil* or *fim*	(C)	Hull et al., 1981; Orndorff and Falkow, 1984a,b; Klemm et al., 1985
	Extraintestinal, human	P fimbriae (F13, F14)	*pap*	(C)	Hull et al., 1981; Normark et al., 1983; High et al., 1988
		($F7_2$)	*fst*	(C)	Van Die et al., 1983, 1984b; Rhen et al., 1983a, 1985b
		($F7_1$)	*fso*	(C)	Van Die et al., 1985; Rhen et al., 1983a, 1985a,b
		(F8,F9)	*fei*	(C)	Hacker et al., 1986b; de Ree et al., 1985
		(F11,F12)	*fei*	(C)	Clegg, 1982; Clegg and Pierce, 1983; de Ree et al., 1985; High et al., 1988
		S fimbriae	*sfa*	(C)	Berger et al., 1982; Hacker et al., 1985; Ott et al., 1987
		Type 1C fimbriae	*foc*	(C)	Van Die et al., 1984a, 1985;

				Riegman et al., 1990
	G fimbriae		(C)	Rhen et al., 1986a
	Dr fimbriae-like hemagglutinin	*dra*	(C)	Nowicki et al., 1987, 1988, 1989, 1990
	Afimbrial-1	*afa*I	(C)	Labigne-Roussel et al., 1984, 1985
	Afimbrial-2	*afa*II	(C)	Labigne-Roussel and Falkow 1988
	Afimbrial-3	*afa*III	(C)	Labigne-Roussel and Falkow, 1988
	M agglutinin	*bma*	(C)	Rhen et al., 1986a
	Nonfimbrial	*nfa*I	(C)	Hales et al., 1988
Dogs (humans)	P-related fimbriae (F13)	*prs* or *pap-2*	(C)	Lund et al., 1988a; Lindstedt et al., 1989
Intestinal, humans	CFA I fimbriae		(P)	Willshaw et al., 1983, 1985
	CFA II fimbriae (CS3)		(P)	Manning et al., 1985; Boylan et al., 1987, 1988
	(469-3) fimbriae		(C)	Hinson et al., 1987
	F1845 fimbriae	*dda*	(C&P)	Bilge et al., 1989
	OMP	*eae*	(C)	Jerse et al., 1990; Jerse and Kaper, 1991
Porcine	K88 fimbriae (K88ab)	*fae*	(P)	Mooi et al., 1979, 1981, 1982, 1984; Shipley et al., 1981; Kehoe et al., 1981, 1983
	(K88ac)	*adh*	(P)	Kehoe et al., 1981, 1983
	987P fimbriae		(P)	DeGraff and Klaasen, 1987; Morrissey and Dougan, 1986; Schifferli et al., 1990
Bovine	F17-A fimbriae	*f17*	(C)	Lintermans et al., 1988
Bovine, porcine	K99 fimbriae	*fan*	(C)	Van Embden et al., 1980; DeGraaf et al., 1984; Roosendaal et al., 1987a,b

continued

Table 9–1. Continued

Bacteria	Source of isolation	Adhesin type (serotype)	Gene abbreviation and location (C=chromosome, P=plasmid)	References
		F41 fimbriae	(C)	Moseley et al., 1986, Anderson and Moseley, 1988
	Rabbit	AF/R1 fimbriae	*afr* (P)	Cheney et al., 1990; Wolf and Boedeker, 1990
Haemophilus influenzae	Respiratory tract, human	Fimbriae (type b)	*hif*B (C)	Gilsdorf et al., 1990; Langermann and Wright, 1990; Van Ham et al., 1989
		LKP fimbriae (untypable)	(C)	Kar et al., 1990
Kelbsiella pneumoniae	Urine, human	Type 1 fimbriae	*fim* (C)	Purcell and Clegg, 1983; Clegg et al., 1985, 1987; Gerlach et al., 1989; Gerlach and Clegg, 1988a,b
		MR/K fimbriae	*mrk* (C)	Gerlach et al., 1988, 1989
Moraxella bovis	Eye, bovine	Q(β) fimbriae	(C)	Marrs et al., 1985, 1988; Fulks et al., 1990; Elleman et al., 1990; Beard et al., 1990
Mycoplasma pneumoniae	Respiratory tract, human	P1 adhesin	(C)	Frydenberg et al., 1987, Schaper et al., 1987; Su et al., 1987, 1988; Trevino et al., 1986; Dallo et al., 1988
Neisseria gonorrhoeae	Urogenital	Fimbriae	*pil* (C)	Meyer et al., 1982; Hagblom et al., 1985; Nicolson et al., 1987
		PII protein	*opa* (C)	Stern et al., 1984, 1986
Neisseria meningitidis	Respiratory tract	Class 1 fimbriae	*pil* (C)	Perry et al., 1988

Porphyromonas gingivalis (Bacteroides)	Oral, human	Fimbriae	*pil*	(C)	Dickinson et al., 1988
Pseudomonas aeruginosa	Respiratory tract	Fimbriae	*pil*	(C)	Nunn et al., 1990
Salmonella typhimurium	Intestinal, human	Type 1 fimbriae	*fim*	(C)	Clegg et al., 1985, 1987; Gerlach et al., 1989
Serratia marcescens	Sputum, human	Type 1 fimbriae	*fim*	(C)	Clegg et al., 1985, 1987; Gerlach et al., 1989
Staphylococcus aureus	Skin, respiratory human	Fibronectin binding protein	*fnb*	(C)	Flock et al., 1987
Streptococcus mutans	Oral, human	Saliva receptor		(C)	Somer et al., 1987; Demuth et al., 1988
		Glucan binding protein	*gbp*	(C)	Russell et al., 1985; Banas et al., 1990
Streptococcus sobrinus	Oral,human	Spa A protein	*spa*A	(C)	Holt et al., 1982
Streptococcus sanguis	Oral, human	Fimbriae		(C)	Fives-Taylor et al., 1987; Fenno et al., 1989; Demuth et
		SSP-5A protein		(C)	al., 1988; Ganeshkumar et al.,
		36 kDa protein		(C)	1988, 1991; Rosan et al.,
		80 kDa complex	*ssaB*	(C)	1989
Streptococcus pyogenes	Respiratory tract, human	M protein			
		(type 6)	*emm*6.1	(C)	Scott and Fischetti 1983;
		(type 12)		(C)	Spanier et al., 1984; Kehoe et
		(type 5)	*smp*5	(C)	al., 1985; Mouw et al., 1988
		(type 24)	*smp*24	(C)	
Vibrio cholerae	Intestinal, human	TCP fimbriae	*tcp*	(C)	Shaw and Taylor 1990; Shaw et al., 1990; Taylor et al., 1988
Yersinia enterocolitica	Intestinal, human	Invasin	inv_{ent}	(C)	Miller and Falkow 1988;
		Ail	*ail*	(C)	Miller et al., 1989, 1990;
		Yad A (YOP1)	*Yad*A	(C)	Balligand et al., 1985; Heesemann and Gruter, 1987
Yersinia pseudotuberculosis	Intestinal, human	Invasin	inv_{pstb}	(C)	Isberg and Falkow, 1985;
		Yad A (YOP1)	*Yad*A	(P)	Isberg, 1989; Paerregaard et al., 1991

molecular biology studies have been performed on laboratory derivatives of fecal or extraintestinal isolates of *E. coli*. Because virtually all *E. coli* strains contain the gene cluster coding for the type 1 fimbrial adhesin complex, the choice of laboratory *E. coli* strains as cloning hosts for cloned-gene manipulation and genetic analysis of the type 1 fimbrial genes should be conducted in only well-defined strains. Thus, lack of expression under any growth conditions of the type 1 fimbrial adhesin by a laboratory strain of *E. coli* could result from mutation and/or deletion of one gene in the type 1 gene cluster of the host strain rather than deletion of the entire gene cluster. Unfortunately, this complexity was not realized until recently (Blomfield et al., 1991), resulting in some conflicting results to be discussed later, especially with respect to the role of the regulatory genes in the type 1 gene cluster. The gene cluster coding for the expression of the type 1 fimbrial adhesin resides on the bacterial chromosome at a position corresponding to 98 minutes on the *E. coli* linkage map (Brinton et al., 1961; Buchanan et al., 1985). Type 1 fimbriae are composed of a major 17,000 mol wt subunit aligned in a right-handed helical array with a diameter of 7 nm (Brinton, 1965) (Figure 9–2). In addition to the major subunit, three minor subunits of mol wt 16,000, 17,500, and 32,000 have been identified (Maurer and Orndorff, 1985, 1987; Abraham et al., 1987, 1988a,b; Klemm and Christiansen, 1987; Hanson and Brinton, 1988; Krogfelt and Klemm, 1988; Minion et al., 1986). Until recently, it was thought that the D-mannose binding site was located in the major 18,000 mol wt structure subunit. However, after the genes required for the synthesis and expression of type 1 fimbriae were cloned and mapped, it soon became clear from insertional and deletional mutational analyses that the D-mannose binding activity is dependent on a minor 28–31,000 mol wt protein encoded by *fim*H (*pilE*) rather than in the major structural subunit encoded by *fim*A. Such mutations in *fim*H result in the formation of fimbriae that lack D-mannose binding activity and the organisms fail to agglutinate guinea pig erythrocytes or mannan-containing yeast cells. Recent results have shown that Fim H is an important determinant of adhesion and colonization. Antibodies against Fim H abolish these activities. Keith et al (1986) demonstrated that insertion mutations in *fim*H abolished the ability of a clinical isolate of *E. coli* to colonize the rat bladder. Abraham et al. (1983, 1985) have shown that polyclonal and monoclonal antibodies against Fim H inhibited adhesion of type 1 fimbriated *E. coli* in vitro. Finally, physical evidence for the mannose binding activity of the Fim H protein was provided by Krogfelt et al. (1990), who showed that of the various fimbrial subunits blotted on nitrocellulose only Fim H reacted with a mannosylated-bovine serum albumin (BSA) neoglycoprotein. This interaction was inhibited by D-mannose.

In addition to their role in biogenesis, other role(s) for the other minor fimbrial proteins encoded by *fim*F (17,0005 mol wt) and *fim*G (16 kDa) remain(s) unclear. It has been suggested that they may act in concert to stabilize the Fim H in the fimbrial structure (Krogfelt and Klemm, 1988). It has been reported that *fim*G

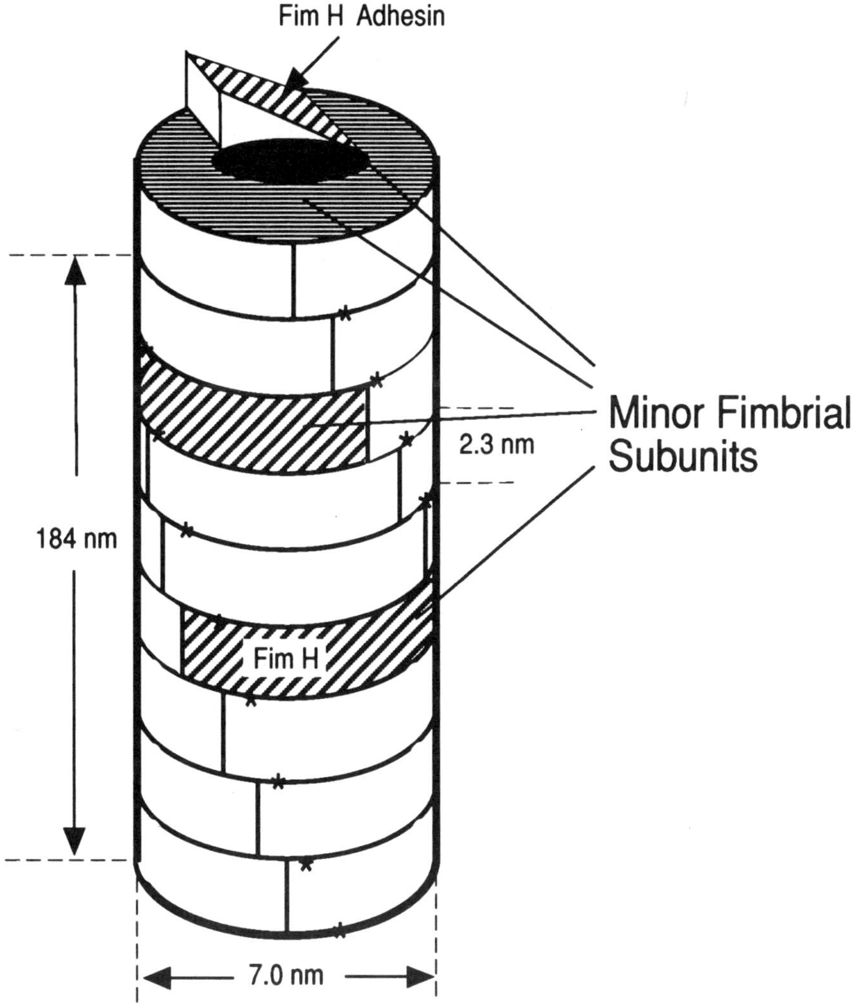

Figure 9–2. Model depicting the helical structure of a type 1 fimbrium. Minor fimbrial subunits are shaded. The Fim H adhesin is at the very tip of the fimbrium. The site indicated by an asterisk is the site where a monoclonal antibody binds. This site represents regular intervals of eight turns or 18.4 nm. (Model derived from Brinton, 1965 and Abraham et al., 1983.)

controls the length of the fimbriae (Maurer and Orndorff, 1985; Klemm and Christiansen, 1987). All three genes are not necessary, however, for the production and expression of the major *fim*A subunit in fimbrial form, because their deletion from a plasmid did not affect the expression of fimbriae by a host cell harboring the plasmid (Krogfelt and Klemm, 1988). In contrast, the expression of these genes seems to require production, assembly, and expression of Fim A in fimbrial form because nonfimbriated phenotypes usually lack MS adhesive properties. Nevertheless, recipient *E. coli* cells, which harbor a recombinant plasmid containing a 6.5-kb insert derived from the type 1 gene cluster, did not produce fimbriae but caused mannose-sensitive hemagglutination (Hultgren et al., 1990). Analysis of the polypeptides encoded by the fragment revealed that it encodes the products of the *fim*D, *fim*H, and probably *fim*F. The results suggest these gene products are required for functional orientation of the MS adhesin on the bacterial surface in the absence of fimbriae. Hultgren et al. (1990) found that the hemagglutinating recombinant strain was not able to adhere to bladder cells and suggested that fimbriae may not be essential for proper orientation of the MS adhesin for the strain to cause hemagglutination but may be required to orient the MS adhesin for binding to other types of animal cells. This reemphasizes the notion that cell–cell interaction is a complex phenomenon and proper orientation of both the adhesin and its receptor of the surface of the cells is required to form a firm attachment that can withstand shearing forces. The minor proteins resist dissociation from the main fimbrial subunit unless the fimbrial preparation is first treated with boiling sodium dodecyl sulfate (SDS). Nevertheless, Hanson and Brinton (1988) were able to fractionate the main fimbrial rod from the three minor proteins. They precipitated the minor protein-free fimbriae with 5% SDS for 24 hours at room temperature and then collected from the supernatant the Fim A-free minor protein components. In this manner the 32,000 mol wt adhesin (Fim H) was purified and its amino acid composition and amino-terminal sequences were determined (Hanson and Brinton, 1988). It was found to be similar to results expected from the DNA sequence of the region containing the three genes coding for the minor fimbrial proteins. There is a limited degree of homology (up to 20%) and there are substantial differences in amino acid composition between Fim H and the major fimbrial subunit (Figure 9–3). However, at the carboxy-terminal regions there are reasonably high degrees of homology between Fim H, Fim A, Fim F, and Fim G. These homologous regions may be involved in interactions between the various subunits (Klemm and Christiansen, 1987). The N-terminal regions of both the Fim F and Fim G proteins show some homology with Fim A protein, suggesting that the proteins are structurally related.

Several studies (Abraham et al., 1987, 1988a; Hanson and Brinton, 1988; Krogfelt et al., 1988) indicate that the Fim H adhesin protein is located at the tips and at the long intervals along the length of the type 1 fimbriae (Figure 9–2). The tip location is also suggested from experiments showing that MS agglutination of erythrocytes (Hanson and Brinton, 1988) or yeast cells (Zafriri et al., 1989) by fimbrial preparations of <1 mg/ml usually requires preliminary aggregation of the latter by some means. The ratio of the Fim H, Fim G and Fim F components to the major subunit is very low (Krogfelt and Klemm, 1988). The 32,000 mol wt tip adhesin, the product of *fim*H, is present in an especially low proportion, estimated to be about six molecules per fimbrial rod. The struc-

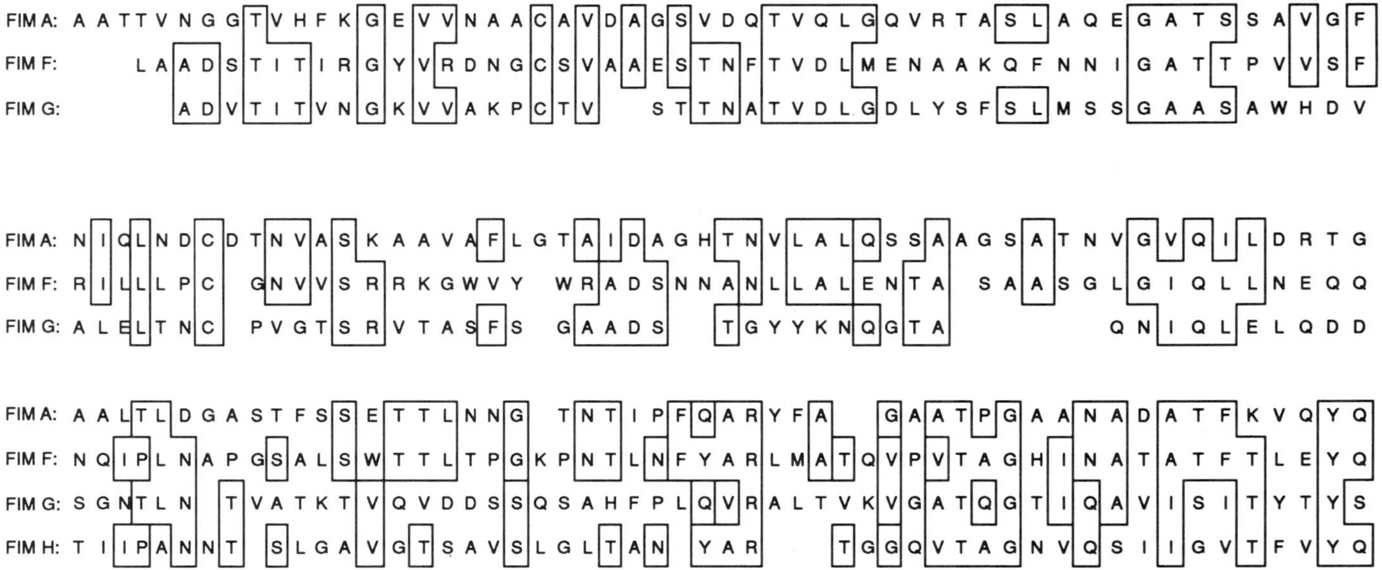

Figure 9–3. Structural homologies of Fim A, Fim F, Fim G, and Fim H proteins. The standard single letter notation is used. Only the carboxy-terminal sequence of Fim H is included. Boxes indicate identical amino acids, whereas gaps were introduced in order to maximize the fit. (Taken from Klemm and Christiansen, 1987, with permission.)

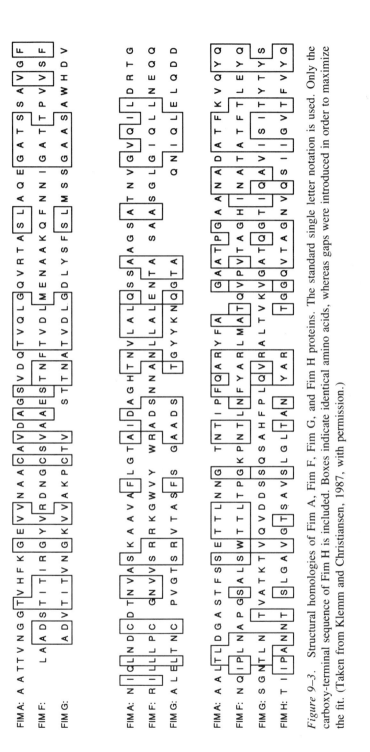

tural features of the tip adhesin protein, either alone or in quaternary structural conformations with one of the other ancillary proteins or with the major Fim A structural subunit, is not known.

Most of the early studies on purified type 1 fimbriae, initiated by Brinton (1965) over 20 years ago, were actually describing the chemical, structural, and immunological properties of Fim A simply because the latter is the major subunit, comprising >90% by weight of the fimbrial suprastructure. Indeed, type 1 fimbrial preparations usually show only a single band of 18,000 mol wt in polyacrylamide gel electrophoresis loaded with moderate amounts of SDS-boiled preparations. Antigenically, the Fim A products vary widely among different genera and species but one or more epitopes seem to be highly conserved among strains of the same species (reviewed in Sharon and Ofek, 1986). For example, type 1 fimbriae (Fim A protein) were found to contain a major *E. coli* type-specific antigen. Those of *Shigella flexneri* contain a major "flexneri-specific antigen" as well as a minor "flexneri-coli antigen." The considerable immunological cross-reactivity between *fim*A product of different strains of the same species and the lack of such cross-reactivity between Fim A of the different enterobacterial species are reflected in the amino acid compositions of fimbrial preparations (Table 9–2). Type 1 fimbriae of different *E. coli* strains have a similar amino acid composition and molecular size. Both these differ markedly from those of type 1 fimbriae of *Klebsiella pneumoniae* and *Salmonella typhimurium*. A high degree of homology has, however, been observed between the N-terminal amino acid sequences of Fim A from *E. coli* (Hermodson et al., 1978), *S. typhimurium* (Waalen et al., 1983), and *K. pneumoniae* (Fader et al., 1982). One possibility is that these homologous N-terminal regions are hidden in the subunits and inaccessible to antibodies. Studies with monoclonal antibodies against type 1 fimbriae of *E. coli* have revealed that certain fimbrial epitopes, probably belonging to the Fim A, react only with fimbrial subunit monomers and dimers, but not with hexamers or higher oligomers, nor with intact fimbriae (Abraham et al., 1983; Eisenstein et al., 1983). These results suggest the antibodies react with hidden or sequestered fimbrial epitopes. In contrast, the reactivity of other monoclonal anti-type 1 fimbrial antibodies is dependent on quaternary structural determinants. The quaternary specific monoclonal antibodies were found by electron microscopy to bind to the fimbriae in a highly discrete, periodic manner (Abraham et al., 1983). The reactivity of these monoclonal antibodies with intact monomeric type 1 fimbriae of other genera of enterobacteria has not been tested. Molecular genetic studies of the relationships among type 1 fimbriae of different genera of enteric bacteria seem to be in line with the immunological studies described above. Buchanan et al. (1985) showed that a DNA probe, including the entire structural gene for the major Fim A subunit and an additional 23,000 mol wt peptide, hybridized with the whole cell DNA of 60 out of 69 *E. coli* isolates, and 6 of 6 *Shigella* isolates, but none of additional genera of *Enterobacteriaceae*. Clegg et al. (1985a,b) found that although genetic complementation occurred

Table 9–2. Amino acid composition of the major fimbrial subunit Fim A of type 1 fimbriae[a]

Amino acid	*Escherichia coli* Bam	*Escherichia coli* K12	*Escherichia coli* 346	*Escherichia coli* Fim A	*Klebsiella pneumoniae*	*Salmonella typhimurium*
Asp	20	18	20	8	27	22
Asn				11		
Thr	20	20	20	19	25	25
Ser	10	9		10	14	23
Glu	13	16	23	3	17	19
Gln				10		
Pro	2	2	3	2	5	11
Gly	17	21	28	16	18	23
Ala	34	34	39	31	30	34
Half-Cys	2	2	2	2	4	nd
Val	13	14	15	15	18	16
Met	0	0	0	0	2	nd
Ile	4	5	5	4	8	7
Leu	10	14	9	10	13	12
Tyr	2	2	2	2	6	4
Phe	8	8	4	7	6	9
His	2	2	1	2	2	3
Lys	3	4	4	3	8	9
Arg	3	2	1	3	5	4
Trp	0	0	0	0	1	0
Total	163	173	156	158	209	221
Mol wt	16,600	17,100	16,600	15,706	21,500	22,100

[a]Composition is given as residues per mole.

Derived from Brinton, 1965; Salit and Gotschlich, 1977; Korhonen et al., 1989; Eshdat et al., 1981; Fader et al., 1982; and Klemm, 1984.

between DNA sequences derived from the same enterobacterial genus, such complementation never occurred between DNA sequences of different genera. The Fim gene cluster can accommodate inserted heterologous DNA sequences and retain function and antigenic identity (Hedegaard and Klemm, 1989).

The *fim*H seems to be more highly conserved, both at the primary structural level and at the gene sequence level, than the other major fimbrial proteins (Krogfelt and Klemm, 1988; Hanson and Brinton, 1988). Abraham et al. (1988b) showed that monoclonal and polyclonal antibodies against *E. coli* Fim H reacted with similar ancillary proteins in western blots of type 1 fimbriae of each isolate of *Enterobacteriaceae* tested (two strains each of *E. coli, Serratia marcescens, Citrobacter freundii,* and *Klebsiella pneumoniae*). Furthermore, a 60-base synthetic oligonucleotide encoding the amino terminus of the mature *E. coli* Fim H hybridized with the whole-cell DNA of most type 1 fimbriated strains, except for *Salmonella typhimurium*. The type 1 fimbriae of the latter may contain a D-mannose binding polypeptide that is distantly related to the *fim*H product because only one of 15 monoclonal antibodies raised against *E. coli* Fim H reacted with a minor 35,000 mol wt protein associated with type 1 fimbriae of *Salmonella*. This was not surprising in view of the fact that the D-mannose combining site of *Salmonella*, as deduced from the inhibitory activity of structurally defined mannose-containing glycosides, differs from that of *E. coli* and *Klebsiella* (see Chapter 5). Whether the 35,000 mol wt polypeptide of *Salmonella* is a tip adhesin analogous to Fim H of *E. coli* and perhaps other enterobacterial genera remains to be determined. Although functional Fim H adhesin can be expressed on the cell surface in *fim*H mutants, the adhesin is intercalated in the fimbrial structure during biogenesis. Mutants that overproduce Fim H appear distorted and fimbriae of *fim*H mutants appear longer (Klemm and Christensen, 1987; Maurer and Orndorff, 1987; Abraham et al., 1988a,b). The insertion of Fim H at long intervals into the fimbrial structure causes increased fragility of the fimbriae, resulting in their fragmentation on freezing and thawing (Ponniah et al., 1991). The hemagglutinating titer of *E. coli* bearing fragmented fimbriae was considerably greater than of bacteria bearing unfragmented fimbriae. If one assumes that fragmentation occurs only at sites of Fim H integration, then the data can be interpreted to suggest that only Fim H at the tip of the flimbrial adhesin complex possesses a functional carbohydrate binding domain (Ponniah et al., 1991). Fim H may also be responsible for the phenomenon of pellicle formation on the surface of static broth cultures because strains deleted of Fim H are fimbriated and do not form the pellicle. Point mutations in *fim*H using muT mutagenesis results in mutants that are capable of forming pellicle but lack the ability to agglutinate erythrocytes (Harris et al., 1990). Because pellicle formation of such mutants is insensitive to mannosides whereas that of wild-type is, it has been postulated that Fim H-Fim H interactions between adjacent cells are responsible for pellicle formation.

In addition to the structural genes, the gene cluster for the type 1 fimbrial adhesin contains genes, designated as *fim*C and *fim*D, responsible for transport and assembly of the fimbriae (Orndorff and Falkow, 1984a,b; Klemm et al., 1985). The assembly of type 1 fimbriae involves three stages: (1) syntheses of precursors; (2) secretion and maturation of the fimbrial subunit; and (3) incorporation of the mature fimbrial subunit into the growing organelle. Immunoelectron microscopy analysis of elongation of type 1 fimbriae revealed that newly synthesized subunits are added to a growing organelle at its base (Lowe et al., 1987) and assembly is dependent on the *sec*A gene product, which is required for normal protein export (Dodd et al., 1984).

Although each of the genes in the cluster can be regulated by the product of a regulatory gene, the *fim*A is the only gene known the transcription of which is regulated by at least two mechanisms involving two distinct genes, the *fim*B and the *fim*E. As originally shown by Brinton (1965), bacteria that possess the type 1 fimbrial gene cluster may spontaneously shift back and forth from a phase of producing fimbriae to a nonproducing phase, a phenomenon known as phase variation (see Chapter 11). The latter has been described in many enterobacterial species and is associated with the adhesive ability of the bacteria (Duguid and Old, 1980). Phase variation is due to the periodic inversion of a 300-bp DNA segment containing the promoter for the fimbrial subunit gene, *fim*A (Abraham et al., 1985; Klemm, 1986) (Figure 9–4). This segment is located immediately upstream from the *fim*A gene. Transactive products of at least two genes, *fim*B and *fim*E, located upstream of *fim*A, mediate the conversion of the element from a position that allows transcription of the *fim*A gene ("on" phase) to a position that does not ("off" phase) and back and forth (Freitag et al., 1985; Klemm, 1986; Eisenstein et al., 1987).

The *fim*B and *fim*E genes control the orientation of the invertible element, presumably because they encode for site-specific recombinases. Fim B and Fim E protein show significant homology with each other and are highly basic, consistent with DNA binding proteins (Klemm, 1986; Dorman and Higgins, 1987; Eisenstein et al., 1987). In attempts to determine the role played by Fim B and Fim A recombinases it was necessary to construct a plasmid-based system that can be used to examine the effects of each regulatory protein on inversion of the invertible element from on to off and from off to on. Studies in two laboratories were carried out using in vivo inversion assays in which a suitable host *E. coli* strain was transformed simultaneously with two compatible plasmids. One of the plasmids contained the entire *fim* gene cluster, including the invertible element but denuded from the *fim*E and *fim*B genes, and the other plasmid contained a DNA element coding either for Fim B or Fim E or for both proteins (Pallesen et al., 1989; Blomfield et al., 1991; McLain et al., 1991). The transformants were allowed to grow in order for recombination events to reach equilibrium and the "switching" activity was measured qualitatively by examining the state of fimbriation of the transformants. The amount of plasmid DNA containing the invertible element at the on and off orientations was determined by Southern hybridization using a DNA probe that hybridizes with the invertible element predigested with a restriction enzyme acting asymmetrically within the element. The latter technique is especially useful to determine the "switching" in a host strain harboring plasmids lacking the *fim* genes but containing the invertible element as a substrate for Fim E or Fim B or both proteins, thereby avoiding complications that might arise from the fimbriated state (Blomfield et al., 1991a,b; McLain et al., 1991). Analyses of the results from both laboratories revealed that (1) the extent of fimbriation was proportional to the relative amount of *fim* DNA invertible element in the on orientation; (2) in the absence of *fim*B and *fim*E, the invertible element remains locked at its initial orientation; and (3) the products of each of the regulatory genes act independently from each other whereby Fim B stimulates DNA inversion from on to off as well as from off to on. Fim E acts preferentially to rearrange the invertible element to the off orientation (McLain et al., 1991). Moreover, it appears that *fim*E mutants, when grown on agar, give rise to two distinct colony types, one of which contains

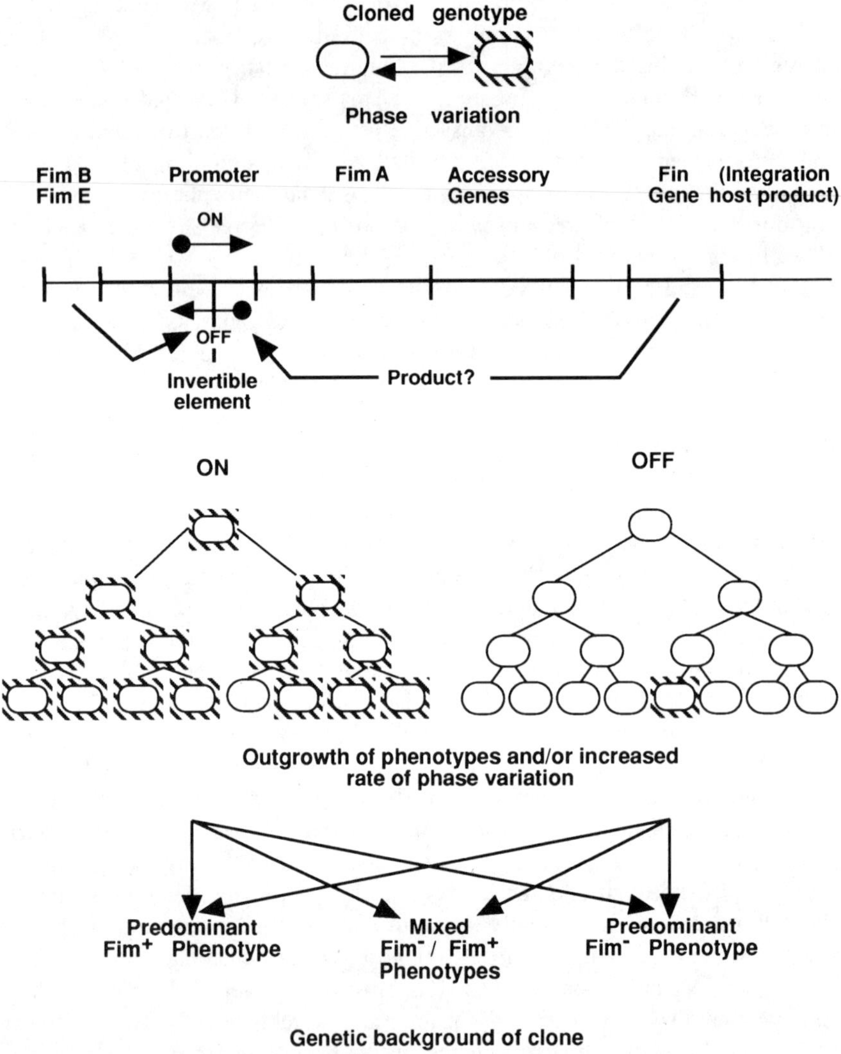

Figure 9–4. Regulation of phase variation in *Escherichia coli*. The *fim* gene cluster is governed by an invertible element of about 300 bp. This region contains the promoter for Fim A, the major structural protein. The "on" and "off" positions are regulated by the products of *fim*B and *fim*E (Figure 9–1 provides a description of the various gene products of the *fim* cluster). Two other proteins, the integration host factor for lambda phage and a product of the *hyp* gene, also modify the level of transcription of the *fim*A gene. (Based on Brinton, 1965 and Abraham et al., 1986.)

mainly fimbriated cells, whereas the other contains mainly nonfimbriated cells (Blomfield et al., 1991), a phenomenon previously described as phase variation (Brinton, 1959; Swaney et al., 1977) in wild strains. In contrast, most of the agar grown strains containing intact *fim*B and *fim*E genes give rise to primarily one type of colony consisting mainly of nonfimbriated cells. This phenomenon probably results from differences in the rate of inversion from on to off and from off to on phases between the *fim*E mutant and its parent strain. The frequency of inversion from off to on was found to be approximately 10^3 per cell per generation in both wild-type and the *fim*E mutant, probably because *fim*E has little influence on rearrangement of the invertible element from off to on. The frequency of inversion from on to off in the *fim*E mutant was markedly higher than that of the parent strain, being $>10^2$ and approximately 10^3 per cell per generation, respectively (Blomfield et al., 1991a,b). It is possible that previous studies showing the phase-variant colonies of *E. coli* on agar employed *fim*E mutant strains, suggesting that such mutants are common in *E. coli*. For example, the *E. coli* strain used to describe the regulatory *hyp* gene, which affects the level of transcription of the major Fim A subunits and fimbriation (Orndorff and Falkow, 1984a,b; Orndorff et al., 1985), now appears to be a *fim*E (*hyp*) mutant (Blomfield et al., 1991a,b). It has been suggested that hyperfimbriation may result by propagating in such *fim*E (or *hyp*) mutants multicopy plasmids containing a *fim* invertible element at the on orientation (Blomfield et al., 1991a,b). Also, the strain CHS50 showing phase-variant colonies employed in earlier studies (Eisenstein, 1981) to determine the frequency of phase variation appears now to contain a copy of the insertion sequence IS*1* in the carboxyl-terminal region coded by *fim*E (Blomfield et al., 1991a,b). This lesion accounts for the observed frequency of phase variation in strain CHS50, similar to the frequency of the *fim*E mutant. It is not clear at this time how prevalent are clinical isolates of *E. coli* with nonfunctional *fim*E, especially because this gene is susceptible to IS*1* insertional mutation. More importantly, such a lesion may affect the course of natural infection because it permits a relatively high frequency of phase variation from off to on in the expression of the type 1 fimbrial adhesin (see Chapter 11). The DNA inversion events within the promoter of *fim*A also require the products of other trans-acting genes mapping at sites distant from the type 1 fimbrial gene cluster. The *him*A and *him*D encode for the integration host factor, a DNA-binding protein that may promote efficient recombination (Dorman and Higgins, 1987; Eisenstein et al., 1987). Another gene, *pil*G (*osm*Z), encodes a histone-like protein that may affect the frequency of inversion (May et al., 1990). In addition, it is possible that environmental factors may affect the frequency of phase variation, believed to be a random process. Brinton (1965) described a growth phase variation, a poorly understood mechanism that may operate by decreasing the rate of production of fimbriae, so that little new synthesis of subunits or net assembly of organelles is occurring by the mid–late exponential phase of growth (Dodd et al., 1984). Due to the phase variation, cultures of individual *E. coli* cells originating from a single clone may yield fimbriate and nonfimbriate cells. Environmental factors clearly cause a change in the generation time of either the type 1 fimbriated or nonfimbriated cells, resulting in outgrowth of one phenotype (phase variant) over the other phenotype (Goochee et al., 1987). For example, growth under static conditions for 24–48 hours may allow time for selection of the predominantly fimbriated phenotype, which grows better under these conditions. This selective growth is abrogated by the addition of glucose to the medium (Eisenstein and Dodd, 1982) or by growing the organisms on agar (Duguid and Old, 1980). In view of the complexity of the selective growth advantage for either the fimbriate or nonfimbriate variants some of the growth conditions discussed above have been examined for their effects on the orientation of the invertible element carried by plasmids lacking the *fim* genes. This approach avoids growth selection of the fimbriate or nonfimbriate states (McLain et al., 1991). None of the growth conditions tested, including bacteria grown in static broth, in well-aerated broth, or in agar, were found to affect the frequency of inversion of the invertible element that controls the phenotypic expression of the type 1 fimbrial adhesin. Nevertheless, the possibility that environmental factors may control phase variation cannot be excluded. This is especially true in *fim*E mutants that now appear to be common among wild-type strains undergoing phase variation in the expression of the fimbrial adhesin due to *fim*B activity (Blomfield et al., 1991a,b). For example, whereas growth in broth invariably selects for type 1 fimbriated bacteria, growth on agar selects two types of categories of strains: the regulated variants and random phase variants (Hultgren et al., 1986). Colonies of the regulated variants contain mostly nonfimbriated cells, whereas those of the random variants consist of a mixture of fimbriated and nonfimbriated cells. It is suspected that the isolates exhibiting random variation during growth on agar represent *fim*E mutants and thus reflect a higher rate of phase variation from the off to on position of the invertible element. As the phenomenon of phase variation clearly takes place in vivo and is important for the infectious process (see Chapter 11), it is important to distinguish all the DNA elements involved in phase variation and especially to determine the environmental factors that directly affect the frequency of phase variation from those that simply confer a growth advantage for the fimbriated or nonfimbriated variants.

A remarkable feature of the mannose-specific fimbrial adhesin is that it is a highly conserved adhesin by many enterobacterial genera (see Table 5–1). A possible evolutionary pressure for this conservation is discussed in Chapter 11. Of special interest is how the gene clusters coding for type 1 fimbriae from various enterobacterial genera relate to each other and to those of *E. coli*. Clegg et al. (1985a,b) cloned the type 1 fimbrial gene cluster of four enterobacterial species (Table 9–1). Examination of proteins encoded by the *fim* gene clusters in minicells showed that the gene products encoded by the cloned inserts of DNA elements differ in number and size for the various enterobacterial genera

(Clegg et al., 1987). Six, four, and three products encoded by DNA elements from *S. typhimurium, K. pneumoniae,* and *S. marcescens,* respectively, were necessary for the phenotypic expression of functional type 1 fimbriae in *E. coli. E. coli* harboring a plasmid, containing the *K. pneumoniae* gene cluster mutated in *fim*A, did not express fimbriae but caused mannose-sensitive hemagglutination (Gerlach et al. 1989). It appears that, like in *E. coli,* the genes coding for the mannose-specific adhesin and other accessory products necessary for transport and expression of mannose binding activity in *K. pneumoniae* are separate entities and may be expressed in nonfimbriated form on the host strain harboring the *fim* gene cluster. Nonfimbriated host strains harboring a plasmid containing the *K. pneumoniae fim* gene cluster with a mutation in the *fim*A gene become fully fimbriated from any of the three enteric genera tested (Gerlach et al., 1989). Collectively, the findings suggest that expression and organization of genes encoding the mannose-specific fimbrial adhesins in enterobacterial genera are similar to those in *E. coli* described above. Nevertheless, DNA hybridization using *fim* DNA elements as probes, showed little nucleotide sequence homology among the *fim* gene clusters of the various enteric genera. Recently, it was found that the sequence of DNA elements coding for the F17-A adhesin complex expressed by bovine isolates of *E. coli* showed extensive homology with *fim*A, *pap*A, and the coding genes for G fimbriae of human isolates (Lintermans et al., 1988).

Type P Fimbrial Adhesin

Most isolates of *E. coli* from pyelonephritis patients express a fimbrial lectin specific for the Galα1,4Gal sequence, usually found in the P blood group and in glycolipids of erythrocytes and other types of animal cells (see Chapter 3). The chromosomal gene cluster coding for the suprastructure of the P-fimbrial adhesin complex was first cloned in 1981 from a pyelonephritogenic isolate of *E. coli* strain J96 (Hull et al., 1981). Most of the molecular biology of the individual genes coding for this pyelonephritis-associated adhesin (the so-called *pap* gene cluster) was performed on a subclone of a 9.8-kb *Eco*R1–*Bam*HI fragment cloned in plasmid PAP5 (Lindberg et al., 1984). *E. coli* K12 strain Hb101, harboring pPAP5, which contains the wild-type *pap* cluster, expresses P fimbriae and causes Galα1,4Gal-specific hemagglutination of erythrocytes. Mapping of the various genes in the *pap* cluster and unraveling the structure and role in regulation or biogenesis of the products encoded by the genes was achieved by DNA sequencing and examination of the polypeptides expressed in minicells harboring pPAP5. The minicells contained the wild-type *pap* cluster or one of a series of Tn5 insertions in the cluster (Baga et al., 1984, 1985, 1987; Norgren et al., 1984, 1987; Lindberg et al., 1986, 1989; Lund et al., 1987; Tennent et al., 1990).

The gene cluster coding for the expression of type P fimbrial adhesins is very similar to that for type 1 fimbriae (Figure 9–1). It contains regulatory genes, a major fibrillin gene, minor fibrillin genes, and genes for secretion and assembly. The major fimbrial subunit Pap A determines the antigenic diversity of the P fimbrial adhesin complex resulting in P fimbrial serotypes F7 through F13 (Table 9–1). Amino acid sequences of the various serotypes of the major P fimbrial subunits reveal that they contain conserved regions important for the biogenesis of the adhesin complex and hypervariable domains that determine the antigenic specificity (Van Die et al., 1987, 1988a). Although the hypervariable regions are needed for the biogenesis of the fimbrial structure, probably to provide correct spacing between the conserved regions, substitution of amino acids within these regions does not affect fimbrial expression (Van Die et al., 1988b). As with type 1 fimbriae, it was thought that the Galα-1,4Gal binding site was located in the major 15,000 mol wt structural subunit. That assumption was based primarily on the observations showing, as with whole bacteria, purified fimbriae adhere to animal cells and the adhesion is specifically inhibited by the digalactoside (Leffler and Svanborg-Eden, 1986; Hoschützky et al., 1989). Furthermore, antibodies raised against purified fimbriae inhibit adhesion and hemagglutination caused by P-fimbriated *E. coli* (Kornhenen et al., 1981). However, after the genes required for the synthesis and expression of type P fimbriae were cloned and mapped, it soon became clear that deletion mutants could be obtained that were fimbriated but not capable of hemagglutinating erythrocytes, or afimbriated, but capable of causing Galα1,4Gal-specific hemagglutination (Lindberg et al., 1984; Uhlin et al., 1985). Further insertional and deletional mutational analyses revealed that the Gal-Gal binding activity was dependent on a minor 36,000 mol wt protein encoded by *pap*G. *E. coli* HB101 harboring the *pap* gene cluster that had an insertional mutation in *pap*G produces normal fimbriae but is not able to cause hemagglutination. Fimbriae isolated from *pap*G mutants are not able to adhere to animal cells (Lindberg et al., 1984; Hoschützky et al., 1989). Direct evidence for the Galα-1,4Gal binding activity of Pap G was obtained in experiments in which the protein complexed with Pap D (the periplasmic transporter protein). The Pap G protein was purified from the periplasmic space by affinity chromatography using Galα1,4Gal-containing beads (Hultgren et al., 1989). Also, Pap G protein purified from P fimbrial preparations and immobilized onto microtiter plates bound erythrocytes, the binding of which was inhibited by Galα-1,4Gal-containing oligosaccharides (Hoschützky et al., 1989).

The role of the other minor fimbrial proteins encoded by *pap*E and *pap*F was examined by analyzing their phenotypic effects on *E. coli* HB101 harboring the pPAP5 plasmid with linker insertional mutations in each of the genes. The number of fimbriae per cell was markedly reduced in *pap*F mutants, suggesting that Pap F may play a role in the assembly of fimbriae (Norgren et al., 1984). Although *pap*E mutants were fimbriated and caused hemagglutination of erythrocytes, their purified fimbriae lost the ability to hemagglutinate (Lindberg et al., 1984). Immunogold labeling of the *E. coli* with antisera made monospecific to each of the minor fimbrial proteins revealed that Pap E, F, and G are located preferentially at the tips of the fimbriae (Lindberg et al., 1987). Because *pap*F and/or *pap*G mutants and their isolated fimbriae lacked digalactoside-binding activities, it was suggested that Pap E forms a composite structure with Pap G and Pap F to anchor these minor fimbrial proteins firmly to the fimbrial structure, whereas Pap F provides a proper orientation and conformation for Pap G to function as a digalactoside-binding protein (Tennent et al., 1990). The association, however, of these minor fimbrial subunits with the fimbrial structure is not essential for the organisms to express all three minor fimbrial subunits on their surfaces in functional form because nonfimbriated *pap*A mutants interact with antibodies against each of the proteins (Lindberg et al., 1987). These mutants are capable of binding Galα1,4Gal-containing structures on erythrocytes, tissue culture cells, and Galα1,4Gal-coated beads (Lindberg et al., 1984; Uhlin et al., 1985). In the wild-type and various other isolates, Pap E, Pap F, and Pap G are invariably associated with the fimbrial structure. It is reasonable to assume that this mode of presentation of the P-fimbrial adhesin complex has an advantage in vivo as a nonfimbriated Galα1,4Gal-specific *E. coli* may not be selected throughout evolution. Furthermore, immunolabeling with antisera specific for each of these proteins has revealed that in three F serotypes examined ($F7_1$, F11, and F13), as well as in HB101 harboring the *pap* cluster, the proteins were found to be located at the tips of the fimbriae (Lindberg et al., 1987; Riegman et al., 1988, 1990). In one F serotype examined (F9) the Pap G was found at the tip and alongside the fimbriae (Riegman et al., 1988), but it is not known whether only the tip-located Pap G possesses Galα1,4Gal binding activity. The latter possibility is supported by the observations that purified P fimbriae do not cause hemagglutination unless they are aggregated or cross-linked (Hoschützky et al., 1989). Recombinant strains harboring the $F7_1$ fimbrial gene cluster deficient in *fso*E and *fso*F adhere poorly to renal tubuli and immobilized fibronectin, compared to nonmutated or *fso*G-mutated recombinant strains (Westerlund et al., 1991). The results suggest that in addition to the digalactoside-specific fimbrial subunits of the P fimbriae complex other minor proteins (E and F) may also serve as adhesins.

E. coli cells possessing a mutation in *pap*H are fimbriated and cause hemagglutination (Baga et al., 1987). Electron microscopic observations have revealed that the fimbriae in these mutants are detached from the cells. The *pap*H mutation could be complemented by intact *pap*H gene in *trans*. Analyses of proteins secreted by minicells harboring the intact pPAP5 plasmid or the plasmid deleted of *pap*H confirmed that *pap*H encodes for a protein of 20,000 mol wt. Pap H protein maintains a high homology with Pap A but contains a unique proline-rich region at its amino-terminus. The role of Pap H could be appreciated only by altering the stoichiometry between Pap A and Pap H. Mutants in which there were an eightfold excess of Pap A produced very long fimbriae. It was postulated that Pap H is embedded in the outer membrane via its proline-rich region and interacts with the fimbrial structure via the Pap A-like region (Tennent et al., 1990). Once such an interaction takes place, the growth of fimbriae is terminated, resulting in firm anchorage of the fimbrial structure to the cell. The relative amounts of Pap H in the cell may therefore determine fimbrial length of the fimbrial rod. Recently, another gene in the pap gene cluster was identified as *pap*J (Tennent et al., 1990). A *pap*J mutant also shed large amounts of fimbrial antigen in the culture medium, but the fimbrial structures were of variable size segments. It was postulated that the Pap J protein functions to ensure correct assembly of the major and minor fimbrial subunits.

The *pap*C gene encodes for a 81,000 mol wt protein that has been determined to be a component of the outer membrane (Norgren et al., 1987). Although *pap*C *E. coli* mutants are not fimbriated, or capable of causing hemagglutination, their cell extracts contain normal amounts of the major subunits of Pap A. In addition, the number of fimbriae per cell are altered in mutants in which the amount of Pap C relative to that of Pap A is manipulated. The results are consistent with the notion that Pap C acts as a membrane channel and as a locus for the initiation of subunit polymerization to form the fimbrial structure.

The 28,000 mol wt protein encoded by *pap*D is found exclusively in the periplasmic space. The "chaperone" role of this protein in the biogenesis of the P fimbriae adhesin complex was deduced from experiments showing that periplasmic extracts of *pap*C mutants contained complexes of Pap D with each of the minor fimbrial subunits Pap E, Pap G, and Pap F, as well as with the major subunit Pap A (Hultgren et al., 1989; Lindberg et al., 1989). Pap D therefore stabilizes other Pap proteins by preventing the subunits from interacting with each other. The lifetime of these Pap D–fimbrial subunit complexes in the periplasmic space must be sufficient for Pap D to enable the delivery of the subunits to the center of fimbrial growth (Tennent et al., 1990). Physicochemical analyses of purified Pap D complexed with Pap G have revealed that the complex is stable and Pap G in the complex maintains carbohydrate binding activity (Kuehn et al., 1991). Although the interaction of Pap D with Pap G is reversible in the sense that complexed Pap D can be displaced by free Pap D, the release of Pap G from the complex in the *E. coli* cytoplasm into the nascent fimbriae is thought to be driven by interaction of the complex with membrane anchor Pap C.

The gene cluster coding for the P-fimbrial adhesin contains a number of genetic elements capable of regulating the expression of the complex on the surface of the organisms. The regulatory genes also confer upon the organisms the ability to alternate between phenotypes, either expressing or not expressing the P-fimbrial complex. At least three different mechanisms of regulation can be distinguished in the P-fimbrial gene cluster. One mechanism is to ensure that each of the gene products involved in the biogenesis of the adhesin complex is transcribed and translated at the relative amounts required for successful assembly of the fimbrial adhesin on the organism's surface. Transcription of *pap*A through *pap*D genes is under the control of one promoter located upstream of the preceding *pap*B (Baga et al., 1985, 1987, 1988). This is in line with earlier studies (Norgren et al., 1984; Uhlin et al., 1985) showing that transcription of cistrons in that region is affected by tranposon insertions, as expected from a polycistronic operon. The most abundant mRNA transcribed from the *pap*B operon is that which leads to the synthesis of Pap A. This mRNA is the product of posttranscriptional cleavage of a primary RNA transcript resulting from the transcription of *pap*B and *pap*A (RNA-BA) (Baga et al., 1988). On the basis of these observations, it was postulated that the stoichiometry of the gene products involved in biogenesis of the fimbrial adhesin complex is the result of attenuated activation of the genes downstream of *pap*A from the upstream promoter and of different half-lives of the mRNAs of the corresponding genes (Tennent et al., 1990). Another regulatory mechanism involves the products of two genes, *pap*B and *pap*I, upstream from *pap*A (Figure 9–1). Transposon-induced mutation in the region containing the *pap*B and *pap*I genes results in a phenotype devoid of fimbriae and adhesiveness (Uhlin et al., 1985). Pap B and Pap I proteins act as *trans*-active positive regulators for the expression of the fimbrial Pap A (Baga et al., 1985; Uhlin et al., 1985). Pap B protein binds to specific regions in the *pap* gene cluster (Forsman et al., 1989) and acts as a positive regulator of high-level transcription of fimbrial genes transcribed from the *pap*B promoter. Overproduction in *trans* of Pap B protein causes a reduction in the expression of products of the genes in the *pap* operon (Göransson et al., 1988; 1989). Transcription of *pap*I is activated by its own promoter. During growth at low temperatures (e.g., 25°C) transcription is considerably reduced, resulting in a nonfimbriated phenotype. The region involved in this thermoregulation of the expression of the P-fimbrial adhesin complex was determined to be in the *pap*I–*pap*B intercistronic region. The intercistronic region also contains a cyclic AMP binding site. The *pap* gene transcription is also dependent on cyclic AMP and its proteinaceous receptor (CRP). The third mechanism of regulation involves the phenomenon of phase variation. Clinical isolates undergo phase variation in the expression of their P-fimbrial adhesin complexes, similar to the expression of type 1 fimbriae (Rhen et al., 1983a; Nowicki et al., 1985). At present, the molecular mechanism involved in switching between the phenotype that expresses the P-fimbrial adhesin complex to one that does not is obscure. Because the phenomenon of phase variation in the expression of the P fimbriae resembles that of type 1 fimbriae it is possible that an invertible element oscillating between "off" and "on" positions at relatively high frequency is responsible. In vitro, *E. coli* phenotypes expressing the P fimbriae seem to overgrow other phenotypes when cultured on agar. The switch frequency of P-fimbrial expression is about 10^{-2}/cell/generation (Rhen et al., 1983b).

Immunological studies have shown that Pap A from various *E. coli* isolates can vary antigenically. These antigenic variants are referred to as F serotypes (Abe et al., 1987; Orskov and Orskov, 1990). Monospecific serum against Pap

F protein reacts with cells harboring a plasmid coding for the expression of three serologically different F serotypes, whereas anti-Pap E serum reacts only with one F serotype (Lund et al., 1988b). The Pap G protein exhibits antigenic heterogeneity within the various F serotypes in spite of the fact that all of the Pap G variants retain their ability to bind digalactosides. Figure 9–5 shows the amino acid sequences of Pap G associated with various F serotypes. The Pap G proteins of F serotypes $F7_2$ and F11 are highly homologous to one another, but both differ from Pap G associated with the F13 serotype. The Pap G proteins of the various F serotypes are probably isolectins. In this sense, Pap G may be similar to legume lectins which are known to possess regions of homology even though the amino acid sequences of their carbohydrate-binding sites are different (Strosberg et al., 1986).

Type S Fimbrial Adhesin

The similarities of the genetic organization of the gene cluster responsible for synthesizing and regulating the expression of the S fimbrial adhesin complex with those of types 1 and P fimbrial adhesins are outlined in Figure 9-1.

The *sfa* determinant codes for three nonfimbrial proteins involved in transport and assembly: Sfa D, Sfa E, and Sfa F, corresponding to 20, 26, and 89,000 mol wt proteins, respectively (Hacker, 1990). The Sfa E is probably analogous to Pap D of P fimbriae and Fim C of type 1 fimbriae. The large Sfa F is the outer membrane protein similar to Pap C and Fim D (Figure 9–1). Whole isolated fimbriae–adhesin complexes at pH 7–8 are dispersed and adhere to animal cells, but are unable to agglutinate erythrocytes. At low pH or in the presence of divalent cations, the fimbriae aggregate and cause hemagglutination of erythrocytes, emphasizing a view that dispersed fimbriae are monovalent with respect to their adhesive functions (Jann and Hoschützky, 1990). Analogous to the organization of the gene cluster coding for P and type 1 fimbrial adhesin complexes, a gene coding for a sialyl-binding protein, but dispensable for fimbriation, was found in the *sfa* genetic determinant (Hacker et al., 1985). Furthermore, neither the recombinant *E. coli* strain harboring a *pap*G mutation nor the fimbriae isolated from the bacteria had any detectable adhesive properties. In contrast to types 1 and P fimbrial adhesins, however, the *sfa*S gene encodes for a 12–14,000 mol wt adhesin and is not located at the extreme 3′ end of the *sfa* gene cluster. It is adjacent to the gene coding for the 32,000 mol wt protein (Schmoll et al., 1989; Hacker, 1990). Physical evidence that Sfa S is a fimbrial protein with sialyl binding properties was obtained by isolating Sfa S from fimbriae–adhesin complexes followed by purifying the adhesin to homogeneity (Moch et al., 1987). When immobilized on the bottom of microtiter plates, the adhesin binds erythrocytes (Jann and Hoschützky, 1990). The binding of erythrocytes by Sfa S adhesin is specifically inhibited by sialyl-containing oligosaccharides, suggesting that the immobilized adhesin maintains the receptor-recognizing specific-

```
PapG F13 MKKWFPAFLF LSLSGGNDAL AGWHNVMFYA FNDYLTTNAG NVKVIDQPQL YIPWNTGSAT ATYYSCSGPE FA-SGVYFQ
PapG F72 ---- L-- --CVSGESS - WN - IV - S LGNVNSYQG- --VITQR - F - ITS - RP - 1 - V-WNQ-N - -DGSWA - YR
PapG F11 --- L___CVSGESS - WN - -IV - S LGDVNSYQG- - VITQR - F - ITS - RP - 1 - V-WNQ - N - -DGFWA-YR

PapG F13 EYLAWMVVPK HVYTNEGFNI FLDVQSKYGW SMENENDKDF YFFVNGYEWD TWTNNGARIC FYPGNMQLN NKFNDLVFRV
PapG F72 -- I --V-F -- K-M-QN-YP - IE - HN - GS - - E - NTG - N-S --LK--K - ERAFDAGNL- QK - ETTR-T E-D-11-K
PapG F11 --I--V - F --K - M - QN - YP -IE - HN - GS - -E -NTG - N - S - LK - K - ERAFDAGNL - QK - EITR - T E - D - II - K

PapG F13 LLPVDLPKGH YNFPVRYIRG IQHHYYDLWQ DHYKMPYDQI KQLPATNTLM LSFDNVGGCQ PSTQVLNIDH GSIVIDRANG
PapG F72 A - A - L - D -SVTIP - TS - I - R - FASY LH ARF - I - NVA -T - RE - EML FL - K - I --R - A - S - E - K - -DLS - NS - N
PapG F11 A- A --L - D -SVKIP - TS - M - R - FASY LH ARF - I - NVA - T - RE - EML FL - K - I --R - A - S - E - K- -DLS - NS - N

PapG F13 NIASQTLSIY CDVPVSVKIS LLRNTPPIY NNNKFSVGLG NGWDSIISLD GVEQSEEILR WYTAGSKTVK IESRLYGEEG
PapG F72 HY - A ---VS --ANIRFM --T - T - S HGK ----H ----V - VN - DTG - TTM - -K - TQNLT - G ---SS
Papg F11 HY - A ---VS ---ANIRFM --T - T - S HGK ----H ----V - VN --DTG - TTM - -K - TQNLT - G ---SS

PapG F13 KRKPGELSGS MTMVLSFP
PapG F72 -IQ - V -- A - LLMIL -
Papg F11 -IQ - V -- A - LLMIL -
```

Figure 9–5. Comparison of amino acid sequence between Pap G-F13, Pap G-$F7_2$, and Pap G-F11. Dashes in Pap G-F13 were introduced in order to attain optimal alignment. Residues in Pap G-$F7_2$ and Pap G-F11 identified to those in Pap G-F13 are shown by the horizontal lines. The first residues of the mature Pap G proteins are denoted by the arrow. Amino acid differences between the Pap G-$F7_2$ and Pap G-F11 proteins are shown by dots. Overlining shows sequence homologies between Pap G-F13 and Prs G-F13. (Adapted from Tennent et al., 1990.)

ity of whole S-fimbriated bacteria or of fimbriae–adhesin complexes isolated from the bacteria. Oligonucleotide-directed, site-specific mutagenesis in *sfa*S has revealed that mutants in which lysine 116 and arginine 118 are replaced by threonine and serine, respectively, are unable to cause hemagglutination, suggesting that these two amino acid residues are important in conferring sialyl-binding activity to Sfa S (Morschauser et al., 1990) (Figure 9–6). Immunogold labeling has shown that the Sfa S adhesin protein is located at the tips of the fimbriae (Moch et al., 1987) similar to the Pap G adhesin.

The phenotypes harboring mutations in *sfa*H cause hemagglutination but the number of fimbriae per cell is considerably reduced, compared to the wild-type cells, suggesting that the function of Sfa H may be similar to that of Pap F and Fim F of types P and 1 fimbriae, respectively. It is reasonable to assume that the functions of the three minor fimbrial proteins encoded by *sfa*H, *sfa*S, and *sfa*G are similar to those encoded by the *pap* and *fim* determinants, although the corresponding genes are located at different positions in the respective gene clusters.

The region that regulates the expression of the type S fimbriae–adhesin complex is located at the proximal (5′) end of the gene cluster (Figure 9–1). Phenotypes harboring transposon-induced mutations in this region are nonfimbriated and do not cause hemagglutination (Hacker et al., 1986a,b). Two gene products, Sfa B and Sfa C, are encoded by the regulatory region analogous to Pap B and Pap I of the P fimbriae–adhesin complex. In fact, sequence studies have revealed that these proteins show a very high degree of homology and are *trans*-complementable to each other (Rhen et al., 1985b; Goransson et al., 1988; Schmoll et al., 1989), suggesting the products of the genes encoded by the proximal end of the cluster share common functions with the corresponding genes in the *pap* cluster, e.g., a positive *trans*-regulatory effect on the expression of the structural gene *sfa*A. A confirmation of this effect was obtained by the construction of a *sfa*A-*pho*A fusion on a plasmid and measuring the amount of Pho A activity as an indicator for the transcriptional activity of *sfa*A (Schneider and Beck, 1986). The data showed that both Sfa A and Sfa C were necessary for the transcription of *sfa*A. In addition, similar to the *pap* cluster, a *sfa*–*lac*Z fusion indicated the presence of two promoters arranged in opposite directions (Hacker et al., 1986a,b).

The expression of S fimbriae–adhesin complexes on bacterial surfaces is under the control of phase variation, similar to that of types 1 and P fimbriae–adhesin complexes. The rate of the shift from S-fimbriated cells to nonfimbriated ones is on the order of one per 100 per cell per generation (Nowicki et al., 1986). The molecular mechanism underlying this phase variation phenomenon is not yet known.

Type 1C Fimbrial Adhesin

Because *E. coli* carrying type 1C fimbriae (F1C) do not cause hemagglutination, the F1C fimbriae are considered not to function as adhesins in spite of the fact

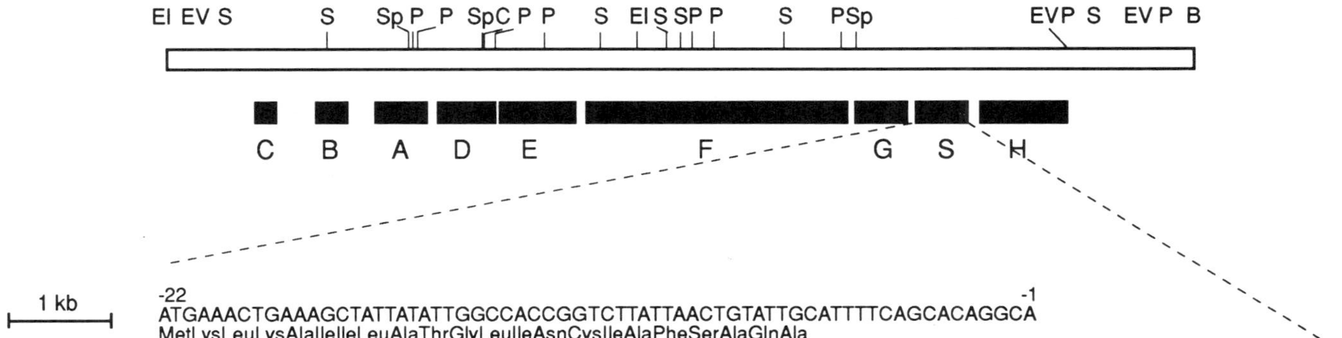

-22 … -1
ATGAAACTGAAAGCTATTATATTGGCCACCGGTCTTATTAACTGTATTGCATTTTCAGCACAGGCA
MetLysLeuLysAlaIleIleLeuAlaThrGlyLeuIleAsnCysIleAlaPheSerAlaGlnAla

+1
GTGGATACGACGATTACTGTGACGGGTAATGTTTTGCAAAGAACATGTAATGTACCAGGGAATGTGGATGTTTCTTTGGGTAATCTGTAT
ValAspThrThrIleThrValThrGlyAsnValLeuGlnArgThrCtsAsnValProGlyAsnValAspValSerLeuGlyAsnLeuTyr

+31
GTATCAGACTTTCCCAATGCAGGAAGTGGATCTCCATGGGTTAATTTTGATCTGTCTCTCACCGGATGCCAGAATATGAATACTGTTCGG
ValSerAspPheProAsnAlaGlySerGlySerProTrpValAsnPheAspLeuSerLeuThrGlyCysGlnAsnMetAsnThrValArg

+61
GCAACATTTAGTGGTACTGCGGATGGGCACACATACTATGCGAATACAGGGATGCTGGCGGTATCAAGATTGAAATTCAGGACAGCCAT
AlaThrPheSerGlyThrAlaAspGlyGlnThrTyrTyrAlaAsnThrGlyAsnAlaGlyAsnAlaGlyGlyIleLysIleGluIleGlnAspArgAsp

+91
GGAAGTAATGCATCATATCACAATGGTATGTTCAAGACGCTTAATGTACAAAATAATAATGCAACCTTTAATCTTAAAGCCCGTGCAGTG
GlySerAsnAlaSerTyrIlisAsnGlyMetPheLysThrLeuAsnValGlnAsnAsnAsnAlaThrPheAsnLeuLysAlaArgAlaVal

+121 … +141
AGTAAAGGCGAGGTTACTCCTGGAAATATCAGTTCTGTTATAACCGTCACCTATACCTATGCG
SerLysGlyGlnValThrProGlyAsnIleSerSerValIleThrValThrTyrThrThrTyrAla

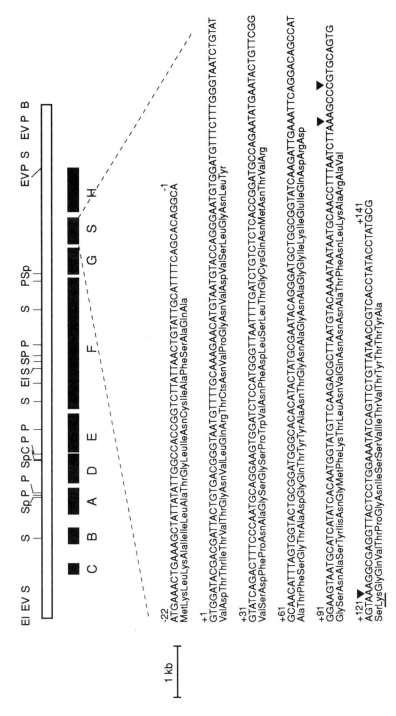

Figure 9–6. The amino acid sequence of Sfa S as deduced from nucleotide sequence. The amino acids used for site-specific mutagenesis are marked by triangles. (Adapted from Morschauser et al., 1990.)

that morphologically these fimbriae resemble the mannose-specific type 1 fimbriae (Klemm et al., 1982). Recently, however, evidence was obtained to suggest that the type 1C fimbriae mediate binding of the bacteria to human kidney cells (Virkola et al., 1988). The *foc* gene cluster coding for type 1C fimbrial adhesin complex (serotype F1C) has been cloned (Van Die et al., 1985). Analyses of *foc* DNA sequences and of proteins encoded by the *foc* gene cluster in minicells and immunoreactivity of the fimbriae have revealed that the F1C fimbrial–adhesin complexes are highly related to those of S fimbriae (Riegman et al., 1990). These results, coupled with the observations showing that the receptor specificity of type 1C fimbrial adhesin resembles that of S fimbriae, both being specific for sialyl residues, led to the conclusion that F1C and S fimbriae belong to the same family (Marre et al., 1990). The organization of the genes in the *foc* gene cluster resemble those coding for other fimbrial adhesins (Figure 9–1). This gene was mapped between *foc*D and *foc*G (see Figure 9–1) (van Die et al., 1991). Seven genes have now been identified in the *foc* cluster (Figure 9–4) encoding for Foc A, the major fimbrial subunit; Foc D and Foc C proteins are probably involved in assembly and transport of minor subunits into fimbrial structure; Foc I is of unknown function; Foc H (previously designated Foc E, Marre et al., 1990) is likely to represent a minor fimbrial subunit which may function as adhesin.

It is possible that Foc F, serves as an adhesin because of sequence homology of the DNA regions encoding for the Foc minor subunits with those encoding for the minor subunits of the *sfa*S (sialic acid-specific) gene cluster (Van Die et al., 1991). The deduced amino acid sequences encoded by *foc*F, G, and H reveal strong homology to *sfa*G, *sfa*S, and *sfa*H, respectively, whereas there is little homology to the minor fimbrial subunits of type 1 and P fimbrial–adhesin complexes.

Colonization Factor Adhesins (CFAs)

The fimbrial adhesins known as colonization factor adhesins (CFAs) are expressed by enterotoxigenic human isolates of *E. coli* (Evans and Evans, 1990). The DNA elements coding for the CFA complex reside on large plasmids in enterotoxigenic *E. coli*. The plasmids usually carry a superfamily of adhesins designated as CFA/I, II, III, and IV. The production of each of these CFAs is associated with isolates belonging to particular serotypes (Evans and Evans, 1990). CFA/I and CFA/III may be part of the same family (Evans and Evans, 1990). The other adhesin superfamilies are composed of a number of antigenically distinct adhesins.

The plasmid genes coding for the CFA/I adhesin complex have been cloned and found to reside in two regions of the plasmid (Willshaw et al., 1983). One region encodes for at least six proteins, in minicells including the major fimbrial subunit, whereas the other region codes for at least three proteins, one of which seems to be essential for assembly (Willshaw et al., 1985). The primary structure of the major fimbrial subunit of the CFA/I adhesin complex lacks cysteine residues, a property that seems to be shared by other adhesins of enterotoxigenic *E. coli,* such as K88ab (Gaastra et al., 1981) and CS2 (Klemm, 1985) and CS3 (Boylan et al., 1988) adhesins of the CFA/II family.

A 15,000 mol wt major fimbrial subunit is the only subunit found in the CFA/I fimbrial structure and therefore bears the adhesin activity. The subunit can be dissociated from the fimbriae by heating in sodium dodecyl sulfate (SDS) or water at 100°C (Bühler et al., 1991). The monomeric subunits are stable and inhibit hemagglutination caused by whole bacteria, suggesting that they maintain a receptor binding activity. Two types of monoclonal anti-CFA/I antibodies have been obtained, one of which reacts with the dissociated subunits and the other with the fimbrial structure. The antisubunit antibodies inhibit hemagglutination only weakly and label the entire length of the fimbrial structure (Bühler et al., 1991). The data suggest that whereas CFA/I fimbriae consist of one subunit, the receptor binding activity is restricted only to subunit(s) at the tip of the fimbrial structure (Bühler et al., 1991). The presence of iron during growth suppresses the formation of CFA/I fimbriae by ETEC. The activity of the CFA/I major subunit promoter is increased upon addition of iron chelators to the growth medium (Karjalainen et al., 1991). The iron regulation appears to involve the product of *fur* (ferric uptake regulation).

The CFA/II complex is composed of three different components, designated coli-surface-associated antigens CS1, CS2, and CS3 (Smyth, 1982). The three components are encoded by genes residing in the same plasmid and differ by their receptor specificities as evidenced from the types of erythrocytes agglutinated by the bacteria expressing the individual adhesins (Mooi and DeGraaf, 1985). Although the plasmid contains all the DNA elements required for expression of each of the CS adhesins, phenotypically strains may express CS3 or CS2 alone or CS3 and CS1 or CS3 and CS2. It is not clear at present what exactly determines which of the adhesins is expressed at any one time. Mobilization of the CFA/II-coding plasmid into various *E. coli* hosts has revealed that the genomic background of the host strain affects the type(s) of adhesins that are phenotypically expressed (Mullany et al., 1983; Smith et al., 1983; Boylan and Smyth, 1985). As with other fimbrial adhesins at least some of the polypeptides encoded by the plasmid are required for the biogenesis of the CS3 adhesin complex. The CFA/II-containing plasmids have been shown to be heterogeneous in size in the various strains and to carry the genes coding for the heat-labile and heat-sensitive enterotoxins (Mullany et al., 1983; Penaranda et al., 1983; Smith et al., 1983; Boylan and Smyth, 1985; Echeverria et al., 1986). Because most strains express two types of fimbriae of the CF/II family, it is difficult to characterize the molecular nature of the individual antigenic variants of CFA/II fimbriae. CS2 fimbriae have been isolated by sonication from a strain producing only CS2 or CS3 and CS2 and purified by gel filtration (Klemm et al., 1985; Sjöberg et al., 1988). The CS2 fimbriae were completely dissociated by saturated guanidine hydrochloride into subunits of mol wt of 16,500. Both the fimbrial structure and its subunits suspended in buffer bind to bovine erythrocytes and the binding is inhibited by sialyllactose, suggesting that the fimbrial subunits bear the sialic acid combining site of the CS2 fimbrial structure (Sjöberg et al., 1988). The amino acid composition of the CS2 fimbriae shows a lack of cysteine and a very low content of aromatic amino acids. It is possible that the CS2 fimbriae of the various strains are not identical (based on animo acid composition, especially in the glycine content between CS2 preparations from two different strains) (Klemm et al., 1989); Sjöberg et al., 1988).

The DNA elements required for the biosynthesis of the CS3 adhesin complex have been cloned in *E. coli* minicells (Manning et al., 1985) and found encode for five polypeptides (Boylan et al., 1987, 1988). The heterogeneity of the plasmids coding for the CFA/II may be responsible for the variations observed in the molecular weights of the peptides encoded by the CS3 gene cluster (Boylan et al., 1987).

Another superfamily, CFA/IV or PCF8775 adhesins, is composed of three antigenically distinct adhesins designated as CS4, CS5, and CS6 (Thomas et al., 1982, 1985, 1987). As with other CFAs, the plasmid carrying the CFA/IV complex also carries genes encoding for the diarrheal toxin. The CFA/IV coding plasmid can be mobilized by conjugation into *E. coli* HB101 (Wolf et al., 1989). The transconjugates expressed CS6 fimbriae only, suggesting that the host strain of the CFA/IV coding plasmid does not allow expression of CS4 fimbrial adhesin. The CS6 apparently is fibrillar, composed of the 15,000 mol wt major subunit and its presence on the bacterial surface is important for colonization of intestinal mucosa (Svennerholm et al., 1988). The molecular weight of the major fimbrial subunit of CS4 was reported to be 17,000 by one study (McConnell et al., 1988) and 22,000 by another (Wolf et al., 1989), whereas that of the CS5 adhesin is 21,000 (McConnell et al., 1988). The CS6 adhesin was found as a single polypeptide with a molecular weight of about 14,500 in one strain of ETEC, whereas other strains positive for CS6 antigen lacked the 14.5-kDa protein (McConnell et al., 1988). Expression of both proteins was dependent on growth temperature (Wolf et al., 1989). Apparently, there are strain variations in the molecular weight of the CFA/IV family of adhesins.

From the foregoing, it is clear that a number of unique features are shared by the CFA-encoding plasmids. They all contain genes coding for enterotoxins, can be mobilized into various host strains of *E. coli* with the aid of conjugative plasmids, and the phenotypic expression of the various adhesins is affected by the genomic background of the strain hosting the CFA-coding plasmid (Evans and Evans, 1990). Very little is known about the regulation of the expression of the CFAs. It is clear that growth in agar favors the expression of CFA (Evans and Evans, 1990). In vitro, the strains may lose the CFA-coding plasmid but retain a toxin-coding plasmid (Evans and Evans, 1990). Because the host strain determines the phenotypic expression of a particular type of CS adhesin in the superfamily CFA, loss and acquisition of CFA-coding plasmids by different strains may represent a mechanism through which the enterotoxigenic *E. coli* strains diversify the specificity and/or antigenicity of their expressed adhesins and at the same time switch from CFA^+ to CFA^- and reverse.

F1845 Fimbriae

The F1845 fimbriae are produced by diarrheal isolates of *E. coli* and are responsible for the diffuse adhesion of the bacteria to HEp-2 cells (Bilge et al., 1989). The chromosomal genes encoding for the fimbrial adhesion complex have been cloned. Five polypeptides are encoded by a 4.7-kb fragment of DNA: Daa A, Daa B, Daa C, Daa D, and Daa E, corresponding to molecular weights of 10,000, 27,000, 95,000, 15,500, and 14,300, respectively (Bilge et al., 1989). The *daa* product was suggested to function in the regulation of the expression of the fimbrial protein. The product of *daa*D gene is not essential for the production of the adhesin mediating diffuse adhesion to the cell line. The major fimbrial subunit of the F1845 fimbrial adhesin complex seems to function as an adhesin. In this respect it resembles other fimbrial adhesins belonging to the K88, K99, and Dr adhesin complexes. Of special interest is the gene cluster coding for the Dr fimbrial adhesin. The organization of the genes encoding the Dr adhesin complex appears to differ from that of the F1845 (see Figure 9–1). The two operons possess genes for structural and functional similarities between the DNA determinants of the two adhesins (see below). Although *daa*-related sequences are found on the chromosome of most strains, the plasmids of about one third of the strains tested are also found to contain homology with a *daa*C probe.

469-3 Fimbriae

The 469-3 fimbrial adhesins are expressed by enteroinvasive *Escherichia coli* (EIEC) responsible for dysentery-like diarrheal disease. The gene(s) encoding for the 469-3 fimbrial adhesins are located on a 16-kb chromosomal DNA fragment derived from strain 469-3 isolated from an infant with enteritis (Hinson et al., 1987). The *E. coli* strain HB101 harboring a plasmid containing the 16-kb fragment expressed 469-3 flexible fimbriae of 2 nm diameter (similar to the parent 369-3 isolate), caused mannose-resistant hemagglutination of human erythrocytes, and adhered to human colonic enterocytes. Southern blotting analysis using a 2.7-kb probe derived from the 16-kb fragment revealed the DNA element encoding for the 469-3 fimbrial adhesin is a single chromosomal copy in the 469-3 strain, as well as in another enteritis isolate called 444-3, which also expresses mannose-resistant adhesion, suggesting that this fimbrial adhesin is shared by other EIEC strains.

Dr Hemagglutinins

The family of the fimbrial adhesins known as Dr fimbriae (previously 075X adhesin) and specific for the Dr blood group antigen are expressed by uropathogenic *E. coli* isolates (Nowicki et al., 1989). A characteristic feature of the hemagglutination caused by strains carrying the Dr fimbrial adhesin complex is its inhibition by chloramphenicol and tyrosine-containing peptides (Nowicki et al., 1988; Kist et al., 1990). A recombinant plasmid, containing an insert of about 6 kb of a DNA fragment obtained from the chromosome of a Dr-positive uropathogenic isolate of *E. coli,* confers Dr-specific hemagglutination upon appropriate host strains. The proteins encoded by *dra* plasmid containing genes and its transposon insertion derivatives were examined by an in vitro transcription–translation method (Nowicki et al., 1989). It appears that the *dra* region contains five genes, designated as *dra*A, *dra*B, *dra*C, *dra*D, and *dra*E encoding for polypeptides of 15,600, 5,000, 18,000, 90,000, and 32,000 mol wt, respectively (Figure 9–1). Four genes—*dra*A, *dra*C, *dra*D, and *dra*E—are required for the recombinant strain to express Dr-specific hemagglutination. The *dra*A encodes the 15,000 major fimbrial subunit and appears to also be responsible for the Dr-specific hemagglutination (Nowicki et al., 1988, 1989). In this respect, the molecular organization of the Dr fimbrial adhesin complex resembles that of K99 and K88 fimbrial adhesins. Interestingly, the morphology of the Dr adhesin is also similar to that of K99 and K88, being of the flexible and thin type fimbriae (Nowicki et al., 1987; Kist et al., 1990; see Table 4–2). Kist et al. (1990) recently isolated and purified the Dr fimbrial adhesin complex from two uropathogenic *E. coli*. The N-terminal amino acid sequences of the major Dr fimbrial subunit of the two isolates were homologous to each other and demonstrated high degrees of homology with AFA/I adhesin. Although the gene order of the Dr gene cluster is somewhat different from that of F1845, the two gene clusters share close genetic and functional relationships (Swanson et al., 1991). Comparison of the nucleotide sequences of the major structural subunits between the two adhesins shows a high degree of homology. Plasmids containing DNA constructs of chimeric elements coding for either accessory genes of one adhesin and genes coding for major structural subunits of the other adhesin (and vice versa) were mobilized into a host strain of *E. coli*. The recombinant phenotypes caused hemagglutination only of Dr-positive erythrocytes and chloramphenicol directly inhibited the hemagglutination caused by recombinants expressing the major Dr fimbrial subunit. The results suggest that the accessory gene products of the two adhesins are functionally interchangeable and the structural major subunit of both adhesins contains the receptor-binding domains. Furthermore, plasmid DNA constructs containing a hybrid gene composed of DNA sequences coding for the amino-terminus of the F1845 fimbrial adhesin and a DNA element coding for the carboxyl-terminus of the Dr adhesin were assayed for hemagglutination activity. The recombinant *E. coli* strain containing the hybrid adhesin gene and a plasmid containing the Dr accessory genes caused chloramphenicol-resistant hemagglutination, suggesting that the chloramphenicol-binding region of the Dr hemagglutinin is contained in the amino-terminal portion of the adhesin.

K88 Fimbriae–adhesin Complex

Strains of *E. coli* expressing the K88 adhesin complex colonize the small intestine of the pig (Moon, 1990). Structurally, the fimbriae belong to the flexible type and are very thin, with a diameter of only 2–5 nm. These appendages have been termed "fibrillar" or filaments (Orskov and Orskov, 1990). Antigenically, they are classified as an F4 serotype. A number of antigenic variants have been identified as K88ab, K88ac, and K88ad (Orskov and Orskov, 1990). The genetic determinant coding for the biosynthesis of the K88ab filament–adhesin complex resides on a large nonconjugative plasmid of approximately 50 MDa which also encodes for raffinose utilization (Shipley et al., 1978). A fragment of 6.5 kb, derived from the wild-type plasmids of K88ac or K88ab and cloned in plasmids pFM205 or pMK005, respectively, was shown to be sufficient for encoding the biosynthesis of a functional K88 fimbrial–adhesin complex on the surface of *E. coli* K-12 harboring the cloned fragments (Mooi et al., 1979; Shipley et al., 1981). Six structural genes designated *fae*C through *fae*H (Figure 9–1) were identified in the K88 gene cluster after analysis of the proteins produced by various deletions and transposon–insertion mutants in *E. coli* minicells (Mooi et al., 1981, 1982, 1984). The *fae*G gene encodes for the major fimbrial subunits identified as a 27,000 mol wt protein. Mutants harboring the K88ab determinant in which *fae*C, *fae*D, or *fae*E have been inactivated are nonfimbriated, do not adhere to animal cells, nor do they cause hemagglutination. They do produce low levels of Fae G, suggesting that they are involved in biosynthesis. The inactivation of *fae*H results in a marked reduction of the fimbriation and ability to cause hemagglutination. Impairment of the *fae*D gene results in an afimbriated mutant that accumulates precursor complexes of other fimbrial subunits in the periplasmic space (Mooi et al., 1982; 1983). Analysis of the structure of *fae*D product of the K88ab gene cluster reveals a large central region in the protein with an alternation of turns and β sheets consistent with a transmembrane protein (Mooi et al., 1986). It is possible that the K88ab cluster contains two or more genes encoding for Fae B and Fae I, the functions of which have not been determined (DeGraaf, 1990). None of the Fae G, Fae C, and Fae H proteins are detected in mutants harboring an inactivated *fae*E gene, suggesting that Fae E may function to stabilize the other fimbrial subunits similar to Pap D. Indeed, complexes of Fae E and Fae G have been detected in cells harboring the K88ab gene cluster (Mooi et al., 1983). Five structural genes, designated *adh*A through *adh*E, have been identified in the K88ac gene cluster by analyses of products in minicells, similar to K88ab (Kehoe et al., 1981, 1983). The major fimbrial subunit is encoded by *adh*D. Expression of all the genes in the cluster is essential for high-level expression of the K88ac fimbriae–adhesin complex, suggesting that the function of these K88ab genes are similar to those of the corresponding K88ab genes (*adh*E, *adh*A, *adh*B, and *adh*C correspond, respectively, to *fae*C, *fae*D, *fae*E, and *fae*D).

Perhaps the most striking difference between the K88 cluster proteins and those of the P, type 1 and S is that evidence for the presence of minor fimbrial subunits with adhesin function has not been obtained. At present, it seems that the major fimbrial subunit functions as the adhesin (Oudega and DeGraaf, 1988). If so, it remains to be seen whether only those subunits at the tip of the fimbrial structure function as adhesins. In support of this possibility are studies showing that cyanogen bromide fragments obtained from the major K88 fimbrial subunit inhibited the adhesion of the K88 carrying *E. coli* to erythrocytes and intestinal epithelial cells (Jacobs et al., 1987a). The shortest fimbriae-derived peptides showing inhibition of adhesion were the tripeptides Ser-Leu-Phe, Ile-Ala-Phe, and Ala-Ile-Phe. The conserved region of all three antigenic variants of the K88 fimbrial adhesin contains the tripeptides. Replacement of Phe residues with Ser by site-directed mutagenesis gives rise to mutants that are capable of causing hemagglutination except for mutants in which the Phe at position 150 was changed to Ser (Jacobs et al., 1987b). The receptor specificity of the K88 fimbrial adhesin may be similar to that of gonococcal fimbriae and B fragment of diphtheria toxin because both of these molecular species contain tripeptide sequences identical to that of the *E. coli* K88 fimbrial adhesin. The foregoing results are surprising, because inhibition was achieved by a small peptide instead of a carbohydrate. It is known, however, that numerous carbohydrate-binding proteins (lectins) can also complex with peptides (reviewed by Barondes, 1988). The possibility that bacterial adhesins may have more than one receptor site has not been considered in the past.

The regulation of the biosynthesis and expression of K88 fimbriae–adhesin complexes on bacterial surfaces is poorly understood. It is possible that the K88 cluster contains two regulatory genes, one of which (*fae*I) is located downstream of *fae*H and the other (*fae*B) is located upstream of *fae*C (Oudega and DeGraaf, 1988; DeGraaf, 1990). The clone of the K88 cluster in pFm205 plasmid contains only part of these genes. Further studies are required to show how the degree of fimbriation is affected in clones that possess the whole K88 gene cluster. As with most MR fimbrial adhesins of *E. coli,* the surface expression of K88 is repressed at a low temperature of growth, perhaps due to temperature-dependent promoter activity as seems to be the case for P fimbriae (DeGraaf and Mooi, 1986). In addition, the optimal amount of fimbria–adhesin complex expression is observed at maximal growth rates (Jacobs and DeGraaf, 1985; Blomberg and Conway, 1991).

K99 fimbriae–adhesin complex

E. coli carrying the K99 fimbriae–adhesin complex are usually isolated from bovine and procine sources and adhere to intestinal epithelial cells of calves, lambs, and pigs. The *E. coli* also cause hemagglutination of sheep and horse erythrocytes (reviewed in DeGraaf and Mooi, 1986). Serologically, they are classified as the F5 serotype based on epitopes of the major fimbrial subunit (Orskov and Orskov, 1990). Structurally, they are flexible, possessing a diameter of about 5 nm (DeGraaf, 1990).

The biosynthesis of the K99 fimbrial–adhesin complex reaches a maximum at the end of exponential growth and gradually decreases thereafter, even though the total bacterial mass in the culture increases three to four times (Jacobs and DeGraaf, 1985). The biogenesis and function of K99 fimbriae–adhesin is under the control of a genetic determinant located on a conjugative plasmid of approximately 52 MDa (Smith and Linggood, 1972; So et al., 1974). A 7.1-kb pair DNA fragment obtained from the parental plasmid was cloned in the pBR325 plasmid (Van Embden et al., 1980). *E. coli* strains harboring this recombinant plasmid express K99 fimbriae, cause sialidase-sensitive hemagglutination, and adhere to the brush border of porcine intestinal epithelial cells, indicating that the fragment contains the gene cluster needed for the expression of the K99 fimbria–adhesin complex. The genetic organization and function of the genes in the K99 cluster was studied by analyzing the products in *E. coli* minicells harboring a recombinant plasmid containing genetically manipulated K99 determinants (e.g., deletions and transposon insertions). The analyses indicated the presence of six structural genes, designated *fan*C-*fam*H (DeGraaf et al., 1984; Roosendaal et al., 1987a,b) (Figure 9–1). The major fimbrial subunit of 16,000 mol wt is encoded by *fan*C and is believed to function as an adhesin that binds the bacteria to carbohydrate residues on animal cells (Oudega and DeGraaf, 1988). Mutations in *fan*D or *fan*E result in nonfimbriated, nonhemagglutinating, and nonadhering phenotypes. Reduced fimbriation and adhesion are characteristics of phenotypes harboring the K99 determinant mutated in *fan*F, *fan*G, and *fan*H. Because all of the proteins encoded by these genes are synthesized with a leader sequence enabling the proteins to cross the periplasmic space, it is assumed that Fan F, Fan G, and Fan H are analogous to fimbrial minor proteins of other fimbriae–adhesin complexes. In support of this is the significant amino acid sequence homology between Fan F and Fan H and the major fimbrial subunit (Roosendaal et al., 1987a,b). It is still not clear which fimbrial subunit serves as an adhesin. Currently, the evidence seems to point out that as with K88, the major fimbrial subunit of the K99 fimbriae–adhesin complex acts as the adhesin (Oudega and DeGraff, 1988). The presence of two additional genes in the K99 cluster can be deduced from analyses of the DNA sequence of the region upstream from *fan*C. These genes, designated *fan*A and *fan*B, encode for proteins of 11,000 mol wt, devoid of leader sequences, suggesting that they reside in the cytoplasm (Roosendaal et al., 1987a,b). There is a high degree of homology between Fan A and Fan B, and interestingly also to Pap B. Inactivation of either of these genes results in about a 10-fold reduction in expression of the K99 fimbriae–adhesin complex. The *fan*A and *fan*B mutations could be complemented in *trans*. The data argue in favor of the hypothesis that Fan A and Fan B are *trans*-acting positive regulators of the biosynthesis of the K99 adhesin complex (Roosendaal et al., 1987a,b). Promoter activity has been detected preceding each of the *fan*A, *fan*B, and *fan*C genes (Roosendaal et al., 1989). The promoters P_1 and P_2 possess high and intermediate activities, respectively, as assayed in an *E. coli* harboring a recombinant plasmid in which the promoter-containing region was cloned next to a promoter-free gene coding for galactokinase (Roosendaal et al., 1989). The *fan*C promoter activity is very low. Three factor-dependent terminators of transcription have been detected in intercistronic regions between *fan*A and *fan*B (T_1), *fan*B and *fan*C (T_2), and *fan*C and *fan*D (T_3) (Roosendaal et al., 1989). It was postulated that the stoichiometry of transcription of the various minor structural proteins, as compared to the major fimbrial subunit, may be the result of attenuation of transcription by T_3 which has an efficiency of about 90% (DeGraaf, 1990).

F41 Fimbriae

The chromosomal gene cluster encoding the F41 fimbriae produced by porcine enterotoxigenic strains of *E. coli* was cloned and found to contain genes encoding for polypeptides of molecular weight 29, 30, 32 and 86,000. In this respect, the gene arrangement of the F41 fimbrial adhesin complex resembles that of other fimbrial adhesins (Anderson and Moseley, 1988). Although the F41 fimbrial adhesin is produced by strains that also produce the K99 fimbrial adhesin (Moon, 1990), the chromosomal DNA elements coding for it have some sequence homology with the plasmid DNA determinant encoding the K88 fimbrial adhesin (Moseley et al., 1986). The biosynthesis of the F41 fimbrial adhesin complex is similar to that of K88 and K99 in that it reaches a maximum at the end of the exponential growth (Jacobs and DeGraaf, 1985).

987P Fimbriae

The adhesin carried by the rigid 987P fimbriae is produced by porcine enterotoxigenic strains of *E. coli* (Moon, 1990). The cloned genetic determinant encoding the 987P adhesin remarkably resembles that of other fimbrial adhesins. It encodes at least five polypeptides with apparent molecular weights of 16.5, 20.5, 28.5, 39, and 81,000 (DeGraaf and Klaasen, 1986, 1987; Morrissey and Dougan, 1986). Recently, it was established that the 987P gene cluster resides on a 35 MDa plasmid in enterotoxigenic strains (Schifferli et al., 1990). An *E. coli* harboring a recombinant plasmid containing a 12-kb insert from the 35-MDa plasmid constitutively expressed the 987P fimbrial adhesin. The same strain harboring the intact 35-MDa plasmid derived from the parental 987P strain neither expressed fimbriae nor synthesized the fimbrial subunits. It was postulated that some negative regulatory DNA determinants acted *trans* in concert with the natural DNA of the parental strain to regulate expression of 987P fimbrial adhesin (Schifferli et al., 1990). This conclusion is supported by the findings that expression of the 987P fimbrial adhesin from recombinant plasmids largely depends on the host *E. coli* strain harboring the plasmid (Morrissey and Dougan, 1986). Furthermore, none of the recombinant strains show phase variation in the expression of the fimbrial adhesin as is the case in the parental strains. Another DNA element that was found to regulate the transcription of the gene encoding for the major fimbrial subunit is an ISI DNA element. The ISI element is usually found in a transposon form containing a heat-stable gene (Klaasen et al., 1990). This toxin-transposon element is found on the chromosome and seems to activate expression of the fimbrial adhesin.

F17 (Fy or Att25) Fimbriae

The genetic element coding for the F17 fimbrial adhesin complex was cloned from the chromosome of an enterotoxigenic bovine isolate (Lintermans et al., 1988). An *E. coli* host harboring a plasmid containing a 8.5 kb derived from the parent strain produced F17 fimbriae, adhered to intestinal calf cells, and the adhesion was specifically inhibited by *N*-acetylglucosamine. The amino acid sequence analysis of the major fimbrial subunit (19,5000 mol wt) deduced from the cDNA revealed an interesting homology with Fim A of type 1 fimbrial adhesin and with Pap A of the P-fimbrial adhesin in which about 40–50 amino acids could be aligned with these proteins. In contrast, less homology was observed with the major fimbrial subunits of K99 and K88a fimbrial adhesins. Even more interesting is the very high homology of F17 fimbrial adhesin with the N-termini regions of the *N*-acetylglucosamine-specific G fimbriae of uropathogenic *E. coli*. These homologies emphasize the notion that the host specificity of the various *E. coli* strains is determined by the receptor specificity of the fimbrial adhesin. The receptor specificity probably arose from multiple random mutations in ancestral fimbrial genes resulting in various degrees of homology in the fimbriae of various strains.

CS31A Fimbriae

The gene cluster for the CS31 fimbriae resides in a plasmid of an *E. coli* isolated from bovine blood (Girardeau et al., 1988). The DNA element coding for the CS31A fimbrial adhesin was cloned from a genomic library of DNA from the parent strain using DNA probes derived from the gene cluster encoding for the F41 fimbrial adhesin complex, exposing a strong relationship between the DNA elements encoding for these fimbrial adhesins (Korth et al., 1991). *E. coli* HI101 harboring plasmid containing cloned CS31A DNA fragments express rigid fimbriae composed of a 28,000 mol wt protein as the major fimbrial subunit and adhere to tissue culture cells. Strong homology is also observed with the K88 fimbrial subunit. Nevertheless, analysis of the protein profile in minicells of the DNA elements encoding for the F41, CS31A, and K88 fimbrial adhesins showed that each determinant encodes distinct protein profiles.

Common Features of Fimbriae–adhesin Complexes

Based on the extensive studies on some of the *E. coli* fimbriae–adhesin complexes discussed above, a number of common features related to function and organization of the genetic determinants in the gene cluster coding for the various types or classes of fimbrial adhesins are apparent (Figure 9–1). At the same time, subtle differences in the function and organization of the genes in the cluster are noticeable. This is not surprising if it is assumed that evolutionary pressures selected clones of *E. coli* with slight variations in the regulation of the expression of their fimbriae–adhesin complexes to best fit colonization of a particular site and/or host. In this section, an attempt is made to emphasize the common features of the molecular biology of fimbriae–adhesin complexes of *E. coli* and to outline the most important differences.

The regulation of the biosynthesis and surface expression of the fimbriae–adhesin complex seems to be at two levels. One is involved with the regulatory genes responsible for the stoichiometry of the various products needed for expressing fimbrial rods with minor and major transport proteins and of the anchoring of the various subunits. This level seems to be regulated by a region that contains one or two promoters. The other is responsible for the phenomenon of phase variation. Although a definite proof arguing that the fimbriae–adhesin gene cluster acts as an operon exists only for K99, S, and P, it is likely that all other clusters function as operons as well. This means that the stoichiometry of the various subunits required for the expression of fimbriae is under genetic control. The mechanism through which this regulation operates, however, may be different for the various fimbrial adhesins.

The organization of the gene cluster seems to follow a certain pattern: A common large gene in all operons or gene clusters encodes for a 85–88,000 mol wt protein that is found in the outer membrane and seems to serve as an anchor for assembled fimbriae and/or plays a role in subunit polymerization. Impairment of this gene results in afimbriated mutants that accumulate precursor complexes of other fimbrial subunits in the periplasmic spaces (Mooi et al., 1982, 1983). Analysis of the structure of this protein in the K88ab cluster has revealed a large central region in the protein with an alternation of turns and β sheets, consistent with a transmembrane protein (Mooi et al., 1986).

A second gene present in all gene clusters is located proximal to the large gene and encodes a smaller protein (23–29,000 kDa mol wt) that is found in the periplasm only (Figure 9–1). Mutation in this gene abolishes both formation of fimbriae and accumulation of subunits in the periplasmic space. It has been postulated that this protein is involved in stabilizing other fimbrial subunits in the periplasmic space by forming complexes with them as a "chaperon." It also may function in transporting other fimbrial subunits from the cytoplasm to the outer membrane and in this sense it was named "periplasmic transporter" protein. The products of *pap*D, *fim*C, *sfa*E, *fae*E, and *fan*E have been postulated to belong to this family of chaperone proteins. Physical evidence, however, for their ability to form stable complexes has been obtained only for Pap D protein (Hultgren et al., 1989). Nevertheless, analysis of the amino acid sequences of Fan E, Fae E, and Sfa E show that all these proteins are 30%–40% identical and 60% similar, especially in the location of hydrophobic amino acid residues within the protein sequence (Hultgren et al., 1991). From this similarity and the crystal structure of the Pap D protein (Holmgren and Bränden, 1989), it was concluded that these chaperone proteins consist of two globular domains in a topology similar to immunoglobulins. Each of the chaperone proteins is predicted to contain a variable region to bind the specific fimbrial subunit in the periplasm. Studies on the stability of the Pap D–Pap G complex show that it is stable in 6 *M* urea and is destroyed only under reducing conditions. The carboxyl-terminal sequence of Pap G adhesin is important for binding Pap D (Hultgren et al., 1991). Site-directed mutagenesis of Pap D in its postulated variable region to generate a mutant that produces Pap D with a glutamic acid at position 167 instead of histidine caused a significant change in the rate of fimbrial assembly. Although not conclusive, the data on Pap D suggest that the family of chaperone proteins have a common shape and are involved in specific interactions with fimbrial subunits, including those with adhesin function. Complex formation is followed by rapid dissociation by an as yet unknown mechanism to allow the fimbrial subunits to form the fimbrial–adhesin suprastructure.

Another group of genes that encode for the so-called minor fimbrial subunits are a common feature of the gene clusters coding for the fimbrial–adhesin complexes. On the basis of their molecular weights and degrees of amino acid homology with each other and with the major fimbrial subunit, the minor fimbrial proteins can be divided into two groups. One group is composed of proteins of molecular weights in the range of 14–20,000 and shows a substantial amino acid homology with the major fimbrial subunit. Their numbers vary between three and four among the various fimbrial types. All gene clusters analyzed also contain a gene coding for a protein or about 30,000 mol wt (Figure 9–1). The similarity between the primary sequence of this protein and the major subunit is not as pronounced as that among the first group of the minor fimbrial subunits. Altogether, the highly conserved regions of the minor fimbrial proteins, especially in the amino and carboxyl termini, are not surprising considering the requirements needed for subunit–subunit interaction for polymerization of the mature fimbriae. In some cases, using specific antibodies and immunoelectron microscopy, the proteins are detected as minor components of fimbriae (Lindberg et al., 1987).

It is likely that each of these fimbrial minor proteins encoded by all gene clusters studied share a common function in the biosynthesis of the fimbriae–adhesin complex. At the same time, each of the different minor fimbrial subunits may possess a distinct and specific role in the gene cluster coding for a particular fimbrial type. For example, the 30,000 mol wt protein encoded by gene clusters in all fimbrial types may share a common function in the biosynthesis as well as a unique biological activity, such as adhesin function ascribed only in type 1 and P fimbriae.

The adhesin function of the fimbrial adhesin complex is mediated by either the major fimbrial subunits, as is the case for the K88, K99, Dr, CS2, F1845, and CFA I, or one of the minor fimbrial subunits, such as for Pap G, Fim H, and Sfa S in the P, type 1, and S fimbriae, respectively. In either case, three types of activities are associated with the fimbrial subunits that serve as adhesins. One is receptor binding activity, another is necessary for incorporation into the mature fimbrial structure and one for forming complexes with the chaperone protein in the periplasmic space. Only limited information is available as to which region(s) of the adhesin are required for each of these activities. As with other minor fimbrial subunits, the assembly into the fimbrial structure is mediated by the conserved carboxyl terminus of the molecule (Hultgren et al., 1989).

All fimbrial operons studied contain regulatory genes located at their proximal (5′) ends and are responsible for regulating *in trans* the expression of the adhesin complexes. The sequences of the regions among the gene clusters coding for the various types of MR fimbrial adhesins (i.e., P, S, and K99 fimbriae) are well conserved (Ott et al., 1987; Goransson et al., 1988) but are different from those involved in regulation of MS fimbrial adhesins.

One probable common role encoded by the genes in the various fimbrial gene clusters is the biogenesis of the fimbriae–adhesin complex. On the basis of the experimental data available, the model depicted in Figure 9–7 is proposed for the biosynthesis of the suprastructural fimbriae–adhesin complex (DeGraaf, 1990; Hultgren et al., 1991). Major and minor fimbrial subunits form a complex with the periplasmic 23–29,000 mol wt chaperone protein after their translocation through the cytoplasmic membrane. The interaction between the chaperone and the fimbrial subunit prevents polymerization and degradation, and at the same time confers upon the polypeptides a conformation favorable to their translocation across the outer membrane. Because most of the minor fimbrial proteins are located at the tip it is assumed that they are the first proteins to interact with the 80,000 mol wt pore protein in the outer membrane. Presumably, they possess the highest affinity for this large outer membrane pore protein. The 30,000 mol wt protein and one other minor protein bind to each other and to the 80,000 mol wt outer membrane protein upon their dissociation from the chaperone protein to form an initiation complex. A second minor fimbrial subunit binds to the initiation complex and to the pore protein causing detachment of the initiation complex from the pore protein. Upon arrival of the major subunit–chaperone

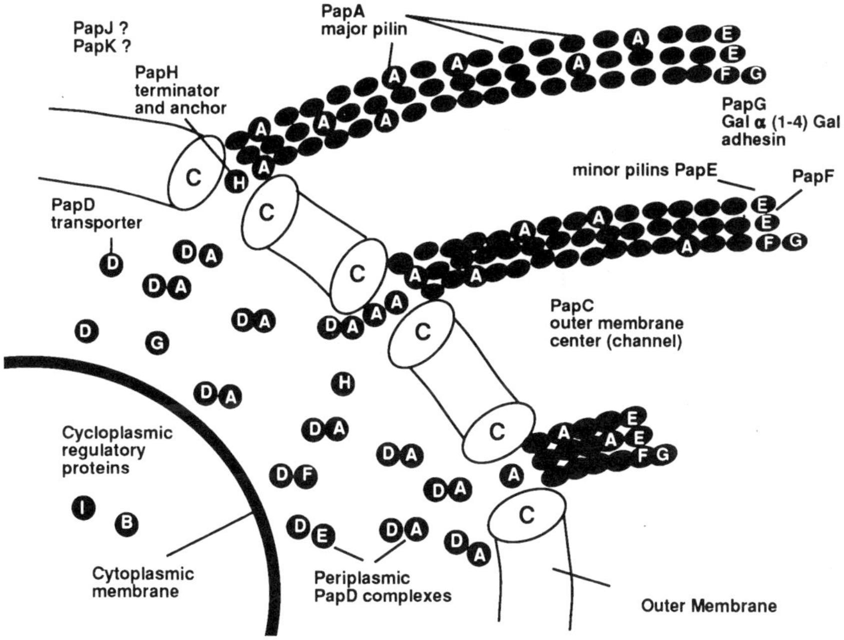

Figure 9–7. Proposed mechanism of biogenesis of the P fimbrium of *Escherichia coli*. The text describes the sequence of reactions leading to the mature fimbrium. Figure 9–1 lists the gene products of the *pap* cluster and their role(s) in synthesis, secretion, and assembly of the fimbrium. (Adapted from Tennent et al., 1990.)

complexes at the pore protein, the major subunit would be dissociated from the periplasmic chaperone protein and incorporated into the growing fimbriae behind the minor fimbrial proteins. As soon as another 30,000 mol wt protein or another minor fimbrial subunit or an initiation complex is delivered to the site of fimbrial growth, further major subunit export is prevented and thus terminates the elongation of the fimbriae.

The above model assumes that newly synthesized subunits are added to the growing fimbriae at the base. Physical evidence for such assembly has been obtained for type 1 fimbriae by using immunoelectron microscopy (Lowe et al., 1987). Another assumption in the proposed model is the distinction between major subunits and minor subunits that serve to regulate the length, number per cell, and adhesive function upon the suprastructure of the fimbriae–adhesin complex. The numbers and lengths of the fimbriae produced depend(s) entirely on the relative ratio of the concentrations of major subunits and each of the minor subunits. Indeed, in the few cases studied, it has been found that mutation in one of the minor fimbrial subunits affects the regulation of the number and length of fimbriae.

In the above discussion the common features of the structural organization and functions of gene products in the gene cluster coding for fimbrial adhesin complexes of *E. coli* were emphasized. In spite of the similarities, it is likely that minor variations have evolved throughout evolution. Some of these minor variations are apparent, especially in the organization of the genes coding for similar functions in the cluster. Random mutations and/or recombination events between homologous regions in one of the structural fimbrial proteins may be responsible for the emergence of receptor binding domains in either the major fimbrial subunit or in one of the minor fimbrial subunits. Recombination events may also be intrastrain following acquisition of fimbrial genes from one strain to another, providing some degree of homology in the DNA of the fimbrial gene cluster. For example, there are homologies between the major subunits of Dr and F1845 (Swanson et al., 1991) and between the minor fimbrial subunits encoded by *fan*F of K99 fimbriae and *fim*H or *fim*G of types 1 and P fimbriae, respectively (Ono et al., 1991). In most cases, only subunits at the tip have functional receptor-binding activities, even in cases where the adhesin is located along the fimbrial side (e.g., Fim H of type 1 fimbriae) or is the major subunit (e.g., CFA I). Thus from the functional point of view it really does not matter where the adhesin resides as long as it is presented in an extended and functional form from the surface.

A- or Nonfimbrial Adhesins of Extraintestinal *Escherichia coli*

The family of nonfimbrial or afimbrial adhesins of *E. coli* causing extraintestinal infections in humans (e.g., urinary tract infections and/or sepsis) constitute a heterogeneous family of adhesins with respect to their receptor specificity, molecular weight, and organization of the gene cluster coding for the adhesin complex. They include a number of classes of adhesins designated as the M agglutinin (Rhen et al., 1986a,b), afimbrial adhesins (AFAs) (Lambigne-Roussel et al., 1984, 1985; Walz et al., 1985), and nonfimbrial adhesins (NFAs) (Goldhar et al., 1984, 1987; Grunberg et al., 1988; Hoschützky et al., 1989). The nonfimbrial adhesins are best expressed during growth on agar at 37°C. Unlike the fimbrial adhesins which require cross-linking or immobilization for hemagglutination or binding of erythrocytes, respectively, the nonfimbrial adhesins in soluble form readily cause hemagglutination of human erythrocytes. Very little is known about the molecular organization of the adhesins on the bacterial surface. It is likely, however, that a number of genes are required for their surface organization because at least in a number of cases examined thus far it has been found that the adhesins are expressed as capsules on bacterial surfaces (Orskov et al., 1985; Kröncke et al., 1990). The genes coding for some of the nonfimbrial adhesins have also been cloned (Table 9–1).

The family of AFA include three types of adhesins: AFA-I, AFA-II, and AFA-III. The detailed organization of the gene cluster has been studied only in the case of the AFA-I adhesin. As with the gene cluster coding for the fimbrial adhesins, the cloned *nfa*-1 gene cluster contains at least five genes designated as *afa*A, *afa*E, *afa*D, *afa*B, and *afa*C encoding polypeptides of molecular weights of 13,000, 16,000, 18,000, 30,000, and 100,000, respectively (Labigne-Roussel et al., 1984, 1985). The five genes belong to the same transcriptional unit and the *afa*E gene was identified as the structural gene for the adhesin. The *afa*B and *afa*C genes are required for functional expression of the adhesin. The *afa*A and *afa*D may function as regulatory genes (Figure 9–2). Thus, the organization of genes encoding for at least one nonfimbrial adhesin remarkably resembles those of the fimbrial adhesins (Figure 9–2).

Four NFAs from different isolates of *E. coli* causing extraintestinal infections have been described (Goldhar et al., 1984, 1987; Grunberg et al., 1988; Hoschützky et al., 1989). They represent a heterogeneous group of structures, which tend to form aggregates of high molecular weight ($>10^6$) composed of protein subunits in the mol wt range of 14,000–21,000. They are readily released from the bacterial surface and are expressed as capsule-like structures on the bacterial surface (Kröncke et al., 1990). The smallest unit size of the DNA fragment required to encode the NFA-1 was shown to be approximately 6.5 kb (Hales et al., 1988). No similarities between the N-terminal amino acid sequence of NFA-1 and the M-agglutinin of AFA-1 have been observed (Hales et al., 1988). As with fimbrial adhesins, expression of the NFA adhesin is temperature regulated.

The *bma* genes, associated with M-agglutinin (Figure 9–1), are also contained in a 6.5-kb chromosomal DNA fragment of pyelonephritogenic strains of *E. coli* (Rhen et al., 1986a). Analysis of proteins encoded by the fragment in minicells revealed that it contains at least five genes encoding for polypeptides of 12.5, 30, 80, 18.5, and 21,000 mol wt (Rhen et al., 1986b). The latter polypeptide was identified as the M-agglutinin, but all five genes are essential for expression of the adhesin on the surface of recombinant strains harboring the 6.5-kb-containing plasmid. It appears, therefore, that the gene organization of the blood group M agglutinin resembles that of *afa*I (Figure 9–1) and perhaps other genes encoding for fimbrial adhesins, but there is no significant sequence homology between the M-agglutinin subunit and that of AFA-I, type 1, Pap, or K99 fimbrillins.

Molecular Epidemiology of *Escherichia coli* Adhesins

The aim of the molecular epidemiology of bacterial adhesins is to determine the frequency of the presence of adhesin-coding DNA sequences in the genome of clinical isolates. Usually, it is achieved by hybridization between the total genomic DNA of the isolates and a radiolabeled DNA probe obtained from the genomic DNA of a representative isolate and that contains a DNA sequence known to be part of the operon or gene cluster required for the expression of the adhesin. Two types of assays can be performed to obtain different types of epidemiological information. One assay employs the dot-blot technique in which the genomic DNA of the clinical isolate is blotted onto a suitable filter paper, followed by hybridization with the probe and by autoradiography. The intensity of the autoradiograph may suggest multiple copies in the genomic DNA of the isolate of DNA elements homologous to the adhesin-encoding DNA probe. The technique is useful to screen large numbers of clinical isolates for positive or negative hybridization, and to suggest the presence or absence, respectively, of adhesin-encoding DNA elements in the genome of the clinical isolates (Table 9–3). The other assay employs restriction fragment polymorphism techniques in which the genomic DNA of the isolates are fragmented by restriction endonucleases, separated by size using agar gel electrophoresis, and blotted with the adhesin-encoding DNA probe. This technique is usually used on a limited number of clinical isolates but allows one to determine polymorphism among the isolates containing adhesin-encoding gene copies and permits assignment of their relative location in the genome by examining the pattern of fragments hybridizing with the DNA probe. The probe is usually excised from a recombinant plasmid in which the adhesin-encoding insert DNA element is cloned. Although both techniques are now widely employed in molecular epidemiology of microorganisms, using DNA probes encoding for specific products, their application to bacterial adhesin complexes bears specific features that must be considered. One feature relates to the choice of the DNA probe and the other to the phenotype–genotype relationship.

Because the biogenesis and the expression of adhesin complexes are encoded by DNA elements containing regulatory genes and genes encoding for products essential for assembly and expression of functional adhesin, the interpretation of hybridization data is already influenced by probe-associated variables, the most important of which are size and strain origin of the probe. The probe may span the entire adhesin-encoding operon or may include only one gene of the operon. Antigenic variations reflected in differences between the adhesin gene clusters of the clinical isolate and the probe may result in poor hybridization. This effect may partially explain the occurrence of a few strains showing poor or no hybridization with adhesin-encoding probe but which phenotypically express the adhesin (Plos et al., 1990). In contrast, positive hybridization may occur in isolates that express an adhesin with a receptor specificity that differs from that of the probe-origin strain. This is due to the homology between the accessory genes or DNA elements present in the adhesin-encoding gene cluster of the isolate and those present on the probe. This effect probably explains the relatively low percentage of strains of which the DNA does not hybridize with a small size probe containing the gene encoding for the *afa* adhesin only but does so with large size probes containing most of the *afa* operon (Labigne-Roussel and Falkow, 1988). In some cases, hybridization with probes encoding various regions of a particular gene of the adhesin complex is useful for screening large numbers of strains for the frequency of occurrence of conserved versus

Table 9–3. Frequency of chromosomal DNA hybridization with adhesin-encoding DNA probe and phenotypic expression of the adhesin in clinical isolates of enterobacteria

Type of adhesin and bacteria		Size and origin of DNA probe	Percent hybrid-positive stains	Percent adhesin-expressing strains	References
Type 1 fimbriae					
E. coli	n = 69	3 kb, pyelonephritic strain	60	34	Buchanan et al., 1985; Mobley et al., 1987
	n = 52		81	69	
Type 1 fimbriae					
K. pneumoniae	n = 20	11, 2, 0.7 kb urinary strain IA565	85	45	Gerlach et al., 1989
P fimbriae					
E. coli	n = 120	3.6 kb,	39	43[a]	Hull et al., 1984; Mobley et al., 1987; Plos et al., 1990; Archambaud et al., 1988a,b Hornick et al., 1991
	n = 52	pyelonephritic strain	60	44	
	n = 217	SH1	61	44	
	n = 102	0.3 kb, pyelonephritic strain J96	78	61	
	n = 26	0.4 kb, *papA*	54	50	
S fimbriae					
E. coli	n = 102	0.8 kb, urinary strain 536	23	3	Archambaud et al., 1988a,b
Afimbrial, AFA I					
E. coli	n = 102	2.5 kb, pyelonephritic strain KS52	12	10	Archambaud et al., 1988a,b; Labigne-Roussel and Falkow, 1988
	n = 138	11+2.5 kb "B" probe	11	10	
		0.8 kb (*afa*E) "C" probe	3	10[b]	
P fimbriae and afimbrial AFA I					
E. coli	n = 102	2.5 and 0.3 kb	9	8	Archambaud et al., 1988a,b
P fimbriae and S fimbriae					
E. coli	n = 102	2.5 and 0.8 kb	17	1	Archambaud et al., 1988a,b
K88 fimbriae					
E. coli	n = 645	6.5 kb, K88ab serotype	23	29	Lanser and Anargyros, 1985
K99 fimbriae					
E. coli	n = 645	7.1 kb, K99 serotype	5	5	Lanser and Anargyros, 1985

[a]The strains have been assayed for mannose-resistant hemagglutination and it is likely that a considerable percentage did not express the P fimbrial adhesin.

[b]These strains caused hemagglutination mediated by non-AFA because they lacked the *afa*E gene.

[c]Includes *E. cloacae, E. aerogenes, K. pneumoniae, K. oxytoca,* and *Serratia* spp.

Escherichia coli

variable regions of the subunits. For example, hybridization with 12 different synthetic 15-base probes corresponding to *pap*A from four P-fimbriated strains revealed considerable heterogeneity of *pap*A sequences among 89 uropathogenic strains, especially in the region encoding the immunodominant hypervariable region of Pap A (Denich et al., 1991).

In spite of the above reservations a large number of strains show positive hybridization with the adhesin-encoding DNA probes but are not able to phenotypically express the adhesin (Table 9–3). A number of possibilities may partially explain these results. One possibility is related to the number of adhesin-gene copies in the various isolates. For example, in a study by Plos et al. (1990), it was found that the presence of multiple *pap* gene copies was associated with pyelonephritogenic, and to a lesser extent with cystic, isolates as compared to fecal isolates. The pyelonephritogenic isolates also show a relatively high degree of correlation between expression of the P-fimbrial adhesin complex and positive hybridization with the *pap* probe. All other possible explanations must assume the existence of differences at the DNA level between strains that do and those that do not phenotypically express the adhesin in vitro, in spite of the fact that both groups of strains are grown under the same conditions and their genomic DNA hybridize equally well with the same adhesin-encoding probe. This assumption is based on studies showing that the clinical origins of strains unable to phenotypically express the adhesin are distinct from those of strains capable of expressing the adhesin. These differences were shown for the *pap*-positive strains between fecal and pyelonephritogenic isolates (Plos et al., 1990) and for *pil*- or *fim*-positive strains isolated from multiple (and one episode) of urinary tract infections (Mobley et al., 1987). A number of differences at the DNA level between expressing and not expressing strains are plausible, such as: (1) presence of genes in nonexpressed loci as found in *N. gonorrhoeae;* (2) mutation in one or more genes essential for expression; and (3) deletion of gene(s) in the adhesin-encoding probe. Especially attractive is the possibility that the differences rely on the DNA elements involved in regulation of the expression of genes essential for the functional expression of the adhesin. *E. coli* strains can be grouped by their ability to express the type 1 fimbrial adhesin on agar into two categories (Hultgren et al., 1986). Whereas all strains are capable of expressing the mannose-specific adhesin after growth in broth, one group of so-called "random phase" variants also express the adhesin on agar. In contrast, the other "regulated" variants do not. Clearly, the control of the rate of phase variation between type 1 fimbriated and nonfimbriated is different between the two groups of strains and this difference is manifested in vitro on agar-grown bacteria. With the realization of subtle differences that affect the expression of adhesins at the DNA level, probes might be developed that can more precisely predict the ability of strains to express or not to express their adhesins.

Another interesting finding from the hybridization studies emerges from results employing DNA probes encoding for more than one type of adhesin. It becomes clear that whereas most *E. coli* clones are positive for the probe encoding for type 1 fimbriae, they also may be genotypically positive for one other adhesin only (Table 9–3). Only few percent of isolates tested contain genomic DNA encoding for three types of adhesins (type 1 fimbriae and two more types) as tested with *afa, sfa,* and *pap* probes (Table 9–3). Evolutionary pressure has no doubt selected clones of *E. coli* with genes encoding for two adhesins one of which is always type 1 fimbriae and the other represents its receptor specificity expected from the various clinical isolates.

Restriction fragment length polymorphism using DNA probe derived from the *pap* gene cluster and restriction digest of genomic DNA from uropathogenic *E. coli* revealed that a considerable number of strains had more than one copy of *pap* homologous sequences (Hull et al., 1988; Arthur et al., 1989a). Furthermore, DNA probes from various regions of the *pap* operon reveal that a number of strains, including fecal isolates, do not express P fimbriae although there are multiple copies of *pap* sequences corresponding to only a part of the *pap* operon (Arthur et al., 1989a,b, 1990). It was suggested that a variety of intra- and interchromosomal recombination events between incomplete and not expressed *pap* gene clusters may give rise to antigenic variants of each of the major and minor fimbrial subunits. Some of these recombination events may occur only in the regions encoding the minor fimbrial subunits giving rise to variants with different receptor specificities, such as strains carrying the Prs fimbrial adhesin, which are often coexpressed with the P-fimbrial adhesin in urinary isolates (see Chapter 10).

Adhesins of *Neisseria gonorrhoeae*

The genomic material of *Neisseria gonorrhoeae,* the causative agent of the sexually transmitted disease gonorrhea, contains information for the expression of a number of adhesins. The molecular biology of two classes of adhesins has been studied the most: the fimbrial and the outer membrane adhesins. The molecular biology of these adhesins is unique in that a single clone uses a novel and complex mechanism that allows the organism to store in its chromosomal DNA different versions of incomplete genetic information coding for each class of adhesin. At any one time one single version of the mature adhesin is expressed on the bacterial surface. In addition, the organisms are capable of switching on and off the synthesis of the adhesins by a mechanism of phase variation. In the following, the genetic mechanisms responsible for variations in fimbriae and outer membrane protein II adhesins are discussed.

The Gonococcal Fimbrial Adhesins

The gonococcal fimbriae are approximately 6 nm in diameter and 1–4 μm in length (Swanson et al., 1971) and contain a major fimbrial protein that has a molecular weight of 18–20,000. Genes coding for fimbrial products in *E. coli* have generally been referred to as *fim, pap, sfa, fae,* and *fan* genes. Throughout this book, these designations have been retained. Equivalent genes in *N. gonorrhoeae,* in contrast, have been designated as *pil,* for pili. This is in spite of the fact that the words fimbriae and pili are interchangeable for *N. gonorrhoeae*. In order to maintain consistency with the literature, *pil* will be employed in the following discussion. The gonococcal pili are required for virulence and mediate the adhesion of the organisms to animal cells in a host- and tissue-specific manner (reviewed in McGee et al., 1983). Like many other Gram-negative fimbriae, they have carbohydrate-binding specificities (see Table 5–1). Physical evidence that proteins other than the major pilus subunit copurify with the fimbrial structure has been obtained (Muir et al., 1988), but genetic evidence for the existence of minor pilus proteins is lacking. Until such evidence is obtained, if at all, it must be assumed that the biogenesis and adhesive function of the pilus structure does not require any other minor pilus-associated protein(s). The gonococcal pilin is synthesized in the cytoplasm as prepilin, consisting of a seven-amino-acid stretch at the amino-terminus which functions as the transport leader peptide (Hermodson et al., 1978; Meyer et al., 1984). During processing, this sequence is nicked and modified to the mature pilin by *N*-methylation at the amino-terminal phenylalanine. This type of processing is a common feature of fimbrial proteins from a number of other species, such as *Pseudomonas aeruginosa* and *Moraxella bovis* (Paranchych, 1990). The mature pilin adhesin consists of three main regions: a conserved amino-terminal hydrophobic region involved in polymerization; a central semivariable region; and a highly variable immunodominant region.

A characteristic feature of the gonococcal pilus adhesin has been related to observations showing that pili subunits from different clinical isolates vary in a number of physicochemical properties, such as size, isoelectric point, and amino acid composition (Lambden, 1982) and are immunologically distinct (Buchanan, 1975). More importantly, they also differ in their receptor specificities (Lambden et al., 1980; Heckels, 1982). Phenotypically, a single clone of the gonococcus undergoes phase and antigenic variations in the expression of its pili at a relatively high frequency. The phase variation phenomenon describes the transition from piliated to nonpiliated variants and vice versa (pilus$^+$, pilus$^-$) and in that sense is similar to the phase variation described above for *E. coli* fimbriae, although as shall be seen, the underlying genetic mechanism is quite different. The term antigenic variation describes transition from one piliated variant to another piliated variant containing one or more distinct epitopes in its variable region. Some of the antigenic pilus variants may also possess distinct receptor specificities, e.g., variants expressing pilus with a receptor specificity different from the parent enabling the former to adhere to substrata that do not bind the latter (Heckels, 1982). The transition between a pilus$^+$-rich population to a pilus-poor population (phase variation) and vice versa is associated with changes in colonial morphology from colony types 1 and 2 to colony types 3 and 4, respectively (Kellogg et al., 1963; Swanson et al., 1971; Ofek et al., 1974).

The molecular biology underlying the phase and antigenic/receptor specificity variations of the gonococcal pilus is remarkable and a number of genetic mechanisms have been suggested to account for this diversity. The gene coding for gonococcal pilin was cloned in a plasmid vector (Meyer et al., 1982). *E. coli* transformed with a plasmid containing 2-kb or 4-kb inserts derived from the gonococcal chromosome produced gonococcal pilins but were not able to assemble these pilin molecules into a pilus structure on its surface. These results suggest that genes responsible for assembly and expression of the gonococcal pilus are not in clusters as is the case for *E. coli* fimbriae. A plasmid containing an 800-bp insert derived from the 2-kb fragment produced pilus protein in minicells. DNA sequencing confirmed the presence of sequences corresponding to a promoter, leader peptide, and pilus protein in the 800-bp fragment. Hybridization experiments with various synthetic oligonucleotide probes specific for various regions of the pilus gene revealed that the chromosomal DNA of a single clone of gonococci contains two loci (reviewed in Seifert and So, 1988; Swanson and Koomey, 1989; Meyer, 1990). One locus is present in most strains as a single copy, designated as *pil*E. This locus contains a promoter and a complete and expressed gene that encodes for the prepilin that undergoes normal processing and assembly into a mature pilus structure. The other locus contains nonexpressed and incomplete pilin genes clustered on either side of *pil*E and is designated *pil*S_1, *pil*S_2, etc. Each *pil*S contains a number of copies of incomplete pilus gene sequences (Meyer et al., 1982; Swanson et al., 1986). Further studies reveal that the incomplete pilus gene copies in the *pil*S loci are found in a tandem array flanked by short, repetitive, and conserved sequences. They also lack the sequences coding for the amino-terminal region of the mature gonococcus pilus as well as the promoter sequences. These copies therefore are not expressed and can be regarded as "silent" gene copies (Figure 9–8). Each silent gene copy contains a number of short regions flanked by highly conserved DNA sequences. Similar distributions of regions are found in the complete and expressed gene copy in the *pil*E locus. DNA sequence analyses show that certain regions of the expressed and nonexpressed *pil* gene copies are identical in all gene copies (conserved regions), but some vary slightly (semivariable regions). The hypervariable region is flanked by nucleotides that encode for cysteine residues. The sequence differences among the various *pil* genes include single nucleotide exchanges, and insertions or deletions of one or more base pairs (Haas and Meyer, 1986).

This unique arrangement of genomic DNA containing complete and incomplete *pil* gene sequences allows recombination between one of the partial gene segments of *pil*S and an expressed and complete pilin gene in *pil*E. The gene arrangement involves the insertion of stretches of different lengths and distinct portions of

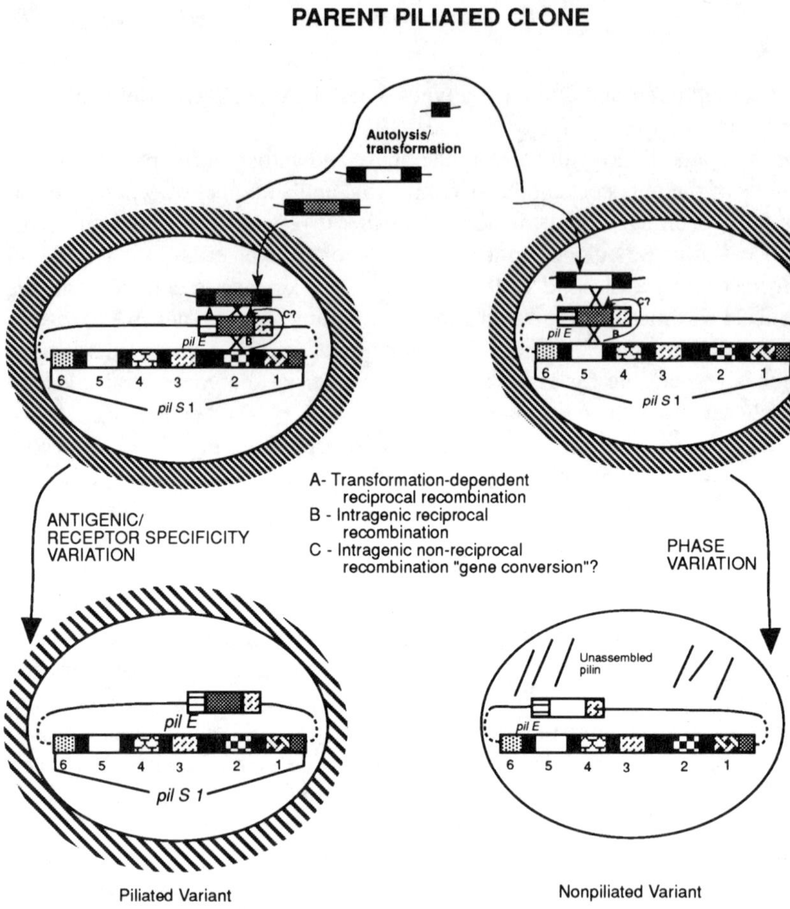

Figure 9–8. Diagrammatic representation of phase and antigenic/receptor specificity variations in pilus adhesins of gonococci. An estimated 12–20 partial nonexpressed pilin genes are clustered at several silent loci (*pil*S) and usually one complete pilin gene in expressed loci (*pil*E). In the diagram is shown one of the silent loci (*pil*S1) containing six partial pilin genes (numbered 1 through 6) truncated by repeat units (black boxes), and a complete pilin gene containing the leader sequence (horizontal dashed box) and the conserved amino terminal sequence (diagonal dashed box) both of which are missing from the silent copies of the incomplete genes. The parent cells can switch the variable region of their complete gene in ?? *pil*E locus with DNA of one of the incomplete gene copies from *pil*S by: (1) uptake (transformation) and reciprocal recombination with free DNA released from cells that have autolyzed in the culture; (2) intracellular and intragenic reciprocal recombination; and (3) intragenic gene conversion and nonreciprocal recombination the contribution of which has been questioned recently (Gibbs et al., 1989). As a result of switching the DNA of the complete gene the cells may either express pili with a variant antigenic and receptor specificity or pilin monomers that are not able to assemble into pilus. (Figure modified from Seifert and So, 1988; Meyer, 1990; Swanson and Koomey, 1989).

one partial pilin gene into the complete pilin gene, resulting in the formation of a new complete pilin gene. The new complete pilin gene may then undergo another round of recombination with stretches of another incomplete gene.

Two major mechanisms of reciprocal recombinations are found to be responsible for such gene rearrangements. One mechanism is intercellular, DNase-sensitive, and dependent on transformation of competent cells with exogenous DNA elements of the *pil*S genes (Norlander et al., 1979; Seifert and So, 1988; Gibbs et al., 1989). Nonpiliated gonococci transform chromosomal DNA at a frequency 3 to 4 log values lower than their piliated variants. Transformation of nonpiliated variants to piliated variants may occur, albeit at a much lower frequency than that of piliated variants (Gibbs et al., 1989). The extensive autolysis of gonococci in culture (Hebeler and Young, 1975) provides an abundant source of *pil*S DNA elements liberated from lysed cells within the population, enabling intercellular transformation-dependent recombination. The other pathway of reciprocal recombination between DNA sequences of *pil*S and *pil*E loci is intracellular and DNase-resistance, and involves intragenic reciprocal recombination between a silent and an expressed locus on the same chromosome. Another mechanism involving nonreciprocal intragenic recombination between the complete pilin gene and the incomplete pilin gene has been suggested, resulting in replacement of the previous pilin gene in the *pil*E locus by one of the incomplete pilin genes in the *pil*S locus of the same chromosome. This type of gene rearrangement has been termed "gene conversion." As pointed out recently, however, the reciprocal recombination between transformed partial pilin gene and complete pilin gene may have the appearance of "gene conversion" because the sequence of the partial *pil* gene in *pil*S locus of the transformant remains intact.

Phenotypically, variants arising after recombinational insertion of different stretches from the partial pilin genes into the parental complete and expressed pilin gene are heterogeneous. The complete pilin gene in the recombinant strains may either encode for pilus adhesin with an antigenic/receptor specificity distinct from the parental strain, or for pilin subunits that are unable to assemble into mature pili. Piliated revertants arise as a result of replacing stretches of the complete pilin genes encoding for missense pilin with a sequence from another partial pilin gene in *pil*S resulting in the synthesis of pilin that assembles into mature pilus adhesin with a distinct antigenic/receptor specificity. It follows that a growing population of a single clone of gonococci is heterogeneous, containing nonpiliated variants and a number of piliated variants, each of which may exhibit a different receptor specificity and antigenically distinct epitopes in the variable regions of the pilin protein (Figure 9–8). It is difficult to assess the frequency of the recombination processes but it is estimated to be as high as 1 in 100 or 1000. Such a high frequency may cause confusion in experiments aiming to determine the receptor specificity or pilin region involved in receptor binding in a particular clone of the gonococcus. This is especially true when high densities of bacteria are tested because the target substrata in the assay system may select for a particular variant expressing a specific pilus adhesin. The conclusion is based mainly on the observations showing that polyclonal antibodies against peptides encompassing residues 41–50 and 69–54 contained in the semivariable region of the pilin are potent inhibitors of adhesion of gonococci to animal cells (Rothbard et al., 1985). Also a large tryptic peptide encompassing residues 31–111 of the pilin molecule binds to endocervical epithelial cells but not to buccal epithelial or Hela cells (Schoolnik et al., 1984). These observations are inconsistent with those showing that gonococci and its purified pilus bind avidly to buccal epithelial cells (Pearce and Buchanan, 1978). It is possible that the receptors on cervical cells are distinct from those on buccal cells. Conversely, a fraction of the population of gonococci and its pili adhesins in the reaction mixture may possess adhesins for the buccal receptors and another fraction may express adhesins for the cervical receptors. Perhaps examination of DNA sequences of pili genes of adherent bacteria detached from a particular substratum in comparison to those detached from another substratum may give a clue to the origin and frequency of variation in receptor specificity.

A number of mutations in the complete pilin gene are found to be responsible for phase variations from piliated to nonpiliated states (Swanson, 1988, 1990). The mutations do not involve recombination processes because they occur in rec^- mutants and include: (1) a single base addition (causing a premature translation termination signal) or its deletion in the complete pilin gene, resulting in nonpiliated or revertant, piliated variants, respectively; (2) a single base revertible substitution in the leader peptide of the complete gene resulting in the synthesis of a prepilin molecule that is not processed in the mature pilin subunit; and (3) a deletion of a large segment of the complete pilin gene resulting in a nonpiliating phenotype incapable of reverting to a piliated phenotype.

Another mechanism of phase variation in the expression of gonococcal fimbriae was found to be dependent on phase variation in the expression of a 110,000 mol wt protein termed Pil C (Jonsson et al., 1991). Only one of two *pil*C genes is expressed by the fimbriated phenotype and inactivation of the gene results in a nonfimbriated phenotype. High frequency of frameshift mutations occurring in a G-containing tract within the region encoding the signal peptide of Pil C appears to be responsible for the phase variation in the expression of Pil C. It was suggested that Pil C forms an outer membrane protein required for assembly of the gonococcal fimbriae and therefore antigenic variation in the fimbriae of revertants that express Pil C is not likely to occur.

The relative contribution of the different mechanisms described above for the phase and antigenic/receptor specificity variations in gonococcal pili is unsettled. In piliated cells the contribution of transformation-mediated recombination seems to be much more important than the intragenomic reciprocal recombination or mutations involving addition/deletion or substitution of a single base (Seifert and So, 1988; Gibbs et al., 1989). In nonpiliated cells the intragenic and intercellular reciprocal recombination processes may be as important as the transformation-mediated recombination. The contribution of "gene conversion" involving intragenic nonreciprocal recombination between partial and complete genes has been questioned. Some investigators argue that it may be the major genetic mechanism responsible for phase and antigenic/receptor specificity variations (Swanson and Koomey, 1989), whereas others stress its importance only in shift from nonpiliated to piliated variants (Seifert and So, 1988). Even others cast doubt on its existence as a mechanism responsible for pilin variations in gonococci (Gibbs et al., 1989). More important are the observations showing that pilin variants arise in vitro during a natural course of infection (Zak et al., 1984) and in infections of human volunteers (Swanson et al., 1987). Which of the various genetic mechanisms found to be responsible for the pilin variations in vitro are preferentially used by gonococci growing at a specific site during a natural course of infection remain to be determined.

The Gonococcal Outer Membrane PII Adhesins

The PII outer membrane proteins (Opa) comprise a family of related molecules (PII_a, PII_b, PII_c, etc.) that share a common transmembrane region and a variable region exposed on the surfaces of the organisms. Their molecular weights range from 29,000 to 36,000 and are distinguished by their electrophoretic mobilities when solubilized at 100°C (Swanson, 1980; Heckels, 1982; Swanson and Barrera, 1983). When expressed, they confer upon agar-grown organisms opaque colonies, become a major component of the outer membrane, and function as adhesins that mediate the binding of the bacteria to each other (Lambden et al., 1979; Virji and Everson, 1981); to various animal cells, including tissue culture cells and epithelial cells (Lambden et al., 1979; James et al., 1980); and, most importantly, to human polymorphonuclear phagocytes (Rest et al., 1982; Virji and Heckels, 1986).

A characteristic feature of the PII adhesins is that a single clone of a gonococcus contains in its chromosome a number of PII genes. The expression of each of the PII genes is under phase variation and the gonococcus can express none or one or more PII adhesins at any given time. Each of the various PII genes encodes for an antigenically distinct PII protein. The receptor specificity of these antigenically distinct PII adhesins may not be the same in that some mediate the binding of the organisms to buccal epithelial cells and others to tissue culture cells (Heckels, 1982). Moreover, even though all the variant PII adhesins tested mediate the interaction of the gonococci with PMNs, the chemiluminescence response of the phagocytic cells following adhesion varies among PII clones. Thus, the family of the PII adhesins is similar to that of the pilus adhesins in the sense that they undergo phase and antigenic/receptor specificity variations. The molecular biology responsible for the variations of PII adhesins, however, is quite different from that of the pilus adhesins. All of the *opa* loci in the

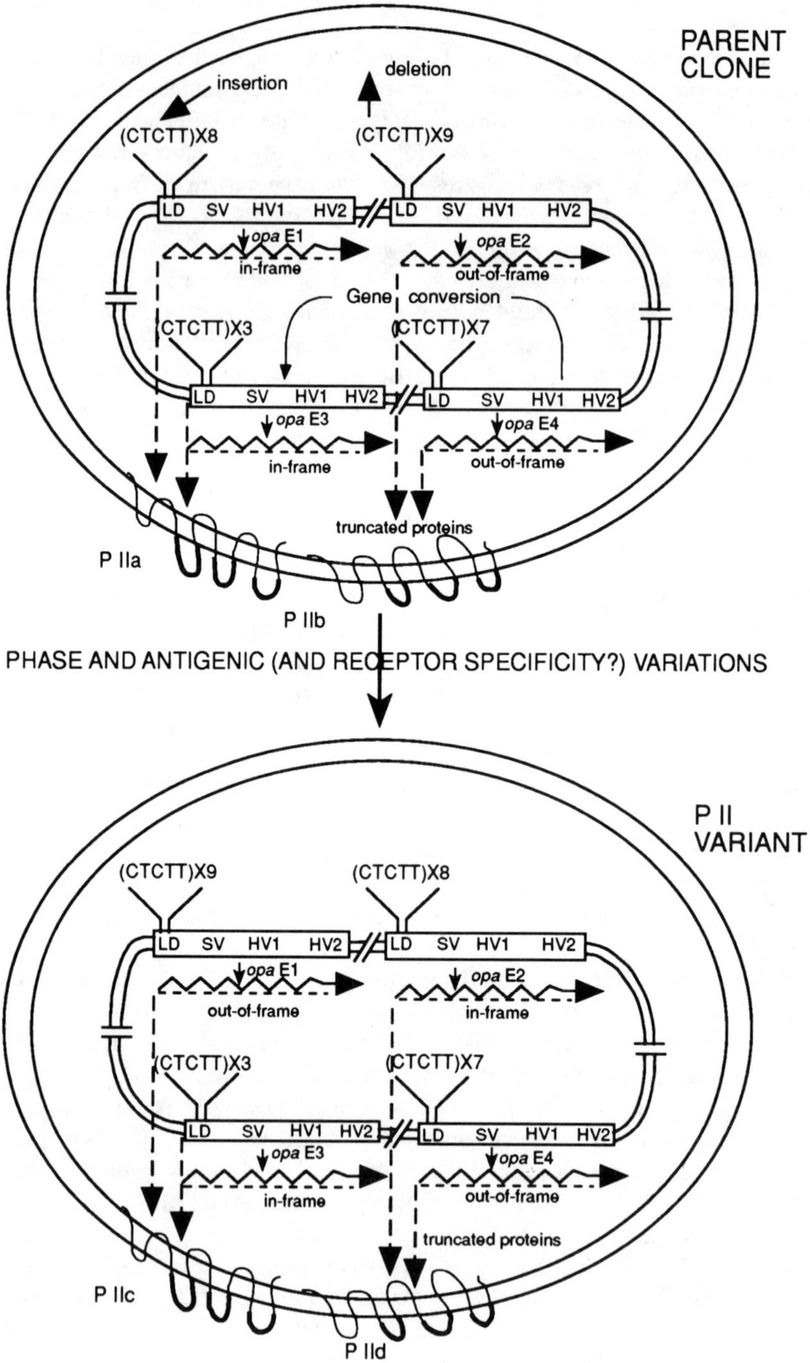

Figure 9–9. Legend on opposite page

chromosome of a single clone contain a functional promoter and a complete PII gene copy. They are scattered in the chromosome and their number varies between 7 and 11 (designated *opa*E_1, *opa*E_2, *opa*E_3, etc.,) according to the test strain or clone of gonococci (Stern et al., 1984; Schwalbe and Cannon, 1986; Muralidharan et al., 1987). They contain DNA sequences in the variable region of their *opa* gene that may give rise to expression of antigenic/receptor specificity PII protein variants or to the lack of expression of the protein altogether (Schwalbe et al., 1985). The *opa* genes in both PII^- and PII^+ phenotypes are transcribed into mRNA and apparently the transcripts in the PII^- phenotypes are not translated into a mature PII outer membrane protein. DNA sequences of a number of PII genes obtained from a PII^- and PII^+ phenotypes have revealed that the phase variation in the regulation of the expression of functional PII protein depends on the numbers of a repeating five-nucleotide CTCTT sequence located within the signal peptide coding region (Figure 9–9) (Stern et al., 1986; Muralidharan et al., 1987; Connell et al., 1988). Depending on the numbers of this repeating peptide, the PII gene would either be in-frame to code for the synthesis of full-length PII protein or out-of-frame resulting in the synthesis of a smaller protein

Figure 9–9. Genetic control of the expression and variation of PII outer membrane adhesins. There are 2–13 estimated numbers of PII gene loci (designated *opa*E1, *opa*E2, etc.) scattered in the chromosome. Most of the loci are consecutively transcribed but the translation of the transcript into mature PII protein is dependent on the numbers of CTCTT repeats in the leader sequence (LD) located in the 5′ region of each of the PII genes. The number of CTCTT copies between the ATG translation codon of LD and the nucleotide triplet that encodes for the amino-terminal alanyl residue of the mature PII protein determines the correct reading frame of individual *opa* genes. Hence, whether mature PII protein is synthesized and expressed in the outer membrane or a truncated protein of unknown biological function is synthesized depends on the numbers of CTCTT copies. The upper figure shows a parent clone that undergoes variations in four of its *opa* loci. Two loci (*opa*E1 and *opa*E3) contain 8 and 3, respectively, CTCTT repeats in their LD regions and the transcribed mRNA is in-frame for the synthesis and expression of two mature PII adhesin variants (PIIa and PIIb). In contrast, the other two *opa* loci (*opa*E2 and *opa*E4) contain 9 and 7, respectively, of CTCTT repeats and the transcribed mRNA is out-of-frame resulting in the synthesis of a truncated protein. At a relatively high frequency the number of CTCTT repeats can change by either inserting or deleting one or more CTCTT repeat. The mechanism for such insertion/deletion events is not known but it is responsible for turning on or off (phase variation) the synthesis of mature PII adhesin as exemplified in the figure of *opa*E1 and *opa*E2. Also, the antigenic diversity (and presumably the receptor specificity) of the PII adhesins can occur by a process of recombination (perhaps gene conversion?). Such recombination enables the rearrangement of genes especially in the variable regions (HV1 and HV2) and to a lesser extent in the semivariable regions (SV). The result is the synthesis of an antigenically distinct PII adhesin expressed in the outer membrane of the new PII variant gonococcus (exemplified in the figure for *opa*E3 and *opa*E4). (The figure is based on Seifert and So, 1988; Swanson and Koomey, 1989; and Meyer, 1990.)

unable to be processed into mature outer membrane protein. The deletion/insertion of the CTCTT units occurs in a spontaneous fashion, affecting the translation of an individual *opa* gene at a relatively high frequency (e.g., 10^{-2} variants per parent cell) and therefore seems to be the major mechanism for the phase variation in the expression of PII adhesins by gonococci as well as in the number of PII variant genes expressed at any one time.

The diversity of antigenic/receptor specificity of the PII adhesins appears to be the result of nonreciprocal intragenic recombination between two *opa genes* affecting only a part of their sequences (Stern et al., 1986; Connell et al., 1988; Swanson, 1990). These recombination processes, resembling gene conversion-like processes, yield PII genes with different combinations in the hypervariable regions of the PII adhesin.

As with the pilus adhesin, phase and antigenic/receptor specificity variations are likely to take place in vivo. For example, gonococci reisolated from humans challenged with a population rich in PII^- variants yielded a population rich in PII^+ variants (Swanson and Koomey, 1989).

Summary

The gonococcus, and to a less extent *E. coli,* is (are) perhaps a champion among human pathogens studied in its ability to diversify its receptor specificity and to regulate the expression of its adhesins. The genetic control and variations of the synthesis of other adhesins expressed by other pathogens have not been studied yet in detail. Studies may reveal similarly complex and remarkable mechanisms to modulate adhesins. Because the synthesis of most bacterial adhesins, however, is neither constitutive nor essential, it is likely that they undergo phase and antigenic/receptor specificity variations as well. As can be seen in Table 9–1, *N. meningitidis* appears to employ mechanisms similar to *N. gonorrhoeae* to diversify the receptor specificity of its adhesins (Stern et al., 1986; Stern and Meyer, 1987; Kawula et al., 1988; Perry et al., 1988). Although the genetic mechanisms of regulation of the fimbrial adhesins of *E. coli* discussed above are quite different, some similarities to those of *N. gonorrhoeae* can be noted. For example, both species produce two types of adhesins, only one of which is capable of mediating the interaction of the organisms to phagocytic cells. Variation in receptor specificity in *E. coli* may not be as high as in *N. gonorrhoeae* but in at least one type of *E. coli* adhesin, the P-fimbrial adhesin encoded by *pap*G, it has been found that the products of the *pap*G gene from two different chromosomal loci have different receptor specificities (Lund et al., 1987).

The bacterial species such as *Pseudomonas aeruginosa, Neisseria meningitidis, Moraxella bovis,* and *Bacteroides nodosus* (*Dichelobacter nodusus*) express flexible fimbriae similar to those of *Neisseriae gonorrhoeae* and collectively these fimbriae have been termed methylphenylalanine fimbriae (Paranchych, 1990).

Based on structural similarities and perhaps also arrangement of genes encoding for these fimbriae, it was suggested to group them as type 4 fimbriae so as to include the fimbrial adhesin of *V. cholerae* (Shaw and Taylor, 1990). The gene cluster coding for the type 4 fimbriae of the above bacterial species has been cloned, and except for the *Bacteriodes* fimbriae, the type 4 fimbriae of all other species has been shown to play a role in the process of adhesion of the organisms to animal cells (Paranchych, 1990; and Table 9–2). At this stage little is known about the receptor specificity of many of these fimbrial adhesins to allow any general conclusions as to possible similarities between the genera and species. It is likely, however, that the biogenesis of the type 4 fimbrial adhesins follows a common pathway for their assembly because the respective gene coding for the major fimbrial subunit is expressed when mobilized into *P. aeruginosa* (Beard et al., 1990; Elleman et al., 1990). Nevertheless, it is expected to find differences in the mechanisms involved in the antigenic and perhaps receptor specificity of these fimbrial adhesins. For example, transition between one fimbrial variant to another in *M. bovis* is the result of an inversion of a 2.1-kb fragment of DNA (Marrs et al., 1988). As more information on the receptor specificity and biogenesis becomes available, it may be possible to conclude whether these fimbriae can be grouped under one family with respect to their adhesion properties. Recently, however, it has been shown that monoclonal antibodies against pili from *P. aeruginosa, M. bovis, N. gonorrhoeae, D. nodosus,* and *V. cholerae* cross-reacted with the various pili (Patel et al., 1991). Residues 1–25 of the amino-terminal region were responsible for the antigenic similarities. Also, the regulation of the expression of each of the two types of adhesins (MR and MS in *E. coli* and the pilus and PII in gonococci) in both species is independent and under phase variation control.

The frequency of turning off and on (phase variation) the expression of each class of adhesin and of the periodical switching of the affinity or receptor specificity of the adhesins by one single clone is quite high (estimated to be in the range of 1 for 100–1000 cells in gonococci). This means that a fairly dense population of bacteria would be predicted to contain a sufficient number of organisms expressing a particular adhesin to enable the bacteria to adhere to a variety of substrata. Such a high number of variants may provide an inoculum sufficient to initiate colonization of substrata containing receptors specific for the adhesin(s). For example, it was possible to isolate from experimentally infected animals a gonococcal clone expressing a pilus adhesin distinct from that of the challenge strain (Lambden et al., 1981). It is clear that bacteria are endowed with different genetic mechanisms for diversifying and expressing different receptors thereby providing for a unique survival potential. There are only meager data revealing how staphylococci, streptococci, pseudomonads, and other bacteria can undergo preprogrammed changes in the genes coding for their adhesins. Based on studies with *E. coli* and *N. gonorrhoeae,* it is likely that each species will possess unique genetic mechanisms allowing for adhesin diversification. The evolutionary goal of adhesin diversification is to enable the bacterium to colonize new substrata that it may encounter (Chapter 11 provides a more detailed discussion on the biological need for adhesin variation).

References

Abe, C., S. Schmitz, I. Moser, G. Boulnois, N.J. High, I. Orskov, F. Orskov, B. Jann, and K. Jann. 1987. Monoclonal antibodies with fimbrial F1C, F12, F13, and F14 specificities obtained with fimbriae from *E. coli* 04:K12H$^-$. *Microb. Pathogen.* **2**:71–77.

Abraham, S.N., D.L. Hasty, W.A. Simpson, and E.H. Beachey. 1983. Antiadhesive properties of a quarternary structure hybridoma antibody against type 1 fimbriae of *Escherichia coli. J. Exp. Med.* **158**:1114–1128.

Abraham, S.N., J. Babu, C. Giampapa, D. Hasty, W. Simpson, and E. Beachey. 1985. Protection against *Escherichia coli*-induced urinary tract infections with hybridoma antibodies directed against type 1 fimbriae or complementary D-mannose receptors. *Infect. Immun.* **48**:625–628.

Abraham, J.M., C.S. Freitag, R.M. Gander, J.R. Clements, V.L. Thomas, and B.I. Eisenstein. 1986. Fimbrial phase variation and DNA rearrangements in uropathogenic isolates of *Escherichia coli. Mol. Biol. Med.* **3**:495–508.

Abraham, S.N., J.D. Goguen, D. Sun, P. Klemm, and E.H. Beachey. 1987. Identification of two ancillary subunits of *Escherichia coli* type 1 fimbriae by using antibodies against synthetic oligopeptides of *fim* gene products. *J. Bacteriol.* **169**:5530–5536.

Abraham, S.N., J.D. Goguen, and E.H. Beachey. 1988a. Hyperadhesive mutant of type-1 fimbriated *Escherichia coli* associated with formation of Fim H organelles (fimbriosomes). *Infect. Immun.* **56**:1023–1029.

Abraham, S.N., D. Sun, J.B. Dale, and E.H. Beachey. 1988b. Conservation of the D-mannose-adhesion protein among type 1 fimbriated members of the family *Enterobacteriaceae. Nature* **336**:682–684.

Anderson, D.G. and S.L. Moseley. 1988. *Escherichia coli* F41 adhesin: genetic organization, nucleotide sequence, and homology with the K88 determinant. *J. Bacteriol.* **170**:4890–4896.

Archambaud, M., P. Courcoux, and A. Labigne-Roussel. 1988a. Detection by molecular hybridization of PAP, AFA, and SFA adherence systems in *Escherichia coli* strains associated with urinary and enteral infections. *Ann. Inst. Pasteur/Microbiol.* **139**:575–588.

Archambaud, M., P. Courcoux, V. Ouin, G. Chabanon, and A. Labigne-Roussel. 1988b. Phenotypic and genotypic assays for the detection and identification of adhesins from pyelonephritic *Escherichia coli. Ann. Inst. Pasteur/Microbiol.* **139**:557–573.

Arthur, M., C. Campanelli, R.D. Arbeit, C. Kim, S. Steinbach, C.E. Johnson, R.H. Rubin, and R. Goldstein. 1989a. Structure and copy number of gene clusters related to the *pap*P-adhesin operon of uropathogenic *Escherichia coli. Infect. Immun.* **57**:314–321.

Arthur, M., C.E. Johnson, R.H. Rubin, R.D. Arbeit, C. Campanelli, C. Kim, S. Steinbach, M. Agarwal, R. Wilkinson, and R. Goldstein. 1989b. Molecular epidemiology of adhesin and hemolysin virulence factors among uropathogenic *Escherichia coli*. *Infect. Immun.* **57**:303–313.

Arthur, M., R.D. Arbeit, C. Kim, P. Beltran, H. Crowe, S. Steinbach, C. Campanelli, R.A. Wilson, R.K. Selander, and R. Goldstein. 1990. Restriction fragment length polymorphisms among uropathogenic *Escherichia coli: pap*-related sequences compared with rrn operons. *Infect. Immun.* **58**:471–479.

Baga, M., S. Normark, J. Hardy, P. O'Hanley, D. Lark, O. Olsson, G. Schoolnik, and S. Falkow. 1984. Nucleotide sequence of the *pap* A gene encoding the *pap* pilus subunit of human uropathogenic *Escherichia coli*. *J. Bacteriol.* **157**:330–333.

Baga, M., M. Göransson, S. Normark, and B.E. Uhlin. 1985. Transcriptional activation of a Pap pilus virulence operon from uropathogenic *Escherichia coli*. *EMBO J.* **4**:3387–3893.

Baga, M., M. Norgren, and S. Normark. 1987. Biogenesis of *E. coli* Pap-pili: Pap H, a minor pilin subunit involved in cell anchoring and length modulation. *Cell* **49**:241–251.

Baga, M., M. Göransson, S. Normark, and B.E. Uhlin. 1988. Processed mRNA with differential stability in the regulation of *E. coli* pilin gene expression. *Cell* **52**:197–206.

Balligand, G., Y. Laroche, and G. Cornelis. 1985. Genetic analysis of virulence plasmid from a serogroup 9 *Yersinia enterocolitica* strain: role of outer membrane protein P1 in resistance to human serum and autoagglutination. *Infect. Immun.* **48**:782–786.

Banas, J.A., R.R.B. Russell, and J.J. Ferretti. 1990. Sequence analysis of the gene for the glucan-binding protein of *Streptococcus mutans* Ingbritt. *Infect. Immun.* **58**:667–673.

Barondes, S. 1988. Bifunctional properties of lectins: lectins redefined. *Trends Biochem. Sci.* **13**:480–482.

Beachey, E.H., C.S. Giampapa, and S.N. Abraham. 1988. Bacterial adherence. Adhesin receptor-mediated attachment of pathogenic bacteria to mucosal surfaces. *Am. Rev. Respir. Dis.* **138**:S45–S48.

Beard, M.K.M., J.S. Mattick, L.J. Moore, M.R. Mott, C.F. Marrs, and J.R. Egerton. 1990. Morphogenetic expression of *Moraxella bovis* fimbriae (pili) in *Pseudomonas aeruginosa*. *J. Bacteriol.* **172**:2601–2607.

Berger, H., J. Jacker, A. Juarez, C. Hughes, and W. Geobel. 1982. Cloning of the chromosomal determinants encoding hemolysin production and mannose-resistant hemagglutination in *Escherichia coli*. *J. Bacteriol.* **152**:1241–1247.

Bilge, S.S., C.R. Clausen, W. Lau, and S.L. Moseley. 1989. Molecular characterization of a fimbrial adhesin, F1845, mediating diffuse adherence of diarrhea-associated *Escherichia coli* to HEp-2 cells. *J. Bacteriol.* **171**:4281–4289.

Blomberg, L. and P.L. Conway. 1991. Influence of raffinose on the relative synthesis rate of K88 fimbria and the adhesive capacity of *Escherichia coli* K88. *Microb. Pathogen.* **11**:143–147.

Blomfield, I.C., M.S. McClain, and B.I. Eisenstein. 1991a. Type 1 fimbriae mutants of *Escherichia coli* K12: characterization of recognized afimbriate strains and construction of new *fim* deletion mutants. *Mol. Microbiol.* **5**:1439–1445.

Blomfield, I.C., M.S. McClain, J.A. Princ, P.J. Calie, and B.I. Eisenstein. 1991b. Type 1 fimbriation and fimE mutants of *Escherichia coli* K12. *J. Bacteriol.* **173**:5298–5307.

Boylan, M. and S.J. Smyth. 1985. Mobilization of CS fimbriae-associated plasmids of enterotoxigenic *Escherichia coli* of serotype 06:K15:H16 or H- into various wild-type hosts. *FEMS Microbiol. Lett.* **29**:83–89.

Boylan, M., D.C. Coleman, and C.J. Smyth. 1987. Molecular cloning and characterization of the genetic determinant encoding CS3 fimbriae of enterotoxigenic *Escherichia coli. Microb. Pathogen.* **2**:195–209.

Boylan, M., C.J. Smyth, and J.R. Scott. 1988. Nucleotide sequence of the gene encoding the major subunit of CS3 fimbriae of enterotoxigenic *Escherichia coli. Infect. Immun.* **56**:3297–3300.

Brinton, C.C., Jr. 1959. Non-flagellar appengates of bacteria. *Nature (Lond.)* **183**:782–786.

Brinton, C.C., Jr. (1965). The structure, function, synthesis and genetic control of bacterial pili and a molecular mechanism for DNA and RNA transport in Gram-negative bacteria. *Trans. NY Acad. Sci.* **27**:1003–1054.

Brinton, C.C., P.C. Gemski, Jr., S. Falkow, and L.S. Baron. 1961. Location of the piliation factor on the chromosome of *Escherichia coli. Biophys. Biochem. Res. Commun.* **5**:293–299.

Brown, D.R. and C.D. Parker. 1987. Cloning of the filamentous hemagglutinin of *Bordetella pertussis* and its expression in *Escherichia coli. Infect. Immun.* **55**:154–161.

Buchanan, T.M. 1975. Antigenic heterogeneity of gonococcal pili. *J. Exp. Med.* **141**:1470–1475.

Buchanan, K., S. Falkow, R.A. Hull, and S.I. Hull. 1985. Frequency among *Enterobacteriaceae* of the DNA sequences encoding type 1 pili. *J. Bacteriol.* **162**:799–803.

Bühler, T., H. Hoschützky, and K. Jann. 1991. Analysis of colonization factor antigen I, an adhesin of enterotoxigenic *Escherichia coli* 078:H11:fimbrial morphology and location of the receptor-binding site. *Infect. Immun.* **59**:3876–3882.

Cheney, C.P., S.B. Formal, P.A. Schad, and E.C. Boedeker. 1990. Genetic transfer of a mucosal adherence factor (R1) from an enteropathogenic *Escherichia coli* strain into a *Shigella flexneri* strain and the phenotypic suppression of this adherence factor. *J. Infect. Dis.* **147**:711–723.

Clegg, S. 1982. Cloning of genes determining the production of mannose-resistant fimbriae in a uropathogenic strain of *Escherichia coli* belonging to serogroup 06. *Infect. Immun.* **38**:739–744.

Clegg, S. and J.K. Pierce. 1983. Organization of genes responsible for the production of mannose-resistant fimbriae of a uropathogenic *Escherichia coli* isolate. *Infect. Immun.* **42**:900–906.

Clegg, S., S. Hull, R. Hull, and J. Pruckler. 1985a. Construction and comparison of recombinant plasmids encoding type 1 fimbriae of members of the family *Enterobacteriaceae*. *Infect. Immun.* **48**:275–279.

Clegg, S., J. Pruckler, and B.K. Purcell. 1985b. Complementation analyses of recombinant plasmids encoding type 1 fimbriae of members of the family *Enterobacteriaceae*. *Infect. Immun.* **50**:338–340.

Clegg, S., B.K. Purcell, and J. Pruckler. 1987. Characterization of genes encoding type 1 fimbriae of *Klebsiella penumoniae, Salmonella typimurium, and Serratia marcescens*. *Infect. Immun.* **55**:281–287.

Connell, T.D., W.J. Black, T.H. Kawula, D.S. Barritt, J.A. Dempsey, K. Kverneland, Jr., A. Stephenson, B.S. Schepart, G.L. Murphy, and J.G. Cannon. 1988. Recombination among protein II genes of *Neisseria gonorrhoeae* generates new coding sequences and increases structural variability in the protein II family. *Mol. Microbiol.* **2**:227–236.

Dallo, S.F., C.J. Su, J.R. Horton, and J.B. Baseman. 1988. Identification of P1 gene domain containing epitope(s) mediating *Mycoplasma pneumoniae* cytadherence. *J. Exp. Med.* **167**:718–723.

DeGraaf, F.K. 1990. Genetics of adhesive fimbriae of intestinal *Escherichia coli*. *Curr. Top. Microbiol. Immunol.* **151**:29–53.

DeGraaf, F.K., and P. Klassen. 1986. Organization and expression of genes involved in the biosynthesis of 987P fimbriae. *Mol. Gen. Genet.* **204**:75–81.

DeGraaf, F.K. and P. Klassen. 1987. Nucleotide sequence of the gene encoding the 987P fimbrial subunit of *Escherichia coli*. *FEMS Microbiol. Lett.* **42**:253–258.

DeGraaf, F.K. and F.R. Mooi. 1986. The fimbrial adhesins of *Escherichia coli*. *Adv. Microbial Physiol.* **28**:65–143.

DeGraaf, F.K., B.E. Krenn, and P. Klaasen. 1984. Organization and expression of genes involved in the biosynthesis of K99 fimbriae. *Infect. Immun.* **43**:508–514.

Delisse-Gathoye, A.M., C. Locht, F. Jacob, M. Raaschou-Nielsen, I. Heron, J.L. Ruelle, M. DeWild, and T. Cabezon. 1990. Cloning, partial sequence, expression, and antigenic analysis of the filamentous hemagglutinin gene of *Bordetella pertussis*. *Infect. Immun.* **58**:2895–2905.

Demuth, D.R., C. Davis, A. Corner, R. Lamont, P. Leboy, and D. Malamud. 1988. Cloning and expression of *Streptococcus sanguis* surface antigen that interacts with a human salivary agglutinin. *Infect. Immun.* **56**:2484–2490.

Denich, K., A. Craiu, H. Rugo, G. Muralidhar, and P. O'Hanley. 1991. Frequency and organization of papA homologous DNA sequences among uropathogenic digalactoside-binding *Escherichia coli* strains. *Infect. Immun.* **59**:2089–2096.

deRee, J.M., P. Schwillens, P. Promes, I. Van Die, H. Bergmans, and J.F. Van den Bosch. 1985. Molecular cloning and characterization of F9 fimbriae from a uropathogenic *Escherichia coli*. *FEMS Microbiol. Lett.* **26**:163–169.

Dickinson, D.P., M.A. Kubiniec, F. Yoshimura, and R.J. Genco. 1988. Molecular cloning and sequencing of the gene encoding the fimbrial subunit protein of *Bacteroides gingivalis*. *J. Bacteriol.* **170**:1658–1665.

Dodd, D.C., P.J. Bassford, Jr., and B.I. Eisenstein. 1984. Dependence of secretion and assembly of type 1 fimbrial subunits of *Escherichia coli* on normal protein export. *J. Bacteriol.* **159**:1077–1079.

Donkersloot, J.A., J.O. Cisar, M.E. Wax, R.J. Harr, and B.M. Chassy. 1990. Expression of *Antinomyces viscosus* antigens in *Escherichia coli:* cloning of a structural gene (*fim*A) for type 2 fimbriae. *J. Bacteriol.* **162**:1075–1078.

Dorman, C.J. and C.F. Higgins. 1987. Fimbrial phase variation in *Escherichia coli:* dependence on integration host factor and homologies with other site-specific recombinases. *J. Bacteriol.* **169**:3840–3843.

Duguid, J.P. and Old, D.C. 1980. Adhesive properties of *Enterobacteriaceae*. In: E.H. Beachey (ed.), *Bacterial Adherence (Receptors and Recognition, Vol. 6).* Chapman and Hall, London, pp. 85–217.

Echeverria P., J. Seriwatana, D.N. Taylor, S. Changchawalit, C.J. Smyth, J. Twohig, and B. Rowe. 1986. Plasmids coding for colonization factor antigens I and II, LT, and ST-A2 in *Escherichia coli. Infect. Immun.* **51**:626–630.

Eisenstein, B.I. 1981. Phase variation of type 1 fimbriae in *Escherichia coli* is under transcriptional control. *Science* **214**:337–339.

Eisenstein, B.I. and D.C. Dodd. 1982. Pseudocatabolite repression of type 1 fimbriae of *Escherichia coli. J. Bacteriol.* **151**:1560–1567.

Eisenstein, B.I., J.R. Clements, and D.C. Dodd. 1983. Isolation and characterization of a monoclonal antibody directed against type 1 fimbriae organelles from *Escherichia coli. Infect. Immun.* **42**:333.

Eisenstein, B.I., D.S. Sweet, V. Vaughn, and D.I. Friedman. 1987. Integration host factor is required for the DNA inversion that controls phase variation in *Escherichia coli. Proc. Natl. Acad. Sci. USA* **84**:6506–6510.

Elleman, T.C., P.A. Hoyne, and A.W.D. Lepper. 1990. Characterization of the pilin gene of *Moraxella bovis* Dalton 2d and expression of pili from *M. bovis in Pseudomonas aeruginosa. Infect. Immun.* **58**:1678–1684.

Eshdat, Y., F.J. Silverblatt, and N. Sharon. 1981. Dissociation and reassembly of *Escherichia coli* type 1 pili. *J. Bacteriol.* **148**:308–314.

Evans, D.J., Jr. and D.G. Evans. 1990. Colonization factor antigens of human pathogens. *Curr. Top. Microbiol. Immunol.* **151**:129–145.

Fader, R.C., L.K. Duffy, C.P. Davis, and A. Kurosky. 1982. Purification and chemical characterization of type-1 pili isolated from *Klebsiella pneumoniae. J. Biol. Chem.* **257**:3301–3305.

Fenno, J.C., D.J. LeBlanc, and P. Fives-Taylor. 1989. Nucleotide sequence analysis of a type 1 fimbrial gene of *Streptococcus sanguis* FW213. *Infect. Immun.* **57**:3527–3533.

Fives-Taylor, P.M., F.L. Macrina, T.J. Pritchard, and S.S. Peene. 1987. Expression of *Streptococcus sanguis* antigens in *Escherichia coli:* cloning of a structural gene for adhesion fimbriae. *Infect. Immun.* **55**:123–128.

Flock, J.I., G. Froman, K. Jonsson, B. Guss, C. Signas, B. Nilsson, G. Raucci, M. Hook, T. Wadström, and M. Lindberg. 1987. Cloning and expression of the gene for a fibronectin-binding protein from *Staphylococcus aureus. EMBO J.* **6**:2351–2357.

Forsman, K., M. Göransson, and B.E. Uhlin. 1989. Autoregulation and multiple DNA interactions by a transcriptional regulatory protein in *E. coli* pili biogenesis. *EMBO J.* **8**:1271–1278.

Freitag, C.S., J.M. Abraham, J.R. Clements, Jr., and B.I. Eisenstein. 1985. Genetic analysis of the phase variation control of expression of type 1 fimbriae in *Escherichia coli. J. Bacteriol.* **162**:668–675.

Frydenberg, J., K. Lind, and P.C. Hu. 1987. Cloning of *Mycoplasma pneumoniae* DNA and expression of P1-epitopes in *Escherichia coli*. Isr. J. Med. Sci. **23**:759–762.

Fulks, K.A., C.F. Marrs, S.P. Stevens, and M.R. Green. 1990. Sequence analysis of the inversion region containing the pilin genes of *Moraxella bovis*. *J. Bacteriol.* **172**:310–316.

Gaastra, W., F.R. Mooi, A.R. Stuitje, and F.K. DeGraaf. 1981. The nucleotide sequence of the gene encoding the K88ab protein subunit of porcine enterotoxigenic *Escherichia coli. FEMS Microbiol. Lett.* **12**:41–46.

Ganeshkumar, N., M. Song, and B.C. McBride. 1988. Cloning of *Streptococcus sanguis* adhesin which mediates binding to saliva-coated hydroxyapatite. *Infect. Immun.* **56**:1150–1157.

Ganeshkumar, N., P.M. Hannam, P.E. Kolenbrander, and B.C. McBride. 1991. Nucleotide sequence of a gene coding for a saliva-binding protein (SsaB) from *Streptococcus sanguis* 12 and possible role of the protein in coaggregation with *Actinomyces*. *Infect. Immun.* **59**:1093–1099.

Gerlach, G.F. and S. Clegg. 1988a. Characterization of two genes encoding antigenically distinct type-1 fimbriae of *Klebsiella pneumoniae*. *Gene* **64**:2321–2340.

Gerlach, G.F. and S. Clegg. 1988b. Identification and characterization of the genes encoding the type 3 and type 1 fimbrial adhesins of *Klebsiella pneumoniae*. *J. Bacteriol.* **171**:1262–1270.

Gerlach, G.F., B.L. Allen, and S. Clegg. 1988. Molecular characterization of the type 3 (MR/K) fimbriae of *Klebsiella pneumoniae*. J. Bacteriol. **170**:3547–3553.

Gerlach, G.F., S. Clegg, N.J. Ness, D. Swenson, B.L. Allen, and W.A. Nichols. 1989. Expression of type 1 fimbriae and mannose-sensitive hemagglutinin by recombinant plasmids. *Infect. Immun.* **57**:764–770.

Gibbs, C.P., B.-Y Reimann, E. Schultz, A. Kaufmann, R. Haas, and T.F. Meyer. 1989. Reassortment of pilin genes in *Neisseria gonorrhoeae* occurs by two distinct mechanisms. *Nature* **338**:651–652.

Gilsdorf, J.R., C.F. Marrs, K.W. McCrea, and L.J. Forney. 1990. Cloning, expression and sequence analysis of the *Haemophilus influenzae* type b strain M43p$^+$ pilin gene. *Infect. Immun.* **58**:1065–1072.

Girardeau, J.P., M. Der Vartanian, J.L. Ollier, and M. Contrepois. 1988. CS31A, a new K88-related fimbrial antigen on bovine enterotoxigenic and septicemic *Escherichia coli* strains. *Infect. Immun.* **56**:2180–2188.

Goldhar, J., R. Perry, and I. Ofek. 1984. Extraction and properties of nonfimbrial mannose resistant hemagglutinin from a urinary isolate of *Escherichia coli*. *Curr. Microbiol.* **11**:49–54.

Goldhar, J., R. Perry, J.R. Golecki, H. Hoschultzky, B. Jann, and K. Jann. 1987. Nonfimbrial, mannose-resistant adhesins from uropathogenic *Escherichia coli* 083:K1:H4 and 014:K?:H11. *Infect. Immun.* **55**:1837–1842.

Goochee, C.F., R.T. Hatch, and T.W. Cadman. 1987. Some observations on the role of type 1 fimbriae in *Escherichia coli* autoflocculation. 1986. *Biotech. Bioeng.* **29**:1024–1034.

Göransson, M., K. Forsman, and B.E. Uhlin. 1988. Functional and structural homology among regulatory cistrons of pili-adhesin determinants in *Escherichia coli. Mol. Gen. Genet.* **212**:412–417.

Göransson, M., K. Forsman, and B.E. Uhlin. 1989. Regulatory genes in the thermoregulation of *Escherichia coli* pili gene transcription. *Genes Dev.* **3**:123–130.

Grunberg, J., R. Perry, H. Hoschutzky, B. Jann, K. Jann, and J. Goldhar. 1988. Nonfimbrial blood group N-specific adhesin (NFA-3) from *Escherichia coli* 020:KX104:H−, causing systemic infection. *FEMS Microbiol. Lett.* **56**:241–246.

Haas, R. and T.F. Meyer. 1986. The repertoire of silent pilus genes in *Neisseria gonorrhoeae:* evidence for gene conversion. *Cell* **44**:107–115.

Hacker, J. 1990. Genetic determinants coding for fimbriae and adhesins of extra-intestinal *Escherichia coli. Curr. Top. Microbiol. Immunol.* **151**:1–27.

Hacker, J., G. Schmidt, C. Hughes, S. Knapp, M. Marget, and W. Goebel. 1985. Cloning and characterization of genes involved in production of mannose-resistant, neuraminidase-susceptible (X) fimbriae from a uropathogenic 06:K15:H31 *Escherichia coli* strain. *Infect. Immun.* **47**:434–440.

Hacker, J., T. Jarchau, S. Knapp, R. Marre, G. Schmidt, T. Schmoll, and W. Goebel. 1986a. Genetic and in vivo studies with S-fimbrial antigens and related virulence determinants of extraintestinal *Escherichia coli* strains. In: D. Lark (ed.). *Protein–Carbohydrate Interactions in Biological Systems.* Academic Press, London, pp. 125–133.

Hacker, J., M. Ott, G. Schmidt, R. Hull, and W. Goebel. 1986b. Molecular cloning of the F8 fimbrial antigen from *Escherichia coli. FEMS Microbiol. Lett.* **36**:139–144.

Hagblom, P., E. Segal, E. Billyard, and M. So. 1985. Intragenic recombination leads to pilus antigenic variation in *Neisseria gonorrhoeae. Nature (Lond.)* **315**:156–158.

Hales, B.A., H. Beverly-Clarke, N.J. High, K. Jann, R. Perry, J. Goldhar, and G.J. Boulnois. 1988. Molecular cloning and characterization of the genes for a non-fimbrial adhesin from *Escherichia coli. Microb. Pathogen.* **5**:9–17.

Hanson, M.S. and C.C. Brinton, Jr. 1988. Identification and characterization of the *Escherichia coli* type 1 pilus tip adhesin protein. *Nature (Lond.)* **322**:265–268.

Harris, S.L., D.A. Elliott, M.C. Blake, L.M. Must, M. Messenger, and P. Orndorff. 1990. Isolation and characterization of mutants with lesions affecting pellicle formation and erythrocyte agglutination by type 1 piliated *Escherichia coli. J. Bacteriol.* **172**:6411–6418.

Hebeler, B.H. and F.E. Young. 1975. Autolysis of *Neisseria gonorrhoeae. J. Bacteriol.* **122**:385–392.

Heckels, J.E. 1982. Structural comparison of *Neisseria gonorrhoeae*. In: Schlessinger, D. (ed.), *Microbiology–1982*. American Society for Microbiology, Washington, pp. 301–304.

Hedegaard, L. and P. Klemm. 1989. Type 1 fimbriae of *Escherichia coli* as carriers of heterologous antigenic sequences. *Gene* **85**:115–124.

Heesemann, J. and L. Gruter. 1987. Genetic evidence that the outer membrane protein YOP1 of *Yersinia enterocolitica* mediates adherence and phagocytosis resistance to human epithelial cells. *FEMS Microbiol. Lett.* **40**:37–41.

Hermodson, M.A., K.C.S. Chen, and T.M. Buchanan. 1978. *Neisseria* pili proteins: amino-terminal sequence and identification of an unusual amino acid. *Biochemistry* **17**:442–445.

High, N.J., B.A. Hales, K. Jann, and G.J. Boulnois. 1988. A block of urovirulence genes encoding multiple fimbriae and hemolysin in *Escherichia coli* 04:K12:H^-. *Infect. Immun.* **56**:513–517.

Hinson, G., S. Knutton, M.K.L. Lam-Po-Tang, A.S. McNeish, and P. H. Williams. 1987. Adherence to human colonocytes of an *Escherichia coli* strain isolated from severe infantile enteritis: Molecular and ultrastructural studies of afibrillar adhesin. *Infect. Immun.* **55**:393–402.

Holmgren, A. and C.I. Bränden. 1989. Crystal structure of chaperone protein Pap D reveals an immunoglobulin fold. *Nature* **342**:248–251.

Holt, R.C., Y. Abiko, S. Saito, M. Smorawinska, J.B. Hansen, and R. Curtiss III. 1982. *Streptococcus mutans* genes that code for extracellular proteins in *Escherichia coli* K-12. *Infect. Immun.* **38**:147–156.

Hornick, D.B., B.L. Allen, M.A. Horn, and S. Clegg. 1991. Fimbrial types among respiratory isolates belonging to the family *Enterobacteriaceae*. *J. Clin. Microbiol.* **29**:1795–1800.

Hoschützky, H., R. Lottspeich, and K. Jann. 1989. Isolation and characterization of the α-galactosyl-1,4-β-galactosyl-specific adhesin (P adhesin) from fimbriated *Escherichia coli*. *Infect. Immun.* **57**:76–81.

Hull, R.A., R.E. Gill, P. Hsu, B.H. Minshew, and S. Falkow. 1981. Construction and expression of recombinant plasmids encoding type 1 or D-mannose-resistant pili from a urinary tract infection *Escherichia coli* isolate. *Infect. Immun.* **33**:933–938.

Hull, R.A., S.I. Hull, and S. Falkow. 1984. Frequency of gene sequences necessary for pyelonephritis-associated pili expression among isolates of *Enterobacteriaceae* from human extraintestinal infections. *Infect. Immun.* **43**:1064–1067.

Hull, S.I., S. Bieler, and R.A. Hull. 1988. Restriction fragment length polymorphisms and multiple copies of DNA sequences homologous with probes for P-fimbriae and hemolysin genes among uropathogenic *Escherichia coli*. *Can. J. Microbiol.* **34**:307–311.

Hultgren, S.J., W.R. Schwan, A.J. Schaeffer, and J.L. Duncan. 1986. Regulation of production of type 1 pili among urinary tract isolates of *Escherichia coli*. *Infect. Immun.* **54**:613–620.

Hultgren, S.J., F. Lindberg, G. Magnusson, J. Kihlberg, J.M. Tennent, and S. Normark.

1989. The Pap G adhesin of uropathogenic *Escherichia coli* contains separate regions for receptor binding and for the incorporation into the pilus. *Proc Natl. Acad. Sci. USA* **86**:4357–4361.

Hultgren, S.J., J.L. Duncan, A.J. Schaeffer, and S.K. Amundsen. 1990. Mannose-sensitive haemagglutination in the absence of piliation of *Escherichia coli. Mol. Microbiol.* **4**:1311–1318.

Hultgren, S.J., S. Normark, and S. Abraham. 1991. Chaperone-assisted assembly and molecular architecture of adhesive pili. *Annu. Rev. Microbiol.* **45**:383–415.

Isberg, R.R. 1989. Determinants for thermoinducible cell binding and plasmid-encoded cellular penetration detected in the absence of the *Yersinia pseudotuberculosis* invasin protein. *Infect. Immun.* **57**:1998–2005.

Isberg, R.R. and S. Falkow. 1985. A single genetic locus encoded by *Yersinia pseudotuberculosis* permits invasion of cultured animal cells by *Escherichia coli* K-12. *Nature (Lond.)* **317**:262–264.

Jacob, F., C. Capiau, A.M. Gathoye, C. Locht, and T. Cabezon. 1988. Molecular cloning of the filamentous hemagglutinin structural gene from *Bordetella pertussis* in *Escherichia coli*. In: Mebel, S., H. Stompe, M. Drescher, and S. Rustenback (eds.). *FEMS Symposium on Pertussis*. Berlin, German Democratic Republic, pp. 62–65.

Jacobs, A.A.C. and F.K. DeGraaf. 1985. Production of K88, K99 and F41 fibrillae in relation to growth phase and a rapid procedure for adhesin purification. *FEMS Microbiol. Lett.* **26**:15–19.

Jacobs, A.A.C., J. Venema, R. Leeven, H. Van Pelt-Heerschap, and F.K. DeGraaf. 1987a. Inhibition of adhesive activity of K88 fibrillae by peptides derived from the K88 adhesin. *J. Bacteriol.* **169**:735–741.

Jacobs, A.A.C., B. Roosendaal, J.F.L. van Breemen, and F.K. DeGraaf. 1987b. Role of phenylalanine 150 in the receptor-binding domain of the K88 fibrillar subunit. *J. Bacteriol.* **169**:4907–4911.

James, J.F., C.J. Lammel, D.L. Draper, and G.F. Brooks. 1980. Attachment of *Neisseria gonorrhoeae* colony phenotype variants to eukaryotic cells and tissues. In: Normark, S. and D. Danielsson (eds.), *Genetics and Immuno-biology of Pathogenic Neisseria*. University of Umea, Sweden, pp. 213–216.

Jann, K. and Hoschützky. 1990. Nature and organization of adhesins. *Curr. Top. Microbiol. Immunol.* **151**:55–70.

Jones, G.W. and R.E. Isaacson. 1983. Proteinaceous bacterial adhesins and their receptors. *Crit. Rev. Microbiol.* **10**:229–260.

Jerse, A.E. and J.B. Kaper. 1991. The *eae* gene of enteropathogenic *Escherichia coli* encodes a 94-kilodalton membrane protein, the expression of which is influenced by the EAF plasmid. *Infect. Immun.* **59**:4302–4309.

Jerse, A.E., J. Yu, B.D. Tall, and J.B. Kaper. 1990. A genetic locus of enteropathogenic *Escherichia coli* necessary for the production of attaching and effacing lesions on tissue culture cells. *Proc. Natl. Acad. Sci. USA* **87**:7839–7843.

Jonsson, A.B., G. Nyberg, and S. Normark. 1991. Phase variation of gonococcal pili by frameshift mutation in *pil*C, a novel gene pilus assembly. *EMBO J.* **10**:477–488.

Kar, S., S.C.-M. To, and C.C. Brinton, Jr. 1990. Cloning and expression in *Escherichia coli* of LKP pilus genes from a nontypeable *Haemophilus influenzae* strain. *Infect. Immun.* **58**:903–908.

Karjalainen, T.K., D.G. Evans, D.J. Evans, Jr., D.Y. Graham, and C.-H. Lee. 1991. Iron represses the expression of CFA/I fimbriae of enterotoxigenic *E. coli*. *Microb. Pathogen.* **11**: 317–323.

Karr, J., B. Nowicki, L. Truong, R. Hull, J. Moulds, and S. Hull. 1990. *pap*-2-encoded fimbriae adhere to the P blood group O related glycosphingolipid stage-specific embryonic antigen 4 in the human kidney. *Infect. Immun.* **58**:4055–4062.

Kawula, T.H., E.L. Aho, D.S. Barritt, D.G. Klapper, and J.G. Cannon. 1988. Reversible phase variation of expression of *Neisseria meningitidis* class 5 outer membrane proteins and their relationship to gonococcal proteins II. *Infect. Immun.* **56**:380–386.

Kehoe, M., R. Sellwood, P. Shipley, and G. Dougan. 1981. Genetic analysis of K88-mediated adhesion of enterotoxigenic *Escherichia coli*. *Nature* **291**:122–126.

Kehoe, M., M. Winther, and G. Dougan. 1983. Expression of a cloned K88ac adhesion antigen determinant: identification of a new adhesion cistron and role of a vector-encoded promoter. *J. Bacteriol.* **155**:1071–1077.

Kehoe, M.A., T.P. Poirer, E.H. Beachey, and K.N. Timmis. 1985. Cloning and genetic analysis of serotype 5 M protein determinant of group A streptococci: evidence for multiple copies of the M5 determinant in the *Streptococcus pyogenes* genome. *Infect. Immun.* **48**:190–197.

Keith, B.R., L. Maurer, P.A. Spears, and P.E. Orndorff. 1986. Receptor-binding function of type I pili effects bladder colonization by a clinical isolate of *Escherichia coli*. *Infect. Immun.* **53**:693–696.

Kellogg, D.S., Jr., W.L. Peacock, Jr., W.E. Deacon, L. Brown, and C.I. Pirkle. 1963. *Neisseria gonorrhoeae*. I. Virulence genetically linked to colonial variation. *J. Bacteriol.* **85**:1274–1279.

Kist, M.L., I.E. Salit, and T. Hofmann. 1990. Purification and characterization of the Dr hemagglutinins expressed by two uropathogenic *Escherichia coli* strains. *Infect. Immun.* **58**:695–702.

Klaasen, P., M.J. Woodward, F.G. van Zijderveld, and F.K. De Graaf. 1990. The 987P gene cluster in enterotoxigenic *Escherichia coli* contains an ST_{pa} transposon that activates 987P expression. *Infect. Immun.* **58**:801–807.

Klemm, P. 1984. The *fim*A gene encoding the type-1 fimbrial subunit of *Escherichia coli* nucleotide sequence and primary structure of the protein. *Eur. J. Biochem.* **143**:395–399.

Klemm, P. 1985. Fimbrial adhesins of *Escherichia coli*. *Rev. Infect. Dis.* **7**:321–340.

Klemm, P. 1986. Two regulatory *fim* genes, *fim*B and *fim*E, control the phase variation of type 1 fimbriae in *Escherichia coli*. *EMBO J.* **5**:1389–1393.

Klemm, P. and G. Christiansen. 1987. Three *fim* genes required for the regulation of length and mediation of adhesion of *Escherichia coli* type 1 fimbriae. *Mol. Gen. Genet.* **208**:439–445.

Klemm, P., I. Orskov, and F. Orskov. 1982. F7 and type-1-like fimbriae from three *Escherichia coli* strains isolated from urinary tract infections: protein chemical and immunological aspects. *Infect. Immun.* **36**:462–468.

Klemm, P., B.J. Jorgensen, I. Van Die, H. De Ree, and H. Bergmans. 1985. The *fim* genes responsible for synthesis of type 1 fimbriae in *Escherichia coli,* cloning and genetic organization. *Mol. General Genet.* **199**:410–414.

Klemm, P., W. Gaastra, M.M. McConnell, and H.R. Smith. 1989. The CS2 fimbrial antigen from *Escherichia coli,* purification, characterization and partial covalent structure. *FEMS Microbiol. Lett.* **26**:207–210.

Knutton, S., A. Phillips, H. Smith, R. Gross, R. Shaw, P. Watson, and E. Price. 1991. Screening for enteropathogenic *Escherichia coli* in infants with diarrhea by the fluorescent-acting staining test. *Infect. Immun.* **59**:365–71.

Korhonen, T.K., K. Lounatmaa, H. Ranta, and N. Kuush. 1980. Characterization of type 1 pili of *Salmonella typhimurium* LT2. *J. Bacteriol.* **144**:800–805.

Kornhonen, T.K., H. Leffler, and C. Svanborg-Eden. 1981. Binding specificity of piliated strains of *Escherichia coli* and *Salmonella typhimurium* to epithelial cells, *Saccharomyces cerevisiae* cells, and erythrocytes. *Infect. Immun.* **32**:796–804.

Korth, M., R. Schneider, and S. Moseley. 1991. An F41-K88-related genetic determinant of bovine septicemic *Escherichia coli* mediates expression of CS31A fimbriae and adherence to epithelial cells. *Infect. Immun.* **59**:2333–2340.

Krogfelt, K.A. and P. Klemm. 1988. Investigation of minor components of *Escherichia coli* type 1 fimbriae: protein, chemical and immunological aspects. *Microb. Pathogen.* **4**:231–238.

Krogfelt, K.A., H. Bergmans, and P. Klemm. 1990. Direct evidence that the Fim H protein is the mannose-specific adhesin of *Escherichia coli* type 1 fimbriae. *Infect. Immun.* **58**:1995–1998.

Kröncke, K.D., I. Orskov, F. Orskov, B. Jann, and K. Jann. 1990. Electron microscopic study of coexpression of adhesive protein capsules and polysaccharide capsules in *Escherichia coli. Infect. Immun.* **58**:2710–2714.

Kuehn, M.J., S. Normark, and S.J. Hultgren. 1991. Immunoglobulin-like Pap D chaperone caps and uncaps interactive surface of nascently translocated pilus subunits. *Proc. Natl. Acad. Sci. USA* **88**:10586–10590.

Labigne-Roussel, A. and S. Falkow. 1988. Distribution and degree of heterogeneity of the afimbrial-adhesin-encoding operon (*afa*) among uropathogenic *Escherichia coli* isolates. *Infect. Immun.* **56**:640–648.

Labigne-Roussel, A., D. Lark, G. Schoolnik, and S. Falkow. 1984. Cloning and expression of an afrimbrial adhesin (AfaI) responsible for blood group-independent, mannose-resistant hemagglutination from a pyelonephritic *Escherichia coli* strain. *Infect. Immun.* **46**:251–259.

Labigne-Roussel, A., M.A. Schmidt, W. Walz, and S. Falkow. 1985. Genetic organization of the afimbrial adhesin operon and nucleotide sequence from a uropathogenic *Escherichia coli* gene encoding an afimbrial adhesin. *J. Bacteriol.* **162**:1285–1292.

Lambden, P.R. 1982. Biochemical comparison of pili from variants of *Neisseria gonorrhoeae*. *J. Gen. Microbiol.* **128**:2105–2111.

Lambden, P.R., J.E. Heckels, L.T. James, and P.J. Watt. 1979. Variations in surface protein composition associated with virulence properties in opacity types of *Neisseria gonorrhoeae*. *J. Gen. Microbiol.* **114**:305–312.

Lambden, P.R., J.N. Robertson, and P.J. Watt. 1980. Biological properties of two distinct pilus types produced by isogenic variants of *Neisseria gonorrhoeae* P9. *J. Bacteriol.* **141**:393–396.

Lambden, P.R., J.E. Heckel, H. McBride, and P.J. Watt. 1981. The identification and isolation of novel pilus types produced by variants of *Neisseria gonorrhoeae* P9 following selection *in vivo*. *FEMS Microbiol. Lett.* **10**:339–341.

Langermann, S. and A. Wright. 1990. Molecular analysis of the *Haemophilus influenzae* type b pili gene. *Mol. Microbiol.* **4**:221–230.

Lanser, J.A. and P.A. Anargyros. 1985. Detection of *Escherichia coli* adhesins with DNA probes. *J. Clin. Microbiol.* **22**:425–427.

Leffler, H. and C. Svanborg-Eden. 1986. Glycolipids as receptors for *Escherichia coli* lectins or adhesins. In: D. Mirelman (ed.), *Microbial Lectins and Agglutinins: Properties and Biological Activity*. Wiley-Interscience, Orlando, pp. 83–111.

Lindberg, F.P., B. Lund, and S. Normark. 1984. Genes of pyelonephritogenic *E. coli* required for digalactoside-specific agglutination of human cells. *EMBO J.* **3**:1167–1173.

Lindberg, F., B. Lund, and S. Normark. 1986. Gene products specifying adhesion of uropathogenic *Escherichia coli* are minor components of pili. *Proc. Natl. Acad. Sci. USA* **83**:1891–1895.

Lindberg, F., B. Lund, L. Johansson, and S. Normark. 1987. Localization of the receptor-binding protein adhesin at the tip of the bacterial pilus. *Nature* **328**:84–87.

Lindberg, F., J.M. Tennent, S.J. Hultgren, B. Lund, and S. Normark. 1989. Pap D, a periplasmic transport protein in P-pilus biogenesis. *J. Bacteriol.* **171**:6052–6058.

Lindstedt, R., N. Baker, P. Falk, R. Hull, S. Hull, J. Karr, H. Leffler, C. Svanborg-Eden, and G. Larson. 1989. Binding specificities of wild-type and cloned *Escherichia coli* strains that recognize globo-A. *Infect. Immun.* **57**:3389–3394.

Lintermans, P., P. Pohl, F. Deboeck, A. Bertels, C. Schlicker, J. Vandekerckhove, J. Van Damme, M. Van Montagu, and H. DeGreve. 1988. Isolation and nucleotide sequence of the F17-A gene encoding the structural protein of the F17 fimbriae in bovine enterotoxigenic *Escherichia coli*. *Infect. Immun.* **56**:1475–1484.

Lowe, M.A., S.C. Holt, and B.I. Eisenstein. 1987. Immunoelectron microscopic analysis of elongation of type 1 fimbriae in *Escherichia coli*. *J. Bacteriol.* **169**:157–163.

Lund, B., F. Lindberg, B.I. Marklund, and S. Normark. 1987. The PapG protein is the α-D-galactopyranosyl-(1→4)-β-D-galactopyranose-binding adhesin of uropathogenic *Escherichia coli*. *Proc. Natl. Acad. Sci. USA* **84**:5898–5902.

Lund, B., B.I. Marklund, N. Strömberg, F. Lindbert, K.A. Karlsson, and S. Normark. 1988a. Uropathogenic *Escherichia coli* can express serologically identical pili of different receptor binding specificitites. *Mol. Microbiol.* **2**:255–263.

Lund, B., F. Lindberg, and S. Normark. 1988b. The structure and antigenic properties of the tip-located P-pilus proteins of uropathogenic *Escherichia coli*. *J. Bacteriol.* **170**:1887–1894.

Manning, P.A., K.N. Timmis, and G. Stevenson. 1985. Colonization factor antigen II (CFA/II) of enterotoxigenic *Escherichia coli:* molecular cloning of the CS3 determinant. *Mol. Genet.* **200**:322–327.

Marre, R., B. Kreft, and J. Hacker. 1990. Genetically engineered S and F1C fimbriae differ in their contribution to adherence of *Escherichia coli* to cultured renal tubular cells. *Infect. Immun.* **58**:3434–3437.

Marrs, C.F., G. Schoolnik, J.M. Koomey, J. Hardy, J. Rothbard, and S. Falkow. 1985. Cloning and sequencing of a *Moraxella bovis* pilin gene. *J. Bacteriol.* **163**:132–139.

Marrs, C.F., W.W. Ruehl, G.K. Schoolnik, and S. Falkow. 1988. Pilin gene variation of *Moraxella bovis* is caused by an inversion of the pilin genes. *J. Bacteriol.* **170**:3032–3039.

Mattei, D., F. Pichot, J. Bellalou, O. Mercereau-Puijalon, and A. Ullmann. 1986. Molecular cloning of a coding sequence of *Bordetella pertussis* filamentous hemagglutinin gene. *FEMS Microbiol. Lett.* **37**:73–77.

Maurer, L. and P.E. Orndorff. 1985. A new locus, *pil*E, required for the binding of type 1 piliated *Escherichia coli* to erythrocytes. *FEMS Microbiol. Lett.* **30**:59–66.

Maurer, L. and P.E. Orndorff. 1987. Identification and characterization of genes determining receptor binding and pilus length of *Escherichia coli* type 1 pili. *J. Bacteriol.* **169**:640–645.

May, G., P. Dersch, M. Haardt, A. Middendorf, and E. Bremer. 1990. The *osm*Z(*bgl*Y) genes encodes the DNA-binding protein H-NS (H1a), a component of the *Escherichia coli* K12 nucleoid. *Mol. Gen. Genet.* **224**:81–90.

McConnell, M.M., L.V. Thomas, G.A. Willshaw, H.R. Smith, and B. Rowe. 1988. Genetic control and properties of coli surface antigens of colonization factor IV (PCF 8775) of enterotoxigenic *Escherichia coli*. *Infect. Immun.* **56**:1974–1980.

McGee, Z.A., D.S. Stephens, L.H. Hoffman, W.F. Schlech, and R.G. Horn. 1983. Mechanisms of mucosal invasion by pathogenic *Neisseria*. *Rev. Infect. Dis.* **5**:S708–S714.

McLain, M., I. Blomfield, and B.I. Eisenstein. 1991. Roles of *fim*B and *fim*E in site specific DNA inversion associated with phase variation of type 1 fimbriae in *Escherichia coli*. *J. Bacteriol.* **173**:5308–5314.

Meyer, T.F. 1990. Variation of pilin and opacity-associated protein in pathogenic *Neisseria* species. In: Iglewski, B.H. and V.L. Clarks (eds.), *Molecular Basis of Bacterial Pathogenesis*. Academic Press, San Diego, pp. 137–153.

Meyer, T.F., N. Mlawer, and M. So. 1982. Pilus expression in *Neisseria gonorrhoeae* involves chromosomal rearrangement. *Cell* **30**:45–52.

Meyer, T.F., E. Billyard, R. Haas, S. Storzbach, and M. So. 1984. Pilus genes of *Neisseria gonorrhoeae:* chromosomal organization and DNA sequence. *Proc. Natl. Acad. Sci. USA* **81**:6110–6114.

Miller, V.L. and S. Falkow. 1988. Evidence for two genetic loci in *Yersinia enterocolitica* that can promote invasion of epithelial cells. *Infect. Immun.* **56**:1242–1248.

Miller, V.L., J.J. Farmer III, W.E. Hill, and S. Falkow. 1989. The *ail* locus is found uniquely in *Yersinia enterocolitica* serotypes commonly associated with disease. *Infect. Immun.* **57**:121–131.

Miller, V.L., J.B. Bliska, and S. Falkow. 1990. Nucleotide sequence of the *Yersinia enterocolitica ail* gene and characterization of the Ail protein product. *J. Bacteriol.* **172**:1062–1069.

Minion, F.C., S.N. Abraham, E.H. Beachey, and J.D. Goguen. 1986. The genetic determinant of adhesive function in type 1 fimbriae of *Escherichia coli* is distinct from the gene encoding the fimbrial subunit. *J. Bacteriol.* **165**:1033–1036.

Mirelman, D. and I. Ofek. 1986. Introduction to microbial lectins and aglutinins. In: Mirelman, D. (ed.), *Microbial Lectins and Agglutinins: Properties and Biological Activity*. Wiley-Interscience, New York, pp. 1–19.

Mobley, H.L., G.R. Chippendale, J.H. Tenney, R.A. Hull, and J.W. Warren. 1987. Expression of type 1 fimbriae may be required for persistence of *Escherichia coli* in the catherized urinary tract. *J. Clin. Microbiol.* **25**:2253–2257.

Moch, T., H. Hoschützky, J. Hacker, K.D. Kröncke, and K. Jann. 1987. Isolation and characterization of the α-sialyl beta-2,3-galactosyl-specific adhesin from fimbriated *Escherichia coli*. *Proc. Natl. Acad. Sci. USA* **84**:3462–3466.

Mooi, F.R. and F.K. DeGraaf. 1985. Molecular biology of fimbriae of enterotoxigenic *Escherichia coli*. *Curr. Top. Microbiol. Immunol.* **118**:119–138.

Mooi, F.R., F.K. DeGraaf, and J.D. van Embden. 1979. Cloning, mapping, and expression of the genetic determinant that encodes for the K88ab antigen. *Nucleic Acids Res.* **6**:849–865.

Mooi, F.R., N. Harms, D. Bakker, and F.K. DeGraaf. 1981. Organization and expression of genes involved in the production of the K88ab antigen. *Infect. Immun.* **32**:1155–1163.

Mooi, F.R., C. Wouters, A. Wijfjes, and F.K. DeGraaf. 1982. Construction and characterization of mutants impaired in the biosynthesis of the K88ab antigen. *J. Bacteriol.* **150**:512–521.

Mooi, F.R., A. Wijfjes, and F.J. DeGraaf. 1983. Identification and characterization of precursors in the biosynthesis of the K88ab fimbria of *Escherichia coli*. *J. Bacteriol.* **154**:41–49.

Mooi, F.R., M. van Buuren, G. Koopman, B. Roosendaal, and F.K. DeGraaf. 1984. A K88ab gene of *Escherichia coli* encodes a fimbria-like protein distinct from the K88ab fimbrial adhesin. *J. Bacteriol.* **159**:482–487.

Mooi, F.R., I. Claassen, D. Bakker, H. Kuipers, and F.K. DeGraaf. 1986. Regulation and structure of an *Escherichia coli* gene coding for an outer membrane protein involved in export of the K88ab fimbrial subunits. *Nucleic Acids Res.* **14**:2443–2457.

Moon, H. 1990. Colonization factor antigens of enterotoxigenic *Escherichia coli* in animals. *Curr. Top. Microbiol. Immun.* **151**:147–166.

Morschhauser, J., H. Hoschutsky, K. Jann, and J. Hacker, 1990. Functional analysis of the sialic acid binding adhesin Sfa S of pathogenic *Escherichia coli* by site-specific mutagenesis. *Infect. Immun.* **58**:2133–2138.

Morrissey, P.M. and G. Dougan. 1986. Expression of a cloned 987P adhesion-antigen fimbrial determinant in *Escherichia coli* K-12 strain HB101. *Gene* **43**:79–84.

Moseley, S.L., G. Dougan, R.A. Schneider, and H.W. Moon. 1986. Cloning of chromosomal DNA encoding the F41 adhesin of enterotoxigenic *Escherichia coli* and genetic homology between adhesins F41 and K88. *J. Bacteriol.* **167**:799–804.

Mouw, A.R., E.H. Beachey, and V. Burdett. 1988. Molecular evolution of streptococcal M protein: cloning and nucleotide sequence of the type 24 M protein gene and relation to other genes of *Streptococcus pyogenes*. *J. Bacteriol.* **170**:676–684.

Muir, L.L., R.A. Strugnell, and J.K. Davis. 1988. Proteins that appear to be associated with pili in *Neisseria gonorrhoeae*. *Infect. Immun.* **56**:1743–1747.

Mullany, P., A.M. Field, M.M. McConnell, S.M. Scotland, H.R. Smith, and B. Rowe. 1983. Expression of plasmids coding for colonization factor antigen II (CFA/II) and enterotoxin production in *Escherichia coli*. *J. Gen. Microbiol.* **129**:3591–3601.

Muralidharan, K., A. Stern, and T.F. Meyer. 1987. The control mechanism of opacity protein expression in the pathogenic neisseriae. *Antonie van Leeuwenhoek* **53**:435–440.

Nicholson, I.J., A.C.F. Perry, M. Virji, J.E. Heckels, and J.R. Saunders. 1987. Localization of antibody-binding sites by sequence analysis of cloned pilin genes from *Neisseria gonorrhoeae*. *J. Gen. Microbiol.* **133**:825–833.

Norgren, M., S. Normark, D. Lark, P. O'Hanley, G. Schoolnik, S. Falkow, C. Svanborg-Eden, M. Baga, and B.E. Uhlin. 1984. Mutations in *E. coli* cistrons affecting adhesion to human cells do not abolish Pap pili fiber formation. *EMBO J.* **3**:1159–1165.

Norgren, M., M. Baga, J. Tennent, and S. Normark. 1987. Nucleotide sequence, regulation and functional analysis of the *pap*C gene required for cell surface localization of Pap pili of uropathogenic *Escherichia coli*. *Mol. Microbiol.* **1**:169–178.

Norlander, L., J. Davies, A. Norqvist, and S. Normark. 1979. Genetic basis for colonial variation in *Neisseria gonorrhoeae*. *J. Bacteriol.* **138**:762–769.

Normark, S., D. Lark and R. Hull. 1983. Genetics of digalactoside-binding adhesin from a uropathogenic *Escherichia coli* strain. *Infect. Immun.* **41**:942–949.

Nowicki, B., M. Rhen, V. Vaisanen-Rhen, A. Pere, and T.K. Korhonen. 1985. Kinetics of phase variation between S and type 1 fimbriae of *Escherichia coli*. *FEMS Microbiol. Lett.* **28**:237–242.

Nowicki, B., J. Vuopio-Varkila, P. Viljanen, T.K. Korhonen, and P.H. Makela. 1986. Fimbrial phase variation and systemic *Escherichia coli* infection studied in the mouse peritonitis model. *Microb. Pathogen.* **1**:335–347.

Nowicki, B., J.P. Barrish, T. Korhonen, R.A. Hull, and S.I. Hull. 1987. Molecular cloning of the *Escherichia coli* 075X adhesin. *Infect. Immun.* **55**:3168–3173.

Nowicki, B., J. Moulds, R. Hull, and S. Hull. 1988. A hemagglutinin of uropathogenic *Escherichia coli* recognizes the Dr blood group antigen. *Infect. Immun.* **56**:1057–1060.

Nowicki, B., C. Svanborg-Eden, R. Hull, and S. Hull. 1989. Molecular analysis and epidemiology of the Dr hemagglutinin of uropathogenic *Escherichia coli*. *Infect. Immun.* **57**:446–451.

Nowicki, B., A. Labigne, S. Moseley, R. Hull, S. Hull, and J. Moulds. 1990. The Dr hemagglutinin, afimbrial adhesins AFA-1 and AFA-III, and F1845 fimbriae of uropathogenic and diarrhea-associated *Escherichia coli* belong to a family of hemagglutinins with Dr receptor recognition. *Infect. Immun.* **58**:279–281.

Nunn, D., S. Bergman, and S. Lory. 1990. Products of three accessory genes, *pil*B, *pil*C and *pil*D, are required for biogenesis of *Pseudomonas aeruginosa* pili. *J. Bacteriol.* **172**:2911–2919.

Ofek, I., E.H. Beachey, and A.L. Bisno. 1974. Resistance of *Neisseria gonorrhoeae* to phagocytosis: relationship to colonial morphology and surface pili. *J. Infect. Dis.* **129**:310–316.

Ono, E., M.F. Lavin, and M. Naiki. 1991. The nucleotide sequence of the genes, *fan*E and *fan*F, of *Escherichia coli* K99 fimbriae. *Jpn. J. Vet. Res.* **39**:1–10.

Orndorff, P.E. and S. Falkow. 1984a. Organization and expression of genes responsible for type 1 piliation in *Escherichia coli*. *J. Bacteriol.* **159**:736–744.

Orndorff, P.E. and S. Falkow. 1984b. Identification and characterization of a gene product that regulates type 1 piliation in *Escherichia coli*. *J. Bacteriol.* **160**:61–66.

Orndorff, P.E., P.A. Spears, D. Schauer, and S. Falkow. 1985. Two modes of control of *pil*A, the gene encoding type 1 pilin in *Escherichia coli*. *J. Bacteriol.* **164**:321–330.

Orskov, I. and F. Orskov. 1990. Serological classification of fimbriae. *Curr. Top. Microbiol. Immun.* **151**:71–90.

Orskov, I., A. Birch Anderson, J.P. Duguid, J. Stenderup, and F. Orskov. 1985. An adhesive protein capsule of *Escherichia coli*. *Infect. Immun.* **47**:191–200.

Ott, M., T. Schmoll, W. Goebel, I. Van Die, and J. Hacker. 1987. Comparison of the genetic determinant coding for the S-fimbrial adhesin (*sfa*) of *Escherichia coli* to other chromosomally encoded fimbrial determinants. *Infect. Immun.* **55**:1940–1943.

Oudega, B. and F.K. DeGraaf. 1988. Genetic organization and biogenesis of adhesive fimbriae of *Escherichia coli*. *Antonie van Leeuwenhoek* **54**:285–299.

Paerregaard, A., F. Espersen, and M. Skurnik. 1991. Role of the *Yersinia* outer membrane protein YadA in adhesion to rabbit intestinal tissue and rabbit intestinal brush border membrane vesicles. *Acta Pathol. Microbiol. Immunol. Scand.* **99**:226–232.

Pallesen, L., O. Madsen, and P. Klemm. 1989. Regulation of the phase switch controlling expression of type 1 fimbriae in *Escherichia coli*. *Mol. Microbiol.* **3**:925–931.

Paranchych, W. 1990. Molecular studies on *N*-methylphenylalanine pili. In: Iglewski, B.H. and V.L. Clark (eds.), *The Bacteria, Vol. XI. Molecular Basis of Bacterial Pathogenesis*. Academic Press, San Diego, pp. 61–78.

Patel, P., C.F. Marrs, J.S. Mattick, W.W. Ruehl, R.K. Taylor, and M. Koomey. 1991. Shared antigenicity and immunogenicity of type 4 pilins expressed by *Pseudomonas aeruginosa, Moraxella bovis, Neisseria gonorrhoeae, Dichelobacter nodosus,* and *Vibrio cholerae*. *Infect. Immun.* **59**:4674–4676.

Pearce, W.A. and T.M. Buchanan. 1978. Attachment role of gonococcal pili. Optimum conditions and quantitation of adherence of isolated pili to human cells *in vitro*. *J. Clin. Invest.* **61**:931–943.

Penaranda, M.E., D.G. Evans, B.E. Murray, and D.J. Evans, Jr. 1983. ST:LT:CFA/II plasmids in enterotoxigenic *Escherichia coli* belonging to serogroups 06, 08, 085a, and 0139. *J. Bacteriol.* **154**:980–983.

Perry, A.C.F., I.J. Nicolson, and J.R. Saunders. 1988. *Neisseria meningitidis* strain C114 contains silent, truncated pilin genes that are homologous to *Neisseria gonorrhoeae pil* sequences. *J. Bacteriol.* **170**:1691–1697.

Plos, K., T. Carter, S. Hull, R. Hull, and C. Svanborg-Eden. 1990. Frequency and organization of *pap* homologous DNA in relation to clinical origin of uropathogenic *Escherichia coli*. *J. Infect. Dis.* **161**:518–524.

Ponniah, S., R.O. Endres, D.L. Hasty, and S.N. Abraham. 1991. Fragmentation of *Escherichia coli* type 1 fimbriae exposes cryptic D-mannose-binding sites. 1991. *J. Bacteriol.* **173**:4195–4202.

Purcell, B.K. and S. Clegg. 1983. Construction and expression of recombinant plasmids encoding type 1 fimbriae of a urinary *Klebsiella pneumoniae* isolate. *Infect. Immun.* **39**:1122–1127.

Relman, D.A., M. Domenighini, E. Tuomanen, R. Rappuoli, and S. Falkow. 1989. Filamentous hemagglutinin of *Bordetella pertussis:* nucleotide sequence and crucial role in adherence. *Proc. Natl. Acad. Sci. USA* **86**:2637–2641.

Rest, R.F., S.H. Fischer, Z.Z. Inghaum, and J.F. Jones. 1982. Interactions of *Neisseria gonorrhoeae* with human neutrophils: effects of serum and gonococcal opacity on phagocyte killing and chemiluminescence. *Infect. Immun.* **36**:737–744.

Rhen, M., J. Knowles, M.E. Penttila, M. Sarvas, and T.K. Korhonen. 1983a. P-fimbriae of *Escherichia coli:* molecular cloning of DNA fragments containing the structural genes. *FEMS Microbiol. Lett.* **19**:119–123.

Rhen, M., P.H. Makela, and T.K. Korhonen. 1983b. P-fimbriae of *Escherichia coli* are subject to phase variation. *FEMS Microbiol. Lett.* **19**:267–271.

Rhen, M., I. Van Die, V. Rhen, and H. Bergmans 1985a. Comparison of the nucleotide sequence of the genes encoding the KS71A and F7 fimbrial antigens of uropathogenic *Escherichia coli*. *Eur. J. Biochem.* **151**:573–577.

Rhen, M., V. Vaisanen-Rhen, A. Pere, and T.K. Korhonen. 1985b. Complementation and regulatory interaction between two cloned fimbrial gene clusters of *Escherichia coli* strain KS71. *Mol. Gen. Genet.* **200**:60–64.

Rhen, M., P. Klemm, and T.K. Korhonen. 1986a. Identification of two hemagglutinins of *Escherichia coli, N*-acetyl-D-glucosamine-specific fimbriae and a blood group M-specific agglutinin, by cloning the corresponding gene in *Escherichia coli* K-12. *J. Bacteriol.* **168**:1234–1242.

Rhen, M., V. Vaisanen-Rhen, M. Saraste, and T.K. Korhonen. 1986b. Organization of genes expressing the blood-group-M-specific hemagglutinin of *Escherichia coli:* identification and nucleotide sequence of the M-agglutinin subunit gene. *Gene* **49**:351–360.

Riegman, N., I. Van Die, J. Leunissen, W. Hoekstra, and H. Bergmans. 1988. Biogenesis of $F7_1$ and $F7_2$ fimbriae of uropathogenic *Escherichia coli:* influence of the FsoF and FsoG proteins and localization of the Fso/FstkE protein. *Mol. Microbiol.* **2**:73–80.

Riegman, N., R. Kusters, H. Van Veggel, H. Bergmans, P.V.B.E. Henegouwen, J. Hacker, and I Van Die. 1990. F1C fimbriae of a uropathogenic *Escherichia coli* strain: genetic and functional organization of the *foc* gene cluster and identification of minor subunits. *J. Bacteriol.* **172**:1114–1120.

Roosendaal, E., M. Boots, and F.K. DeGraaf. 1987a. Two novel genes, *fan*A and *fan*B, involved in the biogenesis of K99 fimbriae. *Nucleic Acids Res.* **15**:5973–5984.

Roosendaal, E., A.A.C. Jacobs, P. Rathman, C. Sondermeyer, F. Stegehuis, B. Oudega and F.K. DeGraaf. 1987b. Primary structure and subcellular localization of two fimbrial subunit-like proteins involved in the biosynthesis of K99 fimbrillae. *Mol. Microbiol.* **1**:211–217.

Roosendaal, B., J. Damoiseaux, W. Jordi, and F.K. DeGraaf. 1989. Transcriptional organization of the DNA region controlling expression of the K99 gene cluster. *Mol. Gen. Genet.* **215**:250–256.

Rosan, B., C.T. Baker, G.M. Nelson, R. Berman, R.J. Lamont, and D.R. Demuth. 1989. Cloning and expression of an adhesin antigen of *Streptococcus sanguis* G9B in *Escherichia coli. J. Gen. Microbiol.* **135**:531–538.

Rothbard, J.B., R. Fernandez, L. Wang, N.N.H. Teng, and G.K. Schoolnik. 1985. Antibodies to peptides corresponding to a conserved sequence of gonococcal pilins block bacterial adhesion. *Proc. Natl. Acad. Sci. USA* **82**:915–919.

Russell, R.R.B., D. Coleman, and G. Dougan. 1985. Expression of a gene for glucan-binding protein from *Streptococcus mutans* in *Escherichia coli. J. Gen. Microbiol.* **131**:295–299.

Salit, I.E. and E.C. Gotschlich. 1977. Hemagglutination by purified type 1 *Escherichia coli* pili. *J. Exp. Med.* **146**:

Schaper, U., J.S. Chapman, and P.C. Hu. 1987. Preliminary indication of unusual codon usage in the DNA coding sequence of the attachment protein of *Mycoplasma pneumoniae. Isr. J. Med. Sci.* **23**:361–367.

Schifferli, D.M., E.H. Beachey, and R.K. Taylor. 1990. The 987P fimbrial gene cluster of enterotoxigenic *Escherichia coli* is plasmid encoded. *Infect. Immun.* **58**:149–156.

Schmoll, T.H. Hoschützky, J. Morschhauser, F. Lottspeich, and K. Jann. 1989. Analysis of genes coding for the sialic acid-binding adhesin and two other minor fimbrial subunits of the S fimbrial adhesin determinant of *Escherichia coli. Mol. Microbiol.* **3**:1735–1744.

Schneider, K. and C.F. Beck. 1986. Promoter probe vectors for the analysis of divergently arranged promoters. *Gene* **42**:37–48.

Schoolnik, G.K., R. Fernandez, J.Y. Tai, J. Rothbard, and E.C. Gotschlich. 1984. Gonococcal pili: primary structure and receptor binding domain. *J. Exp. Med.* **159**:1351–1370.

Schwalbe, R.S. and J.G. Cannon. 1986. Genetic transformation of genes for protein II in *Neisseria gonorrhoeae. J. Bacteriol.* **167**:186–190.

Schwalbe, R.S., P.F. Sparling, and J.G. Cannon. 1985. Variation of *Neisseria gonorrhoeae* protein II among isolates from an outbreak caused by a single gonococcal strain. *Infect. Immun.* **49**:250–252.

Scott, J.R. and V.A. Fischetti. 1983. Expression of streptococcal M protein in *Escherichia coli. Science* **221**:758–760.

Seifert, H. and M. So. 1988. Genetic mechanisms of bacterial antigenic variation. *Microbiol. Rev.* **52**:327–336.

Sharon, N. and I. Ofek. 1986. Mannose-specific bacterial surface lectins. In: Mirelman, D. (ed.), *Microbial Lectins and Agglutinins: Properties and Biological Activity*. Wiley-Interscience, New York, pp. 55–81.

Shaw, C.E. and R.K. Taylor. 1990. *Vibrio cholerae* 0395 tcpA pilin gene sequence and comparison of predicted protein structural features to those of type 4 pilins. *Infect. Immun.* **58**:3042–3049.

Shaw, C.E., K.M. Peterson, J.J. Mekalanos, and R.K. Taylor. 1990. Genetic studies of *Vibrio cholerae* TCP pilus biogenesis. *Adv. Res. Cholera Related Diarrheas* **7**:51–58.

Shipley, P.L., C.L. Gyles, and S. Falkow. 1978. Characterization of plasmids that encode for the K88 colonization antigen. *Infect. Immun.* **20**:559–566.

Shipley, P.L., G. Dougan, and S. Falkow. 1981. Identification and cloning of the genetic determinant that encodes for the K88ac adherence antigen. *J. Bacteriol.* **145**:920–925.

Sjöberg, P., M. Lindahl, J. Porath, and T. Wadstrom. 1988. Purification and characterization of CS2, a sialic acid-specific haemagglutinin of enterotoxigenic *Escherichia coli. Biochem. J.* **255**:105–111.

Smith, H.W. and M.A. Linggood. 1972. Further observations on *Escherichia coli* enterotoxins with particular regard to those produced by atypical piglet strains and by calf and lamb strains. The transmissible nature of these enterotoxins and of a K antigen possessed by calf and lamb strains. *J. Med. Microbiol.* **5**:243–250.

Smith, H.R., S.M. Scotland, and B. Rowe. 1983. Plasmids that encode for production of colonization factor antigen II and enterotoxin production in strains of *Escherichia coli. Infect. Immun.* **40**:1236–1239.

Smyth, C.J. 1982. Two mannose-resistant haemagglutinins on enterotoxigenic *Escherichia coli* or serotype 06:K15:H16 or H-isolated from travellers' and infantile diarrhoea. *J. Gen. Microbiol.* **128**:2081–2096.

So, M., J.F. Crandall, J.H. Crosa, and S. Falkow. 1974. Extrachromosomal determinants which contribute to bacterial pathogenicity. In: Schlessinger, D. (ed.), *Microbiology-1974*. American Society for Microbiology, Washington, pp. 16–20.

Sommer, P., T. Bruyere, J.A. Ogier, J.M. Garnier, J.M. Jeltsch, and J.P. Klein. 1987. Cloning of the saliva-interacting protein gene from *Streptococcus mutans. J. Bacteriol.* **169**:5167–5173.

Spanier, J.G., S.J.C. Jones, and P. Cleary. 1984. Small DNA deletions creating avirulence in *Streptococcus pyogenes. Science* **225**:935–938.

Stern, A. and T.F. Meyer. 1987. Common mechanism controlling phase and antigenic variation in pathogenic neisseriae. *Mol. Microbiol.* **1**:5–12.

Stern, A., P. Nickel, T.F. Meyer, and M. So. 1984. Opacity determinants of *Neisseria gonorrhoeae:* gene expression and chromosomal linkage to the gonococcal pilus gene. *Cell* **37**:447–456.

Stern, A., M. Brown, P. Nickel, and T.F. Meyer. 1986. Opacity genes in *Neisseria gonorrhoeae:* control of phase and antigenic variation. *Cell* **47**:61–71.

Stibitz, S., A.A. Weiss, and S. Falkow. 1988. Genetic analysis of a region of the *Bordetella pertussis* chromosome encoding filamentous hemagglutinin and the pleiotropic regulatory locus vir. *J. Bacteriol.* **170**:2904–2913.

Strosberg, A.D., D. Buffard, M. Lauwereys, and A. Foriers. 1986. Legume lectins: a large family of homologous proteins. In: Liener, I., N. Sharon and I.J. Goldstein (eds.), *The Lectins, Properties, Functions and Applications in Biology and Medicine*. Academic Press, Orlando, pp. 249–264.

Su, C., V.V. Tryon, and J.B. Baseman. 1987. Cloning and sequence analysis of cytadhesin P1 gene from *Mycoplasma pneumoniae. Infect. Immun.* **55**:3023–3029.

Su, C.-J., A. Chavoya, and J.B. Baseman. 1988. Regions of *Mycoplasma penumoniae* cytadhesin P1 structural gene exist as multiple copies. *Infect. Immun.* **56**:3157–3161.

Svennerholm, A.-M., Y.L. Vidal, J. Holmgren, M.M. McConnell, and B. Rowe. 1988. Role of PCF8775 antigen and its coli surface subcomponents for colonization, disease, and protective immunogenicity of enterotoxigenic *Escherichia coli* in rabbits. *Infect. Immun.* **56**:523–528.

Swaney, L.M., Y.P. Liu, C.M. To, C.C. To, K. Ippen-Ihler, and C.C. Brinton, Jr. 1977. Isolation and characterization of *Escherichia coli* phase variants and mutants deficient in type 1 pilus production. *J. Bacteriol.* **130**:495–505.

Swanson, J. 1980. Adhesion and entry of bacteria into cel Infls: a model of the pathogenesis of gonorrhea. In: Smith, H., J.J. Skehel, and M.J. Turner (eds.), *The Molecular Basis of Microbial Pathogenicity*. Verlag Chemi GmbH, Weinheim, Federal Republic of Germany, pp. 17–40.

Swanson, J. 1988. Genetic mechanisms responsible for changes in pilus expression by gonoccocci. *UCLA Symp. Mol. Cell. Biol. New Ser.* **20**:347–363.

Swanson, J. 1990. Pilus and outer membrane protein II variation in *Neisseria gonorrhoeae*. In: Ayoub, E.M., G.H. Cassell, W.C. Branche, Jr., and F.J. Henry, (eds.), *Microbial Determinants of Virulence and Host Response*. American Society for Microbiology, Washington, pp. 197–205.

Swanson, J. and O. Barrera. 1983. Gonococcal pilus subunit size heterogeneity correlates with transitions in colony piliation phenotype, not with changes in colony opacity. *J. Exp. Med.* **158**:1459–1472.

Swanson, J. and M. Koomey. 1989. Mechanisms of variation of pili outer membrane protein II in *Neisseria gonorrhoeae*. In: Berg, D.E. and M.M. Howe (eds.), *Mobile DNA*. American Society for Microbiology, Washington, pp. 743–761.

Swanson, J., S.J. Kraus, and E.C. Gotschlich. 1971. Studies on gonococcus infection. I. Pili and zones of adhesion: their relation to gonococcal growth patterns. *J. Exp. Med.* **134**:886–906.

Swanson, J., S. Bergström, K. Robbins, O. Barrera, D. Corwin and J.M. Koomey.

1986. Gene conversion involving the pilin structural gene correlates with pilus$^+$ pilus$^-$ changes in *Neisseria gonorrhoeae*. *Cell* **47**:267–276.

Swanson, J., K. Robbins, O. Barrera, D. Corwin, J. Boslego, J. Ciak, M. Blake, and J.M. Koomey. 1987. Gonococcal pilin variants in experimental gonorrhea. *J. Exp. Med.* **165**:1344–1357.

Swanson, T.N., S. Bilge, B. Nowicki, and S.L. Moseley. 1991. Molecular structure of the Dr adhesin: nucleotide sequence and mapping of receptor-binding domain by use of fusion constructs. *Infect. Immun.* **59**:261–268.

Taylor, R., C. Shaw, K. Peterson, P. Spears, and J. Mekalanos. 1988. Safe, live *Vibrio cholerae* vaccines? *Vaccine* **6**:151–156.

Tennent, J.M., S. Hultgren, B.-I. Macklund, K. Forsman, M. Goransson, B.E. Uhlin, and S. Normark. 1990. Genetics of adhesin expression in *Escherichia coli*. In: Iglewski, B. and V.L. Clark (eds.), *The Bacteria, Vol. XI, The Molecular Basis of Bacterial Pathogenesis*. Academic Press, San Diego, pp. 79–?.

Tennent, J.M., F. Lindberg, and S. Normark, 1990. Integrity of *Escherichia coli* P pili during biogenesis: properties and role of Pap J. *Mol. Microbiol.* **4**:747–758.

Thomas, L.V., A. Cravioto, S.M. Scotland, and B. Rowe. 1982. New fimbrial antigenic type (E8775) that may represent a colonization factor in enterotoxigenic *Escherichia coli* in humans. *Infect. Immun.* **35**:1119–1124.

Thomas, L.V., M.M. McConnell, B. Rowe, and A.M. Field. 1985. The possession of three novel coli surface antigens by enterotoxigenic *Escherichia coli* strains positive for the putative colonization factor PCF8775. *J. Gen. Microbiol.* **131**:2319–2326.

Thomas, L.V., B. Rowe, and M.M. McConnell. 1987. In strains of *Escherichia coli* 0167 a single plasmid encodes for the coli surface antigens CS5 and CS6 of putative colonization factor PCF8775, heat-stable enterotoxin, and colicin Ia. *Infect. Immun.* **55**:1929–1931.

Trevino, L.B., W.G. Haldenwang, and J.B. Baseman. 1986. Expression of *Mycoplasma pneumoniae* antigens in *Escherichia coli*. *Infect. Immun.* **53**:129–134.

Uhlin, B.E., M. Norgren, M. Baga, and S. Normark. 1985. Adhesion to human cells by *Escherichia coli* lacking the major subunit of a digalactoside-specific pilus-adhesin. *Proc. Natl. Acad. Sci. USA* **82**:1800–1804.

Van Die, I., C. Van den Hondel, H.J. Hamstra, W. Hoekstra, and H. Bergmans. 1983. Studies on the fimbriae of an *Escherichia coli* 06:K2:H1:F7 strain: molecular cloning of a DNA fragment encoding a fimbrial antigen responsible for mannose-resistant hemagglutination of human erythrocytes. *FEMS Microbiol. Lett.* **19**:77–82.

Van Die, I., B. Van Geffen, W. Hoekstra, and H. Bergmans. 1984a. Type 1C fimbriae of a uropathogenic *Escherichia coli* strain: cloning and characterization of the genes involved in the expression of the 1C antigen and nucleotide sequence of the subunit gene. *Gene* **34**:187–196.

Van Die, I., I. Van Megen, W. Hoekstra, and H. Bergmans. 1984b. Molecular organization of the genes involved in the production of $F7_2$ fimbriae, causing mannose-resistant hemagglutination, of a uropathogenic *Eshcerichia coli* 06:K2:H1:F7 strain. *Mol. Gen. Genet.* **194**:528–533.

Van Die, I., G. Spierings, I. Van Megen, E. Zuidweg, W. Hoekstra, and H. Bergmans. 1985a. Cloning and genetic organization of the gene cluster encoding $F7_1$ fimbriae of a uropathogenic *Escherichia coli* and comparison with the $F7_2$ gene cluster. *FEMS Microbiol. Lett.* **28**:329–334.

Van Die, I., B. van Geffen, W. Hoekstra, and H. Bergmans. 1985b. Type 1C fimbriae of a uropathogenic *Escherichia coli* strain: cloning and characterization of the genes involved in the expression of the 1C antigen and nucleotide sequence of the subunit gene. *Gene* **34**:187–196.

Van Die, I., W. Hoekstra, and H. Bergmans. 1987. Analysis of the primary structure of P-fimbrillins of uropathogenic *Escherichia coli*. *Microb. Pathogen.* **3**:149–154.

Van Die, I., N. Riegman, O. Gaykema, I. van Megen, W. Hoekstra, H. Bergmans, H. de Ree, and H. van Bosch. 1988a. Localization of antigenic determinants on P-fimbriae of uropathogenic *Escherichia coli*. *FEMS Microbiol. Lett.* **49**:95–100.

Van Die, I., M. Wauben, I. van Megen, H. Bergmans, N. Riegman, W. Hoekstra, P. Pouwels, and B. Enger-Valk. 1988b. Genetic manipulation of major P-fimbrial subunits and consequences for formation of fimbriae. *J. Bacteriol.* **170**:5870–5876.

Van Die, I., C. Kramer, J. Hacker, H. Bergmans, W. Jongen, and W. Hoekstra. 1991. Nucleotide sequence of the genes coding for minor fimbrial subunits of the F1C fimbriae of *Escherichia coli*. *Res. Microbiol.* **142**:653–658.

Van Embden, J.D.A., F.K. DeGraaf, L.M. Schouls, and J.S. Teppema. 1980. Cloning and expression of a deoxyribonucleic acid fragment that encodes for the adhesive antigen K99. *Infect. Immun.* **29**:1125–1133.

Van Ham, S.M., F.R. Mooi, M.G. Sindhunata, W.R. Maris, and L. van Alphen. 1989. Cloning and expression in *Escherichia coli* of *Haemophilus influenzae* fimbrial genes established adherence to oropharyngeal epithelial cells. *EMBO J.* **8**:3535–3540.

Virji, M. and J.S. Everson. 1981. Comparative virulence of opacity variants of *Neisseria gonorrhoeae* strain P9. *Infect. Immun.* **31**:965–970.

Virji, M. and J.E. Heckels. 1986. The effect of protein II and pili on the interaction of *Neisseria gonorrhoeae* with human polymorphonuclear leucocytes. *J. Gen. Microbiol.* **132**:503–512.

Virkola, R., B. Westerlund, H. Holthofer, J. Parkkinen, M. Kekomaki, and T.K. Korhonen. 1988. Binding characteristics of *Escherichia coli* adhesins in human urinary bladder. *Infect. Immun.* **56**:2615–2622.

Waalen, K., K. Sletten, L.O. Froholm, V. Vaisanen, and T.K. Korhonen. 1983. The N-terminal amino acid sequence of type 1 fimbria (pili) of *Salmonella typhimurium* LT2. *FEMS Microbiol. Lett.* **16**:149–151.

Walz, W., A. Schmidt, A. Labigne-Roussel, S. Falkow, and G. Schoolnik. 1985. AFA-1: a cloned afimbrial X-type adhesin from a human pyelonephritic *Escherichia coli* strain. *Eur. J. Biochem.* **152**:315–321.

Westerlund, B., I. van Die, C. Kramer, P. Kuusela, H. Holthöfer, A.M. Tarkkanen, R. Virkola, N. Riegman, H. Bergmans, W. Hoekstra, and T.K. Korhonen. 1991. Multifunctional nature of P fimbriae of uropathogenic *Escherichia coli:* mutations in *fso*E and *fso*F influence fimbrial binding to renal tubuli and immobilized fibronectin. *Mol. Microbiol.* **5**:2965–2975.

Willshaw, G.A., H.R. Smith, and R. Rowe. 1983. Cloning of regions encoding colonization factor antigen I and heat-stable enterotoxin in *Escherichia coli*. *FEMS Microbiol. Lett.* **16**:101–106.

Willshaw, G.A., H.R. Smith, M.M. McConnell, and B. Rowe. 1985. Expression of cloned plasmid regions encoding colonization factor antigen I (CFA/I) in *Escherichia coli*. *Plasmid* **13**:8–16.

Wolf, M.K. and E.C. Boedeker. 1990. Cloning of the genes for AF/RI pili from rabbit enteroadherent *Escherichia coli* RDEC-1 and DNA sequence of the major structural subunit. *Infect. Immun.* **58**:1124–1128.

Wolf, M.K., G.P. Andrews, B.D. Tall, M.M. McConnell, M.M. Levine, and E.C. Boedeker. 1989. Characterization of CS4 and CS6 antigenic components of PCF8775, a putative colonization factor complex from enterotoxigenic *Escherichia coli* E8775. *Infect. Immun.* **57**:164–173.

Yeung, M.K. and J.O. Cisar. 1988. Cloning and nucleotide sequence of a gene for *Actinomyces naeslundii* WVU45 type 2 fimbriae. *J. Bacteriol.* **170**:3803–3809.

Yeung, M.K. and J.O. Cisar. 1990. Sequence homology between the subunits of two immunologically and functionally distinct types of fimbriae of *Actinomyces* spp. *J. Bacteriol.* **172**:2462–2468.

Yeung, M.K., B.M. Chassy, and J.O. Cisar. 1987. Cloning and expression of a type 1 fimbrial subunit of *Actinomyces viscosus* T14V. *J. Bacteriol.* **169**:1678–1683.

Zafriri, D., I. Ofek, R. Adar, M. Pocino, and N. Sharon. 1989. Inhibitory activity of cranberry juice on adherence of type 1 and type P fimbriated *Escherichia coli* to eucaryotic cells. *Antimicrob. Agents Chemother.* **33**:92–98.

Zak, K., J.L. Diaz, D. Jackson, and J.E. Heckels. 1984. Antigenic variation during infection with *Neisseria gonorrhoeae:* detection of antibodies to surface proteins in sera of patients with gonorrhea. *J. Infect. Dis.* **149**:166–174.

10

Recent Developments in Bacterial Adhesion to Animal Cells

Although at the beginning of this century it was found that bacteria interact with animal cells by causing hemagglutination of erythrocytes (Guyot, 1908), systematic research on bacterial adhesion and hemagglutination reactions started only in the 1950s with the work of Duguid (Duguid et al., 1955). Extensive and uninterrupted studies by many investigators on bacterial adhesion to animal cells started only about two decades ago. These extensive studies may be divided roughly, with some overlap, into three 7-year periods. In addition to the numerous proceedings from symposia held intermittently during these periods, the emerging studies have been summarized in a number of books and reviews. The studies included in the first 7-year period and published up to 1978 are characterized largely by establishment of the role of bacterial adhesion in the infectious process and the specificity of the bacteria–animal cell interactions, but only a few adhesins were defined. The most important studies published during that period have been discussed in the book on bacterial adhesion edited by the late Ed Beachey (Beachey, 1980) and in a review by the same author (Beachey, 1981) as well as in other reviews (Jones, 1977; Gibbons, 1977; Ofek et al., 1978; Duguid et al., 1979; Ofek and Beachey, 1980). Studies included in the second 7-year period and published between 1979 and 1985 continued to assess the role of adhesion in vivo, but focussed on identifying the various bacterial adhesins and their receptor specificities. During this period researchers also began the cloning of the adhesin gene clusters involved in expression and regulation of the adhesins, especially those of *Escherichia coli* and *Neisseria gonorrhoeae*. Most of these studies have been summarized and discussed in books (Sussman, 1985; Savage and Fletcher, 1985; Mirelman, 1986), monographs (Beachey et al., 1982), and reviews (Gaastra and De Graaf, 1982; Swanson, 1983; Jones and Isaacson, 1983; Hirsch, 1985; Klemm, 1985; Mooi and DeGraaf, 1985; Uhlin et al., 1985; Sparling and Cannon, 1986; DeGraaf and Mooi, 1986), all of which contain brief summaries on studies published in the first 7-year period as well. Studies emerging during the third 7-year period and published from 1985 through 1991 represent recent advances in bacterial adhesion. Only those from this period written in English are discussed in this chapter. In general, these studies continued to characterize the adhesins, their receptor specificities, and their molecular biology. The recent studies also employed DNA probes for the identification of the genomic operons coding for the adhesin complexes as well as the distribution of adhesin genes initiating the phenomenon of phase variation. Reviews and books discussing and summarizing some of these studies, especially those dealing with adhesins of *E. coli* (Clegg and Gerlach, 1987; Reid and Sobel, 1987; Oudega and DeGraaf, 1988; Jann and Jann, 1990; Tennent et al., 1990; Hultgren et al., 1991; Johnson, 1991; Krogfelt, 1991), *N. gonorrhoeae* (Murphy and Cannon, 1988; Seifert and So, 1988; Swanson and Koomey, 1989; Meyer, 1990), oral bacteria (Kolenbrander 1989; Handley, 1990), and a limited number of other bacterial species as well (Ofek and Sharon, 1988; Finlay and Falkow, 1989; Baddour et al., 1990; London, 1991), served as an invaluable source for preparing this chapter. A comprehensive review of the foregoing literature citations has revealed that bacterial adhesion is an essential, but is only one of many, processes leading to colonization and disease. Furthermore, it is clear that most, if not all, pathogenic bacterial clones are capable of expressing multiple adhesins, each with a distinct receptor specificity and with an independent mechanism of regulating adhesin expression. The understanding of the precise role in vivo of each of the multiple adhesins expressed by a single bacterial clone is a primary goal of this chapter. The rationale for the necessity for multiple adhesins is discussed separately in Chapter 11. Specific advantages conferred by a particular adhesin to the ability of the bacterial clone to colonize a particular tissue are also summarized in this chapter. It is for this reason that the studies are presented by the bacterial species, its source of isolation, a specific adhesin and/or its receptor specificity, and the substrata tested or examined for adhesion. It is also for this reason that studies dealing strictly with the molecular biology of the adhesin or its fine carbohydrate specificity or the effects of sublethal concentrations of antibiotics on its expression have been largely excluded, as these are dealt with in other chapters. In some cases, a great deal is known about the role of a few adhesins in vivo. In many other cases, there is virtually no information regarding an in vivo role for the adhesin. Therefore, a comprehensive discussion is given for those few adhesins and only a brief description of the data is given for the other adhesins.

On reviewing the voluminous literature on adhesion it became apparent that the amount of studies on adhesion of *Escherichia coli,* as reflected by the number of published articles during each of the 7-year periods, is equal to or greater than those combined on all other bacterial species colonizing animal tissues. For this reason, this chapter is divided in two parts, one of which deals with studies on adhesion of *E. coli* and the other on adhesion of all other bacterial species. It is not surprising, therefore, that the *E. coli* adhesins are the best characterized of all bacterial adhesins. Studies on some of the adhesins have aided understanding of basic principles in bacterial adhesion as discussed in this chapter as well as in Chapter 11. Nevertheless, *E. coli* adhesins have evolved to enable a particular type of pathogen to colonize specific animal tissues. It is likely that if more efforts are devoted to other bacterial species, especially Gram-positive bacteria, new strategies employed by numerous bacterial adhesins will be realized. For example, research on oral bacterial adhesins, which as a group have not been studied as extensively as *E. coli,* has revealed that the adhesins of one bacterial species interact with complementary receptor structures on the surfaces of other bacterial species as a strategy for colonization of the oropharyngeal tissues. Abbreviations used in the following are shown below.

List of Abbreviations Used in this Chapter

BLD	blood
BM	biomaterials
CBP	carbohydrate-binding protein
CSF	cerebrospinal fluid
CS	corneal surface
DAEC	diffuse adhesion
Eae	effacing-attaching factor
EAEC	enteroadherent—*E. coli*
EAF	enteroadherent factor
EC	epithelial cells
ECM	extracellular matrix (fibronectin, collagen, heparin, laminin)
EHEC	enterohemorrhagic *E. coli*
EIEC	enteroinvasive *E. coli*
ENC	endothelial cells
ENT	enterocytes
ENV	environment
EPEC	enteropathogenic *E. coli*
ERT	erythrocytes
ETEC	enterotoxigenic *E. coli*
FIM	fimbriae
HA	hydroxylapatite
IG	immunoglobulin
IM	immobilized mucus (on plastic surface)
IS	immobilized carbohydrate (on beads or on thin layers)
LPS	lipopolysaccharide
LOS	lipooligosaccharide
LRT	lower respiratory tract
LTA	lipoteichoic acid
MC	M-cells
MUC	mucus

OC	organ culture
OMP	outer membrane protein
ORL	oral
PC	phagocytic cell
RT	respiratory tract
SHA	saliva-coated hydroxylapatite
SKN	skin
SS	synthetic surface (e.g., contact lenses, dacron fibers, hydroxylapatite)
TA	teichoic acid
TB	tissue biopsy
TC	tissue culture
TCP	toxin-coregulated pili
TS	tissue surface
URG	urogenital
URT	upper respiratory tract
UTI	urinary tract

PART A: RECENT ADVANCES ON *ESCHERICHIA COLI* ADHESION

Escherichia coli

Clinical sources:	Gut, blood, urinary tract
Adhesin:	Type 1 fimbriae
Receptor specificity:	Mannosides
Substrata:	EC, ECM, ERT, PC, TB, TC

Because the type 1 fimbrial adhesin is expressed by *E. coli* strains from all clinical categories (e.g., ENT, UTI, BLD), many investigators dealing with gastrointestinal, urinary tract, or blood infections have continued to focus their studies on two major issues. One issue is related to the function of type 1 fimbrial adhesin during the infectious process. The other issue is related to the identity of glycoconjugates or oligosaccharides that serve as receptors (on animal cells and in mucus) or act as potent specific inhibitors, respectively, for the fimbrial lectin. In both issues, the hope is still to find a common function for the type 1 fimbrial adhesin so that one or another approach can be designed for inhibiting the adhesion. A goal is to prevent or treat diverse types of infections caused by the genus *E. coli,* taking advantage of adhesin specificity. The approaches employed to reveal the function(s) of the type 1 fimbrial adhesin, especially in initiating infection on mucosal surfaces, include relative infectivity of type 1 bearing *E. coli,* efficacy of type 1 fimbrial vaccines, expression of the fimbrial adhesin during infection, distribution of receptors for the fimbrial adhesin in various tissues, and the phenotypic expression by various isolates of the adhesin during in vitro growth. In Chapter 5 some of the results accumulated over the last decade showing a positive role of type 1 fimbriae in infections caused by various enterobacterial genera were discussed. In the following, the focus is on studies performed during the last half decade on *E. coli* to illustrate the controversial results obtained concerning the role played by the type 1 fimbrial adhesin of *E. coli* in initiating infection on mucosal surfaces.

The relative advantages of type 1 fimbriated *E. coli* to cause mucosal infections have been determined in experimental bladder infections of laboratory animals using recombinant strains expressing type 1 fimbriae as compared with strains that do not express any type of adhesin (O'Hanley et al., 1985a; Keith et al., 1986; Marre and Hacker, 1987). Experiments exploying phase variant strains rich or poor in type 1 fimbriated cells for bladder colonization of mice (Hultgren et al., 1986; Schaeffer et al., 1987; Schwan et al., 1989) or experiments employing type 1-deficient and isogenic type 1 fimbriated mutants of K1 *E. coli* strains in the oropharnyngeal colonization of rats (Bloch and Orndorff, 1990) also support a role for type 1 fimbriae in infection. A dose-dependent advantage for type 1 fimbriae to initiate colonization of the mouse intestine was shown by experiments comparing the infectivity of *E. coli* suspended in D-mannose to inhibit the adhesiveness of type 1 fimbriae to that suspended in sucrose (Goldhar et al., 1986). Enterotoxigenic type 1 fimbriated *E. coli* colonized the pig small intestine, whereas a nonfimbriated variant derivative failed to do so (Nakazawa et al., 1986). A direct role of type 1 fimbrial adhesin in inducing specific tissue damage has been suggested by studies showing that the degree of renal scarring developing in rats after intrarenal injection of *E. coli* is greatest among type 1-bearing organisms (Topley et al., 1989b). The tissue damaging activity has been attributed to the release of oxygen radicals and neutral proteases from phagocytic cells following lectinophagocytosis of the type 1 fimbriated bacteria (see also Chapter 7). In contrast, no advantage for type 1 fimbriated *E. coli* in colonization of the mouse large intestine was observed in infections employing *E. coli* strains expressing type 1 fimbriae as compared with isogenic strains rendered deficient of Fim A or Fim H by transduction (McCormick et al., 1989) or by direct mutagenesis of fimA (Bloch and Orndorff, 1990) in rats. Moreover, nonfimbriated *E. coli* persisted significantly longer periods of time over type 1 fimbriated bacteria in the bladder of human volunteers infected with mixtures of fimbriated and nonfimbriated isogenic mutants, suggesting that type 1 fimbriae confer a disadvantage to the bacteria by enhancing their clearance from the bladder (Andersson et al., 1991). Type 1 fimbriae also are responsible for enhanced clearance of the type 1 fimbriated phenotype of *E. coli* from phagocyte-rich sites because they mediate lectinophagocytosis of the organisms (see Chapter 7).

The second approach to assess the role of type 1 fimbriae in infection is concerned with the efficacy of either passive or active immunization. Although early studies failed to show protective effects of type 1 fimbrial vaccines in humans (reviewed in Sharon and Ofek, 1986) and mice (O'Hanley et al., 1985a) a number of studies demonstrate protective immunity with type 1 fimbriae in other animals (see Table 5–21). Because all these studies employed fimbrial preparations as vaccines, most of the antibodies induced by such vaccines were directed against Fim A, which exhibits considerable antigenic heterogeneity (Sharon and Ofek, 1986). Furthermore, only hybridoma antibodies directed against a quaternary structural epitope of Fim A inhibited the adhesion of type 1 fimbriated *E. coli* to eucaryotic cells (Abraham et al., 1985). Because the type 1 fimbrial adhesin is a heteropolymer consisting of a number of different minor subunits (see Chapter 9), it is clear that the protective activity of vaccines consisting of the minor fimbrial subunit, perhaps the more conserved ones, of the type 1 fimbrial adhesin should be examined. Preliminary studies showed that anti-Fim H, a conserved minor fimbrial subunit that is responsible for binding mannose residues, protected urinary tract infections in mice against challenge with type 1 fimbriated *E. coli* (Madison et al., 1990). Primary stimulated lymphoid cells committed to produce IgA anti-type 1 fimbriae have a tendency to "home" into the mammary gland but after repeated stimulation with type 1 fimbrial antigen, the lymphoid cells home to the intestine (Dahlgren et al., 1990). It is not clear whether this is a unique property of type 1 fimbrial antigens or a more general property of all fimbrial antigens. A somewhat unique type of vaccine consisting of antibodies against anti-type 1 fimbrial idiopathic antibodies has been proposed (Paque et al., 1990). These antiidiopathic antibodies are believed to represent the molecular image of the antigenic determinant against which the idiopathic antibodies are directed and may therefore be used as a vaccine to induce humoral and cellular immunity against desired antigenic determinants of type 1 fimbrial adhesins. Although the antiidiopathic vaccine elicited cellular and humoral immune responses against antigenic determinants associated with 27,000 and 29,000 mol wt proteins of type 1 fimbriae, it failed to protect mice against intraperitoneal infection with type 1 fimbriated *E. coli*.

The third approach includes studies showing, by direct specific staining techniques, the expression of type 1 fimbriae by adherent organisms in tissue biopsies of experimentally infected animals (Dominick et al., 1985; Jayappa et al., 1985; Sherman et al., 1989a,b; Krogfelt et al., 1991). Evidence was also obtained to suggest that type 1 fimbriae are expressed by *E. coli* causing intestinal infections in human volunteers (Karch et al., 1987b). Although in earlier studies it was found that the majority of *E. coli* shed in the urine of patients were devoid of type 1 fimbriae (reviewed in Sharon and Ofek, 1986), recent studies have shown the expression of the fimbrial adhesin occurs in up to 20% of the organisms shed in urine of 5 of 20 (Pere et al., 1987) and 31 of 41 (Kiselius et al., 1989) patients with urinary tract infections. The results showing expression of type 1 fimbriae in vivo have encouraged studies concerned with the construction of recombinant live *E. coli* vaccines by fusing DNA elements coding for small antigenic peptide sequences into a specific region of *fim*A (Hedegaard and Klemm, 1989).

A fourth approach represents attempts to find a possible correlation between the phenotypic expression of the adhesin in a particular culture condition and the source of isolation. There are numerous reports on the proportion of strains expressing type 1 fimbriae among isolates from various clinical sources, especially from patients with UTI (reviewed in Johnson, 1991) and to a lesser extent from patients with septicemia and meningitis (Selander et al., 1986) and intestinal infections (Blanco et al., 1985). Because virtually all *E. coli* isolates, including those from farm animals such as bovines and poultry (Achtman et al., 1986), as well as from rabbit pathogens (Sherman et al., 1985), are genotypically capable of expressing type 1 fimbriae as determined either morphologically or functionally (e.g., MS hemagglutination) and presumably the *fim* gene cluster (see Chapter 9), it has been assumed that the genetic control of the phase variation in the expression of the fimbrial adhesin might be different in various strains during growth in a particular culture condition. Due to differences in patient selection and definition as well as in culture conditions among the various studies it is difficult to draw any universal conclusions. Moreover, even in the same study, no information is given as regards variables that might affect the expression of the fimbrial adhesin for the various strains tested, such as number of passages of the test strain before and after storage or the percent fimbriated bacteria in the bacterial population of the inocula and in the subcultured bacteria. Two studies help to illustrate the complexity involved in the phenotypic expression of type 1 fimbriae by various isolates. In one study (Mobley et al., 1987) the proportion of strains isolated from the urethra of persistently colonized patients and expressing type 1 fimbriae after one broth passage from stock culture was significantly higher than that of transiently colonized patients. In another study that minimized the number of laboratory passages of the test strains, no differences in fimbriation between persistently and transiently colonized patients were found. It was noted, however, that the phenotypic expression of the type 1 fimbrial adhesin changes over time in the same patient (Amundsen et al., 1988). Nevertheless, careful examination of the proportion of organisms in colonies of agar-grown urinary isolates of *E. coli* has revealed that there are two categories of strains. One of these (regulated variants) contains strains that suppress fimbriation during growth on agar, whereas the other (random phase variants) contains strains that produce colonies consisting of a mixed population of type 1 fimbriated and nonfimbriated (Hultgren et al., 1986). These differences appear to reflect differences in the genomic locus within the gene cluster coding for the type 1 fimbrial adhesin (Blomfield et al., 1991; McClain et al., 1991). A useful probe independent of growth conditions or number of laboratory passages may be available to examine the correlation between phenotypic expression and clinical source of the isolate. Type 1 fimbriated *E. coli* differ in their surface hydrophobicity as measured by the salt aggregation test, one group being significantly more hydrophobic than the other. It has been suggested that the relative hydrophobicity of the type 1 fimbriated phenotype may be associated with a particular infection (Sherman et al., 1985; Gonzalez et al., 1988a). Because type 1 fimbriae mediate phagocytosis and cause release of oxygen metabolites from phagocytic cells (see Chapter 7) it was suggested the fimbriae may play a role in the development of renal scarring (Topley et al., 1989b). Because growth conditions may affect the outgrowth of type 1 fimbriated *E. coli,* studies have been initiated to determine if urinary constituents can also affect the growth of the type 1 fimbriated *E. coli* phenotype. It has been found that the type 1 fimbriated phenotype outgrew the nonfimbriated phenotype after growth in broth supplemented with urea (Ofek et al., 1989). Interestingly, the presence of urea in the medium seems to favor growth of type 1 fimbriated phenotype, whereas agar media usually favor growth of non type 1 fimbriated phenotypes (Conventi et al., 1989). There are, however, other high molecular weight constituents in human urine that may cause outgrowth of non type 1 fimbriated *E. coli,* suggesting that the balance between urea concentration and putative high molecular weight constituents in urine may determine the proportion of the type 1 fimbriated phenotype in the *E. coli* population in the bladder of patients with UTI (Ofek et al., 1989). Another host factor affecting the adhesion of type 1-carrying *E. coli* may be estrogen. Vaginal and bladder epithelial cells obtained from estrogen-stimulated rats bound significantly more MS *E. coli* as compared to nonstimulated rats (Sobel and Kay, 1986). The enhanced adhesion to cells from estrogen-induced animals was not restricted to MS type 1 bearing *E. coli*. *E. coli* bearing other classes of adhesins also bound in higher numbers to cells from the estrogen-stimulated rats. The results may have relevance to the susceptibility of females to develop urinary tract infections with *E. coli* during menstruation.

Research has continued on the fine carbohydrate specificity and on the tissue distribution of soluble and cell-bound receptor molecules containing oligomannosaccharides specific for the type 1 fimbrial adhesin. Many types of mucosal cells including buccal and tracheal (Fader et al., 1988; Nogare, 1989); intestinal (Wold et al., 1988); cecal and colonic brush borders (Wadolkowski et al., 1988a,b); bladder mucosa (Yamamoto et al., 1990a); primary cultures of human uroepithelial cells (Hopkins et al., 1990); and vaginal, bladder, and urethral epithelial cells (Fujita et al., 1989; see also Table 11–5), all of which may be encountered by *E. coli* during natural courses of infections, were found to adhere with type 1 fimbriated *E. coli*. Other studies have found that type 1 fimbriae promoted adhesion of ETEC to enterocyte basolateral cell surfaces but not to brush border of human duodenal enterocytes (Knutton et al., 1985) and of EHEC to human and rabbit ileal brush borders and enterocytes (Durno et al., 1989). Interestingly, adhesion of type 1 fimbriated *E. coli* to rabbit intestinal cells was dependent on the age of the rabbits. Adhesion to ileal and colonic cells obtained from rabbits during the first week of life was 13–19% of that of cells from adult animals (Ashkenazi et al., 1991). If type 1 fimbrial adhesion is important in certain strains of *E. coli* known not to be able to cause infection during the first few months (e.g., hemorrhagic colitis caused by EHEC), then age-dependent adhesion mediated by type 1 fimbriae may explain susceptibility to infection. Urinary isolates of *E. coli* carrying type 1 fimbriae bound poorly to formalin-fixed human ileal mucosa (Yamamoto et al., 1990a). A special case is the finding that type 1 fimbriae, but not other adhesins (e.g., CFA/I and CFA/II), mediate adhesion of *E. coli* to M-cells in human ileal lymphoid follicle epithelium, which is responsible for initiating mucosal immunity (Yamamoto et al., 1990a). Examination of the binding of fluorescein-labeled and purified type 1 fimbriae to histologic sections of human kidney tissue revealed the presence of receptors for the fimbrial adhesins on connective tissue layers of blood vessel walls, on muscular layers, and in the cytoplasm and lumenal surfaces of proximal tubular cells, but not on epithelium of the distal tubulus, collecting ducts, glomeruli, and vascular endothelium (Vaisanen-Rhen et al., 1985; Virkola, 1987; Virkola et al., 1988; Korhonen et al., 1990). The results were interpreted to suggest that type 1 fimbrial adhesins do not confer upon the organisms the ability to invade renal parenchyma. In contrast, other studies have shown type 1 fimbrial mediated adhesion of *E. coli* to human and rabbit kidney sections (Smith, 1986a,b). The adhesion to the kidney tissue was inhibited by anti-type 1 fimbriae and by hypertonic salt or urea solutions. Moreover, the Manα-1,2Man dimannoside epitope, which may serve as a receptor for the type 1 fimbrial adhesin, was found on epithelial surfaces of the vagina, bladder, ureter, renal pelvis, and kidney, including the collecting tubules and loop of Henle cells, but not in the glomeruli or fibroblasts of the lamina propria of the urogenital tract (O'Hanley et al., 1985a). It should be noted, however, that not all strains expressing type 1 fimbriae can adhere to a particular animal cell. For example, not all type 1 fimbriated *E. coli* strains tested adhere to HEp-2 cells (Old et al., 1986) or cause hemagglutination of human erythrocytes (see Chapter 2). Differences in the surface hydrophobicity and ability to bind to intestinal cells (Sherman et al., 1985) and rabbit kidney tissue (Smith, 1986a) of type 1 fimbriated bacteria were noted and may explain the heterogeneity in the ability of type 1 fimbriated *E. coli* to bind to certain animal cells. Type 1 fimbriated bacteria also adhere to flat catheter materials, possibly due to surface hydrophobicity of the fimbriated bacteria. Coating of the surface with heparin sulfate prevents this type of adhesion (Ruggieri et al., 1987). Several important conclusions become evident from studies on adhesion of type 1 fimbriated *E. coli* to host cells. One conclusion is that adhesion mediated by type 1 fimbriae to tissue culture cells confers upon the organisms growth advantages and enhanced toxicity (Zafriri et al., 1987). Moreover, the enhanced toxicity of labile toxin secreted by *E. coli* adherent via type 1 fimbriae escapes neutralization by antitoxin antibodies (Ofek et al., 1990). Although these properties are shared by bacteria that adhere via other types of adhesins (Linder et al., 1988). They emphasize the notion that adhesion of *E. coli* mediated by type 1 fimbriae confers advantages similar to that mediated by other adhesins. Another aspect is related to the ability of type 1 fimbriated *E. coli* to adhere to mutant tissue culture cells that are defective in the synthesis of complex *N*-linked oligosaccharide units (Firon et al., 1985). The mutant that accumulated *N*-linked oligomannose units in its glycoproteins bound the largest numbers of type 1 fimbriated *E. coli*, suggesting oligomannose and hybrid units of cell surface glycoproteins serve as preferred receptors for mannose-specific

E. coli. The accessibility of mannooligosaccharide-containing glycoproteins on animal cell surfaces to function as receptors for binding type 1 fimbriated *E. coli* appears to be important for adhesion. This became evident from studies showing poor adhesion of type 1 fimbriated *E. coli* to HEp-2 cells in spite of the fact that these tissue culture cells bound type 1 fimbriated *Salmonella typhimurium* (Tavendal and Old, 1985). The presence of type 1 fimbriae, however, on the surface of *E. coli* markedly enhanced the adhesion of HEp-2 cells of organisms carrying so-called narrow-spectrum, mannose-resistant adhesin, suggesting a role for type 1 fimbriae in stabilizing *E. coli* adhesion mediated by another adhesin (Tavendale and Old, 1985). Fibronectin masks receptors that bind type 1 fimbriated *E. coli* on buccal epithelial cells. This masking effect may partially explain the lack of long-term colonization of the upper respiratory tract by type 1 fimbriated bacteria (Hasty and Simpson, 1987).

Two classes of mannooligosaccharide-containing molecules that interact with type 1 fimbriated *E. coli* are found in host tissues and fluids. One class of molecules includes glycoproteins found as integral components of animal cells that function as receptors that bind the fimbriated bacteria. The second class includes cell-free or mucous-associated glycoproteins. Two types of isoreceptors have been isolated from cell membranes and identified as oligomannose and hybrid *N*-linked units of glycoproteins. One type is a 65,000 mol wt glycoprotein isolated from guinea pig erythrocytes (Giampapa et al., 1988), and the other type includes glycoproteins belonging to the integrin superfamily CD11/CD18 (which are also known as leukocyte adhesion molecules) isolated from human polymorphonuclear leukocytes (Rodriguez-Ortega et al., 1987; Gbarah et al., 1991). Of interest is the finding that although the leukocyte cell membranes contain 10–12 concanavalin A-reactive oligomannose-containing glycoproteins, the mannose-specific type 1 fimbriae preferentially interact only with three glycoproteins belonging to the CD11/CD18 integrin superfamily. In contrast, type 1 fimbriated *E. coli* binds to numerous colonic and cecal brush border glycoproteins but no data are available as to which of these glycoproteins serve as receptor(s) that tether the organisms to the intestinal tract (Wadolkowski et al., 1988a,b). Purified carcinoembryonic antigen, which is normally localized at the apical border of epithelial cells of the large intestine and anchored to the cell membrane via phosphatidylinositol, binds type 1 fimbriated *E. coli* (Leusch et al., 1990, 1991). The binding is abolished by deglycosylation of the glycoprotein (Leusch et al., 1990) and inhibited by purified fimbrial lectin (Leusch et al., 1991), suggesting that it is mediated by mannooligosaccharide units of the glycoprotein. The relative concentrations of mannosides and aromatic mannosides needed to inhibit the binding of the bacteria and its mannose-specific fimbrial lectin to the purified glycoprotein antigen are similar to those needed to inhibit the binding of type 1 fimbriated *E. coli* to intestinal cells (Firon et al., 1987; Leusch et al., 1991). The results can be interpreted to suggest that the glycoprotein, known as carcinoembryonic antigen, may serve as a receptor on intestinal epithelial cells for binding the fimbriated *E. coli*. Another class of molecules that interacts with type 1 fimbriae are constituents of mucous layers. A number of glycoproteins with apparent mol wts of 50.5, 66, 73, and 94 kDa in colonic and cecal mucus of mice have been identified and found to bind type 1 fimbriated *E. coli* (Cohen et al., 1986; Wadolkowski et al., 1988b). A plasmid-cured, type 1 fimbriated derivative, which no longer produces the *E. coli* colicin, however, bound poorly to these mucous glycoproteins. Purified rat mucins bound only one of six type 1 fimbriated EHEC and none of other type 1 fimbriated *E. coli* isolates tested (Sajjan and Forstner, 1990a,b). The EHEC isolate (0157:H7, strain CL-49) that bound the purified mucin was significantly more hydrophobic compared to the other *E. coli* isolates tested. Hydrophobic effect mediators (e.g., *p*-nitrophenol) inhibited the mucin binding by the bacteria, suggesting that hydrophobic interactions stabilize the mannose-specific interaction between the CL-49 strain and rat mucin. The glycopeptide responsible for binding the type 1 fimbriated C1–49 strain was identified as a 118-kDa glycopeptide that contains *N*-linked manno-oligosaccharide side units. Human and rabbit small intestine mucins, pig gastric mucin, and human colonic mucin competitively inhibited the binding of type 1 fimbriated CL-49 to rat mucin, suggesting that the mucins contain similar mannooligosaccharide units specific for the fimbrial lectin. The fine carbohydrate specificity of the fimbrial lectin of strain CL-49 to the mucin is somewhat different from that reported for other *E. coli* strains (see Chapter 5). For example, compared to methyl-α-mannoside, *p*-nitrophenyl-α-mannoside was only seven times more potent as an inhibitor rather than the 30–50 times reported for other *E. coli* strains (see Chapter 5). Also, the dimannosides Manα-1,2-mannose and Manα-1,3-mannose were as effective as inhibitors as the *p*-nitrophenyl-α-mannoside, whereas in other strains they were poor inhibitors. Mucus-associated glycoproteins are not the only types of glycoproteins that interact specifically with type 1 fimbriated *E. coli* in the intestinal tract. The mannooligosaccharide units of secretory immunoglobulins, especially those of the IgA2 subclass, have been found to bind the mannose-specific fimbrial lectins and to inhibit the binding of the type 1 fimbriated *E. coli* to colonic epithelial cells (Wold et al., 1990). Several studies have focussed on Tamm–Horsfall glycoprotein, often referred to as uromucoid or urinary slime and previously shown to interact specifically with type 1 fimbriated *E. coli*. Because the glycoprotein is synthesized and secreted by epithelial cells of the ascending limb of Henle's loop and the distal convoluted tubule of the kidney into the urine, most of the studies have dealt with the possible effect of the glycoprotein–*E. coli* interaction on the pathogenesis of urinary tract infections and to a lesser extent with the specificity of this interaction. The glycoprotein probably interacts with the fimbrial lectin via its oligomannoside side chain because (1) binding of the glycoprotein to type 1 fimbriated bacteria is specifically inhibited by mannosides (Kuriyama and Silverblatt, 1986; Mobley et al., 1987; Reinhart et al., 1990b); (2) the glycoprotein competitively inhibited binding of type 1 fimbriated *E. coli* to immobilized synthetic dimannoside

(O'Hanley et al., 1985a); and (3) purified type 1 fimbriae bound to the immobilized glycoprotein and the binding was specifically inhibited by mannosides (Parkkinen et al., 1988). The concentration of Tamm–Horsfall glycoprotein secreted in urine varies widely during 24 hours, but it is not significantly different in women with recurrent urinary tract infections as compared to controls (Reinhart et al., 1990a). In contrast, the mean concentration of Tamm–Horsfall glycoprotein is significantly lower in infants less than 1 year old who have documented *E. coli* urinary tract infections compared to healthy age-matched controls (Israele et al., 1987). In the same study, however, the mean concentration of the urinary glycoprotein in children (age range 1–15 years) was significantly higher as compared to age-matched controls. Other studies have shown that although the glycoprotein interacts with exfoliated transitional cells in voided urine, cell-bound Tamm–Horsfall glycoprotein could not be detected in formalin-fixed biopsy specimens, suggesting that the interaction of the glycoprotein with the exfoliated cells occurs after their detachment (Fowler et al., 1987). At relatively high concentrations, the glycoprotein inhibited adhesion of type 1 fimbriated bacteria to a transitional carcinoma cell line (Duncan, 1988) and to primary epithelial cell culture of renal tubular origin (Hawthorn et al., 1991) as well as to polymorphonuclear leukocytes (Kuriyama and Silverblatt, 1986), whereas at low concentrations the glycoprotein enhanced adhesion only to the tissue culture cells (Duncan, 1988). It is possible, therefore, that Tamm–Horsfall glycoprotein may have a dual effect on the infectious process by acting as a nonspecific defense mechanism in inhibiting adhesion or it may act to predispose children to urinary tract infections by either inhibiting phagocytosis of the bacteria at high concentrations or promoting adhesion to bladder cells at low concentrations (Hawthorn et al., 1991). Depletion of Tamm–Horsfall glycoprotein from urine caused a considerable drop in the ability of the urine to inhibit *E. coli* adhesion mediated by type 1 fimbriae, suggesting that Tamm–Horsfall is the major glycoconjugate that interacts with type 1 fimbrial adhesin in urine (Sobota and Apicella, 1991). A relatively high concentration of calcium may modulate the ability of the Tamm–Horsfall glycoprotein to inhibit adhesion of type 1 fimbriated *E. coli* (Sobota and Apicella, 1991). Urine, however, contains additional inhibitors of *E. coli* type 1 fimbriae. These inhibitors include mannooligosaccharides consisting of three to five saccharides. These saccharides inhibit hemagglutination caused by type 1 fimbriated *E. coli* at a concentration of about one half of that of the saccharides present in pooled human urine (Parkkinen et al., 1988). The source of these short oligosaccharides in urine is not clear at the moment. *N*-linked mannooligosaccharide units derived from ribonuclease glycoproteins, coupled to lipids and separated on thin-layer chromatography, avidly bound the fimbriated *E. coli* (Rosenstein et al., 1988), whereas those saccharides derived from human IgG or transferrin did not bind to the fimbriated bacteria. A 60-kDa salivary mannooligosaccharide-containing glycoprotein inhibits specifically the adhesion of type 1 fimbriated *E. coli* to buccal cells (Babu et al., 1986). It should be noted that not all glycoproteins containing high-mannose type oligosaccharide units bind the fimbrial lectin. For example, immobilized ovalbumin binds poorly to type 1 fimbriated *E. coli* (Leusch et al., 1991), whereas short oligomannoside-type glycoaspargine derived from this glycoprotein strongly inhibited hemagglutination caused by type 1 fimbriated *E. coli* (Neeser et al., 1986). Studies focussing on fine carbohydrate specificity of the type 1 fimbrial lectin of *E. coli* confirmed earlier studies (see Chapter 5). First, the carbohydrate-combining site of this lectin preferentially recognizes unsubstituted α-1,3-linked mannosyl residues from *N*-linked carbohydrate chains. These structures are abundant in a wide variety of animal and plant glycoproteins (Nesser et al., 1986; Rosenstein et al., 1988). Second, the fimbrial lectin possesses a hydrophobic region in or close to its carbohydrate-combining site as evident from testing the inhibitory activity of additional α-1inked aromatic derivatives of mannose on adhesion of type 1 fimbriated *E. coli* (Firon et al., 1987). Particularly relevant are the findings that 4-methylumbelliferyl-α-mannoside was 1000-fold stronger than was methyl-α-mannoside in inhibiting the adhesion of *E. coli* to ileal epithelial cells and was more efficient than the methyl derivative in removing adherent bacteria from the epithelial cells (Firon et al., 1987). In contrast, fructose is a weak inhibitor, being 10 times weaker than methyl-α-mannoside, but the sugar is normally present in relatively high concentrations in most juices (about 10 times higher than that required for inhibition of type 1 fimbrial adhesion of *E. coli*) (Zafriri et al., 1989). It was postulated that the claimed beneficial effects of cranberry juice in preventing urinary tract infections caused by *E. coli* may be due to the presence in the juice of both fructose as an inhibitor of type 1 fimbrial adhesin and another inhibitor of MR adhesin (Ofek et al., 1991).

The gene cluster coding for the type 1 fimbrial adhesin in *E. coli* has been conserved throughout evolution, suggesting that the adhesin should confer an advantage to this bacterium, the habitat of which is invariably a higher animal species. It has been suggested that the type 1 adhesin might enhance killing of the organisms in deep tissues because it mediates phagocytosis of the organisms by animal and human phagocytes (see Chapter 7). The assessment of its contribution, however, to the infectious process initiated at the phagocyte-free mucosal surfaces is difficult in the face of the conflicting results cited earlier. For example, it is clear that some of the experimental infections suggest an essential role for the adhesin, whereas others do not. The cell-free compounds that bind the fimbrial adhesin may either promote adhesion or survival (e.g., intestinal mucus or uromucoid) or serve as a defense against infection (e.g., low molecular weight inhibitors in urine or S-IgA in the gut). There seems to be little doubt that the adhesin is expressed by the organisms at one or more stages of the infectious process. A plausible explanation for these controversial results may be that the expression of this fimbrial adhesin confers an advantage at a defined stage(s) of the infectious process and may be of a disadvantage at another stage. Moreover, this advantage may be superseded in infections initiated by relatively high doses or at a site unfavorable for the type 1 fimbriated phenotype. The search will undoubtedly continue to explore potent and specific inhibitors of the mannose-specific fimbrial adhesin (Neeser et al., 1986; Firon et al., 1987) to provide a knowledge-based approach to prevent or treat infections caused by *E. coli* (see also Chapters 5 and 11).

Escherichia coli

Clinical source:	ETEC
Adhesins:	CFA
Receptor specificities:	Sialoglycoconjugates
Substrata:	ERT, EC, MC, TB, TC

Four groups of *E. coli* causing intestinal infections have been described: enterotoxigenic *E. coli* (ETEC), enteropathogenic *E. coli* (EPEC), enteroinvasive *E. coli* (EIEC), and enterohemorrhagic *E. coli* (EHEC) (Levine, 1987; Levine et al., 1988). Each category contains strains belonging to certain O serotypes that cause diarrhea in humans by a distinct mechanism (Robins-Browne, 1987). It is not surprising, therefore, that the strains belonging to each of the various categories also exhibit distinct mechanisms of mannose-resistant adhesion to intestinal mucosa in vivo and to tissue culture cells in vitro. The CFA adhesins of ETEC probably have been studied the most. The adhesins characterized as CFA are expressed by ETEC diarrheal isolates that produce the heat-labile (LT) or heat-stable toxin. The CFAs can be distinguished antigenically using monoclonal and absorbed polyclonal antibodies. The most well characterized of the CFAs are CFA/I, CFA/II, and CFA/IV (also known as PCF8775 fimbriae). CFA I is a single fimbrial antigen containing several epitopes (Lopez-Vidal et al., 1988), whereas CFA/II and CFA/IV each consist of three distinct antigens known as *E. coli* surface (CS) antigens. CFA/II comprises the CS1, CS2, and CS3 fimbrial antigens (Lopez-Vidal and Svennerholm, 1990). CFA/IV comprises the fimbrial antigens CS4, CS5, and the nonfimbrial antigen CS6 (Thomas et al., 1985c). Epidemiological surveys revealed that most, if not all, *E. coli* isolates harboring plasmids encoding for the CFA family of adhesins also encode the diarrheagenic toxins (Gothefors et al., 1985; McConnell et al., 1985, 1988; Binsztein et al., 1991). However, isolates harboring plasmids encoding for one type or both types of enterotoxins devoid of CFA have been found in about 25% of ETEC (Danbara et al., 1987; Binsztein et al., 1991). This one-way association between toxin and CFA production was interpreted to suggest that this family of adhesins confers an advantage upon the ETEC to colonize preferentially the small intestine, whereas non-ETEC strains usually colonize the colon (Evans and Evans, 1990). These epidemiological surveys also confirmed earlier studies showing that the CFA/II isolates express either the CS3 antigen alone or CS3 and CS1 or CS2 but no isolates were found to express all three antigens (Binsztein et al., 1991). Similarly, some CFA/IV-positive isolates produced only CS6, whereas the other isolates in addition to CS6 produced either CS5 or CS4, but none of 19 CFA/IV-positive isolates tested produced all three CS antigens. In one study strains producing CS5 only were isolated from diarrheal patients (Manning et al., 1987). Functionally, the adhesive properties could be assessed by hemagglutination reactions of erythrocytes from different animal species. Confirming earlier studies (reviewed in Evans and Evans, 1990) it was found that the CFA/II-positive strains caused hemagglutination of bovine erythrocytes, and none of the isolates reacted with human or guinea pig erythrocytes, whereas the CFA/I-positive isolates caused hemagglutination of both human and bovine erythrocytes but not of guinea pig erythrocytes (Binsztein et al., 1991). In other studies, it was found that chicken erythrocytes also reacted with CFA/II isolates (Ahren et al., 1986). The hemagglutinating properties of both CFA/I- and CFA/II-carrying strains were retained on storage but rapidly lost after 5–10 subcultures (Ahren et al., 1986). It was suggested that this occurred due to the tendency of the ETEC strains to lose the CFA-encoding plasmid during subculture (Evans and Evans, 1990). The CS4- and CS6-carrying strains agglutinated human and bovine erythrocytes, whereas the CS5- and CS6-carrying strains agglutinated human, bovine, and guinea pig erythrocytes. One variant expressing CS6 alone did not cause hemagglutination (Thomas et al., 1985a). Because of the complexity of these CFA-CS adhesins, each with a distinct antigenicity and probably receptor specificity as well, it is difficult to evaluate their relative contributions to the infectivity.

The relative hydrophobicity of CFA/I-carrying strains was stronger than that of CFA/II strains (Gonzalez et al., 1988a; Binsztein et al., 1991), although both groups of strains were isolated from patients with diarrhea in about the same frequency (McConnell et al., 1985; Binsztein et al., 1991). In vitro, ETEC carrying either CS1 and CS3 or CS2 and CS3 or CS3 bound to the brush border of isolated human duodenal enterocytes, whereas a CFA/I-carrying ETEC showed poor adhesion (Knutton et al., 1985). Although both CFA/I- and CFA/II-bearing ETEC bound to formalin-fixed tissue biopsy specimens excised from the terminal ileum or colon or jejunum of adults and children, CFA/II-positive strains adhered considerably better than the CFA/I-positive ones (Yamamoto et al., 1991). The adhesion appears to be specific because the CFA-negative derivatives of the strains did not adhere at all and the CFA-positive strains adhered poorly to human urethral specimens or to porcine or rabbit intestinal specimens. Expression of flagella associated with motility by CFA/I and CFA/II appears to enhance the adhesion of the bacteria to formalin-fixed human ileum (Yamamoto et al., 1990a). Of several tissue culture cell lines tested (HEp-2, Hela, Vero, HRT 18, Hutu 80, MDBK, and MDCB) only the carcinoma cell line Caco-2 bound ETEC carrying CFA/I or CFA/II (CS1 and CS3 antigens) (Darfeuille-Michaud et al., 1990). The adhesion to Caco-2 cells was inhibited by purified adhesins and antiadhesin antibodies. The adhesion was observed only in >6 days of culture and was not observed with derivative strains cured of the plasmid coding for adhesins. The CFA/I- and CFA/II-carrying strains also adhered equally well to the brush border of human enterocytes. Electron micrographs showed that the adhesion of ETEC strains to brush border microvilli was not associated with any lesion (Darfeuille-Michaud et al., 1990). *E. coli* carrying CS4 and CS5 fimbrial adhesins bound to human intestinal enterocytes and cultured intestinal cells, whereas CS6 appeared to lack adhesive activity (Knutton et al., 1989b). Nevertheless, *E. coli* strains carrying CS6 alone colonized rabbit intestine as well as strains carrying CS4 and CS6 or CS5 and CS6, whereas mutants lacking the CS adhesins colonized the rabbit intestine poorly (Svennerholm et al., 1988). The possibility that strains carrying CS6 alone coexpress another unidentified adhesin has not been excluded (Knutton et al., 1989b). The identity of sialoglycoconjugates that act as receptors for CFA/I-carrying ETEC was confirmed in human erythrocytes as sialoglycoproteins with an apparent 26,000 mol wt. The receptors can be isolated from erythrocyte membranes by extraction with lithium diiodosalicylate (Pieroni et al., 1988). Attempts to identify membrane glycoproteins that serve as receptors for CFA-carrying organisms employed binding of either intact bacteria or purified CFAs to blots of electrophoretically separated cell membrane glycoproteins. Three brush border glycoproteins in HT-29 cells grown under conditions that promote their enterocytic differentiation were identified as possible receptors for binding ETEC carrying CS3 or CS1+CS3 adhesins (Nesser et al., 1988a). In other studies, it was found that purified CFA/I, CS2, and CS3 bound to two glycoproteins found in lipid-free cell membrane components of rabbit brush border and HT-29 cells but not in cell membrane components of other cell lines, consistent with the results obtained on adhesion of CFA-carrying *E. coli* described earlier (Wenneras et al., 1990). The attachment site on the glycoproteins involved in binding the ETEC adhesin may involve oligosaccharide side chains because the *Evonymus europaea* plant lectin, a GalNAc-specific lectin, inhibited the adhesion of the ETEC to the differentiated HT-29 cells and exhibited affinity for the same glycoproteins that bound the purified CS3 adhesin. Moreover, the CS3-mediated adhesion to HT-29 cells and hemagglutination caused by CS3-carrying ETEC was inhibited by the same types of complex carbohydrates and glycopeptides and was not affected by pretreating the target cells with sialidase (Nesser et al., 1988a). Hemagglutination was inhibited by lactosamine-containing oligosaccharides and by a glycopeptide mixture derived from human erythrocytes (Nesser et al., 1986, 1988a). In contrast, CS2 adhesin is a sialic acid-specific adhesin because binding of both the fimbrial structure and its subunits by bovine erythrocytes is specifically inhibited by sialyllactose (Sjoberg et al., 1988). Strains expressing CS2 and CS3 of the CFA/II adhesin complex adhere much better to enterocytes than strains expressing type 1 fimbriae or CFA/I fimbriae (Knutton et al., 1985). The expression or accessibility of receptors that bind the CFA adhesins requires differentiation. At this stage of research, it is not clear how valid are the models of adhesion to erythrocytes and tissue culture cell lines for adhesion to and colonization of human small intestine by ETEC carrying the

CFA adhesins. For example, the accessibility of the receptors for CFA/II adhesin on human erythrocyte membrane is unstable when whole blood is stored at 4°C. The accessibility is more common in erythrocytes from African-origin donors and is superior for CS1 and CS3 carrying ETEC as compared to ETEC bearing CS2 and CS3 or CS3 (Evans et al., 1988).

Attempts to examine the contribution of the CFA adhesins to the infectious process leading to diarrhea focussed on two main issues: production of the adhesin in vivo and efficacy of vaccine composed of the purified CFA adhesins. The in vitro production of CFA/I has been studied in detail and found to be regulated by the availability of iron in the medium. Low iron concentrations, likely to be the case in the small intestine, induced expression of CFA/I adhesin even in broth culture (Evans and Evans, 1990). Evidence for the production of the CFA adhesins in vivo is inferred from determination of the immune response to the adhesins during an infection. The majority of patients who have been examined develop serum IgG and IgA antibody as well as intestinal IgA to CFA/I and CFA/II adhesins (Stoll et al., 1986). Experimental infections employing the nonligated rabbit intestine model reveal that ETEC carrying CS3 alone or CS3 and CS1 as well as CFA/I adhesins are capable of inducing diarrhea and at the same time capable of conferring protection against reinfection (Svennerholm et al., 1990; Mynott et al., 1991). Passive immunization with monoclonal anti-CS1, -CS2, or -CS3 antibodies protected against experimental infection with ETEC carrying the corresponding CS adhesin, suggesting that all types of CFA/II adhesins are important in inducing diarrhea in the rabbit model. A vaccine composed of colchicine-killed CFA/I-carrying ETEC protected volunteers against challenge with the homologous strain and induced IgA intestinal anti-CFA/I adhesin antibodies (Evans and Evans, 1990). Because the vaccine also protected volunteers challenged with CFA/II-carrying ETEC, it was not clear whether adhesin-targeted vaccine alone can confer protection against ETEC infections. This is interesting in view of one study reporting that infection of rabbit ileal loop with nontoxigenic CFA/II-carrying *E. coli* caused fluid secretion and diarrhea for 72 hours postinfection (Wanke and Guerrant, 1987). In other studies using the intestinal adult rabbit model, it was concluded that optimum protection against diarrhea caused by ETEC-carrying CFA adhesins may be achieved by inducing both antiadhesin and antitoxin immunity (Ahren and Svennerholm, 1985). Because both the diarrheagenic toxin and the CFA adhesins are encoded by the same plasmid, it was possible to mobilize the CFA/I-encoding plasmid into a host mutant of *Salmonella typhi galE* strain Ty21a resulting in a live attenuated vaccine strain that produces abundant amounts of both CFA/I and heat-labile toxin (Yamamoto et al., 1985).

In a recent survey of diarrheal isolates of ETEC, it was found that about 25% of the strains were negative for CFA/I, II, and IV adhesins but caused mannose-resistant hemagglutination of erythrocytes from different species, suggesting that they express an as yet undefined adhesin or colonization factor (Binsztein et al.,

1991; Dominique et al., 1991). In another survey, it was found that ETEC strains isolated from children with diarrhea adhered to the brush border of the human intestine but did not produce any of the CFAs described above (Darfeuille-Michaud et al., 1987). It is likely, therefore, that new adhesins acting as colonization factors are produced among enterotoxigenic *E. coli*. Indeed, a number of new fimbrial and nonfimbrial adhesins that were proposed to function as colonization factors produced by ETEC diarrheal isolates have been described including: (1) CFA/III, fimbriae of 7 nm diameter composed of a major 18,000 mol wt polypeptide subunit. The fimbriae may be responsible for binding the organisms to brush borders of human enterocytes and the Caco-2 tissue cell line (Darfeuille-Michaud et al., 1990); (2) fimbriae 6–8 nm in diameter composed of a major 19,800 mol wt polypeptide subunit that is encoded by a plasmid and designated PCF 0159:H4 (Tacket et al., 1987); (3) curli fibers 3 nm in diameter observed only on nonhemagglutinating bacteria subcultured from human intestinal mucosa and responsible for the adhesion of the bacteria to human enterocytes (Knutton et al., 1987c); (4) 7.5-nm fimbriae composed of a major polypeptide of 17,500 mol wt and a 15,500 mol wt minor polypeptide encoded by a 100-MDa plasmid and responsible for hemagglutination of bovine erythrocytes and designated CS17 (McConnell et al., 1990); (5) a nonfimbrial antigen, designated 2230, that may be responsible for the observed adhesion of 2230 antigen-carrying ETEC to brush borders of human enterocytes and the Caco-2 tissue cell line (Darfeuille-Michaud et al., 1986, 1990); and (6) a nonfimbrial plasmid-encoded antigen that mediates adhesion of *E. coli* 8786 (isolated from a child with diarrhea) to a distinct receptor found on the Caco-2 cell line but not on HeLa and HEp-2 cultured cells (Aubel et al., 1991).

Escherichia coli

Clinical source:	ENT, EPEC, human
Adhesin:	F1845 Fim
Receptor specificity:	ND
Substrata:	EC, ERT, TC

EPEC bacteria are of particular interest, because based on the mode or pattern of their adhesion to enterocytes and tissue culture cells, they can be subdivided into three distinct subentities (Levine et al., 1988). Strains in one EPEC subentity adhere to Hep–2 or HeLa tissue cultures as discrete clusters of bacteria. The clusters or microcolonies of the bacteria are associated with the tissue culture cells in a patchy pattern termed localized adhesion (Scaletsky et al., 1985; Nataro et al., 1985) (Figure 2–3) (cooperative adhesion may lead to adherent patches of bacteria, Chapter 2). Although localized patterns of adhesion of *E. coli* to uroepithelial cells from patients with acute cystitis were observed (Reid, 1989), the EPEC strains exhibiting the localized adhesion belong to specific EPEC serogroups and are referred to as classic EPEC (Levine et al., 1988). The adhesion of *E. coli* of the second subentity of EPEC to HEp-2 cell monolayers also involves the clustering of the bacteria but the clusters are larger than those associated with localized adhesion and are found in between the cells, giving an aggregative pattern of adhesion (Nataro et al., 1987a; Vial et al., 1988) (Figure 2–3). The strains exhibiting the aggregative form of adhesion have been referred to as enteroadherent-aggregative *E. coli* (EAEC) (Vial et al., 1988). The mannose-resistant adhesion of strains belonging to the third entity of EPEC does not involve clustering or aggregation of the bacteria and has been termed diffuse adhesion (Figure 2–3) (Nataro et al., 1985; Scaletsky et al., 1985). Whereas the mode of adhesion of enteric isolates of *E. coli* to HEp-2 cells serves to identify strains associated with a particular entity of EPEC, the adhesion of the strains to enterocytes in vitro and in vivo as observed in ultrathin sections of adherent bacteria is quite different. This was first noted in biopsy specimens from infants with chronic diarrhea in which the EPEC were adherent to the enterocytes in regions with a disrupted microvillus membrane involving brush border effacement and the formation of pedestal and cup by the lumenal plasma membrane surrounding the attached bacteria. The pattern of intimate association with cup formation of EPEC strains with enterocytes has been termed attaching and effacing adhesion and has been observed in intestinal biopsy specimens from infants with diarrhea caused by EPEC as well as in cultured small intestinal mucosa exposed to EPEC in vitro (Knutton et al., 1987a,b). The attachment/effacement mode of adhesion is a characteristic of EPEC strains exhibiting a localized mode of adhesion to tissue culture cells (Knutton et al., 1987a,b). It has also been observed in intestinal biopsy specimens from infants with diarrhea caused by EPEC exhibiting the diffuse mode of adhesion to HEp-2 cells (Sherman et al., 1989a) and in the intestinal mucosa of rabbits infected with EPEC strains at sites of bacterial adhesion (Embaye et al., 1989). Local alternations of brush border membrane caused by adherent *E. coli* in small bowel biopsy specimens taken from patients with gastrointestinal pathology were also noted (Cerf et al., 1986). Studies employing HEp-2 or human embryonic lung (HEL) cells revealed that the attachment/effacement adhesion is associated with dense concentrations of microfilaments in the apical cytoplasm and actin condensation beneath the adherent bacteria (Knutton et al., 1989a). The concentration of cytoskeletal actin at the site of adhesion is characteristic for strains showing attachment/effacement modes of adhesion. The effacement lesion may represent a novel mechanism through which some EPEC strains cause membrane damage that leads to destruction of the enterocyte membrane and eventually to diarrhea as a direct consequence of adhesion. Many *E. coli* strains, however, isolated from normal individuals were shown to exhibit either localized (Mathewson et al., 1985, 1987; Escheverria et al., 1987a; Levine et al., 1988) or diffuse (Scaletsky et al., 1985; Mathewson et al., 1985, 1987; Echeverria et al., 1987a; Gomes et al., 1989) modes of adhesion to tissue culture cells, suggesting that in order to cause diarrhea the EPEC must possess a number of distinct virulence properties in addition to its ability to exhibit attaching and effacing adhesion to enterocytes (Levine et al., 1988). EPEC strains exhibiting attachment/effacement modes of adhesion confined to the distal halves of the small intestine of infected piglets do not cause diarrhea (Tzipori et al., 1989). In contrast, a number of cases of diarrhea in calves, pigs, lambs, and dogs are associated with infections with *E. coli* that exhibit attachment/effacement lesions (Janke et al., 1989). In humans, the attachment/effacement mode of adhesion to intestinal cells is probably not the only mode of adhesion required for the EPEC strains to cause diarrhea (Sherman et al., 1989a).

There was no association between the production of cytotoxin and the ability of the EPEC strains to exhibit either localized or diffuse modes of adhesion to tissue culture cells (Escheverria et al., 1987b; Karch et al., 1987a). Invasion into the target cells may be another virulence property that follows the adhesion of EPEC to target cells. EPEC strains were found to invade into Henle 407 cells (Miliotis et al., 1989) and HEp-2 cells (Donnenberg et al., 1989). The EPEC strain invading HEp-2 cells hybridizes with a probe specific for strains exhibiting localized adhesion, it exhibits localized adhesion, and it causes attachment/effacement associated with polymerization of actin. Analysis of mutants obtained by transposon insertion revealed that multiple genetic loci in addition to those encoding for adhesion to and induction of actin polymerization in epithelial cells are required by EPEC to invade the epithelial cells (Donnenberg et al., 1990). It appears, therefore, that the unique patterns of adhesion of the various entities of EPEC strains are mediated by factors the expression of which is not coordinated with any other virulence property. The expression of these adhesion factors, however, seems to be a prerequisite, but not the sole determinant, for the EPEC strain to cause diarrhea. Efforts to identify the molecular species responsible for a particular mode of adhesion of EPEC to HEp-2 or Henle cells revealed that they are encoded by a plasmid, as is the case with many other enteric adhesins (Table 9–1). A 60-MDa plasmid encodes for the property of localized pattern of adhesion to HEp-2 cells (Nataro et al., 1985; Levine et al., 1988; Baldini et al., 1986; Wu and Peng, 1991). A probe of a 1-kilobase fragment of DNA from the plasmid hybridized with DNA of the vast majority of strains exhibiting a localized pattern of adhesion (Baldini et al., 1986; Levine et al., 1988; Forestier et al., 1989; Gomes et al., 1989). EPEC isolated from outbreaks of neonatal diarrhea and exhibiting localized adhesion also cause diarrhea in rabbits. Strains cured from their 60-MDa plasmid lost all these activities (Wu and Peng, 1991). Strains with a particular O serotype showing positive localized adhesion and hybridizing with the plasmid DNA probe are recovered in much higher proportions from patients with diarrhea as compared to controls (Scaletsky et al., 1985; Echeverria et al., 1987a,b; Levine et al., 1988; Gomes et al., 1989; Kaper et al., 1985). It has been cautioned that minor modifications in the methodology may result in difficulties in determining microscopically the pattern of adhesion of EPEC strains to tissue culture cells (Levine et al., 1988; Vial et al., 1988;

Mathewson and Cravioto, 1989). Thus any relationship between a specific pattern of adhesion with the ability to cause disease, or to a specific molecular species and DNA elements, as being involved in the adhesion remains to be resolved. For example, of 23 EPEC strains exhibiting adhesion to HEp-2 cells associated with actin polymerization (as assayed by the fluorescein-actin staining test), only 10 hybridized with the EAF DNA probe and showed localized adhesion. All the strains adhered equally well to human small intestinal mucosa (Knutton et al., 1991). *E. coli* isolated from patients with ulcerative colitis in relapses or remission or with Crohn's disease exhibit mannose-resistant hemagglutination and differ from fecal *E. coli* isolates from healthy individuals because they adhered to homologous or heterologous buccal cells or to HeLa cells considerably better and exhibit more hydrophobic surfaces. It is not clear how these *E. coli* isolates should be classified (Burke and Axon, 1986, 1987a,b; 1988).

Attempts to identify the factor(s) responsible for localized and attachment/effacement adhesion patterns employed antibodies absorbed with strains cured from the plasmid encoding for EPEC adhesion. Volunteers infected with EPEC strains exhibiting localized adhesion developed antibodies that identified a 94,000 mol wt outer membrane protein (Levine et al., 1988), but the antibodies did not inhibit localized adhesion of the challenge strain (Chart et al., 1988). Localized adhesion of EPEC strains to the HeLa cell line was inhibited by relatively high concentrations of GalNAc (0.1 *M*) but not by other monosaccharides (Scaletsky et al., 1988). Absorbed antibodies that inhibited localized adhesion reacted with two components (32,000 and 29,000 mol wt proteins) of the outer membrane, one of which, the 32,000 mol wt polypeptide, was predominant. It was suggested that it may be the EPEC adhesion factor that mediates localized adhesion (Scaletsky et al., 1988). The genes encoding for localized adhesion and attachment/effacement to rabbit enterocytes of a strain isolated from a child with gastroenteritis were found to be encoded by a plasmid that did not hybridize with the probe from the 60-MDa plasmid (Fletcher et al., 1990). Localized adhesion and hemagglutination of two diarrheal isolates of EPEC were found to be encoded by 55–57 MDa plasmids (Pal and Ghose, 1990). Absorbed antisera identified two proteins with subunit wts of 18,000 and 14,500 that were present on the surfaces of the two strains and inhibited localized adhesion of the strains to HeLa cells. Localized adhesion was reduced by pretreating the bacteria with pronase and heat (100°C, 10 min). A chromosomal gene, *eae,* necessary for the attachment/effacement mode of adhesion to tissue culture cells in EPEC was characterized. A DNA probe for the gene is now available to screen for EPEC strains (Jerse et al., 1990). The *eae* gene encodes for a 94,999 mol wt outer membrane protein. The presence of the native plasmid encoding for the localized pattern of adhesion caused an increase in the amount of Eae protein, suggesting that the plasmid has a regulatory role for expression of the attachment/effacement mode of adhesion (Jerse et al., 1991; Jerse and Kaper, 1991). Eae protein is probably the 94,000 mol wt protein identified in sera of individuals infected with Eae-positive strains (Levine et al., 1988), suggesting that it is produced during infection. The ability of the strains to exhibit attaching/effacing lesions is dependent on the presence of both plasmid encoding gene(s) and a chromosomal *eae* gene (Donnenberg and Kaper, 1991; Francis et al., 1991). The plasmid gene encodes for an adhesion factor, EAF (Nataro et al., 1985, 1987a,b; Knutton et al., 1987a,b). Studies employing isogenic strains deficient in either EAF or *eae* gene revealed that mutants expressing only EAF protein adhered efficiently to HEp-2 cells, but the presence of both EAF and *eae* was required for efficient effacement of the cultured cells (Francis et al., 1991). Other studies have shown that attachment/effacement lesions develop in the intestine of piglets challenged with either *E. coli* strains isolated from humans with diarrhea or with the plasmid-free derivatives of the strains. This is in spite of reduced adhesion of the plasmid-free derivatives to HEp-2 or 407 intestinal cells (Hall et al., 1990). These studies are consistent with the hypothesis that the attachment/effacement mode of adhesion to intestinal cells may be the result of two stages of adhesion, each mediated by a distinct adhesin complex (Knutton et al., 1987a,b). The first stage includes adhesion to the target cells and is possibly mediated by the fimbrial adhesin complex EAF encoded by a plasmid (EPEC adhesion factor). The second stage is characterized by effacement, actin aggregation, and pedestal formation and requires an outer membrane protein encoded by the chromosomal *eae* gene. The plasmid-encoded adhesin involved in the first stage of adhesion is probably not confined to a particular adhesin because it could be substituted by type 1 fimbrial adhesin or an afimbrial adhesin (Francis et al., 1991). It has been suggested that in the course of infection with wild-type EPEC, the plasmid-encoded EAF adhesin provides tissue specificity for human epithelial cells (Jerse et al., 1991). In assays using tissue culture cells the EAF adhesin can be substituted by adhesins for which receptors are available on the target tissue (Cantey and Moseley, 1991; Francis et al., 1991). The findings show that the development of diarrhea in volunteers infected with strains exhibiting a localized mode of adhesion is dependent on the presence of a plasmid (Levine et al., 1988). The occurrence of strains exhibiting localized adhesion in specimens from healthy individuals supports the notion that strains lacking the factor(s) responsible for either the first stage or the second stage of adhesion to enterocytes are not capable of causing diarrhea. The localized pattern of adhesion may be induced by tissue-derived factors (Vuopio-Varkila and Schoolnik, 1991). The localized adhesion to HEp-2 cells of EPEC strains is associated with de novo synthesis of a 18,500 mol wt polypeptide. The pattern of localized adhesion also implies the adherent organisms are bound adjacent to each other to form a microcolony of adherent bacteria. Indeed, a surface protein complex termed the bundle-forming pilus has been detected in *E. coli* exhibiting localized adhesion (Giron et al., 1991b). The fimbrial protein has a mol wt of 19,500, is encoded by 92-kb plasmid, may be similar to the adhesion-inducible 18,500 mol wt protein described earlier, and may bind individual bacteria together. Its amino-terminal region is similar to that of TCP fimbrial protein of *Vibrio cholera* but it does not react with anti-TCP antisera and a *tcp* DNA probe does not hybridize with DNA prepared from two EPEC strains exhibiting localized adhesion. The most interesting features of this bundle-forming protein (or protein complex) are that its expression is not consecutive and greatly dependent on growth of the organisms on tissue culture cells or on blood agar and it causes hemagglutination of human type O erythrocytes. Thus, it is possible that the bundle-forming protein has a lectin-like activity specific for carbohydrates expressed by EPEC exhibiting localized adhesion. The protein is probably not required for the attachment/effacement lesion but may confer growth advantages under certain specific growth conditions that induce its formation. EPEC strains negative for the EAF probe, but showing localized adhesion associated with typical effacing lesions and accumulation of actin in cultured HEp-2 cells beneath the adherent bacteria, have been isolated from infants with diarrhea, suggesting that plasmid-encoded proteins other than EAF are capable of conferring localized adhesion to EPEC (Scotland et al., 1991a). In attempts to develop a simple test for detecting EPEC strains that exhibit localized adhesion, it has been found that antisera raised against classic EPEC strains and absorbed with its plasmid-cured nonadhering derivative are capable of detecting in enzyme-linked immunosorbent assay (ELISA) all the EPEC strains exhibiting localized adhesion (100% sensitivity). None of the *E. coli* strains, including EPEC strains (100% specificity) that did not exhibit localized adhesion, were unreactive with the antibody probe (Albert et al., 1991).

E. coli exhibiting aggregative patterns of adhesion (EAEC) cause intestinal tissue damage in rabbits and ileal loops, similar to the damage associated with "neurotoxic" syndrome in rabbits infected with live organisms, suggesting involvement of toxin(s) (Vial et al., 1988). EAEC isolates from sporadic cases of diarrhea in children belong to classic EPEC serotypes, hybridize with specific EAEC DNA probes, and cause MR hemagglutination of rat erythrocytes (Baudry et al., 1990; Scotland et al., 1991b). In one study, it was found that the EAEC were isolated at relatively high frequencies from children with persistent diarrhea compared to controls or children with acute diarrhea (Wanke et al., 1991). An *E. coli* diarrheal isolate exhibiting aggregative patterns of adhesion to HeLa cells was investigated for its ability to adhere to various tissue cells (Yamamoto et al., 1991). The strain adheres to colonic, ileal, and urethral epithelial cells as well as to M-cells, but adhesion to jejunal or ileal villi is poor. The molecular mechanisms responsible for the aggregative pattern of adhesion to HEp-2 monolayers are not clear, but they may involve adhesins that bind soluble components released by the tissue culture cells. Mobilization of a 55–65 MDa plasmid from a wild-type EPEC strain exhibiting nonaggregative modes of adhesion was accompanied by the acquisition of smooth lipopolysaccharide, fimbriae, and the property of the aggregative mode of adhesion (Vial et al., 1988). A putative adhesin associated with fine fibrillar appendages may mediate adhesion to erythrocytes and HEp-2 cells (Travendale and Old, 1985).

The diffuse pattern of adhesion lacks any special features and is essentially indistinguishable from a regular pattern of adhesion. It is recognized only in the context of the source of isolation of the strain, e.g., adhesin(s) expressed by EPEC the adhesion of which to tissue culture cell lines is not of the localized or aggregative pattern. Screening of isolates from childhood diarrhea patients has revealed that more than one-half of the strains show the diffuse type of adhesion not mediated by type 1 fimbriae (Giron et al., 1991a). It is possible, therefore, that EPEC showing the diffuse type of adhesion, also termed DAEC, includes a heterogeneous group of strains each expressing an adhesin with a distinct receptor specificity. Two adhesins expressed by such strains have been described. One is the F1845 which is encoded by a plasmid and mediates adhesion to HEp-2 cells (Bilge et al., 1989). Preliminary studies show that this fimbrial adhesin also mediates adhesion to cecal and colonic epithelial cells in piglets. The receptor on erythrocytes that binds the F1845 adhesin is the Dr group antigen which is expressed also by cells of the intestinal tract (Nowicki et al., 1988a, 1990). F1845 expressing E. coli adhere to the differentiated human colon cell lines HT-29 and Caco-2 (Kerneis et al., 1991). Purified F1845 adhesin complex binds preferentially to apical domains of the cell lines. About one-third of the strains isolated from children with diarrhea hybridized with a DNA probe derived from the chromosome of a F1845-carrying strain (Giron et al., 1991a). The other plasmid-encoded adhesin, designated AIDA-I, is carried by EPEC isolated from infantile diarrhea patients and mediates adhesion of the bacteria to HeLa cells (Benz and Schmidt, 1989). The adhesin is nonfimbrial and may be an outer membrane protein, but its ability to mediate adhesion to intestinal cells has not been determined. In addition to the above putative adhesins of *E. coli* causing intestinal infections, it appears that the adhesion to murine intestinal cells is markedly enhanced by introducing the colicin V plasmid in recipient strains (Darken and Savage, 1987). The mechanism of such an effect is not clear, but seems to involve genes coding for the conjugal transfer function of some colicin V plasmids which often are found in intestinal isolates of *E. coli*.

Escherichia coli

Clinical source:	EIEC
Adhesins:	469–3 Fim
Receptor specificities:	ND
Substrata:	EC, ERT, TC

The 469–3 fimbrillar adhesin expressed by EIEC mediates the binding of the bacteria to human epithelial cell lines and to brush borders of human colon and is responsible for the mannose-resistant hemagglutination caused by the bacteria (Hinson et al., 1987). No binding of 469-3-carrying bacteria was noted to duodenal enterocytes [described separately (see Chapter 9) but it seems that at least

one more EIEC isolate (strain 444-3) expresses a closely related adhesin as evident from the Southern blot analysis of total DNA.]

Escherichia coli

Clinical source:	EHEC
Adhesins:	ND
Receptor specificities:	ND
Substrata:	ERT

At least two different MR adhesins that lack hemagglutinating activity are expressed by the EHEC of serotype 0157:H7. One class of adhesin seems to be fimbrial, containing a major fimbrial subunit of 16,000 mol wt, encoded by a plasmid and mediating adhesion of the organisms to Henle 407 intestinal cells and mucins, but not to HEp-2 cells or human animal erythrocytes (Karch et al., 1987b; Sajjan et al., 1990a,b). The other adhesin is nonfimbrial, encoded by a plasmid and laboratory strain of *E. coli* transformed with the plasmid, and exhibits weak adhesion associated with formation of cup-like structures and possible actin polymerization at sites of bacterial adhesion, reminiscent of attachment/effacement described for EPEC earlier (Toth et al., 1990). EHEC belonging to serotype 026 isolated from infants with diarrhea is known to exhibit localized adhesion to HEp-2 cells associated with actin polymerization and hybridizes with a DNA probe derived from the plasmid of an EHEC strain belonging to the 0157 serotype (Scotland et al., 1990). In other studies, nonfimbriated EHEC strains belonging to the 0157 serotype were found to adhere to HEp-2 cells, lack hemagglutinating activity, and exhibit low surface hydrophobicity (Sherman et al., 1987). The adhesion was inhibited by an outer membrane extract or antibodies raised against outer membrane preparations, but not by lipopolysaccharide antigens or isolated flagella or antiflagellar antiserum, suggesting that constituents of the outer membrane of the organisms mediate binding of EHEC 0157:H7 to HEp-2 cells (Sherman and Soni, 1988; Sherman et al., 1991). Antisera against the proteins, especially anti–94,000 mol wt outer membrane protein, inhibited adhesion and protected fluid accumulation in rabbit loops challenged with *E. coli* 0157:H7. The attachment/effacement mode of adhesion of EHEC to intestinal cells was observed in gnotobiotic piglets infected with EHEC (Francis et al., 1986; Tzipori et al., 1986, 1989; Sherman and Soni, 1988). In vitro, it is difficult to observe a specific pattern of adhesion of EHEC to tissue culture cells. Mobilizing a plasmid encoding for F1845 adhesin of an EPEC strain into EHEC strains confer upon the recipients the ability to exhibit a diffuse pattern of adhesion associated with aggregation of cell actin (Cantey and Moseley, 1991). These observations have been interpreted to support the notion postulated for EPEC strains that adhesion followed by actin aggregation and invasion into the target cells of EPEC involves a two-stage adhesion process. The initial adhesion stage may not be confined to a specific adhesin, which acts merely to enable close contact of the organisms to the target cell and thus to permit the induction in the adherent bacteria of products required for the subsequent stage of adhesion.

Escherichia coli

Clinical source:	ETEC, piglets
Adhesin:	K88 Fim
Receptor specificity:	ND
Substrata:	EC, ERT, MC

The K88 fimbrial adhesin carried by ETEC isolated from piglets consists of a family of three distinct antigenic subtypes, designated b, c, and d, all of which are recognized by monoclonal antibodies but share a common determinant designated a (Thorns et al., 1987; Thorns and Roeder, 1988; van Zijderveld et al., 1990). It is possible that the receptor specificity is different for each subtype of K88 fimbrial adhesin because intestinal cells from some pigs bind organisms bearing either one of the three fimbrial subtypes. Cells from some pigs bind organisms bearing one or two antigenic variants, whereas cells from others bind none of the three variants (Rapacz and Hasler-Rapacz, 1986; Moon, 1990). Pigs in which intestinal cells bind the appropriate K88 fimbrial variant are also susceptible to infection initiated by that variant (Sarmiento et al., 1988), confirming earlier studies and suggesting that K88-mediated adhesion plays a major role in infection (reviewed in Moon, 1990). Availability of K88 receptors on small intestinal epithelial cell-brush borders appears, as compared to other cell types, to be the best indicator of the susceptible phenotype of a particular strain of pigs (e.g., large white pigs), but not of other strains of pigs (e.g., Chinese Meishan pigs) (Valpotic et al., 1989; Bertin and Duchet-Suchaux, 1991). This provides the best example of natural immunity to infection based on availability of accessibility on the infected tissue for receptors specific for the adhesin carried by the bacterial pathogen. Availability of receptors for the K88 fimbrial adhesins, however, is not restricted to the pig intestinal brush border because *E. coli* bearing K88 fimbrial adhesins bind to a variety of epithelial cells including intestinal villi, buccal cells, and erythrocytes of weaned piglets (Cox and Houvenaghel, 1987) as well as to mouse intestinal cells (Laux et al., 1986). These results emphasize an earlier notion that adhesion per se is prerequisite, but not sufficient, for colonization because colonization of small intestine by K88-bearing *E. coli* is restricted to pigs. It is not surprising, therefore, that one of the best and first adhesin-based vaccines is the K88 fimbrial vaccine given to dams in order to induce anti-K88 antibodies in the colostrum and thus to protect passively the weaning offspring against diarrhea (Table 5–12). The fimbrial vaccine is now commercially available through recombinant DNA technology (Greenwood et al., 1988). The success of this vaccine is evident from epidemiological observations showing that the frequency of isolation of K88-positive strains (0149) from piglets and weaned pigs has been reduced over the last 20 years in Sweden (Soderlind et al., 1988). Because the diarrhea caused by the K88-bearing organisms is caused by the heat-labile and heat-resistant enterotoxins, studies were performed to examine the plasmid encoding for the adhesins and enterotoxins (Wasteson and Olsvik, 1991). The plasmids of five out of nine K88-bearing strains contained genes encoding for both the adhesin and the labile enterotoxin, whereas none of the heat-stable encoding plasmids contained fimbrial genes. The results suggest that in ETEC of farm animals there is no common environmental pressure for selecting the K88 fimbrial genes and the enterotoxin genes as is the case for CFA-carrying ETEC (see above). In addition to intestinal cells, the K88 adhesins also mediate binding of the organisms to the porcine ileal mucus (Conway et al., 1990). The K88 fimbrial receptors in porcine mucus are sensitive to periodate or proteolytic enzymes, and are specifically removed from the mucus by purified K88 fimbrial preparations which appear to complex with a 40,000–42,000 mol wt glycoprotein in the mucus (Metcalfe et al., 1991). Relatively low or high densities of the receptors are present in the ileal mucus of newly born piglets or 35-day-old piglets, respectively. It was postulated that low amounts of the receptor may facilitate adhesion and render the newborn piglets susceptible to infection, whereas large amounts of the K88 fimbrial receptors in the mucus may actually bind the fimbriae and inhibit the adhesion of the K88-carrying *E. coli* to the intestinal wall, thereby preventing the development of infection (Conway et al., 1990).

Escherichia coli

Clinical source:	ETEC, piglets
Adhesin:	987 Fim
Receptor specificity:	ND
Substrata:	EC, ERT

ETEC strains carrying 987P fimbrial adhesin do not cause hemagglutination of erythrocytes of a number of species tested (Moon, 1990) but they do adhere to brush borders of the intestinal epithelium (Dean et al., 1989). A number of glycoproteins that bind the 987P fimbrial adhesin have been identified in the small intestine of neonatal and older pigs as well as of adult rabbits (Dean and Isaacson, 1985a,b). Antibodies against the receptor from rabbits react with the small-intestine epithelium of both neonatal pigs and older pigs. Whereas the expression of the adhesion in vitro is under phase variation (see Chapter 9), the 987P fimbriated phenotype is selected during growth in the porcine small intestine (Broes et al., 1988), even in pigs inoculated with poorly fimbriated bacteria (Dean et al., 1989). The data collectively suggest that the 987P fimbrial adhesin is important for infectivity. Intestinal infections and diarrhea caused by 987P-bearing ETEC is confined to neonatal pigs, although no differences have been noticed in the availability of 987P fimbrial receptors on intestinal cells or in the rate of growth or the expression of the fimbrial adhesin during intestinal growth between neonatal and older pigs. Other studies have shown that *E. coli* grown to express the 987P adhesin complex failed to adhere to intestinal cells from two strains of adult pigs (Bertin and Duchet-Suchaux, 1991). It has been shown among the various cell-associated glycoproteins that those binding the purified 987P fimbrial adhesin become more readily soluble and trapped in the mucus of older pigs only (Dean, 1990). It has been postulated that a relatively high concentration of such soluble 987P-binding glycoproteins may inhibit the ability of the organisms to bind to and colonize the intestinal tissue (Dean, 1990). This notion is another example of how modulation of adhesion by soluble host constituents can affect age-related resistance to infection.

Escherichia coli

Clinical source:	ETEC, piglets
Adhesins:	F42 Fim
Receptor specificity:	ND
Substrata:	EC, ERT

Enterotoxigenic strains isolated from piglets with diarrhea express a fimbrial adhesin designated as F42 that is antigenically distinct for the K88, K99, and 987P adhesins (Yano et al., 1986). The strains cause mannose-resistant hemagglutination of erythrocytes from humans, guinea pigs, horses, sheep, and chickens, whereas K88-carrying *E. coli* do not react with any of the erythrocytes and K99- and F41-carrying *E. coli* react with erythrocytes of all species, except those from the chicken. The purified F42 fimbriae cause hemagglutination with a pattern similar to that caused by whole bacteria. F42-carrying *E. coli* also adhere to tissue culture cells and to the pig brush border. Anti-F42 sera specifically inhibit hemagglutination and adhesion to HeLa cells and brush border. The results suggest that the receptor specificity of the F42 fimbrial adhesin is distinct from that of fimbrial adhesins expressed by other enterotoxigenic strains. The growth conditions required for the expression of the F42 adhesin, however, are similar to those of other enterotoxigenic *E. coli* in that the adhesin is expressed at 37°C, but not at 18°C, after growth in agar. No information is available as to the frequency of isolation of F42-carrying strains.

Escherichia coli

Clinical source:	ETEC, piglets, calves
Adhesin:	F99 Fim
Receptor specificity:	NeuGcGal
Substrata:	EC, ERT, MC

Early work established that the K99 fimbrial adhesin is the major adhesin that enables the ETEC to colonize and cause diarrhea in neonatal pigs and calves (reviewed in Moon, 1990). It now appears, however, that many K99-carrying isolates frequently produce in addition to the type 1 fimbrial adhesin, other adhesins, such as F41 and/or F17 (FY), each with a distinct receptor specificity (see below). Although these non-K99 adhesins are important for infectivity, the contribution of adhesion mediated by the K99 fimbrial adhesin of ETEC isolates to the infectious process is probably indispensable. This conclusion is based on several observations. First, the restriction of infections caused by K99-bearing ETEC to the neonatal period may be partially due to the lack of K99-specific receptors that bind the organisms to intestinal cells of adult pigs, whereas intestinal cells of piglets contain the NeuGc-Gal-Glc-Cer glycolipid which can bind the K99 adhesin (Teneberg et al., 1990). Second, the passive lacteal immunity provided by the K99 fimbrial vaccine confers protection to suckling piglets and calves against ETEC strains carrying K99 and F41 (and type 1 fimbriae) (Moon et al., 1988). In contrast, F41 vaccine did not confer protection against strains that produce both K99 and F41 fimbrial adhesins (Runnels et al., 1987). Third, loss of K99 genes during the course of infection leads to reduced colonization and the prevalence of such emerging K99-F41^{+} variants in naturally occurring disease is low (Mainil et al., 1987). Attempts to understand the unique and apparently indispensable advantage(s) conferred by the K99 fimbrial adhesin in ETEC strains have focussed on the interaction of the microorganisms with mucous constituents (Mouricout and Julien, 1987). The binding of *E. coli* carrying K99 alone to an immobilized glycoprotein fraction obtained from calf mucus was three to four times that of *E. coli* carrying F41 or F17, whereas binding of the nonfimbriated bacteria was very low to none. The binding isotherm to the mucous glycoproteins by *E. coli* expressing K99 or K99 and F41 or K99, F41, and F17 adhesins was nonlinear, characterized by a typical bell-shaped curve, whereas that of F41- or F17-carrying *E. coli* was either concave or linear, respectively. Hill plot analyses of the binding revealed that the K99-carrying strains showed positive cooperativity, whereas the F41-carrying strains showed negative cooperativity and the F17(FY)-carrying strain showed no cooperativity with one type of binding site (Chapter 2 for a discussion of cooperativity). The unique binding isotherm of K99-carrying *E. coli* may be due to the distinct attachment site of the mucous glycoproteins responsible for binding the bacteria. Intact sialic acids and galactose in the mucous glycoproteins were required for optimal binding of K99-carrying *E. coli,* whereas desialylated glycoproteins preferentially bind the F41-carrying *E. coli*. The relative importance of the K99 fimbrial adhesin and the efficacy of colostrum anti-K99 antibodies in protecting piglets and calves against diarrhea caused by K99-carrying ETEC stimulated research in identifying oligosaccharides that function as K99 receptor analogues. The goal was to prevent infection caused by K99-bearing ETEC by inhibiting the K99-mediated adhesion of the organisms to intestinal cells or mucous constituents.

Purified K99 fimbriae and whole bacteria reacts with *N*-glycolylneuraminyllactosyl ceramide (NeuGclCer) purified from equine erythrocytes. *N*-Glycolyl and NeuGc groups are essential for recognition, suggesting that the receptor for the

K99 adhesin on erythrocytes is a glycolilpid (Ono et al., 1989; see also Chapter 5). The attachment site specific for the K99 fimbrial adhesin in mucous glycoproteins constitutes complex carbohydrate structures, because simple sugars (i.e., sialic acids, galactose, *N*-acetylgalactosamine, glucose, and glucosamine) present in mucin oligosaccharide side chains do not inhibit the adhesion of the K99-carrying microorganisms to immobilized glycoproteins (Mouricout and Julien, 1987). In contrast, glycopeptides isolated from glycoproteins present in bovine plasma inhibit the adhesion of the organisms to both immobilized mucous glycoprotein (Mouricout and Julien, 1987) and to sheep erythrocytes (Mouricout and Julien, 1986). Furthermore, the sialylated glycopeptides bearing the terminal NeuGc (α2,3)Gal sequence strongly inhibited hemagglutination caused by K99^{+} *E. coli*. Oral administration of the plasma-derived glycopeptides to colostrum-deprived newborn calves challenged with K99-carrying ETEC caused a marked reduction of intestinal colonization by the organisms and protected the calves against lethal doses of the challenge strain (Mouricout et al., 1990). In other studies, it has been shown that there was no correlation between the ability of the K99-bearing *E. coli* to adhere to enterocytes and the susceptibility of pigs to infection caused by the K99 *E. coli* (Bertin and Duchett-Suchaux, 1991). It now appears there exist two phenotypes of piglets that can be distinguished by the ability of their enterocytes to bind K99-carrying *E. coli* and the type sialoglycolipids predominating in the cell membrane of their enterocytes (Seignole et al., 1991). The enterocyte of one phenotype binds the K99-expressing bacteria and contains relatively high amounts of NeuGc(α2,3)Gal-containing monosialoglycolipids, which serve as receptors for binding the K99 fimbrial adhesin complex, whereas the enterocytes of the other phenotype of piglets do not bind the K99-bearing *E. coli* and are poor in glycolipids that bind the K99 adhesin. The susceptibility of the two phenotypes to diarrhea caused by K99-bearing *E. coli* remains to be investigated.

Escherichia coli

Clinical source:	Piglet, calves
Adhesin:	F41 Fim
Receptor specificity:	Glycophorin
Substrata:	EC, ERT, MC

Strains carrying only the F41 adhesin bind to intestinal cells from pigs (Bertin and Duchet-Suchaux, 1991). The receptor specificity of F41 fimbrial adhesin is distinct from that of the K99 adhesin, although both adhesins are produced by clones causing diarrhea in newborn piglets and calves. This distinct receptor specificity is apparent from experiments showing that the interaction of F41-carrying *E. coli* with erythrocytes is specifically inhibited by anti-F41 antibodies or GalNAc (Yano et al., 1986; Wadström, 1988). A strong hemagglutination was observed with donors of the MM or MN blood types, but those with the

NN blood types were weakly agglutinated by the F41 adhesin (Brooks et al., 1989). Glycophorin A, which bears the M or N determinants, serves as the F41 receptor because the interactions of both the purified F41 fimbrial adhesin and the F41-carrying *E. coli* are inhibited strongly by MM glycophorin and less strongly by NN glycophorin. Also, the attachment site of the mucous glycoproteins that binds the F41-expressing bacteria weakly, via a negative cooperativity, is distinct form the K99 fimbrial adhesin attachment site (Mouricout and Julien, 1987). It is not clear at present how the F41 fimbrial adhesin may function, especially during the course of natural infection in piglets and calves, because most isolates that produce F41 adhesin also produce the K99 adhesin (Moon, 1990). Experimental infections in mice showed that the production of the F41 adhesin is essential for the ability of the challenge *E. coli* to cause diarrhea (Bertin, 1985). The F41 adhesin is expressed in vivo because piglets infected with F41-fimbriated bacteria mount anti-F41 antibodies (Runnels et al., 1987) and organisms adherent to ileum can be visualized by use of anti-F41 antibodies (Runnels et al., 1987; Thorns and Roeder, 1988). Piglets challenged with enterotoxigenic *E. coli* expressing only the F41 fimbrial adhesin develop diarrhea (Runnels et al., 1987). Piglets suckling from gilts vaccinated with recombinant *E. coli* harboring a plasmid encoding for the F41 fimbrial adhesin are protected against F41-producing challenge *E. coli,* but not against a challenge strain that produces both the F41 and K99 fimbrial adhesins (Runnels et al., 1987). In contrast, infant mice from suckling mothers vaccinated with F41-bearing *E. coli* are protected against diarrhea caused by an enterotoxigenic strain that produces either F41 alone or F41 and K99 fimbrial adhesins (Duchet-Suchaux, 1988). Vaccine strains bearing adhesin only were less protective in this mouse model. Anti-F41 antibodies in colostrum from cows contributed in protecting calves challenged with an enterotoxigenic strain bearing K99 and F41 and F17 (FY) adhesins (Contrepois and Girardeau, 1985). Variants carrying the F41 adhesin only emerged during the infection in piglets with K99- and F41-producing strains, probably due to anti-K99 pressure but it is not clear whether the F41-producing variants are less virulent so as to allow enhanced recovery from clinical disease (Mainil et al., 1987). The composite view of the results is that certain enterotoxigenic strains acquire the ability to express the F41 adhesin, in addition to other adhesins that confer an advantage in colonization of piglet and calve intestines at specific stages of the infectious process.

Escherichia coli

Clinical source	ENT, BLD, piglets, calves
Adhesion:	CS1541 Fim
Receptor specificity:	ND
Substrata:	TC

In spite of the similarities between the CS1541 and K88 or F41 fimbrial adhesin complexes (see Chapter 9), it can be reasonably assumed that the CS1541 adhesin has a distinct receptor specificity. It does not cause hemagglutination (Girardeau et al., 1988), but mediates binding of the bacteria to Henle 407 cells (Korth et al., 1991). Furthermore, although CS1541-carrying *E. coli* do not bind to enterocytes isolated from newborn colostrum-deprived piglets, the organisms cause diarrhea in piglets. The fimbrial adhesin is found on the surface of the bacteria colonizing the intestine (Broes et al., 1989). The adhesin is found in both enterotoxigenic and nonenterotoxigenic bovine isolates and a considerable number of strains also produce the K99 fimbrial adhesin (Contrepois et al., 1989).

Escherichia coli

Clinical source:	ENT, calves
Adhesion:	F17 Fim
Receptor specificity:	GlcNAc
Substrata:	EC

The role of F17 in mediating adhesion of the organisms to intestinal cells has been deduced from genetic manipulation of the gene cluster encoding for the fimbrial adhesin. *E. coli* host harboring a plasmid containing a 8.5-kb insert from the parent strain produced F17 fimbriae, adhered to intestinal calf cells, and the adhesion was specifically inhibited by *N*-acetylglucosamine (Lintermans et al., 1988). Most enterotoxigenic bovine isolates that carry the F17 adhesin also produce other non type 1 adhesins such as K99 (Contrepois et al., 1985; Morris et al., 1985). It is possible that the F17 fimbrial adhesin plays a role in infection because antibodies against F17 (FY) in colostrum contribute to the protection of suckling calves against challenge with a K99, F41, and F17 producing strain (Contrepois and Girardeau, 1985). F17 carrying *E. coli* binds to immobilized, mucus-derived glycoproteins but the magnitude of the binding is considerably less than that of K99-carrying *E. coli* (Mouricout and Julien, 1987). It is possible that the F17 adhesin confers only a marginal advantage to the infectivity of K99-carrying strains because only a small percentage of isolates are $K99^+$ and $F17(FY)^+$ (Contrepois et al., 1985).

Escherichia coli

Clinical source:	ENT, rabbits
Adhesin:	AF/R1 Fim
Receptor specificity:	Sialic acid
Substrata:	EC, MC

The adhesion of an *E. coli* 015 RDEC-1 bearing AF/R1 plasmid encoded fimbrial adhesin to rabbit intestine has been extensively studied, mainly because it plays a major role in the ability of the bacteria to cause diarrhea in rabbits. The diarrhea is related to a mechanism similar to that of EPEC in that it exhibits attachment/effacement to human intestinal cells. The AF/R1 fimbriae mediate adhesion of bacteria to rabbit intestinal cells because purified fimbriae or anti-

AF/R1 fimbriae inhibit adhesion of RDEC-1 to the intestinal cells and strains not expressing the fimbrial adhesin fail to bind to the intestinal cells (Cantey et al., 1989; Rafiee et al., 1991) or to the rabbit ileal microvillus membrane (Drumm et al., 1988). Although the AF/R1-bearing *E. coli* cause attachment/effacement lesions in vivo, the AF/R1 fimbrial adhesin appears not to be directly involved in this process which requires products of distinct DNA elements (Cantey and Moseley, 1991). The role of the AF/R1 fimbrial adhesin may be in conferring an advantage at a particular stage of the infectious process, especially under a natural course of infection initiated by low doses of bacteria. This conclusion is based on studies showing that mutants of RDEC-1 unable to express the AF/R1 fimbrial adhesin have a reduced ability to cause diarrhea in rabbits at inocula 100-fold larger than those of the parent strain (Wolf et al., 1988; Cantey et al., 1989). The mutant failed to adhere to M-cells in Peyer's patches (Cantey et al., 1989), confirming earlier studies showing that within a few hours following inoculation, the AF/R1-bearing RDEC-1 bacteria colonize the Peyer's patch lymphoepithelial cells of rabbits. The interaction with M-cells may be important for inducing mucosal immunity to O antigens of *E. coli* (Axelrod, 1985; Cantey et al., 1987). Milk immune secretory IgA effectively inhibits the adhesion of RDEC-1 strains to rabbit intestinal brush borders (Boedeker et al., 1987). The AF/R1 fimbrial adhesin may also mediate binding of the bacteria to mucous constituents because RDEC-1 organisms are seen associated with mucus in vivo (Hill, 1985; Sherman and Boedeker, 1987). Purified fimbriae or whole RDEC-1 are also aggregated by rabbit mucous glycoproteins (Sherman and Boedeker, 1987). Furthermore, only AF/R1 expressing RDEC-1 bind to immobilized rabbit mucus (Drumm et al., 1988). Excess of mucous constituents capable of interacting with RDEC-1 strains of *E. coli* inhibit adhesion of the organisms to plastic surfaces, presumably by altering the surface hydrophobicity which appears to be associated with adhesive structures (Mack and Sherman, 1991). Whereas the expression of the type 1 fimbrial adhesin by RDEC-1 may abrogate the advantages conferred by the AF/R1 adhesin (Sherman et al., 1985), the AF/R1 adhesin may have a specific role because purified AF/R1 fimbrial vaccines protect against subsequent colonization by RDEC-1 *E. coli* (McQueen et al., 1987). AF/R1 fimbriated EDEC-1 *E. coli* are more hydrophobic and adhere better to rabbit ileal microvillus membranes, mucus, and mucin, as compared to phenotypes lacking this fimbrial adhesin (Drumm et al., 1989). It is possible that during a natural course of infection initiated by low doses of bacteria, transient colonization of either lymphoepithelial cells or mucus mediated by the AF/R1 fimbrial adhesin is essential for the RDEC-1 strain to produce progenies expressing distinct adhesive properties. The progenies enable the infectious process to continue, leading to attachment/effacement lesions and diarrhea. Studies on the identity of the receptor on rabbit small intestinal cells suggest that the AF/R1 fimbrial adhesin may mediate the first stage of adhesion in the process leading to attachment/effacement of intestinal brush border (Rafiee et al., 1991). The receptor was identified as a glycoprotein complex composed of 130,000 and 140,000 mot wt subunits. The binding of AF/R1 adhesin to the receptor is dependent on intact sialic acid residues. The unique feature of this AF/R1 receptor is that it is linked to the cytoskeletal component myosin–1 which is linked to actin. Although mutants not expressing AF/R1 adhesin trigger the development of attachment/effacement lesions, it has been postulated that the AF/R1 adhesin greatly promotes the development of such lesions due to the link between the AF/R1 receptor and cytosolic myosin (Rafiee et al., 1991). Nevertheless, in order to demonstrate the characteristic features of attachment/effacement mode of adhesion in HeLa cells including aggregation of HeLa cell actin at points of adhesin of RDEC-1, it was necessary to mobilize into the rabbit pathogen DNA elements coding for the F1845 adhesin. The HeLa cells presumably lack AF/R1 receptors (Cantey and Moseley, 1991).

Escherichia coli

Clinical source:	ENT, rabbits
Adhesin:	ND
Receptor specificity:	ND
Substrata:	EC, TC

A survey of *E. coli* isolated from weanling diarrheic rabbits reveals many strains that do not produce enterotoxin exhibit plasmid-dependent attachment to rabbit intestinal cells and cause effacement and cup-like projections associated with inflammatory response of the microvillous brush border (Peeters et al., 1988; Licois et al., 1991; Reynaud et al., 1991). Many of the rabbit diarrheal isolates of *E. coli* 0103 adhere to rabbit intestinal villi and to HeLa cells in a diffuse adhesion pattern (Milon et al., 1990) and are hydrophobic (Baloda et al., 1986). A 32,000 mol wt protein extracted from the surface of one strain may function as an adhesin because antisera raised against the protein inhibited adhesion. It appears that the rabbit diarrheagenic strains share many properties with the human EPEC and may serve as a model to study the mechanism through which the attachment/effacement induces diarrhea.

Escherichia coli

Clinical source:	BLD, RT, avian
Adhesin:	AC/I Fim
Receptor specificity:	ND
Substrata:	EC

AC/I fimbriae are expressed by about 50% of *E. coli* 078 strains isolated from blood of avian colisepticemia and mediate adhesion of the organisms to chicken tracheal and intestinal cells (Yerushalmi et al., 1990). The AC/I fimbriae-carrying *E. coli* do not cause hemagglutination of erythrocytes from a number of species and adhere poorly to human buccal cells. Adhesion to tracheal cells was detected after oral inoculation of chickens with AC/I fimbriae bearing *E. coli,* suggesting that the AC/I fimbrial adhesin complex mediates the adhesion of the organisms in vivo. Monoclonal anti-AC/I antibodies directed against the major 18,000 mol wt fimbrial subunit of the AC/I fimbrial adhesin complex inhibit adhesion of the organisms to tracheal epithelial cells. Because avian isolates of *E. coli,* similar to most *E. coli* isolates, also express the type 1 fimbrial adhesin complex, it remains to be seen if a unique advantage is conferred by AC/I fimbriae (the first MR fimbrial adhesins described in avian isolates of *E. coli*) in the ability of the organisms to colonize the chicken lower respiratory tracts.

Escherichia coli

Clinical source:	ENT, human
Adhesin:	ND
Receptor specificity:	ND
Substrata:	MC

The importance of the interaction of *E. coli* with the highly glycosylated mucous layer lining the alimentary tract is reflected by studies showing that the organisms must somehow interact with specific mucous constituents and grow in the mucous layer in order to subsequently penetrate the layer to reach and bind to the underlying target epithelial cells (Cohen et al., 1985b; Wadolkowski et al., 1988a; Embaye et al., 1989). In certain cases, it was found that colonization of the mucous layer with a loose attachment to the brush border can impair electrolyte transport, which may then lead to diarrhea in susceptible hosts (Schlager et al., 1990). Moreover, binding to and colonization of intestinal mucus on epithelial cells may be important in areas (e.g., cecum and colon) where the balance between the rate of microbial multiplication and rate of movement of lumenal content is positive for the microbe (Savage, 1987). Growth in mucus probably requires a number of different traits, but interaction with specific mucous constituents may be essential. A number of studies have focused on identifying the mucus-specific adhesins of *E. coli* and their receptors. Adhesion of the bacteria to mouse colonic mucus is greater than adhesion to bovine serum albumin. *E. coli* lipopolysaccharide (LPS) specifically inhibits the adhesin and purified capsules also inhibit the binding to mucus, but this inhibition was found to be nonspecific (Cohen et al., 1985a). Sialoglycoconjugate-specific adhesins CFA/I, CFA/II (CS2), and K99 of ETEC also seem to interact with sialic acid-rich glycoconjugates in human and pig mucins, respectively (Wadström and Baloda, 1986). In addition, the F41 (Wadström and Baloda, 1986) adhesin also interacts with mucous constituents. The data suggest that at least certain *E. coli* clones capable of colonizing mouse intestinal cells produce a number of adhesins specific for different glycoproteins in mouse mucus. The relative advantage of each of these adhesins in the ability of the bacteria to interact with human mucus is unknown. It is possible that each of the different mucus-specific adhesins confers an advantage at certain stage(s) of the colonization process because binding to rabbit mucus was found to vary by region of the gut and the strain employed (Wanke et al., 1990). Periodate oxidation of side chain oligosaccharide of the mucous glycoproteins reduces binding of the enteropathogenic *E. coli* to immobilized rabbit mucus, suggesting that the binding of the mucus-specific adhesins is dependent on intact vicinal hydroxyl groups in the oligosaccharide chain of the mucous glycoproteins.

Escherichia coli

Clinical source:	ENT, human
Adhesin:	Curli
Receptor specificity:	ND
Substrata:	ECM

In addition to the overlying mucous layer it is likely that enteropathogenic *E. coli* may encounter constituents of the extracellular matrix (ECM) during the infectious process. Fibronectin-binding proteins associated with specific fimbrial adhesins of *E. coli* were discussed separately in another chapter. There are a number of reports on fibronectin-binding proteins expressed by *E. coli,* but the role of these proteins in adhesion is not known. The coiled surface structures called curli bind soluble fibronectin and apparently mediate the binding of fibronectin by *E. coli* that express these structures (Olsen et al., 1989). Curli are composed of a single subunit, but only certain strains of *E. coli* (e.g., about half of *E. coli* strains isolated from bovine mastitis or fecal isolates and one strain of *E. coli* isolated from a human patient with UTI) can assemble the curlin subunits into the structures identified as curli. This is in spite of the fact that all strains examined, including two *E. coli* K12 strains, hybridize with a curli-specific probe. Curli are expressed in agar-grown bacteria only at 26°C and no data are available on their expression by various human isolates of *E. coli*. Nor is it known if the curli can actually mediate adhesion to cell surface fibronectin. It is therefore too early to speculate on the role of the curli structures in the infectious process. In addition, it has been found that about 15 of 50 EPEC, 1 of 8 EIEC, 25 of 104 ETEC, and 1 of 16 meningitis isolates tested avidly bound fibronectin (Wadström and Baloda, 1986; Wadström, 1988). A number of EPEC strains isolated from patients with ulcerative colitis and grown under appropriate conditions bound fibrinogen, collagen, and fibronectin (Wadström and Baloda, 1986; Ljungh and Wadström, 1988). In addition, an ETEC strain was found to bind all types of collagen tested (I–V) and their cyanogen bromide fragments (Visai et al., 1990). Fibronectin and laminin also bind to the ETEC strain and fibronectin competes with the collagen binding sites, suggesting that the putative binding sites for collagen are either the same as those for fibronectin or in close proximity (Visai et al., 1990, 1991). The ETEC strain expresses two fibronectin-binding proteins, one of which, a 17,000 mol wt protein, binds the organisms to the amino-terminal region of fibronectin, and the other, a 55,000 mol wt protein, interacts with the heparin-binding domain of fibronectin. Streptococcal and staphylococcal fibronectin-binding proteins inhibit the binding of fibronectin and its amino-terminal fragment, but not its heparin-binding peptide to the ETEC strain, suggesting that the bacterial proteins bind to the amino-terminal region. The expression of binding sites for various constituents of the extracellular matrix (ECM) on the bacterial surface in vitro was studied in detail employing a diarrheal isolate of *E. coli* (Ljungh et al., 1991). The binding of laminin, vitronectin, fibronectin, and collagen by the organisms was maximal during the stationary phase of growth in broth. The growth conditions are critical for the expression of ECM binding proteins on the surface of the enteropathogenic *E. coli*. For example, growth in medium supplemented with 1% NaCl or deprived of Mg^{2+} and Mn^{2+} ions, respectively, suppressed or increased ECM binding by the organisms. The expression of ECM-binding proteins on the organism's surface is thermoregulated (Ljungh et al., 1991). The results suggest that biosynthesis and expression of receptors for all four matrix proteins is highly regulated and is affected by environmental signal transduction. It is not clear, however, which *in vivo* conditions are required for stimulating the expression of ECM-binding proteins on the surface of the *E. coli*. It is also possible that various enteropathogenic strains are capable of expressing distinct ECM-binding proteins because the ECM-binding characteristics appear to vary in different strains (Visai et al., 1990; Ljungh et al., 1991). Moreover, binding of collagen fibers by diarrheal *E. coli* is probably different from binding of soluble collagen because it takes place only in distilled water. Addition of salts to the reaction mixture removes bacteria attached to the collagen fibers (Campbell et al., 1987). *E. coli* isolated from bovine mastitis also bind fibronectin (Faris et al., 1987). A number of strains exhibit a high capacity of binding to the amino-terminal 29,000 mol wt fragment and adhere to bovine skin fibroblasts in much higher numbers, compared to strains that poorly bind the amino-terminal fragment. One bovine strain binds to fibronectin, but not to its 29,000 mol wt amino-terminal fragment. Growth of the bacteria in a penicillin- or tetracycline-containing medium suppresses the cell surface components responsible for binding to fibronectin and for adhesion to fibroblasts (Faris et al., 1987).

Escherichia coli

Clinical source;	UTI, human
Adhesin:	P Fim
Receptor specificity:	Galα1,4Gal
Substrata:	EC, ECM, ERT

The majority of strains isolated from patients with pyelonephritis contain the gene cluster encoding for the P-fimbrial adhesin (reviewed by Johnson, 1991). P-fimbriated strains are rarely selected to cause other types of extraintestinal *E.*

coli infections, such as neonatal sepsis or meningitis (Korhonen et al., 1985). This association clearly indicates that the P-fimbrial adhesin has a unique role in enabling the *E. coli* to colonize and subsequently cause damage to glomerular tissue. Although a great deal is known at present about the receptor specificity (e.g., Gal-Gal containing glycolipids) and molecular biology of the P-fimbrial adhesin, extensive studies aimed at understanding the specific role of the adhesin in the infectious process leading to pyelonephritis continue to emerge. The agglutination test with Galα-1,4Gal beads and hybridization of genomic DNA with DNA probes of *pap* gene cluster provide a screen for the incidence of the P-fimbrial adhesin of intestinal and extraintestinal isolates (de Man et al., 1987). The incidence of strains producing the P-fimbrial adhesin is high (79%–100%) in uncompromised patients with pyelonephritis (Domingue et al., 1985; Elo et al., 1985; Gander et al., 1985; O'Hanley et al., 1985b; Enerback et al., 1987; Archambaud et al., 1988b; Jacobson et al., 1988a,b; Marild et al., 1988; Westerlund et al., 1988; Dalet et al., 1991); and urosepsis in children, but not bacteremia from other sources (Brauner et al., 1985a,b; Arthur et al., 1990). The incidence decreases considerably in patients with compromised conditions, such as urinary tract abnormalities, urinary tract instrumentation, pregnancy, diabetes, and systemic corticosteroid therapy, all of which appear to render the patients susceptible to pyelonephritis caused by *E. coli* strains that do not produce P fimbriae (Brauner et al., 1985, 1987; Brauner and Oslenson, 1987, 1987a,b; Domingue et al., 1985; Elo et al., 1985; Gander et al., 1985; Dowling et al., 1987; Stenqvist et al., 1987; Johnson, 1988a; Sandberg et al., 1988; Arthur et al., 1989; de Man et al., 1989; Lomberg et al., 1989; Dalet et al., 1991). In spite of the high prevalence of P-fimbriated *E. coli* in pyelonephritis in uncompromised individuals, it is not feasible, as suggested (Dowling et al., 1987), to use the test for P fimbriation by pyelonephritic isolates as a marker for compromised or uncompromised hosts due to significant overlap (Johnson, 1988b). The proportion of *E. coli* strains expressing the P-fimbrial adhesin or harboring the *pap* gene cluster also declines progressively among isolates from cystitis, bacteriuria, and feces (Domingue et al., 1985; Elo et al., 1985; Gander et al., 1985; Jacobson et al., 1985a, 1987; O'Hanley et al., 1985b; Hacker et al., 1986b; Dowling et al., 1987; Enerback et al., 1987; Israele et al., 1987; Lidefelt et al., 1987; Stenqvist et al., 1987; Archambaud et al., 1988a; Marild et al., 1988; Nicolle et al., 1988; Sandberg et al., 1988; Westerlund et al., 1988; Arthur et al., 1989; de Man et al., 1989; Dalet et al., 1991). Like other enterobacteria, the large intestine serves as a reservoir for P-fimbriated strains with no preference over strains that carry other types of adhesins (Kallenius and Mollby, 1988; Kallenius et al., 1985; Jacobson et al., 1987; Tullus et al., 1988), supporting the fecal–urethral transmission hypothesis of uropathogenic bacteria (Stamm et al., 1989). In a prospective study of bacteriuric episodes from 1970 to 1984, it was found that there was a linear relationship between the intensity of the inflammatory response (e.g., body temperature, C-reactive protein, micro-sedimentation rate, and urinary leukocyte) and the proportion of P-fimbriated isolates (de Man et al., 1988). None of 196 diarrheal isolates of *E. coli* examined contained the *pap* gene cluster (Archambaud et al., 1988a). The epidemiological observations, taken together, suggest that P-fimbriated strains of *E. coli* are specifically selected to cause the more clinically severe forms of UTI, especially pyelonephritis, in otherwise healthy individuals. The specific advantage(s), if any, conferred upon *E. coli* by the P-fimbrial adhesin over other MR *E. coli* adhesins in causing pyelonephritis in the uncompromised host remains an enigma.

Perhaps one clue to the problem are the types of urinary tract infections in which the P-fimbriated strains are specifically not selected. The incidence of P-fimbriated strains from renal scarring associated with vesicourethral reflux is very low (Jacobson, 1986; Lomberg et al., 1986a,b; Jacobson et al., 1987; Lomberg and Svanborg-Eden, 1989; de Man et al., 1989; Topley et al., 1989a). Scarring may be the result of extensive interaction of the invading *E. coli* with phagocytic cells to release extracellular tissue damaging agents (Mundi et al., 1991). The inability of P-fimbriated strains to induce renal scarring may be due to their inability to interact and stimulate phagocytic cells, in contrast to the type 1 fimbriated *E. coli* (see Chapter 5). Because most P-fimbriated strains are also capable of producing the type 1 fimbrial adhesin, however, there must be a specific negative role for P fimbriae in impairing the ability of the P fimbriae carrying strains to initiate renal scarring in a compromised host. For example, strains isolated from patients with pyelonephritis associated with renal scarring induce more often the extracellular release of oxygen radicals in human neutrophils, compared to strains isolated from a nonscarred group of patients with pyelonephiritis (Mundi et al., 1991). The molecular mechanisms through which the scarred strains cause release of oxidative metabolism is not known. Interestingly, P-fimbriated strains of *E. coli* were isolated from chicken septicemia but the relatedness of these poultry isolates to those of human is not clear (Achtman et al., 1986).

An issue to be considered concerning the susceptibility of the host to acquire infection with P-fimbriated *E. coli* is the presence of soluble glycoconjugates that can specifically bind to and inhibit adhesion of the organisms to target epithelial cells. So far, urine has been tested and found to be free of any inhibitors of P-fimbrial adhesion (Parkkinen et al., 1988). The effects of sublethal concentrations of antibiotics on the biosynthesis of P fimbriae continue to interest a number of laboratories (see Chapter 11). Another interesting feature of the P-fimbrial adhesin is its expression during infection. Anti-P-fimbrial response following infection with P-fimbriated *E. coli* has been documented, suggesting that the P-fimbrial antigen must have been produced sometime during the infectious process (de Ree and van den Bosch, 1987; Salit et al., 1988; Agata et al., 1989). Due to the phase variation phenomenon, however, it is likely that organisms growing in vitro or in vivo will exhibit a mixed population of P-fimbriated and non-P-fimbriated variants (Gander and Thomas, 1987). Also, the frequency of phase variation appears to be affected by external environmental factors (Blyn et al., 1989). The question is therefore at which stage of the infectious process is there a prevalence of population expressing the P-fimbrial adhesin.

Agglutination of Gal-Gal-coated beads and immunofluorescence have been used to detect P fimbriae on the organism's surface in void urine (Pere et al., 1987; Kisielius et al., 1989). The percent of bacterial cells expressing P fimbriae varies from one patient to another, ranging from no detection to 95% of the cell population. It has been argued that the percentage of P-fimbriated cells may be even more prevalent among *E. coli* adherent to the bladder as compared to bacteria shed in the urine (Johnson, 1988c). It is not clear at this stage whether this simply reflects a small percentage of adherent P-fimbriated cells or whether it reflects a high frequency of switching to the "on" phase under the influence of the close proximity of the bacteria to the uroepithelial cells. Experimental infections in animal models and humans have helped to illuminate the relative importance of P-fimbrial adhesin in the infectious process. The choice of animal model is dependent on the presence of Gal-Gal-containing glycolipids on renal cells accessible for binding the P-fimbriated strain. P-fimbriated *E. coli* cause hemagglutination of erythrocytes from humans, pigs, and horses (Parry and Rooke, 1985). Galα1,4Gal-containing glycolipids in various tissues have not been found in rats, whereas mouse and primate renal tissues contain abundant Galα1,4Gal-containing glycolipids (O'Hanley et al., 1985; Lyerla et al., 1986). The animal models dealt with two major issues, one of which is the relative importance of the P-fimbriated *E. coli* in establishing urinary tract infection, and the other is the efficacy of a P-fimbrial vaccine in preventing infection caused by P-fimbriated *E. coli*. The P-fimbriated *E. coli* colonized and invaded the upper urinary tract with better efficacy as compared to nonfimbriated or type 1 fimbriated *E. coli* in mice and monkeys (O'Hanley et al., 1985; Domingue et al., 1988; Kaack et al., 1988). The P-fimbrial adhesin is capable of presenting endotoxin to the target uroepithelial cells to stimulate local inflammation in LPS responder mice (Linder et al., 1988). The local inflammation was specifically inhibited by soluble digalactoside, suggesting that it was dependent on the digalactoside-specific adhesin Pap G of the P fimbriae. Intravesicular inoculation of P-fimbriated *E. coli* also results in the release of interleukin-6 in the urine (Linder et al., 1991). The interleukin-6 release was dependent only on the presence of Fim G on P fimbriae which binds the digalactoside-containing receptor, whereas S-fimbriated and type 1 fimbriated *E. coli* were ineffective. In summary, it appears that the P-fimbrial adhesin on the surface of the organisms is capable of directly activating mucosal interleukin-6 and indirectly causing local inflammation associated with the influx of polymorphonuclear leukocytes to the site of infection by presenting the endotoxin to the target tissue. In contrast to these animal studies, intravesicular inoculation of, in female volunteers, *E. coli* strains expressing the P-fimbrial adhesin persisted for significantly shorter periods of time compared to nonfimbriated strains (Anderson et al., 1991). The volunteers had a history of recurrent symptomatic urinary tract infections and therefore may represent a population prone to UTI. Clearly, the P-fimbriated strains were selectively eliminated from the bladder, but because type 1 fimbriated *E. coli* also persisted for shorter periods of time, the possibility that any type of adhesin may be a disadvantage to the bacterium in this model of human infection cannot be excluded.

The animal models serve to evaluate the efficacy of P-fimbrial vaccine which continues to interest many investigators. P-fimbrial vaccine or synthetic-derived peptide proved to be effective in preventing the development of urinary tract infection in mice (Hagberg et al., 1985; O'Hanley et al., 1985; Schmidt et al., 1988; Pecha et al., 1989) and in monkeys (Kaack et al., 1988). Studies on inhibition of adhesion of P-fimbriated *E. coli* by polyclonal or monoclonal anti-Pap A antibodies revealed that whereas some monoclonal antibodies inhibit adhesion (Gander and Thomas, 1986; de Ree et al., 1987), most of the monoclonal antibodies and even some polyclonal antibodies poorly inhibit adhesion, even of the homologous P-fimbrial type (Abe et al., 1987; de Ree et al., 1987; Salit et al., 1988). Furthermore, in one study, it was found that the anti-P-fimbrial antibodies developed in patients with pyelonephritis failed to inhibit adhesion of the homologous P-fimbriated *E. coli* (de Ree and van den Bosch, 1987). In another study, it was found that urine samples of patients with UTI did not inhibit the adhesion of a presumably P-fimbriated *E. coli* to uroepithelial cells. In contrast, urine samples from patients with ileocystoplasty did contain adhesion-inhibitory activities associated with relatively high sIgA concentrations (Trinchieri et al., 1990). These somewhat conflicting results probably arise from extensive antigenic heterogeneity of the major P-fimbrial subunit (de Ree et al., 1985, 1986, 1987; Hanley et al., 1985; Moser et al., 1986; Pere, 1986; Old et al., 1987); and from lack of knowledge on the exact epitope(s) of the Pap A subunit needed to be blocked by antibodies in order to bring about efficient inhibition of P-fimbriae-mediated adhesion. It is possible that the best vaccine would be the Fim G tip adhesin that now has been purified (Hoschützky et al., 1989; Lund et al., 1988a). The protective effectiveness of the anti-P-fimbrial immunity has been challenged because of the decreased importance of P-fimbriated strains to cause pyelonephritis in compromised hosts (Kunin, 1986; Johnson et al., 1987). Although the P-fimbrial vaccine affords protection against UTI in experimental models and development of antibacterial antibodies in the urine of volunteers vaccinated intravesicularly with live *E. coli* has been documented (Svanborg-Eden et al., 1990), it has not been shown unequivocally that anti-P-fimbrial antibodies at adhesion-inhibitory concentrations actually reach the urine. Even during urinary tract infection the concentration of urinary sIgA needed to inhibit adhesion of *E. coli* to uroepithelial cells was not high enough to cause a change in the development of symptoms.

The density of P-fimbrial receptors (e.g., P blood group antigens) and the accessibility of these receptors on various cell types for binding P-fimbriated

E. coli have been examined thoroughly, especially as regards susceptibility to infection. The data, however, are complex and conflicting. For example, uroepithelial cells from adults with renal scarring bind higher numbers of P-fimbriated bacteria than control cells, but the frequency of isolation of P-fimbriated strains from individuals with renal scarring as discussed earlier is notably low (Jacobson, 1986b). The ability of uroepithelial cells from nonsecretors to bind P-fimbriated *E. coli* is considerably higher compared to cells from secretors (Lomberg et al., 1986b), suggesting that the digalactoside-containing glycolipids of the P blood group antigen are not accessible for binding P-fimbriated bacteria. It has been suggested that the P blood group antigen is masked by other blood group antigens in secretors, whereas in nonsecretors, the cell surface expression of A, B, and H antigens is impaired. Nonsecretors are more prevalent among patients with renal scarring and women with recurrent UTI as compared with the population at large (Svanborg-Eden et al., 1988; Sheinfeld et al., 1989) but the incidence of infection with P-fimbriated *E. coli* is relatively high only in women with recurrent UTI (Stamm et al., 1989). Furthermore, it has been argued that the observed high risk of healthy individuals with the P1 blood group phenotype to develop pyelonephritis caused by P-fimbriated *E. coli* represents a marker for a UTI-prone subgroup (Svanborg-Eden et al., 1988). The erythrocytes of P1 blood group phenotype are particularly rich in P blood antigens, all of which contain the digalactoside that binds P-fimbriated *E. coli*. Uroepithelial cells, however, from individuals with the P1 group phenotype have no increased capacity to bind P-fimbriated *E. coli,* nor do they have increased density of P-fimbrial receptors over those from individuals with other blood group phenotypes (Jacobson et al., 1985b, 1986a,b; Lomberg et al., 1986a). The accessibility of P-fimbrial receptors on exfoliated uroepithelial cells may also change during UTI caused by Gram-negative bacteria but not by Gram-positive bacteria (Daifuku and Stamm, 1986). The changes of accessibility of P-fimbrial receptors on uroepithelial cells may also be the result of removal of a protective layer that masks the adhesin receptors, such as heparin sulfate (Parsons, 1986; Ruggieri et al., 1987). The density of P-fimbrial receptors and the capacity to bind P-fimbriated bacteria have been examined using histochemical techniques in various cell types of frozen human kidney sections (Nowicki et al., 1986a; Korhonen et al., 1986a; Virkola et al., 1988; Karr et al., 1989; Korhonen et al., 1990). High to moderate densities of P-fimbrial receptors were found on Bowman's capsule, glomerulus, and distal tubulus. A low density was found on epithelium and muscular layer of bladder as well as on urine-sedimented epithelial cells, whereas no detectable P-fimbrial receptors were found on connective tissue of the bladder. Because virtually any glycolipid that contains the digalactoside sequence will bind P-fimbriated *E. coli* in the thin-layer techniques, a host cell may contain a family of potential isoreceptors for the P-fimbriated bacteria. For example, uroepithelial cells, the first type of nucleated cells used to demonstrate binding of *E. coli* carrying P fimbriae, contain a number of glycolipids that bind the P-fimbriated bacteria but it is not clear if all these glycolipids can serve as receptors on the intact cell (Bock et al., 1985). P-fimbriated *E. coli* bind to the adenocarcinoma cell line HT-29 and the binding is specifically inhibited by the soluble digalactosides, suggesting that Gal-Gal-containing glycolipids on HT-29 cells are responsible for binding the organisms (Wold et al., 1988). In contrast to renal and intestinal tissues, polymorphonuclear leukocytes do not bind P-fimbriated bacteria, probably because the Gal-Gal-containing glycolipids present in these cells (Bock et al., 1985) are not accessible for binding the organisms (see Chapter 5). Bladder surface glycosaminoglycans may also mask potential receptors required for bacterial adhesion, but the ability of these molecules to mark specifically P-fimbrial receptors has not been tested thoroughly (Kaufman et al., 1987).

P-fimbriated *E. coli* are found to bind to a loosely associated surface constituent on human colonic cells but the ability of disaccharides to inhibit this binding has not been determined (Wold et al., 1988). In other studies it has been shown that the minor P-fimbrial subunits Fso E and Fso F (and probably Pap E and Pap F) mediate binding of *E. coli* to immobilized intact fibronectin and specifically to the immobilized 32,000 mol wt amino-terminal and 120–140,000 mol wt carboxy-terminal fragments of fibronectin (Westerlund et al., 1989b, 1991). The P-fimbriated bacteria bind weakly to a 40,000 mol wt gelatin-binding fragment and not at all to soluble fibronectin or to immobilized type IV collagen. It is possible that the observed binding of P-fimbriated mutants lacking the digalactoside-specific Fso G subunit to tubular basolateral membranes of human kidney is mediated by the fibronectin-binding proteins Fso E and Fso F (Westerlund et al., 1991). This prospect raises the possibility that the P fimbriae of *E. coli* may mediate binding of the bacteria to host cells in at least two different ways: the digalactoside-dependent adhesion and the fibronectin-specific adhesion. Incorporation of Fim G into the fimbrial structure is required only for the carbohydrate-dependent adhesion. It is not known whether there is any evolutionary pressure to preserve the fibronectin-specific adhesion without the carbohydrate-specific adhesion because no data are available as to the frequency of isolation of P-fimbriated *E. coli* strains deficient of *fim*G. The possession of two receptor specificities on the same P-fimbrial adhesin complex may confer a unique advantage to the organisms in colonizing diverse types of substrata.

Another approach to explain the ability of P-fimbriated *E. coli* to cause the most severe form of UTI (e.g., pyelonephritis) in individuals with unimpaired host defense mechanisms was to employ molecular epidemiology. The techniques of DNA hybridization, using probes from *pap* operon and from genes encoding for other virulence factors, was employed in order to examine a number of issues related to the phenotypic expression of the P adhesins, copy number of P-fimbrial gene clusters, and the linkage of the adhesin genes to genes coding for other virulence factors in various isolates. Several noteworthy findings are relevant to the unique role of the P-fimbrial adhesin: (1) the phenotypic expression of functional P-fimbrial adhesins has now been confirmed by DNA hybridization and is supported by the epidemiological data described above. The incidence of *pap* homologous DNA was the highest in the chromosomes of pyelonephritogenic isolates and gradually decreases in cystitis and asymptomatic bacteriuria (Plos et al., 1989; Arthur et al., 1989); (2) *pap*-positive strains isolated from pyelonephritis expressed phenotypically more often the P-fimbrial adhesin as compared to strains isolated from cystitis or asymptomatic bacteriuria (Plos et al., 1989); (3) the presence of three copies of *pap* gene per chromosome was observed significantly more often in pyelonephritogenic isolates as compared to those from asymptomatic bacteriuria (Plos et al., 1989); (4) the results on the occurrence of *pap*-linked genes encoding for other virulence factors were inconclusive. In some strains, the *pap* gene cluster is genetically linked with the hemolysin gene, and in one strain with genes encoding for other adhesins as well (High et al., 1988; Hull et al., 1988); (5) the copresence of *pap* and *prs* (a *pap*-related operon that encodes for the Prs G adhesin possessing a distinct receptor specificity) (see below) genetic elements in the chromosome and phenotypic coexpression of the two adhesins were found with the highest frequency among *E. coli* isolates from patients with pyelonephritis and gradually less among cystitis and fecal isolates (Arthur et al., 1989); (6) although not necessarily linked to *pap,* the genetic elements coding for hemolysin, aerobactin, and P fimbriae were found at a higher frequency on the chromosome of strains from uncompromised patients with urosepsis than from compromised patients with urosepsis (Johnson et al., 1988); and (7) strains belonging to certain serogroups 04, 06, and 02 have similar P-fimbrial subtypes (Pere, 1986; Pere et al., 1988) but the presence of *pap* genes and P fimbriation are variable within a clonal group of *E. coli* defined by O serotype, outer membrane protein pattern, biotype, and multilocus enzyme type, whereas a number of different clonal groups share common P-fimbrial subtypes (Achtman et al., 1986; Selander et al., 1986; Pere, 1986; Pere et al., 1988; Plos et al., 1989; Arthur et al., 1990). The results suggest that the P-fimbrial genes may have been acquired horizontally between genetically unrelated clones of *E. coli* but vertical transmission within a particular clone of *E. coli* may have occurred as well.

In summary, P-fimbrial adhesins confer upon *E. coli* the ability to initiate a severe form of urinary tract infection, particularly in uncompromised individuals. The adhesion-dependent inflammation model, the availability of Gal-Gal-containing receptors in the renal tissue, and production of other virulence factors (e.g., hemolysin, aerobactin) do not adequately explain the lack of selection of P-fimbriated strains in pyelonephritis or urosepsis in adults with compromised defense mechanisms. In contrast, there are a number of unique features regarding the adhesive properties of P-fimbriated strains. The P adhesin via its Pap G subunit is capable of mediating adhesion of the organisms specifically to Gal-Gal-containing glycolipids and via non-Pap G subunits to fibronectin immobilized on cell surfaces. Furthermore, most isolates from uncompromised pyelonephritis patients contain multiple copies of *pap* and *pap*-related gene clusters, each encoding for adhesins with distinct receptor specificities (see below). The P-fimbrial adhesin complex may possess other unknown adhesive properties because the specificity of the loose binding of P-fimbriated *E. coli* to the cell surface coat of colonic human cells is not known yet. The adhesin may also be involved in internalization of the *E. coli* into human renal tubular epithelial cells by an as yet unknown mechanism because at low inocula only P and type 1 fimbriated pyelonephritogenic strains are internalized by the tubular cells (Warren et al., 1988). It has been argued that the P-fimbriated *E. coli* were selected throughout evolution to inhabit the gastrointestinal tract without provoking any tissue damage. Colonization of the lower and upper urinary tracts may be coincidental (Levin and Svanborg-Eden, 1990). Nevertheless, it is reasonable to assume that these unique receptor specificities of the P-fimbrial adhesin complex are required to overcome certain host defense mechanism(s) to colonize and induce inflammation in the upper urinary tract of otherwise healthy individuals.

Escherichia coli

Clinical source:	UTI, human, canine
Adhesin:	Prs Fim
Receptor specificity:	GalNAc (Gal-Gal)
Substrata:	EC, ERT

A group of uropathogenic *E. coli* was found to contain a *pap*-related gene cluster encoding for a fimbrial adhesin serologically classified as P fimbriae, but the adhesin exhibited receptor specificities distinct from that of the P-fimbrial adhesin. This Pap-related adhesin preferentially recognizes the terminal GalNAcα1,3GalNAc moiety of the Forssman antigen or GalNAcα1,3 bound to Galα,4Gal disaccharide of globo-A glycolipids (Senior et al., 1988; Stromberg et al., 1990). *E. coli* carrying this fimbrial adhesin strongly agglutinates sheep erythrocytes or P1 erythrocytes of the A blood group (Lund et al., 1988a; Senior et al., 1988; Karr et al., 1989; Lindstedt et al., 1989). The terms Prs for P-related sequence (Lund et al., 1988b) or Pap–2 for a second Pap-related sequence (Lindstedt et al., 1989) or ONAP for agglutination of O-negative, A-positive P1 erythrocytes (Senior et al., 1988) or F for Forssman antigen (Arthur et al., 1989) were coined to describe this fimbrial adhesin. For clarity, this book refers to the adhesin as the Prs fimbrial adhesin complex, GalNAc-(Gal-Gal) as its receptor specificity, and Prs G as the minor fimbrial subunit possessing the receptor-binding activity (Lund et al., 1988b; Stromberg et al., 1990). The importance of the Prs fimbrial adhesin complex may be inferred from its occurrence among various uropathogenic strains of *E. coli* and the distribution of the Prs adhesin receptor in various tissues. In humans the frequency of strains coexpressing both the P and Prs fimbrial adhesins is highest among *E. coli* isolated from patients with pyelonephritis and decreases considerably among strains isolated from cystitis or among fecal isolates (Arthur et al., 1989). In contrast, the proportion of strains expressing only the Prs adhesin was the highest among fecal isolates and much lower among cystitis isolates and virtually absent among pyelonephritis isolates. In another study, it was found that the magnitude of inflammatory response of patients with UTI caused by *E. coli* expressing only the Prs adhesin was significantly lower than that caused by *E. coli* expressing the Gal-Gal specific adhesin. P fimbriae-like antigens are also expressed by *E. coli* causing urinary tract infections in dogs (Garcia et al., 1988; Low et al., 1988). A number of these dogs' uropathogenic isolates were found to express the Prs adhesin only (Senior et al., 1988; Stromberg et al., 1990). The distribution of Prs adhesin receptors in various tissues was studied with Prs fimbrial preparations purified from recombinant *E. coli* harboring a plasmid containing the *prs* gene cluster. Purified Prs fimbrial adhesin complex bound to Bowman's capsule, glomerulus, lumen of some collecting ducts, smooth muscles from the bladder wall and bladder, and uroepithelia of renal pelvis and bladder, but weakly to tubules of the kidney (Karr et al., 1989, 1990). These data are consistent with the studies showing that the human kidney contains the Forssman antigen (a major glycolipid of sheep erythrocytes) and Prs-specific receptor sequences (Breimer et al., 1985). Glycolipids that bind Prs-expressing *E. coli* have also been detected in extracts from human uroepithelial cells (Lindstedt et al., 1991). The expression of the Prs adhesin receptor by uroepithelial cells was found to be restricted to individuals with blood group A and positive secretor state. A urinary isolate that expresses only Prs fimbrial adhesin complex bound only to uroepithelial cells of such individuals (Lindstedt et al., 1991). Interestingly, all six patients examined with UTI caused by *E. coli* expressing only Prs fimbrial adhesin were blood group A-positive, a statistic much higher than in the population at large. The results strongly suggest that the Prs fimbrial adhesin plays an important role in conferring an advantage for the colonization of the human upper urinary tract by uropathogenic *E. coli* also expressing the P-fimbrial adhesin. Strains that express Prs fimbrial adhesin as the only non type 1 fimbrial adhesin appear to be less virulent and are selected to cause UTI with reduced inflammatory response by individuals with secretor status and who are blood group A-positive (Lindstedt et al., 1991). Less clear is the situation in dogs. With a limited number of strains examined it appears that about one-third of the uropathogenic dog isolates express only the Prs fimbrial adhesin but no information is given as to the severity of the infection caused by these Prs-bearing strains. It was suggested that the distinct receptor specificity of Prs G adhesin enables *E. coli* strains bearing the Prs fimbrial adhesin to colonize specifically the urinary tract of dogs because (1) MDCK II tissue culture cells derived from dog kidney are rich in the Forssman antigen, a glycolipid complex accessible for avidly binding with *E. coli* bearing the Prs adhesin, but inaccessible for binding Pap G-bearing *E. coli* (Stromberg et al., 1990); and (2) all urinary isolates examined bearing Prs fimbrial adhesin did not express the Gal-Gal-specific P-fimbrial adhesin (Senior et al., 1988; Stromberg et al., 1990). The findings that human intestinal and urinary tracts may also become colonized by strains bearing only Prs fimbrial adhesin (Arthur et al., 1989; Lindstedt et al., 1991) suggest that the Prs G receptors of certain individuals may also become specifically accessible for Prs fimbriae-bearing organisms to render these individuals susceptible to a less severe form of UTI. It remains to be seen how severe is the UTI caused by *E. coli* strains bearing only Prs fimbrial adhesin in dogs. Such knowledge will be of value in relating the receptor specificity of the Prs G adhesin to a particular type of UTI infection in dogs.

Escherichia coli

Clinical source:	BLD, human
Adhesin:	S Fim
Receptor specificity:	SialylGal
Substrata:	EC, TB, TC

Although the *sfa* gene cluster coding for the S-fimbrial adhesin was originally cloned from *E. coli* isolated from urine of patients with UTI (Hacker et al., 1985), it soon became clear that the frequency of strains that either express S-fimbrial adhesin or contain the *sfa* gene cluster is low in patients with UTI. The *sfa*$^+$ strains were found mainly among isolates from patients with neonatal sepsis and meningitis (Korhonen et al., 1985; Ott et al., 1986). Moreover, most of the *sfa*-containing urinary strains also contain the gene cluster encoding for the P-fimbrial adhesin complex, suggesting that the S fimbriae alone are not sufficient for colonization of the urinary tract (Archambaud et al., 1988a). The epidemiological surveys showing that not all of the strains of *E. coli* from meningitis specimens contain the *sfa* gene cluster (Ott et al., 1986) and most of the strains expressing the S-fimbrial adhesin are homogeneous, belonging to a limited number of serotypes and electrophoretic types (Korhonen et al., 1985; Selander et al., 1986), suggest the S-fimbriae-bearing strains form a genetically related clonal group capable of causing neonatal meningitis and sepsis. Studies aimed to assess the contribution of the receptor specificity of the S-fimbrial adhesin to the ability of certain *E. coli* clones to colonize brain tissues include experiments in the molecular biology of the linkage between the *sfa* gene cluster and other virulence genes, accessibility of S adhesin receptors, presence of specific S-fimbrial inhibitors in various tissues, and expression of the S adhesin in vivo in experimental models of infections. As with other fimbrial adhesins, the S-fimbrial adhesins are subject to a relatively high rate of phase variation under either broth or agar growth conditions (Nowicki et al., 1985a,b, 1986b). The exact molecular mechanism at the DNA level responsible for the "on" and "off" switch in the expression of the S-fimbrial adhesin is poorly understood. A DNA element contained in a region coding for α-hemolysin II and Prs fimbrial adhesin functions as a positive regulator in *trans* to promote the transcription of *sfa* gene cluster (Knapp et al., 1986). Noncoding DNA sequences lacking the *sfa* gene cluster at both sides also seem to represent a unique property of virulent strains and have been termed the "virulence gene block" (Ott et al., 1986; Hacker, 1990). Exactly how these flanking DNA elements affect the expression of the S-fimbrial adhesin is not known. It is possible the expression of the S-fimbrial adhesin is linked with the expression of other virulence factors and that this linkage is unique to certain adhesins in certain clones of *E. coli,* such as the S-fimbrial producing clones (Hughes et al., 1987; Hacker, 1990). Although sialylgalactosides are common in glycoproteins of animal cell membranes, the accessibility of these residues for binding the S-fimbriated *E. coli* is shared by a rather limited number of cells. Glycophorin A on the erythrocyte surface is probably the major S adhesin receptor. The protein contains *O*-linked sialylgalactoside residues that bind the S-fimbriated bacteria (Parkkinen et al., 1986). S-fimbriated bacteria form a complex with purified glycophorin A immobilized on the bottom of microtiter plates and in blots of erythrocyte membrane lysates. The adhesion can be inhibited by sialic acid-containing oligosaccharides (Parkkinen et al., 1986). A number of other different cell types bind the S-fimbrial adhesin. Purified S-fimbrial adhesin or a recombinant *E. coli* expressing S adhesin adheres to various tissue sites of frozen sections of human kidney (Korhonen et al., 1986b). Except for the connective tissue that bound the P-fimbrial adhesin, but not the S-fimbrial adhesin, all other tissue sites of the kidney as well as the bladder epithelium and uroepithelial cells complex with the S-fimbriated *E. coli* and its fimbrial adhesin (Korhonen et al., 1986b, 1990; Virkola et al., 1988). Whereas this pattern of adhesion of S-fimbriated bacteria is similar to that of P-fimbriated or type 1 fimbriated bacteria, cultured human endothelial cells bind only the S-fimbriated bacteria (Parkkinen et al., 1989). S-fimbriated *E. coli* and its isolated S-fimbrial adhesin also adhere to the lumenal surfaces of the vascular endothelium and of the epithelium lining the choroid plexuses as well as brain ventricles of frozen sections of rat brain, whereas P-fimbriated *E. coli* do not bind to the brain tissues (Parkkinen et al., 1988). Although Sfa S is the minor sialylgalactoside specific fimbrial adhesin, the major Sfa A fimbrial subunit and the minor Sfa G fimbrial subunit contribute to the adhesion of the recombinant S-fimbriated *E. coli* to cultured renal tubular cells derived from pig kidneys (Marre et al., 1990). Whenever tested, sialyllactose inhibited adhesion of S-fimbriated bacteria or its S-fimbrial adhesin to the various tissues, suggesting that the binding was mediated by the sialylgalactose specific Sfa S adhesin. Although S-fimbriated bacteria interact poorly, if at all, with phagocytic cells (Nowicki et al., 1986b; Konig et al., 1986), release of mediators of inflammation (e.g., leukotriene, oxygen radicals, lysosomal enzymes) from human polymorphonuclear leukocytes by hemolysin-producing strains is markedly potentiated by the presence of S fimbriae on the surface of the organisms (Konig et al., 1986). It is not clear at this stage whether this potentiation effect is specific for the S-fimbrial dependent adhesion.

A number of animal models have been employed to assess the role of the S-fimbrial adhesin in vivo. In one set of experiments the S-fimbriated bacteria were grown in broth and a subpopulation of bacteria enriched with phenotypes bearing type 1 or S fimbriae or nonfimbriated phenotypes were obtained by adsorption on erythrocytes or yeast, followed by elution. The various subpopulations were injected intraperitoneally in mice (Nowicki et al., 1986b) or infant rats (Saukkonen et al., 1988). In both animal models the type S fimbriated phenotype was the most virulent as judged by mortality and persistence in body fluids, with no major shift to other phenotypes. The type 1 fimbriated phenotype was eliminated rapidly, probably because of lectinophagocytosis (see Chapter 7). In contrast, inoculation with a population rich in the nonfimbriated phenotype resulted in a shift within a short period of time to a population rich in the S-fimbriated phenotype in the peritoneal cavity or the blood. This shift appears to be the result of outgrowth of the S-fimbriated phenotype in media supplemented with mouse serum or peritoneal fluid. In the mouse model, the S-fimbriated phenotype spread to the blood and anti-S-fimbrial serum protected the mice from a lethal dose of S-fimbriated phenotype. In the infant rat model, the organisms also spread to the cerebrospinal fluid (CSF). The number of nonfimbriated bacteria in the CSF was equal to that of S-fimbriated in animals inoculated with a population of bacteria enriched with the S phenotype. The data were interpreted to suggest that the nonfimbriated phenotype confers an advantage for the population to penetrate the endothelium, thereby reaching the CSF where the S-fimbriated phenotype overgrows the nonfimbriated phenotype (Saukkonen et al., 1988). *E. coli* strains constructed to express S fimbriae colonized the kidney in higher numbers than isogenic strains expressing type 1 fimbriae, whereas the latter colonized the bladder in higher numbers than the former 1 week after intravesicular injection of the recombinant strains (Marre and Hacker, 1987). S fimbrial adhesin contributed to the infectivity (e.g., nephropathogenicity) of the wild-type uropathogenic strain given intravenously in mice, but its expression enhanced the infectivity of the strain in intranasal-induced lung toxicity or in a subcutaneous-induced sepsis model in mice (Hacker et al., 1986a). Specific inhibition of S-fimbrial adhesin by sialyloligosaccharide chains of the Tamm–Horsfall glycoprotein in human urine may partially explain the low ability of S-fimbriated *E. coli* to cause UTI (Parkkinen et al., 1988).

In summary, in spite of the accessibility of S adhesin receptors in kidney and bladder cells, S-fimbriated bacteria rarely cause urinary tract infections, which reemphasizes the notion that accessibility of bacterial adhesin receptors is not sufficient for colonization. S fimbriae provide *E. coli* the ability to colonize endothelial and brain tissue cells. Their expression is genetically linked to the expression of other virulence factors in wild-type strains. Virulence in animal models of S-fimbriated bacteria is dependent on the site of inoculation. There is a rapid population shift from nonfimbriated phenotypes to S-fimbriated phenotypes in vivo. These S fimbriae-associated properties enable certain *E. coli* clones to initiate blood infection and meningitis in neonates and to a much lesser extent urinary tract infections. The gene cluster encoding for the S-fimbrial adhesin apparently undergoes mutations resulting in strains bearing closely related fimbriae, some of which retain adhesin function with slightly different receptor specificities, such as type 1C fimbrial adhesin (see below), whereas others may lose adhesin function. For example, a blood isolate of *E. coli* has been shown to express fimbriae termed Sfr (Pawelzik et al., 1988). Recombinant strains of *E. coli* harboring a plasmid containing the DNA element coding for Sfr fimbriae express the Sfr fimbrial antigen, but fail to cause hemagglutination of erythrocytes from different species, including human and bovine erythrocytes, which normally react with S-fimbriated *E. coli*. At this stage, the Sfr fimbriae have been designated as nonadhesive (Hacker, 1990). Because the closely related type 1C fimbrial adhesin also does not cause hemagglutination, but mediates adhesion of *E. coli* to a number of host cells, it remains to be determined if the adhesion of Sfr-bearing recombinant strains to various host tissues is truly reflective of nonadhesive fimbriae.

Escherichia coli

Clinical source:	UTI,
Adhesin:	Type 1C Fim
Receptor specificity:	ND
Substrata:	TB, TC

The type 1C fimbrial adhesion complex, which shares a structural similarity with type 1 fimbriae, was originally thought to lack any adhesin function because it did not cause hemagglutination of erythrocytes from a number of species. It is now clear that it is serologically distinct from other fimbrial adhesins (Pere et al., 1985; Schmitz et al., 1986). It mediates binding of *E. coli* to various tissues (Virkola et al., 1988; Marre et al., 1990) and is related genetically, immunologically, as well as functionally to the S-fimbrial adhesin (Marre et al., 1990; Van Die et al., 1991, Chapter 9). None of 50 fecal isolates tested expressed the type 1C fimbrial adhesin, but about 17–20% of urinary isolates and 15% of blood and meningeal isolates express the type 1C fimbrial adhesin. Most of these strains also express the P-fimbrial adhesin (Pere et al., 1985, 1987; Selander et al., 1986). Organisms expressing type 1C fimbrial adhesin were detected in two urine sediment samples and in none of the urine samples out of 20 examined from patients with UTI (Pere et al., 1987). Agar-grown *E. coli* isolated from four patients expressed the type 1C fimbriae, suggesting that as with other fimbrial adhesins, the expression of type 1C adhesin undergoes phase variation. Type 1C-carrying strains do not bind to uroepithelial cells but they do bind to epithelial cells of the distal nephron, collecting duct, and endothelial cells of vessels in human kidney tissue sections (Korhonen et al., 1990; Virkola et al., 1988) and to human buccal cells (Schmitz et al., 1986). *E. coli* strains engineered to express type 1C fimbrial adhesin bind to cultured renal tubular cells from pig kidney. This adhesion is inhibited strongly by *N*-acetyllactosamine and poorly by other sialylgalactoside-containing oligosaccharides which normally strongly inhibit the adhesion of S-carrying *E. coli* to the cells (Marre et al., 1990). The results suggest that the receptor specificity of the type 1C fimbrial adhesin differs slightly from that of S-fimbrial adhesin, as expected from the high percent homology between the Sfa S adhesin and the Foc G minor fimbrial subunit of the type 1C fimbrial adhesin complex (Van Die et al., 1991). This difference is apparently responsible for the lack of hemagglutinating activity of strains carrying type 1C fimbrial adhesin. It has been postulated because the type 1 fimbrial adhesin is mostly produced by P-fimbriated strains that cause pyelonephritis and other severe form of *E. coli* infections, the type 1C fimbrial adhesin with its unique receptor specificity may confer upon certain P-fimbriated strains the ability to overcome specific host defense mechanisms. The defense mechanisms normally operate in a limited number of hosts that select for strains expressing type 1C and P-fimbrial adhesins (Korhonen et al., 1990).

Escherichia coli

Clinical source:	UTI, human
Adhesin:	Dr Fim
Receptor specificity:	Dr blood group
Substrata:	EC, ERT, TB

The Dr fimbrial adhesin complex was originally described as 075X adhesin because it was expressed exclusively by *E. coli* strains of O-group 75 (Korhonen et al., 1986c; Nowicki et al. 1987). When the erythrocyte receptor that binds these 075 *E. coli* was identified as the Dr blood group antigen the adhesin was designated as the Dr fimbrial adhesin complex or Dr hemagglutinin (Nowicki et al., 1988b). Further studies using hybridization with a DNA probe, *dra*D, derived from the Dr gene cluster, revealed that although a high proportion of *dra*D-positive *E. coli* strains were found among 075, a high proportion of strains belonging to 012 and a low proportion of strains belonging to other uropathogenic O serotypes also hybridized with the *dra*D probe (Nowicki et al., 1989). The probe hybridized with about one-fourth of strains isolated from patients with UTI. The Dr DNA probe hybridized with about 6% of the strains isolated from patients with either symptomatic bacteriuria or pyelonephritis and with 15% of fecal strains, but with a significantly higher proportion (28%) of strains isolated from patients with cystitis. It appears that the Dr fimbrial adhesin partially enables *E. coli* to cause inflammatory disease of the lower urinary tract.

In attempts to understand the type of contribution of the Dr fimbrial adhesin, and especially the receptor specificity of this adhesin, to the infectivity of the *E. coli,* a number of studies have been carried out to determine the distribution of the Dr receptors in various tissues. The Dr^{a} antigen is contained in the decay-accelerating factor, a cell membrane protein that is involved in regulating the complement cascade to prevent complement-mediated lysis of cells (Nowicki et al., 1988a). Various forms of the Dr fimbrial adhesin complex, including those expressed by urinary isolates (Nowicki et al., 1986b) and by recombinant strains that contain the Dr gene cluster (Nowicki et al., 1987, 1988b), and fimbrial adhesin immobilized on microparticles (Westerlund, 1991), were employed to locate Dr adhesin receptors in frozen sections of human tissues from various organs. The Dr fimbrial adhesins bind to several sites of the urinary tract, including the renal interstitium, Bowman's capsule, and tubular basement membrane, but not epithelial cells, urethral transitional epithelial cells, or to uroepithelial cells exfoliated in urine. They also bind to different parts of the digestive, respiratory, urinary, and genital tracts, and skin, as well as to colonic, bronchial, and endometrial glands (Nowicki et al., 1988a). Dr adhesin receptors are not restricted to human tissue because the purified Dr fimbrial adhesin also binds to various parts of the entire urinary tract of dogs in a manner similar to its adhesion to human tissues (Westerlund et al., 1987). Although the Dr antigen is a receptor for other adhesins expressed by uropathogenic (AFA-I and AFA-III afimbrial adhesins) and diarrheal (F1845 adhesin) strains of *E. coli,* the attachment site that specifically binds the Dr fimbrial adhesin is distinct (Nowicki et al., 1990). For example, hemagglutination caused by Dr-fimbriated *E. coli* is not affected by pretreating the erythrocytes with trypsin or papain and is abolished by chymotrypsin or pronase pretreatment. In contrast, hemagglutination caused by *E. coli* carrying AFA-I is completely abolished by pretreating the erythrocytes with chymotrypsin and pronase and partially abolished with papain, whereas that caused by AFA-III is abolished by either chymotrypsin or pronase. That caused by the F1845 adhesin is completely abolished by chymotrypsin and only partially by pronase. Furthermore, a monoclonal anti-Dr group antigen inhibited hemagglutination caused only by Dr fimbrial adhesin complex. The receptor analogues, chloramphenicol and acetylated tyrosine, inhibit the hemagglutination caused only by Dr fimbrial adhesin as well as the binding of *E. coli* expressing the Dr adhesin to tubular basement membrane and Bowman's capsule of the human kidney, suggesting that the attachment site for binding Dr adhesin contains tyrosine-like residues (Nowicki et al., 1988a, 1990). The Dr antigen is probably not the only Dr adhesin receptor, especially in the basement membrane, because *E. coli* expressing the Dr fimbrial adhesin complex as well as the purified adhesin strongly bind type IV collagen and its 7S domain (Westerlund et al., 1989a). Binding of the Dr adhesin to type IV collagen was inhibited by chloramphenicol, and was not affected by pretreating the collagen with endoglycosidase-H or periodate, suggesting that the adhesion involves protein–protein interactions. The purified adhesin also binds weakly to laminin, fibronectin, and other types of collagens (Westerlund et al., 1987, 1989a). Perhaps the specific attachment site on the Dr antigen and its accessibility for binding the Dr carrying *E. coli* in various host cell, together with collagen binding activity of the Dr adhesin, contributes to the ability of strains expressing the Dr fimbrial adhesins to cause cystitis.

Escherichia coli

Clinical source:	UTI, human
Adhesin:	AFA-I
Receptor specificity:	ND
Substrata:	EC, ERT, TC

The AFA-I adhesin complex belongs to a family of adhesins that are not in fimbrial form, but are expressed by about 10% of urinary or blood isolates of *E. coli*. AFA-I carrying *E. coli* cause MR hemagglutination of human erythrocytes and adhere to uroepithelial cells and to HEp-2 cells (Labigne-Roussel and Falkow, 1988). The Afa E subunit of the afimbrial adhesin-I (AFA-I) complex is probably responsible for these adhesive activities (Labigne-Roussel et al., 1985; Labigne-Roussel and Falkow, 1988). Hybridization experiments with a DNA probe derived from the *afa*-I gene cluster revealed that about 10% of strains isolated from urine of patients with urinary tract infections and 5% of fecal isolates hybridized with a probe overlapping most of the genes contained in the *afa* operon, but only four strains hybridized with a probe containing a DNA sequence encoding the last 83 C-terminal residues of the *afa*E adhesin. DNA elements corresponding to the *afa* operon from each of two strains that did not hybridize with the *afa*-I adhesin probe were cloned into *E. coli*. The recombinant *E. coli* caused MR hemagglutination, adhered to uroepithelial cells, and lacked DNA sequences of the AFA-I adhesin-encoding gene, suggesting that they encode an afimbrial adhesin closely related to that of AFA-I but containing immunologically distinct adhesins. Although the AFA adhesin complex of these two strains was designated AFA-II and AFA-III, it is not clear how distinct are their receptor specificities, compared to that of the AFA-I adhesin complex, to warrant a different adhesin designation. Preliminary studies showing that pretreatment of erythrocytes with trypsin did not affect hemagglutination caused by AFA-III and partially abolished hemagglutination caused by AFA-I are not conclusive evidence that the receptor specificities of the two AFA adhesins are very different from each other.

The AFA adhesin complex and its gene cluster is carried by about 10% of pyelonephritogenic strains, most of which also express the P-fimbrial adhesin complex and contain the *pap* gene cluster, suggesting that strains containing only the *afa* gene cluster are rarely selected as uropathogens (Archambaud et al., 1988a,b). Interestingly, the *afa* gene cluster is carried by about 10% of EPEC strains which probably also contain gene clusters encoding for additional MR adhesins (Archambaud et al., 1988b, and above). The epidemiological data suggest that the AFA adhesin complex alone is not sufficient to endow upon *E. coli* the ability to colonize the urinary or the intestinal tracts. The epidemiological data emphasize the premise that *E. coli* capable of expressing multiple adhesins, such as P fimbriae and AFA adhesin, with distinct receptor specificities, are selected by certain hosts to cause a severe form of UTI because they are adapted to overcome host defense mechanisms at particular stages of the infectious process.

Escherichia coli

Clinical source:	UTI, BLD, human
Adhesin:	NFA-1,2 Fim
Receptor specificity:	ND
Substrata:	ERT, PC, TC

An *E. coli* strain (083:K1-H4), isolated from urine and blood of an elderly patient, was found to express a nonfimbrial adhesin designated as NFA-1 (Goldhar et al., 1987). The NFA-1 carrying isolate, recombinant *E. coli* harboring a plasmid containing NFA-1-encoding genes, and high molecular weight aggregates of purified soluble NFA-1 adhesin complex cause hemagglutination of human erythrocytes only, and bind to human kidney tissue culture cells as well as to human polymorphonuclear leukocytes (Goldhar et al., 1987, 1991; Hales et al., 1988). The adhesion of the NFA-1 expressing *E. coli* is inhibited by the purified adhesin, suggesting that the NFA-1 adhesin complex mediates the adhesion of the organisms to the cells. The adhesin is best expressed by agar-grown bacteria, released in the medium by broth-grown bacteria, and is present on the bacterial surface as a capsule-like layer. Its expression is probably under phase variation control because the agar-grown bacteria can be fractionated by hemadsorption into NFA expressing and nonexpressing subpopulations. An *E. coli* strain (014:K?-H11) isolated from urine of a patient with UTI was found to express an antigenically distinct nonfimbrial adhesin designated as NFA-2 (Goldhar et al., 1987). Although NFA-2 is not expressed as a capsule-like layer on the surface of the organisms, it probably mediates the binding of the bacteria to a closely related receptor on erythrocytes and human kidney cells. It is difficult at this stage to appreciate the role of the NFA-2 adhesins in the infectious process as no data are available about their frequency of isolation in various types of *E. coli* infections.

Escherichia coli

Clinical source:	BLD, human
Adhesin:	NFA-3
Receptor specificity:	Glycophorin
Substrata:	ERT, PC

An *E. coli* (020:KX104:H) isolated from the blood of a patient with sepsis was found to express a nonfimbrial adhesin designated as NFA-3 (Grunberg et al., 1988). Human erythrocytes of blood group NN are preferentially agglutinated by NFA-3 carrying *E. coli* and its purified NFA-3 adhesin complex, which like other NFA adhesins, is also in the form of high molecular weight aggregates. Interestingly, NFA-3 adhesin preferentially mediates the binding of the bacteria to human PMNs from donors with blood group NN (Grunberg et al., 1992). Glycophorin A^{NN}-like determinants appear to serve as receptors on PMN for binding to and subsequently ingestion and killing by the phagocytic cells of the NFA-3-carrying *E. coli*. No data are available on the frequency of isolation of *E. coli* expressing the NFA-3 adhesin.

Escherichia coli

Clinical source:	UTI, human
Adhesin:	NFA-4
Receptor specificity:	Glycophorin A^{MM}
Substrata:	ERT

An *E. coli* (07:K98:H6) isolated from a patient with UTI caused hemagglutination of chicken and human erythrocytes. Human erythrocytes belonging to human blood group NN are preferentially agglutinated (Hoschützky et al., 1989). A purified nonfimbrial soluble hemagglutinin, designated as NFA-4, in the form of high molecular weight aggregates analogous to other NFA adhesins, also preferentially agglutinated erythrocytes of blood group NN. The hemagglutination was readily inhibited by glycophorin A^{MM}. Like NFA-1, the NFA-4 adhesin also is expressed on the *E. coli* surface as a capsule-like layer (Kröncke et al., 1990). Glycophorin A^{MM} also serves as a receptor for a nonfimbrial adhesin designated M adhesin expressed by a urinary isolate strain of *E. coli* 02 serotype and possibly by another strain designated KS300 (Rhen et al., 1986). The M agglutinin or adhesin is specifically inhibited by L-serine and L-threonine and is antigenically distinct from NFA-4 but it is not clear whether the two adhesins bind to the same attachment site on glycophorin A^{MM}. Because both the NFA-4- and M-carrying strains of *E. coli* were isolated from patients with UTI and the two adhesins are nonfimbrial and exist in the form of high molecular weight aggregates in solution, it is preferred at this stage to classify the adhesins under the same category of receptor specificity. Perhaps when more data as to the frequency of isolation of *E. coli* strains expressing blood group MM-specific adhesins become available it will be possible to determine the importance of this group of adhesins in vivo and whether they represent a heterogeneous group of adhesins with respect to their receptor specificities. In a preliminary study, it was found that only one out of 102 strains of *E. coli* isolated from patients with pyelonephritis expressed the M adhesin (Archambaud et al., 1988b).

Escherichia coli

Clinical source:	UTI
Adhesin:	G Fim
Receptor specificity:	GlcNAc
Substrata:	ERT

An *E. coli* 02 urinary isolate that expresses the M adhesin described above also expresses a GlcNAc-specific fimbrial adhesin. Both the fimbrial adhesin and a recombinant *E. coli* harboring the G fimbriae-encoding genes cause hemagglutination of erythrocytes treated with endo-β-galactosidase to expose internal GlcNAc residues (Rhen et al., 1986). Kidney tissue avidly binds the plant lectin wheat germ agglutinin, suggesting that it contains accessible GlcNAc residues (Rhen et al., 1986). It is difficult, however, to determine at this stage whether G-fimbriated strains are associated with a particular form of urinary tract infection as only one strain expressing the GlcNAc specific fimbrial adhesin has been reported. Because the G fimbriae-carrying strain agglutinates only endogalactosidase-treated erythrocytes, it is possible that isolates expressing this adhesin escape detection.

Escherichia coli

Clinical source:	UTI, dog
Adhesin:	ND
Receptor specificity:	Sialic acid?
Substrata:	ERT

Hemagglutination caused by canine isolates is frequently sensitive. The DNA of the isolates contain sequences that hybridize with probes obtained from DNA coding for the P-fimbrial adhesin of human isolates (Low et al., 1988). The results suggest that the receptor specificity of the adhesin of canine isolates is different from that of the Gal-Gal-specific adhesin of pyelonephritogenic isolates, although both adhesins are carried by closely related fimbrial structures.

Escherichia coli

Clinical source:	UTI, human
Adhesin:	NFA-6, ND
Receptor specificity:	ND
Substrata:	ERT, PC, TC

Because of the phenomenon of phase variation the assay for detecting the phenotypic expression of adhesins expressed by *E. coli* may miss strains that contain the gene cluster encoding for at least one adhesin complex but are in the "off" phase. Hybridization of colonies with DNA probes derived from gene cluster(s) encoding for three or four *E. coli* adhesins (S, type 1C, P- and S-fimbrial adhesins) "genotypic assay," revealed that about 10% of the strains isolated from the urine of patients with pyelonephritis had no homology with any of the different probes, nor did they adhere to uroepithelial cells (Archambaud et al., 1988b). In another study, it was found that about 30% of the strains isolated from patients with UTI, including pyelonephritis, cystitis, and bacteriuria, lacked the adhesin-coding gene cluster of the AFA adhesin or the P, S, type 1C fimbrial adhesin gene clusters (Archambaud et al., 1988a). Although probes derived from NFAs or M or G encoding gene clusters were not used in the genotypic assay, it is possible that there are many other unidentified adhesins expressed by uropathogenic *E. coli*. For example, adhesion of two isolates of *E. coli* to HEp-2 cells was dependent of the presence on the bacterial surface of both the type 1 fimbrial adhesin and an unidentified MR adhesin (Tavendale and Old, 1985). This type 1 and MR adhesin was not inhibited by mannose, suggesting the adhesion is mediated by an unidentified adhesin and that the contribution of type 1 fimbrial adhesin complex to the ability of the strain to adhere to HEp-2 cells does not involve the mannose-binding site of Fim H.

A poorly defined adhesin, termed MIAT, was found to mediate adhesion of 7 out of 250 urinary strains of *E. coli* tested to human uroepithelial and buccal cells (Pruzzo et al., 1989b). This adhesin is a receptor for coliphage T7 because irradiated phage specifically inhibits the adhesion of the *E. coli* strains to the epithelial cells and spontaneous T7 phage-resistant mutants bind more poorly to the epithelial cells as compared to their T7 phage-susceptible derivatives. The coliphage T7 receptor-adhesin, termed MIAT, is expressed by *E. coli* strains that either do or do not express other adhesins, such as the type 1 fimbrial adhesin or other MR adhesins. The presence of MIAT adhesin on the surface of the organisms seems to interfere with type 1 mediated attachment of the bacteria to metabolically active human PMN.

An *E. coli* strain (021:K4:H4) isolated from the stool of a patient with diarrhea was found to express an immunologically distinct nonfimbrial adhesin, termed NFA-6 or Z1, which is expressed on the bacterial surface in the form of a capsule (Orskov et al., 1985; Kroncke et al., 1990). This poorly defined adhesin mediates the binding of the organisms to uroepithelial and buccal cells.

The surface hydrophobicity of the bacteria may contribute greatly to the ability of the organisms to adhere to animal cells (Doyle and Rosenberg, 1990). Although early studies attempted to correlate hydrophobicity of fimbriated *E. coli* and adhesion there is no conclusive evidence that a particular hydrophobin on the bacterial surface is involved in the adhesion process. A positive correlation between the hydrophobicity of the surfaces of various *E. coli* isolates and their ability to cause hemagglutination or to adhere to host cells (e.g., foreskin and rabbit intestinal cells) or mucus has been observed (Fussell et al., 1988; Drumm et al., 1989; Herpay et al., 1991) but no such correlation has been observed in other studies employing buccal cells as target substrata for adhesion (Sainz et al., 1987). No correlation has been found between expression of P-fimbriae and by *E. coli* and cell surface hydrophobicity although P-fimbriated pyelonephritogenic isolates were more hydrophobic than fecal isolates (Jacobson et al., 1987). In one study, it was found that the binding of a urinary isolate of *E. coli* to immobilized glycolipids in thin-layer chromatograms is unrelated to the presence of fimbriae and is dependent on both the saccharide and the covalently bound lipid moiety, suggesting the involvement of hydrophobic interaction in the binding process (Rosenstein et al., 1988).

PART B: RECENT ADVANCES ON ADHESION OF NON-ESCHERICHIA COLI BACTERIA

Actinomyces israelii

Clinical source:	ORL
Adhesin:	Sialic acid and L-rhamnose containing polysaccharide, type 1 and/or 2 fimbriae?
Receptor specificity:	Carbohydrate binding protein; collagen; proline-rich salivary protein
Substrata:	Oral bacteria, HA (collagen coated), EC (sialidase treated)
References:	Weiss et al., 1987a; Gibbons, 1989; London et al., 1989; Childs and Gibbons, 1990; Liu et al., 1991; Tables 8–6 and 8–7.

Actinomyces naeslundii

Clinical source:	ORL
Adhesins:	Type 2 Fim
Receptor specificity:	β-Galactoside; GalNAcβ1,3Gal
Substrata:	Oral streptococci, ERT, EC, PC, SHA

Actinomyces naeslundii expresses the β-galactoside-specific fimbrial adhesin, type 2 fimbriae, through which the organisms interact with various types of cells encountered in the oral cavity, including oral epithelial cells and human polymorphonuclear leucocytes, and other oral bacteria (e.g., *S. sanguis;* Table 8–6) (Mergenhagen et al., 1987). The genes coding for the type 2 fimbrial adhesin consisting of 54,000 mol wt mature subunits have been cloned (Table 9–1). The purified fimbriae cause haemagglutination only after crosslinking (Cisar, 1986), suggesting that they act as monovalent structures similar to the type 1 fimbrial adhesin (Chapter 9). The receptor site bearing β-galactoside residues on the eucaryotic cells is, however, cryptic, because adhesion is greatly enhanced by enzymatic desialylation of cells (Childs and Gibbons, 1988, 1990). The relationship between the ability of *A. naeslundii* to adhere to and colonize various surfaces during health and disease has been examined by quantitating the amount of sialic acid residues on epithelial cells obtained from healthy individuals and from patients with gingivitis (Gibbons et al., 1990). The results show that the amount of sialic acid released by sialidase from the patients' cells was significantly less than that released from healthy individuals, suggesting that during gingivitis the epithelial cells become desialylated and presumably possess increased amounts of galactoside-containing receptors accessible for binding the *A. naeslundii*. It has been suggested that sialidase (as well as other enzymes) may be released from inflammatory host cells, epithelial cells, or plaque bacteria in individuals with altered oral hygiene, enhancing adhesion of the *Actinomyces* to oral cells and increasing the amount of the type 2-bearing bacteria colonizing the oral cavity. This hypothesis is consistent with the shift in the oral microbiota during periodontal disease (Table 8–1). Using specific fluorescein labeled antibodies, it has been shown that oral epithelial cells obtained from healthy individuals contain mostly adherent *A. naeslundii* bearing type 2 fimbriae (Cisar, 1986; Cisar et al., 1985; 1988). Because both *A. naeslundii* and *A. viscosus* express type 2 fimbrial adhesins with the same receptor specificity the results suggest that factors other than adhesion (e.g., growth conditions) permit colonization of oral epithelial cells preferentially by type 2 fimbriae carrying *A. naeslundii*.

A. naeslundii also binds to GalNAcβ1,3Gal-containing glycolipids, but there are no data to suggest that glycolipids can serve as receptors for the bacteria on mammalian cells (Brennan et al., 1987). In contrast, a 160,000 mol wt glycoprotein on the surface of oral epithelial cells was found to serve as a receptor that binds *A. naeslundii* via its type 2 fimbrial lectin (Brennan et al., 1986). The 160,000 mol wt glycoprotein was readily extracted by *n*-octylglucoside, suggesting that it is loosely bound to the epithelial cells. *A. naeslundii* also interacts with human salivary mucin (Koop et al., 1990). A 180,000 mol wt salivary glycoprotein has been found to serve as a receptor that binds *A. naeslundii* to buccal epithelial cells (Babu et al., 1991). The enhanced adhesion of the bacteria to epithelial cells coated with the salivary glycoprotein is sensitive to lactose, suggesting that it is mediated by the type 2 fimbrial adhesin complex. It is possible, therefore, that desialylated and β-galactoside-containing salivary glycoproteins become preferentially adsorbed onto oral epithelial cells, thus serving as receptors for *A. naeslundii*. The coaggregation of *A. naeslundii* with many strains of *S. milleri* (Eifuku et al., 1991) and other oral bacteria (Table 8–6) is not inhibited by lactose, suggesting that the *Actinomyces* express adhesins other than type 2 fimbriae. Also, interaction of *A. naeslundii* with *Prevotella denticola* is inhibited by chlorhexidine, whereas that with *S. oralis* is not, suggesting that *A. naeslundii* possesses two distinct molecular mechanisms of adhesion, one of which participates with the streptococci and the other with the Gram-negative oral bacteria (Smith et al., 1991).

Actinomyces viscosus

Clinical source:	ORL
Adhesins:	Type 2 Fim, type 1 Fim
Receptor specificities:	β-Galactoside; GalNAcβ1,3Gal; proline-rich salivary protein (PRP); statherin
Substrata:	Oral bacteria, ERT, SHA, EC, PC

A viscosus is capable of expressing two fimbrial adhesins. One adhesin is the β-galactoside-specific (type 2 fimbriae), which is immunologically and functionally similar to the type 2 fimbriae of *A. naeslundii* described above (Yeung and Cisar, 1990). Whenever tested, the receptor specificity of *A. viscosus* type 2 fimbrial adhesins and the types of substrata that bind the type 2 fimbriated *A. viscosus* are the same as those of type 2 fimbriated *A. naeslundii* (Gabriel et al., 1985; Cisar, 1986). The other fimbrial adhesin is type 1 fimbriae and is produced only by *A. viscosus*. The type 1 fimbrial adhesin was cloned and found to consist of about 54,000 mol wt mature major subunit (Yeung et al., 1987; Yeung and Cisar, 1990). It possesses 33% and 34% sequence identity with the type 2 fimbrial adhesins of *A. viscosus* and *A. naeslundii*, respectively. Functionally, the type 1 fimbrial adhesin of *A. viscosus* is distinct from that of the type 2 fimbrial adhesin. Studies with mutants and specific antibodies have confirmed earlier findings showing that the type 1 fimbrial adhesin is responsible for binding the organisms to saliva-coated hydroxylapatite, whereas type 2-carrying *Actinomyces* bind poorly to the apatites (Clark et al., 1986; Cisar et al., 1988, 1991). Additional studies have confirmed that human salivary acidic proline-rich proteins and statherin are the principal salivary constituents responsible for promoting the adhesion of type–1-bearing *A. viscosus* to saliva-coated apatitic surfaces (Gibbons and Hay, 1988; Gibbons et al., 1988; Clark et al., 1989; Leung et al., 1990). A purified salivary proline-rich glycoprotein from parotid saliva adsorbed onto gold particles or Latex beads or hydroxylapatite adhered specifically to bacteria expressing type 1 fimbriae but did not bind with strains lacking type 1 fimbriae. The type 1 fimbriae did not interact with Latex beads treated with other proteins, such as salivary amylase, salivary histidine-rich polypeptides, or laminin. Soluble proline-rich proteins did not inhibit the interaction of the *Actinomyces* with the immobilized salivary protein, suggesting that the salivary protein bound to apatite undergoes conformational changes to expose cryptic receptor sites that are recognized by the type 1 fimbriae. It was suggested further that evolutionary pressures selected the type 1 bearing *A. viscosus* so that the organisms can participate in the buildup of the dental plaque as an initial colonizer in the presence of saliva, which contains the soluble form of the cryptic receptor (Gibbons, 1989; Gibbons et al., 1990). The type 1 fimbriae appear also to be responsible for binding the *Actinomyces* to human type I or type III collagen in fiber form or immobilized on apatite surface. The binding is probably due to structural similarities between collagen and the carboxy-terminal region of the salivary proline-rich protein (Gibbons, 1989; Liu et al., 1991). The importance of binding to collagen in the buildup of plaque is not clear but a role in promoting invasion of the bacteria into gingival connective tissue has been suggested (Gibbons, 1989).

A viscosus binds to pellicle, probably via the two fimbrial adhesins, because immunoelectron microscopy has revealed that the type 2 adhesin is expressed in a similar pattern of distribution as that of its polysaccharide receptor (McIntire et al., 1988) on *S. sanguis* in samples of human dental plaques. These results were obtained by use of specific fluorescein or rhodamine-labeled antibodies (Cisar et al., 1985; Mergenhagen et al., 1987). Furthermore, the two fimbrial adhesins are probably coexpressed by the *Actinomyces*, because both anti-type

2 and anti-type 1 fimbrial antibodies react with bacteria in plaque samples (Cisar, 1986) and in suspension after in vitro growth (Cisar et al., 1991). Also *A. viscosus* was exhibited very densely on almost 100% of the cells examined whereas *A. naeslundii,* which carries only type 2 fimbriae, exhibited very sparse fimbriae on about 90% of the cells (Handley, 1990). *A viscosus* probably is capable of producing other adhesins because some coaggregation reactions with oral bacteria are not inhibited with lactose and are heat-resistant (Eifuku et al., 1991; Tables 8–6 and 8–7).

The central role of β-galactoside-specific type 2 fimbrial lectin in adhesion of *A. viscosus* to various oral substrata has prompted studies aimed at identifying potent inhibitors of the fimbrial lectin (McIntire et al., 1988). Using desialylated erythrocytes as a target substratum, it was found that desialylated bovine caseinoglycomacropeptides (8 μg/ml of galactose residues) strongly inhibited hemagglutination (Neeser et al., 1988b). Inhibition is probably obtained by GalNAcβ1,3Gal sequences contained in these glycolipids. A number of proteinaceous macromolecules (albumin, desialylated bovine caseinoglycomacropeptide, and mucin type glycoproteins) were potent inhibitors of adhesion of *A. viscosus* to polystyrene surfaces. The same compounds inhibited the adhesion of other oral bacteria (*S. sanguis, S. mutans*) to polystyrene surfaces, suggesting a nonspecific inhibition. Neither related mono- nor disaccharide compounds contained in the glycopeptide inhibited adhesion to polystyrene. Highly positively or negatively charged polypeptides and polysaccharides were weak inhibitors. The results suggest that the polystyrene–A *viscosus* interaction is promoted by the hydrophobic effect.

Aeromonas hydrophila

Clinical source:	ENT
Adhesin:	Fimbriae
Receptor specificities:	Fibronectin (29,000 mol wt amino-terminal region), collagen types I and IV, laminin, vitronectin, fucose
Substrata:	ECM, ERT, TC, TB
References:	Burke et al., 1986; Atkinson et al., 1987; Carrello et al., 1988; Elbashir and Millership, 1989; Ascencio et al., 1990a,b, 1991; Hokama et al., 1990; Honma and Nakasone, 1990; Hokama and Iwanga, 1991; Nishikawa et al., 1991

Aeromonas sobria

Clinical source:	ENT
Adhesin:	Fimbriae
Receptor specificities:	Collagen type I and vitronectin, galactose, mannose
Substrata:	ECM, ERT, TB
References:	Ascencio et al., 1991; Hokama and Iwanaga, 1991

Bacillus thuringiensis

Clinical source:	Insects
Adhesin:	Fimbriae?
Receptor specificity:	ND
Substrata:	ERT
Reference:	Smirnova et al., 1991

Bacteroides fragilis

Clinical source:	ENT
Adhesins:	Fimbria? 70,000 mol wt protein
Receptor specificity:	α-D-glucosamine, D-galactosamine
Substrata:	ERT, EC, TC, IM, *E. hirae* (coaggregation)
References:	Rogemond and Guinet, 1986; van Doorn et al., 1987; Pruzzo et al., 1989a; Brook and Myhal, 1991

Bacteroides melaninogenicus

Clinical source:	URG
Adhesins:	ND
Receptor specificity:	ND
Substrata:	EC
Reference:	Cook et al., 1989

Bordetella avium

Clinical source:	RT
Adhesins:	ND
Receptor specificity:	ND
Substrata:	EC, ERT, TB
References:	Arp, 1986; Arp and Brooks, 1986; Arp and Hellwig, 1988; Arp et al., 1988

Bordetella bronchiseptica

Clinical source:	URT, pigs
Adhesins:	ND
Receptor specificity:	Sialyl glycoconjugates, mucins
Substrata:	EC, ERT
References:	Ishikawa and Isayama, 1987a,b, 1988, 1990; Jacques et al., 1988; Nakai et al., 1988; Chung et al., 1990

Bordetella parapertussis

Clinical source:	URT
Adhesin:	ND
Receptor specificity:	Gal(3S04)β1, asialogangliosides
Substratum:	TC
References:	Ewanowich et al., 1989a,b; Brennan et al., 1991

Bordetella pertussis

Clinical source:	URT
Proposed adhesins:	Filamentous hemagglutinin; toxin; pertactin
Receptor specificities:	Heparin and sulfatides; lactose; sialylated glycoconjugates; integrins
Substrata:	EC, PC, ERT, TC

Bordetella pertussis inhabits the human upper respiratory tract and is capable of expressing a number of adhesins that mediate adhesion of the bacteria to ciliated and nonciliated cells as well as to macrophages (Weiss and Hewlett, 1986; see also Chapter 7). Extensive studies on the molecular mechanisms of *Bordetella* adhesion have revealed a number of unique features of the adhesins and their receptors on animal cells. One adhesin is the filamentous hemagglutinin, 1 2-nm filamentous structure, found in a cell-associated form outside the outer membrane as a microcapsule (Parker and Armstrong, 1988) or in a soluble form secreted into the culture medium during growth. Studies showing inhibition of adhesion of *B. pertussis* to tissue culture cells by antihemagglutinin antibodies (Gorringe et al., 1985; Redhead, 1985; Urisu et al., 1986; Relman et al., 1989) and reduced adhesion to ciliated cells and tissue culture cells of *B. pertussis* mutants with insertional mutations in or deleted of the *fha* structural hemagglutinin gene (Tuomanen and Weiss, 1985; Urisu et al., 1986; Relman et al., 1989) have confirmed earlier studies suggesting a central role for the filamentous hemagglutinin in mediating adhesion of *Bordetella* to animal cells. Additional studies show that even the soluble hemagglutinin can function as an adhesin (Tuomanen and Weiss, 1985; Urisu et al., 1986). The microorganisms bind the secreted hemagglutinin and their adhesion to human ciliated cells as well as their hemagglutinating activity increase concomitantly. *Bordetella* mutants deficient in production of the hemagglutinin lack the ability to bind to epithelial cells but are capable of binding the soluble hemagglutinin which restores adhesion to the ciliated cells or to WiDr tissue culture cells. The hemagglutinin neither binds to nor increases the adhesion of *S. pneumoniae* or *S. aureus* to cilia (Tuomanen, 1986). The *Bordetella* hemagglutinin is capable of binding to the ciliated cells in a way that it can enhance the adhesion of *B. pertussis* as well as of *H. influenzae* to the cells (Tuomanen, 1986; Tuomanen et al., 1988). The filamentous hemagglutin, therefore, contains a region that binds avidly to constituents on the bacterial surface and one or more regions that bind(s) to animal cell constituents to promote adhesion. While the molecular mechanism of the hemagglutinin–bacteria interaction is poorly understood, studies have shown that the filamentous hemagglutinin contains at least three domains with distinct receptor specificities. One domain of the filamentous hemagglutinin is specific for galactose and lactosamine and especially for lactosyl-containing glycolipids (Tuomanen et al., 1988). Studies employing antihemagglutinin or antilactosyl containing glycolipid antibodies or soluble lactose-containing carbohydrates as inhibitors of adhesion have confirmed that the filamentous hemagglutinin binds *B. pertussis* to lactosyl-containing glycolipids readily accessible on human ciliary tufts. The cloning and sequencing of the major fimbrial subunit of the filamentous hemagglutinin (see Table 9–1) have revealed that within the 220,000 mol wt protein there are two sites containing the peptide triplets Arg-Gly-Asp (RGD) (Relman et al., 1989). One of the RGD-containing sites was shown to function as a domain that mediates nonopsonic attachment and phagocytosis of *B. pertussis* to human macrophages (Chapter 7) and as such it may confer adhesin/invasin properties to the filamentous hemagglutinin. Because *B. pertussis* has been shown to attach to and invade HeLa cells in vitro and antifilamentous hemagglutinin inhibits invasion (Ewanowich et al., 1989a) the possibility that integrins may also serve as receptors for filamentous hemagglutinin-mediated endocytosis by nonprofessional phagocytes cannot be excluded as shown for *Yersinia.* The filamentous hemaggluutinin also binds heparin, a component of the extracellular matrix coating of many types of animal cells. Heparin is composed of sulfated β1,4-linked glucosamine and glucuronic acid residues (Menozzi et al., 1991). Sulfated polysaccharides, such as D-Gal–6-sulfate, dextran sulfate, and heparin, at relatively low concentrations, inhibit the adhesion of *B. pertussis* to WiDr tissue culture cells and to hamster tracheal cells (Brennan et al., 1991) suggesting that heparin or a sulfated galactose binding domain of the filamentous hemagglutinin participates in the process of adhesion of *Bordetella* to the animal cells.

Pertussis toxin may also serve as an adhesin that binds *B. pertussis* to ciliated respiratory cells. Mutants deficient in production of pertussis toxin bind poorly to ciliated cells (Tuomanen and Weiss, 1985; Relman et al., 1989). The toxin is found transiently in relatively small amounts on the bacterial surface (Parker and Armstrong, 1988) but it is mostly secreted into the culture medium (Weiss and Hewlett, 1986). The secreted toxin is capable of binding to the *Bordetella* surface via one region and to the surface of the cilia via other region(s), therefore bridging between the two cell types (Tuomanen and Weiss, 1985; Tuomanen et al., 1988). The receptor specificity of the toxin is the same as that of the filamentous hemagglutinin because the toxin binds the same lactose-containing glycolipids on TLC. While glycolipids from Chinese hamster ovary cells do not bind with the *Bordetella* toxin, the toxin does bind to a NeuNAcGalβ1,4GlcNAc oligosaccharide sequence present in a 165,000 mol wt glycoprotein extracted from the tissue culture cells (Brennan et al., 1988). Of the two distinct oligomers of the toxin, only the β-oligomer binds the sialylated but not the desialylated glycoprotein. The sialylated glycoprotein apparently is not accessible on the tissue culture cells for binding the *Bordetella* via its toxin because toxin-deficient mutants bind as well to the cells as the parent strain (Relman et al., 1989). In contrast, ciliated epithelial cells interact poorly with the toxin-deficient mutant. Binding is restored by adding exogenous toxin either to the epithelial cells or to the bacterial surface, but this toxin-mediated adhesion is lactose-specific (Tuomanen et al., 1988) and different from that described for the glycoprotein. Because the β oligomer of the toxin consists of four subunits, it is possible that the lactose binding site of *Bordetella* toxin resides in subunit(s) distinct from that specific for the sialylated oligosaccharides. It is likely that in other cell types certain sialylated oligosaccharide sequences of either glycoproteins or glycolipids are accessible for mediating adhesion of *B. pertussis* via its toxin.

Bordetella pertussis expresses on its surface a 69,000 mol wt protein termed pertactin, which contains the tripeptide Arg-Gly-Asp sequence (Leininger et al., 1991). Chinese hamster ovary cells bind the purified protein and the binding is inhibited by Arg-Gly-Asp-containing synthetic peptides, suggesting that integrins on the surface of the tissue cells are involved in interaction with pertactin. It is possible that pertactin can serve as an adhesin that mediates adhesion of *B. pertussis* to the tissue culture cells because pertactin mutants bind significantly less to Chinese hamster cells and HeLa cells than the parent strain. Although *B. pertussis* expresses rigid fimbriae and antifimbrial serum inhibits adhesion to tissue culture cells (Gorringe et al., 1985), a role for the fimbrial antigen in adhesion to human WiDr tissue culture cells could not be confirmed (Urisu et al., 1986).

In summary, although *B. pertussis* appears to be able to express three potential adhesins, in vivo there is evidence only for the filamentous adhesin function. The role of the filamentous hemagglutinin in mediating adhesion in vivo is inferred from studies showing that mice immunized with filamentous hemagglutinin and challenged with virulent *Bordetella* have reduced bacterial colonization in the lung and trachea (Kimura et al., 1990). Mutants deficient in filamentous hemagglutinin fail to colonize the upper respiratory tract, but do colonize the lung following aerosol challenge of mice with the mutant. It is possible therefore that the filamentous adhesin is important in mediating adhesion at the initial stages of the infection. Progenies may reach the lower respiratory tract to colonize the lung via other types of adhesins.

Borrelia burgdorferi

Clinical source:	BLD
Adhesins:	Osp B protein
Receptor specificity:	ND
Substrata:	EC, TC
References:	Comstock and Thomas, 1989; Garcia-Monco et al., 1989; Thomas and Comstock, 1989; Szcepanski et al., 1990

Branhamella catarrhalis

Clinical source:	LRT
Adhesin:	ND
Receptor specificity:	ND
Substratum:	EC

References:	Carr et al., 1989; Ahmed et al., 1990; McGaki et al., 1987

Campylobacter coli

Clinical source	ENT
Adhesins:	ND
Receptor specificity:	ND
Substrata:	TB, TC
References:	Campbell et al., 1987; Fauchere et al., 1986, 1989

Campylobacter jejuni

Clinical source:	ENT
Adhesins:	26,000–30,000 mol wt OMP, flagella?, Lipopolysaccharide?
Receptor specificity:	Fucose, mannose
Substrata:	TC, MC, IM

The recognition of *C. jejuni* as a major causative agent of acute diarrheal disease in humans (Walker et al., 1986) has stimulated studies on the properties of its adhesion to animal cells. *C. jejuni* binds to a number of tissue culture cell lines. Several characteristic features have emerged from these studies. First, there appear to be strain variations in the magnitude of *Campylobacter* adhesion to the cell lines INT 407 and HEp-2 as well as cell line variations in their capacity to bind a particular clinical isolate of *C. jejuni* (McSweegan and Walker, 1986; Neman-Simhi and Megraud, 1988; Konkel and Joens, 1989). In one study, it was found that *C. jejuni* strains isolated from patients with diarrhea or fever bound in higher numbers to Hela cells than strains isolated from patients without either one of these symptoms (Fauchere et al., 1986). Also, the capacity of *C. jejuni* to adhere to HEp-2 cells is probably under phase-like control because laboratory passages give rise to nonadhesive variants (Konkel and Joens, 1989). Another property is that *C. jejuni* adherent to HEp-2 or HELa cell lines were internalized by the cell lines (Fauchere et al., 1986; de Melo and Pechere, 1988; de Melo et al., 1989; Konkel and Joens, 1989; Konkel et al., 1990). Whenever tested, the invasion process was dependent on an intact active network of microfilaments of the animal cells. In one study, it was noted that *C. jejuni* internalized by HEp-2 cells was digested after phagolysosome fusion, suggesting that the invasiveness may not be a mechanism of survival for *Campylobacter* (de Melo and Pechere, 1990). As with adhesion, there were strain variations in the capacity of *Campylobacter* adherent to HEp-2 cell line to undergo endocytosis (Konkel and Joens, 1989; Lindblom et al., 1990). Variations in the ability of the HEp-2 cell line to bind and internalize *C. jejuni* were also noted in virus-infected cells. HEp-2 cells infected with echovirus or coxsackievirus bound and internalized considerably more *Campylobacter* than noninfected cells or cells infected by poliovirus or porcine enterovirus (Konkel and Joens, 1990). Because the adhesion of other *Campylobacter* species (*C. hyointestinalis* and *C. mucosalis*) was not affected by the viruses, it was concluded that the virus-dependent adhesion represents a specific interaction between components on the surface of both types of cells.

Histological examination of rabbit intestinal tissue in ileal loops infected with *C. jejuni* revealed that the spiral-shaped bacteria adhered to and invaded into M-cells, suggesting a mechanism of translocating the *Campylobacter* from the intestine to the lymphatic tissues (Walker et al., 1988). Controversial results have been obtained on the capacity of intestinal mucus to interact with *C. jejuni*. Following experimental infection of antibiotic-treated mice with *C. jejuni,* it was observed that most of the spiral-shaped bacteria were associated with the mucus overlying the intestinal cells (Lee et al., 1986). Pretreatment of HEp-2 cells with mucin enhanced both the adhesion and the invasion of *C. jejuni* into the cell line (de Melo and Pechere, 1988). In another study, rabbit small intestinal mucus alone had no effect on the adhesion of *Campylobacter* to the intestinal cell line INT 407. Presence of anti-*Campylobacter* sIgA antibodies in the mucus caused aggregation of the bacteria and inhibited their adhesion to the cells (McSweegen et al., 1987). Finally, considerable intrastrain variations were observed in the capacity of the *C. jejuni* to bind to immobilized rabbit small and intestinal mucus (McSweegan and Walker, 1986). It is possible that as with adhesion, the source of the intestinal mucus and the strain employed may determine the capacity of the spiral-shaped bacteria to interact with mucus.

There is evidence that the oligosaccharide moiety of LPS may serve as an adhesin (or receptor) that binds *C. jejuni* to ill-defined constituents coating the surface of the intestinal cells. Compared to other spiral-shaped bacteria (e.g., *H. pylori*), *C. jejuni* exhibits a weakly hydrophobic surface (Pruul et al., 1990). Bacteria with a high negative charge and a relatively weak hydrophobic surface bind in significantly higher numbers to human intestinal cell line HT-29 than *C. jejuni* strains exhibiting a more hydrophobic surface and weaker negative charge (Walan and Kihlstrom, 1988). Pretreatment of *C. jejuni* with proteolytic enzymes or glutaraldehyde reduced adhesion to INT 407 cells but failed to completely inhibit adhesion (McSweegan and Walker, 1986). In contrast, both glutaraldehyde and proteolytic treatments of the INT 407 intestinal cells abolished almost completely their ability to bind the spiral-shaped bacteria. Whole cell lysate partially inhibited the adhesion of *Campylobacter* to HEp-2 cells and this inhibitory activity was abolished by pretreating the lysate with sodium metaperiodate, but not with proteolytic enzymes or heating to 100°C for 30 minutes. Fucose and mannose inhibited partially the adhesion of *Campylobacter* to INT 407 cells. The binding of *C. jejuni* LPS to the intestinal cells was also inhibited by fucose or by pretreating the intestinal cells with glutaraldehyde or the polysaccharides with sodium metaperiodate. Finally, *C. jejeuni* LPS at 500 μg per well completely inhibited the adhesion of the bacteria to the INT 407 cell line. The data suggest that the LPS, properly exposed on the surface of *C. jejeuni,* confers hydrophilic properties to the bacteria and promotes their adhesion to intestinal cell lines by combining with a proteinaceous component possessing lectin-like activity specific for mannose and fucose. Interestingly, studies employing interaction of 29 different strains of *C. jejuni* with lectins indicate that neither fucose nor mannose was found on the surface of the bacteria. Most of the strains reacted with lectins capable of binding the GlcNAc and GalNAc/Gal groups (Wong et al., 1985). Also, a fucose- and/or mannose-specific lectin on the HEp-2 tissue culture cell line has not been found yet. It is clear, therefore, that the exact carbohydrate sugar specificity of the *C. jejuni* surface polysaccharide and the tissue cell line require further studies. *C. jejuni* also adheres to the porcine intestinal brush border but the involvement of LPS in the adhesion process could not be confirmed (Naess et al., 1988).

Further studies indicate that *C. jejuni* is capable of expressing additional proteinaceous adhesin(s) that may also promote invasiveness. Flagella were implicated as adhesins or carriers of adhesins because nonflagellated mutant strains adhered much less than the flagellated parent strain and isolated flagella preparations bound to a monolayer of INT 407 cells (McSweegan and Walker, 1986). In other studies, however, flagella bound poorly to murine small intestine cells as compared to other *C. jejuni* derived proteins (Moser and Hellmann, 1989). Whole cell lysates from a highly invasive variant strain of *C. jejuni* inhibited the internalization of the homologous and heterologous strains by HEp-2 cells, whereas the adhesion of the bacteria to the cell line was not affected (Konkel et al., 1990). In contrast, lysates prepared from a noninvasive derivative of the same *C. jejeuni* strain were without effect. Monoclonal anti-*C. jejuni* antibodies partially inhibited the invasiveness of the *Campylobacter* and reacted in western blots with a number of proteins present in whole cell lysates obtained from both the invasive and noninvasive strains. Lysates from only the invasive strain, however, contained antigen(s) in the range of 42,000–38,000 mol wt that reacted with the monoclonal antibodies. Because the monoclonal antibodies also reacted with epitopes present on the flagella and the outer surface of the bacteria, it is possible that some of the surface proteins become associated with the flagella. In other studies, a somewhat different approach was used to identify *C. jejuni* adhesins. In these studies, protein extracts from invasive and noninvasive strains were incubated with the HEp-2 cells and the proteins bound to the cells were analyzed by lysing the HEp-2 cells monolayer and electroblotting with hyperimmune rabbit anti-*C. jejuni* sera after electrophoretic separation of the proteins on sodium dodecyl sulfate (SDS)-polyacrylamide gels. The results indicate that the HEp-2 cells bound outer membrane proteins of 28,000 and 32,000 mol wt that were found in the outer membrane lysate of only the invasive strains (de Melo and Pechere, 1990). In other studies using a similar technique, it was found that HeLa cells bound 26,000 and 30,000 mol wt proteins present in outer membrane extract of invasive strains of *C. jejuni* (Fauchere et al., 1989; Fauchere and Blaser, 1990). Antiserum raised against these HeLa cell-bound proteins inhibited the adhesion of homologous *C. jejuni* to the cell line, whereas antibodies against an unrelated 92,000 mol wt outer membrane protein were without effect. In addition, outer membrane proteins in the 30,000 mol wt range are readily extracted by mild acid treatment (Pei et al., 1991). Anti-*C. jejuni* antibodies identified in the acid extract of the invasive strain three proteins of 28,000, 32,000, and 42,000 mol wt that bound to the HEp-2 cells (de Melo and Pechere, 1990). The 32,000 mol wt protein is of special interest because it was found in both the acid and outer membrane extracts of invasive and noninvasive strains of *C. jejuni,* but HEp-2 and Hela cells bound only the 32,000 mol wt protein from the extracts of the invasive strain. In contrast, animal cell lines MDCK and BHK, which did not internalize the invasive strain of *C. jejuni,* also did not bind the 32,000 mol wt protein, suggesting the involvement of the protein in the adhesion/invasion process.

In summary, it is possible that one or more proteins in the 30,000 mol wt range, belonging to the outer membrane or cell surface (e.g., S. layer) of *C. jejuni,* function as adhesin/invasin that promote(s) interaction with intestinal cells. Some of these adhesins may become associated with the flagella for better orientation to function as adhesins. Although there is circumstantial evidence implicating the oligosaccharide moiety of LPS in the adhesion/invasion process, the large heterogeneity in the polysaccharide composition of the various *C. jejuni* strains as revealed by antilipopolysaccharide antibodies and interaction with plant lectins (Wong et al., 1985) complicates the studies.

Capnocytophaga gingivalis

Clinical source:	ORL
Adhesins:	150,000 mol wt outer membrane protein
Receptor specificity:	Sialic acid
Substrata:	Oral bacteria
References:	Kagermeier and London, 1986; Tempro et al., 1989; Tables 8–6 and 8–7

Capnocytophaga ochracea

Clinical source:	ORL
Adhesins:	155,000 mol wt L-rhamnose binding protein
Receptor specificities:	L-rhamnose, D-fucose, β-galactosides
Substrata:	Oral bacteria
References:	Weiss et al., 1987b, 1990; Kolenbrander et al., 1989; Tables 8–6 and 8–7

Chlamydia trachomatis

Clinical source:	URG
Adhesins:	18 and 32,000 mol wt glycoproteins?, 40,000 mol wt major outer membrane protein (VD IV region)

Receptor specificities: Phosphatidylethanolamine; GalNAcβ1,4Galβ1,4Glc
Substrata: EC, TC, ERT
References: Hackstadt and Caldwell, 1985; Bose and Goswami, 1986; Hackstadt, 1986; Wenman and Meuser, 1986; Maslow et al., 1988; Wyrick et al., 1989; Swanson and Kuo, 1990; Su et al., 1990b; Krivan et al., 1991; Schmiel et al., 1991

Clostridium difficile
Clinical source: ENT
Adhesion: ND
Receptor specificity: ND
Substratum: TB
Reference: Borriello et al., 1988

Corynebacterium diphtheriae (diphtheroids)
Clinical source: ENT
Adhesin: ND
Receptor specificities: Mannose, GlcNAc
Substratum: TC, EC
Reference: Romero-Steiner et al., 1990

Corynebacterium pilosum
Clinical source: URG, bovine
Adhesins: Fimbriae
Receptor specificity: ND
Substrata: EC
References: Kudo et al., 1987; Hayashi et al., 1985

Corynebacterium renale
Clinical source: URG, bovine
Adhesins: Fimbriae
Receptor specificity: ND
Substrata: EC
References: Hayashi et al., 1985; Fukuoka and Yanagawa, 1987; Kudo et al., 1987

Eikenella corrodens
Clinical source: ORL
Adhesins: ND
Receptor specificity: GalNAc
Substrata: ERT, PC, oral bacteria
References: Miki et al., 1987; Yamazaki et al., 1988; Ebisu et al., 1988; Tables 8–6 and 8–7

Enterobacter aerogenes

Clinical source:	URT
Adhesins:	Type 1 fimbriae, MR/P (nonfimbrial), MS thin fimbriae, KR/K fimbriae
Receptor specificity:	Mannosides
Substrata:	ERT
References :	Adegbola and Old, 1985; Gerlach et al., 1989; Hornick et al., 1991

Enterobacter cloacae

Clinical source:	URT
Adhesins:	Type 3 fimbriae (MR/K)
Receptor specificity:	Spermidine
Substrata:	ERT, PC
References:	Gerlach et al., 1989; Hornick et al., 1991; Hirakata et al., 1991

Erysipelothrix rhusiopathiae

Clinical source:	SKN
Adhesin:	ND
Receptor specificity:	Neuraminyllactose
Substrata:	ENC
Reference:	Nakato et al., 1987

Fusobacterium nucleatum

Clinical source:	ORL
Adhesins:	39,500 mol wt galactose-binding lectin
Receptor specificity:	β-Galactosides, lactose
Substrata:	ERT, EC, oral bacteria
References:	Murray et al., 1988; Kaufman and DiRienzo, 1989; Kolenbrander and Andersen, 1989; Kolenbrander et al., 1989; Tables 8–6 and 8–7

Fusobacterium necrophorum

Clinical source:	ENT, bovine
Adhesins:	ND
Receptor specificity:	ND
Substrata:	ERT, EC, TC
References:	Kanoe et al., 1985; Kanoe and Iwaki, 1987

Gardnerella vaginalis

Clinical source:	URG
Adhesin:	Fimbriae, fimbrillae

Receptor specificity:	Galactose, lactose
Substrata:	EC, ERT, TC
References:	Scott et al., 1987, 1989; Scott and Smyth, 1987; Cook et al., 1989

Hafnia alvei

Clinical source:	ENT
Adhesin:	ND
Receptor specificity:	ND
Substrata:	TB
Reference:	Albert et al., 1991a

Haemophilus influenzae

Source	URT
Proposed adhesins:	Fimbriae, nonfimbrial protein
Receptor specificity:	Anton blood group, GalNAcβ1,4Gal
Substrata:	EC, OC, TC, ERT

Haemophilus influenzae adheres to animal cells by at least two mechanisms, one of which includes fimbrial adhesins and the other involves one or more classes of nonfimbrial adhesins. At least two classes of fimbrial adhesin complexes have been described and cloned: one is associated with type b encapsulated strains and the other with nontypable strains of *H. influenzae* (see Table 9–1). The DNA and the derived amino acid sequences of the major fimbrial subunit of the nontypable strains are 77% and 68%, respectively, identical to those of the type b strains (Coleman et al., 1991). It is not clear at this stage whether the adhesin function of the fimbrial adhesin complex resides on the major or a minor fimbrial subunit.

The type b fimbrial adhesin has an apparent molecular size of approximately 24,000 and contains conserved and nonconserved epitopes among different strains of *H. influenzae* type b strains (Guerina et al., 1985; LiPuma and Gilsdorf, 1988; Gilsdorf et al., 1990). The focus on the role of adhesion, especially that mediated by the type b fimbrial adhesin, in *H. influenza* infections is driven by the desire to develop a fimbriae-based vaccine that will induce protective antiadhesive antibodies (Brinton et al., 1989; Tanpowpong et al., 1989). In vitro, there is good evidence that the type b fimbrial adhesin complex promotes adhesion of the organisms to oropharyngeal epithelial cells and is responsible for causing hemagglutination of human erythrocytes. *E. coli* harboring type b fimbrial genes acquire the ability to bind epithelial cells and cause hemagglutination (van Ham et al., 1989). Inactivation of the fimbrial genes causes significant decreases in adhesion to buccal cells (Weber et al., 1991). Nonfimbriated strains bind more poorly to buccal cells than type b fimbriae-carrying strains (Sable et al., 1985; van Alphen et al., 1987; Patrick et al., 1989). Monoclonal antifimbrial antibodies specifically inhibit both hemagglutination caused by the bacteria and adhesion of the homologous and heterologous strains of *H. influenzae* to oropharyngeal cells (van Alphen et al., 1988; Sterk et al., 1991). Polyclonal antifimbrial antibodies inhibit adhesion of homologous and heterologous strains of *H. influenzae,* suggesting that fimbrial epitopes shared by fimbrial adhesins of many *H. influenzae* type b strains may be a prospect for the antiadhesive vaccine (LiPuma and Gilsdorf, 1988; Brinton et al., 1989). Moreover, purified fimbriae caused hemagglutination of human erythrocytes and bind to oropharyngeal cells (van Alphen et al., 1988). In addition to buccal cells, type b fimbriae also mediate adhesion to adenoid tissues (Loeb et al., 1988) and to trachea in organ culture (Smith et al., 1989). Like other fimbrial adhesins, the type b fimbrial adhesin undergoes phase variation at a frequency of about 7–3 $\times$ 10^{-4} per bacterium per generation for fimbriated and hemagglutinating to nonfimbriated and nonhemagglutinating phenotypes and vice versa (Farley et al., 1990). Buccal epithelial cells appear to select fimbriated variants from a population of predominantly nonfimbriated cells by causing overgrowth of fimbriated subpopulations (Patrick et al., 1989). Unlike the nonfimbrial adhesin (see below) capsule formation does not interfere with the ability of fimbriated *H. influenzae* type b to adhere to human buccal epithelial cells (LiPuma and Gilsdorf, 1987). In vivo, the type b fimbrial adhesin is probably expressed during a course of infection because a rise of antifimbrial antibodies occurs following infection with *H. influenzae* type b (Erwin et al., 1985). Type b fimbrial adhesin complex confers an advantage in colonization of the upper respiratory tract as evidenced from experiments showing that mutationally inactivated fimbrial genes exhibit a significant decrease in both adhesion to buccal cells and nasal colonization of yearling Rhesus monkeys (Weber et al., 1991). *H. influenzae* type b, however, undoubtedly undergoes phase variation in vivo as it does in vitro. Fimbriated variants adhering to buccal epithelial cells were selected from a population of mixed variants and shown to be more effective than the nonfimbriated variants in the colonization of rats following intranasal inoculation (Anderson et al., 1985). Cultures obtained from nasal washes of infected rats, however, contained predominantly nonfimbriated variants. As for *E. coli* fimbrial adhesins, the results may be interpreted to suggest that the fimbriae confer an initial advantage for a transient colonization of oral cells, giving rise to nonfimbriated progeny more suitable to survive on and probably adhere to other types of cells (Anderson et al., 1985; see also Chapter 11). This notion is supported by studies showing that most *H. influenzae* type b strains isolated from the nasopharynx of patients with systemic infections contain predominantly nonfimbriated populations of bacteria (Mason et al., 1985).

H. influenzae, type b expresses a poorly defined nonfimbrial adhesin with a distinct receptor specificity as described in studies showing that human foreskin fibroblasts and HEp-2 tissue culture cells bind more nonfimbriated than fimbriated variants (Sable et al., 1985). It is not clear whether fimbrial receptors are available on HEp-2 or HeLa tissue cells because a low level of adhesion observed in tissue cells exposed to a population consisting primarily of fimbriated variants may be due to the presence of small numbers of nonfimbriated bacteria in the population originating from the phase variation phenomenon (Karam-Sarkis et al., 1991). In one study, it was found that human adenoidal cells in organ culture system bound equally well the fimbriated and the nonfimbriated variants of a single clone, but the two variants did not compete for the same receptors (Loeb et al., 1988). The nonfimbrial adhesin may be important for conferring invasiveness to adherent bacteria. Adhesion associated with invasion of human tissue culture cells by nonfimbriated variants of *H. influenzae* type b was noted by a number of investigators (Farley et al., 1990; St.-Geme and Falkow, 1991, Virji et al., 1991a). The capsule interfered with adhesion and invasion. It was postulated that the relatively high frequency of capsule loss in *H. influenzae* type b results in nonencapsulated variants that can adhere and invade cells via a nonfimbrial adhesin functioning as an invasin (St. Geme and Falkow, 1991).

Although *H. influenzae* type b accounts for 95% of cases of systemic *H. influenzae* infections, nonencapsulated and nontypable strains account for a considerable number of local diseases, including otitis media, sinusitis, bronchitis, and conjunctivitis, as well as of nasopharyngeal carriage (Maxon, 1990). In one study, it was noted that all nontypable *H. influenzae* isolates were fimbriated, but there was a high degree of variability in the number of fimbriae per bacterium as well as in the percentage of fimbriated bacteria in the population, ranging from <10% fimbriated bacteria to 100% (Bakaletz et al., 1988a). Adhesion experiments have shown that there was no correlation between the degree of fimbriation and the magnitude of adhesion of the various isolates to human oropharyngeal cells or chinchilla tracheal cells in organ culture. Neither was there a correlation between the strength of hemagglutination of human erythrocytes caused by the various isolates and the degree of fimbriation. Although this lack of correlation between fimbriation and adhesion of hemagglutination may suggest that the fimbriae of nontypable *H. influenzae* do not function as adhesins, it is possible that factors other than fimbriae on the surface of the isolates interfere with the presentation of the fimbrial adhesin. Fimbriation, however, appears to promote adhesion of nontypable isolates of *H. influenzae* to human buccal epithelial cells but not to human respiratory cells in organ culture (Read et al., 1991). Moreover, *E. coli* harboring a gene cluster encoding for the fimbriae from nontypable strains of *H. influenzae* cause mannose-resistant hemagglutination (Kar et al., 1990; Coleman et al., 1991) and antibody directed against the fimbrial adhesin of a nontypable *H. influenzae* strain is protective in an experimental model of otitis media (Karasic et al., 1989). The relationship between the receptor of the type b fimbrial adhesin to that of the nontypable strains is not clear, although both types of fimbriae appear to mediate adhesion preferentially to human buccal cells, are responsible for mannose-resistant hemagglutination in vitro, and share common epitopes. Compared to type b strains, the nontypable strains adhere in greater numbers to isolated nasal, nasopharyngeal, and buccal epithelial cells (Porras et al., 1985; Harada et al., 1990) and to ciliated cells in whole organ perfusion assays (Bakaletz et al., 1988b). It is possible that nontypable *H. influenzae* strains produce another type of fimbrial adhesin that does not mediate hemagglutination but is expressed as thin fimbriae (Bakaletz et al., 1989).

The nontypable *H. influenzae* are also capable of expressing nonfimbrial adhesins that mediate binding of the organisms to tissue culture cells (Jallat et al., 1991). Studies employing tissue culture cells derived from normal human conjunctiva epithelium and nontypable and nonfimbriated *H. influenzae* revealed that adhesion increased over time, peaking by 4–6 hours of incubation to give 150 bacteria per epithelial cell (St. Geme and Falkow, 1990). This adhesion was dependent on actual growth of the bacteria and was associated with de novo synthesis of proteins by the adherent bacteria. Because adherent bacteria have growth advantages over nonadherent bacteria in a number of systems (see Chapter 11) the de novo synthesis of proteins may simply represent more metabolically active bacteria. Like the nonfimbriated *H. influenzae* type b, the nontypable and nonfimbriated *H. influenzae* entered, albeit in small numbers, the epithelial cells in a process that involved actin filaments and microtubules. In summary, type b and untypable *H. influenzae* strains appear to be able to express fimbrial and nonfimbrial adhesins in which the nonfimbrial adhesin mediates adhesion and entry under certain conditions into respiratory epithelial cells in organ culture, whereas the fimbrial adhesins mediate binding to buccal epithelial cells and are responsible for hemagglutination reactions. It is likely that *Haemophilus* produces other types of fimbrial or nonfimbrial adhesins. The expression of the adhesins in functional form and the choice of type of animal cell employed may allow the distinction between the various adhesins. It is also possible that strains may, under certain growth conditions, coexpress two classes with distinct receptor specificities. The receptor specificity of the fimbrial adhesin of type b *H. influenzae* has been studied by employing erythrocytes and buccal epithelial cells from donors with various group antigens. Initial studies showed that the presence of Anton or Wj antigen (AnWj) on erythrocytes was essential for hemagglutination. Anti-Wj sera inhibit hemagglutination caused by the fimbriated bacteria (van Alphen et al., 1986). The AnWj antigen is always expressed on buccal epithelial cells in functional form for binding the *H. influenzae* in AnWj blood group-positive phenotypes, but its expression on cells in AnWj-negative blood group phenotypes depends on the genetic background or physiological state of the individuals. For example, in neonates up to 40 days after birth or in some AnWj-blood group negative individuals the receptor for binding the fimbriated *H. influenzae* is expressed on the buccal cells but not on the erythrocytes (van Alphen et al., 1987, 1990). In contrast, in some individuals the AnWj gene, which has a recessive mode of inheritance, is not expressed at all and therefore the receptor for binding the *H. influenzae* is not expressed on either type of cell (van Alphen et al., 1990). It would be of interest to see whether such individuals are genetically immune against infections with type b *H. influenzae*. Although anti-AnWj antibodies inhibit hemagglutination caused by the fimbriated bacteria, but not adhesion of the organisms to buccal epithelial cells, it is possible that the chemical structure required for binding the fimbrial adhesin complex of *H. influenzae* is carried by two isoreceptors, one of which is expressed on erythrocytes and the other on buccal epithelial cells and possibly on other cell types. The attachment site on erythrocytes and buccal cells specific for the fimbrial adhesin apparatus of *H. influenzae* is a complex structure of oligosaccharides contained in the gangliosides GM1, GM2, GM3, and GD1a, all of which inhibit, in nanomolar concentrations, adhesion to oropharyngeal cells of the fimbriated bacteria as well as the hemagglutination caused by whole bacteria (van Alphen et al., 1991). Similar inhibition patterns have been observed in hemagglutination and adhesion assays employing recombinant *E. coli* expressing the *H. influenzae* fimbrial adhesin, whereas the ganglioside GM1 does not inhibit hemagglutination caused by *H. aegypticus,* suggesting that the target for the inhibitory carbohydrate is specific for the fimbrial adhesin of *H. influenzae*. The asialo derivative of GM1 or sialoglycoproteins or 30 different mono- and disaccharides did not inhibit adhesion or hemagglutination. The minimal structure, therefore, that binds the *H. influenzae* fimbrial adhesin is sialyllactose ceramide. The receptor that contains this structure is probably glycolipid only on both the erythrocytes and epithelial cells, because this structure is usually not found in glycoproteins and trypsinization of buccal cells has no effect on the ability of the treated cells to bind *H. influenzae*. In other studies, it has been shown that *H. influenzae* binds to GalNAcβ1,4Gal-containing glycolipids (Krivan et al., 1988). It has been suggested that GalNAcβ1,4Gal may be specific for the nonfimbrial adhesin of type b *H. influenzae* (van Alphen et al., 1991). *H. influenzae* adherent to ciliated and nonciliated nasal cells has been observed in specimens obtained from guinea pigs infected intransally with type b or nontypable *H. influenzae,* suggesting that adhesion of the microorganisms takes place in vivo (Harada et al., 1989). The adhesion-related host factors that render individuals susceptible to *H. influenzae* infections have been studied by a number of investigators. Nasopharyngeal cells obtained from children, the primary target population for *H. influenzae* infections, bound more nontypable *H. influenzae* than cells of adults (Shimamura et al., 1990). Further, *Haemophilus* adheres in greater numbers to nasopharyngeal cells from patients with otitis media or chronic sinusitis than to cells for normal children (Harada et al., 1990; Shimamura et al., 1990). Buccal epithelial cells bind less efficiently with *H. influenzae* than nasal or nasopharyngeal cells which are the normal reservoirs and the source of the pathogen. It is possible that at the port of entry *H. influenzae* first encounters and adheres transiently to buccal epithelial cells, thereby gaining growth advantages until progeny reach the pharyngeal area, the preferred environment for adhesion and colonization.

The interaction of soluble host factors with *H. influenzae* may also modulate the infectious process. The concentrations of soluble Anton blood group determinants in saliva have been found to be too low to inhibit the in vitro adhesion of *Haemophilus* to buccal cells or hemagglutination caused by the bacteria (van Alphen et al., 1989). Saliva also contains secretory IgA anti-*H. influenzae* antibodies. The higher the concentration of such antibodies in saliva of patients with otitis media the lower the ability of the patient's oral epithelial cells to bind to *H. influenzae* in vitro (Shimamura et al., 1990; Taylor et al., 1990). Saliva also contains non-IgA inhibitors of adhesion of the bacteria to buccal cells and the level of the inhibitory activity is higher in patients with histories of recurrent bronchitis than in patients who do not suffer from such disease (Taylor et al., 1990). Milk contains inhibitors of nontypable *H. influenzae* adhesion to buccal epithelial cells (Anderson et al., 1986; Aniansson et al., 1990). The anti-adhesive activity resides in the casein fraction of human milk as the free oligosaccharides are without effect. *H. influenzae* also binds avidly to IgD via an outer membrane protein, termed protein D, with an apparent mol wt of 42,000 (Janson et al., 1991). Protein D in functional form is expressed by all *H. influenzae* strains, including encapsulated strains belonging to serotypes a through f and nonencapsulated (nontypable) strains as well as by *H. haemolyticus* and *H. aegyptius* strains (Akkoyunlu et al., 1991). It is not clear how IgD or consumption of milk can modulate adhesion of *H. influenzae* in vivo to affect the infectious process. Nevertheless, adhesion of *H. influenzae* to various types of cells of the upper respiratory tract appears to be essential for successful colonization of the host prone to develop an infection caused by this pathogen.

Haemophilus influenzae (biogroup aegypticus)

Clinical source:	EYE
Adhesin:	Fimbriae, OMP?
Receptor specificity:	ND
Substrata:	ERT, TC
Reference:	St. Geme et al., 1991

Haemophilis parainfluenzae

Clinical source:	ORL
Adhesin:	ND
Receptor specificity:	ND
Substrata:	Oral bacteria, SHA, TS
References:	Liljemark et al., 1986, 1988

Helicobacter pylori

Clinical source:	ENT
Adhesin:	Fibrillar hemagglutinin
Receptor specificities:	*N*-acetylneuraminyllactose, collagen, laminin
Substrata:	ERT, TB, TC, PC, EC, IM

The studies showing that *H. pylori* is responsible for chronic gastric inflammation, including duodenal and gastric ulceration (Blaser, 1990), have stimulated research aimed at identifying the *Helicobacter* adhesins and their receptors on gastric mucosa (Wadström et al., 1991). The importance of *Helicobacter* adhesion to gastric cells in the ulceration process is evident from ultrastructural observations of biopsy specimens showing increased levels of *H. pylori* adherent to gastric mucosa from patients with histological epithelial damage as compared to low levels of adherent bacteria in biopsy specimens without epithelial damage (Hessey et al., 1990). In other studies it was observed that the spiral-shaped *Helicobacter* is in close contact with the gastric epithelium in a manner reminiscent of the pedestal or "cup formation" seen in the adhesion of enteropathogenic *E. coli* (Goodwin et al., 1986; Bode et al., 1988). In vitro, *Helicobacter* adheres in close association in the form of cup-like structures to HEp-2 and INT-407 tissue culture cell lines (Neman-Simha and Megraud, 1988) as well as to Y-1 mice adrenal cells (Evans et al., 1989a). *H. pylori* also adheres to the KATO-III cell line, an epithelial cell line derived from human gastric carcinoma, with the "attachment and effacement" type of adhesion associated with morphological alterations and occasional cell invasion (Wyle et al., 1990; Dunn et al., 1991; Hemalatha et al., 1991). Adhesion to KATO-III cells was best at 37°C and was enhanced in the presence of sublethal concentrations of erythromycin. Finally, human gastric tissue biopsies that either contained adherent *Helicobacter* or were exposed to low numbers of *Helicobacter* allowed growth of the adherent bacteria with time, suggesting adhesion to gastric cells may confer a growth advantage upon the adherent *Helicobacter* (Rosberg et al., 1991).

Attempts to identify the *Helicobacter* adhesins and their receptors employed mostly erythrocytes as target cells for adhesion. Some strains of *H. pylori* cause hemagglutination of erythrocytes from one species, whereas other strains cause hemagglutination of erythrocytes from a number of species including humans (Emody et al., 1988; Evans et al., 1988; Huang et al., 1988; Nakazawa et al., 1989). The data suggest that *H. pylori* is capable of expressing a number of hemagglutinins with distinct receptor specificities. In one study, it was found that the *Helicobacter* hemagglutinin is specific for neuraminyl residues on erythrocytes because pretreatment of bovine and human erythrocytes with sialidase abolished the hemagglutination by *Helicobacter*, and neuraminyllactose was the best inhibitor of hemagglutination (Evans et al., 1988). Of several glycoproteins tested, glycophorin A and fetuin also inhibited hemagglutination, whereas asialofetuin was without effect. It was possible to isolate from the bacteria by octylglucoside a hemagglutinin that has a fibrillar structure, binds to immobilized fetuin, and, when coated onto Latex beads, causes hemagglutination of bovine erythrocytes. Further studies show that fetuin, orosomucoid, and *N*-acetylneuraminyllactose and monosialogangliosides, but not asialogangliosides, inhibited hemagglutination between *Helicobacter* and erythrocytes from a number of species, including human, bovine, and rabbits (Emody et al., 1988; Slomiany et al.,

1989). Adhesion of *H. pylori* to Y-1 adrenal mouse cell line was abolished by pretreating the cell line with sialidase and was inhibited by fetuin, suggesting that the sialic acid specific hemagglutinin promotes the adhesion of the bacteria to the adrenal cells (Evans et al., 1989a). The sialic acid-specific fibrillar adhesin is probably expressed during a natural course of colonization of the stomach with the bacteria because most of the patients with gastric and duodenal ulcers or volunteers infected with *H. pylori* develop antibodies to the hemagglutinin extracted from one single strain (Evans et al., 1989b). Certain strains of *H. pylori* release the sialic acid-specific hemagglutinin which recognizes both NeuNAc2,3Gal and NeuNAc2,6Gal structures during growth in broth (Robinson et al., 1990). Examination of the adhesion of the strains to HEp-2 cells showed that strains producing soluble hemagglutinin adhered to the tissue culture cells in a manner involving capping of the plasma membrane, whereas strains not producing soluble hemagglutinin adhered but did not induce capping (Armstrong et al., 1991).

H. pylori binds to neuraminyllactose ceramide extracted from human gastric mucosa and separated by thin-layer chromatography (Saitoh et al., 1991). The *H. pylori* sialic acid-specific hemagglutinin is resistant to the gastric enzymes pepsin and trypsin, but sensitive to pronase and papain and may function as an adhesin that promotes adhesion of the spiral-shaped bacteria to gastric epithelial cells. Although *E. coli* exhibiting sialic acid-specific adhesion are hydrophobic, *H. pylori* expresses in vitro a relatively hydrophilic surface (Emody et al., 1988; Pruul et al., 1990; Smith et al., 1990). Depending on the test used to determine hydrophobicity, however, it is possible to demonstrate a relatively higher hydrophobic surface in *H. pylori* than in other Gram-negative pathogens, such as *Campylobacter jejuni* (Pruul et al., 1990).

H. pylori also binds to the sulfated glycolipid, galactose-sulfate ceramide (Saitoh et al., 1991). The SO_3-galactose-specific binding of the bacteria to the chromatogram-bound lipid was abolished by sonication of the bacteria, whereas the neuraminyl-specific binding was not affected, suggesting that the *Helicobacter* is capable of producing two adhesins with distinct receptor specificities. Sulfated galactosyl ceramide and lactoceramides were potent inhibitors of the hemagglutination of human erythrocytes by *H. pylori,* whereas desulfated derivatives were without effect (Slomiany et al., 1989). The relative inhibitory activity of lactosylceramide sulfate and GM3 ganglioside was additive, consistent with the notion that the population of *H. pylori* contains two distinct hemagglutinins. Interestingly, the drug sucrose octasulfate, an anti-ulcer agent, also inhibited the *Helicobacter* hemagglutinin, as well as the adhesion of the bacteria to HEp-2 cells (Slominary et al., 1989; Armstrong et al., 1991), suggesting that part of the therapeutic effect of the drug may be due to antiadhesive activity. The relative level of glycolipids that bind *H. pylori* in thin-layer chromatography is estimated to be higher in human stomach antrum than in fundus and in adults compared with infant tissues (Lingwood et al., 1989; Saitoh et al., 1991). Fetuin or sialic acid does not inhibit adhesion of *H. pylori* to HeLa cells, suggesting that adhesion to these tissue culture cells is promoted by a different class of adhesins (Fauchere and Blaser, 1990). Surface-associated material that contains a major 60,000 mol wt protein may be another class of adhesin that binds *H. pylori* to HeLa cells (Fauchere and Blaser, 1990). The adhesins bind to HeLa cell membranes and inhibit the adhesion of *Helicobacter* to the tissue culture cells. Microscopic observations indicate that at certain stages of the gastric ulceration, spiral-shaped *Helicobacter* are seen within the gastric mucous layer overlying the epithelial cells (Tricottet et al., 1986). In vitro, *Helicobacter* binds to human gastric mucin immobilized on plastic (Tzouvelekis et al., 1991). Sialidase and sodium metaperiodate treatment of the immobilized mucin partially abolished the adhesion of the bacteria to the mucin, suggesting that the sialic acid hemagglutinin described above may promote the adhesion of the bacteria. Because mucin also inhibited the adhesion of the *Helicobacter* to HEp-2 cells, it is possible that *H. pylori* interacts with mucin to gain a growth advantage permitting progenies to penetrate the mucous layer and reach the gastric epithelium. The ulceration process induced by *H. pylori* adherent to gastric mucosa exposes underlying basement membrane proteins that may serve as substrata for binding the bacteria. This notion is supported by the finding showing high-affinity binding of collagen type IV and laminin to *H. pylori* (Trust et al., 1991).

Klebsiella oxytoca

Clinical source:	URT
Adhesin:	Type 3 fimbriae (MR/K), type 1 fimbriae
Receptor specificity:	Mannosides
Substrata:	ERT
References :	Podschun et al., 1987; Gerlach et al., 1989; Hornick et al., 1991

Klebsiella pneumoniae

Clinical source:	URT, URG, SKN
Adhesin:	Type 3 fim (MR/K), type 1 fimbriae
Receptor specificity:	Mannosides, GalNAcβ1,4Gal, type V collagen
Substrata:	BM, EC, ERT, PC, TC
References:	Old et al., 1985a; Daifuku and Stamm 1986; Bhat and Panhotra, 1987; Guevara et al., 1987; Podschun et al., 1987; Athamna and Ofek, 1988; Chandra, 1988; Fader et al., 1988; Krivan et al., 1988; Lata and Panhotra, 1987; Gerlach et al., 1989; Tarkkanen et al., 1990; Wurker et al., 1990; Athamna et al., 1991; Hornick et al., 1991; Hirakata et al., 1991

Lactobacillus fermentum

Clinical source:	ENT, pigs, mice
Adhesin:	12,000–13,000 mol wt protein, LTA?
Receptor specificity:	ND
Substrata:	EC, TB, SS
References:	Sherman and Savage, 1986; Wadström et al., 1987; Conway and Kjelleberg, 1989; Conway and Adams, 1989; Pedersen and Tannock, 1989; Henriksson et al., 1991

Leptospira interrogans

Clinical source:	BLD, URG, bovine, human
Adhesin:	ND
Receptor specificity:	ND
Substrata:	TC
Reference:	Thomas and Higbie, 1990

Listeria monocytogenes

Clinical source:	BLD, CSF, URG
Adhesin:	ND
Receptor specificity:	ND
Substrata:	SS
Reference:	Mafu et al., 1991

Mobiluncus curtissii

Clinical source:	URG
Adhesin:	ND
Receptor specificity:	ND
Substrata:	EC
References:	De Boer and Plantema, 1988; Cook et al., 1989

Moraxella bovis

Clinical source:	Eye, bovine
Adhesin:	Q (b) Fimbriae
Receptor specificity:	ND
Substrata:	ERT, TC
Reference:	Annuar and Wilcox, 1985

Morganella (Proteus) morganii

Clinical source:	URT, BLD
Adhesin:	Type 3 fimbriae (MR/K)
Receptor specificity:	Spermidine
Substrata:	ERT
Reference:	Gerlach et al., 1989

Mycobacterium bovis (bacillus Calmette-Guerin)

Clinical source:	URT
Adhesin:	ND
Receptor specificity:	Fibronectin, heparin?
Substrata:	EC
References:	Ratliff et al., 1987, 1988; Kavoussi et al., 1990; Hudson et al., 1991

Mycoplasma genitalium

Clinical source:	URG
Adhesin:	P1-like protein
Receptor specificity:	ND
Substrata:	ERT
References	Clyde and Hu, 1986; Hu et al., 1987; Morrison-Plummer et al., 1987

Mycoplasma mycoides

Clinical source:	LRT, bovine sheep
Adhesin:	ND
Receptor specificity:	ND
Substrata:	ENC, ERT
References:	Valdivieso-Garcia et al., 1989

Mycoplasma pneumoniae

Clinical source:	URG
Adhesin:	P1 protein
Receptor specificities:	α2,3-Linked sialic acid, Gal($3SO_4$) β1 (sulfated glycolipids); laminin
Substrata:	TC, EC, ERT

The adhesion of *M. pneumoniae* has attracted scientific interest focusing mainly on the previously described P1 adhesin and the receptor specificity of the adhesin (previous studies have been summarized by Razin, 1985a,b). The P1 gene has not been cloned and sequenced (Su et al., 1987; Inamine et al., 1988). The mol wts of the mature protein and its leader sequence containing precursor are estimated at 169,000 and 176,000, respectively. The protein has no homology to known proteinaceous adhesins, and lacks cystine residues, suggesting that it is not stabilized by disulfide bridges (Razin and Yogev, 1989; Jacobs, 1991). The mycoplasma chromosome contains one single copy of the whole P1 gene but some regions of the P1 gene exist as multiple copies (Su et al., 1988). Recombination events between the multiple copy regions and the structural P1 gene may be responsible for observed restriction fragment polymorphism and sequence divergency in the P1 adhesin gene among clinical isolates of *M. pneumoniae* (Dallo et al., 1990b; Su et al., 1990a). The P1 gene, however, contains conserved regions which are likely to code for determinants that bind to the receptor membrane anchoring region and perhaps signal leader amino acid sequences. Attempts to express the P1 protein in recombinant *E. coli* result invariably in the production of truncated fusion proteins (Trevino et al., 1986; Frydenberg et al., 1987; Schaper et al., 1987; Dallo et al., 1988); probably because mycoplasma use the universal stop codon (Dallo et al., 1988; Razin and Yogev, 1989). Because monoclonal P1 antibodies inhibit adhesion of *M. pneumoniae* to erythrocytes (Morrison-Plummer et al., 1986), studies were initiated aimed at characterizing P1 region(s) with adhesin function. The strategy was to employ monoclonal antibodies to various regions of the P1 adhesin molecule as inhibitors of mycoplasma adhesion to sheep erythrocytes (Dallo et al., 1988; Gerstenecker and Jacobs, 1990). After topological mapping of the P1 regions that react with antibodies inhibiting mycoplasma adhesion, it was concluded that three protein regions are important for P1 protein to function as an adhesin. One region maps at the amino-terminal region, the other approximately in the middle of the P1 molecule flanked by two hydrophobic regions, and the third region is close to the carboxylate terminal (Jacobs, 1991). It was suggested that the P1 molecule transverses the mycoplasma membrane several times in a manner that the three P1 regions involved in adhesion protrude on the outer surface to form an adhesin active domain. The P1 protein is highly immunogenic and is one of the major antigens against which *M. pneumoniae* patients develop antibodies (Jacobs et al., 1986; Kanamori et al., 1987; Hu et al., 1985; Vu et al., 1987), suggesting that the P1 adhesin is expressed in vivo. As expected from a protein molecule as large as P1 adhesin, it contains multiple epitopes that readily induce polyclonal antibodies in mice (Hu et al., 1985). Sera from patients with respiratory disease caused by *M. pneumoniae* react with synthetic peptides corresponding to the amino-terminal and middle regions of the P1 molecule, suggesting that at least two regions of the molecule with adhesin function are immunogenic in man (Jacobs et al., 1987, 1989). Furthermore, a test based on inhibition of adhesion of mycoplasma to erythrocytes by antisera from patients may be of diagnostic value because false-positive reactions have not been detected in patients free from infection with *M. pneumoniae* (Jacobs et al., 1985). Nevertheless, the findings that antibodies to other surface components, such as glycolipids, actually enhance the adhesion of mycoplasmas to erythrocytes (Morrison-Plummer et al., 1986) may give rise to misinterpretation of results when based solely on the adhesion-inhibition test. The proximity of T-cell epitope to B-cell epitopes involved in the adhesin function in the P1 molecule may explain the immunodominance and enhanced antiadhesive antibody production in humans or experimental animals infected with *M. pneumoniae* (Jacobs et al., 1988, 1990).

Previous studies have shown that mycoplasmas undergo a poorly understood phase-like variation in their capacity to interact with erythrocytes. It was also shown that in adherent organisms the P1 protein undergoes clustering at the tip of the mycoplasmas (Razin, 1985a,b). It now appears that the population of growing mycoplasmas contains hemadsorption-negative and -positive variants in about equal numbers in spite of the fact that P1 antigen was detected in both the adhesive and nonadhesive variants (Kahane et al., 1985). It is possible, therefore, that for the P1 protein to function as an adhesin it must be able to move freely in the membrane and to interact in some way with the cytoskeletal elements adjacent to the inner side of the membrane permitting it to be densely clustered at the tip region of the mycoplasma. This complex assembly of the P1 adhesin is likely to require a number of accessory proteins. For example, the expression of a number of low and high molecular weight proteins was found to be missing in hemadsorption-negative variants of *M. pneumoniae* (Baseman et al., 1987; Stevens and Krause, 1990; Krause and Lee, 1991). Of special interest is the 32,000 mol wt protein, termed P30, expressed on the surface of adhesive variants (Baseman et al., 1987). Anti-P30 antibodies inhibit adhesion of the bacteria to erythrocytes. The P30 gene has been cloned and sequenced and found to share substantial homology with the carboxy-terminal region of P1 adhesin and to contain a proline-rich region at its carboxy-terminal region (Dallo et al., 1990a).

Both glycoproteins and glycolipids with α2,3-sialylated sequences may serve as isoreceptors for binding *M. pneumoniae* (Loomes et al., 1985). *M. pneumoniae,* however, does not bind gangliosides and neutral glycolipids separated on thin-layer chromatograms (Krivan et al., 1989). Because sialyl-containing glycolipids strongly inhibit the adhesion of mycoplasma to erythrocytes (Loomes et al., 1985) it is possible that presentation of the sialic acid in glycolipids immobilized on the plates is inappropriate for binding the bacteria, emphasizing the limitations of the adhesion of bacteria to glycolipids separated by thin-layer chromatography. Alternatively, it was suggested that gangliosides inhibit adhesion of mycoplasmas by masking other receptors that mediate adhesion (Krivan et al., 1989), but no evidence for this effect is available. In other studies, it was found that treatment of animal cells with sialidase does not result in diminished adhesion of mycoplasmas (Izumikawa et al., 1986; Geary and Gabridge, 1987). In one case, it was found that a glycoprotein of 100,000 mol wt with terminal *N*-acetylglucosamine may also serve as a receptor on fibroblast tissue culture cells for binding mycoplasmas (Geary and Gabridge, 1987). Because of the structural similarity of the *N*-acetylglucosamine to sialic acids (they share common *N*-acetamido groups with neighboring hydroxyl groups), as exemplified by the ability of wheat germ agglutinin to interact with both carbohydrate structures, it has been suggested that *N*-acetylglucosamine residues in glycoconjugates on the surface of certain animal cells may serve as receptor sites for adhesion of mycoplasma. *M. pneumoniae* also binds avidly to a number of α2,3-sialylated glycoproteins, including extracellular matrix laminin, fetuin, and human chorionic gonadotropin when immobilized on plastic (Roberts et al., 1989). In addition, the bacteria bind to sulfated galacto-lipides on thin-layer chromatograms (Krivan et al., 1989). The Gal($3SO_4$)β1-specific adhesion is probably mediated by an adhesin distinct from that specific for the sialylated glycoconjugates. The relative importance of the Gal(SO_4)β1 and sialyl-containing receptors in promoting the adhesion of *M. pneumoniae* was examined by determining the relative ability of sialyl lactose and dextran sulfate to inhibit adhesion of the bacteria to the adenocarcinoma cell line WiDr. Each of these compounds partially inhibited the mycoplasma adhesion, whereas in combination they reduced adhesion by 90%, suggesting that both types of receptors participate in the adhesion of mycoplasma to the tissue culture cells.

Few studies have focused on the consequences of adhesion of mycoplasma to animal cells or to soluble glycoconjugates bearing receptor sites that bind to the bacterial surfaces. The specific interaction of mycoplasma with sialoglycolipids of the I antigen present in sera of several mammalian species results in the deposition of the glycolipid on the surface of bacteria growing in serum-supplemented medium (Uemura et al., 1988). The stereospecific interaction of I antigen with mycoplasma during a natural course of infection was offered as a possible explanation for triggering the immune system to produce anti-I autoantibodies (cold agglutinins) which predominate in sera of patients with pneumoniae caused by mycoplasma. Adhesion of *M. pneumoniae* to hamster tracheal cells and HeLa tissue culture cells allows growth of the bacteria in nonpermissive medium (Chen and Krause, 1988; Krause and Chen, 1988). The growth advantages are associated with loss of viability of the tissue culture cells, but the transformed HeLa cells are more susceptible to the adherent mycoplasma than the tracheal cells. Adhesion was also suggested to enhance tissue damage caused by superoxide anions released by the mycoplasma adherent to cultured fibroblasts (Almagor et al., 1986).

Neisseria gonorrhoeae

Source:	URG
Proposed adhesins:	PII protein; PI (Opa I)?; fimbriae; carbohydrate binding proteins.
Receptor specificities:	Glycolipids (GalNAcβ1,4Galβ1,4Glc and GlcNAcβ1,3Galβ1,4Glc)
Substrata:	OC, TC, PC, IS, EC

Earlier studies have established that *N. gonorrhoeae* is capable of expressing two major classes of adhesins (Sparling, 1988; Swanson, 1990). One class includes the membrane protein known as PII and the other classes are the fimbrial adhesin complexes. It is possible that gonococci express other classes of adhesins as discussed below. Gonococci are also shown to adhere to cells from the female genital tract, red blood cells, buccal cells, sperm, polymorphonuclear leukocytes (PMNs), and fallopian tube organ explants. Gonococcal metabolism sensitive to KCN and other respiratory chain inhibitors, but not to protein synthesis inhibitors, is required for optimal adhesion of the organisms to phagocytic cells and HeLa cells, suggesting that the adhesion is energy-dependent (Weber et al., 1989). Attempts to understand the role of the adhesins, especially the PII and fimbrial adhesins, in the infectious process have been difficult because the two classes of adhesins exhibit both phase (on/off) and antigenic/receptor specificity variations at high frequency within a single strain (Chapter 9). Moreover, because each of the adhesins is under a distinct control mechanism, a large number of variants can be generated from a single strain. Most of the earlier studies on gonococcal adhesion employed organisms that were undefined with respect to the PII or fimbrial adhesins, resulting in apparent contradictions in the results. The availability of monoclonal antibodies that distinguish variants of PII proteins (Barritt et al., 1987) and of fimbrial adhesin complexes (Virji and Heckels, 1985) has made it possible to understand better the role of the adhesin variants in the infectious process.

The family of outer membrane PII proteins exhibit different apparent molecular weights in the range of 24–30,000, are recognized by antibodies specific for either the variable region or the conserved region of the proteins, and display a characteristic temperature-dependent behavior in polyacrylamide gel electrophoreses (an increase in apparent molecular weight after solubilization at 100°C) (Blake, 1985; Barritt et al., 1987). Colonies with opaque appearance invariably express one or more PII variant proteins (also called Opa proteins), whereas transparent colonies may or may not express a particular PII variant (Barritt et al., 1987; Swanson, 1990). Follow-up of urethritis induced in three males by intraurethral administration of predominantly PII nonexpressing gonococci revealed that virtually all gonococci isolated from the infected men express the PII adhesin (Swanson et al., 1988). The preferential selection of the PII-carrying phenotypes or variants during the infection clearly suggest that the PII proteins confer advantages in resisting host defense mechanisms. Although the involvement of PII proteins in adhesion of gonococci to vaginal epithelial cells was suggested (Hagman and Danielsson, 1989), most studies employ tissue cell lines to examine the binding properties of gonococcal PII variants. Strains of nonfimbriated gonococci expressing either PIIa or PIIb or PIIac adhesin variants bind to HeLa cells, whereas phenotypes not expressing PII proteins adhere poorly to the cells (Bessen and Gotschlich, 1986). Following attachment and replacement of the medium containing nonadherent bacteria with fresh medium, the PIIac-bearing gonococci resisted detachment during 1-hour incubation, whereas the number of PIIc bearing gonococci remaining adherent decreased markedly, suggesting that high-affinity adhesion is mediated by the PIIac adhesin variant. The PII-mediated adhesion to epithelial cells in vitro is associated with a number of consequences relevant to events likely to occur during an infection. PIIa-mediated adhesion to HeLa cells confers a growth advantage to the adherent gonococci (Bessen and Gotschlich, 1986). Adhesion-dependent growth effects are similar to those described for *E. coli* (Figure 11–5). In other studies it has been shown that adhesion of nonfimbriated gonococci bearing PIIc or PIId adhesins to human fallopian tube tissues causes damage to the epithelial cells, characterized by a decrease in the number of microvilli on nonciliated cells and invagination and sloughing off of ciliated cells (Dekker et al., 1990). In contrast, adhesion of gonococci via PIIb or PIIa adhesins is associated with little damage to the fallopian cells. The ability of the gonococci to enter mucosal cells as a direct consequence of attachment to the cells is probably a major characteristic of gonococcal infections (Stephens, 1989). PII adhesins appear to contribute to this effect as evident from several studies. Interaction of gonococci with tissue culture cell lines, including HeLa, BHK–21 (baby hamster kidney cells), HecIB (human adenocarcinoma cell line), Chang (conjunctiva epithelial cells) cells, and human cornea in organ culture was found to involve entry of the bacteria in the cytoplasm (Bessen and Gotschlich, 1986; Tjia et al., 1988; Shaw and Falkow, 1988; Richardson and Sadoff, 1988; Makino et al., 1991; Weel et al., 1991). Whenever studied, the internalization required viable organisms, and was dependent on movement of microtubules and microfilaments of the tissue culture cells. In two studies employing HeLa and HecIB cells it was observed that nonfimbriated PII-bearing gonococci were also internalized by the tissue culture cells (Bessen and Gotschlich, 1986; Shaw and Falkow, 1988) and in two other studies employing Chang cells, the participation of PII adhesin in the process of internalization was implicated (Makino et al., 1991; Weel et al., 1991a,b). The extent of invasiveness into Chang cells by gonococcal strains expressing various PII proteins was examined by immunolabeling of intracellular bacteria in situ and by analyzing the PII proteins produced by internalized and adherent bacteria with monoclonal antibodies specific for the various PII protein variants. Strains expressing either 30,000 or 27,500 mol wt PII proteins exhibit high-affinity adhesion and invade the cells in increased numbers, compared to gonococci expressing 31,000 or 28,000 and 29,500 mol wt PII adhesin variants (Makino et al., 1991; Weel et al., 1991a,b). The correlation between high-affinity adhesion and increased invasiveness is consistent with the notion that for an adhesin to function as an invasin mediating internalization of the bacteria a requirement is accessibility of receptors on the animal cells capable of binding the adhesin with a high affinity (see discussion on *Yersinia*). Although the affinity of isolated PII variant adhesins for several cell types has not been studied, it remains to be seen which of the various PII variants interact with high-affinity receptors on urogenital epithelial cells triggering internalization of the organisms. Attempts to characterize the PII receptors on HeLa cells have employed a PII adhesin variant associated with an opaque colony. The PII variant has a molecular weight of approximately 29,000 in the heat-modified form (Bessen and Gotschlich, 1987). This PII protein binds to a number of glycoproteins on nitrocellulose blots after they have been solubilized from HeLa cells and separated by electrophoresis. The binding of PII protein is not inhibited by monosaccharides and evidence has been obtained to suggest that the interaction of HeLa cell macromolecules with PII protein involves a specific protein–protein interaction. The PII adhesins expressed by gonococci exhibiting opaque colonies are involved in two other adhesive activities. One activity relates to the study showing that intergonococcal adhesion is mediated by specific interaction between certain PII variant proteins and carbohydrate moieties of lipopolysaccharide molecules expressed on the gonococcal surface which give rise to the opaque colony (Blake, 1985). The other activity relates to studies showing that PII adhesin mediates binding of the bacteria to PMNs (Shafer and Rest, 1989; see also Chapter 7). The bound bacteria stimulate oxidative bursts and are readily ingested and killed by the phagocytic cells. The receptors involved in these processes can be up-regulated by treating the PMNs with various agents (Farrel and Rest, 1990). Although the stimulation of PMNs by gonococci bearing a distinct PII variant has been found to be inhibited by the monosaccharides D-glucosamine, *N*-acetylneuraminic acid, and D-mannose (Rest et al., 1985), the nature of the receptor(s) and the PMN attachment site involved in the phagocytic process has not been defined, nor is it clear whether different receptors are required for different PII variants. Gonococci bearing only certain PII variants are phagocytized by PMNs. The HV2 region of PIIb variant molecules participate in the process (Shafer and Rest, 1989). The magnitude of stimulation of oxidative bursts varies in PMNs exposed to gonococci bearing different PII variants (Virji and Heckels, 1986). Given the diversity of the PII family of adhesins and the high frequency of shifting from one PII variant to another (Chapter 9) it is difficult at this stage to define which PII variant molecule is capable of functioning as an adhesin only, or as an adhesin with invasin activity, or with lectin activity that recognizes specific carbohydrates on animal cells or on the surface of the organisms. If it is assumed that all or most of PII variants can be expressed, one or more at a time, by a single strain, as the genetic data suggest, then the family of PII proteins can explain a number of events known to occur during a course of infection. These include adhesion of the gonococcus to urogenital epithelial cells which then progressively engulf the bound organisms, zones of interaction between the bacteria, and finally PMNs packed with gonococci and extracellular bacteria seen in purulent exudates of infected patients.

Suppression of fimbrial biosynthesis by sublethal concentrations of antibiotics results in decreased adhesion of the nonfimbriated gonococci to epithelial cells (Gorby and McGee, 1990), confirming an earlier proposed role of fimbriae of *N. gonorrhoeae* on adhesion to epithelial cells (Sparling, 1988; Swanson, 1990). Further studies on properties of fimbriae-mediated adhesion of gonococci have been slow. There may be two major reasons to account for this. One reason has to do with difficulties in designing a fimbrial vaccine to induce antiadhesive immunity because of antigenic variations of the fimbrial adhesins expressed by a single strain (see Chapter 9). Further studies in this direction have employed synthetic peptides corresponding to conserved regions of the major fimbrial subunit in order to stimulate antipeptide antibodies that otherwise are weakly developed when whole fimbriae are used as immunogens (Rothbard et al., 1985). Sera obtained by immunizing rabbits with peptides conjugated to a carrier protein and corresponding to residues 41–50 and 69–84 inhibit the adhesion of homologous and heterologous strains of *N. gonorrhoeae* to a human endometrial carcinoma cell line. Other studies, however, have shown that epitopes contained in the conserved regions of the gonococcal fimbrial adhesin complex are inaccessible on native fimbriae (Robinson et al., 1989). Regions of three-dimensional structure of the gonococcal fimbriae, which belong to the type 4 class fimbriae, have been suggested as candidates for vaccines, but information on the antiadhesive capacity of antibodies against such epitopes are lacking (Getzoff et al., 1988). The inability of the gonococcal fimbrial adhesin complex to induce endocytosis of the bacteria by epithelial cells may be another reason for the apparent lack of interest in exploring their adhesive properties. Although fimbriae may facilitate adhesion of gonococci to tissue culture cells, their presence on the bacterial surface was found either to not affect (Shaw and Falkow, 1988) or to impede (Makino et al., 1991) the PII-mediated internalization of the bacteria by the epithelial cells. Although gonococci lacking the PII adhesin, but expressing the fimbrial adhesin, bound in considerably higher numbers than the bacteria not expressing both adhesins, the adherent gonococci were more readily detached from the epithelial cell surface than adherent PII-bearing organisms (Bessen and Gotschlich, 1986). The results suggest that fimbriae mediate weak, reversible adhesion of gonococci to epithelial cells. Phagocytic cells probably lack fimbrial receptors because none of three PII-gonococci strains, each bearing a different fimbrial variant, were killed by or stimulated an oxidative burst in human PMNs (Virji and Heckels, 1986).

Gonococci bind to certain lactose-containing glycolipids separated by thin-layer chromatography or immobilized on the bottoms of microtiter plates (Stromberg et al., 1988a; Nyberg et al., 1990; see Chapter 7). The gene encoding the gonococcal lectin specific for the lactosyl-containing glycolipid was cloned. Transformant *E. coli* harboring the lectin gene binds to the same glycolipids as those bound by the gonococcal cells (Paruchuri et al., 1990). The lactose-specific lectin has a mol wt of 36,000, is expressed on the surface of the organisms devoid of PII or fimbrial adhesins, and its gene is duplicated in one strain tested. Although lactose ceramide and lactose-containing gangliotriosyl ceramide were found to be present on human endocervical cells in culture, the adhesin function of the lactose-specific lectin of gonococci has not yet been shown. Gonococci have been found to bind to maltose and glucosamine immobilized on beads, suggesting the presence of a maltose- and glucosamine-specific lectin on the bacterial surface (Perrollet and Guinet, 1986). The maltose-specific lectin has an apparent mol wt of 440,000 consisting of several identical subunits of 65,000 mol wt. Antibodies against the lectin protect chick embryos infected with gonococci, but the adhesin function of the lectin has not been studied. It appears that gonococcal infection is initiated by low-affinity and reversible attachment of the invading gonococci to mucosal cells lining the urogenital or the upper respiratory tract. This type of weak adhesion may be mediated by the fimbrial adhesin complex or the lactose-specific lectin. Some of the progeny of the bound organisms express one or more PII variant adhesins which bind the organisms to the epithelial cells with a high affinity leading to endocytosis of the bound bacteria. The process of strong and intimate attachment is associated with apparent zones of adhesion formed between the gonococci and the cell membrane of the epithelial cells (Weel et al., 1991a,b). Endocytosis may also involve the pore-forming activity of the PI outer membrane protein, which can translocate from the bacterial surface to the host cell membrane (Elkins and Sparling, 1990). Following invasion, progeny of intracellular gonococci are exocytosed into the subepithelial space (Sparling, 1988). The exocytosed population may consist of gonococci bearing a PII variant adhesin which mediates phagocytosis of the organisms by PMNs and of gonococci either lacking PII molecules or expressing PII variant molecules which allow the gonococci to resist phagocytosis and eventually multiply. The processes of phagocytosis that result in release of tissue damaging agents and of gonococcal proliferation may cause release of toxic products (such as lipopolysaccharides), setting the stage for acute inflammatory responses characterized by a purulent discharge of exudate so characteristic of local gonococcal infections.

Neisseria meningitidis

Clinical source:	ORL
Adhesin:	Class 2 (and 1?); class 5 outer membrane proteins
Receptor specificity:	ND
Substrata:	EC, TC, PC, ERT

During infection with *N. meningitidis,* the bacteria interact with mucosal cells of the upper respiratory tract, and following invasion to the blood and meninges, they also interact with phagocytic cells and endothelial cells. Because the meningococci are capable of colonizing the upper respiratory tract for relatively long periods of time, early and recent studies have focused on adhesion to buccal epithelial cells and nasopharyngeal organ culture as well as erythrocytes as a convenient substratum to monitor adhesin activity (Stephens, 1989). Most virulent strains are encapsulated, suggesting that the meningococcal adhesin(s) may have to promote adhesion of hydrophilic bacteria. Indeed, no correlation between the hydrophobicity of the meningococcal cells and the ability of the bacteria to bind to buccal epithelial cells has been found (Criado et al., 1986; Ferreiros et al., 1986).

In all cases, the fimbriae were implicated to facilitate the meningococcal adhesion. It is now clear that there are two morphologically and antigenically distinct fimbriae expressed by *N. meningitidis* (Greenblatt et al., 1988). One class of fimbriae, termed β-type pili, are expressed by most strains isolated from healthy persons (e.g., carrier state) and appear as aggregated bundles on the bacterial surface. The other class, termed α-type pili, are expressed mostly by strains isolated from patients with meningitis and appear as single filaments on the bacterial surface. Strains carrying β-type fimbriae cause hemagglutination of human erythrocytes and adhere in high numbers to human buccal epithelial cells. In contrast, strains expressing α-type fimbriae adhere poorly to human buccal epithelial cells and most of them fail to cause hemagglutination of erythrocytes. The source of isolation, however, is a poor predictor of the ability of the strain to express a particular fimbrial type or to adhere to buccal cells (Criado et al., 1989). Nonfimbriated variants do not cause hemagglutination nor do they adhere to buccal epithelial cells, confirming earlier results implicating adhesive function for β-type fimbriae. Other studies described two classes of meningococcal fimbriae, one of which reacts with anti-gonococcal fimbriae antibodies and termed class I, whereas the other, termed class II, does not react with the antibodies (Heckels, 1989; Virji et al., 1989). Not all strains carrying class I or class II fimbriae are able to interact with erythrocytes or buccal epithelial cells (Pinner et al., 1991). The data are somewhat confusing but tend to suggest that meningococci express different types of fimbrial adhesins with distinct receptor specificities. It can also be assumed that the presence of any type or class of fimbriae on the bacterial surface confers upon the meningococci adhesive properties to one or more types of animal cells. Molecular genetics of meningococcal fimbriae reveal antigenic similarities to those of gonococcal fimbriae but considerable intra- and interstrain heterogeneity with respect to the sizes of the major fimbrial subunits (in the range of 17,000–21,000 mol wt) and antigenicity (Olafson et al., 1985; Stephens et al., 1985; Perry et al., 1987). In particular, the class I fimbrial genes have been cloned and sequenced and the relatedness of this class of fimbriae to the gonococcal fimbriae was documented (Potts and Saunders, 1987). Because the chromosomal DNA of class II-bearing meningococci was found to contain silent truncated fimbrial genes homologous to DNA sequences coding for the gonococcal pili (Perry et al., 1988), variants expressing fimbrial adhesin with a distinct antigenic/receptor specificity are expected to rise by recombination at relatively high frequency, similar to events occurring in gonococci (see Chapter 9). Indeed, such phase transition during a meningococcal infection has been described (Tinsley and Heckels, 1986). Examination of strains isolated from the blood, CSF, and nasopharynges of individual patients revealed that meningococci carrying fimbriae reacting with the monoclonal antigonococci pili (presumably class I fimbriae) gave rise to variants carrying fimbriae lacking the cross-reactive epitope class II fimbriae). As with gonococcal pili, the ability of meningococci to diversify the antigenic and receptor specificity of its fimbrial adhesin confers an advantage to colonize different types of surfaces during the infectious process (Meyer and van Putten, 1989; Meyer et al., 1990). Attempts to localize the region of the major fimbrial subunit responsible for the receptor specificity have been unsuccessful (Heckels, 1989). The ability of the fimbriae to promote adhesion does not involve epitopes common to all meningococcal fimbrial adhesin complexes because antibodies against these regions do not inhibit the meningococcal adhesion to buccal epithelial cells (Stephens et al., 1988). Fimbriae, possibly particular variants, may also promote adhesion for meningococci to human umbilical vein endothelial cells in culture and the presence of capsule interferes with the adhesion process (Virji et al., 1991b).

Another class of adhesins expressed by *N. meningitidis* may be the family of class 5 outer membrane proteins (Stephens, 1989). This class of meningococcal outer membrane proteins shares several properties with the gonococcal outer membrane PII adhesin. Examination of the DNA sequence of the cloned class 5 gene shows a high degree of homology with the PII gene sequences, especially in the sequences coding for the amino-terminal regions of the two outer membrane proteins (Kawula et al., 1988). The class 5 outer membrane gene contains the CTCTT repeat sequence which is important in the regulation of gonococcal PII expression and presumably for the class 5 protein as well. Similar to the family of PII proteins, antigenic variations (and presumably receptor specificity variations) have also been noted among the class 5 proteins. In one study it was possible to identify eight class 5 protein variants from a single clone obtained over a 4-year period (Achtman et al., 1988). Finally, both the gonococcal PII and the meningococcal class 5 proteins are heat modifiable (protein solubilized at 37°C migrates faster than protein solubilized at 100°C in SDS-PAGE). It has also been noted that urogenital isolates of meningococci bearing heat-modifiable protein adhere in higher numbers to vaginal epithelial cells and are phagocytosed more readily by human polymorphonuclear leucocytes than meningococci lacking this protein (Hagman and Danielsson, 1989).

Microscopic observations of human nasopharyngeal organ cultures exposed to live *N. meningitidis* show adhesion of the bacteria to nonciliated cells, whereas heat-killed bacteria adhere poorly (Stephens et al., 1986). The adhesion is associated with ciliar damage but a role for fimbriae or other outer membrane proteins in this process has not been confirmed, suggesting a distinct mechanism of adhesion to nasopharyngenal cells in organ culture. The receptor specificities of the meningococcal adhesins are not known. It is likely, however, that in spite of the similarity between the gonococcal and meningococcal adhesins, distinct receptor specificities of *N. meningitidis* adhesins are involved in adhesion to nasopharyngeal cells and eventually in translocating the bacteria to the blood and meninges.

Neisseria subflava

Clinical source:	URT
Adhesin:	Sia–1 protein
Receptor specificity:	NeuNAcα2,3Galβ1,4Glc
Substrata:	ERT
Reference:	Nyberg et al., 1990

Pasteurella multocida

Clinical source:	URT, LRT, pigs
Adhesin:	Fimbriae (serotype D)
Receptor specificity:	ND
Substrata:	ERT, EC TB
References:	Jacques et al., 1988; Nakai et al., 1988; Trigo and Pijoan, 1988; Chung et al., 1990; Pijoan and Trigo, 1990; Vena et al., 1991

Porphyromonas (Bacteroides) gingivalis

Clinical source:	ORL
Adhesins:	HA-Ag2, fimbriae, β-galactosides, sialic acid
Receptor specificities:	Carbohydrate binding proteins on oral bacteria, histidine-rich salivary protein
Substrata:	EC, ERT, oral bacteria

The periodontal pathogen *P. gingivalis* is capable of expressing a number of adhesins with distinct receptor specificities. One adhesin is the fimbrial complex consisting of a major polypeptide subunit of an apparent mol wt of 43,000, but a size of 35,924 as determined by DNA sequencing of the fimbrial gene (Dickinson et al., 1988). The fimbrial protein has no significant homology to other fimbrial adhesins, but exhibits considerable size, amino-terminal sequence, and antigenic heterogeneity among various strains of *P. gingivalis* (Suzuki et al., 1988; Lee et al., 1991). The fimbrial adhesin promotes attachment to two types of surfaces associated with the oral cavity. One oral substratum is the buccal epithelial cells. It is known that monoclonal antibodies directed against one antigenic variant inhibit adhesion of the *Porphyromonas* to epithelial cells (Isogai et al., 1988). The fimbrial receptors on many of the scraped buccal cells are probably cryptic, because buccal cells pretreated with proteolytic enzymes or sialidase bind several-fold more bacteria, compared to nontreated cells (Childs and Gibbons, 1988, 1990; Gibbons et al., 1990). The fimbriae also promote adhesion of *Porphyromonas* to *Actinomyces viscosus*. Purified fimbrial monomers and antifimbrial antibodies inhibit *Porphyromonas–Actinomyces* coaggregation (Goulbourne and Ellen, 1991). Purified fimbriae also bind specifically to *Actinomyces viscosus* cells. *P. gingivalis* adheres avidly to *A. viscosus* bound to saliva-coated hydroxylapatite (Schwarz et al., 1987; Li and Ellen, 1989) or to hydrophobic surfaces (Rosenberg et al., 1991), suggesting that the fimbrial adhesin may be responsible for the preferential localization of *P. gingivalis* in dental plaques during normal hygiene. It has been postulated that during poor oral hygiene elevated levels of proteolytic enzymes may expose cryptic fimbrial receptors on epithelial cells to cause a shift in the amount of *Porphyromonas* colonizing the epithelial gingival tissue (Gibbons et al., 1990). Protease-containing vesicles excreted by the bacteria may also contribute to exposing cryptic receptor sites on epithelial or *A. viscosus* substrata (Li et al., 1991). The fimbrial adhesin may also play a direct role in enhancing inflammatory responses during periodontal disease, because in purified form it stimulates production of thymocyte-activating factor by human gingival fibroblasts, and induces in macrophages the expression of interleukin-1 (Hanazawa et al., 1988, 1991). Although the majority of the *P. gingivalis* isolates express the fimbrial adhesin during growth in vitro (Suzuki et al., 1988), not all cells in the population express fimbriae (Handley and Tipler, 1986; Handley, 1990), suggesting that some degree of phase variation occurs in the expression of the fimbrial adhesin. *P. gingivalis* produces two classes of hemagglutinins. One class is soluble and secreted in the culture supernatant. The identity of the component responsible for hemagglutinating activity, however, is not clear. In one case using sheep erythrocytes, it was postulated to be a lipid component associated with a 40,000 mol wt protein (Okuda et al., 1986) and in another case using human erythrocytes it was suggested that the secreted hemagglutinin may involve interaction of the hemagglutinin with arginine residues on the erythrocyte surface (Inoshita et al., 1986). The function of the soluble hemagglutinins as adhesins has not been confirmed. The other class of hemagglutinin is a nonfimbrial surface protein complex, designated HA-Ag2, isolated from the bacteria and shown to be responsible for hemagglutination of formalin-treated human erythrocytes (Mouton et al., 1989, 1991). The cell-associated hemagglutinin is an outer membrane component, consisting of two polypeptides of 43,000 and 49,000 mol wt when extracted as extracellular vesicles. When extracted in the presence of EDTA, 33,000 and 38,000 mol wt peptides were recovered, suggesting that an additional polypeptide of 10,000 mol wt is associated with each of the large polypeptides in an EDTA-sensitive linkage. The protein complex binds to constituents present in membrane lysates obtained from human non-formalin-treated erythrocytes, but its role in adhesion to other eucaryotic cells has not been established. Although in one study it was shown that purified fimbriae and synthetic fimbrial peptides corresponded to two regions of the fimbrial subunit agglutinating rabbit erythrocytes (Ogawa et al., 1991b), a role for fimbriae as a hemagglutinin using formalin-treated human erythrocytes (Isogai et al., 1988) could not be confirmed. Systematic studies employing erythrocytes of various sources and defined *P. gingivalis*-derived components are clearly needed to resolve the controversy relating to the types and numbers of hemagglutinins produced by the *Porphyromonas*. The expression of hemagglutinins probably is under the control of phase variation because the expression of at least one type of hemagglutinin responsible for binding *Porphyromonas* to sheep erythrocytes is greatly enhanced in hemin-starved bacteria (Smalley et al., 1991). *P. gingivalis* vesicles derived from the outer membrane may carry nonfimbrial adhesins for *A. viscosus* surfaces (Ellen and Grove, 1989). Isolated vesicles also enhance the adhesion of other oral bacteria to each other or of streptococci to hydroxylapatite (Grenier and Mayrand, 1987; Singh et al., 1989). The involvement of vesicles in the adhesion process may be complex because these organelles may carry adhesins that bind to one oral bacterium and adhesins specific for other oral bacteria, as well as adhesins specific for particular salivary constituents on tooth surfaces. *P. gingivalis* binds to histidine-rich peptides (histatin) with relatively low affinity (K_d of 1.5×10^6 M^{-1}; Murakami et al., 1991b). The function of histatin-binding component as an adhesin, which may tether the *Porphyromonas* to saliva-coated surfaces, has not been established. Salivary histatin, however, may interfere with the function of other *Porphyromonas* adhesins because the synthetic histidine-rich peptides were found to inhibit the adhesion of *P. gingivalis* to erythrocytes and to *Streptococcus mitis* (Murakami et al., 1990, 1991a). *P. gingivalis* interacts specifically with soluble host-derived fibrinogen (Lantz et al., 1986). It was suggested that the fibrinogen–*Porphyromonas* interaction is likely to contribute to the ability of the periodontal pathogen to adhere to and colonize tissues but no evidence for such adhesion promoting activity is available.

Carbohydrates on *P. gingivalis* surfaces may also serve as adhesins (or receptors) that bind the bacteria to other oral bacteria, promoting either the process of active destructive periodontal disease or formation of dental plaques. For example, *Porphyromonas* interacts with *Fusobacterium nucleatum*. The interaction is sensitive to lactose and is abolished by treating the *Fusobacterium* with heat or proteolytic enzymes, suggesting that β-galactoside-containing polysaccharides on the *Porphyromonas* surface are responsible for this coaggregation (Kolenbrander and Andersen, 1989; Table 8–6). Moreover, although previous studies failed to observe interaction of *P. gingivalis* with oral streptococci, following immobilization of the streptococci onto beads, it was possible to detect binding of the *Porphyromonas* to *S. sanguis* (Stinson et al., 1991). The *Porphyromonas–Streptococcus* interaction was abolished by pretreating the *Porphyromonas* with sialidase, suggesting that a sialic acid binding protein of the *S. sanguis* surface interacts with sialic acid residues on the *P. gingivalis* surface.

Some of the *P. gingivalis* adhesins may function in concert at a particular stage of the infectious process caused by this successful periodontal pathogen. Receptor sites may become available to other *Porphyomonas* adhesins (e.g., fibrinogen-binding adhesin and fimbriae or hemagglutinins) at different stages of the infectious process, especially during periodontal disease involving invasion and gingival bleeding. At this time it is not clear toward which adhesin should an antiadhesin vaccine be directed to prevent *P. gingivalis* infections. Extensive efforts are being devoted to the fimbrial adhesin as a candidate for a vaccine to induce secretory IgA antiadhesion antibodies in the oral cavity. Fimbrial antigens of various strains can be divided into only two immunoreactive groups (Ogawa et al., 1991a,b). The eliciting of secretory IgA antifimbrial antibodies in experimental animals requires muramyl dipeptides as adjuvants (Ogawa et al., 1989) and is also restricted to the H-2 haplotype (Shimauchi et al., 1991). These limited studies have promising prospects because the fimbrial adhesin is involved in promoting adhesion of *P. gingivalis* to different types of oral substrata.

Prevotella (Bacteroides) loescheii

Clinical source:	ORL
Adhesins:	Fimbriae
Receptor specificity:	GalNAc
Substrata:	ERT, oral bacteria

Two fimbrial adhesins with distinct receptor specificities have been described in *P. loescheii*. Both fimbrial adhesins appear to resemble fibrils or fibrillae in structure and to form a peritrichous array of surface appendages (Weiss et al., 1987a). One fimbrial adhesin is a lectin that promotes the interaction of the *Prevotella* with *Streptococcus oralis* strain 34. The streptococcus-specific adhesin is also responsible for the hemagglutination of sialidase-treated erythrocytes from various species caused by whole bacteria. Compared to other carbohydrates, GalNAc (0.2–1.6 m*M*), and to a lesser extent D-galactosamine (1.6–12.5 m*M*), were the most effective inhibitors of the coaggregation with *S. sanguis* and hemagglutination of sialidase-treated human erythrocytes. (Weiss et al., 1987a, 1989). In contrast, the *Prevotella* fimbrial adhesin responsible for *Prevotella–Actinomyces* intractions appears to recognize specific proteinaceous materials on the surface of the *Actinomyces* cells. It is possible to generate monoclonal antibodies to each of the two fimbrial adhesins or polyclonal antifimbrial antibodies which specifically inhibit the interactions between *P. loescheii* and its streptococcal and *Actinomyces* partners (Weiss et al., 1988a,b). Immunoblot analyses have helped to identify a 75,000 mol wt subunit for the galactosamine-specific fimbrial adhesin and a 43,000 mol wt subunit for the *Actinomyces*-specific fimbrial adhesin. The antiadhesin antibodies were used to localize the immunoreactive adhesin for both types of fimbrial adhesins in the distal portion of the fibrillar structure, suggesting that the antigen possessing the adhesin function is associated with a minor fimbrial subunit, presumably localized at the tip of the fimbrial adhesin complexes. Moreover, gold particles of two different sizes coated with antibodies directed against one type of adhesin revealed that all of the *Prevotella* cells coexpressed the two types of adhesins but different fibrillar clumps (London et al., 1989). The total numbers of the two types of adhesins on the cell surface are about equal (ca. 300–400 adhesin molecules per cell) as evident from Scatchard plot analysis of the binding of radiolabeled adhesin-specific monoclonal antibodies by the bacteria (Weiss et al., 1988a,b). Affinity chromatography, using monoclonal adhesin-specific antibodies, has also helped to purify and identify the *N*-acetylgalactosamine-specific adhesin molecule complex as a hexamer consisting of 75,000 mol wt monomers (London and Allen, 1990). The streptococcus- and erythrocyte-specific adhesin is a basic protein with a pI of 8.0–8.4 that aggregates at neutral and basic pH values. The soluble hexameric form of the adhesin inhibits the interactions of streptococci and sialidase-treated erythrocytes and *P. loescheii,* whereas the aggregated form of the *Prevotella* adhesin agglutinates both the streptococci and the erythrocytes. Galactosides specifically inhibit the aggregation activities caused by the purified adhesin complex, suggesting that the soluble form of the lectin acts as a hapten, whereas the aggregated form acts as a multivalent lectin-like adhesin. *P. loescheii* probably produces other adhesins. In a preliminary survey, it was found that *P. loescheii* binds avidly to hydroxylapatite coated with proline-rich proteins (Gibbons, 1989). The two types of fibrillar adhesins, as well as the putative adhesin specific for the proline-rich salivary protein, may function in concert to promote adhesion to and colonization of supergingival and subgingival plaques. In another study, it was noted that chlorhexidine, cetylpyridinium chloride, and octenidine dihydrochloride interfered with coaggregation of *P. loescheii* with streptococci as well as with *Actinomyces* (Smith et al., 1991). The results are interpreted to suggest that these antimicrobial agents bind to fimbrillar structures bearing both the streptococcal and the *Actinomyces* types of adhesins in a manner that interferes with their adhesive functions.

Prevotella intermedia (Bacteriodes intermedius)

Clinical source:	ORL
Adhesins:	Type C "appendages"
Receptor specificity:	β-Galactoside?
Substrata:	ERT, EC, oral bacteria
References:	Gibbons, 1989; Leung et al., 1989; Okuda et al., 1989; Childs and Gibbons, 1990; Gibbons et al., 1990 Tables 8–6 and 8–7

Propionibacterium freundenreichii

Clinical source:	Milk
Adhesin:	ND
Receptor specificity:	Lactosylceramide
Substratum:	ERT
Reference:	Stromberg et al., 1988b

Proteus mirabilis

Clinical source:	UTI
Adhesin:	MR/P fimbriae; type 3 fimbriae (MR/K)
Receptor specificity:	Tamm–Horsfall glycoprotein
Substratum:	EC, ERT, TB
References:	Daifuku and Stamm, 1986; Cellini et al., 1987, 1988; Mobley and Chippendale, 1990; Sareneva et al., 1990; Bahrani et al., 1991; Hawthorne et al., 1991; Hornick et al., 1991

Providencia (Proteus) stuartii

Clinical source:	SKN, URG
Adhesin:	Type 3 fimbriae (MR/K)
Receptor specificity:	Spermidine

Substratum:	EC, ERT, SS
References:	Mobley et al., 1986, 1988; Gerlach et al., 1989

Pseudomonas aeruginosa

Clinical sources:	ENV, URT, URG, ENT, LRT
Proposed adhesins:	Alginate, fimbriae, hydrophobins, carbohydrate-binding proteins, exoenzyme S
Receptor specificities:	Sialic acid, Galβ1,3GlcNAc, lactose, fucose, galactose, *N*-acetylmannosamine, asialogangliosides, GalNAcβ1,3Gal
Substrata:	EC, CS, PC, TC, BM, SS, IM, ERT

Pseudomonas aeruginosa can survive and multiply in numerous types of niches, including aqueous solutions, hand creams, hospital humidifiers, and biomaterials. As an opportunistic pathogen, *P. aeruginosa* is also capable of colonizing various sites of animal bodies, including the skin, urinary tract, ear, eye, lung, and CSF, causing serious localized diseases. Generalized systemic infections tend to occur only in injured, immunodeficient, or otherwise compromised patients. It is not surprising, therefore, that this genus has developed a number of adhesion mechanisms, each being specific for a particular type of substratum. Various model systems have been established for *P. aeruginosa* adhesion including the frequently colonized contact lens and tracheal target surfaces and wounds (Mertz et al., 1987; Miller et al., 1991; Singh et al., 1990; Niederman et al., 1986; Plotkowski et al., 1989, 1991; Stern and Zam, 1986; Todd et al., 1989). The organisms can even interact with saliva and endogenous oral bacteria (Komiyama, 1987; Komiyama et al., 1988). In spite of a considerable number of studies on the various mechanisms of adhesion, little is known on the association between an adhesion mechanism and the ability of the bacterium to colonize and cause a disease at a defined site in the body. In the following, the adhesion mechanisms will be reviewed and their possible contribution to a particular type of infection caused by *Pseudomonas* will be discussed.

One mechanism of adhesion involves one or more hydrophobins on the bacterial surface. The importance of hydrophobic interactions is evident from studies showing a positive correlation between the degree of hydrophobicity of various strains and the ability of the organisms to bind to polystyrene (van Loosdrecht et al., 1987a,b) or phagocytic cells (Speert et al., 1986). Also, contact lenses composed of nonionic polymers bind more bacteria, compared to the less hydrophobic surfaces (Miller and Ahearn, 1987). The interaction of the organisms with erythrocytes is readily inhibited by hydrophobic compounds (Garber et al., 1985; Glick et al., 1987). Because phagocytic cells or erythrocytes probably express receptors specific for *Pseudomonas* adhesins, it is possible that, similar to other bacteria (see Chapter 11), surface hydrophobins provide a first step adhesin to overcome repulsive forces between the bacteria and the substrata, permitting specific adhesins to interact with appropriate receptors.

There is evidence that *Pseudomonas* fimbriae is an adhesin that binds the organism to a number of substrata, including human buccal and ciliated tracheal epithelial cells (Paranchych et al., 1986), mouse epidermal cells (Sato and Okinaga, 1987), cultured bovine tracheal cells (Saiman et al., 1989, 1990) and to immobilized human tracheobronchial mucin (Biesbrock et al., 1991; Kubeceh et al., 1988; Ramphal et al., 1987). The *Pseudonomas* fimbriae belong to type 4 fimbriae (also known as *N*-methylphenylalanine pili), which are shared by *Neisseria, Moraxella, Bacteroides,* and *Vibrio,* and are characterized by a fimbrial subunit containing methylated phenylalanine at its amino terminus (Paranchych, 1990; Patel et al., 1991). The major fimbrial subunits of all these genera possess a conserved N-terminal region that is very hydrophobic (Sastry et al., 1985). Inhibition experiments using monoclonal antibodies directed against various regions of the fimbrillin have revealed that the disulfide-bridged C-terminal region is involved in mediating the adhesion of the organisms to human buccal epithelial cells (Lee et al., 1989; Doig et al., 1989, 1990). A synthetic peptide containing the sequence identical to that of fimbrillin residues 128–144, which form a disulfide bridge, binds to both buccal and tracheal epithelial cells and competitively inhibits the binding of intact fimbriae to the cells (Irvin et al., 1989). In spite of the evidence for fimbriae-mediated adhesion, it is clear that mucoid as well as nonmucoid strains of *Pseudomonas* are capable to adhering to various substrata via nonfimbrial adhesins (Ramphal et al., 1991b).

Another adhesin that binds *Pseudomonas* to tracheal cells and tracheobronchial mucins may be the mucoid exopolysaccharide alginate. The polysaccharide binds to buccal epithelial and tracheal cells, as well as to bronchotracheal mucin. Antipolysaccharide inhibits binding of the organisms to the tracheal cells (Ramphal and Pier, 1985; Doig et al., 1987; Ramphal et al., 1987; Baker et al., 1990). Also, mucoid strains of *Pseudomonas* adhere much better to tracheal cells compared to nonmucoid strains or compared to alginate-producing bacteria grown in an antibiotic medium to reduce alginate production (Geers and Baker, 1987). Alginate appears to play a role in the adhesion of *Pseudomonas* to contact lenses (Slusher et al., 1987). Microscopic evidence suggests that mucoid strains may adhere to ciliated tracheal cells and to inert surfaces, including contact lenses, by polysaccharide-like materials (Butrus and Klotz, 1987; Marcus and Baker, 1985; Miller and Ahearn, 1987). The presence of alginate in the reaction mixture causes an increase in the number or bacteria adherent to tracheal cells or immobilized mucin (Ramphael et al., 1987). It has been suggested that the alginate may act to trap and tether the organisms to the substrata, thereby allowing other adhesins, such as the fimbrial adhesin complex, to bind to specific receptors (Baker et al., 1990). If the target animal cell or substrata lacks the ability to bind alginate, then the presence of an exopolysaccharide coat on the bacterial surface may actually impede the ability of the organisms to attach to such animal cells. This may explain the reduced ability of the mucoid strains to bind to phagocytic cells or to primary cultures of ciliated epithelial cells. The phagocytic cells lack the ability to bind alginate (Cabral et al., 1987; Grant et al., 1991). The genetic regulation of alginate expression is complex and involves a number of regulatory genes some of which require signal transduction for expression (Ohman and Goldberg, 1990). Because virtually all cystic fibrosis isolates of *P. aeruginosa* are mucoid it was postulated that the lung environment of cystic fibrosis patients provides the trigger required to turn on the production of the alginate adhesin.

Pseudomonas produces a number of different lectins or lectin-like molecules (Gilboa-Garber, 1988). There is good evidence that some of these lectins mediate adhesion of the organisms to animal cells. N-Acetylneuraminic acid specifically inhibits the binding of the organisms to immature ocular epithelia (Hazlett et al., 1986, 1991); to mouse lung, liver, to kidney thin sections (Ko et al., 1987); and to the tracheal epithelium (Baker et al., 1990). Sialogangliosides specifically inhibit binding of the bacteria to scarified adult mouse cornea, as well as inhibit the aggregation of the bacteria by human saliva (Komiyama et al., 1987). Adhesion of *Pseudomonas* to rabbit bladder epithelial cells is reduced by pretreating the mucosal cells with sialidase (Chiarini et al., 1989). The organisms bind to sialic acid-containing glycosphingolipids on thin layers (Baker et al., 1990). The data, taken together, suggest that a sialic acid-specific lectin mediates adhesion of *Pseudomonas* to various substrata, but it is not clear what constitutes that molecular nature of the lectin. The fimbrial adhesin complex and its C-terminal region appear to recognize periodate-sensitive oligosaccharide side chains of glycoproteins on buccal epithelial cells (Doig et al., 1987, 1989). The binding of the fimbriae to blots of glycoproteins obtained from buccal epithelial cells is inhibited by L-fucose and *N*-acetylneuraminic acid, but it is uncertain whether the postulated sialic acid-binding lectin is actually contained in the fimbrial adhesin complex.

There are a number of reports showing that adhesion of *P. aeruginosa* to various substrata may involve lectins specific for carbohydrates other than sialic acid. For example, adhesion of the bacteria to scarified adult and uninjured newborn mouse corneal epithelia is inhibited by *N*-acetylmannosamine (Hazlett et al., 1987), whereas adhesion to the tracheal epithelium is inhibited by D-galactose and *N*-acetyl-D-glucosamine (Baker et al., 1990). Moreover, putative nonfimbrial adhesins produced by both mucoid and nonmucoid strains of *Pseudomonas,* specific for Galβ1,4GlcNAc or Galβ1,3GlcNAc sequences, may be involved in mediating binding of the organisms to human bronchial mucins (Ramphal et al., 1991a). Finally, although not directly related to adhesion, *Pseudomonas* binds to a number of structures contained in neutral glycosphingolipid molecules or neoglycolipids or asialogangliosides. In one study, it was found that the minimal structure recognized on glycolipids is GalNAcβ1,4Gal (Krivan et al., 1988), whereas in another study, the minimal structure required for binding the organisms was identified as Galβ1,4Glc (lactosyl) (Karlsson, 1989; Baker et al., 1990). Moreover, at least five types of neutral glycosphingolipids and asialo GM1 obtained from rabbit corneal epithelia cells were found to bind *Pseudomonas* (Panjwani et al., 1990). Exoenzyme S located on *Pseudomonas* surfaces may function as an adhesin that binds the organisms to GalNAcβ1,4Gal, or Galβ1,4Glc structures on buccal epithelial cells because the exoenzyme binds to glycosphingolipid containing these sequences. The adhesion of the bacteria to buccal cells is inhibited by the exoenzyme as well as by monoclonal antiexoenzyme antibodies (Baker et al., 1991).

Pseudomonas aeruginosa also produces two well-characterized intracellular lectins, one of which is specific for galactose and the other for fucose and mannose (Gilboa-Garber, 1988). It was observed that lectins released from lysing *Pseudomonas* and mixed with intact *P. aeruginosa* caused the bacteria to bind to rabbit corneal epithelial culture cells. The adhesion was specifically inhibited by D-galactose and D-mannose, suggesting that the adhesion is mediated by the lectins released from the lysed bacteria (Wentworth et al., 1991). In other studies, it was found that adhesion of laboratory strains of *P. aeruginosa* to HEp-2 tissue culture cells was markedly enhanced by factors released in the culture of broth-grown bacteria (Ogaard et al., 1985). The data are interpreted to suggest that when some members of a population of *P. aeruginosa* undergo lysis during anaerobic growth, they release lectins that bind to surviving intact bacteria in a manner that allows the carbohydrate combining site of the lectins to mediate adhesion of the organisms to galactose- or mannose-containing glycoconjugates (Wentworth et al., 1991). Adhesion of *P. aeruginosa* to traumatized corneal tissues appears to be enhanced (Klotz et al., 1989, 1990). The involvement of mannosyl or glucosyl residues on corneal epithelia as receptors that bind *Pseudomonas* is supported by the findings showing concanavalin A, a mannose- and glucose-specific lectin, inhibits binding of the bacteria to corneal epithelia (Blaylock et al., 1990). The proposed mechanism (Wentworth et al., 1991) for lectin-dependent adhesion of *P. aeruginosa* in which dying cells in the microcolony confer adhesive properties to living offspring is new and resembles that proposed for *B. pertussis* (see above).

Pseudomonas cepacia

Clinical source:	LRT
Adhesin:	Fimbriae
Receptor specificity:	GalNAcβ1,4Gal
Substratum:	EC
Reference:	Eaves and Doyle, 1988; Krivan et al., 1988; Saiman et al., 1990b

Salmonella choleraesuis

Clinical source:	ENT
Adhesin:	Type 1 fimbriae

Receptor specificity: Mannosides
Substrata: TC
References: Finlay et al., 1988a,b; 1989a,b

Salmonella derby
Clinical source: ENT
Adhesin: 41,000 mol wt outer membrane protein?
Receptor specificity: ND
Substrata: TC
Reference: Budiarti et al., 1991

Salmonella dublin
Clinical source: ENT, bovine and human
Adhesin: Type 1 fimbriae
Receptor specificity: Mannosides, fibronectin
Substrata: ERT
Reference: Kristiansen et al., 1987

Salmonella enteriditis
Clinical source: ENT
Adhesin: Type 1 fimbriae, type 3 fimbriae (MR/K), SEF 17 fimbriae?
Receptor specificities: Mannosides, fibronectin
Substrata: EC, ERT
References: Baloda et al., 1985, 1988; Feutrier et al., 1986; Halula and Stocker, 1987; Aslanzadeh and Paulissen, 1990; Baloda, 1988; Collinson et al., 1991; Mulley et al., 1991

Salmonella salinatis
Clinical source: ENT
Adhesin: Type 1 fimbriae
Receptor specificities: Mannosides, manooligosaccharides in carcinoembryonic antigen family
Substrata: ERT, MC
References: Kohbata et al., 1986; Leusch et al., 1991; Yakubu et al., 1989

Salmonella typhi
Clinical source: ENT
Adhesin: Type 1 fimbriae
Receptor specificities: Mannosides, manooligosaccharides in carcinoembryonic antigen family

Substrata: ERT, MC
References: Kohbata et al., 1986; Leusch et al., 1991

Salmonella typhimurium
Clinical source: ENT
Adhesin: Type 1 fimbriae
Receptor specificities: Mannosides, fibronectin, collagen
Substrata: EC, TC, TB, ERT, SS, ECM

S. typhimurium has attracted much interest with the hope that studies gained on the ability of the bacteria to interact with host cells and their extracellular matrix may be used to understand the initial events of salmonellosis in man. The studies on adhesion and its relation to infectivity have focused on three aspects. One aspect is concerned with the role of the type 1 fimbriae expressed by many strains of *Salmonella typhimurium* (as well as human isolates of *Salmonella*) in the infectious process. Type 1 fimbriated strains of *S. typhimurium* adhere to rat (Lindquist et al., 1987; Moustafa Abd El Monem et al., 1988; Omoike et al., 1990) and human (Baloda et al., 1988) enterocytes in higher numbers than nonfimbriated strains. The mannose-specific adhesion to the enterocytes was followed by structural damage to the epithelial cells. Addition of chloramphenicol did not inhibit the adhesion but it did prevent tissue damage. Dead bacteria also adhered to the enterocytes but were unable to induce tissue damage. The results were interpreted to suggest that tissue damage caused by the *Salmonella* involves a sequential multiphasic process in which adhesion to mannooligosaccharide-containing receptors is the first step to enable close proximity of the *Salmonella* to the enterocytes. The adhesion of nonfimbriated *Salmonella* to intestinal cells could be enhanced by including the plant lectin concanavalin A in the reaction mixture, which probably acts to bridge between the bacteria and the rat enterocytes by binding to mannooligosaccharide containing glycoconjugates present on the surfaces of both types of cells (Abud et al., 1989). In rats the availability of mannooligosaccharides on enterocytes for binding the type 1 fimbriated *S. typhimurium* is under developmental control, increasing gradually with age and reaching adult levels at weaning (Moustafa Abd El Monem et al., 1988). Although the change in diets does not affect the ability of the enterocytes of preweaning rats to bind type 1 fimbriated *Salmonella,* other studies have shown that adhesion to intact mucosa in vivo is higher in well-fed rats than in malnourished rats (Omoike et al., 1990). Studies on adhesion of *S. typhimurium* to poultry skin cells show that the mannose-specific type 1 fimbriae promote adhesion only to skin tissue biopsy specimens immersed into the bacterial suspension but not when the bacterial suspension was spread over the skin tissue cells (Lillard, 1986), emphasizing the role of type 1 fimbrial adhesin in overcoming shearing forces to allow other adhesins to function. Type 1 fimbriae confer hydrophobic properties upon the *Salmonella* surface and thus facilitate adhesion of the bacteria to a number of mineral particles (biotite, felspar, magnetite, and quartz) (Stenstrom and Kjelleberg, 1985).

Another aspect of interest is the identity of nontype 1 fimbrial adhesin expressed by *S. typhimurium*. Mannose-resistant hemagglutination was readily demonstrated in many strains of *Salmonella,* including *S. typhimurium,* confirming earlier results (Halula and Stocker, 1987). The hemagglutinin could be detected by using glutaraldehyde-fixed erythrocytes from sheep and goat whereas none of the strains tested reacted with erythrocytes from nine other species. The hemagglutinin is apparently loosely bound to the bacterial surface because it could be released in soluble form from the bacterial surface by washing and it is also present in abundant amounts in the culture medium. Treatments of erythrocytes with a number of enzymes, including sialidase or protease, did not affect their ability to hemagglutinate, and among the various compounds tested for inhibition of hemagglutination, only a component(s) in heart infusion broth was active. This finding may resolve some of the contradictory results obtained in earlier studies in detecting the hemagglutinin activity of broth-grown bacteria. *S. typhimurium* also binds to collagen and fibronectin (Baloda et al., 1985, 1988, 1991; Gonzalez et al., 1988b). The binding of *Salmonella* to the component of the extracellular matrix is under tight genetic control, depending on the medium employed to grow the bacteria and in the case of fibronectin it appears to be affected by the physical state of the fibronectin (e.g., soluble or immobilized on surface). Growth of *Salmonella* in agar appears to be superior in stimulating the expression of surface constituents that bind both the fibronectin and its aminoterminal region. *S. typhimurium* is likely to interact with mucus constituents to gain a growth advantage because the organisms are able to penetrate mucous layers immobilized on microtiter plates (Nevola et al., 1987). LPS-deficient mutants have a poor ability to penetrate mouse cecal mucus (McCormick et al., 1988). This may be related to the observation made in another study showing greater numbers of type 1 fimbriae on rough strains of *Salmonella* (Baloda et al., 1988). The type 1 fimbriae may be responsible for trapping of the organisms within the mucous layer. The loosely bound mucus overlying the intestinal cells forms a barrier and only *Salmonella* that can bind to and penetrate through the mucous layer can reach the underlying epithelial cells. This notion is supported by the findings showing higher adhesion of *S. typhimurium* to washed than to unwashed chicken ceca (McHan et al., 1988). *S. typhimurium* expresses thin 3-nm fimbriae and an extracellular mucopolysaccharide-like capsule, but the role of these structures in adhesion has not been determined (Grund and Weber, 1988; Grund, 1991).

The ability of pathogenic *Salmonella* strains to invade epithelial cells has become the focus of interest of a number of investigators (Peterson and Niesel, 1988; Finlay and Falkow, 1989). *S. typhimurium* is capable of invading a number of tissue culture cell lines, including Hela, MDCK, and HEp-2 cells. A number of interesting findings related to the adhesion of the organisms have emerged from these studies. Stable adhesion to the tissue culture cells is required prior to invasion (Finlay et al., 1989a,b). The stable adhesion is dependent on metabolically active *Salmonella* (both *S. typhimurium* and *S. choleraesuis*) and is associated with de novo synthesis of several proteins. The induction of synthesis of proteins is time dependent, paralleling the kinetics of adhesion of the bacteria to the monolayer and occurring only in bacteria adherent to dead, fixed tissue culture cells and is enhanced by pretreating the cell monolayer with trypsin and sialidase. The data were interpreted to suggest that interaction of *Salmonella* with specific receptors, which may be somewhat cryptic on HEp-2 or MDCK cells, triggers the synthesis of the bacterial proteins required for firm adhesion and subsequent invasion into the tissue culture cells. The *Salmonella* invasion process is dependent on functional microfilaments. It is greatly enhanced under anaerobic conditions with *Salmonella* harvested from the exponential phase of growth (Ernst et al., 1990). Even though the synthesis of proteins is required for the adhesion/invasion process, *S. typhimurium* must first attach, perhaps loosely, to the cells in order for de novo protein synthesis to be initiated. The type 1 fimbrial adhesin may be responsible for such initial low-affinity, less stable, adhesion of *S. typhimurium* to tissue culture cells because nonfimbriated cells are less invasive (Ernst et al., 1990). Nevertheless, mutant strains that are unable to synthesize the adhesion-induced proteins are avirulent in mice, suggesting that loose adhesion is not sufficient for virulence.

Simonsiella spp.

Clinical source:	ORL
Adhesin:	Short fibrils
Receptor specificity:	ND
Substrata:	EC
Reference:	Pankhurst et al., 1988

Serratia liquefaciens

Clinical source:	RT
Adhesin:	Type 3 fimbriae (MR/K), type 1 fimbriae
Receptor specificities:	Spermidine, mannosides
Substrata:	ERT
Reference:	Gerlach et al., 1989

Serratia marcescens

Clinical source:	URT, URG
Adhesin:	Type 3 fimbriae (MR/K), type 1 fimbriae
Receptor specificities:	Spermidine, mannosides
Substrata:	ERT, EC
References:	Daifuku and Stamm, 1986; Gerlach et al., 1989; Hornick et al., 1991; Obana et al., 1991

Shigella dysenteriae

Clinical source:	ENT
Adhesin:	ND
Receptor specificity:	*N*-acetylneuraminic acid
Substrata:	ERT, SS, TC
References:	Kabir et al., 1985; Watanabe and Nakamura, 1985; Qadri et al., 1989; Sen et al., 1990

Shigella flexneri

Clinical source:	ENT
Adhesin:	Congo red binding protein?
Receptor specificity:	ND
Substrata:	ERT, TC, SS
References:	Kabir et al., 1985; Maurelli et al., 1985; Watanabe and Nakamura, 1985; Dinari et al., 1986; Reijntjens et al., 1986; Clerc and Sansonetti, 1987; Daskaleros and Payne, 1987; Qadri et al., 1989

Staphylococcus aureus

Clinical source:	URT, SKN, ENT, BLD
Adhesins:	LTA, TA, hydrophobins, fibronectin-binding protein, laminin-binding protein
Receptor specificities:	ECM (fibronectin, laminin, collagen)
Substrata:	Immobilized ECM proteins, EC, SS, BM, EC, ERT

See Chapter 6

Staphylococcus epidermidis
(and coagulase-negative staphylococci)

Clinical source:	SKN
Adhesins:	Hydrophobins, slimectin polysaccharide
Receptor specificity:	ECM (fibronectin, laminin, collagen)
Substrata:	BM, SS, ERT

See Chapter 6

Staphylococcus saprophyticus

Clinical source:	URG
Adhesins:	LTA, lectin
Receptor specificities:	β-D-Galactosyl, sialic acid, and GalNAc
Substrata:	EC, ERT, PC, TC

See Chapter 6

Streptococcus agalactiae
Clinical source: URG, URT
Adhesins: LTA, hydrophobins
Receptor specificities: Fibronectin, GlcNAc
Substrata: EC, TC
See Chapter 6

Streptococcus bovis
Clinical source: BLD, ENT, URG, human and ruminant
Adhesins: LTA
Receptor specificity: ND
Substratum: EC
Reference: von Hunolstein et al., 1987

Streptococcus cricetus
Clinical source: ORL
Adhesin: Glucan-binding protein
Receptor specificity: Glucan (6–10 α-1,6 glucose residues)
Substrata: SS (glass), dental plaque?
See Chapter 8

Streptococcus defecticus
Clinical source: ORL
Adhesin: ND
Receptor specificity: ND
Substrata: ECM
Reference: Tart and van de Rijn, 1991

Streptococcus dysgalactiae
Clinical source: Mammary gland, bovine
Adhesin: ND
Receptor specificity: Vitronectin (S protein), fibronectin (210,000) mol wt carboxy-terminal fragment), RGD-containing peptide (Gly-Arg-Gly-Asp-Ser)
Substratum: EC
References: Valentin-Weigand et al., 1988; Filippsen et al., 1990

Streptococcus equi
Clinical source: URT, horse
Adhesin: ND
Receptor specificity: Fibronectin (carboxy-terminal fragment)
Substratum: EC
Reference: Valentin-Weigand et al., 1988

Streptococcus faecalis (Enterococcus hirae)

Habitat:	ENT, URG
Adhesins:	ND
Receptor specificity:	ND
Substrata:	EC, TC, PC, ECM
Reference:	Guzman et al., 1989; Schollin, 1988; Schollin and Danielsson, 1988; Tart and van de Rijn, 1991; Hirakata et al., 1991

Streptococcus gordonii (formerly subtype of *S. sanguis*)

Clinical source:	ORL
Adhesin:	38,000 mol wt protein
Receptor specificity:	Proline-rich proteins
Substrata:	SHA, EC, oral bacteria
References:	Kolenbrander and Andersen, 1990; Jenkinson and Easingwood, 1990; Gibbons et al., 1991; Smith et al., 1991 (see also Chapter 8)

Streptococcus milleri (anginosus)

Clinical source:	ORL
Adhesins:	β-Galactoside-specific protein? LTA
Receptor specificity:	Lactose, laminin, fibronectin
Substrata:	SS (glass), oral bacteria
References:	Switalski et al., 1987; Hogg and Manning, 1988; Eifuku et al., 1990, 1991a,b

Streptococcus mitis (formerly *S. sanguis C5*)

Clinical source:	ORL
Adhesins:	Sialic acid-specific protein
Receptor specificities:	Sialic acid, galactose, laminin, fibronectin
Substrata:	ERT, EC

S. mitis, like *S. sanguis,* is among the early colonizers of the tooth surface (see Chapter 8). The organism expresses lectins and hydrophobins on its surface that serve as adhesins. In addition to the galactose-binding protein described in earlier studies, a lectin specific for *N*-acetylneuraminic acid (α2,3) was purified and characterized from *S. mitis* (Murray et al., 1986). The sialic acid-specific lectin is bivalent, consisting of 65 mol wt subunit and two other subunits of 96,000 and 70,000 mol wt linked by disulfide bonds. Epithelial cells pretreated with sialidase bind considerably fewer streptococci (Childs and Gibbons, 1988, 1990). The sialic acid-dependent adhesion of *S. mitis* to epithelial cells is saturable and follows a Langmuir isotherm consistent with one binding site. It is possible, therefore, that the sialic acid-binding protein is the major adhesin that binds the *S. mitis* to oral epithelial cells. The sialic acid-containing receptors on epithelial cells are also susceptible to proteolytic cleavage caused by either trypsin, papain, or lysates of polymorphonuclear leukocytes (Gibbons, 1989; Gibbons et al., 1990). It has been suggested that during periodontal disease or poor oral hygiene when enzymatic activity in the oral cavity increases, streptococcal receptors, such as *S. mitis,* may be cleaved to allow periodontal pathogens to colonize the enzymatically modified oral substrata. A number of salivary glycoproteins can also interact via terminal sialic acids contained in their oligosaccharide side chains. It has been posed that proline-rich salivary glycoproteins can potentially interact with *S. mitis* via terminal sialic acid residues (and galactose) on their triantennary oligosaccharide side chains (Murray et al., 1986). Nevertheless, proline-rich glycoprotein immobilized on hydroxylapatite (Gibbons, 1989) or enamel coated with whole saliva (Pratt-Terpstra et al., 1989) poorly bind *S. mitis* cells, suggesting that conformational changes in the immobilized salivary glycoprotein may render the sialic acid containing receptor sites cryptic for binding the streptococci. In other studies, it has been found that salivary histatin does not bind *S. mitis* cells (Murakami et al., 1991a,b). It is possible, therefore, that interaction between the sialic acid-specific lectin with sialic acid-containing oligosaccharides of salivary glycoproteins may promote clearance of *S. mitis* from tooth surfaces. This is in contrast to *Actinomyces* bacteria which bind only to immobilized proline-rich salivary glycoproteins and thus are protected from clearance mechanisms (Gibbons and Hay, 1988).

S. mitis interacts with a number of Gram-negative and Gram-positive oral bacteria, but the specificity of these interactions has not been determined (Tables 8–6 and 8–7). The interaction with *S. mutans* is enhanced in the presence of saliva, suggesting that some of the *S. mitis*–oral bacteria interactions in vivo may be promoted by bridging salivary molecules (Lamont and Rosan, 1990). *S. mitis* also binds to extracellular matrix derived from tissue cultures (Tart and van de Rijn, 1991), to laminin (Switalski et al., 1987), and to immobilized fibronectin (Hogg and Manning, 1988). The interaction with fibronectin may involve LTA-like hydrophobic amphiphiles. It is possible, therefore, that *S. mitis* is less cariogenic than other oral streptococci, but may be more pathogenic in triggering streptococcal subacute endocarditis by virtue of its ability to colonize damaged tissues with exposed components of the extracellular matrix.

Streptococcus mutans

Clinical source:	ORL
Adhesins:	CBP, hydrophobins, LTA
Receptor specificity:	Salivary glycoproteins, fibronectin
Substrata:	SHA, EC SS (glass)

See Chapter 8

Streptococcus pneumoniae

Clinical source:	URT
Adhesins:	ND

Receptor specificity:	Galβ1,4GlcNAc
Substrata:	EC, ERT
See Chapter 6	

Streptococcus pyogenes

Clinical source:	URT
Adhesins:	LTA, fibronectin binding protein, M-protein, vitronectin-binding proteins
Receptor specificities:	Fibronectin, galactose, and fucose? fibrinogen, vitronectin
Substrata:	EC, PC, TC, ENC
See Chapter 6	

Streptococcus rattus

Clinical source:	ORL
Adhesins:	ND
Receptor specificities:	Fibronectin, collagen, mucin, sialic acid, glucan
Substrata:	Oral bacteria, SHA
References:	Babu and Dabbous, 1986; Koop et al., 1990; Lamont and Rosan, 1990

Streptococcus salivarius

Clinical source:	ORL
Adhesins:	Fibrils (AgB and AgC), glucan? LTA
Receptor specificities:	Fibronectin, laminin
Substrata:	SHA, EC, TC, ECM, oral bacteria

S. salivarius is abundant in the human oral cavity and preferentially colonizes the soft tissue of the tongue. *S. salivarius* interacts in vitro with other oral bacteria (Tables 8–6 and 8–7), laminin of basement membrane (Switalski et al., 1987), and to immobilized fibronectin (Hogg and Manning, 1988). Evidence has been obtained showing that the interaction of the streptococci with the Gram-negative cocci of the genus *Veillonella* is important in in vivo colonization. Only veillonellae isolated from the tongue coaggregated with *S. salivarius,* whereas subgingival veillonellae isolates fail to coaggregate with the streptococci (Hughes et al., 1988). Studies have focused, therefore, on the mechanism of adhesion to epithelial cells and to members of the genus *Veillonella* (Handley et al., 1990). Mutants deficient in their ability either to adhere to buccal epithelial cells or to aggregate with veillonellae have enabled the identification of two adhesins with distinct receptor specificities emanating from the surface of the streptococci as fibrils. One fibrillar adhesin, designated antigen B, mediates coaggregation with *V. parvula*. The *Veillonella*-binding protein is associated with fibrils of about 91 nm long and consists of protein with a mol wt of 380,000 (Weerkamp et al.,

1986a,b). The other adhesin, designated antigen C, is associated with 72-nm long fibrils, and promotes adhesion of *S. salivarius* to buccal epithelial cells, HeLa tissue culture cells, and saliva-coated hydroxylapatite (Weerkamp et al., 1986a,b, 1987). It has a mol wt of 250,000 and contains significant amounts of carbohydrate, but no evidence for covalent linkage of the carbohydrate to the proteinaceous adhesin exists. Both fibrillar adhesins contribute to the hydrophobicity of the streptococcal surface, although other bacterial surface components also appear to contribute to the hydrophobicity. In isolated form, the fibrillar adhesins appear flexible. They are longer in length than the cell-bound form, being 87 nm long for the epithelial-specific adhesin and 184 nm long for the *Veillonella*-specific adhesin. They are thin rods with thickened globular ends. Only the rod structure is destroyed by proteases, suggesting that similar to fimbrial adhesins, the fibrillar adhesins are actually a complex of several different molecules. It has been suggested that the antigen C adhesin is actually a complex of three adhesins with distinct receptor specificities, each involved in promoting adhesion of the *S. salivarius* to different types of substrata: one for hemagglutination reactions, one for binding to salivary components, and one for epithelial cell adhesion (Handley, 1990). It was postulated further that the adhesive function may reside at the tip, perhaps associated with the protease-resistant globular structures. The presence of other surface constituents markedly affects the ability of the fibrillar antigens to function as adhesins. For example, the presence of ruthenium red staining material on the bacterial surface was found to increase the streptococcal hydrophobicity, which may indirectly affect the adhesive capacity of the streptococci (Handley et al., 1988; Busscher et al., 1989). The expression of the fibrillar adhesin appears to be phenotypically stable, whereas the thickness of the ruthenium red staining material and the hydrophobicity of the bacterial surface are phenotypically variable during growth in vitro (Harty and Handley, 1989). It is possible, therefore, that although there is no evidence for phase variation in the expression of the fibrillar adhesins (Handley, 1990), *S. salivarius* may modulate its ability to adhere to epithelial cells by varying the expression of other surface components that indirectly affect its adhesion, similar to the strategy employed by *S. pyogenes*.

About one-half of the *S. salivarius* isolates lack Lancefield K antigen and antigens C and B adhesins but express rigid peritrichous fimbriae. The K-negative fimbriated streptococci adhere poorly to epithelial cells, suggesting that fimbriae either lack adhesin function or act as adhesins to other substrata (Handley et al., 1987). Based on preliminary studies showing interaction of mutant streptococci not expressing any fibrillar adhesins with certain *Veillonella* strains, it was postulated that *S. salivarius* is probably capable of expressing as yet unidentified nonfibrillar adhesins specific for certain *Veillonella* (Handley, 1990). In addition, *S. salivarius* binds, albeit in relatively low numbers, to mycelial (germ tube) *Candida albicans* cells (Jenkinson et al., 1990). Finally, although *S. salivarius* is not always found in plaque, it may provide products such as glucosyltransferases that allow other oral streptococci to adhere and colonize plaque surfaces (Giffard et al., 1991).

Streptococcus sanguis

Clinical source:	ORL
Adhesins:	Sialic acid binding protein, β-galactoside specific lectin, fibrils, class I and II antigens, proteinaceous hydrophobins, LTA
Receptor specificities:	Salivary glycoproteins, fibronectin, sialic acid, β-galactosides, GPIa
Substrata:	SHA, EC, platelets

See Chapter 8

Streptococcus sobrinus

Clinical source:	ORL
Adhesin:	Glucan-binding protein; hydrophobins
Receptor specificity:	Glucan
Substrata:	SS (glass), dental plaque?

See Chapter 8

Streptococcus suis

Clinical source:	URT, LTR, pigs
Adhesins:	ND
Receptor specificity:	ND
Substrata:	TB
Reference:	Gottschalk et al., 1991

Treponema denticola

Clinical source:	ORL
Adhesins:	ND
Receptor specificities:	Galactose, mannose, fibronectin, laminin, collagen
Substrata:	SHA, EC, TC
References:	Reijntjens et al., 1986; Cimasoni and McBride, 1987; Dawson and Ellen, 1990; Weinberg and Holt, 1990; Haapasalo et al., 1991

Treponema hyodysenteriae

Clinical source:	URG, ENT, pigs
Adhesins:	*N*-acetylneuraminic acid residues?
Receptor specificity:	ND
Substratum:	TC
Reference:	Bowden et al., 1989

Treponema pallidum

Clinical source:	URG
Adhesins:	12,000 mol wt peptide (derived from P1, P2, and P3 proteins)
Receptor specificity:	Fibronectin, endothelial 90,000 mol wt protein
Substrata:	EC, PC, TC
References:	Thomas et al., 1985a,b, 1988, 1989; Baughn, 1986; Steiner et al., 1987

Ureaplasma urealyticum

Clinical source:	URG
Adhesin:	ND
Receptor specificities:	*N*-acetylneuraminic acid, dextran sulfate
Substrata:	ERT, TC
References:	Kotani and McGarrity, 1986; Robertson and Sherburne, 1991; Saada et al., 1991

Vibrio cholerae

Source:	ENT
Proposed adhesins:	Hemagglutinin/protease, cell-associated hemagglutinins, toxin coregulated pili (fimbriae) (TCP), outer membrane protein
Receptor specificity:	L-Fucose, mannose
Substrata for adhesion assays:	TB, EC

Similar to enterotoxigenic *E. coli, V. cholerae* adheres to and colonizes the small intestine, causing diarrhea by elaborating a potent enterotoxin. Early studies have established that *V. cholerae* is capable of expressing a number of adhesins, some of which are strain-specific and others that are not. In vitro adhesion assays have employed formalin-fixed intestinal tissue biopsy specimens and the scanning electron microscope to evaluate the adhering capacity of *Vibrio* to various types of intestinal cells (Nakasone and Iwanaga, 1987; Yamamoto et al., 1988). Although many types of cells of the lymphoid follicles bind the *Vibrios,* of special interest are the M-cells which bind the most and internalize the *V. cholerae* 01 (Owen et al., 1986; Yamamoto and Yokota, 1989b). Further studies have focused on two major classes of *V. cholerae* adhesins: the toxin-coregulated pili and the hemagglutinins. There are two subclasses of hemagglutinins produced by various *V. cholerae* biotypes. One subclass is the soluble hemagglutinin secreted in the culture medium during late exponential growth of the organisms. The soluble hemagglutinin is actually a zinc-dependent metalloprotease that is structurally and immunologically related to *P. aeruginosa* elastase (Hase and Finkelstein, 1990). Functionally, the two metalloenzymes are also related in the sense that both enzymes cause hemagglutination of chicken erythrocytes, the hemagglutinating activity of which is inhibited by zinc chelators and unaffected by the presence of fucose or mannose. The adhesin function of the *Vibrio* hemagglutinin/protease is not clear. It is expressed by virtually all strains of *V. cholerae,* including less virulent non-O group 1 strains and strains belonging to O group 1 of biotypes Classic and El Tor (Booth and Finkelstein, 1986). The receptor specificity of the hemagglutinin of non-01 *V. cholerae* strains, however, may be different from that of the 01 strain because the hemagglutinating activity of one non-01 strain examined was not inhibited by a zinc chelator and was sensitive to serum constituents (Honda et al., 1987). Because of its wide distribution among pathogenic and less pathogenic strains of *V. cholerae* the hemagglutinin/protease may be needed by all strains to function at one or more stages of the infectious process.

The other subclass of hemagglutinins expressed by *V. cholerae* is cell-associated. Screening of large numbers of *V. cholerae* 01 and non-01 strains revealed that many strains cause hemagglutination of human and chicken erythrocytes (Booth and Finkelstein, 1986; Yamamoto et al., 1988; Jonson et al., 1989). The hemagglutination was either inhibited by mannose only, or by fucose only, or by both sugars or by neither sugar, suggesting that the subclass of cellular hemagglutinins constitutes a heterogeneous group of hemagglutinins with distinct receptor specificities. The expression of the cellular hemagglutinins on the bacterial surface is best during exponential growth, suggesting that it is under phase variation control (Yamamoto et al., 1988). The diversity of the receptors of the various cell-associated hemagglutinins has been difficult to resolve because the test bacterial culture may contain a heterogeneous population of *Vibrios* with respect to the receptor specificity of the hemagglutinin expressed by the organisms. For example, in one study (Jonson et al., 1989) it was noted that some *Vibrio* strains of the E1 Tor biotype caused hemagglutination of chicken and human erythrocytes, but only the hemagglutination of chicken erythrocytes was inhibited by mannose. In another study (Booth and Finkelstein, 1986), the hemagglutination caused by some Classic and El Tor *V. cholerae* 01 strains was only partially inhibited by either mannose or fucose, suggesting that the test *Vibrio* suspensions in both studies consisted of at least two subpopulations, each expressing a hemagglutinin with a distinct specificity. Strains capable of aggregating mannose-coated beads usually are inhibited only by mannose, whereas strains capable of aggregating fucose-coated beads are inhibited only by fucose, suggesting that *Vibrio* is capable of expressing either mannose- or fucose-specific hemagglutinins. At least some of the strains are probably capable of producing other types of hemagglutinins (Jonson et al., 1989). Because the expression of the hemagglutinins is probably under phase variation control, it is not certain whether all of the *Vibrio* strains are genotypically capable of producing all types of hemagglutinins or whether some of the hemagglutinins are biotype-specific. It was noted that the mannose-specific hemagglutinin is phenotypically expressed by most E1 Tor strains, whereas the fucose-specific hemagglutinin is mostly expressed by the classical biotype. A few strains of either biotype are able to express phenotypically both types of hemagglutinins (Jonson et al., 1989). The mannose-specific, cell-associated hemagglutinin of O group strain of El Tor biotype is expressed on the surface of the *Vibrios* in the form of thin and flexible fimbriae composed of major 17,000 mol wt subunits (Jonson et al., 1991a). Isolated El Tor fimbriae cause mannose-sensitive hemagglutination of chicken erythrocytes and anti-fimbrial antibodies inhibit the hemagglutination caused by El Tor fimbriae-bearing bacteria. Purified fimbriae with a similar molecular weight extracted from four El Tor strains also caused hemagglutination of human erythrocytes, but the carbohydrate specificity of the fimbrial hemagglutinin was not determined (Iwanaga et al., 1989). Immunologically, virtually all El Tor strains react with monoclonal antibodies against the fimbriae, confirming the observations that most El Tor strains exhibit mannose-specific hemagglutinin (Jonson et al., 1991a). Hemagglutinins not inhibited by L-fucose or mannose and expressed by *V. cholerae* non-01 may also be in the form of thin flexible fimbriae consisting of a 16,000 mol wt major subunit because the purified fimbriae cause mannose- and fucose-resistant hemagglutination and because the hemagglutinating activity of whole bacteria is inhibited by antifimbrial antiserum (Nakasone and Iwanaga, 1990a). About 25% of *V. cholerae* of 01 and non-01 strains express fimbriae that immunologically cross-react with the mannose- and fucose-resistant fimbriae of non-01 *V. cholerae*. Fimbriae of similar morphology and molecular weight as the major subunit have been identified in strains of *V. cholerae* non-01 exhibiting mannose- and fucose-sensitive hemagglutination (Ehara et al., 1987). The data suggest that probably many types of cell-associated hemagglutinins with diverse receptor specificities may be in the form of fimbriae on the surface of *V. cholerae*.

A number of studies have attempted to examine the ability of the cellular hemagglutinins to mediate adhesion of *V. cholerae* to intestinal cells. Suspensions of *V. cholerae* (El Tor biotype) prepared directly from the intestines of rabbits infected with the *Vibrio* express mannose-specific hemagglutinin, suggesting that the hemagglutinin is expressed in vivo (Jonson et al., 1990). In addition, the ability of cell-associated hemagglutinin-deficient mutants of *V. cholerae* El Tor to colonize ileal tissue is significantly lower than that of the parent strain (Finn et al., 1987). In other studies, a role for fimbrial hemagglutinin in mediating adhesion of *Vibrio* bacteria to formalized rabbit intestinal tissue in vitro could not be confirmed. Nonadherent strains and adherent El Tor strains expressed the fimbrial hemagglutinin (Iwanaga et al., 1989). Neither the isolated fimbrial hemagglutinins nor the anti-fimbrial antibodies inhibited adhesion of *V. cholerae* strains exhibiting mannose- and fucose-resistant hemagglutinin to the rabbit intestine (Nakasone and Iwanaga, 1990a,b). Furthermore, *V. cholerae* E1 Tor exhibiting mannose-resistant hemagglutination and nonhemagglutinating classic biotype strain adhered equally well to intestinal cells of adult rabbits after inoculation of the bacteria into ligated small intestinal loops (Teppema et al., 1987). The iting mannose-resistant hemagglutination and nonhemagglutinating classic biotype strain adhered equally well to intestinal cells of adult rabbits after inoculation of the bacteria into ligated small intestinal loops (Teppema et al., 1987). The data suggest that the receptors for a number of cell-associated hemagglutinins of *V. cholerae* are not accessible on mucosal surfaces of the excised and formalin-treated rabbit intestinal biopsy materials. In contrast, the cellular hemagglutinins appear to be involved in the adhesion of *V. cholerae* of both 01 and non-01 groups to intestinal cells and to the mucous blanket overlying the intestinal cells when assayed employing intact or formalin-treated small intestine segments excised from humans (Yamamoto et al., 1988; Yamamoto and Yokota, 1988a,b). In general, the adhesion observed by scanning electron microscopy is more prominent to and in much greater numbers to the mucous coat overlying the intestinal cells than to the underlying intestinal cells in the tissue specimen. The involvement of the cellular hemagglutinins in this adhesion process is deduced from experiments showing that: (1) There is good correlation between the extent of adhesion to the intestinal tissue and mucus and the hemagglutinating titer of the test *Vibrio* suspension. Bacteria grown for 20 hours to suppress expression of the hemagglutination adhered poorly if at all; (2) Heating of the bacteria to abolish hemagglutinin activity also diminished adhesion of the treated bacteria; (3) L-Fucose inhibited the adhesion of vibrios bearing fucose-specific hemagglutinin, whereas the adhesion of vibrios expressing fucose- and/or mannose-resistant hemagglutinin was not inhibited by L-fucose.

The other adhesin(s) expressed by many strains of *V. cholerae* 01 of both classic and El Tor biotypes is expressed on the bacterial surface as rigid fimbriae belonging to the type 4 class of fimbriae, also termed *N*-methylphenyalanine fimbriae (Shaw and Taylor, 1990; Jonson et al., 1991b). The most remarkable feature of these fimbriae is that their biosynthesis and expression is under the control of a regulatory transmembrane protein, Tox R, which also controls the expression of cholera toxin (Miller et al., 1987; Herrington et al., 1988) and other virulence factors (Peterson and Mekalanos, 1988). The role of the toxin-coregulated fimbriae, termed TCP, in mediating adhesion to intestinal cells or mucous layer is inferred from studies showing that colonization of human volunteers with mutants deleted of either *tox*R or of *tcp*A genes resulted in marked decrease in the ability of the mutants to colonize the intestine (Herrington et al., 1988; Taylor et al., 1988). Furthermore, anti-TCP antibodies protect infant mice challenged with TCP and toxin producing *V. cholerae* and anti-TCP antibodies inhibit adhesion of *V. cholerae* to epithelial cells in vitro (Sun et al., 1990). Among various monoclonal antibodies directed against the major fimbrial subunit Tcp A only those reacting with the carboxy half of the Tcp A fimbrial subunit protect mice against *Vibrio* infection, suggesting that the hypothetical domain of Tcp A important for intestinal colonization and presumably needed for mediating adhesion to intestinal cells may be similar to that of type 4 fimbriae of *Pseudomonas aeruginosa* (Sun et al., 1991; see also above discussion on *Pseudomonas*).

opment of infection in rabbit ileal loop after *Vibrio* challenge (Sengupta et al., 1989). All 10 strains of *V. cholerae* 01 tested bound fibronectin, the binding of which was specific for the carbohydrate-containing region of the fibronectin molecule (Wiersma et al., 1987). Type II collagen and fibrinogen were bound weakly or not at all. The nature of the fibronectin binding constituent on the surface of the bacteria is not known, but it appears that cellular hemagglutinins are not involved. *V. cholerae,* like other enteropathogens, is capable of producing a number of adhesins that are probably required for different stages of the infectious process. Adhesion to mucous coatings of intestinal cells and subsequently to the underlying epithelial cells as well as to the cells of the lymphoid tissue, especially to M-cells, are required to mount secretory immune responses. In addition, *V. cholerae,* like many other *Vibrio* species, inhabits water where it must survive and multiply for relatively long periods of time before it again encounters humans. Probably some of the adhesins described above are also needed to survive in an aquatic environment. In this regard, it is of interest that the marine *Vibrio furnissii* expresses carbohydrate-binding proteins that are used as nutrium sensorium (Yu et al., 1987, 1991), suggesting that some of the hemagglutinins expressed by *V. cholerae* may also have the same function. It appears that some of the adhesins are strain or biotype specific. This is not surprising, considering the evolutionary view that the classic and E1 Tor biotypes of *V. cholerae* developed in different areas as human pathogens (reviewed in Beachey, 1980).

Vibrio parahaemolyticus

Clinical source:	ENT
Adhesin:	Ha7 fimbriae
Receptor specificity:	ND
Substrata:	ERT, MC, TB
References:	Honda et al., 1988; Levett and Daniel, 1988; Yamamoto and Yokota, 1989b; Nakasone and Iwanaga, 1990b, 1991; Yamamoto et al., 1990b

Vibrio vulnificus

Clinical source:	BLD, SKN
Adhesin:	ND
Receptor specificities:	Fructose, L-fucose, and mannose
Substratum:	TC
Reference	Gander and LaRocco, 1989

Yersinia enterocolitica

Source:	ENT
Proposed adhesins:	Outer membrane proteins: Ail, Invasin, Yad A
Receptor specificity:	Not defined
Substrata:	EC, ERT, ECM, IM

Yersinia enterocolitica is a member of an enterobacterial group capable of invading into epithelial cells of the gut, eventually reaching lymphoid tissue, causing gastrointestinal syndromes (Cover and Aber, 1989). It is closely related to the enteric pathogen *Y. pseudotuberculosis*. Both *Yersinia* species share common properties related to their interaction with animal cells. These *Yersinia* species, whether dead or alive, penetrate into eucaryotic animal cells immediately after they have adhered to the cells, suggesting that unlike other invasive enteropathogens (e.g., salmonellae and shigellae) or neisseriae (see above), active metabolism in *Yersinia* is not required for adhesion to and invasion (Miller et al., 1988b; Finlay and Falkow, 1989). There are several molecular mechanisms through which *Yersinia* interacts with animal cells, some of which are shared by the two *Yersinia* species and others are unique to each species.

Y. enterocolitica is capable of expressing at least two classes of adhesins that individually function as invasins mediating the internalization of the organisms into animal cells. The genes encoding for both classes of adhesins/invasins have been cloned. The contribution of their products, which have been identified and characterized, to the infectious process has been studied (Miller and Falkow, 1988; Miller et al., 1989, 1990; Pepe and Miller, 1990). One invasin (invasin$_{ent}$) encoded by the *inv*$_{ent}$ gene on the *Y. enterocolitica* chromosome is an outer membrane protein of 92,000 mol wt. It shares extensive homology with the invasin$_{pstb}$ of *Y. pseudotuberculosis*. Recombinant *E. coli* expressing invasin$_{ent}$ exhibit a relatively high level of invasion into HEp-2 and CHO cells and like the parent *Y. enterocolitica* strain, they are found within a membrane-bound vacuole. Although invasin$_{ent}$ shares homology with invasin$_{pstb}$, recombinant *E. coli* expressing the latter are several-fold more invasive, especially in MDCK tissue culture cells, than are *E. coli* expressing the *Y. enterocolitica*. It has been suggested that a deletion of about 100 amino acids in the *Y. enterocolitica* invasin may account for the difference in invasiveness between the two recombinant *E. coli* strains. The results, however, strongly suggest that the *Y. enterocolitica* invasin is capable of mediating both adhesion to and invasion into animal cells, albeit to a limited degree. The LD_{50} of the *inv* mutant in orally infected mice is comparable to that of the parent strain, suggesting that molecules other than invasin$_{ent}$ may compensate the *inv* mutation. One such other molecule may be the second adhesin/invasin which is encoded by *ail* (for attachment invasion locus). The *ail* product has an apparent molecular weight of 17,000 and is probably an outer membrane protein with six or seven membrane-spanning domains. *E. coli* transformed with a plasmid containing the *ail* gene, and expressing the *ail* adhesin on their surface, exhibit low levels of invasion to HEp-2 cells and high levels of invasion of CHO cells, despite efficient adhesion of the recombinant *E. coli* to both types of cell lines. Ail mutants of *Y. enterocolitica* exhibit three- to eightfold decrease in invasion of CHO and HEp-2 cells, respectively, but are one order of magnitude higher in LD_{50} of orally infected mice than their parent strain. A clue as to the possible role of each adhesin/invasin of *Y. enterocolitica* in infectivity may be deduced from studies on the invasiveness and distribution of *inv* and *ail* genes in nonpathogenic commensal *Yersinia* species and pathogenic strains belonging to *Y. enterocolitica, Y. pseudotuberculosis,* and *Y. pestis*. The ability of the strains to invade tissue culture cell lines and the presence of sequences that hybridize with DNA probes corresponding to the *inv* and *ail* loci in DNA fragments derived from the strains were examined (Miller et al., 1988b, 1989). All *Yersinia* strains examined hybridized with the *inv* probe, but the pattern of hybridizing DNA fragments derived from invasive and pathogenic strains was distinct from that of fragments derived from noninvasive and nonpathogenic strains. Furthermore, DNA fragments derived from invasive strains of the *Y. enterocolitica* and invasive strains of *Y. pseudotuberculosis* and *Y. pestis* hybridized with the *ail* probe, whereas noninvasive strains of *Y. enterocolitica* or nonpathogenic and noninvasive *Yersinia* species did not hybridize with the *ail* probe. The results strongly argue for a critical role of Ail protein in mediating adhesion/invasion of *Y. enterocolitica* into animal cells. Although the *ail* and inv_{ent} products confer upon *E. coli* the ability to attach and invade into tissue culture cells, clinical isolates of *Y. enterocolitica* probably express on their surface additional adhesins that may or may not function as invasins. For example, antibodies raised against whole formalized *Y. enterocolitica* do not inhibit adhesion but strongly inhibit invasion into HeLa cells (Schiemann and Nelson, 1988). In other studies, it has been noted that *Y. enterocolitica* grown at 35°C is more hydrophobic and adheres better to tissue culture cells than organisms grown at 25°C (Schiemann et al., 1987). Plasmid encoding products were found to be important for the intestinal colonization of mice orally infected with *Y. enterocolitica* (Kapperud et al., 1987) and for adhesion of the bacteria to rabbit and human intestinal cells (Paerregaard et al., 1990). Adhesion of plasmid-cured strains to HeLa or fetal intestinal epithelial cells is equally efficient to that of the plasmid-containing parent strain, suggesting that tissue culture cells may not allow the detection of potential adhesins because receptors for such adhesins are not expressed. Further studies have shown that the temperature-inducible and plasmid-encoded outer membrane protein yadA (formerly called Yop 1) is responsible for autoagglutination of the organisms and mediates adhesion of *Y. enterocolitica* to erythrocytes (Kapperud et al., 1985, 1987) as well as to intestinal tissue and brush border membranes (Mantle et al., 1989; Paerregaard et al., 1991b). Furthermore, *E. coli* transformed with a plasmid containing the Yad A gene exhibited an enhanced adhesion to intestinal tissue and brush border vesicles. Yad A probably does not function as an invasin, because conjugal mobilization of a plasmid-containing DNA fragment encoding the outer membrane protein into nonadherent and noninvasive *Y. enterocolitica* strain conferred upon the transconjugants the ability to adhere without invasion into HEp-2 cells (Heesemann and Gruter, 1987). When the Yad A encoding plasmid was mobilized into plasmidless *Y. enterocolitica* strains capable of adhering and invading HEp-2 cells, the transconjugants lost their invasiveness but not their ability to adhere to the HEp-2 cells. The results were interpreted to suggest that Yad A does not function as an invasin but its presence on the bacterial surface actually interferes with the ability of invasins (e.g., the chromosomally encoded invasin$_{ent}$ or Ail protein) to mediate endocytosis of the organisms (Heesemann and Gruter, 1987). Similar biochemical and genetic studies show that the Yad A plasmid adhesin also mediates binding of *Y. enterolitica* to intestinal mucus and purified mucin (Mantle et al., 1989) as well as to types I, II, and IV collagen (Emody et al., 1989). The *Yersinia* binds in significantly greater numbers to immobilized human mucin than to rabbit mucin, whereas rat mucin binds the fewest numbers of bacteria. The binding of mucus by the *Yersinia* inhibits the ability of the organisms to adhere to intestinal brush border membranes. The adhesion is not affected by the presence of monosaccharides in the reaction mixture or by pretreating the mucus with periodate, suggesting that mucous oligosaccharides are not involved. It is possible that mucous constituents remain tightly associated with the brush border membranes and as such serve as Yad A receptors belonging to the extracellular coat or matrix compartment of the intestinal cells (Chapter 3). Yad A-mediated adhesion to collagen was suggested to be important in the colonization of the lamina propria invaded by *Y. enterocolitica* (Emody et al., 1989).

Y. enterocolitica, similar to other invasive bacteria, is capable of expressing a number of adhesins, some of which also function as invasins. Together with other virulence factors, the adhesins facilitate colonization and enable the organisms to cause symptomatic infection. The Yad A adhesin confers hydrophobic properties upon the bacterial surface and the ability to bind to various substrata, including mucous constituents that coat the intestinal epithelial cells. Progenies arising from the growing *Y. enterocolitica* and expressing adhesins/invasins eventually penetrate through the mucous barrier to reach, bind to, and penetrate into the underlying epithelial intestinal cells.

Yersinia frederidsenii

Clinical source:	ENT
Proposed adhesins:	ND
Receptor specificity:	ND
Substrate for adhesin assays:	ERT
Reference:	Old et al., 1985b

Yersinia pseudotuberculosis

Clinical source:	ENT
Proposed adhesins:	Outer membrane proteins: Ail? Invasin, Yad A
Receptor specificity:	Integrins
Substrata for adhesion assays:	EC, ERT, ECM

Yersinia pseudotuberculosis produces a number of plasmid and chromosomally encoded classes of adhesins/invasins with distinct receptor specificities (Isberg, 1989). One class may be the fibrillae-associated outer membrane protein Yad A encoded by a 40–48 MDa plasmid (Kapperud et al., 1985). A number of plasmid-containing strains of *Y. pseudotuberculosis* tested and grown at 37°C adhere significantly better to HEp-2 cells than their plasmid-cured derivatives (Kawaoka et al., 1988). Unlike that of the Yad A of *Y. enterolitica,* however, the role of Yad A adhesin in colonization in vivo is controversial. Insertional mutagenesis of Yad A does not affect the colonization of mice orally infected with the Yad A-inactivated *Y. pseudotuberculosis* (Kapperud et al., 1987), but markedly reduces the ability of the latter to adhere to intestinal tissue and brush border membranes as well as to polystyrene surfaces, suggesting loss of surface hydrophobicity (Paerregaard et al., 1991a,b). Moreover, mobilization of Yad A gene into *E. coli* confers upon the transformants the ability to adhere to the intestinal cells while mobilization of the Yad A gene from *Y. pestis* is without effect. Similar genetic approaches have provided evidence that Yad A mediates binding of the *Y. pseudotuberculosis* to types I and II collagen (Emody et al., 1989). Strains of *Y. pseudotuberculosis* hybridize with the *ail* probe derived from *Y. enterocolitica* as discussed above (Miller et al., 1989). It was postulated that the *ail* locus may encode the thermoinducible adhesin that mediates binding of *Y. pseudotuberculosis* to tissue culture cells (Isberg, 1990).

The chromosomally encoded invasin of *Y. pseudotuberculosis* (invasin$_{pstb}$) is probably the major adhesin/invasin that mediates a high level of endocytosis of the bacteria by a number of cultured cell lines in vitro (Isberg and Falkow, 1985; Isberg, 1990). Recombinant *E. coli* containing the *inv* locus and expressing invasin$_{pstb}$ is specifically enriched after penetrating and growing into tissue culture cells infected with *E. coli* strains harboring a chromosomal library of *Y. pseudotuberculosis* (Isberg and Falkow, 1985). Further genetic biochemical and immunochemical studies, including insertional mutations, mobilization of the invasin$_{pstb}$ gene deleted of various regions into *E. coli,* and use of antiinvasin antibodies have confirmed that an outer membrane protein of 103,000 mol wt functions as an adhesin/invasin mediating endocytosis of *Y. pseudotuberculosis* by tissue culture cells (Isberg et al., 1987; Isberg and Leong, 1988). HEp-2 cells also bind to invasin-containing bands on blots after fractionation of invasin preparations on SDS-polyacrylamide gels, suggesting that the invasin retains its receptor specificity. The biosynthesis of invasin$_{pstb}$ is thermoregulated, but unlike plasmid-encoded products, it is depressed approximately fourfold at 37°C as compared to 25°C (Isberg et al., 1988). The depression of invasin biosynthesis is associated with about a 10-fold decrease in the ability of *Y. pseudotuberculosis* growth at 37°C to invade tissue culture cells. It appears, therefore, that *Y. pseudotuberculosis* is capable of expressing a number of adhesins, including Yad adhesins, best expressed at 37°C (e.g., plasmid-encoded Yad A, Ail, and possibly other chromosomally encoded adhesins) and adhesins best expressed at 25°C (e.g., chromosomally encoded invasin$_{pstb}$ and at least another plasmid-encoded product). It has been postulated that each of these adhesins is expressed one at a time by a single bacterium in order to enhance the chances of the organism to survive in various niches encountered by the bacterium during infection (Isberg, 1990). Accordingly, the invading *Y. pseudotuberculosis* probably expresses invasin$_{pstb}$ because it grows at low temperature in foodstuffs. The invasin-expressing phenotype probably confers an advantage for the organisms in surviving the initial stages of the infectious process by allowing the organisms to penetrate into cells and escape a deleterious environment. At later stages of the infection, progenies growing at physiological temperatures express other adhesins/invasins, although these are less efficient in mediating invasion into deep tissues (Isberg, 1990). In vivo experiments support this view because they show that the onset of death of mice infected orally with *inv* mutants of *Y. pseudotuberculosis* is delayed, compared to the parent strain. The observations suggest that *inv* mutations could be compensated given enough time for progenies expressing other adhesin/invasins to follow the infectious process (Rosqvist et al., 1988). Perhaps the most remarkable advancement in understanding of the molecular mechanism involved in the invasive process of *Y. pseudotuberculosis* (and possibly other invasive enteropathogens) is the identity of the receptor on tissue culture cells. The cell binding domain of the 986-amino acid invasin molecule was identified as the carboxy-terminal 192-amino-acid region (Leong et al., 1990) because (1) antiinvasin monoclonal antibodies directed against the carboxy-terminal 192 amino acids only inhibited adhesion of HEp-2 cells to invasin immobilized on the bottom of microtiter plates, and (2) HEp-2 cells bind to immobilized lysates of recombinant *E. coli* transformed with plasmids containing restriction DNA fragments of the *inv* gene corresponding to the carboxy-terminal 192-amino-acid region on the invasin, and (3) deletion derivatives of the invasin molecule that bound HEp-2 cells and purified from lysates of the recombinant *E. coli* inhibit invasion of *Y. pseudotuberculosis* into the tissue culture cells. The carboxy-terminal region fused with maltose-binding protein retains its ability to bind HEp-2 cells (Leong et al., 1990) and is sufficient to confer invasiveness of bacteria into HEp-2 cells (Isberg, 1991). The results led to the conclusion that invasin$_{pstb}$ receptor on the surface of the animal cells must actively function to internalize the invasin-bearing bacteria.

Having proven that the *inv* product is the adhesin/invasin ligand on the surface of *Y. pseudotuberculosis,* further studies have been carried out to identify the receptor for this gland on mammalian cultured cells. Affinity chromatography on immobilized invasin-agarose columns and solubilized surface-labeled membrane glycoproteins of cultured cell lines were employed to isolate and identify the β_1 chain integrin family of cell adhesion molecules as the invasin$_{pstb}$ receptors (Isberg and Leong, 1990). This family of integrins, like other members of the integrin superfamily, consists of membrane heterodimeric glycoproteins, each composed of different α subunits (e.g., α_1, α_2, α_3, α_4, α_5, α_6) and the same β_1 subunit to form unique heterodimers (e.g., $\alpha_1\beta_1$, $\alpha_2\beta_1$, $\alpha_3\beta_1$, $\alpha_4\beta_1$, $\alpha_5\beta_1$, $\alpha_6\beta_1$) on the surface of various type of animal cells (Hynes, 1987). They are especially suited to mediate endocytosis because they are bound to cytoskeletal components inside the animal cells and extend through the membrane to serve as receptors for various ligands, such as fibronectin or complement receptors. It appears, therefore, that integrins are involved in internalization of bacteria by both professional phagocytes, as shown for type 1 fimbriated bacteria (Chapter 7) and by nonprofessional phagocytes. To the extent studied, only the dimers, $\alpha_3\beta_1$, $\alpha_4\beta_1$, $\alpha_5\beta_1$, $\alpha_6\beta_1$, each expressed by different tissue culture cell lines, have been found to serve as invasin$_{pstb}$ receptors. Integrins containing another β subunit do not function as invasin receptors (Isberg and Leong, 1990). Because the β_1 family of integrins also serve as ligands for various glycoproteins of the extracellular matrix, especially for collagen and fibronectin, it was of interest to examine the similarities in the attachment sites on the β_1 integrins involved in binding invasin$_{pstb}$ and fibronectin. First, monoclonal antibodies against fibronectin receptors inhibit attachment and invasion of *Y. pseudotuberculosis* to the mammalian cells (Isberg and Leong, 1990; Isberg, 1990). Second, in a series of studies (van Nhieu and Isberg, 1991) using a hybrid molecule consisting of maltose-binding protein fused with the last 479 amino acids of the carboxy-terminal of the invasin, as an invasin$_{pstb}$ derivative (MBP-Inv) it was shown that (1) the cell attachment domain of fibronectin competitively inhibits the binding of the invasin derivative to $\alpha_5\beta_1$ dimers isolated from human placenta and immobilized on filter paper; (2) the affinity of the binding of the invasin derivative to $\alpha_5\beta_1$ integrin is considerably higher than the binding of fibronectin with a two order of magnitude difference in the dissociation constant ($K\bar{d}$ of 5×10^{-9} M and 8×10^{-7} M for invasin and fibronectin, respectively); and (3) although invasin$_{patb}$ lacks an Arg-Gly-Asp sequence that is contained in fibronectin and is recognized by the integrins, a hexapeptide containing this tripeptide inhibits binding of both fibronectin and the invasin derivative to the immobilized integrin. The results suggest that invasin and fibronectin bind to identical or closely located sites on the integrin molecule (Nhieu and Isberg, 1991). In view of the fact that bacteria coated with fibronectin bind to but are not internalized by animal cells, whereas invasin$_{pstb}$-coated bacteria adhere and invade into the cells, it has been postulated that the difference in the binding affinity between the two ligands may explain the findings. Thus, proper orientation of the invasin on the bacterial surface and the ability of the invasin to interact with a relatively high affinity to a receptor that links to the cytoskeletal elements of the animal cell may be necessary for triggering invasion of the bacteria into nonprofessional phagocytic cells, such as intestinal epithelial cells.

References

Abe, C., S. Schmitz, I. Moser, G. Boulnois, N.J. High, I. Orskov, F. Orskov, B. Jann, and K. Jann. 1987. Monoclonal antibodies with fimbrial F1C, F12, F13 and F14 specificities obtained with fimbriae from *E. coli* 04:k12:H7. *Microb. Pathogen.* **2**:71–77.

Abraham, S.N., J. P. Babu, C. S. Giampapa, D. L. Hasty, W. A. Simpson, and E. H. Beachey. 1985. Protection against *Escherichia coli*-induced urinary tract infections with hybridoma antibodies directed against type 1 fimbriae or complementary D-mannose receptors. *Infect. Immun.* **48**:625–628.

Abud, R.L., B.L. Lindquist, R.K. Ernst, J.M. Merrick, E. Lebenthal, and P.C. Lee. 1989. Concanavalin A promotes adherence of *Salmonella typhimurium* to small intestinal mucosa of rats. *Proc. Soc. Exp. Biol. Med.* **192**:81–86.

Achtman, M., M. Heuzenroeder, B. Kusecek, H. Ochman, D. Caugant, R.K. Selander, V. Vaisanen-Rhen, T.K. Korhonen, S. Stuart, F. Ørskov, and I. Ørskov. 1986. Clonal analysis of *Escherichia coli* 02:K1 isolated from diseased humans and animals. *Infect. Immun.* **51**:268–276.

Achtman, M., M. Neibert, B.A. Crowe, W. Strittmatter, B. Kusececk, E. Weyse, M.J. Walsh, B. Slawig, G. Morelli, A. Moll, and M. Blake. 1988. Purification and characterization of eight class 5 outer membrane protein variants from a clone of *Neisseria meningitidis* sero group A. *J. Exp. Med.* **168**:507–525.

Adegbola, R.A. and D.C. Old. 1985. Fimbrial and non-fimbrial haemagglutinins in *Enterobacter aerogenes. J. Med. Microbiol.* **19**:35–43.

Agata, N., M. Ohta, H. Miyazawa, M. Mori, N. Kido, and N. Kato. 1989. Serological response to P-fimbriae of *Escherichia coli* in patients with urinary tract infections. *Eur. J. Clin. Microbiol. Infect. Dis.* **8**:156–159.

Ahmed, K., K. Matsumoto, N. Rikitomi, T. Nagatake, T. Yoshida, and K. Watanabe. 1990. Effects of ampicillin, cefmetazole and minocycline on the adherence of *Branhamella catarrhalis* to pharyngeal epithelial cells. *Tohoku J. Exp. Med.* **161**:1–7.

Ahren, C.M. and A.M. Svennerholm. 1985. Experimental enterotoxin-induced *Escherichia coli* diarrhea and protection induced by previous infection with bacteria of the same. *Infect. Immun.* **50**:255–261.

Ahren, C.M., L. Gothefors, B.J. Stoll, M.A. Salek, and A.M. Svennerholm. 1986. Comparison of methods for detection of colonization factor antigens on enterotoxigenic *Escherichia coli. Clin. Microbiol.* **23**:586–591.

Akkoyunlu, M., M. Ruan, and A. Forsgren. 1991. Distribution of protein D, an immunoglobulin D-binding protein, in *Haemophilus* strains. *Infect. Immun.* **59**:1231–1238.

Albert, M.J., K. Alam, M. Islam, J. Montanaro, A.S. Rahaman, K. Haider, M.A. Hossain, A.K. Kibriya, and S. Tzipori. *Hafnia alvei*, a probable cause of diarrhea in humans. 1991a. *Infect. Immun.* **59**:1507–1513.

Albert, M.J., M. Ansaruzzaman, S. M. Faruque, P. K. Neogi, K. Heider, and S. Tzipori. 1991b. An ELISA for the detection of localized adherent classic enteropathogenic *Escherichia coli* serogroups. *J. Infect. Dis.* **164**:906–909.

Almagor, M., I. Kahane, C. Gilon, and S. Yatziv. 1986. Protective effects of the glutathione redox cycle and vitamin E on cultured fibroblasts infected by *Mycoplasma pneumoniae. Infect. Immun.* **52**:240–244.

Amundsen, S., C.C. Wang, W.R. Schwan, J.L. Duncan, and A.J. Schaeffer. 1988. Role of *Escherichia coli* adhesins in urethal colonization of catheterized patients. *Urology* **140**:651–655.

Andersson, B., O. Porras, L.A. Hanson, T. Lagergard, and C. Svanborg-Eden. 1986. Inhibition of attachment of *Streptococcus pneumoniae* and *Haemophilus influenzae* by human milk and receptor oligosaccharides. *J. Infect. Dis.* **153**:232–238.

Anderson, P.W., ME. Pichichero, and E.M. Connor. 1985. Enhanced nasopharyngeal colonization of rats by piliated *Haemophilus influenzae* type b. *Infect. Immun.* **48**:565–568.

Andersson, P., I. Engberg, G. Lidin-Janson, K. Lincoln, R. Hull, S. Hull, and C. Svanborg. 1991. Persistence of *Escherichia coli* bacterium is not determined by bacterial adherence. *Infect. Immun.* **59**:2915–2921.

Aniansson, G., B. Andersson, R. Lindstedt, and C. Svanborg. 1990. Anti-adhesive activity of human casein against *Streptococcus pneumoniae* and *Haemophilus influenzae. Microb. Pathogen.* **8**:315–323.

Annuar, B.O. and G.E. Wilcox. 1985. Adherence of *Moraxella bovis* to cell cultures of bovine origin. *Res. Vet. Sci.* **39**:241–246.

Archambaud, M., P. Courcoux, and A. Labigne-Roussel. 1988a. Detection by molecular hybridization of PAP, AFA, and SFA adherence systems in *Escherichia coli* strains associated with urinary and enteric infections. *Ann. Inst. Pasteur/Microbiol.* **139**:575–588.

Archambaud, M., P. Courcoux, V. Ouin, G. Chabanon, and A. Labigne-Roussel. 1988b. Phenotypic and genotypic assays for the detection and identification of adhesins from pyelonephritic *Escherichia coli. Ann. Inst. Pasteur/Microbiol.* **139**:557–573.

Armstrong, J.A., M. Cooper, C.S. Goodwin, J. Robinson, S.H. Wee, M. Burton, and V. Burke. 1991. Influence of soluble haemagglutinins on adherence of *Helicobacter pylori* to HEp-2 cells. *J. Med. Microbiol.* **34**:181–187.

Arp, L.H. 1986. Adherence of *Bordetella avium* to turkey tracheal mucosa: effects of culture conditions. *Am. J. Vet. Res.* **47**:2618–2620.

Arp, L.H. and E.E. Brooks. 1986. An *in vivo* model for the study of *Bordetella avium* adherence to tracheal mucosa in turkeys. *Am. J. Vet. Res.* **47**:2614–2617.

Arp, L.H. and D.H. Hellwig. 1988. Passive immunization versus adhesion of *Bordetella avium* to the tracheal mucosa of turkeys. *Avian Dis.* **32**:494–500.

Arp, L.H., R.D. Leyh, and R.W. Griffith. 1988. Adherence of *Bordetella avium* to tracheal mucosa of turkeys: correlation with hemagglutination. *Am. J. Vet. Res.* **49**:693–696.

Arthur, M., C.E. Johnson, R.H. Rubin, R.D. Arbeit, C. Campanelli, C. Kim, S. Steinbach, M. Agarwal, R. Wilkinson, and R. Goldstein. 1989. Molecular epidemiology of adhesin and hemolysin virulence factors among uropathogenic *Escherichia coli. Infect. Immun.* **57**:303–313.

Arthur, M., R.D. Arbeit, C. Kim, P. Beltran, H. Crowe, S. Steinback, C. Campanelli, R.A. Wilson, R.K. Selander, and R. Goldstein. 1990. Restriction fragment length polymorphisms among uropathogenic *Escherichia coli* isolates: *pap*-related sequences compared with *rrn* operons. *Infect. Immun.* **58**:471–479.

Ascencio, F., P. Aleljung, O. Olusanya and T. Wadström. 1990a. Types I and IV collagen and fibrinogen binding to *Aeromonas* species isolated from various infections. *Int. J. Med. Microbiol.* **273**:186–194.

Ascencio, F., P. Aleljung and T. Wadström. 1990b. Particle agglutination assays to identify fibronectin and collagen surface receptors and lectins in *Aeromonas* and *Vibrio* species. *Appl. Environ. Microbiol.* **56**:1926–1931.

Ascencio, F., A. Ljungh, and T. Wadström. 1991. Comparative study of extracellular matrix protein binding to *Aeromonas hydrophila* isolated from diseased fish and human infection. *Microbios* **65**:135–146.

Ashkenazi, S., L. May, M. LaRocco, E.L. Lopez, and T.G. Cleary. 1991. The effect of postnatal age on the adherence of enterohemorrhagic *Escherichia coli* to rabbit intestinal cells. *Pediatr. Res.* **29**:14–19.

Aslanzadeh, J. and L.J. Paulissen. 1990. Adherence and pathogenesis of *Salmonella enteritidis* in mice. *Microbiol. Immunol.* **34**:885–893.

Athamna, A. and I. Ofek. 1988. Enzyme-linked immunosorbent assay for quantitation of attachment and ingestion stages of bacterial phagocytosis. *J. Clin. Microbiol.* **26**:62–66.

Athamna, A., I. Ofek, Y. Keisari, S. Markowitz, G.G.S. Dutton, and N. Sharon. 1991. Lectinophagocytosis of encapsulated *Klebsiella pneumoniae* mediated by surface lectins of guinea pig alveolar macrophages and human monocyte-derived macrophages. *Infect. Immun.* **59**:1673–1682.

Atkinson, H.M., D. Adams, R.S. Savvas, and T.J. Trust. 1987. *Aeromonas* adhesin antigens. *Experientia* **43**:372–374.

Aubel, D., A. Darfeuille-Michaud and B. Joly. 1991. New adhesive factor (antigen 8786) on a human enterotoxigenic *Escherichia coli* 0117:H4 strain isolated in Africa. *Infect. Immun.* **59**:1290–1299.

Axelrod, D.A. 1985. Primary and secondary *in vitro* immune response of the rabbit Peyer's patch and spleen to RDEC-1 pili. *Clin. Immunol. Immunopathol.* **37**:124–134.

Babu, J.P. and M.K. Dabbous. 1986. Interaction of salivary fibronectin with oral streptococci. *J. Dent. Res.* **65**:1094–1098.

Babu, J.P., S.N. Abraham, M.K. Dabbous, and E.H. Beachey. 1986. Interaction of a 60-kilodalton D-mannose-containing salivary glycoprotein with type 1 fimbriae of *Escherichia coli*. *Infect. Immun.* **54**:104–108.

Babu, J.P., M.K. Dabbous, and S.N. Abraham. 1991. Isolation and characterization of a 180-kiloDalton salivary glycoprotein which mediates the attachment of *Actinomyces naeslundii* to human buccal epithelial cells. *J. Periodont. Res.* **26**:97–106.

Baddour, L.M., G.D. Christensen, W.A. Simpson, and E.H. Beachey. 1990. Microbial adherence. In: Mandell, G.L., R.G. Douglas, Jr., and J.E. Bennett (eds.), *Principles and Practice Infectious Diseases,* 3rd ed., Churchill-Livingstone, New York, pp. 9–25.

Bahrani, F.K., D.E. Johnson, D. Robbins, and H.L.T. Mobley. 1991. *Proteus mirabilis* flagella and MR/P fimbriae: isolation, purification, N-terminal analysis, and serum antibody response following experimental urinary tract infection. *Infect. Immun.* **59**:3574–3580.

Bakaletz, L.O., B.M. Tallan, T. Hoepf, T.F. DeMaria, H.G. Birck, and D.J. Lim. 1988a. Frequency of fimbriation of nontypable *Haemophilus influenzae* and its ability to adhere to chinchilla and human respiratory epithelium. *Infect. Immun.* **56**:331–335.

Bakaletz, L.O., T.M. Hoepf, T.F., DeMaria, and D.J. Lim. 1988b. The effect of antecedent influenza A virus infection on the adherence of *Haemophilus influenzae* to chinchilla tracheal epithelium. *Am. J. Otolaryngol.* **9**:127–134.

Bakaletz, L.O., B.M. Tallan, W.J. Andrzejewski, T.F. DeMaria, and D.J. Lim. 1989. Immunological responsiveness of chinchillas to outer membrane and isolated fimbrial proteins of nontypable *Haemophilus influenzae*. *Infect. Immun.* **57**:3226–3229.

Baker, N., G.C. Hansson, H. Leffler, G. Riise, and C. Svanborg-Eden. 1990. Glycosphingolipid receptors for *Pseudomonas aeruginosa*. *Infect. Immun.* **58**:2361–2366.

Baker, N.R., V. Minor, C. Deal, M.S. Shahrabadi, D.A. Simpson, and D.E. Woods. 1991. *Pseudomonas aeruginosa* exoenzyme S is an adhesin. *Infect. Immun.* **59**:2859–2863.

Baldini, M.M., J.P. Nataro, and J.B. Kaper. 1986. Localization of a determinant for HEp-2 adherence by enteropathogenic *Escherichia coli*. *Infect. Immun.* **52**:334–336.

Baloda, S.B. 1988. Characterization of fibronectin binding to *Salmonella enteritidis* strain 27655R. *FEMS Microbiol. Lett.* **49**:483–488.

Baloda, S.B., A. Faris, G. Froman, and T. Wadström. 1985. Fibronectin binding to *Salmonella* strains. *FEMS Microbiol. Lett.* **28**:1–5.

Baloda, S.B., G. Froman, J.E. Peeters, and T. Wadström. 1986. Fibronectin binding and cell-surface hydrophobicity of attaching effacing enteropathogenic *Escherichia coli* strains isolated from newborn and weanling rabbits with diarrhoea. *FEMS Microbiol. Lett.* **34**:225–229.

Baloda, S.B., A. Faris, and K. Krovacek. 1988. Cell surface properties of enterotoxigenic and cytotoxic *Salmonella enteritidis* and *Salmonella typhimurium*: studies on hemagglutination, cell-surface hydrophobicity, attachment to human intestinal cells and fibronectin binding. *Microbiol. Immunol.* **32**:447–459.

Baloda, S.B., R. Dyal, E.A. Gonzalez, J. Blanco, L. Hajdu and I. Mansson. 1991. Fibronectin binding by *Salmonella* strains: evaluation of a particle agglutination assay. *J. Clin. Microbiol.* **29**:2824–2830.

Barritt, D.S., R.S. Schwalbe, D.G. Klapper and J.G. Cannon. 1987. Antigenic and structural differences among six proteins II expressed by a single strain of *Neisseria gonorrhoeae*. *Infect. Immun.* **55**:2026–2031.

Baseman, J.B., J. Morrison-Plummer, D. Drouillard, B. Puleo-Scheppke, V.V. Tryon, and S.C. Holt. 1987. Identification of a 32-kilodalton protein of *Mycoplasma pneumoniae* associated with hemadsorption. *Isr. J. Med. Sci.* **23**:474–479.

Baudry, B., Savarino, S.J., P. Vial, J.B. Kaper, and M.M. Levine. 1990. A sensitive and specific DNA probed to identify enteroaggregative *Escherichia coli*, a recently discovered diarrheal pathogen. *J. Infect. Dis.* **161**:1249–1251.

Baughn, R.E. 1986. Antibody-independent interactions of fibronectin, C1q, and human neutrophils with *Treponema pallidum*. *Infect. Immun.* **54**:456–464.

Beachey, E.H. (ed). 1980. *Bacterial adherence. (Receptors and Recognition, Vol. 6)*, Chapman and Hall, New York.

Beachey, E.H. 1981. Bacterial adherence: adhesin–receptor interactions mediating the attachment of bacteria to mucosal surfaces. *J. Infect. Dis.* **143**:325–345.

Beachey, E.H., B.I. Eisenstein, and I. Ofek. 1982. Bacterial adherence in infectious diseases. *Current Concepts*, Upjohn Company, Kalamazoo, MI.

Benz, I. and M.A. Schmidt. 1989. Cloning and expression of an adhesin (AIDA-I) involved in diffuse adherence of enteropathogenic *Escherichia coli*. *Infect. Immun.* **57**:1506–1511.

Bertin, A. 1985. F42 Antigen as a virulence factor in the infant mouse model of *Escherichia coli* diarrhoea. *J. Gen. Microbiol.* **131**:3037–3045.

Bertin, A.M. and M.F. Duchet-Suchaux. 1991. Relationship between virulence and adherence of various enterotoxigenic *Escherichia coli* strains to isolated intestinal epithelial cells from chinese Meishan and European large white pigs. *Am. J. Vet. Res.* **52**:45–49.

Bessen, D. and E.C. Gotschlich. 1986. Interactions of gonococci with HeLa cells: attachment, detachment, replication, penetration, and the role of protein II. *Infect. Immun.* **54**:154–160.

Bessen, D. and E.C. Gotschlich. 1987. Chemical characterization of binding properties of opacity-associated protein II from *Neisseria gonorrhoeae*. *Infect. Immun.* **55**:141–147.

Bhat, S. and B.R. Panhotra. 1987. Adherence of *Klebsiella* to human buccal epithelial cells. *Indian J. Med. Res.* **86**:433–436.

Biesbrock, A.R., M.S. Reddy, and M.J. Levine. 1991. Interaction of a salivary mucin-secretory immunoglobulin A complex with mucosal pathogens. *Infect. Immun.* **59**:3492–3497.

Bilge, S.S., C.R. Clausen, W. Lau, and S.L. Moseley. 1989. Molecular characterization of a fimbrial adhesin, F1845, mediating diffuse adherence of diarrhea-associated *Escherichia coli* to HEp-2 cells. *J. Bacteriol.* **171**:4281–4289.

Binsztein, N., M.J. Jouve, G.I. Viboud, L. Lopez Moral, M. Rivas, I. Orskov, C. Ahren, and A.M. Svennerholm. 1991. Colonization factors of enterotoxigenic *Escherichia coli* isolated from children with diarrhea in Argentina. *J. Clin. Microbiol.* **29**:1893–1898.

Blake, M.S. 1985. Functions of the outer membrane proteins of *Neisseria gonorrhoeae*. In: Jackson, G.G. and H. Thomas (eds.) *The Pathogenesis of Bacterial Infections*. Springer-Verlag, Berlin, pp. 51–66.

Blanco, J., E.A. Gonzalez, and R. Anadone. 1985. Colonization and hemagglutination patterns of human *Escherichia coli*. *Eur. J. Clin. Microbiol.* **4**:316–326.

Blaser, M.J. 1990. *Helicobacter pylori* and the pathogenesis of gastroduodenal inflammation. *J. Infect. Dis.* **161**:626–633.

Blaylock, W.K., B.Y. Yue, and J.B. Robin. 1990. The use of concanavalin A to competitively inhibit *Pseudomonas aeruginosa* adherence to rabbit corneal epithelium. *CLAO. J.* **16**:223–227.

Bloch, C.A. and P.E. Orndorff. 1990. Impaired colonization by and full invasiveness of *Escherichia coli* K1 bearing a site-directed mutation in the type 1 pilin-gene. *Infect. Immun.* **58**:275–278.

Blomfield, I.D., M.S. McClain, J.A. Princ, P.J. Calie, and B.I. Eisenstein. 1991. Type 1 fimbriation and *fim*E mutants of *Escherichia coli* K–12. *J. Bacteriol.* **173**:5298–5307.

Blyn, L.B., B.A. Braaten, C.A. White-Ziegler, D.H. Rolfson, and D.A. Low. 1989. Phase-variation of pyelonephritis-associated pili in *Escherichia coli*: evidence for transcriptional regulation. *EMBO J.* **8**:613–620.

Bock, K., M.E. Breimer, A. Brignote, G.C. Hannsson, K.A. Karlsson, G. Larson, H. Leffler, B.E. Samuelsson, N. Stromberg, C. Svanborg Eden, and J. Thurin. 1985. Specificity of binding of a strain of uropathogenic *Escherichia coli* to Galα1,4Gal-containing glycosphingolipids. *J. Biol Chem.* **260**:8545–8551.

Bode, G., P. Malfertheiner, and H. Ditschuneit. 1988. Pathogenetic implications of ultrastructural findings in *Campylobacter pylori* related gastroduodenal disease. *Scand. J. Gastroenterol.* **142**:25–39.

Boedeker, E.C., C.P. Cheney, and J.R. Cantey. 1987. Inhibition of enteropathogenic *Escherichia coli* (strain RDEC-1) adherence to rabbit intestinal brush broders by milk immune secretory immunoglobulin A. *Adv. Exp. Med. Biol.* **216B**:919–930.

Booth, R.A. and R.A. Finkelstein. 1986. Presence of hemagglutinin/protease and other potential virulence factors in 01 and non-01 *Vibrio cholerae. J. Infect. Dis.* **154**:183–186.

Borriello, S.P., A.R. Welch, F.E. Barclay, and H.A. Davies. 1988. Mucosal association by *Clostridium difficile* in the hamster gastrointestinal tract. *J. Med. Microbiol.* **25**:191–196.

Bose, S.K. and P.C. Goswami. 1986. Enhancement of adherence and growth of *Chlamydia trachomatis* by estrogen treatment of HeLa cells. *Infect. Immun.* **53**:646–650.

Bowden, C.A., L.A. Joens, and L.M. Kelley. 1989. Characterization of the attachment of *Treponema hyodysenteriae* to Henle intestinal epithelial cells *in vitro. Am. J. Vet. Res.* **50**:1481–1485.

Brauner, A. and C.G. Ostenson. 1987. Bacteremia with P-fimbriated *Escherichia coli* in diabetic patients: correlation between proteinuria and non-P-fimbriated strains. *Diabetes Res.* **6**:61–65.

Brauner, A., M. Leissner, B. Wretlind, I. Julander, S.B. Svenson, and G. Kallenius. 1985. Occurrence of P-fimbriated *Escherichia coli* in patients with bacteremia. *Eur. J. Clin. Microbiol.* **4**:566–569.

Brauner, A., J.M. Boeufgras, S.H. Jacobson, B. Kaijser, G. Kallenius, S.B. Svenson, and B. Wretlind. 1987. The use of biochemical markers, serotype and fimbriation in the detection of *Escherichia coli* clones. *J. Gen. Microbiol.* **133**:2825–2834.

Breimer, M.E., G.C. Hansson, and H. Leffler. 1985. The specific glycosphingolipid composition of human urethal epithelial cells. *J. Biochem.* **98**:1169–1180.

Brennan, M.J., J.O. Cisar, and A.L. Sandberg. 1986. A 160 kilodalton epithelial cell surface glycoprotein recognized by plant lectins that inhibits the adherence of *Actinomyces naeslundii*. *Infect. Immun.* **52**:840–845.

Brennan, M.J., R.A. Joralmon, J.O. Cisar, and A.L. Sandberg. 1987. Binding of *Actinomyces naeslundii* to glycosphingolipids. *Infect. Immun.*. **55**:487–489.

Brennan, M.J., J.L. David, J.G. Kenimer, and C.R. Manclark. 1988. Lectin-like binding of pertussis toxin to a 165-kilodalton chinese hamster ovary cell glycoprotein. *J. Biol. Chem.* **263**:4895–4899.

Brennan, M.J., J.H. Hannah, and E. Leininger. 1991. Adhesion of *Bordetella pertussis* to sulfatides and to the GalNAc beta1,4Gal sequence found in glycosphingolipids. *J. Biol. Chem.* **266**:18827–18831.

Brinton, C.C. Jr., M.J. Carter, D.B. Derber, S. Kar, J.A. Kramarik, A.C.-C. To, S.C.-M. To, and S. Wood. 1989. Design and development of pilus vaccines for *Haemophilus influenzae* diseases. *Pediatr. Infect. Dis. J.* **8**:S54–S61.

Broes, A., J.M. Fairbrother, S. Lariviere, M. Jacques, and H. Johnson. 1988. Virulence properties of enterotoxigenic *Escherichia coli* 08:KX105 strains isolated from diarrheic piglets. *Infect. Immun.* **56**:241–246.

Broes, A., J.M. Fairbrother, M. Jacques, and S. Lariviere. 1989. Requirement for capsular antigen KX105 and fimbrial antigen CS1541 in the pathogenicity of porcine enterotoxigenic *Escherichia coli* 08:KX105 strains. *Am. J. Vet. Res.* **53**:43–47.

Brook, I. and M.L. Myhal. 1991. Adherence of *Bacteroides fragilis* group species. *Infect. Immun.* **59**:742–744.

Brooks, D.E., J. Cavanagh, D. Jayroe, J. Janzen, R. Snoek, and T.J. Trust. 1989. Involvement of the MN blood group antigen in shear-enhanced hemagglutination induced by the *Escherichia coli* F41 adhesin. *Infect. Immun.* **57**:377–383.

Budiarti, S., Y. Hirai, J. Minami, S. Katayama, T. Shimizu, and A. Okabe. 1991. Adherence to HEp-2 cells and replication in macrophages of *Salmonella derby* of human origin. *Microbiol. Immunol.* **35**:111–123.

Burke, D.A. and A.T. Axon. 1986. Adhesive *Escherichia coli* in inflammatory bowel disease and infective diarrhoea. *Br. Med. J.* **297**:102–104.

Burke, D.A. and A.T. Axon. 1987a. Ulcerative colitis and *Escherichia coli* with adhesive properties. *J. Clin. Pathol.* **40**:782–786.

Burke, D.A. and A.T. Axon. 1987b. HeLa cell and buccal epithelial cell adhesion assays for detecting intestinal *Escherichia coli* with adhesive properties in ulcerative colitis. *J. Clin. Pathol.* **40**:1402–1404.

Burke, D.A. and A.T. Axon. 1988. Hydrophobic adhesion of *E. coli* in ulcerative colitis. *Gut* **29**:41–43.

Burke, V., M. Cooper, and J. Robinson. 1986. Haemagglutination patterns of *Aeromonas* spp. related to species and source of strains. *Aust. J. Exp. Biol. Med. Sci.* **64**:563–570.

Busscher, H.J., A.H. Weerkamp, H.C. Van der Mei, D. Van Steenberghe, M. Quirynen, I.H. Pratt, M. Marechal, and P.G. Rouxhet. 1989. Physico-chemical properties of oral streptococcal cell surfaces and their relation with adhesion to solid substrata in vitro and in vivo. *Colloids Surfaces* **42**:345–353.

Butrus, S.I. and S.A. Klotz. 1987. The adherence of *Pseudomonas aeruginosa* to soft contact lenses. *Ophthalmology* **94**:1310–1314.

Cabral, D.A., B.A. Loh, and D.P. Speert. 1987. Mucoid *Pseudomonas aeruginosa* resists nonopsonic phagocytosis by human neutrophils and macrophages. *Pediatr. Res.* **22**:429–431.

Campbell, S., S. Duckworth, C.J. Thomas, and T.A. McMeekin. 1987. A note on adhesion of bacteria to chicken muscle connective tissue. *J. Appl. Bacteriol.* **63**:67–71.

Cantey, J.R. and S.L. Moseley. 1991. HeLa cell adherence, actin aggregation, and invasion by nonenteropathogenic *Escherichia coli* possessing the *eae* gene. *Infect. Immun.* **59**:3924–3929.

Cantey, J.R., A.G. Pinson, and R.K. Blake. 1987. Mucosal and systemic immune response to O-antigen after colonization and diarrhea due to a Peyer's patch-adherent *Escherichia coli* (strain RDEC-1). *J. Infect. Dis.* **156**:1022–1025.

Cantey, J.R., L.R. Inman, and R.K. Blake. 1989. Production of diarrhea in the rabbit by a mutant of *Escherichia coli* (RDEC-1) that does not express adherence (AF/R1) pili. *J. Infect. Dis.* **160**:136–141.

Carr, B., J.B. Walsh, D. Coakley, T. Scott, E. Mulvihill, and C. Keane. 1989. Effect of age on adherence of *Branhamella catarrhalis* to buccal epithelial cells. *Gerontology* **35**:127–129.

Carrello, A., K.A. Silburn, J.R. Budden, and B.J. Chang. 1988. Adhesion of clinical and environmental *Aeromonas* isolates to HEp-2 cells. *J. Med. Microbiol.* **26**:19–27.

Cellini, L., R. Piccolomini, N. Allocati and G. Ravagnan. 1987. Adhesive properties of *Proteus* genus related to antimicrobial agents resistance. *Microbiologica* **10**:291–299.

Cellini, L., R. Piccolomini, N. Allocati, A. Di-Girolamo, and G. Catamo. 1988. Effect of pefloxacin on adherence of *Proteus and Providencia* spp. to squamous epithelial cells. *Chemioterapia* **7**:298–301.

Cerf, M., B. Gaudin, A. Cazier, J. Barge, J. Bizet and E. Bergogne-Berezin. 1986. Bacterial adhesion in human upper gastrointestinal tract. *Diagn. Microbiol. Infect. Dis.* **5**:282–291.

Chandra, R.K. 1988. Increased bacterial binding to respiratory epithelial cells in vitamin A deficiency. *Br. Med. J.* **297**:834–835.

Chanteloup, N.K., M. Dho-Moulin, E. Esnault, A. Bree, and J.P. Lafout. 1991. Serological conservation and location of the adhesin of avian *Escherichia coli* type 1 fimbriae. *Microb. Pathogen.* **10**:271–280.

Chart, H., S.M. Scotland, G. A. Willshaw and B. Rowe. 1988. HEp-2 adhesion and the expression of a 94kDa outer-membrane protein by strains of *Escherichia coli* belonging to enteropathogenic serogroups. *J. Gen. Microbiol.* **134**:1315–1321.

Chen, Y.Y. and D.C. Krause. 1988. Parasitism of hamster trachea epithelial cells by *Mycoplasma pneumoniae*. *Infect. Immun.* **56**:570–576.

Chiarini, F., P. Mastromarino, G.B. Orsi, and T. Riscaldati. 1989. Adhesiveness of *Pseudomonas aeruginosa* to rabbit vesical mucosa: effect of glycosidases on cellular binding. *Annali Di Igiene* **1**:399–408.

Childs, W.C. III and R.J. Gibbons. 1988. Methods of Percoll density gradients for studying attachment of bacteria to oral epithelial cells. *J. Dent. Res.* **67**:826–830.

Childs, W.C. III and R.J. Gibbons. 1990. Selective modulations of bacterial attachment to oral epithelial cells by enzyme activities associated with poor oral hygiene. *J. Periodont. Res.* **25**:172–178.

Chung, W.B., M.T. Collins, and L.R. Backstrom. 1990. Adherence of *Bordetella bronchiseptica* and *Pasteurella multocida* to swine nasal ciliated epithelial cells *in vitro*. *Acta Pathol. Microbiol. Immunol. Scand.* **98**:453–461.

Cimasoni, G. and B.C. McBride. 1987. Adherence of *Treponema denticola* to modified hydroxyapatite. *J. Dent. Res.* **66**:1727–1729.

Cisar, J.O. 1986. Fimbrial lectins of the oral actinomyces. In: Mirelman, D. (ed.), *Microbial Lectins and Agglutinins*, John Wiley & Sons, New York, pp. 183–196.

Cisar, J.O., M.J. Brennan, and A.L. Sandberg. 1985. Lectin-specific interaction of *Actinomyces* fimbriae with oral streptococci. In: Mergenhagen, S.E. and B. Rosan (eds.), *Molecular Basis of Oral Microbial Adhesion*. American Society for Microbiology, Washington, pp. 159–163.

Cisar, J.O., A.E. Vatter, W.B. Clark, S.H. Curl, S. Hurst-Calderone, and A.L. Sandberg. 1988. Mutants of *Actinomyces viscosus* T14V lacking type 1, type 2, or both types of fimbriae. *Infect. Immun.* **56**:2984–2989.

Cisar, J.O., E.L. Barsumian, R.P. Siraganian, W.B. Clark, M.K. Yeung, S.D. Hsu, S.H. Curl, A.E. Vatter, and A.L. Sandberg. 1991. Immunochemical and functional studies of *Actinomyces viscosus* T14V type 1 fimbriae with monoclonal and polyclonal antibodies directed against the fimbrial subunit. *J. Gen. Microbiol.* **137**:1971–1979.

Clark, W.B., T.T. Wheeler, D.D. Lane, and J.O. Cisar. 1986. *Actinomyces* adsorption mediated by type–1 fimbriae. *J. Dent. Res.* **65**:1166–1168.

Clark, W.B., J.E. Beem, W.E. Nesbitt, J.O. Cisar, C.C. Tseng, and M.J. Levine. 1989. Pellicle receptors for *Actinomyces viscosus* type 1 fimbriae in vitro. *Infect. Immun.* **57**:3003–3008.

Clegg, S. and G.F. Gerlach. 1987. Enterobacterial fimbriae. *J. Bacteriol.* **169**:934–938.

Clerc, P. and P.J. Sansonetti. 1987. Entry of *Shigella flexneri* into HeLa cells: evidence for directed phagocytosis involving actin polymerization and myosin accumulation. *Infect. Immun.* **55**:2681–2688.

Clyde, Jr., W.A. and P.C. Hu. 1986. Antigenic determinants of the attachment protein of *Mycoplasma pneumoniae* shared by other pathogenic *Mycoplasma* species. *Infect. Immun.* **51**:690–692.

Cohen, P.S., J.C. Arruda, T.J. Williams, and D.C. Laux. 1985a. Adhesion of a human fecal *Escherichia coli* strain to mouse colonic mucus. *Infect. Immun.* **48**:139–145.

Cohen, P.S., R. Rossoll, V.J. Cabelli, S.L. Yang, and D.C. Laux. 1985b. Relationship between the mouse colonizing ability of a human fecal *Escherichia coli* strain and its ability to bind a specific mouse colonic mucous gel protein. *Infect. Immun.* **40**:62–69.

Cohen, P.S., E.A. Wadolkowski, and D.C. Laux. 1986. Adhesion of a human fecal *Escherichia coli* strain to a 50.5-kDa glycoprotein receptor present in mouse colonic mucus. *Microecol. Ther.* **16**:231–241.

Coleman, T., S. Grass, and R. Munson Jr. 1991. Molecular cloning, expression, and sequence of the pilin gene from nontypable *Haemophilus influenzae* M37. *Infect. Immun.* **59**:1716–1722.

Collinson, S.K., L. Emody, K.H. Muller, T.J. Trust, and W.W. Kay. 1991. Purification and characterization of thin, aggregative fimbriae from *Salmonella enteritidis*. *J. Bacteriol.* **173**:4773–4781.

Comstock, L.E. and D.D. Thomas. 1989. Penetration of endothelial cell monolayers by *Borrelia burgdorferi*. *Infect. Immun.* **57**:1626–1628.

Contrepois, M.G. and J. Girardeau. 1985. Additive protective effects of colostral antipili antibodies in calves experimentally infected with enterotoxigenic *Escherichia coli*. *Infect. Immun.* **50**:947–949.

Contrepois, M., J.L. Martel, C. Bordas, F. Hayers, A. Millet, J. Ramisse, and R. Sendral. 1985. Frequence des pili FY et K99 parmi des souches de *Escherichia coli* isolies de veaux diarrheiques en France. *Ann. Rech. Vet.* **16**:25–28.

Contrepois, M., J.M. Fairbrother, Y.K. Kaura, and J.P. Girardeau. 1989. Prevalence of CS31A and F165 surface antigens in *Escherichia coli* isolates from animals in France, Canada and India. *FEMS Microbiol. Lett.* **59**:319–324.

Conventi, L., G. Errico, S. Mastroprimiano, R. Delia, and F. Busolo. 1989. Characterization of *Escherichia coli* adhesins in patients with symptomatic urinary tract infections. *Genitourin. Med.* **65**:183–186.

Conway P.L. and R.F. Adams. 1989. Role of erythrosine in the inhibition of adhesion of *Lactobacillus fermentum* strain 737 to mouse stomach tissue. *J. Gen. Microbiol.* **135**:1167–1173.

Conway, P.L. and S. Kjelleberg. 1989. Protein-mediated adhesion of *Lactobacillus fermentum* strain 737 to mouse stomach squamous epithelium. *J. Gen. Microbiol.* **135**:1175–1186.

Conway, P.L., A. Welin, and P.S. Cohen. 1990. Presence of K88-specific receptors in porcine ileal mucus is age dependent. *Infect. Immun.* **58**:3178–3182.

Cook, R.L., G. Reid, D.G. Pond, C.A. Schmitt, and J.D. Sobel. 1989. Clue cells in bacterial vaginosis: immunofluorescent identification of the adherent gram-negative bacteria as *Gardnerella vaginalis*. *J. Infect. Dis.* **160**:490–496.

Cover, T.L. and R.C. Aber. 1989. *Yersinia enterocolitica*. *N. Engl. J. Med.* **321**:16–24.

Cox, E. and A. Houvenaghel. 1987. In vitro adhesion of K88ab-, K88ac-, and K88ad-positive *Escherichia coli* to intestinal villi, to buccal cells and to erythrocytes of weaned piglets. *Vet. Microbiol.* **15**:201–207.

Criado, M.T., C.M. Ferreiros, and V. Sainz. 1986. Studies on the implication of surface hydrophobicity in the adherence of *Neisseria meningitidis* to buccal epithelial cells. *Med. Microbiol. Immunol.* **175**:27–34.

Criado, M.T., V. Sainz, M.C. del-Rio, C.M. Ferreiros, J. Criado, J. Carballo, and M.J. Souto. 1989. Failure of adherence to buccal cells and surface hydrophobicity as virulence markers in *Neisseria meningitidis*. *Med. Microbiol. Immunol.* **178**:53–59.

Dahlgren, U.I., A.E. Wold, L.A. Hanson, and T. Midtvedt. 1990. The secretory antibody resonse in milk and bile against fimbriae and LPS in rats monocolonized or immunized in the Peyer's patches with *Escherichia coli*. *Immunology* **71**:295–300.

Daifuku, R. and W.E. Stamm. 1986. Bacterial adherence to bladder uroepithelial cells in catheter-associated urinary tract infection. *N. Engl. J. Med.* **314**:1208–1213.

Dalet, F., T. Segovia, and G. Del Rio. 1991. Frequency and distribution of uropathogenic *Escherichia coli* adhesins: a clinical correlation of over 2,000 cases. *Eur. Urol.* **19**:295–303.

Dallo, S.F., C.J. Su, J.R. Horton, and J.B. Baseman. 1988. Identification of P1 gene domain contianing epitope(s) mediating *Mycoplasma pneumoniae* cytadherence. *J. Exp. Med.* **167**:718–723.

Dallo, S.F., A. Chavoya, and J.B. Baseman. 1990a. Characterization of the gene for a 30-kilodalton adhesin-related protein of *Mycoplasma pneumoniae*. *Infect. Immun.* **58**:4163–4165.

Dallo, S.F., J.R. Horton, C.J. Su, and J.B. Baseman. 1990b. Restriction fragment length polymorphism in the cytadhesin P1 gene of human clinical isolates of *Mycoplasma pneumoniae*. *Infect. Immun.* **58**:2017–2020.

Danbara H., K. Komase, Y. Kirii, M. Shinohara, H. Arita, S. Makino, and M. Yoshikawa. 1987. Analysis of plasmids of *Escherichia coli* 0148:H28 from travelers diarrhea. *Microb. Pathogen.* **3**:269–278.

Darfeuille-Michaud, A., C. Forestier, B. Joly, and R. Cluzel. 1986. Identification of nonfimbrial adhesive factor of an enterotoxigenic *Escherichia coli*. *Infect. Immun.* **52**:468–475.

Darfeuille-Michaud, C. Forrestier, R. Masseboeuf, C. Rich, S. M'Boup, B. Joly, and F. Denis. 1987. Multiplicity of serogroups and adhesins in enteropathogenic and enterotoxigenic *Escherichia coli* isolated from acute diarrhea in Senegal. *J. Clin. Microbiol.* **25**:1048–1051.

Darfeuille-Michaud, A., D. Aubel, G. Chauviere, C. Rich, M. Bourges, A. Servin, and B. Joly. 1990. Adhesion of enterotoxigenic *Escherichia coli* to human colon carcinoma cell line Caco-2 in culture. *Infect. Immun.* **58**:893–902.

Darken, J. and D. Savage. 1987. Influences of conjugal genetic transfer functions of colicin V plasmids on adhesion of *Escherichia coli* to murine intestinal tissue. *Infect. Immun.* **55**:2483–2489.

Daskaleros, P.A. and S.M. Payne. 1987. Congo red binding phenotype is associated with hemine binding and increased infectivity of *Shigella flexneri* in the HeLa cell model. *Infect. Immun.* **55**:1393–1398.

Dawson, J.R. and R.P. Ellen. 1990. Tip-oriented adherence of *Treponema denticola* to fibronectin. *Infect. Immun.* **58**:3924–3928.

De-Boer, J.M. and F.H. Plantema. 1988. Ultrastructure of the *in situ* adherence of *Mobiluncus* to vaginal epithelial cells. *Can. J. Microbiol.* **34**:757–766.

de Man, P., B. Cedergren, S. Enerback, A.C. Larsson, H. Leffler, A.L. Lundell, B. Nilsson, and C. Svanborg-Eden. 1987. Receptor-specific agglutination tests for detection of bacteria that bind globoseries glycolipids. *J. Clin. Microbiol.* **25**:401–406.

de Man, P., U. Jodal, K. Lincoln, and C. Svanborg-Eden. 1988. Bacterial attachment and inflammation in the urinary tract. *J. Infect. Dis.* **158**:29–35.

de Man, P., I. Claeson, I.M. Johanson, U. Jodal, and C. Svanborg-Eden. 1989. Bacterial attachment as a predictor of renal abnormalities in boys with urinary tract infections. *J. Pediatr.* **115**:915–922.

de Melo, M.A. and J.C. Pechere. 1988. Effect of mucin on *Campylobacter jejuni* association and invasion on HEp-2 cells. *Microb. Pathogen.* **5**:71–76.

de Melo, M.A. and J.C. Pechere. 1990. Identification of *Campylobacter jejuni* surface proteins that bind to eucaryotic cells *in vitro*. *Infect. Immun.* **58**:1749–1756.

de Melo, M.A., G. Gabbiani, and J.C. Pechere. 1989. Cellular events and intracellular survival of *Campylobacter jejuni* during infection of HEp-2 cells. *Infect. Immun.* **57**:2214–2222.

de Melo, M.A., G. Gabbiani, and J.C. Pechere. 1989. Cellular events and intracellular survival of *Campylobacter jejuni* during infection of HEp-2 cells. *Infect. Immun.* **57**:2214–2222.

de Ree, J.M. and J.F. van den Bosch. 1987. Serological response to the P fimbriae of uropathogenic *Escherichia coli* in pyleonephritis. *Infect. Immun.* **55**:2204–2207.

de Ree, J.M., P. Schwillens, and J.F. van den Bosch. 1985. Monoclonal antibodies that recognize the P fimbriae F7, F7, F9 and F11 from uropathogenic *Escherichia coli*. *Infect. Immun.* **50**:900–904.

de Ree, J.M., P. Schwillens and J.F. van den Bosch. 1986. Monoclonal antibodies for serotyping the P fimbriae of uropathogenic *Escherichia coli*. *J. Clin. Microbiol.* **24**:121–125.

de Ree, J.M., P. Schwillens and J.F. van den Bosch. 1987. Monoclonal antibodies raised against Pap fimbriae recognize minor component(s) involved in receptor binding. *Microb. Pathogen.* **2**:113–121.

Dean, E.A. 1990. Comparison of receptors for 987P pili of enterotoxigenic *Escherichia coli* in the small intestines of neonatal and older pigs. *Infect. Immun.* **58**:4030–4035.

Dean, E.A. and R.E. Isaacson. 1985a. Purification and characterization of a receptor for the 987P pilus of *Escherichia coli*. *Infect. Immun.* **47**:98–105.

Dean, E.A. and R.E. Isaacson. 1985b. Location and distribution of a receptor for the 987p pilus of *Escherichia coli* in small intestines. *Infect. Immun.* **47**:345–348.

Dean, E.A., S.C. Whipp, and H.W. Moon. 1989. Age-specific colonization of porcine intestinal epithelium by 987-P-piliated enterotoxigenic *Escherichia coli*. *Infect. Immun.* **57**:82–87.

DeGraaf, F.K. and F.R. Mooi. 1986. The fimbrial adhesins of *Escherichia coli*. *Adv. Microb. Physiol.* **28**:65–143.

Dekker, N.P., C.J. Lammel, R.E. Mandrell, and G.F. Brooks. 1990. Opa (protein II) influences gonococcal organization in colonies, surface appearance, size and attachment to human fallopian tube tissues. *Microb. Pathog.* **9**:19–31.

Dickinson, D.P., M.A. Kubiniec, F. Yoshimura, and R.J. Genco. 1988. Molecular cloning and sequencing of the gene encoding the fimbrial subunit protein of *Bacteroides gingivalis*. *J. Bacteriol.* **170**:1658–1665.

Dinari, G., T.L. Hale, O. Washington, and S.B. Formal. 1986. Effect of guinea pig or monkey colonic mucus on *Shigella* aggregation and invastion of HeLa cells by *Shigella flexneri* 1b and 2a. *Infect. Immun.* **51**:975–978.

Doig, P., N.R. Smith, T. Todd, and R.T. Irvin. 1987. Characterization of the binding of *Pseudomonas aeruginosa* alginate to human epithelial cells. *Infect. Immun.* **55**:1517–1522.

Doig, P., W. Paranchych, P.A. Sastry, and R.T. Irvin. 1989. Human buccal epithelial cell receptors of *Pseudomonas aeruginosa*: identification of glycoproteins with pilus binding activity. *Can. J. Microbiol.* **35**:1141–1145.

Doig, P., P.A. Sastry, R.S. Hodges, K.K. Lee, W. Paranchych, and R.T. Irvin. 1990. Inhibition of pilus-mediated adhesion of *Pseudomonas aeruginosa* to human buccal epithelial cells by monoclonal antibodies directed against pili. *Infect. Immun.* **58**:124–130.

Dominick, M.A., M.J. Schmerr, and A.E. Jensen. 1985. Expression of type 1 pili by *Escherichia coli* strains of high and low virulence in the intestinal tract of gnotobiotic turkeys. *Am. J. Vet. Res.* **46**:270–275.

Dominigue, A., A. Darfeuille-Michaud, and B. Joly. 1991. New adhesive factor (antigen 8786) on a human enterotoxigenic *Escherichia coli* 0117:H4 strain isolated in Africa. *Infect. Immun.* **59**:1290–1299.

Dominigue, G.J., J.A. Roberts, R. Laucirica, M.H. Ratner, D.P. Bell, G.M. Suarez, G. Kallenius, and S. Svenson. 1985. Pathogenic significance of P-fimbriated *Escherichia coli* in urinary tract infections. *J. Urol.* **133**:983–989.

Dominigue, G.J., R. Laucirica, P. Baglia, S. Covington, J.A. Robledo, and S.C. Li. 1988. Virulence of wild-type *Escherichia coli* uroisolates in experimental pyelonephritis. *Kidney Int.* **34**:761–765.

Donnenberg, M.S. and J.B. Kaper. 1991. Construction of an *eae* deletion mutant of enteropathogenic *Escherichia coli* by using a positive-selection suicide vector. *Infect. Immun.* **59**:4310–4317.

Donnenberg, M.S., A. Donohue-Rolfe, and G.T. Keusch. 1989. Epithelial cell invasion: An overlooked property of enteropathogenic *Escherichia coli* (EPEC) associated with the EPEC adherence factor. *J. Infect. Dis.* **160**:452–459.

Donnenberg, M.S., S.B. Calderwood, A. Donohue-Rolfe, G.T. Keusch, and J.B. Kaper. 1990. Construction and analysis of TnphoA mutants of enteropathogenic *Escherichia coli* unable to invade HEp-2 cells. *Infect. Immun.* **58**:1565–1571.

Dowling, K., J.A. Roberts, and M.B. Kaack. 1987. P-fimbriated *Escherichia coli* urinary tract infection: a clinical correlation. *South. Med. J.* **80**:1533–1536.

Doyle, R.J. and M. Rosenberg (eds). 1990 *Microbial Cell Surface Hydrophobicity*. American Society for Microbiology, Washington.

Drumm, B., A.M. Robertson, and P.M. Sherman. 1988. Inhibition of attachment of *Escherichia coli* RDEC-1 to intestinal microvillus membranes by rabbit ileal mucus and mucin *in vitro*. *Infect. Immun.* **56**:2437–2442.

Drumm, B., A.W. Neumann, Z. Policova, and P.M. Sherman. 1989. Bacterial cell surface hydrophobicity properties in the mediation of in vitro adhesion by the rabbit enteric pathogen *Escherichia coli* strain RDEC-1. *J. Clin. Invest.* **84**:1588–1994.

Duchet-Suchaux, M. 1988. Protective antigens against enterotoxigenic *Escherichia coli* 0101:K99, F41 in the infant mouse diarrhea model. *Infect. Immun.* **56**:1364–1370.

Duguid, J.P. I.W. Smith, G. Dempster, and P.N. Edmunds. 1955. Non-flagellar filamentous appendages ("fimbrial") and hemagglutinating activity in *Bacterium coli*. *J. Pathol. Bacteriol.* **70**:335–348.

Duguid, J.P., S. Clegg, and M.I. Wilson. 1979. The fimbrial and non-fimbrial haemagglutinins of *Escherichia coli*. *J. Med. Microbiol.* **12**:213–227.

Duncan, J.L. 1988. Differential effect of Tamm-Horsfall protein on adherence of *Escherichia coli* to transitional epithelial cells. *J. Infect. Dis.* **158**:1379–1382.

Dunn, B.E., M. Altmann, and G.P. Campbell. 1991. Adherence of *Helicobacter pylori* to gastric carcinoma cells: analysis by flow cytometry. *Rev. Infect. Dis.* **8**:S657–S664.

Durno, C., R. Soni, and P. Sherman. 1989. Adherence of vero cytotoxin-producing *Escherichia coli* serotype 0157:H7 to isolated epithelial cells and brush border membranes in vitro: role of type 1 fimbriae (pili) as a bacterial adhesin expressed by strain CL-49. *Clin. Invest. Med.* **12**:194–200.

Eaves, D.J. and R.J. Doyle. 1988. Surface characteristics of *Pseudomonas cepacia*. *Microbios* **53**:119–128.

Ebisu, S., H. Nakae, and H. Okada. 1988. Coaggregation of *Eikenella corrodens* with oral bacteria mediated by bacterial lectin-like substance. *Adv. Dent. Res.* **2**:323–327.

Echeverria, P., D.N. Taylor, A. Donohue-Rolfe, K. Supawat, O. Ratchtrachenchai, J. Kaper, and G.T. Keusch. 1987a. HeLa cell adherence and cytotoxin production by enteropathogenic *Escherichia coli* isolated from infants with diarrhea in Thailand. *J. Clin. Microbiol.* **25**:1519–1523.

Echeverria, P., D.N. Taylor, K.A. Bettelheim, A. Chatkaeomorakot, S. Changchawalit, A. Thongcharoen, and U. Leksomboon. 1987b. HeLa cell-adherent enteropathogenic *Escherichia coli* in children under 1 year of age in Thailand. *J. Clin. Microbiol.* **25**:1472–1475.

Ehara, M., M. Ishibashi, Y. Ichinose, M. Iwanaga, S. Shimotori, and T. Naito. 1987. Purification and partial characterization of pili of *Vibrio cholerae* 01. *Vaccine* **5**:283–288.

Eifuku, H., T. Yakushiji, J. Mizuno, and N. Kudo. 1990. Cellular coaggregation of oral *Streptococcus milleri* with actinomyces. *Infect. Immun.* **58**:163–168.

Eifuku, H., K. Kitada, T. Yakushiji, and M. Inoue. 1991a. Lactose-sensitive and -insensitive cell surface interactions of oral *Streptococcus milleri* strains and actinomyces. *Infect. Immun.* **59**:460–463.

Eifuku-Koreeda, H., T. Yakushiji, K. Kitada, and M. Inoue. 1991b. Adherence of oral

"*Streptococcus milleri*" cells to surface in broth cultures. *Infect. Immun.* **59**:4103–4109.

Elbashir, A.M. and S.E. Millership. 1989. Haemagglutinating activity of *Aeromonas* spp. from different sources; attempted use as a typing system. *Epidemiol. Infect.* **102**:221–229.

Elkins, C. and P.F. Sparling. 1990. Outer membrane proteins of *Neisseria gonorrhoeae*, p. 207–217. In: E.M. Ayoub, G.H. Cassell, Jr., F.J. Henry, (ed.), American Society for Microbiology, Washington, D.C.

Ellen, R.P. and D.A. Grove. 1989. *Bacteroides gingivalis* vesicles bind to and aggregate *Actinomyces viscosus*. *Infect. Immun.* **57**:1618–1620.

Elo, J., L.G. Tallgren, V. Vaisanen, T.K. Korhonen, S.B. Svenson, and P.H. Makela. 1985. Association of P and other fimbriae with clinical pyelonephritis in children. *Scand. J. Urol. Nephrol.* **19**:281–284.

Embaye, H., R.M. Batt, J.R. Saunders, B. Getty, and C.A. Hart. 1989. Interaction of enteropathogenic *Escherichia coli* 0 1 1 1 with rabbit intestinal mucosa in vitro. *Gastroenterology* **96**:1079–1086.

Emody, L., A. Carlsson, A. Ljungh, and T. Wadstrom. 1988. Mannose-resistant haemagglutination by *Campylobacter pylori*. *Scand. J. Infect. Dis.* **20**:353–354.

Emody, L., J. Heesemann, H. Wolf-Waltz, M. Skurnik, G. Kapperud, P. O'Toole, and T. Wadstrom. 1989. Binding to collagen by *Yersinia enterocolitica* and *Yersinia pseudotuberculosis*: evidence for yopA-mediated and chromosomally encoded mechanisms. *J. Bacteriol.* **171**:6674–6679.

Enterback, A., A.C. Larsson, H. Leffler, A. Lundell, P. de Man, B. Nilsson, and C. Svanborg-Eden. 1987. Binding to galactose α1–4galactoseβ-containing receptors as potential diagnostic tool in urinary tract infection. *J. Clin. Microbiol.* **25**:407–411.

Ernst, R.K., D.M. Dombroski, and J.M. Merrick. 1990. Anaerobiosis, type 1 fimbriae, and growth phase are factors that affect invasion of HEp-2 cells by *Salmonella typhimurium*. *Infect. Immun.* **58**:2014–2016.

Erwin, A.L., G.E. Kenny, A.L. Smith, and T.L. Stull. 1985. Human antibody response to outer membrane proteins and fimbriae of *Haemophilus influenzae* type b. *Can. J. Microbiol.* **34**:723–729.

Evans, Jr., D.J. and D.G. Evans. 1990. Colonization factor antigens of human pathogens. *Curr. Top. Microbiol. Immun.* **151**:129–145.

Evans, D.G., D.J. Evans, Jr., J.J. Moulds, and D.Y. Graham. 1988a. *N*-Acetylneuraminyllactose-binding fibrillar hemagglutinin of *Campylobacter pylori*: a putative colonization factor antigen. *Infect. Immun.* **56**:2896–2906.

Evans, Jr., D.J., D.G. Evans, S.R. Diaz, and D.Y. Graham. 1988b. Mannose-resistant hemagglutination of human erythrocytes by enterotoxigenic *Escherichia coli* with colonization factor antigen II. *J. Clin. Microbiol.* **26**:1626–1629.

Evans, D.G., D.J. Evans, Jr., and D.Y. Graham. 1989a. Receptor-mediated adherence of *Campylobacter pylori* to mouse Y-1 adrenal cell monolayers. *Infect. Immun.* **57**:2272–2278.

Evans, Jr., D.J., D.G. Evans, K.E. Smith, and D.Y. Graham. 1989b. Serum antibody responses to the *N*-acetylneuraminyllactose-binding hemagglutinin of *Campylobacter pylori*. *Infect. Immun.* **57**:664–667.

Ewanowich, C.A., A.R. Melton, A.A. Weiss, R.K., Sherburne, and M.S. Peppler. 1989a. Invasion of HeLa 229 cells by virulent *Bordetella pertussis*. *Infect. Immun.* **57**:2698–2704.

Ewanowich, C.A., R.K. Sherburne, S.F.P. Man, and M.S. Peppler. 1989b. *Bordetella parapertussis* invasion of HeLa 229 cells and human respiratory epithelial cells in primary culture. *Infect. Immun.* **57**:1240–1247.

Fader, R.C., K. Gondesen, B. Tolley, D.G. Ritchie, and P. Moller. 1988. Evidence that *in vitro* adherence of *Klebsiella pneumoniae* to ciliated hamster tracheal cells is mediated by type 1 fimbriae. *Infect. Immun.* **56**:3011–3013.

Faris, A., K. Krovacek, G. Froman, and T. Wadström. 1987. Binding of fibronectin to *Escherichia coli* isolated from bovine mastitis from different geographical regions. *Vet. Microbiol.* **16**:129–136.

Farley, M.M., D.S. Stephens, S.L. Kaplan, and E.O. Mason, Jr. 1990. Pilus-and non-pilus-mediated interactions of *Haemophilus influenzae* type b with human erythrocytes and human nasopharyngeal mucosa. *J. Infect. Dis.* **161**:274–280.

Farrell, C.F. and R.F. Rest. 1990. Up-regulation of human neutrophil receptors for *Neisseria gonorrhoeae* expressing PII outer membrane proteins. *Infect. Immun.* **58**:2777–2784.

Fauchere, J.L. and M.J. Blaser. 1990. Adherence of *Helicobacter pylori* cells and their surface components to HeLa cell membranes. *Microb. Pathogen.* **9**:427–439.

Fauchere, J.L., A. Rosenau, M. Veron, E.N. Moyen, S. Richard, and A. Pfister. 1986. Association with HeLa cells of *Campylobacter jejuni* and *Campylobacter coli* isolated from human feces. *Infect. Immun.* **54**:283–287.

Fauchere, J.L., M. Kervella, A. Rosenau, and M. Veron. 1989. Adhesion to HeLa cell of *Campylobacter jejuni* and *C. coli* outer membrane components. *Res. Microbiol.* **140**:379–392.

Ferreiros, C.M., M.T. Criado, V. Sainz, J. Carballo, C. del-Rio, and B. Suarez. 1986. Evaluation of hydrophobicity and adherence of *Neisseria meningitidis* strains and a study of their correlation by analysis of alterations induced by antibiotics. *Ann. Inst. Pasteur Microbiol.* **137**:37–45.

Feutrier, J., W.W. Kay, and T.J. Trust. 1986. Purification and characterization of fimbriae from *Salmonella enteritidis*. *J. Bacteriol.* **168**:221–227.

Filippsen, L.F., P. Valentin-Weigand, H. Blobel, K.T. Preissner, and G.S. Chhatwal. 1990. Role of complement S protein (vitronectin) in adherence of *Streptococcus dysgalactiae* to bovine epithelial cells. *Am. J. Vet. Res.* **51**:861–865.

Finlay, B.B. and S. Falkow. 1989. Common themes in microbial pathogenicity. *Microbiol. Rev.* **53**:210–230.

Finlay, B.B., B. Gumbiner, and S. Falkow. 1988a. Penetration of *Salmonella* through a polarized Madin-Darby canine kidney epithelial cell monolayer. *J. Cell. Biol.* **107**:221–230.

Finlay, B.B., M.N. Starnbach, C.L. Francis, B.A. Stocker, S. Chatfield, G. Dougan, and S. Falkow. 1988b. Identification and characterization of TnphoA mutants of *Salmonella* that are unable to pass through a polarized MDCK epithelial cell monolayer. *Mol. Microbiol.* **2**:757–766.

Finlay, B.B., J. Fry, E.P. Rock, and S. Falkow. 1989a. Passage of *Salmonella* through polarized epithelial cells: role of the host and bacterium. *J. Cell Sci. Suppl.* **11**:99–107.

Finlay, B., F. Heffron, and S. Falkow. 1989b. Epithelial cell surfaces induce *Salmonella* proteins required for bacterial adherence and invasion. *Science* **243**:940–943.

Finn, T.M., J. Reiser, R. Germanier, and S.J. Cryz, Jr. 1987. Cell-associated hemagglutinin-deficient mutant of *Vibrio cholerae*. *Infect. Immun.* **55**:942–946.

Firon, N., D. Duksin, and N. Sharon. 1985. Mannose-specific adherence of *Escherichia coli* to BHK cells that differ in their glycosylation patterns. *FEMS Microbiol. Lett.* **27**:161–165.

Firon, N., S. Ashkenazi, D. Mirelman, I. Ofek, and N. Sharon. 1987. Aromatic alpha-glycosides of mannose are powerful inhibitors of the adherence of type 1 fimbriated *Escherichia coli* to yeast and intestinal epiethelial cells. *Infect. Immun.* **55**:472–476.

Fletcher, J.N., J.R. Saunders, R.M. Batt, H. Embaye, B. Getty, and C.A. Hart. 1990. Attaching effacement of the rabbit enterocyte brush border is encoded on a single 96.5-kilobase-pair plasmid in an enteropathogenic *Escherichia coli* 0111 strain. *Infect. Immun.* **58**:1316–1322.

Forestier, C., A Darfeuille-Michaud, E. Wasch, C. Rich, E. Petat, F. Denis, and B. Joly. 1989. Adhesive properties of enteropathogenic *Escherichia coli* isolated from infants with acute diarrhea in Africa. *Eur. J. Clin. Microbiol. Infect. Dis.* **8**:979–983.

Fowler, Jr., J. E., M. Mariano, and J.L.T. Lau. 1987. Interaction of urinary Tamm-Horsfall protein with transitional cells and transitional epithelium. *J. Urol.* **138**:446–448.

Francis, D.H., J.E. Collins, and J.R. Duimstra. 1986. Infection of gnotobiotic pigs with an *Escherichia coli* 0157:H7 strain associated with an outbreak of hemorrhagic colitis. *Infect. Immun.* **51**:953–956.

Francis, C.L., A.E. Jerse, J.B. Kaper, and S. Falkow. 1991. Characterization of interactions of enteropathogenic *Escherichia coli* 0127:H6 with mammalian cells *in vitro*. *J. Infect. Dis.* **164**:693–703.

Franklin, A.L., T. Todd, G. Gurman, D. Black, P.M. Mankinen-Irvin, and R.T. Irvin. 1987. Adherence of *Pseudomonas aeruginosa* to cilia of human tracheal epithelial cells. *Infect. Immun.* **55**:1523–1525.

Frydenberg, J., K. Lind, and P.C. Hu. 1987. Cloning of *Mycoplasma pneumoniae* DNA and expression of P1-epitopes in *Escherichia coli*. *Isr. J. Med. Sci.* **23**:759–762.

Fujita, K., T. Yamamoto, T. Yokota, and R. Kitagawa. 1989. *In vitro* adherence of type 1-fimbriated uropathogenic *Escherichia coli* to human uretheral mucosa. *Infect. Immun.* **57**:2574–2579.

Fukuoka, T. and R. Yanagawa. 1987. Population shift from piliated to non-piliated bacteria in kidneys, bladder and urine of mice infected with *Corynebacterium renale* strain no. 115 piliated bacteria. *Jpn. J. Vet. Sci.* **49**:1073–1079.

Fussell, E.M., M.B. Kaack, R. Cherry, and J.A. Roberts. 1988. Adherence of bacteria to human foreskins. *J. Urol.* **140**:997–1001.

Gaastra, W. and F.K. DeGraaf. 1982. Host-specific fimbrial adhesins of noninvasive enterotoxigenic *Escherichia coli* strains. *Microbiol. Rev.* **46**:129–161.

Gabriel, O., M.J. Heeb, and M.B. Hinrichs. 1985. Interaction of the surface adhesins of the oral *Actinomyces* spp. with mammalian cells. In: Mergenhagen, S. and B. Rosan (eds.), *Molecular Basis of Oral Microbial Adhesion.* American Society for Microbiology, Washington, pp. 45–52.

Gander, R.M. and M.T. LaRocco. 1989. Detection of pilus-like structures on clinical and environmental isolates of *Vibrio vulnificus. J. Clin. Microbiol.* **27**:1015–1021.

Gander, R.M. and V.L. Thomas. 1986. Utilization of anion-exchange-chromatography and monoclonal antibodies to characterize multiple pilus types on a uropathogenic *Escherichia coli* 06 isolate. *Infect. Immun.* **51**:385–393.

Gander, R.M. and V.L. Thomas. 1987. Distribution of type 1 and P pili on uropathogenic *Escherichia coli* 06. *Infect. Immun.* **55**:293–297.

Gander, R.M., V.L. Thomas, and M. Forland. 1985. Mannose-resistant hemagglutination and P receptor recognition of uropathogenic *Escherichia coli* isolated from adult patients. *J. Infect. Dis.* **151**:508–513.

Garber, N., N. Sharon, D. Shohet, J.S. Lam, and R.J. Doyle. 1985. Contribution of hydrophobicity to hemagglutination reactions of *Pseudomonas aeruginosa. Infect. Immun.* **50**:336–377.

Garcia, E., H.E.N. Bergmans, J.F. Van Den Bosch, I. Orskov, B.A.M. Van Der Zeijst, and W. Gaastra. 1988. Isolation and characterisation of dog uropathogenic *Escherichia coli* strains and their fimbriae. *Antonie van Leeuwenhoek* **54**:149–163.

Garcia-Monco, J.C., B. Gernandez-Villar, and J.L. Benach. 1989. Adherence of the lyme disease spirochete to glial cells and cells of glial origin. *J. Infect. Dis.* **160**:497–506.

Gbarah, A., C.G. Gahmberg, I. Ofek, U. Jacobi, and N. Sharon. 1991. Identification of the leukocyte adhesion molecules CD11 and CD18 as receptors for type 1-fimbriated (mannose-specific) *Escherichia coli. Infect. Immun.* **59**:4524–4530.

Geary, S.J. and M.G. Gabridge. 1987. Characterization of a human lung fibroblast receptor site for *Mycoplasma pneumoniae. Isr. J. Med. Sci.* **23**:462–468.

Geers, T.A. and N.R. Baker. 1987. The effect of sublethal concentrations of aminoglycosides on adherence of *Pseudomonas aeruginosa* to hamster tracheal epithelium. *J. Antimicrob. Chemother.* **19**:561–568.

Gerlach, G.F., B.L. Allen, and S. Clegg. 1989. Type 3 fimbriae among enterobacteria and the ability of spermidine to inhibit MR/K hemagglutination. *Infect. Immun.* **57**:219–224.

Gerstenecker, B. and E. Jacobs. 1990. Topological mapping of the P1-adhesin of *Mycoplasma pneumoniae* with adherence-inhibiting monoclonal antibodies. *J. Gen. Microbiol.* **136**:471–476.

Getzoff, E.D., H.E. Parge, D.E. McRee, and J.A. Tainer. 1988. Understanding the structure and antigenicity of gonococcal pili. *Rev. Infect. Dis.* **10**:S296–S299.

Giampapa, S. N. Abraham, T.M. Chiang, and E.H. Beachey. 1988. Isolation and characterization of a receptor for type 1 fimbriae of *Escherichia coli* from guinea pig erythrocytes. *J. Biol. Chem.* **263**:5362–5367.

Gibbons, R.J. 1982. Adherence of bacteria to host tissue. In: Schlessinger D. (ed.), *Microbiology–1982*. American Society for Microbiology, Washington, pp. 395–406.

Gibbons, R.J. 1989. Bacterial adhesion to oral tissues: a model for infectious diseases. *J. Dent. Res.* **68**:750–760.

Gibbons, R.J. and D.I. Hay. 1988. Human salivary acidic proline-rich proteins and statherin promote the attachment of *Actinomyces viscosus* LY7 to apatitic surfaces. *Infect. Immun.* **56**:439–445.

Gibbons, R.J., D.I. Hay, J.O. Cisar, and W.B. Clark. 1988. Adsorbed salivary proline-rich protein 1 and statherin: receptors for type 1 fimbriae of *Actinomyces viscosus* T14V-J1 on apatitic surfaces. *Infect. Immun.* **56**:2990–2993.

Gibbons, R.J., D.I. Hay, W.C. Childs III, and G. Davis. 1990. Role of cryptic receptors (cryptitopes) in bacterial adhesion to oral surfaces. *Arch. Oral Biol.* **35**:107–114.

Gibbons, R.J., D.I. Hay, and D.H. Schlesinger. 1991. Delineation of a segment of adsorbed salivary acidic proline-rich proteins which promotes adhesion of *Streptococcus gordonii* to apatitic surfaces. *Infect. Immun.* **59**:2948–2954.

Giffard, P.M., C.L. Simpson, C.P. Milward, and N.A. Jacques. 1991. Molecular characterization of a cluster of at least two glucosyltransferase genes in *Streptococcus salivarius* ATCC 25975. *J. Gen. Microbiol.* **137**:2577–2593.

Gilboa-Garber, N. 1988. *Pseudomonas aeruginosa* lectins as a model for lectin production, properties, applications and functions. *Zbl. Bakt. Hyg.* **270**:3–15.

Gilsdorf, J.R., K. McCrea, and L. Forney. 1990. Conserved and nonconserved epitopes among *Haemophilus influenzae* type b pili. *Infect. Immun.* **58**:2252–2257.

Girardeau, J.P., M.D. Vartanian, J.L. Ollier, and M. Contrepois. 1988. CS31A, a new K88-related fimbrial antigen on bovine enterotoxigenic and septicemic *Escherichia coli* strains. *Infect. Immun.* **56**:2180–2188.

Giron, J.A., A.S.Y. Ho, and G.K. Schoolnik. 1991a. An inducible bundle-forming pilus of enteropathogenic *Escherichia coli*. *Science* **254**:710–713.

Giron, J.A., T. Jones, F. Millan-Velasco, E. Costra-Monoz, L. Zarate, J. Fry, G. Frankel, S.L. Moseley, B. Baudry, J.B. Kaper, G.K. Schoolnik, and L.W. Riley. 1991b. Diffuse-adhering *Escherichia coli* (DAEC) as a putative cause of diarrhea in Mayan children in Mexico. *J. Infect. Dis.* **163**:507–513.

Glick, J., N. Garber, and D. Shohet. 1987. Surface haemagglutinating activity of *Pseudomonas aeruginosa*. *Microbios* **50**:69–80.

Goldhar, J., A. Zilberberg, and I. Ofek. 1986. Infant mouse model of adherence and colonization of intestinal tissues by enterotoxigenic strains of *Escherichia coli* isolated from humans. *Infect. Immun.* **52**:205–208.

Goldhar, J., R. Perry, J.R. Golecki, H. Hoschützky, B. Jann, and K. Jann. 1987.

Nonfimbrial, mannose-resistant adhesins from uropathogenic *Escherichia coli* 083:K1:H4 and 014:K?:H11. *Infect. Immun.* **55**:1837–1842.

Goldhar, J., M. Yavzori, Y. Keisari, and I. Ofek. 1991. Phagocytosis of *Escherichia coli* mediated by mannose resistant nonfimbrial haemagglutinin (NFA-1). *Microb. Pathogen.* **11**:171–178.

Gomes, T.A.T., P.A. Blake, and L.R. Trabulsi. 1989. Prevalence of *Escherichia coli* strains with localized, diffuse, and aggegative adherence to HeLa cells in infants with diarrhea and matched controls. *J. Clin. Microbiol.* **27**:266–269.

Gonzalez, E.A., J. Blanco, S.B. Baloda, and T. Wadström. 1988a. Relative cell surface hydrophobicity of *Escherichia coli* strains with various recognized fimbrial antigens and without recognized fimbriae. *Zbt. Bakt. A* **269**:218–236.

Gonzalez, E.A., S.B. Baloda, J. Blanco, and T. Wadström. 1988b. Growth conditions for the expression of fibronectin and collagen binding to *Salmonella. Zbt. Bakt. A* **269**:437–446.

Goodwin, C.S., J.A. Armstrong, and B.J. Marshall. 1986. *Campylobacter pylori*, gastritis, and peptic ulceration. *J. Clin. Pathol.* **39**:353–365.

Gorby, G.L. and Z.A. McGee. 1990. Antimicrobial interference with bacterial mechanisms of pathogenicity: effect of sub-MIC azithromycin on gonococcal piliation and attachment to human epithelial cells. *Antimicrob. Agents Chemother.* **34**:2445–2448.

Gorringe, A.R., L.A.E. Ashworth, L.I. Irons, and A. Robinson. 1985. Effect of monoclonal antibodies on the adherence of *Bordetella pertussis* to vero cells. *Infect. Immun.* **36**:782–794.

Gothefors, L., C. Ahren, B. Stoll, D.K. Barua, F. Orskov, M.A. Salek, and A. Svennerholm. 1985. Presence of colonization factor antigens on fresh isolates of fecal *Escherichia coli*: A prospective study. *J. Infect. Dis.* **152**:1128–1133.

Gottschalk, M., S. Petitbois, R. Higgins, and M. Jacques. 1991. Adherence of *Streptococcus suis* capsular type 2 to porcine lung sections. *Can. J. Vet. Res.* **55**:302–304.

Goulbourne, P.A. and R.P. Ellen. 1991. Evidence that *Porphyromonas (Bacteroides) gingivalis* fimbriae function in adhesion to *Actinomyces viscosus. J. Bacteriol.* **173**:5266–5274.

Grant, M.M., M.S. Niederman, M.A. Poehlman, and A.M. Fein. 1991. Characterization of *Pseudomonas aeruginosa* adherence to cultured hamster tracheal epithelial cells. *Am J. Respir. Mol. Biol.* **5**:563–570.

Greenblatt, J.J., K. Floyd, M.E. Philipps, and C.E. Frasch. 1988. Morphological differences in *Neisseria meningitidis* pili. *Infect. Immun.* **56**:2356–2362.

Greenwood, P.E., S.J. Clark, A.D. Cahill, J. Trevallyn-Jones, and S. Tzipori. 1988. Development and protective efficacy of a recombinant-DNA derived fimbrial vaccine against enterotoxic colibacillosis in neonatal piglets. *Vaccine* **6**:389–392.

Grenier, D. and D. Mayrand. 1987. Functional characterization of extracellular vesicles produced by *Bacteroides gingivalis. Infect. Immun.* **55**:111–117.

Grunberg, J., R. Perry, H. Hoschützky, B. Jann, K. Jann, and J. Goldhar. 1988. Nonfimbrial blood group N-specific adhesin (NFA-3) from *Escherichia coli* 020:KX104:H, causing systemic infection. FEMS *Microbiol. Lett.* **56**:241–246.

Grunberg, J., I. Ofek, and J. Goldhar. 1992. The involvement of nonfimbrial hemagglutinins of *Escherichia coli* in mediating nonopsonic phagocytosis by human polymorphonuclear leukocytes. *Isr. J. Med. Sci.* **28**:74.

Grund, S. 1991. Slime capsule and fimbriae on *Salmonella typhimurium var. cop.*-electron microscopic study. *J. Vet. Med.* **38**:545–551.

Grund, S. and A. Weber. 1988. A new type of fimbriae on *Salmonella typhimurium*. *J. Vet. Med.* **35**:779–782.

Guerina, N.G., S. Langermann, G.K. Schoolnik, T.W. Kessler, and D.A. Goldmann. 1985. Purification and characterization of *Haemophilus influenzae* pili and their structural and serological relatedness to *Escherichia coli* P and mannose-sensitive pili. *J. Exp. Med.* **161**:145–159.

Guevara, J.A., G. Zuccaro, A. Trevisan, and C.D. Denoya. 1987. Bacterial adhesion to cerebrospinal-fluid shunts. *J. Neurosurg.* **67**:438–445.

Guyot, G. 1908. Ueber die bakterielle Hamagglutination. *Zbl. Bakt. Parasitol. Infect.* **47**:640–653.

Guzman, C.A., C. Pruzzo, G. LiPira, and L. Calegari. 1989. Role of adherence in pathogenesis of *Enterococcus faecalis* urinary tract infection and endocarditis. *Infect. Immun.* **57**:1834–1838.

Haapasalo, M., U. Singh, B.C. McBride, and V.J. Uitto. 1991. Sulfhydryl-dependent attachment of *Treponema denticola* to laminin and other proteins. *Infect. Immun.* **59**:4230–4237.

Hacker, J. 1990. Genetic determinants coding for fimbriae and adhesins of extraintestinal *Escherichia coli*. *Curr. Top. Microbiol. Immunol.* **151**:1–27.

Hacker, J., G. Schmidt, C. Hughes, S. Knapp, M. Marget, and W. Goebel. 1985. Cloning and characterization of genes involved in production of mannose-resistant, neuraminidase-susceptible (X) fimbriae from a uropathogenic 06:K15:H31 *Escherichia coli* strain. *Infect. Immun.* **47**:434–440.

Hacker, J., A Schrettenbrunner, G. Schroter, H. Duvel, G. Schmidt, and W. Goebel. 1986b. Characterization of *Escherichia coli* wild-type strains by means of agglutination with antisera raised against cloned P-, S- and MS-fimbriae antigens, hemagglutination, serotyping and hemolysin production. *Zentralbl. Bakteriol. Hyg. A* **261**:219–231.

Hacker, J., H. Hof, L. Emody, and W. Goebel. 1986a. Influence of cloned *Escherichia coli* hemolysin genes, S-fimbriae and serum resistance on pathogenicity in different animal models. *Microb. Pathogen.* **1**:533–547.

Hackstadt, T. 1986. Identification and properties of chlamydial polypeptides that bind eucaryotic cell surface components. *J. Bacteriol.* **165**:13–20.

Hackstadt, T. and H.D. Caldwell. 1985. Effect of proteolytic cleavage of surface-exposed proteins on infectivity of *Chlamydia trachomatis*. *Infect. Immun.* **48**:546–551.

Hagberg, L., H. Leffler, and C. Svanborg-Eden. 1985. Non-antibiotic prevention of urinary tract infection. *Infection* **13** (Suppl. 2):S196–S200.

Hagman, M. and D. Danielsson. 1989. Increased adherence to vaginal epithelial cells and phagocytic killing of gonococci and urogenital meningococci associated with heat modifiable proteins. *Acta Pathol. Microbiol. Immunol. Scand.* **97**:839–844.

Hales, B.A., H. Beverley-Clarke, N.J. High, K. Jann, R. Perry, J. Goldhar, and G.J. Boulnois. 1988. Molecular cloning and characterization of the genes for a non-fimbrial adhesin from *Escherichia coli*. *Microb. Pathogen.* **5**:9–17.

Hall, G.A., C.R. Dorn, N. Chanter, S.M. Scotland, H.R. Smith, and B. Rowe. 1990. Attaching and effacing lesions in vivo and adhesion to tissue culture cells of vero-cytotoxin-producing *Escherichia coli* belonging to serogroups 05 and 0103. *J. Gen. Microbiol.* **136**:779–786.

Halula, M.C. and B.A.D. Stocker. 1987. Distribution and properties of the mannose-resistant hemagglutinin produced by *Salmonella species*. *Microb. Pathogen.* **3**:455–459.

Hanazawa, S., K. Hirose, Y. Ohmori, S. Amano, and S. Kitano. 1988. *Bacteroides gingivalis* fimbriae stimulate production of thymocyte-activating factor by human gingival fibroblasts. *Infect. Immun.* **56**:272–274.

Hanazawa, S., Y. Murakami, K. Hirose, S. Amano, Y. Ohmori, H. Higuchi, and S. Kitano. 1991. *Bacteroides (Porphyromonas) gingivalis* fimbriae activate mouse peritoneal macrophages and induce gene expression and production of interleukin-1. *Infect. Immun.* **59**:1972–1977.

Handley, P.S. 1990. Structure, composition and functions of surface structures on oral bacteria. *Biofouling* **2**:239–264.

Handley, P.S. and L.S. Tipler. 1986. An electron microscopy survey of the surface structures and hydrophobicity of oral and non-oral species of the bacterial genus *Bacteroides*. *Arch. Oral. Biol.* **31**:325–335.

Handley, P.S., D.W. Harty, J.E. Wyatt, C.R. Brown, J.P. Doran, and A.C. Gibbs. 1987. A comparison of the adhesion, coaggregation and cell-surface hydrophobicity properties of fibrillar and fimbriate strains of *Streptococcus salivarius*. *J. Gen. Microbiol.* **133**:3207–3217.

Handley, P.S., J. Hargreaves, and D.W. Harty. 1988. Ruthenium red staining reveals surface fibrils and a layer external to the cell wall in *Streptococcus salivarius* HB and adhesion deficient mutants. *J. Gen. Microbiol.* **134**:3165–3172.

Hanley, J., I.E. Salit, and T. Hofmann. 1985. Immunochemical characterization of P pili from invasive *Escherichia coli*. *Infect. Immun.* **49**:581–586.

Harada, T., T. Shimuzu, K. Nishimoto, and Y. Sakakura. 1989. *Haemophilus influenzae* adherence to and absorption from the nasal mucosa of guinea pigs. *Arch. Otorhinolaryngol.* **246**:218–221.

Harada, T., Y. Sakakura, and C.S. Jin. 1990. Adherence of *Haemophilus influenzae* to nasal, nasopharyngeal and buccal epithelial cells from patients with otitis media. *Eur. Arch. Otorhinolaryngol.* **24**:122–124.

Harty, D.W.S. and P.S. Handley. 1989. Expression of the surface properties of the fibrillar *Streptococcus salivarius* HB and its adhesion deficient mutants grown in continuous culture under glucose limitation. *J. Gen. Microbiol.* **135**:2611–2621.

Hase, C.C. and R.A. Finkelstein. 1990. Comparison of the *Vibrio cholerae* hemagglutinin/protease and the *Pseudomonas aeruginosa* elastase. *Infect. Immun.* **58**:4011–4015.

Hasty, D.L. and W.A. Simpson. 1987. Effects of fibronectin and other salivary macromolecules on the adherence of *Escherichia coli* to buccal epithelial cells. *Infect. Immun.* **55**:2103–2109.

Hawthorn, L.A., A.W. Bruce, and G. Reid. 1991. Ability of uropathogens to bind to Tamm Horsfall protein-coated renal tubular cells. *Urol. Res.* **19**:301–304.

Hayashi, A., R. Yanagawa, and H. Kida. 1985. Adhesion of *Corynebacterium renale* and *Corynebacterium pilosum* to epithelial cells of bovine vulva. *Am. J. Vet. Res.* **46**:409–411.

Hazlett, L.D., M.M. Moon, M. Strejc, and R.S. Berk. 1987. Evidence for *N*-acetylmannosamine as an ocular receptor for *P. aeruginosa* adherence to scarified cornea. *Invest. Ophthalmol. Vis. Sci.* **28**:1978–1985.

Hazlett, L.D., M. Moon, and R.S. Berk. 1986. *In vivo* identification of sialic acid as the ocular receptor for *Pseudomonas aeruginosa*. *Infect. Immun.* **51**:687–689.

Hazlett, L., R. Barrett, C. Klettner, R. Berk, and A. Singh. 1991. *Pseudomonas aeruginosa*: a probe to detail change with age in corneal receptor(s). *Ophthalm. Res.* **23**:141–146.

Heckels, J.E. 1989. Structure and function of pili of pathogenic *Neisseria* species. *Clin. Microbiol. Rev.* **2**:S66–S73.

Hedegaard, L. and P. Klemm. 1989. Type 1 fimbriae of *Escherichia coli* as carriers of heterologous antigenic sequences. *Gene* **85**:115–124.

Heesemann, J. and L. Gruter. 1987. Genetic evidence that the outer membrane protein YOP1 of *Yersinia enterocolitica* mediates adherence and phagocytosis resistance to human epithelial cells. *FEMS Microbiol. Lett.* **40**:37–41.

Hemalatha, S.G., B. Drumm, and P. Sherman. 1991. Adherence of *Helicobacter pylori* to human gastric epithelial cells in vitro. *J. Med. Microbiol.* **35**:197–202.

Henriksson, A., R. Szewzyk, and P.L. Conway. 1991. Characteristics of the adhesive determinants of *Lactobacillus fermentum* 104. *Appl. Environ. Microbiol.* **57**:499–502.

Herpay, M., E. Czirok, E. Szollosy, J. Fekete, I. Gado, and H. Milch. 1991. Relative surface hydrophobicity, antigen K1 and haemagglutinating activity are associated in *Escherichia coli*. *Acta Microbiol. Hungaria* **38**:17–28.

Herrington, D.A., R.H. Hall, G. Losonsky, J.J. Mekalanos, R.K. Taylor, and M.M. Levine. 1988. Toxin, toxin-coregulated pili, and the toxR regulon are essential for *Vibrio cholerae* pathogenesis in humans. *J. Exp. Med.* **168**:1487–1492.

Hessey, S.J., J. Spencer, J.I. Wyatt, G. Sobala, B.J. Rathbone, A.T. Axon, and M.F. Dixon. 1990. Bacterial adhesion and disease activity in *Helicobacter* associated chronic gastritis. *Gut* **31**:134–138.

High, N.J., B.A. Hales, K. Jann, and G.J. Boulnois. 1988. A block of urovirulence genes encoding multiple fimbriae and hemolysin in *Escherichia coli* 04:K12:H^-. *Infect. Immun.* **56**:513–517.

Hill, R.H. 1985. Prevention of adhesion by indigenous bacteria to rabbit cecum epithelium by a barrier of microvesicles. *Infect. Immun.* **47**:540–543.

Hinson, G., S. Knutton, M.K.L. Lam Po Tang, A.S. McNeish, and P.H. Williams. 1987. Adherence to human colonocytes of an *Escherichia coli* strain isolated from severe infantile enteritis: molecular and ultrastructural studies of a fibrillar adhesin. *Infect. Immun.* **55**:393–402.

Hirakata, Y., K. Tomono, K. Tateda, T. Matsumoto, N. Furuya, K. Shimoguchi, M. Kaku and K. Yamaguchi. 1991. Role of bacterial association with Kupffer cells in occurrence of endogenous systemic bacteremia. *Infect. Immun.* **59:**289–294.

Hirsch, D.C. 1985. Fimbriae: relation of intestinal bacteria and virulence in animals. *Adv. Vet. Sci. Comp. Med.* **29**:207–238.

Hogg, S.D. and J.E. Manning, 1988. Inhibition of adhesion of viridans streptococci to fibronectin-coated hydroxyapatite beads by lipoteichoic acid. *J. Appl. Bacteriol.* **65**:483–489.

Hokama, A. and M. Iwanaga. 1991. Purification and characterization of *Aeromonas sobria* pili, a possible colonization factor. *Infect. Immun.* **59**:3478–3483.

Hokama, A., Y. Honma, and N. Nakasone. 1990. Pili of an *Aeromonas hydrophila* strain as a possible colonization factor. *Microbiol. Immunol.* **34**:901–915.

Honda, T., A.B.A. Booth, M. Boesman-Finkelstein, and R.A. Finkelstein. 1987. Comparative study of *Vibrio cholerae* non-01 protease and soluble hemagglutinin with those of *Vibrio cholerae* 01. *Infect. Immun.* **55**:451–454.

Honda, T., M. Arita, E. Ayala, and T. Miwatani. 1988. Production of pili on *Vibrio parahaemolyticus*. *Can. J. Microbiol.* **34**:1279–1281.

Honma, Y. and N. Nakasone. 1990. Pili of *Aeromonas hydrophila*: purification, characterization and biological role. *Microbiol. Immunol.* **34**:83–98.

Hopkins, W.J., C.A. Reznikoff, T.D. Oberley, and D.T. Uehling. 1990. Adherence of uropathogenic *E. coli* to differentiated human uroepithelial cells grown *in vitro*. *J. Urol.* **143**:146–149.

Hornick, D.B., B.L. Allen, M.A. Horn, and S. Clegg. 1991. Fimbrial types among respiratory isolates belonging to the family *Enterobacteriaceae*. *J. Clin. Microbiol.* **29**:1795–1800.

Hoschützky, H., F. Lottspeich, and K. Jann. 1989. Isolation and characterization of the α-galactosyl-1, 4-β-galactosyl-specific adhesin (P adhesin) from fimbriated *Escherichia coli*. *Infect. Immun.* **57**:76–81.

Hu, P.C., C.H. Huang, Y.S. Huang, A.M. Collier, and W.A. Clyde Jr. 1985. Demonstration of multiple antigenic determinants on *Mycoplasma pneumoniae* attachment protein by monoclonal antibodies. *Infect. Immun.* **50**:292–296.

Hu, P.C., U. Schaper, A.M. Collier, W.A. Clyde, Jr., M. Horikawa, Y.S. Huang, and M.F. Barile. 1987. A *Mycoplasma genitalium* protein resembling the *Mycoplasma pneumoniae* attachment protein. *Infect. Immun.* **55**:1126–1131.

Huang, J., C.J. Smyth, N.P. Kennedy, J.P. Arbuthnott, and P.W.N. Keeling. 1988. Hemagglutinating activity of *Campylobacter pylori*. *FEMS Microbiol. Lett.* **56**:109–112.

Hudson, M., E.J. Brown, J.K. Ritchey, and T.L. Ratliff. 1991. Modulation of fibronectin-mediated *Bacillus Calmette-Guerin* attachment to murine bladder mucosa by drugs influencing the coagulation pathways. *Cancer Res.* **51**:3726–3732.

Hughes, C., J. Hacker, H. Duvel, and W. Goebel. 1987. Chromosomal deletions and rearrangements cause coordinate loss of hemolysin, fimbriation and serum resistance in a uropathogenic strain of *Escherichia coli*. *Microb. Pathogen.* **2**:227–230.

Hull, S.I., S. Bieler, and R.A. Hull. 1988. Restriction fragment length polymorphism and multiple copies of DNA sequences homologous with probes for P-fimbriae and hemolysin genes among uropathogenic *Escherichia coli*. *Can. J. Microbiol.* **34**:307–311.

Hultgren, S., W.R. Schwan, A.J. Schaeffer, and J.L. Duncan. 1986. Regulation of production of type 1 pili among urinary tract isolates of *Escherichia coli*. *Infect. Immun.* **54**:613–620.

Hultgren, S.J., S. Normark, and S.N. Abraham. 1991. Chaperone-assisted assembly and molecular architecture of adhesive pili. *Annu. Rev. Microbiol.* **45**:383–415.

Hynes, R.O. 1987. Integrins, a family of cell surface receptors. *Cell* **48**:549–555.

Inamine, J.M., T.P. Denny, S. Loechel, M. Schaper, C.H. Huang, K.F. Bott, and P.C. Hu. 1988. Nucleotide sequence of the P1 attachment-protein gene of *Mycoplasma pneumoniae*. *Gene* **64**:217–229.

Inoshita, E., A. Amano, T. Hanioka, H. Tamagawa, S. Shizukuishi, and A. Tsunemitsu. 1986. Isolation and some properties of exohemagglutinin from the culture medium of *Bacteroides gingivalis* 381. *Infect. Immun.* **52**:421–427.

Irvin, R.T., P. Doig, K.K. Lee, P.A. Sastry, W. Paranchych, T. Todd, and R.S. Hodges. 1989. Characterization of the *Pseudomonas aeruginosa* pilus adhesin: confirmation that the pilin structural protein subunit contains a human epithelial cell-binding domain. *Infect. Immun.* **57**:3720–3726.

Isberg, R.R. 1989. Determinants for themoinducible cell binding and plasmid-encoded cellular penetration detected in the absence of the *Yersinia pseudotuberculosis* invasin protein. *Infect. Immun.* **57**:1998–2005.

Isberg, R.R. 1990. Pathways for penetration of enteroinvasive *Yersinia* into mammalian cells. *Mol. Biol. Med.* **7**:73–82.

Isberg, R.R. 1991. Discrimination between intracellular uptake and surface adhesion of bacterial pathogens. *Science* **252**:934–938.

Isberg, R.R. and S. Falkow. 1985. A single genetic locus encoded by *Yersinia pseudotuberculosis* permits invasion of cultured animal cells by *Escherichia coli* K–12. *Nature (Lond.)* **317**:262–264.

Isberg, R.R. and J.M. Leong. 1988. Cultured mammalian cells attach to the invasin protein of *Yersinia pseudotuberculosis*. *Proc. Natl. Acad. Sci. USA* **85**:6682–6686.

Isberg, R.R. and J.M. Leong. 1990. Multiple β_1 chain integrins are receptors for invasin, a protein that promotes bacterial penetration into mammalian cells. *Cell* **60**:861–871.

Isberg, R.R., D.L. Voorhis, and S. Falkow. 1987. Identification of invasin: a protein that allows enteric bacteria to penetrate cultured mammalian cells. *Cell* **50**:769–778.

Isberg, R.R., A. Swain, and S. Falkow. 1988. Analysis of expression and thermoregulation of the *Yersinia pseudotuberuclosis inv* gene with hybrid *Infect. Immun.* **56**:2133–2138.

Ishikawa, H. and Y. Isayama. 1987a. Effect of antigenic modulation and phase variation on adherence of *Bordetella bronchiseptica* to procine nasal epithelial cells. *Am. J. Vet. Res.* **48**:1689–1691.

Ishikawa, H. and Y. Isayama. 1987b. Evidence for sialyl glycoconjugates as receptors for *Bordetella bronchiseptica* on swine nasal mucosa. *Infect. Immun.* **55**:1607–1609.

Ishikawa, H. and Y. Isayama. 1988. Bovine erythorcyte-agglutinin as a possible adhesin of *Bordetella bronchiseptica* responsible for binding to porcine nasal epithelium. *J. Med. Microbiol.* **26**:205–209.

Ishikawa, H. and Y. Isayama. 1990. Hydrophobic surface properties of *Bordetella bronchiseptica* X-mode cells and their possible role in adherence to porcine nasal mucosa. *Can. J. Vet. Res.* **54**:296–298.

Isogai, H., E. Isogai, F. Yoshimura, T. Suzuki, W. Kagota, and K. Takano. 1988. Specific inhibition of adherence of an oral strain of *Bacteroides gingivalis* 381 to epithelial cells by monoclonal antibodies against the bacterial fimbriae. *Arch. Oral Biol.* **33**:479–85.

Israele, V., A. Darabi, and G. McCracken, Jr. 1987. The role of bacterial virulence factors and Tamm-Horsfall protein in the pathogenesis of *Escherichia coli* urinary tract infection in infants. *Am. J. Dis. Child.* **141**:1230–1234.

Iwanaga, M., N. Nakasone, and M. Ehara. 1989. Pili of *Vibrio cholerae* 01 biotype E1 Tor: a comparative study on adhesive and non-adhesive strains. *Microbiol Immunol.* **33**:1–9.

Izumikawa, K., D.F. Chandler, and M.F. Barile. 1986. *Mycoplasma pneumoniae* attachment to glutaraldehyde treated human WiDr cell cultures. *Proc. Soc. Exp. Biol. Med.* **181**:507–511.

Jacobs, E. 1991. *Mycoplasma pneumoniae* virulence factors and the immune response. *Rev. Med. Microbiol.* **2**:83–90.

Jacobs, E., K. Schopperle, and W. Bredt. 1985. Adherence inhibition assay: a specific serological test for detection of antibodies to *Mycoplasma pneumoniae*. *Eur. J. Clin. Microbiol.* **4**:113–118.

Jacobs, E., K. Fuchte, and W. Bredt. 1986. A 168-kilodalton protein of *Mycoplasma pneumoniae* used as antigen in a dot enzyme-linked immunosorbent assay. *Eur. J. Clin. Microbiol.* **5**:435–440.

Jacobs, E., K. Fuchte, and W. Bredt. 1987. Amino acid sequence and antigenicity of the amino-terminus of the 168 kDa adherence protein of *Mycoplasma pneumoniae*. *J. Gen. Microbiol.* **133**:2233–2236.

Jacobs, E., M. Drews. A. Stuhlert, C. Buttner, P.J. Klein, M. Kist, and W. Bredt. 1988. Immunological reaction of guinea-pigs following intranasal *Mycoplasma pneumoniae* infection and immunization with the 168 kDa adherence protein. *J. Gen. Microbiol.* **134**:473–479.

Jacobs, E., B. Gerstenecker, B. Mader, C.H. Huang, P.C. Hu, R. Halter, and W. Bredt. 1989. Binding sites of attachment-inhibiting monoclonal antibodies and antibodies from patients on peptide fragments of the *Mycoplasma pneumoniae* adhesin. *Infect. Immun.* **57**:685–688.

Jacobs, E., R. Rock, and L. Dalehite. 1990. A B cell, T cell-linked epitope located on the adhesin of *Mycoplasma pneumoniae*. *Infect. Immun.* **58**:2464–2469.

Jacobson, S.H. 1986. P-fimbriated *Escherichia coli* in adults with renal scarring and pyelonephritis. *Acta Med. Scand. Suppl.* **73**:1–64.

Jacobson, S.H., L.E. Lins, S.B. Svenson, and G. Kallenius. 1985a. P-fimbriated *Escherichia coli* in adults with acute pyelonephritis. *J. Infect. Dis.* **152**:426–427.

Jacobson, S.H., L.E. Lins, S.B. Svenson, and G. Kallenius. 1985b. Lack of correlation of P blood group pheontype and renal scarring. *Kidney Int.* **28**:797–800.

Jacobson, S., A. Carstensen, G. Kallenius, and S. Svenson. 1986a. Fluorescence-activated cell analysis of P-fimbriae receptor accessibility on uroepithelial cells of patients with renal scarring. *Eur. J. Clin. Microbiol.* **5**:649–654.

Jacobson, S.H., G. Kallenius, L.-E. Lins, and S.B. Svenson. 1986b. P-fimbriae receptors in patients with chronic pyelonephritis. *J. Urol.* **139**:900–903.

Jacobson, S.H., G. Kallenius, L.E. Lins, and S.B. Svenson. 1987. Symptomatic recurrent urinary tract infections in patients with renal scarring in relation to fecal colonization with P-fimbriated *Escherichia coli*. *J. Urol.* **137**:693–696.

Jacobson, S.H., M. Hammerlind, K.J. Lidefeldt, E. Osterberg, K. Tullus, and K. Brauner. 1988a. Incidence of aerobactin positive *Escherichia coli* strains in patients with symptomatic urinary tract infection. *Eur. J. Clin. Microbiol. Infect. Dis.* **7**:630–634.

Jacobson, S.H., K. Tullus, and A. Brauner. 1989. Hydrophobic properties of *Escherichia coli* causing acute pyelonephritis. *J. Infect.* **19**:17–23.

Jacobson, S.H., K. Tullus, B. Wretlind, and A. Brauner. 1988b. Aerobactin-mediated uptake of iron by strains of *Escherichia coli* causing acute pyelonephritis and bacteremia. *Infect. Immun.* **16**:147–152.

Jacques, M., N. Parent, and B. Foiry. 1988. Adherence of *Bordetella bronchiseptica* and *Pasteurella multocida* to porcine nasal and tracheal epithelial cells. *J. Vet. Res.* **52**:283–285.

Jallat, C., A.M. Darfeuille, and B. Joly. 1991. Adhesive properties of *Haemophilus influenzae* to different human cells. *Pathol. Biol.* **39**:140–146.

Janke, B.H., D.H. Francis, J.E. Collins, M.C. Libal, D.H. Zeman, and D.D. Johnson. 1989. Attaching and effacing *Escherichia coli* infections in calves, pigs, lambs and dogs. *J. Vet. Diagn. Invest.* **1**:6–11.

Jann, K. and B. Jann (eds). 1990. *Current Topics in Microbiology and Immunology*: Springer-Verlag: Berlin, Vol. 151.

Janson, H., L.O. Heden, A. Grubb, M. Ruan, and A. Forsgren. 1991. Protein D, an immunoglobulin D-binding protein of *Haemophilus influenzae*: cloning, nucleotide sequence and expression in *Escherichia coli*. *Infect. Immun.* **59**:119–125.

Jayappa, H.J., R.A. Goodnow, and S.J. Geary. 1985. Role of *Escherichia coli* type 1 pilus in colonization of porcine ileum and its protective nature as a vaccine antigen in controlling Colibacillosis. *Infect. Immun.* **48**:350–354.

Jenkinson, H.F. and R.A. Easingwood. 1990. Insertional inactivation of the gene encoding a 76-kilodalton cell surface polypeptide in *Streptococcus gordonii* Challis has a pleiotropic effect on cell surface composition and properties. *Infect. Immun.* **58**:3689–3697.

Jenkinson, H.F., H.C. Lala, and M.G. Shepherd. 1990. Coaggregation of *Streptococcus sanguis* and other streptococci with *Candida albicans*. *Infect. Immun.* **58**:1429–1436.

Jerse A.E. and J.B. Kaper. 1991. The eae gene of enteropathogenic *Escherichia coli* encodes a 94-kilodalton membrane protein, the expression of which is influenced by the EAF plasmid. *Infect. Immun.* **59**:4302–4309.

Jerse A.E., J. Yu, B.D. Tall, and J.B. Kaper. 1990. A genetic locus of enteropathogenic *Escherichia coli* necessary for the production of attaching and effacing lesion on tissue culture cells. *Proc. Natl. Acad. Sci. USA* **87**:7839–7843.

Jerse A.E., K.G. Gicquelais, and J.B. Kaper. 1991. Plasmid and chromosomal elements involved in the pathogenesis of attaching and effacing *Escherichia coli*. *Infect. Immun.* **59**:3869–3875.

Johnson, J.R. 1988a. Asymptomatic bacteriuria in elderly women. *J. Infect. Dis.* **158**:493.

Johnson, J.R. 1988b. P-fimbriated *E. coli* urinary tract infection. *South. Med. J.* **81**:1070(Lett.)

Johnson, J.R. 1988c. Expression of P fimbriae in urine. *J. Infect. Dis.* **158**:495.

Johnson J.R. 1991. Virulence factors in *Escherichia coli* urinary tract infection. *Clin. Microbiol. Rev.* **4**:80–128.

Johnson, J.R., P.L. Roberts, and W.E. Stamm. 1987. P-fimbriae and other virulence factors in *Escherichia coli* urosepsis: association with patients' characteristics. *J. Infect. Dis.* **156**:225–229.

Johnson, J.R., S.L. Moseley, P.L. Roberts, and W.E. Stamm. 1988. Aerobactin and other virulence factor genes among strains of *Escherichia coli* causing urosepsis: association with patient characteristics. *Infect. Immun.* **56**:405–412.

Jones, G.W. 1977. The attachment of bacteria to the surface of animal cells. In: J.L. Reissig (ed). *Microbial Interactions. (Receptors and Recognition, Vol. 3)*. Chapman and Hall, New York, pp. 139–176.

Jones, G.W. and R.E. Isaacson. 1983. Proteinaceous bacterial adhesins and their receptors. *CRC Crit. Rev. Microbiol.* **10**:229–260.

Jonson, G., J. Sanchez, and A.-M. Svennerholm. 1989. Expression and detection of different biotype-associated cell-bound haemagglutinins of *Vibrio cholerae* 01. *J. Gen. Microbiol.* **135**:111–120.

Jonson, G., A.-M. Svennerholm and J. Holmgren. 1990. Expression of virulence factors by classical and El Tor *Vibrio cholerae in vivo* and *in vitro*. *FEMS Microbiol. Ecol.* **74**:221–228.

Jonson, G., J. Holmgren, and A.M. Svennerholm. 1991a. Epitope differences in toxin-coregulated pili produced by classical and El Tor *Vibrio cholerae* 01. *Microbi. Pathogen.* **11**:179–188.

Jonson, G., J. Holmgren, and A.M. Svennerholm. 1991b. Identification of a mannose-binding pilus on *Vibrio cholerae* El Tor. *Microb. Pathogen.* **11**:433–441.

Kaack, M.B., J.A. Roberts, G. Baskin, and G.M. Patterson. 1988. Maternal immunization with P fimbriae for the prevention of neonatal pyelonephritis. *Infect. Immun.* **56**:1–6.

Kabir, S., S. Ali, and Q. Akhtar. 1985. Ionic, hydrophobic, and hemagglutinating properties of *Shigella* species. *J. Infect. Dis.* **151**:194.

Kagermeier, A. and J. London. 1986. Identification and preliminary characterization of a lectin-like protein from *Capnocytophaga gingivalis* (emended). *Infect. Immun.* **51**:490–494.

Kahane, I., S. Tucker, and J.B. Baseman. 1985. Detection of *Mycoplasma pneumoniae* adhesin (P1) in the nonhemadsorbing population of virulent *Mycoplasma pneumoniae*. *Infect. Immun.* **49**:457–458.

Kallenius, T.G. and R. Mollby. 1988. Faecal colonization with P-fimbriated *Escherichia coli* between 0 and 18 months of age. *Epidem. Infect.* **100**:185–191.

Kallenius, G., S.H. Jacobson, K. Tullus, and S.B. Svenson. 1985. P fimbriae studies on the diagnosis and prevention of acute pyelonephritis. *Infection* **13**:159–162.

Kanamori, M., T. Katsura, N. Ishiyama, S. Ogata, and O. Kitamoto. 1987. Immune responses in *Mycoplasma pneumoniae* infection of infant mouse and man. *Isr. J. Med. Sci.* **23**:568–573.

Kanoe, M. and K. Iwaki. 1987. Adherence of *Fusobacterium necrophorum* to bovine ruminal cells. *J. Med. Microbiol.* **23**:69–73.

Kanoe, M., S. Nagi, and M. Toda. 1985. Adherence of *Fusobacterium necrophorum* to vero cells. *Zbl. Bakt. Mikrobiol. Hyg.* **260**:100–107.

Kaper, J.B., R.E. Black, M.L. Clements, M.M. Levine, J.P. Nataro, H. Karch, M.M. Baldini, and A.D. O'Brien. 1985. The diarrhoea response of humans to some classic serotypes of enteropathogenic *Escherichia coli* is dependent on a plasmid encoding an enteroadhesiveness factor. *J. Infect. Dis.* **152**:550–559.

Kapperud, G., E. Namork, and H.-J. Skarpeid. 1985. Temperature-inducible surface fibrillae associated with the virulence plasmid of *Yersinia enterocolitica* and *Yersinia pseudotuberculosis*. *Infect. Immun.* **47**:561–566.

Kapperud, G., E. Namork, M. Skurnik, and T. Nesbakken. 1987. Plasmid-mediated surface fibrillae of *Yersinia pseudotuberculosis* and *Yersinia enterocolitica*: relationship to the outer membrane protein YOP1 and possible importance for pathogenesis. *Infect. Immun.* **55**:2247–2254.

Kar, S., S.C. To, and C.C. Brinton, Jr. 1990. Cloning and expression in *Escherichia coli* of LKP pilus genes from a nontypable *Haemophilus influenzae* strain. *Infect. Immun.* **58**:903–908.

Karam-Sarkis, D., M. German-Fattal, and P. Bourlioux. 1991. Effect of fusafungine on adherence of *Haemophilus influenzae* type b to human epithelial cells in vitro. *Biomed. Pharmacother.* **45**:301–306.

Karasic, R.B., D.J. Beste, S.C. To, W.J. Doyle, S.W. Wood, M.J. Carter, A.C. To, K. Tanpowpong, C.D. Bluestone, and C.C. Brinton, Jr. 1989. Evaluation of pilus vaccines for prevention of experimental otitis media caused by nontypeable *Haemophilus influenzae*. *Pediatr. Infect. Dis. J.* **8**:S62–S65.

Karch H., J. Heesemann, R. Laufs, A.D. O'Brien, C.O. Tacket, and M.M. Levine. 1987a. A plasmid of enterohemorrhagic *Escherichia coli* 0157:H7 is required for expression of a new fimbrial antigen and for adhesion to epithelial cells. *Infect. Immun.* **55**:455–461.

Karch H., J. Heesemann, R. Laufs, H.P. Kroll, J.B. Kaper, and M.M. Levine. 1987b. Serological response to type 1-like somatic fimbriae in diarrheal infection due to classical enteropathogenic *Escherichia coli*. *Microb. Pathogen.* **2**:425–434.

Karlsson, K.A. 1989. Animal glycosphingolipids as membrane sites for bacteria. *Annu. Rev. Biochem.* **58**:309–350.

Karr, J.F., B. Nowicki, L.K. Troung, R.A. Hull, and S.I. Hull. 1989. Purified P fimbriae from two cloned gene clusters of a single pyelonephritogenic strain adhere to unique structures in the human kidney. *Infect. Immun.* **57**:3594–3600.

Karr J.F., B.J. Nowicki, L.D. Truong, R.A. Hull, J.J. Moulds, and S.I. Hull. 1990. Pap–2-encoded fimbriae adhere to the P-blood group-related glycosphingolipid stage-specific embryonic antigen 4 in the human kidney. *Infect. Immun.* **58**:4055–4062.

Kaufman, J. and J.M. DiRienzo. 1989. Isolation of a corncob (coaggregation) receptor polypeptide from *Fusobacterium nucleatum*. *Infect. Immun.* **57**:331–337.

Kaufman J.E., K. Anderson, and C.L. Parsons. 1987. Inactivation of antiadherence effect of bladder surface glycosaminoglycans as a possible mechanism for carcinogenesis. *Urology* **30**:255–258.

Kavoussi, L.R., E.J. Brown, J.K. Ritchey, and T.L. Ratliff. 1990. Fibronectin-mediated Calmette-Guerin bacillus attachment to murine bladder mucosa. Requirement for the expression of an antitumor response. *J. Clin. Invest.* **85**:62–67.

Kawaoka, Y., T. Mitani, K. Hoshina, K. Otsuki, and M. Tsubokura. 1988. Lack of correlation between haemagglutination and adherence to epithelial cells in *Yersinia pseudotuberculosis*. *J. Med. Microbiol.* **25**:175–181.

Kawula, T.H., E.L. Aho, D.S. Barritt, D.G. Klapper, and J.G. Cannon. 1988. Reversible phase variation of expression of *Neisseria meningitidis* class 5 outer membrane proteins and their relationship to gonococcal proteins II. *Infect. Immun.* **56**:380–386.

Keith, B.R., L. Maurer, P.A. Spears, and P.E. Orndorff. 1986. Receptor-binding function of type 1 pili effects bladder colonization by a clinical isolate of *Escherichia coli*. *Infect. Immun.* **53**:693–696.

Kerneis S., S.S. Bilge, V. Fourel, G. Chauviere, M.H. Coconnier, and A.L. Servin. 1991. Use of purified F1845 fimbrial adhesin to study localization and expression of receptors for diffusely adhering *Escherichia coli* during enterocytic differentiation of human colon carcinoma cell lines HT-29 and Caco-2 in culture. *Infect. Immun.* **59**:4013–4018.

Kimura, A., K.T. Mountzouros, D.A. Relman, S. Falkow, and J.L. Cowell. 1990. *Bordetella pertussis* filamentous hemagglutinin: evaluation as a protective antigen and colonization factor in a mouse respiratory infection model. *Infect. Immun.* **58**:7–16.

Kiselius, P.V., W.R. Schwan, S.K. Amundsen, J.L. Duncan, and A.J. Schaeffer. 1989. *In vivo* expression and variation of *Escherichia coli* type 1 and P pili in the urine of adults with acute urinary tract infections. *Infect. Immun.* **57**:1656–1662.

Klemm, P. 1985. Fimbrial adhesins of *Escherichia coli*. *Rev. Infect. Dis.* **7**:321–340.

Klotz, S.A., Y.K. Au, and R.P. Misra. 1989. A partial thickness epithelial defect increases the adherence of *Pseudomonas aeruginosa* to the cornea. *Invest. Ophthalmol. Vis. Sci.* **30**:1069–1074.

Klotz, S.A., R.P. Misra, and S.I. Butrus. 1990. Contact lens wear enhances adherence of *Pseudomonas aeruginosa* and binding of lectins to the cornea. *Cornea* **9**:266–270.

Knapp, S., J. Hacker, T. Jarchau, and W. Goebel. 1986. Large, unstable inserts in the chromosome affect virulence properties of uropathogenic *Escherichia coli* 0 6 strain 536. *J. Bacteriol.* **168**:22–30.

Knutton, S., D.R. Lloyd, D.C.A. Candy, and A.S. McNeish. 1985. Adhesion of enterotoxigenic *Escherichia coli* to human small intestinal enterocytes. *Infect. Immun.* **48**:824–831.

Knutton, S., D.R. Lloyd, and A.S. McNeish. 1987a. Adhesion of enteropathogenic *Escherichia coli* to human intestinal enterocytes and cultured human mucosa. *Infect. Immun.* **55**:69–77.

Knutton, S., D.R. Lloyd, and A.S. McNeish. 1987b. Identification of a new fimbrial structure in enterotoxigenic *Escherichia coli* (ETEC) serotype 0148:H28 which adheres to human intestinal mucosa: a potentially new human ETEC colonization factor. *Infect. Immun.* **55**:86–92.

Knutton, S., M.M. Baldini, J.B. Kaper, and A.S. McNeish. 1987c. Role of plasmid-encoded adherence factors in adhesion of enteropathogenic *Escherichia coli* to HEP-2 cells. *Infect. Immun.* **55**:78–85.

Knutton, S. T. Baldwin, P.H. Williams, and A.S. McNeish. 1989a. Actin accumulation at sites of bacterial adhesion to tissue culture cells: Basis of a new diagnostic test for enteropathogenic and enterohemorrhagic *Escherichia coli*. *Infect. Immun.* **57**:1290–1298.

Knutton, S., M.M. McConnell, B. Rowe, and A.S. McNeish. 1989b. Adhesion and ultrastructural properties of human enterotoxigenic *Escherichia coli* producing colonization factor antigens III and IV. *Infect. Immun.* **57**:3364–3371.

Knutton, S., A.D. Phillips, H.R. Smith, R.J. Gross, R. Shaw, P. Watson, and E. Price. 1991. TI screening for enteropathogenic *Escherichia coli* in infants with diarrhea by the fluorescent-actin staining test. *Infect. Immun.* **59**:365–371.

Ko, H.L., J. Beuth, J. Solter, H. Schroten, G. Uhlenbruck, and G. Pulverer. 1987. In vitro and in vivo inhibition of lectin mediated adhesion of *Pseudomonas aeruginosa* by receptor blocking carbohydrates. *Infection* **15**:237–240.

Kohbata, S., H. Yokoyama, and E. Yabuuchi. 1986. Cytopathogenic effect of *Salmonella typhi* GIFU 10007 on M cells of murine ileal Peyer's patches in ligated ileal loops: an ultrastructural study. *Microbiol. Immunol.* **30**:1225–1237.

Kolenbrander, P.E. 1989. Surface recognition among oral bacteria: multigeneric coaggregations and their mediators. *CRC Crit. Rev. Microbiol.* **17**:137–159.

Kolenbrander, P.E. and R.N. Andersen. 1989. Inhibition of *coaggregation* between *Fusobacterium nucleatum* and *Porphyromonas (Bacteroides) gingivalis* by lactose and related sugars. *Infect. Immun.* **57**:3204–3209.

Kolenbrander, P.E. and R.N. Andersen. 1990. Characterization of *Streptococcus gordonii* (S. sanguis) PK488 adhesin-mediated coaggregation with *Actinomyces naeslundii* PK606. *Infect. Immun.* **58**:3064–3072.

Kolenbrander, P.E., R.N. Andersen, and L.V.H. Moore. 1989. Coaggregation of *Fusobacterium nucleatum, Selenomonas flueggei, Selenomonas infelix, Selenomonas noxia* and *Selenomonas sputigena* with strains from 11 genera of oral bacteria. *Infect. Immun.* **57**:3194–3203.

Komiyama, K. 1987. Interbacterial adhesion between *Pseudomonas aeruginosa* and indigenous oral bacteria isolated from patients with cystic fibrosis. *Can. J. Microbiol.* **33**:27–32.

Komiyama, K., B.F. Habbick, and S.K. Tumber. 1987. Role of sialic acid in saliva-mediated aggregation of *Pseudomonas aeruginosa* isolated from cystic fibrosis patients. *Infect. Immun.* **55**:2364–2369.

Komiyama, K., B.F. Habbick, and S.K. Tumber. 1989. Whole, sumandibular and parotid saliva-mediated aggregation of *Pseudomonas aeruginosa* in cystis fibrosis patients. *Infect. Immun.* **57**:1299–1304.

Konig, B., W. Konig, J. Scheffer, J. Hacker, and W. Goebel. 1986. Role of *Escherichia coli* alphahemolysin and bacterial adherence in infection: requirement for release of inflammatory mediators from granulocytes and mast cells. *Infect. Immun.* **54**:886–892.

Konkel, M.E. and L.A. Joens. 1989. Adhesion to and invasion of HEp-2 cells by *Campylobacter spp. Infect. Immun.* **57**:2984–2990.

Konkel, M.E. and L.A. Joens. 1990. Effect of enteroviruses on adherence to and invasion of HEp-2 cells by *Campylobacter* isolates. *Infect. Immun.* **58**:1101–1105.

Konkel, M.E., F. Babakhani, and L.A. Joens. 1990. Invasion related antigens of *Campylobacter jejuni. J. Infect. Dis.* **162**:888–895.

Koop, H.M., M. Valentijn-Benz, A.V. Nieuw-Amerongen, P.A. Roukema, and J. deGraaff. 1990. Aggregation of oral bacteria by human salivary mucins in comparison to salivary and gastric mucins of animal origin. *Antonie van Leeuwenhoek* **58**:255–263.

Korhonen, T.K., M.V. Valtonen, J. Parkkinen, V. Vaisanen-Rhen, J. Finne, F. Orskov, S.B. Svenson, and P.H. Makela. 1985. Serotypes, hemolysin production, and receptor recognition of *Escherichia coli* strains associated with neonatal sepsis and meningitis. *Infect. Immun.* **48**:486–491.

Korhonen, T.K., R. Virkola, and H. Holthofer. 1986a. Localization of binding sites for purified *Escherichia coli* P fimbriae in the human kidney. *Infect. Immun.* **54**:328–332.

Korhonen, T.K., J. Parkkinen, J. Hacker, J. Finne, A. Pere, M. Rhen, and H. Holthofer. 1986b. Binding of *Escherichia coli* S fimbriae to human kidney epithelium. *Infect. Immun.* **54**:322–327.

Korhonen, T.K., R. Virkola, V. Vaisanen-Rhen, and H. Holthofer. 1986c. Binding of purified *Escherichia coli* 075X adhesin to frozen sections of human kidney. *FEMS Microbiol. Lett.* **35**:313–318.

Korhonen T.K., R. Virkola, B. Westerlund, H. Holthofer, and J. Parkkinen. 1990. Tissue

Korhonen, T.K., J. Parkkinen, J. Hacker, J. Finne, A. Pere, M. Rhen, and H. Holthofer. 1986b. Binding of *Escherichia coli* S fimbriae to human kidney epithelium. *Infect. Immun.* **54**:322–327.

Korhonen, T.K., R. Virkola, V. Vaisanen-Rhen, and H. Holthofer. 1986c. Binding of purified *Escherichia coli* 075X adhesin to frozen sections of human kidney. *FEMS Microbiol. Lett.* **35**:313–318.

Korhonen T.K., R. Virkola, B. Westerlund, H. Holthofer, and J. Parkkinen. 1990. Tissue tropism of *Escherichia coli* adhesins in human extraintestinal infections. *Curr. Top. Microbiol. Immunol.* **151**:115–127.

Korth, M.J., R.A. Schneider, and S.L. Moseley. 1991. An F41-K88-related genetic determinant of bovine septicemic *Escherichia coli* mediates expression of CS31A fimbriae and adherence to epithelial cells. *Infect. Immun.* **59**:2333–2340.

Kotani, H. and G.J. McGarrity. 1986. *Ureaplasma* infection of cell cultures. *Infect. Immun.* **52**:437–444.

Krause, D.C. and Y.Y. Chen. 1988. Interaction of *Mycoplasma pneumoniae* with Hela cells. *Infect. Immun.* **56**:2054–2059.

Krause, D.C. and K.K. Lee. 1991. Juxtaposition of the genes encoding *Mycoplasma pneumoniae* cytadherence-accessory proteins HMW1 and HMW3. *Gene* **107**:83–89.

Kristiansen, K., S.B. Baloda, J.L. Larsen, and T. Wadström. 1987. Toxins, putative cell adhesins and fibronectin binding properties of *Salmonella dublin*. *Acta Pathol. Microbiol. Immunol. Scand.* **95**:57–63.

Krivan, H., V. Ginsburg, and D.D. Roberts. 1988. *Pseudomonas aeruginosa* isolated from cystic fibrosis patients bind specifically to gangliotetraosylceramide (asialo GM1) and gangliotriaosylceramide (asialo GM2). *Arch. Biochem. Biophys.* **260**:493–496.

Krivan, H.C., L.D. Olson, M.F. Barile, V. Ginsburg, and D.D. Roberts. 1989. Adhesion of *Mycoplasma pneumoniae* to sulfated glycolipids and inhibition by dextran sulfate. *J. Biol. Chem.* **264**:9283–9288.

Krivan, H., B. Nilsson, C.A. Lingwood, and H. Ryu. 1991. *Chlamydia trachomatis* and *Chlamydia pneumoniae* bind specifically to phosphatidylethanolamine in Hela cells and to GalNAcβ1–4Galβ1–4Glc sequences found in asialo-GM_1 and asialo -GM_2. Biochem. *Biophys. Res. Commun.* **175**:1082–1089.

Krogfelt, K.A. 1991. Bacterial adhesion: genetics, biogenesis, and role in pathogenesis of fimbrial adhesins of *Escherichia coli*. *Rev. Infect. Dis.* **13**:721–725.

Krogfelt, K.A., B.A. McCormick, R.L. Burghoff, D.C. Laux, and P.S. Cohen. 1991. Expression of *Escherichia coli* F–18 type 1 fimbriae in the streptomycin-treated mouse large intestine. *Infect. Immun.* **59**:1567–1568.

Kroncke, K.D., I Orskov, F. Orskov, B. Jann, and K. Jann. 1990. Electron microscopic study of coexpression of adhesive protein capsules and polysaccharide capsules in *Escherichia coli*. *Infect. Immun.* **58**:2710–2714.

Kubesch, P., M. Lingner, D. Grothues, M. Wehsling, and B. Tummler. 1988. Strategies of *Pseudomonas aeruginosa* to colonize and to persist in the cystic fibrosis lung. *Scand. J. Gastroenterol.* **143**:S77–S80.

Labigne-Roussel, A. and S. Falkow. 1988. Distribution and degree of heterogeneity of the afimbrial-adhesin-encoding operon (*afa*) among uropathogenic *Escherichia coli* isolates. *Infect. Immun.* **56**:640–648.

Labigne-Roussel, A., M.A. Schmidt, W. Walz, and S. Falkow. 1985. Genetic organization of the afimbrial adhesin operon and nucleotide sequence from a uropathogenic *Escherichia coli* gene encoding an afimbrial adhesin. *J. Bacteriol.* **162**:1285–1292.

Lamont, R.J. and B. Rosan. 1990. Adherence of mutans streptococci to other oral bacteria. *Infect. Immun.* **58**:1738–1743.

Lantz, M.S., R.W. Roland, L.M. Switalski, and M. Hook. 1986. Interactions of *Bacteroides gingivalis* with fibrinogen. *Infect. Immun.* **54**:654–658.

Lata, B.S. and B.R. Panhotra. 1987. Adherence of *Klebsiella* to human buccal epithelial cells. *Indian J. Med. Res.* **86**:433–436.

Laux, D.C., E.F. McSweegan, T.J. Williams, E.A. Wadolkowski, and P.S. Cohen. 1986. Identification and characterization of mouse small intestine mucosal receptors for *Escherichia coli* K–12 (K88ab). *Infect. Immun.* **52**:18–25.

Lee, A., J.L. O'Rourke, P.J. Barrington, and T.J. Trust. 1986. Mucus colonization as a determinant of pathogenicity in intestinal infection by *Campylobacter jejuni*: a mouse cecal model. *Infect. Immun.* **51**:536–546.

Lee, K.K., P. Doig, R.T. Irvin, W. Paranchych, and R.S. Hodges. 1989. Mapping the surface regions of *Pseudomonas aeruginosa* PAK pilin: the importance of the C-terminal region for adherence to human buccal epithelial cells. *Mol. Microbiol.* **3**:1493–1499.

Lee, J.Y., H.T. Sojar, G.S. Bedi, and R.J. Genco. 1991. *Porphyromonas (Bacteroides) gingivalis* fimbrillin: size, amino-terminal sequence, and antigenic heterogeneity. *Infect. Immun.* **59**:383–389.

Leininger, E., M. Roberts, J.G. Kenimer, I.G. Charles, N. Fairweather, P. Novotny, and M.J. Brennan. 1991. Pertactin, an Arg-Gly-Asp-containing *Bordetella pertussis* surface protein that promotes adherence of mammalian cells. *Proc. Natl. Acad. Sci.* **88**:345–349.

Leong, J.M., R.S. Fournier, and R.R. Isberg. 1990. Identification of the integrin binding *Yersinia pseudotuberculosis* invasin protein. *EMBO J.* **9**:1979–1989.

Leung, K.P. H. Fukushima, H. Sagawa, C.B. Walker, and W.B. Clark. 1989. Surface appendages, hemagglutination, and adherence to human epithelial cells of *Bacteroides intermedius. Oral Microbiol. Immunol.* **4**:204–210.

Leung, K.P., W.E. Nesbitt, W. Fischlschweiger, D.I. Hay, and W.B. Clark. 1990. Binding of colloidal gold-labeled salivary proline-rich proteins to *Actinomyces viscosus* type 1 fimbriae. *Infect. Immun.* **58**:1986–1991.

Leusch, H.G., S.A. Hefta, Z. Drzeniek, K. Hummel, Z. Markos-Pusztai, and C. Wagener. 1990. *Escherichia coli* of human origin binds to carcinoembryonic antigen (CEA) and non-specific crossreacting antigen (NCA). *FEBS Lett.* **261**:405–409.

Leusch, H.G., Z. Drzeniek, Z. Markos-Pusztai, and C. Wagener. 1991. Binding of *Escherichia coli* and *Salmonella* strains to members of the carcinoembryonic antigen family: differential binding inhibition by aromatic α-glycosides of mannose. *Infect. Immun.* **59**:2051–2057.

Levett, P.N. and R.R. Daniel. 1988. Attachment of a *Vibrio parahaemolyticus* strain to rabbit brush border membranes. Zbl. *Bkt. Mikrobiol. Hyg.* **268**:248–250.

Levin, B.R. and C. Svanborg-Eden. 1990. Selection and evolution of virulence in bacteria: an ecumenical excursion and modest suggestion. *Parasitology* **100**:S103–S115.

Levine, M.M. 1987. *Escherichia coli* that cause diarrhea: enterotoxigenic, enteropathogenic, enteroinvasive, enterohemorrhagic and enteroadherent. *J. Infect. Dis.* **155**:377–389.

Levine, M.M., V. Prado, R.R. Browne, H. Lior, J.B. Kaper, S.L. Moseley, K. Gicquelais, J.P. Nataro, P. Vial, and B. Tall. 1988. Use of DNA probe and HEp-2 cell adherence assay to detect diarrheagenic *Escherichia coli*. *J. Infect. Dis.* **158**:224–228.

Li, J. and R.P. Ellen. 1989. Relative adherence of *Bacteroides* species and strains to *Actinomyces viscosus* on saliva-coated hydroxyapatite. *J. Dent. Res.* **68**:1308–1312.

Li, J., R.P. Ellen, C.I. Hoover, and J.R. Felton. 1991. Association of proteases of *Porphyromonas (Bacteroides) gingivalis* with its adhesion to *Actinomyces viscosus*. *J. Dent. Res.* **70**:82–86.

Licois, D., A. Reynaud, M. Federighi, B. Gaillard-Martinie, J.F. Guillot, and B. Joly. 1991. Scanning and transmission electron microscopic study of adherence of *Escherichia coli* 0103 enteropathogenic and/or enterohemorrhagic strain GV in enteric infection in rabbits. *Infect. Immun.* **59**:3796–3800.

Lidefelt, K.J., I. Bollgren, G. Kallenius, and S.B. Svenson. 1987. P-fimbriated *Escherichia coli* in children with acute cystitis. *Acta Paediatr. Scand.* **76**:775–780.

Liljemark, W.F., L.J. Fenner, and C.G. Bloomquist. 1986. *In vivo* colonization of salivary pellicle by *Haemophilus, Actinomyces*, and *Streptococcus* species. *Caries Res.* **20**:481–497.

Liljemark, W.F., C.G. Bloomquist, M.C. Coulter, L.J. Fenner, R.J. Skopek, and C.F. Schachtele. 1988. Utilization of a continuous streptococcal surface to measure interbacterial adherence *In vitro* and *in vivo*. *J. Dent. Res.* **67**:1455–1460.

Lillard, H.S. 1986. Role of fimbriae and flagella in the attachment of *Salmonella typhimurium* to poultry skin. *J. Food Sci.* **51**:54–56.

Lindblom, G.B., L.E. Cervantes, E. Sjogren, B. Kaijser, and G.M. Ruiz-Palacios. 1990. Adherence, enterotoxigenicity, invasiveness and serogroups in *Campylobacter jejuni* and *Campylobacter coli* strains from adult humans with acute enterocolitis. Acta Path. *Microbiol. Immunol. Scand.* **98**:179–184.

Linder, H., I. Engberg, I. Mattsby-Baltzer, K. Jann, and C. Svanborg-Eden. 1988. Induction of inflammation by *Escherichia coli* on the mucosal level: requirement for adherence and endotoxin. *Infect. Immun.* **56**:1309–1313.

Linder, H., I Engberg, H. Hoschützky, I. Mattsby-Baltzer, and C. Svanborg. 1991. Adhesion-dependent activation of mucosal interleukin-6 production. *Infect. Immun.* **59**:4357–4362.

Lindquist, B. L., E. Lebenthal, P.C. Lee, M.W. Stinson, and J.M. Merrick. 1987. Adherence of *Salmonella typhimurium* to small-intestinal enterocytes of the rat. *Infect. Immun.* **55**:3044–3050.

Lindstedt, R., N. Baker, P. Falk, R. Hull, S. Hull, J. Karr, H. Leffler, C. Svanborg Eden, and G. Larson. 1989. Binding specificies of wild-type and cloned *Escherichia coli* strains that recognize globo-A. *Infect. Immun.* **57**:3389–3394.

Lindstedt, R., G. Larson, P. Falk, U. Jodal, H. Leffler, and C. Svanborg. 1991. The receptor repertoire defines the host range for attaching *Escherichia coli* strains that recognize globo-A. *Infect. Immun.* **59**:1086–1092.

Lingwood, C.A., H. Law, A. Pellizzari, P. Sherman, and B. Drumm. 1989. Gastric glycerolipid as a receptor for *Campylobacter pylori*. *Lancet* **ii**:238–241.

Lintermans P., P. Pohl, F. Deboeck, A. Bertels, C. Schlicker, J. Vandekerckhove, J. Van Damme, M. Van Montagu, and H. De Greve. 1988. Isolation and nucleotide sequence of the F17-A gene encoding the structural protein of the F–17 fimbriae in bovine enterotoxigenic *Escherichia coli*. *Infect. Immun.* **56**:1475–1484.

LiPuma, J.J. and J.R. Gilsdorf. 1987. Role of capsule in adherence of *Haemophilus influenzae* type b to human buccal epithelial cells. *Infect. Immun.* **55**:2308–2310.

LiPuma, J.J. and J.R. Gilsdorf. 1988. Structural and serological relatedness of *Haemophilus influenzae* type b pili. *Infect. Immun.* **56**:1051–1056.

Liu, T., R.J. Gibbons, D.I. Hay, and Z. Skobe. 1991. Binding of *Actinomyces viscosus* to collagen: association with the type 1 fimbrial adhesin. *Oral Microbiol. Immunol.* **6**:1–5.

Ljungh, A. and T. Wadström. 1988. Subepithelial connective tissue protein binding of *Escherichia coli* isolated from patients with ulcerative colitis. In: Macdermoth, R.P. (ed.), *Inflammatory Bowel Diseases: Current Status and Future Approach*. Elsevier Science Publishers, Amsterdam, pp. 571–575.

Ljungh, A., L. Emody, P. Aleljung, O. Olusanya, and T. Wadström. 1991. Growth conditions for the expression of fibronectin, collagen type I, vitronectin, and laminin binding to *Escherichia coli* strain NG7C. *Curr. Microbiol.* **22**:97–102.

Loeb, M.R., E. Connor, and D. Penney. 1988. A comparison of the adherence of fimbriated and nonfimbriated *Haemophilus influenzae* type b to human adenoids in organ culture. *Infect. Immun.* **56**:484–489.

Lomberg, H. and C. Svanborg-Eden. 1989. *Escherichia coli* virulence and renal scarring. *J. Infect. Dis.* **160**:1082(Lett.)

Lomberg, H., M. Hellstrom, U. Jodal, and C. Svanborg-Eden. 1986a. Renal scarring and non-attaching *Escherichia coli*. *Lancet* **ii**:1341(Lett.)

Lomberg, H., B. Cedergren, H. Leffler, B. Nilsson, A.S. Carlstrom, and C. Svanborg-Eden. 1986b. Influence of blood group on the availability of receptors for attachment of uropathogenic *Escherichia coli*. *Infect. Immun.* **51**:919–926.

Lomberg, H., M. Hellstrom, U. Jodal, I. Orskov, and C. Svanborg-Eden. 1989. Properties of *Escherichia coli* in patients with renal scarring. *J. Infect. Dis.* **159**:579–582.

London, J. 1991. Bacterial adhesins. *Annu. Rep. Med. Chem.* 26:239–264.

London, J. and J. Allen. 1990. Purification and characterization of a *Bacteroides loescheii* adhesin that interacts with procaryotic and eucaryotic cells. *J. Bacteriol.* **172**:2527–2534.

London, J., A.R. Hand, E.I. Weiss, and J. Allen. 1989. *Bacteroides loescheii* PK1295 cells express two distinct adhesins simultaneously. *Infect. Immun.* **57**:3940–3944.

Loomes, L.M., K.I. Uemura, and T. Feizi. 1985. Interaction of *Mycoplasma pneumoniae* with erythrocyte glycolipids of I and i antigen types. *Infect. Immun.* **47**:15–20.

Lopez-Vidal, Y. and A.M. Svennerholm. 1990. Monoclonal antibodies against the different subcomponents of colonization factor antigen II of enterotoxigenic *Escherichia coli*. *J. Clin. Microbiol.* **28**:1906–1912.

Lopez-Vidal Y., P. Klemm, and A.M. Svennerholm. 1988. Monoclonal antibodies against different epitopes on colonization factor antigen I of enterotoxin-producing *Escherichia coli*. *J. Clin. Microbiol.* **26**:1967–1972.

Low, D.A., B.A. Braaten, G.V. Ling, D.L. Johnson, and A.L. Ruby. 1988. Isolation and comparison of *Escherichia coli* strains from canine and human patients with urinary tract infections. *Infect. Immun.* **56**:2601–2609.

Lund, B., F. Lindberg, B.I. Marklund, and S. Normark. 1988a. Tip proteins of pili associated with pyelonephritis: new candidates for vaccine development. *Vaccine* **6**:110–112.

Lund, B., B.I. Marklund, N. Stromberg, F. Lindberg, K.A. Karlsson, and S. Normark. 1988b. Uropathogenic *Escherichia coli* can express serologically identical pili of different receptor binding specificities. *Mol. Microbiol.* **2**:255–263.

Lyerla, T.A., S.K. Gross, and R.H. McCluer. 1986. Glycosphingolipid patterns in primary mouse kidney cultures. *J. Cell Physiol.* **129**:390–394.

Mack, D.R. and P.M. Sherman. 1991. Mucin isolated from rabbit colon inhibits *in vitro* binding of *Escherichia coli* RDEC-1. *Infect. Immun.* **59**:1015–1023.

Madison, B., E.H. Beachey, and S.N. Abraham. 1989. Protection against *E. coli* induced urinary tract infection (UTI) with antibodies directed at the Fim H protein of type 1 fimbriae. Abstracts, Annual Meeting, American Society for Microbiology, p. 38.

Mafu, A.A., D. Roy, J. Goulet, and L. Savoie. 1991. Characterization of physicochemical forces involved in adhesion of *Listeria monocytogenes* to surfaces. *Appl. Environ. Microbiol.* **57**:1969–1973.

Mainil, J.G., P.L. Sadowski, M. Tarsio, and H.W. Moon. 1987. In vivo emergence of enterotoxigenic *Escherichia coli* variants lacking genes for K99 fimbriae and heat-stable enterotoxin. *Infect. Immun.* **55**:3111–3116.

Makino, S., J.P. van-Putten, and T.F. Meyer. 1991. Phase variation of the opacity outer membrane protein controls invasion by *Neisseria gonorrhoeae* into human epithelial cells. *EMBO J.* **10**:1307–1315.

Manning, P.A., G.D. Higgins, R. Lumb, and J.A. Lanser. 1987. Colonization factor antigens and a new fimbrial type, CFA/V, on 0115:H40 and H- strains of enterotoxigenic *Escherichia coli* in central Australia. *J. Infect. Dis.* **156**:841–844.

Mantle, M., L. Basaraba, S.C. Peacock, and D.G. Gall. 1989. Binding of *Yersinia enterocolitica* to rabbit intestinal brush border membranes, mucus, and mucin. *Infect. Immun.* **57**:3292–3299.

Marcus, H. and N.R. Baker. 1985. Characterization of adherence of *Pseudomonas aeruginosa* to hamster tracheal epithelium. *Infect. Immun.* **47**:723–729.

Marild, S., B. Wettergren, M. Hellstrom, U. Jodal, K. Lincoln, I. Orskov, F. Orskov, and C. Svanborg-Eden. 1988. Bacterial virulence and inflammatory response in infants with febrile urinary tract infection or screening bacteriuria. *J. Pediatr.* **112**:348–354.

Marre, R. and J. Hacker. 1987. Role of S- and common-type-fimbriae of *Escherichia coli* in experimental upper and lower urinary tract infection. *Microb. Pathogen.* **2**:223–226.

Marre, R., B. Kreft, and J. Hacker. 1990. Genetically engineered S and F1C fimbriae differ in their contribution to adherence of *Escherichia coli* to cultured renal tubular cells. *Infect. Immun.* **58**:3434–3437.

Maslow, A.S., C.H. Davis, J. Choong, and P.B. Wyrick. 1988. Estrogen enhances attachment of *Chlamydia trachomatis* to human endometrial epithelial cells *in vitro*. *Am J. Obstet. Gynecol.* **159**:1006–1014.

Mason, E.O., Jr., S.L. Kaplan, B.L. Wiedermann, E.P. Norrod, and W.A. Stenback. 1985. Frequency and properties of naturally occurring adherent piliated strains of *Haemophilus influenzae* type b. *Infect. Immun.* **49**:98–103.

Mathewson, J.J. and A. Cravioto. 1989. HEp-2 cell adherence as an assay for virulence among diarrheagenic *Escherichia coli*. *J. Infect. Dis.* **159**:1057–1060.

Mathewson, J.J., P.C. Johnson, H.L. Dupont, D.R. Morgan, S.A. Thornton, L.V. Wood, and C.D. Ericsson. 1985. A newly recognized cause of travelers' diarrhea: enteroadherent *Escherichia coli*. *J. Infect. Dis.* **151**:471–475.

Mathewson, J.J., R.A. Oberhelman, H.L. Dupont, F.J. De la Cabada, and E.V. Garibay. 1987. Enteroadherent *Escherichia coli* as a cause of diarrhea among children in Mexico. *J. Clin. Microbiol.* **25**:1917–1919.

Maurelli, A.T., B. Baudry, H. d'Hauteville, T.L. Hale, and P.J. Sansonetti. 1985. Cloning of plasmid DNA sequences involved in invasion of HeLa cells by *Shigella flexneri*. *Infect. Immun.* **49**:164–171.

Maxon, E.R. 1990. *Haemophilus influenzae*, In: Mandrell, G.L., R.G. Douglas, Jr.., and J.E. Benne (eds.), *Principles and Practice of Infectious Diseases*, 3rd ed. Churchill Livingstone, New York, pp. 1722–1729.

Mbaki, N. Rikitomi, T. Nagatake and K. Matsumoto. 1987. Correlation between *Branhamella catarrhalis* adherence to oropharyngeal cells and seasonal incidence of lower respiratory tract infections. *Tohoku J. Exp. Med.* **153**:111–121.

McClain, M., I.C. Blomfield, and B.I. Eisenstein. 1991. Roles of *fim*B and *fim*E in site-specific DNA inversion associated with phase variation of type 1 fimbriae in *Escherichia coli*. *J. Bacteriol.* **173**:5308–5314.

McConnell, M.M., L.V. Thomas, N.P. Day, and R. Rowe. 1985. Enzyme-linked immunosorbent assays for the detection of adhesion factor antigens of enterotoxigenic *Escherichia coli*. *J. Infect. Dis.* **152**:1120–1127.

McConnell, M.M., L.V. Thomas, G.A. Willshaw, H.R. Smith, and B. Rowe. 1988. Genetic control and properties of coli surface antigens of colonization factor antigen IV (PCF8775) of enterotoxigenic *Escherichia coli*. *Infect. Immun.* **56**:1974–1980.

McConnell, M.M., M. Hibberd, A.M. Field, H. Chart, and B. Rowe. 1990. Characterization of a new putative colonization factor (CS17) from a human enterotoxigenic *Esche-*

richia coli of serotype 0114:H21 which produces only heat-labile enterotoxin. *J. Infect. Dis.* **161**:343–347.

McCormick, B.A., B.A.D. Stocker, D.C. Laux, and P.S. Cohen 1988. Roles of motility, chemotaxis, and penetration through and growth in intestinal mucus in the ability of an avirulent strain of *Salmonella typhimurium* to colonize the large intestine of streptomycin-treated mice. *Infect. Immun.* **56**:2209–2217.

McCormick, B.A., D.P. Franklin, D.C. Laux, and P.S. Cohen. 1989. Type 1 pili are not necessary for colonization of the streptomycin-treated mouse large intestine by type 1-piliated *Escherichia coli* F–18 and E. coli K–12. *Infect. Immun.* **57**:3022–3029.

McEachran, D.W., and R.T. Irvin. 1985. Adhesion of *Pseudomonas aeruginosa* to human buccal epithelial cells: evidence for two classes of receptors. *Can. J. Microbiol.* **31**:563–569.

McHan, F. N.A. Cox, J.S. Bailey, L.C. Blankenship, and N.J. Stern. 1988. The influence of physical and environmental variables on the in vitro attachment of *Salmonella typhimurium* to the ceca of chickens. *Avian Dis.* **32**:215–219.

McIntire, F.C., L.K. Crosby, A.E. Vatter, J.O. Cisar, M.R. McNeil, C.A. Bush, S.S. Tjoa, and P.V. Fennessey. 1988. A polysaccharide from *Streptococcus sanguis* 34 that inhibits coaggregation of *S. sanguis* 34 with *Actinomyces viscosus* T14V. *J. Bacteriol.* **170**:2229–2235.

McQueen, C.E., H. Shoham, and E.C. Boedeker. 1987. Intragastric inoculation with an *E. coli* pilus attachment factor (AF/R1) protects against subsequent colonization by an enteropathogen (*E. coli* strain RDEC-1). *Adv. Exp. Med. Biol.* **216B**:911–918.

McSweegan, E., and R.I. Walker. 1986. Identification and characterization of two *Campylobacter jejuni* adhesins for cellular and mucous substrates. *Infect. Immun.* **53**:141–148.

McSweegan, E., D.H. Burr, and R.I. Walker. 1987. Intestinal mucus gel and secretory antibody are barriers to *Campylobacter jejuni* adherence to INT 407 cells. *Infect. Immun.* **55**:1431–1435.

Menozzi, F.D., C. Gantiez, and C. Locht. 1991. Interaction of the *Bordetella pertussis* filamentous hemagglutinin with heparin. *FEMS Microbiol. Lett.* **78**:59–64.

Mergenhagen, S.E., A.L. Sandberg, B.M. Chassy, M.J. Brennan, M.K. Yeung, J.A. Donkersloot, and J.O. Cisar. 1987. Molecular basis of bacterial adhesion in the oral cavity. *Rev. Infect. Dis.* **S5**:S467–S474.

Mertz, P.M., J.M. Patti, J.J. Marcin, and D.A. Marshall. 1987. Model for studying bacterial adherence to skin wounds. *J. Clin. Microbiol.* **25**:1601–1604.

Metcalfe, J.W., K.A. Krogfelt, H.C. Krivan, P.S. Cohen, and D.C. Laux. 1991. Characterization and identification of a porcine small intestine mucus receptor for the K88ab fimbrial adhesin. *Infect. Immun.* **59**:91–96.

Meyer, T.F. 1990. Variation of pilin and opacity-associated protein in pathogenic *Neisseria* species. In: Iglewski, B.H. and V.L. Clark (eds.), *Molecular Basis of Bacterial Pathogenesis.* Academic Press, New York, pp. 137–155.

Meyer, T.F. and J.P.M. van Putten. 1989. Genetic mechanisms and biological implications of phase variation in pathogenic neisseriae. *Clin. Microbiol. Rev.* **2**:S139–S145.

Meyer, T.F., C.P. Gibbs, and R. Haas. 1990. Variation and control of protein expression in *Neisseria*. *Annu. Rev. Microbiol.* **44**:451–477.

Miki, Y., S. Ebisu, and H. Okada. 1987. The adherence of *Eikenella corrodens* to guinea pig macrophages in the absence and presence of anti-bacterial antibodies. *J. Periodont. Res.* **22**:359–365.

Miliotis, M.D., H.J. Koornhof, and J.I. Phillips. 1989. Invasive potential of noncytotoxic enteropathogenic *Escherichia coli* in an *in vitro* Henle 407 cell model. *Infect. Immun.* **57**:1928–1935.

Miller, M.J. and D.G. Ahearn. 1987. Adherence of *Pseudomonas aeruginosa* to hydrophilic contact lenses and other substrata. *J. Clin. Microbiol.* **25**:1392–1397.

Miller, M.J., L.A. Wilson, and D.G. Ahearn. 1988a. Effects of protein, mucin and human tears on adherence of *Pseudomonas aeruginosa* to hydrophilic contact lenses. *J. Clin. Microbiol.* **26**:513–517.

Miller, V.L. and S. Falkow. 1988. Evidence for two genetic loci in *Yersinia enterocolitica* that can promote invasion of epithelial cells. *Infect. Immun.* **56**:1242–1248.

Miller, V.L., R.K. Taylor, and J.J. Mekalanos. 1987. Cholera toxin transcriptional activator toxR is a transmembrane DNA binding protein. *Cell* **48**:271–279.

Miller, V.L., B.B. Finlay, and S. Falkow. 1988b. Factors essential for the penetration of mammalian cells by *Yersinia*. *Curr. Top Microbiol. Immunol.* **138**:15–39.

Miller, V.L., J.J. Farmer III, W.E. Hill, and S. Falkow. 1989. The *ail* locus is found uniquely in *Yersinia enterocolitica* serotypes commonly associated with disease. *Infect. Immun.* **57**:121–131.

Miller, V.L., J.B. Bliska, and S. Falkow. 1990. Nucleotide sequence of the *Yersinia enterocolitica ail* gene and characterization of the *ail* protein product. *J. Bacteriol.* **172**:1062–1069.

Miller, M.J., L.A. Wilson, and D.G. Ahearn. 1991. Adherence of *Pseudomonas aeruginosa* to rigid gas-permeable contact lenses. *Arch. Ophthalmol.* **109**:1447–1448.

Mirelman, D. (ed.) 1986. *Microbial Lectins and Agglutinins: Properties and Biological Activity*; John Wiley and Son, New York.

Milon A., J. Esslinger, and R. Camguilhem. 1990. Adhesion of *Escherichia coli* strains isolated from diarrheic weaned rabbits to intestinal villi and HeLa cells. *Infect. Immun.* **58**:2690–2695.

Mobley, H.L.T. and G.R. Chippendale. 1990. Hemagglutinin, urease, and hemolysin production by *Proteus mirabilis* from clinical sources. *J. Infect. Dis.* **161**:525–530.

Mobley, H.L.T., G.R. Chippendale, J.H. Tenney, and J.W. Warren. 1986. Adherence to uroepithelial cells of *Providencia stuartii* isolated from the catheterized urinary tract. *J. Gen. Microbiol.* **132**:2863–2872.

Mobley H.L.T., G.R. Chippendale, J.H. Tenney, R.A. Hull, and J.W. Warren. 1987. Expression of type 1 fimbriae may be required for persistence of *Escherichia coli* in the catheterized urinary tract. *J. Clin. Microbiol.* **25**:2253–2257.

Mobley, H.L.T., G.R. Chippendale, J.H. Tenney, A.R. Mayrer, L.J. Crisp, J.L. Penner, and J.W. Warren. 1988. MR/K hemagglutination of *Providencia stuartii* correlates with adherence to catheters and with persistence in catheter-associated bacteriuria. *J. Infect. Dis.* **157**:264–271.

Mooi, F.R. and F.K. DeGraaf. 1985. Molecular biology of fimbriae of enterotoxigenic *Escherichia coli*. *Curr. Top. Microbiol. Immunol.* **118**:119–138.

Moon, H.W. 1990. Colonization factor antigens of enterotoxigenic *Escherichia coli* in animals. *Curr. Top. Microbiol. Immunol.* **151**:147–165.

Moon, H.W., D.G. Rogers, and R. Rose. 1988. Effect of a live oral pilus vaccine on duration of lacteal immunity to enterotoxigenic *Escherichia coli* in swine. *Am. J. Vet. Res.* **49**:2068–2077.

Morris, J.A., W.J. Sojka, and R.A. Ready. 1985. Serological comparison of the *Escherichia coli* prototype strains for the F(Y) and AH25 adhesins implicated in neonatal diarrhoea in calves. *Res. Vet. Sci.* **38**:246–247.

Morrison-Plummer, J., D.K. Leith, and J.B. Baseman. 1986. Biological effects of anti-lipid and anti-protein monoclonal antibodies on *Mycoplasma pneumoniae*. *Infect. Immun.* **53**:398–403.

Morrison-Plummer, J., A. Lazzell, and J.B. Baseman. 1987. Shared epitopes between *Mycoplasma pneumoniae* major adhesin protein P1 and a 140-kilodalton protein of *Mycoplasma genitalium*. *Infect. Immun.* **55**:49–56.

Moser, I., I. Orskov, J. Hacker, and K. Jann. 1986. Characterization of a monoclonal antibody against the fimbrial F8 antigen of *Escherichia coli*. *FEMS Microbiol. Lett.* **34**:329–334.

Moser, I. and E. Hellmann. 1989. *In vitro* binding of *C. jejuni* surface proteins to murine small intestine cell membranes. *Med. Microbiol. Immunol.* **178**:217–228.

Mouricout, M. and R. Julien. 1986. Inhibition of mannose-resistant haemagglutination of sheep erythrocytes by enterotoxigenic *Escherichia coli* in the presence of plasma glycoprotein glycans. *FEMS Microbiol. Lett.* **37**:145–149.

Mouricout, M.A. and R.A. Julien. 1987. Pilus-mediated binding of bovine enterotoxigenic *Escherichia coli* to calf small intestinal mucins. *Infect. Immun.* **55**:1216–1223.

Mouricout, M., J.M. Petit, J.R. Carias, and R. Julien. 1990. Glycoprotein glycans that inhibit adhesion of *Escherichia coli* mediated by K99 fimbriae: treatment of experimental colibacillosis. *Infect. Immun.* **58**:98–106.

Moustafa Abd El Monem, A., M. Saad, B.L. Lindquist, P.C. Lee, R. Abud, J.M. Merrick, and E. Lebenthal. 1988. The effect of postnatal development on the adherence of nonfimbriated and fimbriated *Salmonella typhimurium* to isolated small intestinal enterocytes. *Pediatr. Res.* **24**:508–511.

Mouton, C., D. Bouchard, M. Deslauriers, and L. Lamonde. 1989. Immunochemical identification and preliminary characterization of nonfimbrial hemagglutinating adhesin of *Bacteroides gingivalis*. *Infect. Immun.* **57**:566–573.

Mouton, C., D. Ni-Eidhin, M. Deslauriers, and L. Lamy. 1991. The hemagglutinating adhesin HA-Ag2 of *Bacteroides gingivalis* is distinct from fimbrillin. *Oral Microbiol. Immunol.* **6**:6–11.

Muller, K.H., S.K. Collinson, T.J. Trust, and W.W. Kay. 1991. Type 1 fimbriae of *Salmonella enteritidis*. *J. Bacteriol.* **173**:4765–4772.

Mundi, H., B. Bjorksten, C. Svanborg, L. Ohman, and C. Dahlgren. 1991. Extracellular release of reactive oxygen species from human neutrophils upon interaction with *Escherichia coli* strains causing renal scarring. *Infect. Immun.* **59**:4168–4172.

Murakami, Y., T. Takeshita, S. Shizukuishi, A. Tsunemitus, and S. Aimoto. 1990. Inhibitory effects of synthetic histidine-rich peptides on haemagglutination by *Bacteroides gingivalis* 381. *Arch. Oral Biol.* **35**:775–777.

Murakami, Y., H. Nagata, A. Amano, M. Takagaki, S. Shizukuishi, A. Tsunemitus, and S. Aimoto. 1991a. Inhibitory effects on human salivary histatins and lysozyme on coaggregation between *Porphyromonas gingivalis* and *Streptococcus mitis*. *Infect. Immun.* **59**:3284–3286.

Murakami, Y., S. Shizukuishi, A. Tsunemitus, Y. Kato, and S. Aimoto. 1991b. Binding of a histidine-rich peptide to *Porphyromonas gingivalis*. *FEMS Microbiol. Lett.* **82**:253–256.

Murphy, G.L. and J.G. Cannon. 1988. Genetics of surface protein variation in *Neisseria gonorrhoeae*. *Bioessays* **9**:7–11.

Murray, P.A., M.J. Levine, M.S. Reddy, L.A. Tabak, and E.J. Bergey. 1986. Preparation of a sialic acid-binding protein from *Streptococcus mitis* KS32AR. *Infect. Immun.* **53**:359–365.

Murray, P.A., D.G. Kern, and J.R. Winkler. 1988. Identification of a galactose-binding lectin on *Fusobacterium nucleatum* FN–2. *Infect. Immun.* **56**:1314–1319.

Naess, V., A.C. Johannessen, and T. Hofstad. 1988. Adherence of *Campylobacter jejuni* and *Campylobacter coli* to porcine intestinal brush border membranes. *Acta Pathol. Microbiol. Immunol. Scand.* **96**:681–687.

Mynott, T.L., D.S. Chandler, and R.K.J. Luke. 1991. Efficacy of enteric coated proteases in preventing attachment of enterotoxigenic *Escherichia coli* and diarrheal diseases in Ritard model. *Infect. Immun.* **59**:3708–3714.

Nakai, T., K. Kume, H. Yoshikawa, T. Oyamada, and T. Yoshikawa. 1988. Adherence of *Pasteurella multocida* or *Bordetella bronchiseptica* to the swine nasal epithelial cell *in vitro*. *Infect. Immun.* **56**:234–240.

Nakasone, N. and M. Iwanaga. 1987. Quantitative evaluation of colonizing ability of *Vibro cholerae* 01. *Microbiol. Immunol.* **31**:753–761.

Nakasone, N. and M. Iwanaga. 1990a. Pili of a *Vibrio parahaemolyticus* strain as a possible colonization factor. *Infect. Immun.* **58**:61–69.

Nakasone, N. and M. Iwanaga. 1990b. Pili of *Vibrio cholerae* non 01. *Infect. Immun.* **58**:1640–1646.

Nakasone, N. and M. Iwanaga. 1991. Purification and characterization of pili isolated from *Vibrio parahaemolyticus* Na2. *Infect. Immun.* **59**:726–728.

Nakato, H., K. Shinomiya, and H. Mikawa. 1987. Adhesion of *Erysipelothrix rhusiopathiae* to cultured rat aortic endothelial cells. Role of bacterial neuraminidase in the induction of arteritis. *Pathol. Res. Pract.* **182**:255–260.

Nakazawa, M., M. Haritani, C. Sugimoto, and M. Kashiwazaki. 1986. Colonization of enterotoxigenic *Escherichia coli* exhibiting mannose-sensitive hemagglutination to the small intestine of piglets. *Microbiol. Immunol.* **30**:485–489.

Nakazawa, T., M. Ishibashi, H. Konishi, T. Takemoto, M. Shigeeda, and T. Kochiyama. 1989. Hemagglutination activity of *Campylobacter pylori*. *Infect. Immun.* **57**:989–991.

Nataro, J.P., I.C.A. Scaletsky, J.B. Kaper, M.M. Levine, and L.R. Trabulsi. 1985. Plasmid-mediated factors conferring diffuse and localized adherence of enteropathogenic *Escherichia coli*. *Infect. Immun.* **48**:378–383.

Nataro J.P., J.B. Kaper, R. Robins-Brown, V. Prado, P. Vial, and M.M. Levine. 1987a. Patterns of adherence of diarrheagenic *Escherichia coli* to HEp-2 cells. *Pediatr. Infect. Dis. J.* **6**:829–831.

Nataro J.P., K.O. Maher, P. Mackie, and J.B. Kaper. 1987b. Characterization of plasmids encoding the adherence factor of enteropathogenic *Escherichia coli*. *Infect. Immun.* **55**:2370–2377.

Neeser, J.–R., B. Koellreutter, and P. Wuersch. 1986. Oligomannoside-type glycopeptides inhibiting adhesion of *Escherichia coli* strains mediated by type 1 pili: preparation of potent inhibitors from plant glycoproteins. *Infect. Immun.* **52**:428–436.

Neeser, J.–R., A. Chambaz, K.Y. Hoang, and H. Link-Amster. 1988a. Screening for complex carbohydrates inhibiting hemagglutinations by CFA/I- and CFA/II-expressing enterotoxigenic *Escherichia coli* strains. *FEMS Microbiol. Lett.* **49**:301–307.

Neeser, J.–R., A. Chambaz, S.D. Vedovo, M.J. Prigent, and B. Guggenheim. 1988. Specific and nonspecific inhibition of adhesion of oral actinomyces and streptococci to erythrocytes and polystyrene by caseinoglycopeptide derivatives. *Infect. Immun.* **56**:3201–3208.

Neman-Simha, V. and F. Megraud. 1988. *In vitro* model for *Campylobacter pylori* adherence properties. *Infect. Immun.* **56**:3329–3333.

Nhieu, G.T.V. and R.R. Isberg. 1991. The *Yersinia pseudotuberculosis* invasin protein and human fibronectin bind to mutually exclusive sites on the $\alpha_5\beta_1$ integrin receptor. *J. Biol. Chem.* **266**:24367–24375.

Nicolle, L.E., P. Muir, G.K.M. Harding, and M. Norris. 1988. Localization of urinary tract infection in elderly, institutionalized women with asymptomatic bacteriuria. *J. Infect. Dis.* **157**:65–70.

Niederman, M.S., W.W. Merrill, L.M. Polomski, H.Y. Reynolds, and J.B.L. Gee. 1986. Influence of sputum IgA and elastase on tracheal cell bacterial adherence. *Am. Rev. Respir. Dis.* **133**:255–260.

Nishikawa, Y., T. Kimura, and T. Kishi. 1991. Mannose-resistant adhesion of motile *Aeromonas* to INT407 cells and the differences among isolates from humans, food and water. *Epidemiol. Infect.* **107**:171–179.

Nogare, A.R.D. 1989. Type 1 pili mediate gram negative bacterial adherence to tracheal epithelium. *Amer. Rev. Resp. Dis.* **139:**A39.

Nowicki, B., M. Rhen, V. Vaisanen-Rhen, A. Pere, and T.K. Korhonen. 1985a. Fractionation of a bacterial cell population by absorption to erythrocytes and yeast cells. *FEMS Microbiol. Lett.* **26**:35–40.

Nowicki, B., M. Rhen, V. Vaisanen-Rhen, A. Pere, and T.K. Korhonen. 1985b. Kinetics of phase variation between S and type–1 fimbriae of *Escherichia coli*. *FEMS Microbiol. Lett.* **28**:237–242.

Nowicki, B., H. Holthofer, and T. Saraneva. 1986a. Location of adhesion sites for P fimbriated and for 075X-positive *Escherichia coli* in the human kidney. *Microb. Pathogen.* **1**:169–180.

Nowicki, B., J. Vuopio-Varkila, P. Viljanen, T.K. Korhonen, and P.H. Makela. 1986b. Fimbrial phase variation and systemic *Escherichia coli* infection studied in the mouse peritonitis model. *Microb. Pathogen.* **1**:335–347.

Nowicki, B., J.P. Barrish, T. Korhonen, R.A. Hull, and S.I. Hull. 1987. Molecular cloning of the *Escherichia coli* 075X adhesin. *Infect. Immun.* **55**:3168–3173.

Nowicki, B., J. Moulds, R. Hull, and S. Hull. 1988a. A hemagglutinin of uropathogenic *Escherichia coli* recognizes the Dr blood group antigen. *Infect. Immun.* **56**:1057–1060.

Nowicki, B., L. Truong, J. Moulds, and R. Hull. 1988b. Presence of the Dr receptor in normal human tissues and its possible role in the pathogenesis of ascending urinary tract infection. *Am. J. Pathol.* **133**:1–4.

Nowicki, B., C. Svanborg-Eden, R. Hull, and S. Hull. 1989. Molecular analysis and epidemiology of the Dr hemagglutinin of uropathogenic *Escherichia coli*. *Infect. Immun.* **57**:446–451.

Nowicki, B., A. Labigne, S. Moseley, R. Hull. S. Hull, and J. Moulds. 1990. The Dr hemagglutinin, afimbrial adhesins AFA-1 and AFA-III and F1845 fimbriae of uropathogenic and diarrhea-associated *Escherichia coli* belong to a family of hemagglutinins with Dr receptor recognition. *Infect. Immun.* **58**:279–281.

Nyberg, G., N. Stromberg, A. Jonsson, K.A. Karlsson, and S. Normark. 1990. Erythrocyte gangliosides act as receptors for *Neisseria subflava*: identification of the Sia–1 adhesin. *Infect. Immun.* **58**:2555–2563.

Obana, Y., K. Shibata, and T. Nishino. 1991. Adherence of *Serratia marcescens* in the pathogenesis of urinary tract infections in diabetic mice. *J. Med. Microbiol.* **35**:93–97.

Ofek, I. and E.H. Beachey. 1980. Bacterial adherence. *Adv. Intern. Med.* **25**:503–532.

Ofek, I. and N. Sharon. 1988. Lectinophagocytosis: A molecular mechanism of recognition between cell surface sugars and lectins in the phagocytosis of bacteria. *Infect. Immun.* **56**:539–547.

Ofek, I., E.H. Beachey, and N. Sharon. 1978. Surface sugars of animal cells as determinants of recognition in bacterial adherence. *Trends Biochem. Sci.* **3**:159–160.

Ofek, I., S. Hadar, R. Heiber, D. Zafriri, and M. Maayan. 1989. Effect of urea and urine dialysate on the growth of type 1-fimbriated and nonfimbriated phenotypes of isolates of *Escherichia coli*. In: Kass, E.H. and C.S. Eden (eds.), *Host–Parasite Interactions in Urinary Tract Infections*. The University of Chicago Press, Chicago, pp. 122–128.

Ofek, I., D. Zafriri, J. Goldhar, and B.I. Eisenstein. 1990. Inability of toxin inhibitors to neutralize enhanced toxicity caused by bacteria adherent to tissue culture cells. *Infect. Immun.* **58**:3737–3742.

Ofek, I., J. Goldhar, D. Zafriri, H. Lis, R. Adar, and N. Sharon. 1991. Anti-*Escherichia coli* adhesin acitivity of cranberry and blueberry juices. *N. Engl. J. Med.* **324**:1599.

Ogaard, A.R., K. Bjoror, G. Bukholm, and B.P. Berdal. 1985. Correlation between adhesion of *Pseudomonas aeruginosa* bacteria to cell surfaces and the presence of some factors related to virulence. *Acta Pathol. Microbiol. Immunol. Scand.* **93**:211–216.

Ogawa, T., H. Shimauchi, and S. Hamada. 1989. Mucosal and systemic immune responses in BALB/c mice to *Bacteroides gingivalis* fimbriae administered orally. *Infect. Immun.* **57**:3466–3471.

Ogawa, T., T. Mukai, K. Yasuda, H. Shimauchi, Y. Toda, and S. Hamada. 1991a. Distribution and immunochemical specificities of fimbriae of *Porphyromonas gingivalis* and related bacterial species. *Oral Microbiol. Immunol.* **6**:332–340.

Ogawa, T., Y. Kusumoto, H. Uchida, S. Nagashima, H. Ogo, and S. Hamada. 1991b. Immunobiological activities of synthetic peptide segments of fimbrial protein from *Porphyromonas gingivalis. Biochem. Biophys. Res. Commun.* **180**:1335–1341.

O'Hanley, P., D. Lark, S. Falkow, and G. Schoolnik. 1985a. Molecular basis of *Escherichia coli* colonization of the upper urinary tract in BALB/c mice. *J. Clin. Invest.* **75**:347–360.

O'Hanley, P., D. Low, I. Romero, D. Lark, K. Vosti, S. Falkow, and G. Schoolnik. 1985b. Gal-Gal binding and hemolysin phenotypes and genotypes associated with uropathogenic *Escherichia coli. N. Engl. J. Med.* **313**:414–420.

Ohman, D.E. and J. Goldberg. 1990. Genetics of alginate biosynthesis in *Pseudomonas aeruginosa*. In: Gacesa, P., and N.J. Nicholas (eds.), Pseudomonas *Infection and Alginates*. Chapman and Hall, New York, pp. 206–220.

Okuda, K., A. Yamamoto, Y. Naito, I. Takazoe, J. Slots, and R.J. Genco. 1986. Purification and properties of hemagglutinin from culture supernatant of *Bacteroides gingivalis. Infect. Immun.* **54**:659–665.

Okuda, K., M. Ono, and T. Kato. 1989. Neuraminidase-enhanced attachment of *Bacteroides intermedius* to human erythrocytes and buccal epithelial cells. *Infect. Immun.* **57**:1635–1637.

Olafson, R.W., P.J. McCarthy, A.R. Bhatti, J.S.G. Dooley, J.E. Heckels, and T.J. Trust. 1985. Structural and antigenic analysis of meningococcal piliation. *Infect. Immun.* **48**:336–342.

Old, D.C., A. Tavendale, and B.W. Senior. 1985a. A comparative study of the type–3 fimbriae of *Klebsiella species. J. Med. Microbiol.* **20**:203–214.

Old, D.C., B. Klaassen, and P.B. Crichton. 1985b. A novel non-fimbrial mannose-sensitive and eluting (MSE) haemagglutinin of *Yersinia frederiksenii* strains. *FEMS Microbiol. Lett.* **28**:81–84.

Old, D.C., A.I. Roy, and A. Tavendale. 1986. Differences in adhesiveness among type 1 fimbriae strains of *Enterobacteriaceae revealed by* an *in vitro* HEp2 cell adhesion model. *J. Appl. Bacteriol.* **61**:563–568.

Old, D.C., D.E. Yakubu, and P.B. Crichton. 1987. Demonstration by immuno-electron-microscopy of antigenic heterogeneity among P-fimbriae of strains of *Escherichia coli. J. Med. Microbiol.* **23**:247–253.

Olsen, A., A. Johnson, and S. Normark. 1989. Fibronectin binding mediated by a novel class of surface organelles on *Escherichia coli. Nature* **338**:652–655.

Omoike, I., B. Lindquist, R. Abud, J. Merrick, and E. Lebenthal. 1990. The effect of protein energy malnutrition and refeeding on the adherence of *Salmonella typhimurium* to small intestinal mucosa and isolated enterocytes in rats. *J. Nutr.* **120**:404–411.

Ono, E., K. Abe, M. Nakazawa, and M. Naiki. 1989. Ganglioside epitope recognized by K99 fimbriae from enterotoxigenic *Escherichia coli. Infect. Immun.* **57**:907–911.

Orskov, I., A. Birch-Andersen, J.P. Duguid, J. Stenderup, and F. Orskov. 1985. An adhesive protein capsule of *Escherichia coli. Infect. Immun.* **47**:191–200.

Ott, M., J. Hacker, T. Schmoll, T. Jarchau, T.K. Korhonen, and W. Goebel. 1986. Analysis of the genetic determinants coding for the S-fimbrial adhesin (sfa) in different *Escherichia coli* strains causing meningitis or urinary tract infections. *Infect. Immun.* **54**:646–653.

Oudega, B. and F.K. DeGraaf. 1988. Genetic organization and biogenesis of adhesive fimbriae of *Escherichia coli. Antonie van Leeuwenhoek* **54**:285–299.

Owen, R.L., N.F. Pierce, and R.T. Apple. 1986. M cell transport of *Vibrio cholerae* from the intestinal lumen into Peyer's patches: a mechanism for antigen sampling and for microbial transepithelial migration. *J. Infect. Dis.* **153**:1108–1118.

Paerregaard, A., F. Espersen, J. Hannover-Larsen, and N. Hiby. 1990. Adhesion of *Yersinia enterocolitica* to human epithelial cell lines and to rabbit human small intestinal tissue. *Acta Pathol. Microbiol. Immunol. Scand.* **98**:53–60.

Paerregaard, A., F. Espersen, and M. Skurnik. 1991a. Role of the *Yersinia* outer membrane protein Yad A in adhesion to rabbit intestinal tissue and rabbit intestinal brush border membrane vesicles. *Acta Microbiol. Pathol. Immunol. Scand.* **99**:226–232.

Paerregaard, A., F. Espersen, O.M. Jensen, and M. Skurnik. 1991b. Interactions between *Yersinia enterocolitica* and rabbit ileal mucus: growth, adhesion, penetration, and subsequent changes in surface hydrophobicity and ability to adhere to ileal brush border membrane vesicles. *Infect. Immun.* **59**:253–260.

Pal, R. and A.C. Ghose. 1990. Identification of plasmid-encoded mannose-resistant hemagglutinin and HEp-2 and HeLa cell adherence factors of two diarrheagenic *Escherichia coli* strains belonging to an enteropathogenic serogroup. *Infect. Immun.* **58**:1106–1113.

Panjwani, N., T.S. Zaidi, J.E. Gigstad, F.B. Jungalwala, M. Barza, and J. Baum. 1990. Binding of *Pseudomonas aeruginosa* to neutral glycosphingolipids of rabbit corneal epithelium. *Infect. Immun.* **58**:114–118.

Pankhurst, C.L., D.W. Auger, and J.M. Hardie. 1988. An ultrastructural study of adherence to buccal epithelial cells of *Simonsiella* sp. *Appl. Microbiol.* **6**:125–128.

Paque R.E., R. Miller, and V. Thomas. 1990. Polyclonal anti-idiotypic antibodies exhibit antigenic mimicry of limited type 1 fimbrial proteins of *Escherichia coli. Infect. Immun.* **58**:680–686.

Paranchych, W. 1990. Molecular studies on *N*-methylphenylalanine pili. In: Iglewski, B. and V.L. Clark (eds.), *Molecular Basis of Bacterial Pathogenesis*: Academic Press, San Diego, pp. 61–78.

Paranchych, W., P.A. Sastry, K. Volpel, B.A. Loh, and D.P. Speert. 1986. Fimbriae (pili): molecular basis of *Pseudomonas aeruginosa* adherence. *Clin. Invest. Med.* **9**:113–118.

Parker, C.D. and S.K. Armstrong. 1988. Surface proteins of *Bordetella pertussis*. *Rev. Infect. Dis.* **10** (*Suppl* 2):S327–S330.

Parkkinen, J., G.N. Rogers, T. Korhonen, W. Dahr, and J. Finne. 1986. Identification of the O-linked sialyloligosaccharides of glycophorin A as the erythrocyte receptors for S-fimbriated *Escherichia coli*. *Infect. Immun.* **54**:37–42.

Parkkinen, J., R. Virkola, and T.K. Korhonen. 1988. Identification of factors in human urine that inhibit the binding of *Escherichia coli* adhesins. *Infect. Immun.* **56**:2623–2630.

Parkkinen, J., A. Ristimaki, and B. Westerlund. 1989. Binding of *Escherichia coli* S fimbriae to cultured human endothelial cells. *Infect. Immun.* **57**:2256–2259.

Parry, S.H. and D.M. Rooke. 1985. Adhesins and colonization factors of *Escherichia coli*. In: Sussman, M. (ed.), *Virulence of Escherichia coli*. Academic Press, London, pp. 79–155.

Parsons, C.L. 1986. Pathogenesis of urinary tract infections. Bacterial adherence, bladder defense mechanisms. *Urol. Clin. North Am.* **13**:563–568.

Paruchuri, D.K., H.S. Seifert, R.S. Ajioka, K.A. Karlsson, and M. So. 1990. Identification and characterization of a *Neisseria gonorrhoeae* gene encoding a glycolipid-binding adhesin. *Proc. Natl. Acad. Sci. USA* **87**:333–337.

Patel, P., C.F. Marrs, J.S. Mattick, W.W. Ruehl, R.K. Taylor, and M. Koomey. 1991. Shared antigenicity and immunogenicity of type 4 pilins expressed by *Pseudomonas aeruginosa, Moraxella bovis, Neisseria gonorrhoeae, Dichelobacter nodosus*, and *Vibrio cholerae*. *Infect. Immun.* **59**:4674–4676.

Patrick, C.C., G.S. Patrick, S.L. Kaplan, J. Barrish, and E.O. Mason, Jr. 1989. Adherence kinetics of *Haemophilus influenzae* type b to eucaryotic cells. *Pediatr. Res.* **26**:550–553.

Pawelzik, M., J. Heesemann, J. Hacker, and W. Opferkuch. 1988. Cloning and characterization of a new type of fimbria (S/F1C-related fimbriae) expressed by an *Escherichia coli* 075:K1:H7 blood culture isolate. *Infect. Immun.* **56**:2918–2924.

Pecha, B. D. Low, and P. O'Hanley. 1989. Gal-Gal pili vaccines prevent pyelonephritis by piliated *Escherichia coli* in a murine model. Single-component Gal-Gal pili vaccines prevent pyelonephritis by homologous and heterologous piliated *Escherichia coli* strains. *J. Clin. Invest.* **83**:2102–2108.

Pedersen, K. and G.W. Tannock. 1989. Colonization of the porcine gastrointestinal tract by lactobacilli. *Appl. Environ. Microbiol.* **55**:279–283.

Peeters, J.E., R. Geeroms, and F. Orskov. 1988. Biotype, serotype and pathogenicity of attaching and effacing enteropathogenic *Escherichia coli* strains isolated from diarrheic commercial rabbits. *Infect. Immun.* **56**:1442–1448.

Pei, Z., R.T. Ellison, III, and M.J. Blaser. 1991. Identification, purification, and characterization of major antigenic proteins of *Campylobacter jejuni*. *J. Biol. Chem.* **266**:16363–16369.

Pepe, J.C. and V.L. Miller. 1990. The *Yersinia enterocolitica inv* gene product is an outer membrane protein that shares epitopes with *Yersinia pseudotuberculosis J. Bacteriol.* **172**:3780–3789.

Pere, A. 1986. P fimbriae on uropathogenic *Escherichia coli* 016:K1 and 018 strains. *FEMS Microbiol. Lett.* **37**:19–26.

Pere, A., M. Leinonen, V. Vaisanen-Rhen, M. Rhen, and T.K. Korhonen. 1985. Occurrence of type–1C fimbriae on *Escherichia coli* strains isolated from human extraintestinal infections. *J. Gen. Microbiol.* **131**:1705–1711.

Pere, A., B. Nowicki, H. Saxen, A. Siitonen, and T.K. Korhonen. 1987. Expression of P, type–1 and type 1c fimbriae of *Escherichia coli* in the urine of patients with acute urinary tract infection. *J. Infect. Dis.* **156**:567–574.

Pere, A., R.K. Selander, and T.K. Korhonen. 1988. Characterization of P fimbriae on 01, 07, 075, rough and nontypable strains of *Escherichia coli. Infect. Immun.* **56**:1288–1294.

Perrollet, H. and R.M. Guinet. 1986. Gonococcal cell wall lectin-adhesin with vaccine potential. *Lancet* **31**:1269–1270.

Perry, C.F., C.A. Hart, I.J. Nicolson, J.E. Heckels, and J.R. Saunders. 1987. Inter-strain homology of pilin gene sequences in *Neisseria meningitidis* isolates that express markedly different antigenic pilus types. *J. Gen. Microbiol.* **133**:1409–1418.

Perry, A.C.F., I.J. Nicolson, and J.R. Saunders. 1988. *Neisseria meningitidis* C114 contains silent, truncated pilin genes that are homologous to *Neisseria gonorrhoeae* pil sequences. *J. Bacteriol.* **170**:1691–1697.

Peterson, K.M. and J.J. Mekalanos. 1988. Characterization of the *Vibrio cholerae* toxR regulon: identification of novel genes involved in intestinal colonization. *Infect. Immun.* **56**:2822–2829.

Peterson, J.W. and D.W. Niesel. 1988. Enhancement by calcium of the invasiveness of *Salmonella* for Hela-cell monolayers. *Rev. Infect. Dis.* **10**:S319–S322.

Pieroni, P., E.A. Worobec, W. Paranchych, and G.D. Armstrong. 1988. Identification of a human erythrocyte receptor for colonization factor antigen I pili expressed by H10407 enterotoxigenic *Escherichia coli. Infect. Immun.* **56**:1334–1340.

Pijoan, C. and F. Trigo. 1990. Bacterial adhesion to mucosal surfaces with special reference to *Pasteurella multocida* isolates form atrophic rhinitis. *Can. J. Vet. Res.* **54**:S16–S21.

Pinner, R.W., P.A. Spellman, and D.S. Stephens. 1991. Evidence for functionally distinct pili expressed by *Neisseria meningitidis. Infect. Immun.* **59**:3169–3175.

Plos, K., S.I. Hull, B.R. Levin, I. Orskov, F. Orskov, and C. Svanborg-Eden. 1989. Distribution of the P-associated-pilus (*pap*) region among *Escherichia coli* from natural sources: evidence for horizontal gene transfer. *Infect. Immun.* **57**:1604–1611.

Plotkowski, M.C., G. Beck, J.M. Tournier, M. Bernardo-Filho, E.A. Marques, and E. Puchelle. 1989. Adherence of *Pseudomonas aeruginosa* to respiratory epithelium and the effect of leucocyte elastase. *J. Med. Microbiol.* **30**:285–293.

Plotkowski, M.C., M. Chevillard, D. Pierrot, D. Altemayer, J.M. Zahm, G. Colliot, and E. Puchelle. 1991. Differential adhesion of *Pseudomonas aeruginosa* to human respiratory epithelial cells in primary culture. *J. Clin. Invest.* **87**:2018–2028.

Podschun, R., P. Heineken, and H.G. Sonntag. 1987. Haemagglutinins and adherence properties to HeLa and intestine cells of *Klebsiella pneumoniae* and *Klebsiella oxytoca* isolates. *Zbl. Bakteriol. Parasitenkd. Infektionskr. Hyg. Abt. 1 Orig. Reihe* **A263**:585–593.

Porras, O., C. Svanborg-Eden, T. Lagergard, and L.A. Hanson. 1985. Methods for testing adherence of *Haemophilus influenzae* to human buccal epithelial cells. *Eur. J. Clin. Microbiol.* **4**:310–315.

Potts, W. and J.R. Saunders. 1987. Nucleotide sequence of the structural gene for class 1 pilin from *Neisseria meningitidis*: homologies with the pil E locus of *Neisseria gonorrhoeae*. *Mol. Microbiol.* **1**:647–654.

Pratt-Terpstra, I.H., A.H. Weerkamp, and H.J. Busscher. 1989. The effects of pellicle formation on streptococcal adhesion to human enamel and artificial substrata with various surface free-energies. *J. Dent. Res.* **68**:463–467.

Pruul, H., C.S. Goodwin, P.J. McDonald, G. Lewis, and D. Pankhurst. 1990. Hydrophobic characterisation of *Helicobacter (Campylobacter) pylori*. *J. Med. Microbiol.* **32**:93–100.

Pruzzo, C., C.A. Guzman, and B. Dainelli. 1989a. Incidence of hemagglutination activity among pathogenic and non-pathogenic *Bacteroides fragilis* strains and role of capsule and pili in HA and adherence. *FEMS Microbiol. Lett.* **50**:113–118.

Pruzzo, C., G. Cisani, and G. Satta. 1989b. Mannose-inhibitable adhesin/T7 receptor-mediated interactions with the host in strains of *Escherichia coli* isolated from the urinary tract. In: Kass, E.H. and C. Svanborg-Eden (eds.), *Host–Parasite Interactions in Urinary Tract Infections*. The University of Chicago Press, Chicago, pp. 127–131.

Qadri, F., S. Haq, and I. Ciznar. 1989. Hemagglutinating properties of *Shigella dysenteriae* type 1 and other *Shigella* species. *Infect. Immun.* **57**:2909–2911.

Rafiee, P., H. Leffler, J.C. Byrd, F.J. Cassels, E.C. Boedeker, and Y.S. Kim. 1991. A sialoglycoprotein complex linked to the microvillus cytoskeleton acts as a receptor for pilus (AF/R1) mediated adhesion of enteropathogenic *Escherichia coli* in rabbit small intestine. *J. Cell Biol.* **115**:1021–1029.

Ramphal, R. and G.B. Pier. 1985. Role of *Pseudomonas aeruginosa* mucoid exopolysaccharide in adherence to tracheal cells. *Infect. Immun.* **47**:1–4.

Ramphal, R., C. Guay, and G.B. Pier. 1987. *Pseudomonas aeruginosa* adhesins for tracheobronchial mucin. *Infect. Immun.* **55**:600–603.

Ramphal, R., N. Houdret, L. Koo, G. Lamblin, and P. Roussel. 1989. Differences in adhesion of *Pseudomonas aeruginosa* to mucin glycopeptides from sputa of patients with cystic fibrosis and chronic bronchitis. *Infect. Immun.* **57**:3066–3071.

Ramphal, R., C. Carnoy, S. Fievre, J.C. Michalski, N. Houdret, G. Lamblin, G. Strecker, and P. Roussel. 1991a. *Pseudomonas aeruginosa* recognizes carbohydrate chains containing type 1 (Galβ1–3GlcNAc) or type 2 (Galβ1–4GlcNAc) disaccharide units. *Infect. Immun.* **59**:700–704.

Ramphal, R., L. Koo, K.S. Ishimoto, P.A. Totten, J.C. Lara, and S. Lory. 1991b.

Adhesion of *Pseudomonas aeruginosa* pilin-deficient mutants to mucin. *Infect. Immun.* **59**:1307–1311.

Rapacz, J. and J. Hasler-Rapacz. 1986. Polymorphism and inheritance of swine small intestinal receptors mediating adhesion of three serological variants of *Escherichia coli* producing K88 pilus antigen. *Anim. Genet.* **17**:305–321.

Ratliff, T.L., J.O. Palmer, J.A. McGarr, and E.J. Brown. 1987. Intravesicular bacillus Calmette-Guerin therapy for murine bladder tumors: initiation of the response by fibronectin-mediated attachment of bacillus Calmette-Guerin *Cancer Res.* **47**:1762–1766.

Ratliff, T.L., J.A. McGarr, C. Abou-Zeid, G.A. Rook, J.L. Stanford, J. Asalanzadeh, and E.J. Brown. 1988. Attachment of mycobacteria to fibronectin-coated surfaces. *J. Gen. Microbiol.* **134**:1307–1313.

Razin, S. 1985a. Molecular biology and genetics of mycoplasmas (Mollicutes). *Microbiol. Rev.* **49**:419–455.

Razin, S. 1985b. Mycoplasma adherence. In: Razin S. and M.F. Berile (eds.), *Mycoplasmas, Mycoplasma pathogenicity, Vol. 4.* Academic Press, Orlando, pp. 161–202.

Razin, S. and D. Yogev. 1989. Molecular approaches to characterization of mycoplasmal adhesins. In: Switalski, L., M. Hook, and E.H. Beachey (eds.), *Molecular Mechanisms of Microbial Adhesion.* Springer-Verlag, Berlin, pp. 52–76.

Read, R.C., R. Wilson, A. Rutman, V. Lund, H.C. Todd, A.P. Brain, P.K. Jeffery, and P.J. Cole. 1991. Interaction of nontypable *Haemophilus influenzae* with human respiratory mucosa in vitro. *J. Infect. Dis.* **163**:549–558.

Redhead, K. 1985. An assay of *Bordetella pertussis* adhesion to tissue cells. *J. Med. Microbiol.* **19**:99–108.

Reid, G. 1989. Local and diffuse bacterial adherence on uroepithelial cells. *Curr. Microbiol.* **18**:93–97.

Reid, G. and J.D. Sobel. 1987. Bacterial adherence in the pathogenesis of urinary tract infection: a review. *Rev. Infect. Dis.* **9**:470–484.

Reijntjens, F.M.J., F.H.M. Mikx, J.M.L. Wolters-Lutgerhorst, and J.C. Maltha. 1986. Adherence of oral treponemes and their effect on morphological damage and detachment of epithelial cells in vitro. *Infect. Immun.* **54**:642–647.

Reinhart, H., N. Obedeanu, T. Hooton, W. Stamm, and J. Sobel. 1990a. Urinary excretion of Tamm-Horsfall protein in women with recurrent urinary tract infections. *J. Urol.* **144**:1185–1187.

Reinhart, H.H., N. Obedeanu and J.D. Sobel. 1990b. Quantitation of Tamm-Horsfall protein binding to uropathogenic *Escherichia coli* and lectins. *J. Infect. Dis.* **162**:1335–1340.

Relman, D.A., M. Domenighini, E. Tuomanen, R. Rappuoli, and S. Falkow. 1989. Filamentous hemagglutinin of *Bordetella pertussis*: nucleotide sequence and crucial role in adherence. *Proc. Natl. Acad. Sci. USA* **86**:2637–2641.

Rest, R.F., N. Lee, and C. Bowden. 1985. Stimulation of human leukocytes by protein II^+ gonococci is mediated by lectin-like gonococcal components. *Infect. Immun.* **50**:116–122.

Reynaud, A. M. Federighi, D. Licois, J.F. Guillot, and B. Joly. 1991. R plasmid in *Escherichia coli* 0103 coding for colonization of the rabbit intestinal tract. *Infect. Immun.* **59**:1888–1892.

Rhen, M., P. Klemm, and T.K. Korhonen. 1986. Identification of two new hemagglutinins of *Escherichia coli*, N-acetyl-D-glucosamine-specific fimbriae and a blood group M-specific agglutinin, by cloning the corresponding genes in *Escherichia coli* K–12. *J. Bacteriol.* **168**:1234–1242.

Richardson, W.P. and J.C. Sadoff. 1988. Induced engulfment of *Neisseria gonorrhoeae* by tissue culture cells. *Infect. Immun.* **56**:2512–2514.

Roberts, D.D., L.D. Olson, M.F. Barile, V. Ginsburg, and H.C. Krivan. 1989. Sialic acid-dependent adhesion of *Mycoplasma pneumoniae* to purified glycoproteins. *J. Biol. Chem.* **264**:9289–9293.

Robertson, J.A. and R. Sherburne. 1991. Hemadsorption by colonies of *Ureaplasma urealyticum*. *Infect. Immun.* **59**:2203–2206.

Robins-Browne, R.M. 1987. Traditional enteropathogenic *Escherichia coli* of infantile diarrhea. *Rev. Infect. Dis.* **9**:28–53.

Robinson Jr., E.N., C.M. Clemens, G.K. Schoolnik, and Z.A. McGee. 1989. Probing the surface of *Neisseria gonorrhoeae*: immunoelectron microscopic studies to localize cyanogen bromide fragment 2 in gonococcal pili. *Mol. Microbiol.* **3**:57–64.

Robinson, J., C.S. Goodwin, M. Cooper, V. Burke, and B.J. Mee. 1990. Soluble and cell-associated haemagglutinins of *Helicobacter (Campylobacter) pylori*. *J. Med. Microbiol.* **33**:277–284.

Rodriguez-Ortega, M., I. Ofek, and N. Sharon. 1987. Membrane glycoproteins of human polymorphonuclear leukocytes that act as receptors for mannose-specific *Escherichia coli*. *Infect. Immun.* **55**:968–973.

Rogemond, V. and R.M.F. Guinet. 1986. Lectin-like adhesins in the *Bacteroides fragilis* group. *Infect. Immun.* **53**:99–102.

Romero-Steiner, S., T. Witek, and E. Balish. 1990. Adherence of skin bacteria to human epithelial cells. *J. Clin. Microbiol.* **28**:27–31.

Rosberg, K., T. Berglindh, S. Gustavsson, R. Hubinette, and W. Rolfsen. 1991. Adhesion of *Helicobacter pylori* to human gastric mucosal biopsy specimens cultivated in vitro. *Scand. J. Gastroenterol.* **26**:1179–1187.

Rosenberg, M., I.A. Buivids, and R.P. Ellen. 1991. Adhesion of *Actinomyces viscosus to Porphyromonas (Bacteroides) gingivalis*-coated hexadecane droplets. *J. Bacteriol.* **173**:2581–2589.

Rosenstein, I.J., T. Mizuochi, E. Hounsell, M.S. Stoll, and R.A. Childs. 1988. New type of adhesive specificity revealed by oligosaccharide probes in *Escherichia coli* from patients with urinary tract infection. *Lancet* 1327–1330.

Rosqvist, R., M. Skurnik, and H. Wolf-Watz. 1988. Increased virulence of *Yersinia pseudotuberculosis* by two independent mutations. *Nature (Lond.)* **334**:522–525.

Rothbard, J.B., R. Fernandez, L. Wang, N.N.H. Teng, and G.K. Schoolnik. 1985. Antibodies to peptides corresponding to a conserved sequence of gonococcal pilins block bacterial adhesion. *Proc. Natl. Acad. Sci. USA* **82**:915–919.

Ruggieri, M.R., P.M. Hanno, and R.M. Levin. 1987. Reduction of bacterial adherence to catheter surface with heparin. *J. Urol.* **138**:423–430.

Runnels, P.L., S.L. Moseley, and H.W. Moon. 1987. F41 pili as protective antigens of enterotoxigenic *Escherichia coli* that produce F41, K99, or both pilus antigens. *Infect. Immun.* **55**:555–558.

Saada, A.B., Y. Terespolski, A. Adoni, and I. Kahane. 1991. Adherence of *Ureaplasma urealyticum* to human erythrocytes. *Infect. Immun.* **59**:467–469.

Sable, N.S., E.M. Connor, C.B. Hall, and M.R. Loeb. 1985. Variable adherence of fimbriated *Haemophilus influenzae* type b to human cells. *Infect. Immun.* **48**:119–123.

Saiman, L., J. Sadoff, and A. Prince. 1989. Cross-reactivity of *Pseudomonas aeruginosa* antipilin monoclonal antibodies with heterogenous strains of *P. aeruginosa* and *Pseudomonas cepacia. Infect. Immun.* **57**:2764–2770.

Saiman, L., K. Ishimoto, S. Lory, and A. Prince. 1990a. The effect of piliation and exoproduct expression on the adherence of *Pseudomonas aeruginosa* to respiratory epithelial monolayers. *J. Infect. Dis.* **161**:541–548.

Saiman, L., G. Cacalano, and A. Prince. 1990b. *Pseudomonas cepacia* adherence to respiratory epithelial cells is enhanced by *Pseudomonas aeruginosa. Infect. Immun.* **58**:2578–2584.

Sainz, V., C.M. Ferreiros, and M.T. Criado. 1987. Role of R-plasmids in adherence and hydrophobicity in *Escherichia coli. J. Hosp. Infect.* **9**:175–181.

Saitoh, T., H. Natomi, W.L. Zhao, K. Okuzumi, K. Sugano, M. Iwamori, and Y. Nagai. 1991. Identification of glycolipid receptors for *Helicobacter pylori* by TLC-immunostaining. *FEBS Lett.* **282**:385–387.

Sajjan, S.U. and J.F. Forstner. 1990a. Characteristics of binding of *Escherichia coli* serotype 0157:H7 strain CL-49 to purified intestinal mucin. *Infect. Immun.* **58**:860–867.

Sajjan, S.U. and J.F. Forstner. 1990b. Role of the putative "link" glycopeptide of intestinal mucin in binding of piliated *Escherichia coli* serotype 0157:H7 strain CL-49. *Infect. Immun.* **58**:868–873.

Salit, I.E., J. Hanley, L. Clubb, and S. Fanning. 1988. The human antibody response to uropathogenic *Escherichia coli*: a review. *Can. J. Microbiol.* **34**:312–318.

Sandberg, T., B Kaijser, G. Lidin-Janson, K. Lincoln, F. Orskov, I. Orskov, E. Stokland, and C. Svanborg-Eden. 1988. Virulence of *Escherichia coli* in relation to host factors in women with asymptomatic urinary tract infection. *J. Clin. Microbiol.* **26**:1471–1476.

Sareneva, T., H. Holthofer, and T.K. Korhonen. 1990. Tissue-binding affinity of *Proteus mirabilis* fimbriae in the human urinary tract. *Infect. Immun.* **58**:3330–3336.

Sarmiento J.I., T.A. Casey, and H.W. Moon. 1988. Post weaning diarrhea in swine: experimental model of enterotoxigenic *Escherichia coli* infection. *Am. J. Vet. Res.* **49**:1154–1158.

Sastry, P.A., J.R. Pearlstone, L.B. Smillie, and W. Paranchych. 1985. Studies on the primary structure and antigenic determinants of pilin isolated form *Pseudomonas aeruginosa* K. *Can. J. Biochem. Cell. Biol.* **63**:284–291.

Sato, H. and K. Okinaga. 1987. Role of pili in the adherence of *Pseudomonas aeruginosa* to mouse epidermal cells. *Infect. Immun.* **55**:1774–1778.

Saukkonen, K.M.J., B. Nowicki, and M. Leinonen. 1988. Role of type 1 and S fimbriae in the pathogenesis of *Escherichia coli* 018:K1 bacteremia and meningitis in the infant rat. *Infect. Immun.* **56**:892–897.

Savage, D.C. 1987. Microorganisms associated with epithelial surfaces and stability of the indigenous gastrointestinal microflora. *Nahrung* **31**:383–395.

Savage, D.C. and M. Fletcher (eds.) 1985. *Bacterial Adhesion: Mechanisms and Physiological Significance*: Plenum Press, New York.

Scaletsky, I.C.A., M.L.M. Silva, M.R.F. Toledo, B.R. Davis, P.A. Blake, and L.R. Trabulsi. 1985. Correlation between adherence to HeLa cells and serogroups, serotypes, and bioserotypes of *Escherichia coli*. *Infect. Immun.* **49**:528–532.

Scaletsky, I.C.A., S.R. Milani, L.R. Trabulsi, and L.R. Travassos. 1988. Isolation and characterization of the localized adherence factor of enteropathogenic *Escherichia coli*. *Infect. Immun.* **56**:2979–2983.

Schaeffer, A.J., W.R. Schwan, S.J. Hultgren, and J.L. Duncan. 1987. Relationship of type 1 pilus expression in *Escherichia coli* to ascending urinary tract infections in mice. *Infect. Immun.* **55**:373–380.

Schaper, U., J.S. Chapman, and P.C. Hu. 1987. Preliminary indication of unusual codon usage in the DNA coding sequence of the attachment protein of *Mycoplasma pneumoniae*. *Isr. J. Med. Sci.* **23**:361–367.

Schiemann, D.A. and C.M. Nelson. 1988. Antibody inhibition of HeLa cell invasion by *Yersinia enterocolitica*. *Can. J. Microbiol.* **34**:52–57.

Schiemann, D.A., M.R. Crane, and P.J. Swanz. 1987. Surface properties of *Yersinia* species and epithelial cell interactions in vitro by a method measuring total associated attached and intracellular bacteria. *J. Med. Microbiol.* **24**:205–218.

Schlager, T.A., C.A. Wanke, and R.L. Guerrant. 1990. Net fluid secretion and impaired villous function induced by colonization of the small intestine by nontoxigenic colonizing *Escherichia coli*. *Infect. Immun.* **58**:1337–1343.

Schmidt, A., P. O'Hanley, D. Lark, and G.K. Schoolnik. 1988. Synthetic peptides corresponding to protective epitopes of *Escherichia coli* digalactoside-binding pilin prevent infection in murine pyelonephritis model. *Proc. Natl. Acad. Sci. USA* **85**:1247–1251.

Schmiel, D.H., S.T. Knight, J.E. Raulston, J. Choong, C.H. Davis, and P.B. Wyrick. 1991. Recombinant *Escherichia coli* clones expressing *Chlamydia trachomatis* gene products attach to human endometrial epithelial cells. *Infect. Immun.* **59**:4001–4012.

Schmitz, S., C. Abe, I. Moser, I. Orskov, F. Orskov, B. Jann, and K. Jann. 1986. Monoclonal antibodies against the nonhemagglutinating fimbrial antigen 1C (pseudotype 1) of *Escherichia coli*. *Infect. Immun.* **51**:54–59.

Schollin, J. 1988. Adherence of alpha-hemolytic streptococci to human endocardial endothelial and buccal cells. *Acta Paediatr. Scand.* **77**:705–710.

Schollin, J. and D. Danielsson. 1988. Bacterial adherence to endothelial cells from rat heart, with special regard to alpha-hemolytic streptococci. *Acta Pathol. Microbiol. Immunol. Scand.* **96**:428–432.

Schwan, W.R., S.J. Hultgren, S.K. Amundsen, J.L. Duncan, and A.J. Schaeffer. 1989. Contribution of *Escherichia coli* type 1 pili to ascending urinary tract colonization in mice. In: Kass, E.H. and Svanborg-Eden (eds.), *Host-Parasite Interactions in Urinary Tract Infections*. University of Chicago Press, Chicago, pp. 341–347.

Schwarz, S., R.P. Ellen, and D.A. Grove. 1987. *Bacteroides gingivalis-Actinomyces viscosus* cohesive interactions as measured by a quantitative binding assay. *Infect. Immun.* **55**:2391–2397.

Scotland, S.M., G.A. Willshaw, H.R. Smith, and B. Rowe. 1990. Properties of strains of *Escherichia coli* 026:H11 in relation to their enteropathogenic or enterohemorrhagic classification. *J. Infect. Dis.* **162**:1069–1074.

Scotland, S.M., H.R. Smith, and B. Rowe. 1991a. *Escherichia coli* 0128 strains from infants with diarrhea commonly show localized adhesion and positivity in the fluorescent-actin staining test but do not hybridize with an enteropathogenic *Escherichia coli* adherence factor probe. *Infect. Immun.* **59**:1569–1571.

Scotland, S.M., H.R. Smith, B. Said, G.A. Willshaw, T. Cheasty, and B. Rowe. 1991b. Identification of enteropathogenic *Escherichia coli* isolated in Britain as enteroaggregative or as members of a subclass of attaching-and-effacing *Escherichia coli* not hybridizing with the EPEC adherence-factor probe. *J. Med. Microbiol.* **35**:278–283.

Scott, T.G. and C.J. Smyth. 1987. Haemagglutination and tissue culture adhesion of *Gardnerella vaginalis*. *J. Gen. Microbiol.* **133**:1999–2005.

Scott, T.G., C.J. Smyth, and C.T. Keane. 1987. *In vitro* adhesiveness and biotype of *Gardnerella vaginalis* strains in relation to the occurrence of clue cells in vaginal discharges. *Genitourin. Med.* **63**:47–53.

Scott, T.G., B. Curran, and C.J. Smyth. 1989. Electron microscopy of adhesive interactions between *Gardnerella vaginalis* and vaginal epithelial cells, McCoy cells and human red blood cells. *J. Gen. Microbiol.* **135**:475–480.

Seifert, H.S. and M. So. 1988. Genetic mechanisms of bacterial antigenic variation. *Microbiol. Rev.* **52**:327–336.

Seignole, D., M. Mouricout, Y. Duval-Iflah, B. Quintard, and R. Julien. 1991. Adhesion of K99 fimbriated *Escherichia coli* to pig intestinal epithelium: correlation of adhesive and non-adhesive phenotypes with the sialoglycolipid content. *J. Gen. Microbiol.* **137**:1591–1601.

Selander, R.K., T.K. Korhonen, V. Vaisanen-Rhen, P.H. Williams, P.E. Pattison, and D.A. Caugant. 1986. Genetic relationships and clonal structure of strains of *Escherichia coli* causing neonatal septicemia and meningitis. *Infect. Immun.* **52**:213–222.

Sen, A., M.A. Leon, and S. Palchaudhuri. 1990. Comparative study of attachment to and invasion of epithelial cell lines by *Shigella dysenteriae*. *Infect. Immun.* **58**:2401–2403.

Sengupta, D., K. Datta-Roy, K. Banerjee, and A.C. Ghose. 1989. Identification of some antigenically related outer-membrane proteins of strains of *vibrio cholerae* 01 and non-

01 serovars involved in intestinal adhesion and the protective role of antibodies to them. *J. Med. Microbiol.* **29**:33–39.

Senior, D., N. Baker, B. Cedergren, P. Falk, G. Larson, R. Lindstedt, and C. Svanborg-Eden. 1988. Globo-A: a new receptor specificity for attaching *Escherichia coli*. *FEBS Lett.* **237**:123–127.

Shafer, W.M. and R.F. Rest. 1989. Interactions of gonococci with phagocytic cells. *Annu. Rev. Microbiol.* **43**:121–145.

Sharon, N. and I. Ofek. 1986. Mannose-specific bacterial surface lectins. In: Mirelman, D. (ed.): *Microbial Lectins and Agglutinins*. John Wiley and Sons, New York, pp. 55–81.

Shaw, J.H. and S. Falkow. 1988. Model for invasion of human tissue culture cells by *Neisseria gonorrhoeae*. *Infect. Immun.* **56**:1625–1632.

Shaw, C.E. and R.K. Taylor. 1990. *Vibrio cholerae* 0395 *tcpA* pilin gene sequence and comparison of predicted protein structural features to those of type 4 pilins. *Infect. Immun.* **58**:3042–3049.

Sheinfeld, J., A.J. Schaeffer, C. Cordon-Cardo, A. Rogatko, and W.R. Fair. 1989. Association of the Lewis blood-group phenotype with recurrent urinary tract infections in women. *N. Engl. J. Med.* **320**:773–777.

Sherman, L.A. and D.C. Savage. 1986. Lipoteichoic acids in *Lactobacillus strains* that colonize the mouse gastric epithelium. *Appl. Envir. Microbiol.* **52**:302–304.

Sherman, P.M. and E.C. Boedeker. 1987. Regional differences in attachment of enteroadherent *Escherichia coli* strain RDEC-1 to rabbit intestine: luminal colonization but lack of mucosal adherence in jejunal self-filling blind loops. *J. Pediatr. Gastroenterol. Nutr.* **6**:439–444.

Sherman, P.M. and R. Soni. 1988. Adherence of vero cytotoxin-producing *Escherichia coli* of serotype 0157:H7 to human epithelial cells in tissue culture: role of outer membranes as bacterial adhesins. *J. Med. Microbiol.* **26**:11–17.

Sherman, P.M., W.L. Houston, and E.C. Boedeker. 1985. Functional heterogeneity of intestinal *Escherichia coli* strains expressing type 1 somatic pili (fimbriae): assessment of bacterial adherence to intestinal membranes and surface hydrophobicity. *Infect. Immun.* **49**:797–804.

Sherman, P., R. Soni, M. Petric, and M. Karmali. 1987. Surface properties of the vero cytotoxin-producing *Escherichia coli* 0157:H7. *Infect. Immun.* **55**:1824–1829.

Sherman, P., B. Drumm, M. Karmali, and E. Cutz. 1989a. Adherence of bacteria to the intestine in sporadic cases of enteropathogenic *Escherichia coli*-associated diarrhea in infants and young children: A prospective study. *Gastroenterology* **96**:86–94.

Sherman P., R. Soni, and E. Boedeker. 1989b. Role of type 1 somatic pili (fimbriae) in mucosal attachment of the enteroadherent *Escherichia coli*, strain RDEC-1, in rabbits. *J. Pediatr. Gastroenter. Nutr.* **7**:594–601.

Sherman, P., F. Cockerill III, R. Soni, and J. Brunton. 1991. Outer membranes are competitive inhibitors of *Escherichia coli* 0157:H7 adherence to epithelial cells. *Infect. Immun.* **59**:890–899.

Shimamura, K., H. Shigemi, Y. Kurono, and G. Mogi. 1990. The role of bacterial adherence in otitis media with effusion. *Arch. Otolaryngol.-Head-Neck-Surg.* **116**:1143–1146.

Shimauchi, H., T. Ogawa, and S. Hamada. 1991. Immune response gene regulation of the humoral immune response to *Porphyromonas gingivalis* fimbriae in mice. *Immunology* **74**:362–364.

Singh, A., L.D. Hazlett, and R.S. Berk. 1990. Characterization of *Pseudomonas aeruginosa* adherence to mouse corneas in organ culture. *Infect. Immun.* **58**:1301–1307.

Singh, U., D. Grenier, and B.C. McBride. 1989. *Bacteroides gingivalis* vesicles mediate attachment of streptococci to serum-coated hydroxyapatite. *Oral Microbiol. Immunol.* **4**:199–203.

Sjoberg, P., M. Lindahl, J. Porath, and T. Wadström. 1988. Purification and characterization of CS2, a sialic acid-specific haemagglutinin of enterotoxigenic *Escherichia coli. Biochem. J.* **255**:105–111.

Slomiany, B.L., J. Piotrowski, A. Samanta, K. VanHorn, V.L.N. Murty, and A. Slomiany. 1989. *Campylobacter pylori* colonization factor shows specificity for lactosylceramide sulfate and GM3 ganglioside. *Biochem. Int.* **19**:929–936.

Slusher, M.M., Q.N. Myrvik, J.C. Lewis, and A.G. Gristina. 1987. Extended wear lenses, biofilm, and bacterial adhesion. *Arch. Ophthalmol.* **105**:110–115.

Smalley, J.W., A.J. Birss, A.S. McKee, and P.D. Marsh. 1991. Haemin-restriction influences haemin-binding, haemagglutination and protease activity of cells and extracellular membrane vesicles of *Porphyromonas gingivalis* W50. *FEMS Microbiol. Lett.* **90**:63–68.

Smirnova, T.A., L.I. Kulinich, M.Yu. Galperin, and R.R. Azizbekyan. 1991. Subspecies-specific haemagglutination patterns of fimbriated *Bacillus thuringiensis* spores. *FEMS Microbiol. Lett.* **90**:1–4.

Smith, A.L., P. Jenny, L. Langley, P. Verdugo, M. Villanlon, J. Unadkat, and D. Luchtel. 1989. Kinetic measurement of bacterial adherence to eukaryotic cells. *J. Microbiol. Meth.* **46**:787–796.

Smith, J.I., B. Drumm, A.W. Neumann, Z. Policova, and P.M. Sherman. 1990. In vitro surface properties of the newly recognized gastric pathogen *Helicobacter pylori. Infect. Immun.* **58**:3056–3060.

Smith, J.W. 1986a. Adherence to kidney tissue of *Escherichia coli* from humans with urinary tract infection. *Curr. Microbiol.* **14**:91–94.

Smith, J.W. 1986b. Microbial and host factors that influence adherence of *Escherichia coli* to kidney epithelium. *Am. J. Kidney Dis.* **7**:368–374.

Smith, R.N., R.N. Andersen, and P.E. Kolenbrander. 1991. Inhibition of intergeneric coaggregation among oral bacteria by cetylpyridinium chloride, chlorhexidine digluconate and octenidine dihydrochloride. *J. Periodont. Res.* **26**:422–428.

Sobel, J.D. and D. Kay. 1986. Enhancement of *Escherichia coli* adherence to epithelial cells derived from estrogen-stimulated rats. *Infect. Immun.* **53**:53–56.

Sobota, A.E. and L.L. Apicella. 1991. Reduction in the anti-adherence activity of Tamm-Horsfall protein with increasing concentration of calcium. *Urol. Res.* **19**:177–180.

Soderlind, O., B. Thafvelin, and R. Mollby. 1988. Virulence factors in *Escherichia coli* strains isolated from Swedish piglets with diarrhea. *J. Clin. Microbiol.* **26**:879–884.

Sparling, P.F. 1988. Adherence of *Neisseria gonorrhoeae*: how little we really know. In: M.H. Schrinner, G. Richmond, and U. Schwartz (eds.), *Surface Structures of Microorganisms and Their Interactions with Mammalian Host*. VCH Publishers, Veinheim, Germany, pp. 197–208.

Sparling, P.F. and J.G. Cannon. 1986. Phase and antigenic variation of pili and outer membrane protein II of *Neisseria gonorrhoeae*. *J. Infect. Dis.* **153**:196–201.

Speert, D.P., B.A. Loh, D.A. Cabral, and I.E. Salit. 1986. Nonopsonic phagocytosis of nonmucoid *Pseudomonas aeruginosa* by human neutrophils and monocyte-derived and macrophages is correlated with bacterial piliation and hydrophobicity. *Infect. Immun.* **53**:207–212.

St. Geme, III, J.W. and S. Falkow. 1990. *Haemophilus influenzae* adheres to and enters cultured human epithelial cells. *Infect. Immun.* **58**:4036–4044.

St. Geme, III, J.W. and S. Falkow. 1991. Loss of capsule expression by *Haemophilus influenzae* type b results in enhanced adherence to and invasion of human cells. *Infect. Immun.* **59**:1325–1333.

St. Geme, III, J.W., J.R. Gilsdorf, and S. Falkow. 1991. Surface structures and adherence properties of diverse strains of *Haemophilus influenzae* biogroup aegyptius. *Infect. Immun.* **59**:3366–3371.

Stamm, W.E., T.M. Hooton, J.R. Johnson, C. Johnson, A. Stapleton, P.L. Roberts, and S.D. Fihn. 1989. Urinary tract infections: from pathogenesis to treatment. *J. Infect. Dis.* **159**:400–408.

Steiner, B.M., S. Sell, and R.F. Schell. 1987. *Treponema pallidum* attachment to surface and matrix proteins of cultured rabbit epithelial cells. *J. Infect. Dis.* **155**:742–748.

Stenqvist, K., T. Sandberg, G. Lidin-Janson, F. Orskov, I. Orskov, and C. Svanborg-Eden. 1987. Virulence factors of *Escherichia coli* in urinary isolates from pregnant women. *J. Infect. Dis.* **156**:870–877.

Stenstrom, T.A. and S. Kjelleberg. 1985. Fimbriae mediated nonspecific adhesion of *Salmonella typhimurium* to mineral particles. *Arch. Microbiol.* **143**:6–10.

Stephens, D.S. 1989. Gonococcal and meningococcal pathogenesis as defined by human cell, cell culture, and organ culture assays. *Clin. Microbiol. Rev.* **2**:S104–S111.

Stephens, D.S., A.M. Whitney, J. Rothbard, and G.K. Schoolnik. 1985. Analysis of structure and investigation of structural and antigenic relationships to gonococcal pili. *J. Exp. Med.* **161**:1539–1553.

Stephens, D.S., A.M. Whitney, M.A. Melly, L.H. Hoffman, M.M. Farley, and C.E. Frasch. 1986. Analysis of damage to human ciliated nasopharyngeal epithelium by *Neisseria meningitidis*. *Infect. Immun.* **51**:579–585.

Stephens, D.S., A.M. Whitney, G.K. Schoolnik, and W.D. Zollinger. 1988. Common epitopes of pilin of *Neisseria meningitidis*. *J. Infect. Dis.* **158**:332–342.

Sterk, L.M., L. Van Alphen, L. Geelen van den Broek, H.J. Houthoff, and J. Dankert. 1991. Differential binding of *Haemophilus influenzae* to human tissues by fimbriae. *J. Med. Microbiol.* **35**:129–138.

Stern, G.A. and Z.S. Zam. 1986. The pathogenesis of contact lenses-associated *Pseudomonas aeruginosa* corneal ulceration.I. The effect of contact lens coatings on adherence of *Pseudomonas aeruginosa* to soft contact lenses. *Cornea* **5**:41–45.

Stern, G.A. and Z.S. Zam. 1987. The effect of enzymatic contact lenses on adherence of *Pseudomonas aeruginosa* to soft contact lenses. *Ophthalmology* **94**:115–119.

Stevens, M.K. and D.C. Krause. 1990. Disulfide-linked protein associated with *Mycoplasma pneumoniae* cytadherence phase variation. *Infect. Immun.* **58**:3430–3433.

Stinson, M.W., K. Safulko, and M.J. Levine. 1991. Adherence of *Porphyromonas (Bacteroides) gingivalis* to *Streptococcus sanguis* in vitro. *Infect. Immun.* **59**:102–108.

Stoll, B.J., A.M. Svennerholm, L. Gothefors, D. Barua, S. Huda, and J. Holmgren. 1986. Local and systemic antibody responses to naturally acquired enterotoxigenic *Escherichia coli* diarrhea in an endemic area. *J. Infect. Dis.* **133**:527–532.

Stromberg, N. and K.A. Karlsson. 1990. Characterization of the binding of *Propionibacterium granulosum* to glycosphingolipids adsorbed on surfaces. *J. Biol. Chem.* **265**:11244–12250.

Stromberg, N., C. Deal, G. Nyberg, S. Normark, M. So, and K.A. Karlsson. 1988a. Identification of carbohydrate structures that are possible receptors for *Neisseria gonorrhoeae*. *Proc. Natl. Acad. Sci. USA* **85**:4902–4906.

Stromberg, N., M. Ryd, A.A. Lindberg, and K.A. Karlsson. 1988b. Two species of *Propionibacterium* apparently recognize separate epitopes on lactose of lactosylceramide. *FEBS Lett.* **232**:193–198.

Stromberg, N., B.-I. Marklund, B. Lund, D. Ilver, A. Hamers, W. Gaastra, K.A. Karlsson, and S. Normark. 1990. Host-specificity of uropathogenic *Escherichia coli* depends on differences in binding specificity to Galα1–4Gal-containing isoreceptors. *EMBO J.* **9**:2001–2010.

Su, C.J., V.V. Tryon, and J.B. Baseman. 1987. Cloning and sequence analysis of cytadhesin P1 gene from *Mycoplasma pneumoniae*. *Infect. Immun.* **55**:3023–3029.

Su, C.J., A. Chavoya, and J.B. Baseman. 1988. Regions of *Mycoplasma pneumoniae* cytadhesin P1 structural gene exist as multiple copies. *Infect. Immun.* **56**:3157–3161.

Su, C.J., A. Chavoya, S.F. Dallo, and J.B. Baseman. 1990a. Sequence divergency of the cytadhesin gene of *Mycoplasma pneumoniae*. *Infect. Immun.* **58**:2669–2674.

Su, H., N.G. Watkins, Y.X. Zhang, and H.D. Caldwell. 1990b. *Chlamydia trachomatis*-host cell interactions: role of the chlamydial major outer membrane protein as an adhesin. *Infect. Immun.* **58**:1017–1025.

Sun, D.X. J.J. Mekalanos, and R.K. Taylor. 1990. Antibodies directed against the toxin-coregulated pilus isolated from *Vibrio cholerae* provide protection in the infant mouse experimental cholera model. *J. Infect. Dis.* **16**:1231–1236.

Sun, D., J.M. Seyer, I. Kovari, R.A. Sumrada, and R.K. Taylor. 1991. Localization of protective epitopes within the pilin subunit of the *Vibrio cholerae* toxin-coregulated pilus. *Infect. Immun.* **59**:114–118.

Sussman, M. (ed.) 1985. *Virulence of Escherichia coli: Reviews and methods*, Academic Press, London.

Suzuki, Y., F. Yoshimura, K. Takahashi, H. Tani, and T. Suzuki. 1988. Detection of fimbriae and fimbrial antigens on the oral anaerobe *Bacteroides gingivalis* by negative staining and serological methods. *J. Gen. Microbiol.* **134**:2713–2720.

Svanborg-Eden, S. Hausson, U. Jodal, G. Lidin-Janson, K. Lincoln, H. Linder, H. Lomberg, P. de Man, S. Marild, J. Martinell, K. Plos, T. Sandberg, and K. Stenqvist. 1988. Host-parasite interaction in the urinary tract. *J. Infect. Dis.* **157**:421–426.

Svanborg-Eden, C., B. Anderson, G. Aniansson, R. Lindstedt, P. de Man, A. Nielsen, H. Leffler, and A. Wold. 1990. Inhibition of bacterial attachment: examples from the urinary and respiratory tracts. *Current Top. Microbiol. Immun.* **151**:167–186.

Svennerholm, A.M., Y. Lopez-Vidal, J. Holmgren, M.M. McConnell, and B. Rowe. 1988. Role of PCF8775 antigen and its coli surface subcomponents for colonization, disease, and protective immunogenicity of enterotoxigenic *Escherichia coli* in rabbits. *Infect. Immun.* **56**:523–528.

Svennerholm, A.M., C. Wenneras, J. Holmgren, M.M. McConnell, and B. Rowe. 1990. Roles of different coli surface antigens of colonization factor antigen II in colonization by and protective immunogenicity of enterotoxigenic *Escherichia coli* in rabbits. *Infect. Immun.* **58**:341–346.

Swanson, A.F. and C. C. Kuo. 1990. Identification of lectin-binding proteins in *Chlamydia* species. *Infect. Immun.* **58**:502–507.

Swanson, J. 1983. Gonococcal adherence: selected topics. *Rev. Infect. Dis.* **5**:5678–5684.

Swanson, J. 1990. Pilus and outer membrane protein II variation in *Neisseria gonorrhoeae*. In: Ayoub, E.M., G.H. Cassell, W.C. Branche, Jr., and F.J. Henry (eds.), *Microbial determinants of virulence and host response*. American Society for Microbiology, Washington, pp. 197–205.

Swanson, J. and J.M. Koomey. 1989. Mechanisms for variations of pili and outer membrane protein II in *Neisseria gonorrhoeae*. In: Berg, D.E. and M. Howe (eds.), *Mobile DNA*. American Society for Microbiology, Washington, pp. 743–761.

Swanson, J., O. Barrera, J. Sola, and J. Boslego. 1988. Expression of outer membrane protein II by gonococci in experimental gonorrhea. *J. Exp. Med.* **168**:2121–2129.

Switalski, L.M, H. Murchison, R. Timpl, R. Curtiss III, and M. Hook. 1987. Binding of laminin to oral and endocarditis strains of viridans streptococci. *J. Bacteriol.* **169**:1095–1101.

Szcepanski, A., M.B. Furie, J.L. Benach, B.P. Lane, and H.B. Fleit. 1990. Interaction between *Borrelia burgdorferi* and endothelium *in vitro*. *J. Clin. Invest.* **85**:1637–1647.

Tacket, C.O., D.R. Maneval, and M.M. Levine. 1987. Purification, morphology, and genetics of a new fimbrial putative colonization factor of enterotoxigenic *Escherichia coli* 0159:H4. *Infect. Immun.* **55**:1063–1069.

Tanpowpong, K., C.D. Bluestone, and C.C. Brinton Jr. 1989. Evaluation of pilus vaccines for prevention of experimental otitis media caused by nontypable *Haemophilus influenzae*. *Pediatr. Infect. Dis. J.* **8**:62–65.

Tarkkanen, A.-M., B. Allen, B. Westerlund, H. Holthofer, P. Kuusela, L. Ristelli, S.

Clegg, and T.K. Korhonen. 1990. Type V collagen as target for type–3 fimbriae, enterobacterial adherence organelles. *Mol. Microbiol.* **4**:1353–1361.

Tart, R.C. and I. van de Rijn. 1991. Analysis of adherence of *Streptococcus defectivus* and endocarditis-associated streptococci to extracellular matrix. *Infect. Immun.* **59**:857–862.

Tavendale, A. and D.C. Old. 1985. Haemagglutinins and adhesion of *Escherichia coli* to HEp–2 epithelial cells. *J. Med. Microbiol.* **20**:345–353.

Taylor, D.C., A.W. Cripps, and R.L. Clancy. 1990. Inhibition of adhesion of *Haemophilus influenzae* to buccal cells by respiratory secretions. *Immunol. Cell. Biol.* **68**:335–342.

Taylor, R., C. Shaw, K. Peterson, P. Spears, and J. Mekalanos. 1988. Safe, live *Vibrio cholerae* vaccines? *Vaccine* **6**:151–154.

Tempro, P., F. Cassels, R. Siraganian, A.R. Hand, and J. London. 1989. Use of adhesins-specific monoclonal antibodies to identify and localize an adhesin on the surface of *Capnocytophaga gingivalis* DR2001. *Infect. Immun.* **57**:3418–3424.

Teneberg, S., P. Willemsen, F.K. DeGraaf, and K.A. Karlsson. 1990. Receptor-active glycolipids of epithelial cells of the small intestine of young and adult pigs in relation to susceptibility to infection with *Escherichia coli* K99. *FEBS Microbiol. Lett.* **263**:10–14.

Tennent, J.M., S. Hultgren, B.-I. Marklund, K. Forsman, M. Göransson, B.E. Uhlin and S. Normark. 1990. *In*: Iglewski, B.H. and V.L. Clark: *Molecular Basis of Bacterial Pathogenesis*, Academic Press, San Diego, pp. 79–110.

Teppema, J.S., P.A.M. Guinee, A.A. Ibrahim, M. Paques, and E.J. Ruitenberg. 1987. In vivo adherence and colonization of *Vibrio cholerae* strains that differ in hemagglutinating activity and motility. *Infect. Immun.* **55**:2093–2102.

Thomas, D.D. and E. Comstock. 1989. Interaction of lyme disease spirochetes with cultured eucaryotic cells. *Infect. Immun.* **57**:1324–1326.

Thomas, D.D. and L.M. Higbie. 1990. *In vitro* association of leptospires with host cells. *Infect. Immun.* **58**:581–585.

Thomas, D.D., J.B. Baseman, and J.F. Alderete. 1985a. Putative *Treponema pallidum* cytadhesins share a common functional domain. *Infect. Immun.* **49**:833–835.

Thomas, D.D., J.B. Baseman, and J.F. Alderete. 1985b. Fibronectin tetrapeptide is target for syphilis spirochete cytadherence. *J. Exp. Med.* **162**:1715–1719.

Thomas, L.V., M.M. McConnell, B. Rowe, and A.M. Field. 1985c. The possession of three enovel coli surface antigens by enterotoxigenic *Escherichia coli* strains positive for the putative colonization factor PCF8775. *J. Gen. Microbiol.* **131**:2319–2326.

Thomas, D.D., M. Navab, D.A. Haak, A.M. Fogelman, J.N. Miller, and M.A. Lovett. 1988. *Treponema pallidum* invades endothelial cell monolayers by intracellular penetration. *Proc. Natl. Acad. Sci. USA* **85**:3608–3612.

Thomas, D.D., A.M. Fogelman, J.N. Miller, and M.A. Lovett. 1989. Interactions of *Treponema pallidum* with endothelial cell monolayers. *Eur. J. Epidemiol.* **5**:15–21.

Thorns, C.J. and P.L. Roeder. 1988. Use of monoclonal antibody to detect the F41 fimbrial adhesin of enterotoxigenic *Escherichia coli*. *Res. Vet. Sci.* **44**:394–395.

Thorns, C.J., C.D.H. Boarer, and J.A. Morris. 1987. Production and evaluation of monoclonal antibodies directed against the K88 fimbrial adhesin produced by *Escherichia coli* enterotoxigenic for piglets. *Res. Vet. Sci.* **43**:233–238.

Tinsley, C.R. and J.E. Heckels. 1986. Variation in the expression of pili and outer membrane protein by *Neisseria meningitidis* during the course of meningococcal infection. *J. Gen. Microbiol.* **132**:2483–2490.

Tjia, K.F., J.P. van Putten, E. Pels, and H.C. Zanen. 1988. The interaction between *Neisseria gonorrhoeae* and the human cornea in organ culture. An electron microscopic study. *Graefes Arch. Clin. Exp. Ophthalmol.* **226**:341–345.

Todd, T.R., A. Franklin, I.P. Mankinen, G. Gurman, and R.T. Irvin. 1989. Augmented bacterial adherence to tracheal epithelial cells is associated with gram-negative pneumonia in an intensive care unit population. *Am. Rev. Respir. Dis.* **140**:1585–1589.

Topley, N., R.K. Mackenize, R. Steadman, and J.D. Williams. 1989a. *Escherichia coli* virulence and renal scarring. *J. Infect. Dis.* **160**:1081–1082.

Topley, N., R. Steadman, R. Mackenize, J.M. Knowlden, and J.D. Williams. 1989b. Type 1 fimbriae strains of *Escherichia coli* initiate renal parenchymal scarring. *Kidney Int.* **36**:609–616.

Toth, I., M.L. Cohen, H.S. Rumschlag, L.W. Riley, E.H. White, J.H. Carr, W.W. Bond, and I.K. Wachsmuth. 1990. Influence of the 60-megadalton plasmid on adherence of *Escherichia coli* 0157:H7 and genetic derivatives. *Infect. Immun.* **58**:1223–1231.

Trevino, L.B., W.G. Haldenwang, and J.B. Baseman. 1986. Expression of *Mycoplasma pneumoniae* antigens in *Escherichia coli*. *Infect. Immun.* **53**:129–134.

Tricottet, V., P. Bruneval, O. Vire, J.P. Camiller, F. Bloch, N. Bonte, and J. Roge. 1986. *Campylobacter*-like organisms and surface epithelium abnormalities in active, chronic gastritis in humans: an ultrastructural study. *Ultrastruct. Pathol.* **10**:113–122.

Trigo, E. and C. Pijoan. 1988. Presence of pili in *Pasteurella multocida* strains associated with atrophic rhinitis. *Vet. Rec.* **122**:19.

Trinchieri, A., L. Braceschi, D. Tiranti, S. Dell'Acqua, A. Mandressi, and E. Pisani. 1990. Secretory immunoglobulin A and inhibitory activity of bacterial adherence to epithelial cells in urine from patients with urinary tract infections. *Urol. Res.* **18**:305–308.

Trust, T.J., P. Doig, L. Emody, Z. Kienle, T. Wadström, and P. O'Toole. 1991. High-affinity binding of the basement membrane proteins collagen type IV and laminin to the gastric pathogen *Helicobacter pylori*. *Infect. Immun.* **59**:4398–4404.

Tullus, K., B. Fryklund, B. Berglund, G. Kallenius, and L.G. Burman. 1988. Influence of age on faecal carriage of P-fimbriated *Escherichia coli* and other gram-negative bacteria in hospitalized neonates. *J. Hosp. Infect.* **11**:349–356.

Tuomanen, E. 1986. Piracy of adhesins: attachment of superinfecting pathogens to respiratory cilia by secreted adhesins of *Bordetella pertussis*. *Infect. Immun.* **54**:905–908.

Tuomanen, E. and A. Weiss. 1985. Characterization of two adhesins of *Bordetella pertussis* for human ciliated respiratory-epithelial cells. *J. Infect. Dis.* **152**:118–125.

Tuomanen, E., H. Towbin, G. Rosenfelder, D. Braun, G. Larson, G.C. Hansson, and R. Hill. 1988. Receptor analogs and monoclonal antibodies that inhibit adherence of *Bordetella pertussis* to human ciliated respiratory epithelial cells. *J. Exp. Med.* **168**:267–277.

Tzipori, S., K. Nachsmuth, C. Chapman, and R. Birner. 1986. The pathogenesis of hemorrhagic colitis caused by *Escherichia coli* 0175:H7 in gnotobiotic piglets. *J. Infect. Dis.* **4**:712–716.

Tzipori, S., R. Gibson, and J. Montanaro. 1989. Nature and distribution of mucosal lesions asociated with enteropathogenic and enterohemorrhagic *Escherichia coli* in piglets and the role of plasmid-mediated factors. *Infect. Immun.* **57**:1142–1150.

Tzouvelekis, L.S., A.F. Mentis, A.M. Makris, C. Spiliadis, C. Blackwell, and D.M. Weir. 1991. In vitro binding of *Helicobacter pylori* to human gastric mucin. *Infect. Immun.* **59**:4252–4254.

Uemura, K.I., L.M. Loomes, R.A. Childs, and T. Feizi. 1988. Evidence for occurrence of passively adsorbed I antigen activity on a cultured strain of *Mycoplasma pneumoniae*. *Infect. Immun.* **56**:3015–3020.

Uhlin, B.E., M. Norgren, M. Baga, and S. Normark. 1985. Adhesion to human cells by *Escherichia coli* lacking the major subunit of a digalactoside-specific, pilus adhesin. *Proc. Natl. Acad. Sci.* (USA) **82**:1800–1804.

Urisu, A., J.L. Cowell, and C.R. Manclark. 1986. Filamentous hemagglutinin has a major role in mediating adherence of *Bordetella pertussis* to human WiDr cells. *Infect. Immun.* **52**:695–701.

Vaisanen-Rhen, V., M. Rhen, E. Linder, and T.K. Korhonen. 1985. Adhesion of *Escherichia coli* to human kidney cryostat sections. *FEMS Microbiol. Lett.* **27**:179–182.

Valdivieso-Garcia, A., S. Rosendal, and S. Serebrin. 1989. Adherence of *Mycoplasma mycoides subspecies mycoides* to cultured endothelial cells. *Int. J. Med. Microbiol.* **272**:210–215.

Valentin-Weigand, P., G.S. Chhatwal, and H. Blobel. 1988. Adherence of streptococcal isolates from cattle and horses to their respective host epithelial cells. *Am. J. Vet. Res.* **49**:1485–1488.

Valpotic, I., P.L. Runnels, and H.W. Moon. 1989. In vitro adhesion of K88ac *Escherichia coli* to Peyer's patch and peripheral blood lymphocytes, buccal and rectal epithelial cells or intestinal epithelial brush borders of weaned pigs. *J. Vet. Microbiol.* **20**:357–368.

van Alphen, L., J. Poole, and M. Overbeeke. 1986. The anton blood group antigen is the erythrocyte receptor for *Haemophilus influenzae*. *FEMS Microbiol. Lett.* **37**:69–71.

van Alphen, L., J. Poole, L. Geelen, and H.C. Zanen. 1987. The erythrocyte and epithelial cell receptors for *Haemophilus influenzae* are expressed independently. *Infect. Immun.* **55**:2355–2358.

van Alphen, L., N. van den Berghe, and L. Geelen van den Broek. 1988. Interaction of *Haemophilus influenzae* with human erythrocytes and oropharyngeal epithelial cells is mediated by a common fimbrial epitope. *Infect. Immun.* **56**:1800–1806.

van Alphen, L., M. van Ham, L. Geelen van den Broek, and T. Pieters. 1989. Relationship between secretion of the anton blood group antigen in saliva and adherence of *Haemophilus influenzae* to oropharynx epithelial cells. *FEMS Microbiol. Immunol.* **1**:357–362.

van Alphen, L., C. Levene, L.G. Broek, J. Poole, M. Bennett, and J. Dankert. 1990. Combined inheritance of epithelial and erythrocyte receptors for *Haemophilus influenzae*. *Infect. Immun.* **58**:3807–3809.

van Alphen, L., Geelen van den Broek, L. Blaas, M. van Ham, and Dankert-J. 1991. Blocking of fimbria-mediated adherence of *Haemophilus influenzae* by sialyl gangliosides. *Infect. Immun.* **59**:4473–4477.

Van Die, I., C. Kramer, J. Hacker, H. Bergmans, W. Jongen, and W. Hoekstra. 1991. Nucleotide sequence of the genes coding for minor fimbrial subunits of the F1C fimbriae of *Escherichia coli*. *Res. Microbiol.* **142**:653–658.

van Doorn, J. F.R. Mooi, A.M. van Vught, and D.M. MacLaren. 1987. Characterization of fimbriae from *Bacteroides fragilis*. *Microbiol. Pathog.* **3**:87–95.

van Ham, S.M., F.R. Mooi, M.G. Sindhunata, W.R. Maris, and L. van Alphen. 1989. Cloning and expression in *Escherichia coli* of *Haemophilus influenzae* fimbrial genes establishes adherence to oropharyngeal epithelial cells. *EMBO J.* **8**:3535–3540.

van Loosdrecht, M.C.M., J. Lyklema, W. Norde, G. Schraa, and A.J.B. Zehnder. 1987a. Electrophoretic mobility and hydrophobicity as a measure to predict the initial steps of bacterial adhesion. *Appl. Environ. Microbiol.* **53**:1898–1901.

van Loosdrecht, M.C.M., J. Lyklema, W. Norde, G. Schraa, and A.J.B. Zehnder. 1987b. The role of bacterial cell wall hydrophobicity in adhesion. *Appl. Environ. Microbiol.* **53**:1893–1897.

van Zijderveld, F.G., J. Anakotta, R.A.M. Brouwers, A.M. van Zijderveld, D. Bakker, and F.K. De Graaf. 1990. Epitope analysis of the F4 (K88) fimbrial antigen complex of enterotoxigenic *Escherichia coli* by using monoclonal antibodies. *Infect. Immun.* **58**:1870–1878.

Vena, M.M., B. Blanchard, D. Thomas, and M. Kobisch. 1991. Adherence of *Pasteurella multocida* isolated from pigs and relationship with capsular type and dermonecrotic toxin production. *Ann. Rech. Vet.* **22**:211–218.

Vial, P.A., R. Robins-Browne, H. Lior, V. Prado, J.B. Kaper, J.P. Nataro, D. Maneval, A. Elsayed, and M.M. Levine. 1988. Characterization of enteroadherent *Escherichia coli*, a putative agent of diarrheal disease. *J. Infect. Dis.* **158**:70–79.

Virji, M. and J.E. Heckels. 1985. Role of anti-pilus antibodies in host defense against gonococcal infection studied with monoclonal anti-pilus antibodies. *Infect. Immun.* **49**:621–628.

Virji, M. and J.E. Heckels. 1986. The effect of protein II and pili on the interaction of *Neisseria gonorrhoeae* with human polymorphonuclear leucocytes. *J. Gen. Microbiol.* **132**:503–512.

Virji, M., J.E. Heckels, W.J. Potts, C.A. Hart, and J.R. Saunders. 1989. Identification of epitopes recognized by monoclonal antibodies SM1 and SM2 which react with all pili of *Neisseria gonorrhoeae* but which differentiate between two structural classes of pili expressed by *Neisseria meningitidis* and the distribution of their encoding sequences in the genomes of *Neisseria* spp. *J. Gen. Microbiol.* **135**:3239–3251.

Virji M., H. Kayhty, D.J. Ferguson, C. Alexandrescu, and E.R. Moxon. 1991a. Interactions of *Haemophilus influenzae* with cultured human endothelial cells. *Microb. Pathogen.* **10**:231–245.

Virji, M., H. Kayhty, D.J.P. Ferguson, C. Alexandrescu, J.E. Heckels, and E.R. Moxon. 1991b. The role of pili in the interaction of pathogenic *Neisseria* with cultured human endothelial cells. *Mol. Microbiol.* **5**:1831–1841.

Virkola, R. 1987. Binding characteristics of *Escherichia coli* type 1 fimbriae in the human kidney. *FEMS Microbiol. Lett.* **40**:257–262.

Virkola, R., B. Westerlund, H. Holthofer, J. Parkkinen, M. Kekomaki, and T.K. Korhonen. 1988. Binding characteristics of *Escherichia coli* in human urinary bladder. *Infect. Immun.* **56**:2615–2622.

Visai, L., P. Speziale, and S. Bozzini. 1990. Binding of collagens to an enterotoxigenic strain of *Escherichia coli*. *Infect. Immun.* **58**:449–455.

Visai, L., S. Bozzini, T.E. Petersen, L. Speciale, and P. Speziale. 1991. Binding sites in fibronectin for an enterotoxigenic strain of *E. coli* B342289c. *FEBS Lett.* **290**:111–114.

von Hunolstein, C., M.L. Ricci, R. Scenati, and G. Orefici. 1987. Adherence of *S. bovis* to adult buccal epithelial cells. *Microbiologica* **10**:385–392.

Vu, A.C., H.M. Foy, F.D. Cartwright, and G.E. Kenny. 1987. The principal proteins of isolates of *Mycoplasma pneumoniae* measured by levels of immunoglobulin G in human serum are stable in strains collected over a 10-year period. *Infect. Immun.* **55**:1830–1836.

Vuopio-Varkila, J. and G.K. Schoolnik. 1991. Localized adherence by enteropathogenic *Escherichia coli* is an inducible phenotype associated with the expression of new outer membrane proteins. *J. Exp. Med.* **174**:1167–1177.

Wadolkowski, E.A., D.C. Laux, and P.S. Cohen. 1988a. Colonization of the streptomycin-treated mouse large intestine by a human fecal *Escherichia coli* strain: role of growth in mucus. *Infect. Immun.* **56**:1030–1035.

Wadolkowski, E.A., D.C. Laux, and P.S. Cohen. 1988b. Colonization of the streptomycin-treated mouse large intestine by a human fecal *Escherichia coli* strain: role of adhesion to mucosal receptors. *Infect. Immun.* **56**:1036–1043.

Wadström, T. 1988. Adherence traits and mechanisms of microbial adhesion in the gut. *Bailliere's Clin. Trop. Med. Communicable Dis.* **3**:417–433.

Wadström, T. and S.B. Baloda. 1986. Molecular aspects on small bowel colonization by enterotoxigenic *Escherichia coli*. *Microecol. Ther.* **16**:243–255.

Wadström, T., K. Andersson, M. Sydow, L. Axelsson, S. Lindgren, and B. Gullmar. 1987. Surface properties of lactobacilli isolated from the small intestine of pigs. *J. Appl. Bacteriol.* **62**:513–520.

Wadström, T., J.L. Guruge, S. Wei, P. Aleljung, and A. Ljungh. 1991. The pathogenic mechanisms of *Helicobacter pylori*: a short overview. In: Wadström, T., P.M. Kakela,

A.-M. Svennerholm, and H. Wolf Watz (eds.), *Molecular Pathogenesis of Gastrointestinal Infections*. Plenum, New York, pp. 249–255.

Walan, A. and E. Kihlstrom. 1988. Surface charge and hydrophobicity of *Campylobacter jejuni* strains in relation to adhesion to epithelial HT–29 cells. *Acta Pathol. Microbiol. Immunol. Scand.* **96**:1089–1096.

Walker, R.I., M.B. Caldwell, E.C. Lee, P. Guerry, T.J. Trust, and G.M. Ruiz-Palacios. 1986. Pathophysiology of *Campylobacter* enteritis. *Microbiol. Rev.* **50**:81–94.

Walker, R.I., C.E.A. Schmauder, J.L. Parker, and D. Burr. 1988. Selective association and transport of *Campylobacter jejuni* through M cells of rabbit Peyer's patches. *Can. J. Microbiol.* **34**:1142–1147.

Wanke, C.A. and R.L. Guerrant. 1987. Small-bowel colonization alone is a cause of diarrhea. *Infect. Immun.* **55**:1924–1926.

Wanke, C.A., S. Cronan, C. Goss, K. Chadee, and R.L. Guerrant. 1990. Characterization of binding of *Escherichia coli* strains which are enteropathogens to small-bowel mucin. *Infect. Immun.* **58**:794–800.

Wanke, C.A., J.B. Schorling, L.J. Barrett, M.A. Desouza, and R.L. Guerrant. 1991. Potential role of adherence traits of *Escherichia coli* in persistent diarrhea in an urban Brazilian slum. *J. Pediatr. Infect. Dis.* **10**:746–751.

Warren, J.W., H.L.T. Mobley, and A.L. Trifillis. 1988. Internalization of *Escherichia coli* into human renal tubular epithelial cells. *J. Infect. Dis.* **158**:221–223.

Wasteson, Y. and O. Olsvik. 1991. Specific DNA fragments coding for ST1 and LT1 toxins, and K88 (F4) adhesin in enterotoxigenic *Escherichia coli*. *J. Vet. Med. B.* **38**:445–452.

Watanabe, H. and A. Nakamura. 1985. Large plasmids associated with virulence in *Shigella* species have a common function necessary for epithelial cell penetration. *Infect. Immun.* **48**:260–262.

Weber, A., K. Harris, S. Lohrke, L. Forney, and A.L. Smith. 1991. Inability to express fimbriae results in impaired ability of *Haemophilus influenzae* b to colonize the nasopharynx. *Infect. Immun.* **59**:4724–4728.

Weber, R.D., B.E. Britigan, T. Svendsen, and M.S. Cohen. 1989. Energy is required for maximal adherence of *Neisseria gonorrhoeae* to phagocytic and nonphagocytic cells. *Infect. Immun.* **57**:785–790.

Weel, J.F., C.T. Hopmann, and J.P. van Putten. 1991a. Bacterial entry and intracellular processing of *Neisseria gonorrhoeae* in epithelial cells: immunomorphological evidence for alterations on the major outer membrane protein P.IB. *J. Exp. Med.* **174**:705–715.

Weel, J.F., C.T. Hopmann, and J.P. van Putten. 1991b. In situ expression and localization of *Neisseria gonorrhoeae* opacity proteins in infected epithelial cells: apparent role of Opa proteins in cellular invasion. *J. Exp. Med.* **173**:1395–1405.

Weerkamp, A.H., H.C. van der Mei, and R.S.B. Liem. 1986a. Structural properties of fibrillar proteins isolated from the cell surface and cytoplasm of *Streptococcus salivarius* (K^+) cells and nonadhesive mutants. *J. Bacteriol.* **165**:756–762.

Weerkamp, A.H., P.S. Handley, A. Baars, and J.W. Slot. 1986b. Negative staining and immunoelectron microscopy of adhesion deficient mutants of *Streptococcus salivarius* reveal that the adhesive protein antigens are separate classes of cell surface fibril. *J. Bacteriol.* **165**:746–755.

Weerkamp, A.H., H.C. van der Mei, and J.W. Slot. 1987. Relationship of cell surface morphology and composition of *Streptococcus salivarius* K^+ to adherence and hydrophobicity. *Infect. Immun.* **55**:438–445.

Weinberg, A. and S.C. Holt. 1990. Interaction of *Treponema denticola* TD–4, GM–1, and MS25 with human gingival fibroblasts. *Infect. Immun.* **58**:1720–1729.

Weiss, A.A. and E.L. Hewlett. 1986. Virulence factors of *Bordetella pertussis*. *Annu. Rev. Microbiol.* **40**:661–686.

Weiss, E.I., J. London, P.E. Kolenbrander, A.S. Kagermeier, and R.N. Andersen. 1987a. Characterization of lectin like surface components on *Capnocytophaga ochraceae* ATCC 33596 that mediate coaggregation with gram-positive oral bacteria. *Infect. Immun.* **55**:1198–1202.

Weiss, E.I., P.E. Kolenbrander, J. London, A.R. Hand, and R.N. Andersen. 1987b. Fimbria-associated adhesins of *Bacteroides loescheii* PK 1295 mediate intergeneric coaggregations. *J. Bacteriol.* **169**:4215–4222.

Weiss, E.I., J. London, P.E. Kolenbrander, R.N. Andersen, C. Fischler, and R.P. Siraganian. 1988a. Characterization of monoclonal antibodies to fimbria-associated adhesins of *Bacteroides loescheii* PK 1295. *Infect. Immun.* **56**:219–224.

Weiss, E.I., J. London, P.E. Kolenbrander, A.R. Hand, and R. Siraganian. 1988b. Localization and enumeration of fimbria-associated adhesins of *Bacteroides loescheii*. *J. Bacteriol.* **170**:1123–1128.

Weiss, E.I., J. London, P.E. Kolenbrander, and R.N. Andersen. 1989. Fimbria-associated adhesin of *Bacteroides loescheii* that recognizes receptors on procaryotic and eucaryotic cells. *Infect. Immun.* **57**:2912–2913.

Weiss, E.I., I. Eli, B. Shenitzki, and N. Smorodinsky. 1900. Identification of the rhamnose-sensitive adhesin of *Capnocytophaga ochracea* ATCC 33596. *Arch. Oral Biol. Suppl.* **35**:S127–S130.

Wenman, W.M. and R.U. Meuser. 1986. *Chlamydia trachomatis* elemetary bodies possess proteins which bind to eucaryotic cell membranes. *J. Bacteriol.* **165**:602–607.

Wenneras, C., J. Holmgren, and A.M. Svennerholm. 1990. The binding of colonization factor antigens of enterotoxigenic *Escherichia coli* to intestinal cell membrane proteins. *FEMS Microbiol. Lett.* **54**:107–112.

Wentworth, J.S., F.E. Austin, N. Garber, N. Gilboa-Garber, C.A. Paterson, and R.J. Doyle. 1991. Cytoplasmic lectins contribute to the adhesion of *Pseudomonas aeruginosa*. *Biofouling* **4**:99–104.

Westerlund, B. 1991. Fluorescent microparticles as a rapid tool in bacterial adherence studies. *J. Microbiol. Meth.* **13**:135–143.

Westerlund, B., J. Merenmies, H. Rauvala, A. Miettinen, A.K. Jarvinen, R. Virkola, H. Holthofer, and T.K. Korhonen. 1987. The 075X adhesin of uropathogenic *Escherichia coli*: receptor-active domains in the canine urinary tract and *in vitro* interaction with laminin. *Microb. Pathogen.* **3**:117–127.

Westerlund, B., A. Siitonen, J. Elo, P.H. Williams, T.K. Korhonen, and P.H. Makela. 1988. Properties of *Escherichia coli* isolates from urinary tract infections in boys. *J. Infect. Dis.* **158**:996–1002.

Westerlund, B., P. Kuusela, J. Risteli, L. Risteli, T. Vartio, H. Rauvala, R. Virkola, and T.K. Korhonen. 1989a. The 075X adhesin of uropathogenic *Escherichia coli* is a type IV collagen-binding protein. *Mol. Microbiol.* **3**:329–337.

Westerlund, B., P. Kuusela, T. Vartio, I. van Die, and T.K. Korhonen. 1989b. A novel lectin independent interaction of P fimbriae of *Escherichia coli* with immobilized fibronectin. *FEBS Lett.* **243**:199–204.

Westerlund, B., I. van Die, C. Kramer, P. Kuusela, H. Holthofer, A.M. Tarkkanen, R. Virkola, N. Riegman, H. Bergmans, W. Hoekstra, and T.K. Korhonen. 1991. Multifunctional nature of P fimbriae of uropathogenic *Escherichia coli*: mutations in *fso*E and *fso*F influence fimbrial binding to renal tubuli on immobilized fibronectin. *Mol. Microbiol.* **5**:2965–2967.

Wiersma, E.J., G. Froman, S. Johansson, and T. Wadström. 1987. Carbohydrate specific binding of fibronectin to *Vibrio cholerae* cells. *FEMS Microbiol. Lett.* **44**:365–369.

Wold, A.E., M. Thorssen, S. Hull, and C. Svanborg-Eden. 1988. Attachment of *Escherichia coli* via mannose- or Galα1–4Galβ-containing receptors to human colonic epithelial cells. *Infect. Immun.* **56**:2531–2537.

Wold, A.E., J. Mestecky, M. Tomana, A. Kobata, H. Ohbayashi, T. Endo, and C. Svanborg-Eden. 1990. Secretory immunoglobulin A carries oligosaccharide receptors for *Escherichia coli* type 1 fimbrial lectin. *Infect. Immun.* **58**:3073–3077.

Wolf, M.K., G.P. Andrews, D.L. Fritz, R.W. Sjogren, Jr., and E.C. Boedeker. 1988. Characterization of the plasmid from *Escherichia coli* RDEC–1 that mediates expression of adhesin AF/R1 and evidence that AF/R1 pili promote but are not essential for enteropathogenic disease. *Infect. Immun.* **56**:1846–1857.

Wong, K.H., S.K. Skelton, and J.C. Feeley. 1985. Interaction of *Campylobacter jejuni* and *Campylobacter coli* with lectins and blood group antibodies. *J. Clin. Microbiol.* **22**:134–135.

Wu, S.X. and R.Q. Peng. 1991. Studies on adherence and outer membrane protein of enteropathogenic *Escherichia coli* 0127:H6 and their related plasmids. *Acta Paediatr. Scand.* **80**:1019–1024.

Wurker, M., J. Beuth, H.L. Ko, A. Przondo-Mordarska, and G. Pulverer. 1990. Type of fimbriation determines adherence of *Klebsiella* bacteria to human epithelial cells. *Int. J. Med. Microbiol.* **274**:239–245.

Wyle, F.A., A. Tarnawski, W. Dabros, and H. Gergely. 1990. *Campylobacter pylori* interactions with gastric cell tissue culture. *J. Clin. Gastroenterol.* **12** (*Suppl.*):S99–S103.

Wyrick, P.B., J. Choong, C.H. Davis, S.T. Knight, M.O. Royal, A.S. Maslow, and C.R. Bagnell. 1989. Entry of genital *Chlamydia trachomatis* into polarized epithelial cells. *Infect. Immun.* **57**:2378–2389.

Yakubu, D.E., B.W. Senior, and D.C. Old. 1989. A novel fimbrial. haemagglutinin produced by a strain of *Salmonella* of serotype salinatis. *FEMS Microbiol. Lett.* **57**:29–34.

Yamamoto, T. and T. Yokota. 1988a. *Vibrio cholerae* non–01: production of cell-associated hemagglutinins and in vitro adherence to mucus coat and epithelial surfaces of the villi and lymphoid follicles of human small intestines treated with formalin. *J. Clin. Microbiol.* **26**:2018–2024.

Yamamoto, T. and T. Yokota. 1988b. Electron microscopic study of *Vibrio cholerae* 01 adherence to the mucus coat and villus surface in the human small intestine. *Infect. Immun.* **56**:2753–2759.

Yamamoto, T. and T. Yokota. 1989a. Adherence targets of *Vibrio parahaemolyticus* in human small intestines. *Infect. Immun.* **57**:2410–2419.

Yamamoto, T. and T. Yokota. 1989b. *Vibrio cholerae* 01 adherence to human small intestinal M cells in vitro. *J. Infect. Dis.* **160**:168–169.

Yamamoto, T., Y. Tamura, and T. Yokota. 1985. Enteroadhesion fimbriae and enterotoxin of *Escherichia coli*: genetic transfer to a streptomycin-resistant mutant of the *gal*E oral-route live-vaccine *Salmonella typhi* ty21a. *Infect. Immun.* **50**:925–928.

Yamamoto, T., T. Kamano, M. Uchimura, M. Iwanaga, and T. Yokota. 1988. *Vibrio cholerae* 01 adherence to villi and lymphoid follicle epithelium: in vitro model using formalin-treated human small intestine and correlation between adherence and cell-associated hemagglutinin levels. *Infect. Immun.* **56**:3241–3250.

Yamamoto, T., K. Fujita, and T. Yokota. 1990a. Adherence characteristics to human small intestinal mucosa of *Escherichia coli* isolated from patients with diarrhea or urinary tract infections *J. Infect. Dis.* **162**:896–908.

Yamamoto, T., K. Fujita, and T. Yokota. 1990b. Piliated *Vibrio parahaemolyticus* adherent to human ureteral mucosa. *J. Infect. Dis.* **161**:361–362.

Yamamoto, T., S. Endo, T. Yokota, and P. Echeverria. 1991. Characteristics of adherence of enteroaggregative *Escherichia coli* to human and animal mucosa. *Infect. Immun.* **59**:3722–3739.

Yamazaki, Y., S. Ebisu, and H. Okada. 1988. Partial purification of a bacterial lectin-like substance from *Eikenella corrodens*. *Infect. Immun.* **56**:191–196.

Yano, T., D. da Silva Leite, I.J. Barsanti de Camargo, and A.F. Pestana de Castro. 1986. A probable new adhesive factor (F42) produced by enterotoxigenic *Escherichia coli* isolated from pigs. *Microbiol. Immun.* **30**:495–508.

Yerushalmi, Z., N.I. Smorodinsky, M.W. Naveh, and E.Z. Ron. 1990. Adherence pili of avian strains of *Escherichia coli* 078. *Infect. Immun.* **58**:1129–1131.

Yeung, M.K. and J.O. Cisar. 1990. Sequence homology between the subunits of two immunologically and functionally distinct types of fimbriae of *Actinomyces* spp. *J. Bacteriol.* **172**:2462–2468.

Yeung, M.K., B.M. Chassy, and J.O. Cisar. 1987. Cloning and expression of a type 1 fimbrial subunit of *Actinomyces viscosus* T14V. *J. Bacteriol.* **169**:1678–1683.

Yu, C., A.M. Lee, and S. Roseman. 1987. The sugar-specific adhesion/deadhesion of the marine bacterium *Vibrio furnissii* is a sensorium that continuously monitors nutrient levels in the environment. *Biochem. Biophys. Res. Commun.* **149**:86–92.

Yu, C., A.M. Lee, B.L. Bassler, and S. Roseman. 1991. Chitin utilization by marine bacteria. A physiological function for bacterial adhesion to immobilized carbohydrates. *J. Biol. Chem.* **266**:24260–24267.

Zafriri, D., Y. Oron, B.I. Eisenstein, and I. Ofek. 1987. Growth advantage and enhanced toxicity of *Escherichia coli* adherent to tissue culture cells due to restricted diffusion of products secreted by the cells. *J. Clin. Invest.* **79**:1210–1216.

Zafriri, D., I. Ofek, R. Adar, M. Pocino, and N. Sharon. 1989. Inhibitory activity of cranberry juice on adherence of type 1 and type p fimbriated *Escherichia coli* to eucaryotic cells. *Antimicrob. Agents Chemother.* **33**:92–98.

11

Common Themes in Bacterial Adhesion

Research in bacterial adhesion has yielded several common themes associated with bacteria–substratum interactions. Different organisms have evolved separate and distinct mechanisms of adhesion but many seem to require two mechanisms in order to survive on at least two different substrata. Physiological changes of bacteria or damage to substrata caused by the bacteria may be different in the various systems but appear to be events tightly associated with adhesion phenomena. Finally, interest in modulating bacterial adhesion revealed that there are several ways to interfere in the adhesion process in order to prevent or promote colonization on a particular substratum.

Multiple Mechanisms of Adhesion and Their Role In Vivo

It now seems that individual clones of most bacterial species can express more than one type of adhesin. Chemically, these adhesins may consist of proteins, polysaccharides, lipids, or phosphodiester-containing polymers (teichoic acids). Functionally, they may be lectins, hydrophobins, or polyelectrolytes. Ultrastructurally, adhesins may reside in fimbriae, fibrillae, cell wall, outer membrane, or as loosely attached peripheral components. The adhesins produced by a particular bacterial clone may differ in terms of their chemistry, cellular function, or ultrastructure (Table 11–1). For example, the functional domains of hydrophobins may be due to either lipids or proteins. Lectins that serve as adhesins are always proteins but they may be structurally associated with either fimbriae or cell wall peripheral components. In contrast, some clones may express adhesins which are chemically, functionally, and ultrastructurally indistinguishable, but with different specificities. The adhesin–receptor relationships of bacterial adhesion may exist in a number of forms as depicted in Fig. 11–1.

The chemical, functional, and ultrastructural diversity of adhesins expressed

Table 11–1. Chemical, functional, and ultrastructural identities of multiple adhesins within selected bacterial clones

			Characteristics			
Bacterial clone	Source of isolation	Adhesin designation	Chemical	Functional	Ultrastructural	References
E. coli	Pyelonephritis	Type P	Protein	Lectin	Fimbriae	Leffler and Svanborg-Eden, 1986
		Type 1	Protein	Lectin	Fimbriae	Sharon and Ofek, 1986
S. saprophyticus	Urinary	Gal-GlcNAc	Protein	Lectin	Peripheral	Beuth et al., 1988
		Lipoteichoic acid	Fatty acids	Hydrophobin	Fibrillae	
E. coli	Urosepsis	NFA 1	Protein	Lectin	Peripheral	Goldhar et al., 1984
		Type 1	Protein	Lectin	Fimbriae	Sharon and Ofek, 1986
S. sanguis	Dental plaque	No designation	Protein	Hydrophobin	Peripheral	Nesbitt et al., 1982a,b
		Sialic acid specific lectin	Protein	Lectin	Peripheral	Levine et al., 1978

by a single bacterial clone (see Table 11–1) undoubtedly endows the organisms with the ability to colonize a definite range of substrata. It seems reasonable that evolutionary pressures have selected organisms that are capable of demonstrating more than one mechanism of adhesion. Indeed, as shown in Table 11–2, many bacterial species have been found to express more than one type of adhesin, and in some cases, more than 10 have been described. Traditionally, investigators have emphasized studies on a particular phenotype expressing one type of adhesin. Evidence now has emerged to show that most bacterial clones may exhibit two or more phenotypes, each of which expresses an adhesin of a distinct specificity. The detection of such phenotypes is dependent on the choice of the target substratum possessing receptors specific for the adhesin and on growth conditions that allow proliferation of the phenotype. Although there is no proof that all the adhesins ascribed to any bacterial species shown in Table 11–2 are expressed by a single genotype, it now seems likely that two or more adhesins originate from a single genotype.

In order to understand how multiple adhesins may function two major mechanisms seem plausible. One mechanism requires a multiple-step kinetic adhesion reaction, whereas the other depends on expression of adhesin(s) at predetermined frequencies. the advantage(s) of each of these mechanisms is (are) discussed below. The multistep kinetic model is based on studies with oral streptococci and group A streptococci. Adhesion of *S. sanguis* to saliva-coated hydroxylapatite was found to depend on multiple, interacting sites (Nesbitt et al., 1982a,b; Cowan et al., 1986; 1987a,b). It was suggested that adhesion to saliva-coated hydroxylapatite (SHA) was cooperative in that the binding of a few bacteria tended to promote the binding of other bacteria. This cooperative adhesion was dependent on the hydrophobic effect and on nonhydrophobic interactions. Other work has supported the view that *S. sanguis* adheres to SHA via multiple sites (Gibbons et al., 1983; Morris and McBride, 1984; Fives-Taylor and Thompson, 1985; Morris et al., 1985). The *S. sanguis*–SHA complex seems to proceed by two distinct kinetic steps, the first of which is readily reversible and is mediated by the hydrophobic effect. The second step is mediated primarily by nonhydrophobic interactions and follows the first step. Once the second reaction has taken place, the cell becomes firmly bound and is only very slowly reversible. Detachment would be expected only when all adhesin–receptor interactions are coordinately reversible, the probability of which is low. The findings that agents interfering with ionic or hydrogen bonds (salts, high temperatures) or with the hydrophobic effect (chaotropes, low temperature) inhibit adhesion supports the validity of the multiple-site model (Nesbitt et al., 1982a,b; Cowan et al., 1986, 1987a,b; Zhang et al., 1990). Furthermore, it has been shown that adhesion of *S. sanguis* occurs in two distinct kinetic steps, whereas desorption occurs in a single step (Cowan et al., 1986, 1987a). This behavior is also consistent with the multiple-site model (Hasty et al., 1992). The adhesion of group A streptococci may also obey the two-step kinetic model. One step would involve hydrophobic interactions between

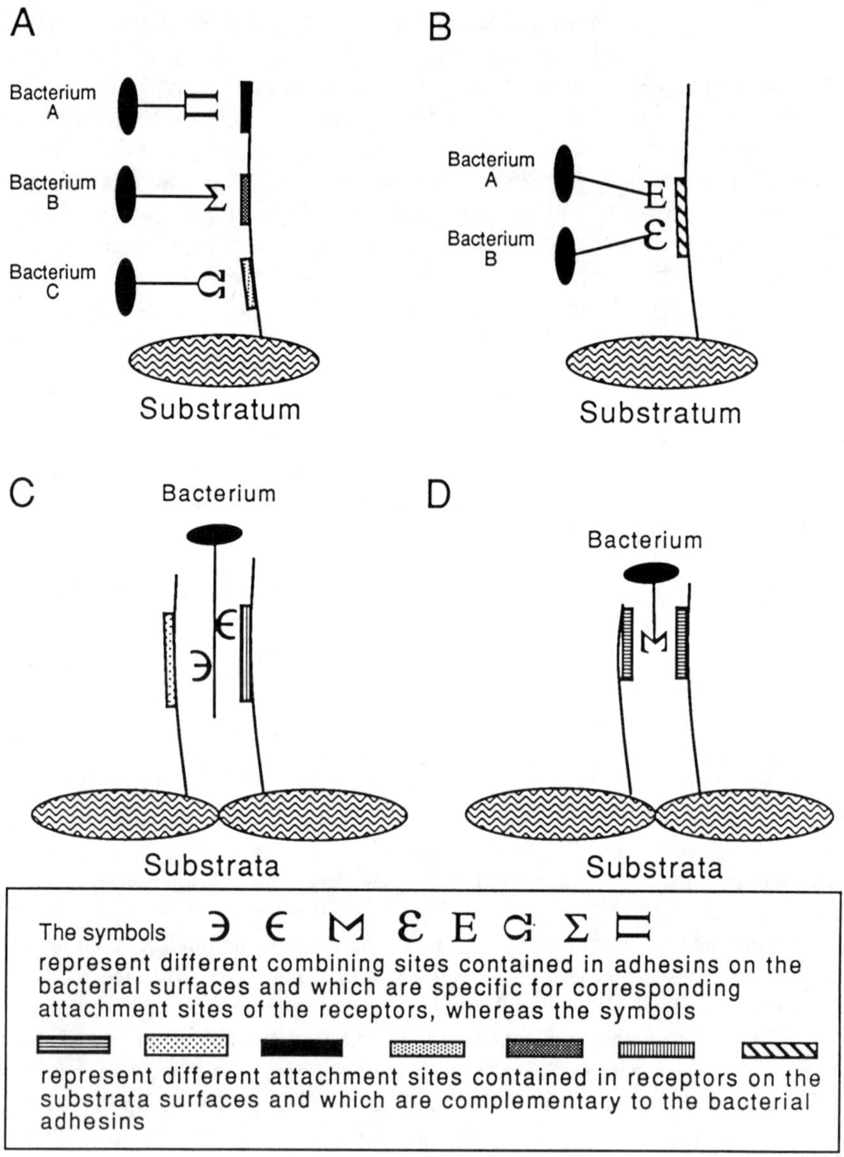
A
B
Bacterium A
Bacterium B
Bacterium C
Substratum
Bacterium A
Bacterium B
Substratum
C
Bacterium
D
Bacterium
Substrata
Substrata
The symbols
represent different combining sites contained in adhesins on the bacterial surfaces and which are specific for corresponding attachment sites of the receptors, whereas the symbols
represent different attachment sites contained in receptors on the substrata surfaces and which are complementary to the bacterial adhesins

Figure 11–1.

LTA and the fatty acid-binding domains of epithelial cells. The correlation between hydrophobicity of the streptococci and the ability of the bacteria to adhere to epithelia supports this premise (Chapter 6 discusses the role of hydrophobicity in the adhesion of pyogenic cocci). A time-dependent, virtually irreversible, adhesion of group A streptococci to fibronectin was observed by Speziale et al. (1984). They found that proteolysis of the bacteria reduced the adhesion to fibronectin. Preparations of a tryptic digest of the bacteria that contained very low amounts of lipoteichoic acid (LTA) were capable of inhibiting adhesion. It is tempting to speculate that the adhesion of group A streptococci to epithelial cells follows a two-step reaction sequence. The first step would be mediated by LTA binding to fibronectin and the second step would be mediated by a fibronectin-binding streptococcal protein. Similarly, a two-step model for the adhesion of enteropathogenic *E. coli* to human intestinal enterocytes and cultured intestinal mucosal cells has been postulated by Knutton et al. (1987). It was suggested that the bacteria interact with intact microvili in an initial dissociable step. The following second step of adhesion involves rearrangement of the microvili and is accompanied by a firm interaction resulting in the virtual irreversible union of the bacterium and the mucosal cell. If this hypothesis is valid, then two-step kinetics of adhesion to enterocytes would be expected. Minion et al. (1984) also reasoned that adhesion of *Mycoplasma pulmonis* to erythrocytes involved a rearrangement of surface molecules on the red cells following an initial adhesion

Figure 11–1. Illustrations of various forms of receptor–adhesin relationships in bacterial adhesion. Four forms designated A-D are shown and describe the following: (A) A substratum receptor bears multiple receptor sites specific for distinct adhesins expressed by different bacterial strains. This form of relationship was found for the Dr blood group antigen that serves as a receptor on erythrocytes and contains three distinct receptor sites specific for the adhesins AFA II, Dr fimbriae, and F 1845 fimbriae produced by different strains of *E. coli*. (B) One receptor site for two distinct adhesins expressed by two different bacteria. This form of relationship was found for the animal cell fibronectin and its amino-terminus region which interacts with streptococcal LTA and staphylococcal FBP (fibronectin-binding protein) adhesins. (C) Different subunits or regions of an adhesin complex with distinct receptor specificities. This form of relationship was found in two bacterial species. In *E. coli,* Fso G and Fso F/H subunits of the P fimbrial adhesin complex are specific for Gal-Gal-containing receptors and fibronectin, respectively. In *Bordetella pertussis,* the filamentous hemagglutinin contains a carbohydrate binding region and an RGD sequence that bind the organism to two different receptors: galactose-containing receptors and integrins, respectively, on animal cells. (D) Isoreceptors on different substrata bearing the same receptor site specific for a bacterial adhesin. This form of relationship was found in many cases, especially in bacteria-producing lectins that function as adhesins capable of binding the organisms to oligosaccharides carried by different types of glycolipids or glycoproteins on various animal cells. An example is the type 1 fimbrial adhesin specific for manooligosaccharides present on different types of glycoproteins expressed by different animal cells (e.g., erythrocytes and phagocytes).

Table 11–2. Examples of multiple adhesins in bacteria

Bacterium	Adhesin	Reference(s) or review(s)
Actinomyces spp.	Glucan-specific lectin	Bourgeau and McBride, 1976
	Gal-specific lectin	Kolenbrander, 1982, 1989
	Sialic acid-specific lectin	McIntire et al., 1983
	Teichoic acid	Ciardi et al., 1977
	Hydrophobin	McBride et al., 1984
	Sialic acid-specific lectin	Levine et al., 1978
Capnocytophaga ochracea	Lectins for sialic acid, β-galactosides, fucose, rhamnose	Weiss et al., 1987
Enterobacter aerogenes	MS and MR hemagglutinins	Adegbola and Old, 1985
	Type 1 fimbriae	Korhonen et al., 1983
	Hemagglutinin	Old and Adegbola, 1983
Escherichia coli	Type 1 fimbriae	DeGraaf, 1990
	K88 fimbriae	Hacker, 1990
	K99 fimbriae	Baddour et al., 1989
	Dr-hemagglutinin	
	G fimbriae	
	M-agglutinin	
	Nonfimbrial adhesin(nfa)	
	P fimbriae (Galα-1,4Gal)	
	CFA/I fimbriae	
	CFA/II fimbriae (CS 1,2,3,)	
	CFA/III fimbriae	
	987P fimbriae	
	F41 fimbriae	
	E8775 fimbriae	
	AF/R1	
	PCF8775(CFA4,5,6)	
	S	
Klebsiella pneumoniae	MS lectin, MR/K agglutinin	Duguid and Old, 1980
Pseudomonas aeruginosa	MS Lectin	Mirelman and Ofek, 1986
	Sialic acid lectin	Hazlett et al., 1986
	Hydrophobin	Garber et al., 1985
	Mucoid exopolysaccharide	Ramphal and Pier, 1985
	Exotoxin S	Baker et al., 1991
	Ganglioside-binding protein	Krivan et al., 1988
	Cytoplasmic lectins	Wentworth et al., 1991
	PA-I and PA-II	
	N-Acetylmannosamine-specific lectin	Hazlett et al., 1987
Serratia marcescens	Type 1 fimbriae	Mirelman and Ofek, 1986
	Hydrophobin	Rosenberg, 1984

continued

Table 11–2. Continued

Bacterium	Adhesin	Reference(s) or review(s)
Staphylococcus aureus	LTA	Carruthers and Kabat, 1983
	Protein	Maxe et al., 1986
	Wall TA	Aly and Levit, 1987
	Collagen binding protein	Holderbaum et al., 1985
Staphylococcus epidermidis	LTA	Chugh et al., 1990
	Hydrophobin	Hogt et al., 1983
	Exopolysaccharide	Christensen et al., 1982
	Fibronectin and collagen binding protein	Wadström et al., 1987
	Wall TA	
Staphylococcus saprophyticus	LTA	Beuth et al., 1988
	Galα-1,4GlcNAc-specific lectin	Gunnarsson et al., 1984
Streptococcus agalactiae	Wall TA	Mattingly and Johnston, 1987
	LTA	Teti et al., 1985
	GlcNAc-specific lectin	Bagg et al., 1982
Streptococcus mutans	Glucosyltransferase	Nalbandian et al., 1974
Streptococcus pyogenes	LTA	Beachey et al., 1983
	M-protein	Tylewska et al., 1988
	S-protein binding adhesin	Valentin-Weigand et al., 1988
Streptococcus sanguis	Hydrophobin	Nesbitt et al., 1982a,b
	Sialic-acid-specific lectin	Levine et al., 1978; Murray et al., 1982
	LTA	DeRienzo et al., 1985
	Fimbrial protein	Fachon-Kalweit et al., 1985
	Platelet-binding protein	Herzberg et al., 1990
Streptococcus sobrinus	Glucan-binding lectin	Drake et al., 1988
	Hydrophobin	Nesbitt et al., 1982a,b

step. Their results are consistent with multiple-step kinetics of adhesion. It now seems clear that adhesion of different kinds of bacteria to various substrata involves distinct adhesins, one interacting after the other. Alternatively, a change in either the surface of the substratum or the bacterium may also result in multiple-step kinetics. To date, the multiple-step or complex kinetics has not been generally appreciated, possibly because of the difficulties of treatment of data or experimental design.

Predetermined frequencies of the expression of multiple adhesion can be achieved only by complex regulatory systems. Unfortunately, very little is known regarding the genetic and physiological regulation of multiple adhesins. It seems unlikely that a single regulatory mechanism would govern the activity of all of the adhesins. A comprehensive understanding of the advantages of such regulation can be achieved by knowing the ecology of the bacterial clone in question and the molecular biology of the adhesins it expresses. Frequently, only the ecology or the molecular biology of one adhesin is known. In many cases, only sparse information is available. The bacteria belonging to the *Enterobacteriaceae* constitute the largest group of bacteria studied with respect to structure–function relationships and the molecular biology of their adhesins. Many members of this family are capable of producing at least two adhesins (reviewed in Duguid and Old, 1980). Of the various members of the *Enterobacteriaceae* a considerable amount of information has been generated on *E. coli*. Therefore, to obtain insight as regards the evolutionary advantages of expressing multiple adhesins, future discussion will be focused on clones belonging to this species.

Genotypically, most *E. coli* isolates (and many other enterobacterial isolates as well) contain the MS type 1 fimbrial protein (Buchanan et al., 1985; Mirelman and Ofek, 1986). Phenotypically, the MS lectin activity associated with type 1 fimbriae is expressed after one or several broth passages (Duguid and Old, 1980). The expression of the type 1 fimbrial lectin is readily detectable down to a sensitivity of 10^7 cells/ml by haemagglutination reactions of whole bacteria employing guinea pig erythrocytes (Duguid and Old, 1980) or mannan-containing yeast cells (Ofek and Beachey, 1978). Duguid and his colleagues (reviewed in Duguid and Old, 1980) noticed the *E. coli* isolates, after growth on agar, were capable of expressing fimbrial and nonfimbrial lectins specific for carbohydrates other than mannose. These lectins were designated MR and are common among the enterobacteria. We now know that most of the MR lectins are specific for complex carbohydrates which may be found in accessible form only on certain types of erythrocytes or tissues (see Chapter 3). It follows that the detection of the various phenotypes depends largely on growth conditions and the types of erythrocytes or tissues tested. This can be illustrated by comparing two separate studies performed on urinary isolates of *E. coli* as shown in Table 11–3. It is clear that the percentage of isolates capable of expressing type 1 fimbriae increases as the number of serial broth passages of the isolate increases, whereas the number of MR-expressing isolates increases with the number of different test erythrocytes. The result is that the percentage of isolates capable of expressing two fimbrial adhesins is widespread among *E. coli* isolates (and probably other enterobacterial species). In fact, just in recent years, several new MR adhesins expressed by numerous isolates of *E. coli* have been reported (Table 9–3; Honda et al., 1984; Karch et al., 1987; Knutton et al., 1987; Rhen et al., 1986; Tacket

Table 11–3. Expression of MS and MR adhesins by urinary isolates (asymptomatic bacteriuria) of Escherichia coli[a]

Growth conditions	Source of erythrocytes tested[b]	Percent strains genotypically capable of expressing adhesins			
		MS	MS and MR	MR	Neither
One broth passage	GP, HU	55	14	3	28
Serial broth passages	F,G,H,HU,O,P,S	47	40	4	9

[a]Derived from Crichton and Old, 1980; Leffler and Svanborg-Eden, 1981.

[b]GP, guinea pig; HU, human group A; F, avian; G, goat; H, horse; O, ox; P, pig; S, sheep.

et al., 1987). It is anticipated that even additional MR adhesins will be found, as growth conditions are varied and new cellular substrata are employed.

From the foregoing, it is tempting to speculate that most *E. coli* isolates (and perhaps other enterobacterial species) are genotypically capable of expressing at least two types of adhesins, one of which is the MS lectin and the other of which may be an adhesin specific for complex carbohydrates or other receptors. The MR lectin may or may not be associated with fimbriae, and its detection also depends on growth conditions and the types of target cells employed in adhesion or agglutination assays. Empirically, it was found that in some isolates the percent organisms expressing the type 1 phenotype may become detectable after one broth passage, whereas that of others requires more broth passages (Fein, 1981; Ofek et al., 1982). Whether or not this difference reflects a distinct property among the various *E. coli* isolates (and perhaps other type 1 fimbriae-producing enterobacterial species) as suggested recently by Hultgren et al. (1986) requires further studies.

A key question relevant to this chapter is why throughout evolution has *E. coli* (and probably other enterobacterial species) conserved two sets of genes coding for the surface expression of at least two adhesins. Why is it that regardless of the source or origin of the infection, the *E. coli* isolate is always capable of expressing the type 1 fimbrial lectin, whereas the other adhesins appear to be associated with isolates from a particular site of infection (Table 11–4)? To examine this question, focus will be on pyelonephritogenic isolates of *E. coli* because these isolates have been studied the most with respect to their detailed

Table 11–4. Multiple adhesins of some Escherichia coli *clones*[a]

	Specificity of adhesins expressed by each clone and associated with	
Source of clones	Type 1 fimbriae MS	Other surface structures MR
UT1, pyelonephritis		Galα-1,4Gal (P fimbriae)
	M	
UT1, others	A	Not defined (e.g., nonfimbrial, NFA,
	N	specific for blood group antigens)
Meningitis	N	Sialylgalactose (S fimbriae)
Diarrhea, human	O	Sialyllactose (CFA/I,II fimbriae)
	S	
Diarrhea, piglets	I	Sialic acid (K99 fimbriae)
	D	
Diarrhea, others	E	Not defined (e.g., CFA/III)
	S	
Fecal microbiota		None?

[a]Adapted from Baddour et al., 1989.

Abbreviation: MS, mannose-specific or sensitive; MR, mannose-resistant.

carbohydrate specificities, the identities of their carbohydrate-containing receptor molecules, and their roles in urinary tract infections.

The detailed molecular mechanisms of the regulation of the genes coding or the type 1 fimbrial complex and for the P-fimbrial complex were described in Chapter 9. Abraham et al. (1986) found that the regulations of the expression of the individual fimbrial adhesins were independent of each other. It follows that phenotypically, due to the phase variation phenomenon, a bacterial culture may be composed of four possible phenotypes: P-fimbriated, type 1 fimbriated, afimbriated, and a composite of types 1 and P-fimbriated. Immunogold and immunofluorescence double-labeling fimbriae confirm this prediction (Nowicki et al., 1984; Gander and Thomas, 1987). The results also show that P and type 1 fimbria occur mostly on separate cells with fewer than 9% of fimbriated cells coexpressing both types of fimbrial lectins in equal numbers under the growth conditions tested (Nowicki et al., 1984; Gander and Thomas, 1987). It seems that genotypically it is difficult for the bacteria to coexpress both types of fimbriae. This notion is supported by similar studies on *E. coli* strains that are capable of expressing type 1 and type S fimbrial lectins (Nowicki et al., 1985a). The percentage of each phenotype in the cell population in a given environment is the result of the dynamic relationship between cell division and of elimination of the phenotype by environmental factor(s).

Assuming that each of the phenotypes is important at a particular stage and site during the infectious process, the following factors are to be considered: (1) the rate of phase variation from one phenotype to another; (2) growth factors that affect the rate of proliferation of each phenotype; (3) the distribution and/or accessibility of glycoconjugates containing carbohydrates specific for each of the fimbrial lectins on cell membranes, especially phagocytic and mucosal cells; and (4) the presence of lectin-reactive glycoconjugates in soluble form in fluids surrounding these cells.

The rate of phase variation in vitro from fimbriated to afimbriated phenotypes varies among the different isolates of *E. coli*. For example, the phase switch for type 1 fimbriae averages one unidirectional change per thousand cells per generation in an *E. coli* K12 derivative (Eisenstein, 1981), but that of a blood isolate is much higher, being 16 cells per thousand (Nowicki et al., 1986). In pyelonephritogenic isolates of *E. coli* the rate is 5 per thousand for P fimbriae (Nowicki et al., 1984). A shift from one fimbrial phase to another may occur either directly or via a nonfimbriate phase. The shift between types 1, P, and afimbriate phases is faster than that of flagellar phases in *Salmonella* (Simon et al., 1980) or fimbriae pili in *Neisseria gonorrhoeae* (Swanson and Barrera, 1983). Studies suggest that phase variation is not affected by such environmental factors as temperature, glucose, or cyclic AMP (Eisenstein, 1981; Eisenstein and Dodd, 1982). To date, therefore, no environmental factors have been found that affect the fimbrial phase variation frequency. Furthermore, the expression of fimbrial protein does not appear to be coordinated with the expression or repression of particular metabolic pathways (Swaney et al., 1977). It follows that the rate of phase variation in fimbrial protein expression is genetically controlled in different *E. coli* isolates (see also Chapter 9).

Although the frequency of phase variation is not affected by environmental factors, the latter may select a phenotype that has switched its phase because that phenotype is best suited to survive in that particular environment. In vitro, the percentage of each of the phenotypes in growing culture is determined by the type and composition of the growth medium. In vivo, the selection is the outcome of the rate of elimination by the host defense mechanisms vs the rate of proliferation of each of the phenotypes at a particular site (Figure 9–4 provides a review of the mechanisms of phase variation in *E. coli*).

Glycoproteins containing mannooligosaccharide side chains are ubiquitous and abundant in living cells, where they serve as attachment sites for the type 1 fimbriated phenotype (Table 11–5). Studies with concanavalin A and other mannose-specific lectins confirm the ubiquity of mannose-containing receptors. In contrast, only human and primate (and possibly the mouse) cells are known to contain the requisite Gal-Gal-containing glycolipids that serve as receptors for binding the P-fimbriated phenotype (details are provided in Chapter 10). Not even all tissues appear to possess receptors for fimbriae. Most importantly, phagocytic cells lack receptors for this fimbrial lectin (see Chapter 7). This difference in distribution of tissue receptors for the two fimbrial lectins is remarkable considering the fact that chemically, ultrastructurally, and functionally the two fimbrial lectins are indistinguishable (Table 11–1).

There is also a difference in the abundance of Gal-Gal-containing glycolipids and mannooligosaccharide-containing glycoproteins in various body fluids. The

Table 11–5. Cells reactive with MS lectins of Enterobacteriaceae.

Kingdom		Cell type	Reference(s)
Man and animals (mouse, rabbit, dog, horse, guinea pig, monkey, pig)	Epithelial	Buccal, tracheal Intestinal Urogenital Endometrial	Ofek and Beachey, 1980; Duguid and Old, 1980; Isaacson et al., 1978; Lindquist et al., 1987; Schaeffer et al., 1979; Nishikawa and Baba, 1985
	Blood	Erythrocytes Lymphocytes Phagocytes Mammary glands	Crichton and Old, 1980; Ponniah et al., 1989; Bar-Shavit et al., 1980 Harper et al., 1978
Avian (turkeys, chicken)	Epithelial	Intestinal	Dominick et al., 1985; Lillard, 1986
Plants (poa, prantnsis)	Grass roots		Haahtela et al., 1985
Parasites	*Schistosoma* Amoeba		Melhem and Loverde, 1984; Mirelman, 1987
Fungi	Yeasts		Duguid and Old, 1980

former are rarely found in soluble form, whereas the latter are commonly found as evidenced from interactions with mannose-specific concanavalin A (Lis and Sharon, 1986). More specifically, while urine is known to lack any inhibitor of P fimbriae (Parkkinen et al., 1988), it does contain inhibitors of type 1 fimbriae. The type 1 fimbrial lectin reacts with Tamm–Horsfall glycoprotein in urine (Orskov et al., 1980) and with glycoprotein preparations obtained from porcine and rabbit gastric mucus as well as from bovine submaxillary mucin (Sherman and Boedeker, 1987). It has been suggested that, depending on the rate of synthesis of mucus along the intestine and the rate of bacterial multiplication, the mucus may serve in clearing enteropathogens from the underlying intestinal villus surface or may serve as a site for bacterial division (Sherman and Boedeker, 1987). Furthermore, the ability of *E. coli* and *Salmonella typhimurium* to interact with mucus glycoproteins was found to enhance the ability of the organisms to colonize the intestine of the mouse (Cohen et al., 1985; Nevola et al., 1987). The same argument may apply to Tamm–Horsfall glycoprotein. The effect on the infectious process of some of the manooligosaccharide-containing glycoproteins (e.g., Tamm–Horsfall glycoprotein) was discussed in Chapters 4 and 5.

There is now ample evidence that *E. coli* undergoes phase variation in vivo similar to that observed in vitro although the exact rate of the phase switch is difficult to ascertain (Maayan et al., 1985; Alkan et al., 1986; Nowicki et al., 1986; Pere et al., 1987; Schaeffer et al., 1987). Different protocols of experimental infections in various animals, coupled with epidemiological observations, show that both P-fimbriated and type 1 fimbriated phenotypes are important for infectivity. In one set of experiments it was shown that the infectivity of each of the phenotypes was prevented or reduced by injecting the organisms in the presence of specific inhibitors of adhesion. Anti-type 1 fimbriae, soluble mannose (and its derivatives), or anti-mannose-containing glycoproteins, anti-P fimbriae, soluble Galα1,4Gal disaccharide, or anti-Galα1,4-Gal-containing glycolipids prevented infections initiated by types 1 and P-fimbriated phenotypes, respectively. In another set of experiments, it was shown that the infectivity of fimbriated phenotypes, as compared to types 1 or P-fimbriated phenotypes, was reduced after intravascular injections. A composite view of the results suggests that both types 1 and P fimbriae are important for the development of urinary tract infections but they do not tell much about how or at which stage of the infectious process they are important.

The following theory, based on two assumptions made earlier, is presented in order to emphasize the importance of multiple adhesins in infections. One assumption was that adhesion to tissue cells confers a growth advantage upon the organisms. Furthermore, in the case of attachment to mucosa, adhesion also enables the bacteria to withstand cleansing mechanisms (see discussion in next section). The other assumption was that each of the phenotypes of the pyelonephritogenic clone of *E. coli* resulting from phase variation confers a survival benefit to the *E. coli* clone during a natural course of infection. It is likely that at the beginning of the infectious process the type 1 fimbriated phenotype contributes much more than the P-fimbriated phenotype to the survival of the bacterial population because it has a better chance to find a readily accessible mannooligosaccharide-containing receptor on epithelial cells. The adherent bacterium would be able to withstand cleansing mechanisms and acquire growth advantages. This survival value mediated by type 1 fimbrial lectin has been conserved by many enterobacterial species, and in that sense, it is a universal mechanism of adhesion by virtue of the fact that the cellular receptors involved in binding the organisms are widely distributed in living cells. This survival value, however, may become dispensible if the phenotype presenting a non-MS-mediating adhesion mechanism reaches the site of infection at a high density.

It follows that infectivity depends largely on the percentage of type 1 fimbriated phenotype in the bacterial population initiating the infection and on the time lapse between the initial dose until colonization can be detected. The following experiment (Goldhar et al., 1986) supports this premise. Infant mice were infected per os with a mixed population of *E. coli* consisting, as a result of the phase variation, of type 1 fimbriated and MR adhesin-bearing phenotypes. The cells were suspended in buffer lacking or containing mannose to block the ability of the type 1 fimbriated phenotype to adhere to tissue cells. The results (Table 11–6) show that in the absence of mannose, colonization was detected as early as 3 hours after infection. The number of mice with detectable colonization also increased as the infective dose increased. In contrast, in the presence of mannose, which reduces the ability of type 1 fimbriated phenotype to adhere to target cells, early colonization was detected only with high infective doses or as late as 7 days with low infective doses. The results were interpreted to suggest that in both cases growth advantages are gained by the type 1 fimbriated phenotype which has a better chance of finding a cell expressing mannooligosaccharide-

Table 11–6. Infectivity in mice of Escherichia coli *801 in the presence and absence of* D-*mannose and sucrose in the inoculum*[a]

	No. of infected mice at indicated times after inoculation with strain 801.					
Inoculating dose/mouse (viable counts)	3 h		24 h		72 h	
	Sucrose	D-Mannose	Sucrose	D-Mannose	Sucrose	D-Mannose
5×10^1	2	0	4	0	10	10
5×10^2	4	0	7	3	10	10
5×10^3	8	3	8	6	10	10
5×10^4	10	6	10	10	10	10
Infectious dose	7×10^2	2.7×10^4	2.0×10^2	2.0×10^3	50	50

[a]Adapted from Goldhar et al., 1986.

Ten mice were tested per group. Carbohydrate concentrations were 20 mg/ml.

containing receptors. In the presence of mannose only a few type 1 fimbriated bacteria could adhere to cells and hence it took a longer period of time for the bacteria to reach a detectable level, as compared to infection in the absence of mannose where more bacteria could adhere to the target cells. In a natural course of adhesion these time-wise and dose-wise differences between types 1 and MR adhesin-bearing phenotypes may be critical in initiating infection.

As the infection proceeds to the upper urinary tract, cleansing mechanisms become less important and infiltration of phagocytic cells becomes the major defense mechanism. Because Galα-1,4Gal-containing receptors are found only on kidney and bladder cells, whereas mannooligosaccharides are found on both bladder cells and phagocytic cells, it is likely that the survival of the bacterial population is now largely dependent on the P-fimbriated phenotype. Unlike the type 1 phenotype, the Gal-Gal-specific P-fimbriated cells escape recognition by the phagocytic cells. The results obtained in the distribution of the two phenotypes between the kidney and bladder after infection with a mixed phenotype population support this view (Table 7–4). In fact, the predominant population in phagocyte-rich sites are phenotypes that either lack detectable fimbrial adhesins or are rich in P-fimbriated bacteria, whereas the population on mucosal surfaces is rich in type 1-fimbriated bacteria.

The presence of mannooligosaccharide-containing glycoconjugates may have a dual effect as discussed above. They may inhibit the MS lectin-mediated interaction between the bacteria and the animal cells and consequently enhance either their elimination at mucosal surfaces due to the effective cleansing of the urine (Orskov et al., 1980), or promote their survival in deep tissues due to their ability to block recognition between the bacteria and the phagocytes (Kuriyama and Silverblatt, 1986). All of these effects are dependent on the concentration of the Tamm–Horsfall glycoprotein at a particular site. Recently, Israele et al. (1987) determined the amount of secreted Tamm–Horsfall glycoprotein by normal infants and patients suffering from urinary tract infections. Although they found that the patients had significantly less Tamm–Horsfall glycoprotein, a firm conclusion regarding the role of this glycoprotein in infection could not be reached. Undoubtedly, further studies are required to establish a definite role for soluble glycoconjugates that bear carbohydrates specific for bacterial lectins.

The association of *E. coli* clones with pyelonephritis may be related to the ability of the strains to alternate between types 1, P, and afimbriated phenotypes. This notion may be explained by assuming that a bacterial inoculum invading deep tissue, such as the kidney, consists of a mixture of the above mentioned phenotypes. The P-fimbriated and afimbriated phenotypes escape phagocytosis and may proliferate. This is especially pronounced in the case of the P-fimbriated phenotype as it can adhere to cells and acquire growth advantages (Zafriri et al., 1987). In contrast, the type 1 fimbriated phenotype undergoes lectinophagocytosis (refer to Chapter 7 and Figure 11–2). The process of phagocytosis is aimed at eliminating the invading organisms but at the same time it is associated with

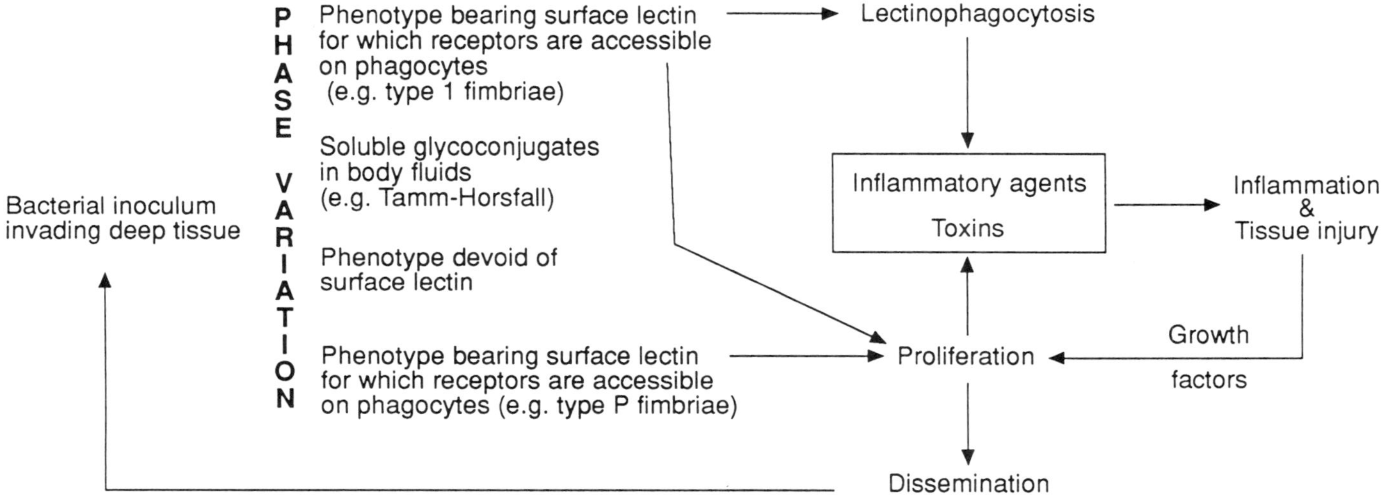

Figure 11–2. Postulated fate of a bacterium in vivo during a natural course of infection. Successful colonization of the bacterium depends on phase variations to yield new surface phenotypes and on the effects of opposing host factors, such as lectinophagocytosis and inflammation. Phenotypes that resist phagocytosis may gain a growth advantage by adhering to tissues.

oxygen bursts and degranulation resulting in the elaboration of tissue damaging factors (H_2O_2, HO· and hydrolytic enzymes). The growing bacteria may also secrete toxic products which become more effective when the bacteria are adherent (Zafriri et al., 1987; Linder et al., 1988). The dynamic balance (Figure 11–2) between the rate of killing by phagocytic cells and the rate of bacterial division may determine whether there will be a prompt elimination of the invading organisms or tissue damage and inflammation. The latter may persist until the elimination of the invading bacteria has been ultimately achieved by opsonophagocytosis.

It is possible that the above suggested roles of the two mechanisms of adhesion of pyelonephritogenic isolates of *E. coli* apply to other *E. coli* isolates listed in Table 11–4 as they each produce the "universal" MS adhesin and they diversify by the other MR mechanisms of adhesion. The MR adhesins, similar to P fimbriae, probably represent site and species specific adaptations (Parry et al., 1983), and most importantly, do not mediate interaction with phagocytes (Sussman et al., 1983; Nowicki et al., 1986). This is supported by the recent findings of Nowicki et al. (1986), who infected mice intraperitoneally with a mixture of phenotypes containing types 1 and S fimbriae. They found that in this phagocyte-rich environment the survival of the clone was associated with overgrowth of the type S fimbriated phenotype. Because an increasing number of reports suggest that other type 1 fimbriae-bearing enterobacteria produce at least one other mechanism of adhesion (Table 11–2) it is plausible that similar considerations apply to all these MS- and MR-producing bacteria.

The concept that one of the multiple mechanisms of adhesion is "universal" and conserved whereas the other(s) is more site and/or species specific may apply to other groups of organisms. For example, LTA, for which receptors are accessible on many types of animal cells (Wicken, 1980), may be the "universal" adhesin of several Gram-positive cocci, although each member of this group of bacteria may produce other distinct adhesins (Table 11–2). Also, similar to type 1 fimbriae, LTA interacts with soluble proteins in body fluids, such as albumin (Simpson et al., 1980). It is not known if the survival value of multiple mechanisms of adhesion in these bacteria is at the level of phenotypic variation.

The phenomenon of phase variation is probably not unique to fimbrial adhesins of enterobacteria. It is known to occur in *N. gonorrhoeae* (Lambden et al., 1979; Swanson and Berrera, 1983), *Bordetella pertussis* (Weiss and Falkow, 1984; Stibitz et al., 1988), and in *Moraxella bovis,* a cattle pathogen (Bovre and Froholm, 1972; Marrs et al., 1988). It may also occur in *Vibrio cholerae,* as those bacteria are known to detach from host intestinal epithelia (Finkelstein and Hanne, 1982). Christensen et al. (1987) described phase variations in *S. epidermidis* for adhesins that bind the organisms to plastic substrata. The production of extracellular polysaccharide by the marine bacterium *Pseudomonas atlantica* also seems to be under a genetic regulation that involves a phase transition (Bartlett et al., 1988). Even the fungal pathogen *Candida albicans* exhibits surface phase variations (Klotz and Penn, 1987) which may account for the multiple adhesins found in this organism.

Phase variation may account for the divergence of results obtained in various laboratories studying adhesion mechanisms of particular bacterial isolates. In some cases colonial morphology reflects a phase transition, as adhesin-producing phenotypes frequently display different colony types than phenotypes lacking adhesins. This phenomenon is commonly observed in *N. gonorrhoeae* (Swanson and Koomey, 1988).

It is tempting to conclude that phase variation and adhesin expression is a property shared by most bacteria. The rate and mechanism of phase transition may vary from bacterium to bacterium. For example, phase variation in expression of the fimbrial adhesin by *Moraxella bovis* is about 1 per 10,000 (Bovre and Froholm, 1972; Marrs et al., 1988), whereas that of type 1 fimbriae of *E. coli* K12 is 1 per 1000 (Eisenstein, 1981; Eisenstein and Dodd, 1982) and that of P fimbriae of pyelonephritogenic isolates of *E. coli* is 1 per 200 (Nowicki et al., 1985b). A new phase variant provides a new opportunity to adhere and colonize a surface different from the one inhabited by its mother. Each bacterial clone is assumed to give rise to two or more phenotypes. Each phenotype would then be able to adhere to a complementary surface. The same property of phase variation would permit the adherent phenotype to give rise to a nonadherent phenotype which in turn may bind to another surface. The switching phenomenon endows the bacterium with the ability to diversify its habitat. A bacterium may be able to adhere avidly to a substratum but factors such as nutritional availability and favorable physical and chemical environments are critical determinants for colonization as well. Due to evolutionary pressures the bacterium conserves at least two sets of genes coding for two or more adhesins in order to survive on at least two different substrata. A composite representation of phase variation in a bacterial clone and its role in adhesion to and colonization of various substrata is presented in Figure 11–3.

Some Consequences of Adhesion

Adhesion of bacteria to surfaces may result in immediate or delayed changes to the bacteria or to the substrata. The delayed changes occur because of colonization following the adhesion reaction and will not be further discussed in this chapter. The following paragraphs review some of the immediate consequences of bacterial adhesion to various substrata. Both the adherent bacteria and the target substrata in a particular microenvironment may be modified as a result of adhesion. Table 11–7 summarizes a few of these immediate changes.

Changes reported to occur in the bacteria upon the adhesion to inert and noninert surfaces may be reflected in the physiology or morphology of the adherent bacteria. Dramatic morphological changes occur in bacteria adherent

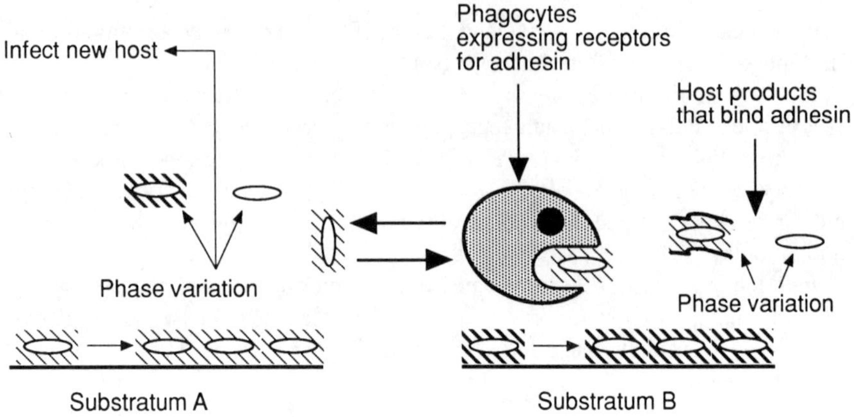

Figure 11–3. Possible role of multiple bacterial adhesins and phase variation in the colonization of two different substrata. An assumption in the model is that adhesion is a prerequisite for the ability of the bacterium to attain nutrients and to resist cleansing mechanisms of the host. The model also states that a unique adhesin is required for adhesion to the two substrata.

to glass under starvation: from bacilli to ovoid shape in order to increase the surface area bound to the substratum (Marshall, 1986). Physiological changes include de novo synthesis of proteins, enhanced growth, and enhanced resistance to antibiotics and other antimicrobials. Finlay et al. (1989) found that when *Salmonella typhimurium* adheres to an epithelial cell line, the bacteria begins to synthesize several new proteins. These new proteins could not be identified in broth cultures. It is possible that these newly induced proteins confer survival traits. At this time the exact function of none of these proteins has been defined. Nevertheless, attachment of the bacteria to either inert or living surfaces endows the adherent bacteria with growth advantages. There is no doubt that the attachment of bacteria to inert surfaces such as rocks, clays, or glass beads confers survival traits upon the bacteria (reviewed in Fletcher, 1985; Pearl, 1985). In the case of marine bacteria a microzone containing the bacteria and surrounding extracellular polymers is probably the main reason for trapping nutrients (Costerton et al., 1985). The growth advantages observed in bacteria adherent to tissue culture cells in vitro (Bessen and Gotschlich, 1986; Zafriri et al., 1987) and to the peritoneal linings (Alkan et al., 1986) or intestinal tissue (Tanaka, 1982) in vivo (described in Figure 11–4) are probably due to high concentrations of nutrients formed by an unstirred layer of ruffle structure on the living cell membrane and by the lid of the adherent bacteria (Figure 11–5). Interestingly enough, a microzone containing bacterial nutrients is thought to exist very close to environmental surfaces, such as submerged objects. This microzone, analogous to the cell surface of a tissue culture cell or an animal cell, promotes growth of bacteria at a rate faster than nutrients in the bulk media (Costerton et al., 1985).

Table 11–7. Changes that may occur as a result of bacterial adhesion to surfaces

Bacterium	Substratum	Result(s)	Reference(s)
Salmonella typhimurium	MDCK epithelial cell monolayers	De novo synthesis of several proteins occurs in the bacteria upon adhesion.	Finlay et al., 1989
Escherichia coli	Human epithelial cell line	Adherent bacteria demonstrate shorter lag times and are more toxic than nonadherent *E. coli*.	Zafriri et al., 1987
Marine pseudomonad	Cellulose dialysis membrane	Adherent cells retain cellular volumes better than nonadherent cells during starvation.	Humphrey et al., 1983
Salmonella typhimurium	Rat intestinal enterocytes	Adherent Fim^+ phenotype produces more damage to the enterocytes than the Fim^- phenotype.	Lindquist et al., 1987
Shigella flexneri	HeLa cells	Triggering of endocytosis.	Clerc and Sansonetti, 1987
Neisseria gonorrhoeae	Human cell	Cellular invasion follows adhesion.	Shaw and Falkow 1988; Swanson, 1980
Escherichia coli	Mouse bladder	Bacteria induce uroepithelial shedding.	Aronson et al., 1988
Pseudomonas aeruginosa	Catheter material	Adherent bacteria are more resistant to tobramycin than are nonadherent *P. aeruginosa*.	Nickel et al., 1985
Rumen bacteria	Rumen of cattle	Adherent bacteria demonstrate a higher rate of urease activity than nonadherent cells.	Javorsky et al., 1987
Achromobacter spp.	Chitin colloids	Chitin-bound bacteria exhibit higher rates of respiration than freely suspended cells.	Jannasch and Pritchard, 1972
Escherichia coli	Anion-exchange resin	Growth is stimulated when the cells become attached to the resin.	Hattori and Hattori, 1963, 1981
Coagulase-negative	Catheter material	Slime-producing strains are more resistant to nafcillin when adherent.	Sheth et al., 1985
Yersinia pseudotuberulosis		Surface protein promotes cellular invasion	Isberg et al., 1987
Escherichia coli	Mouse bladder	Adherent bacteria target LPS to induce inflammation.	Linder et al., 1988

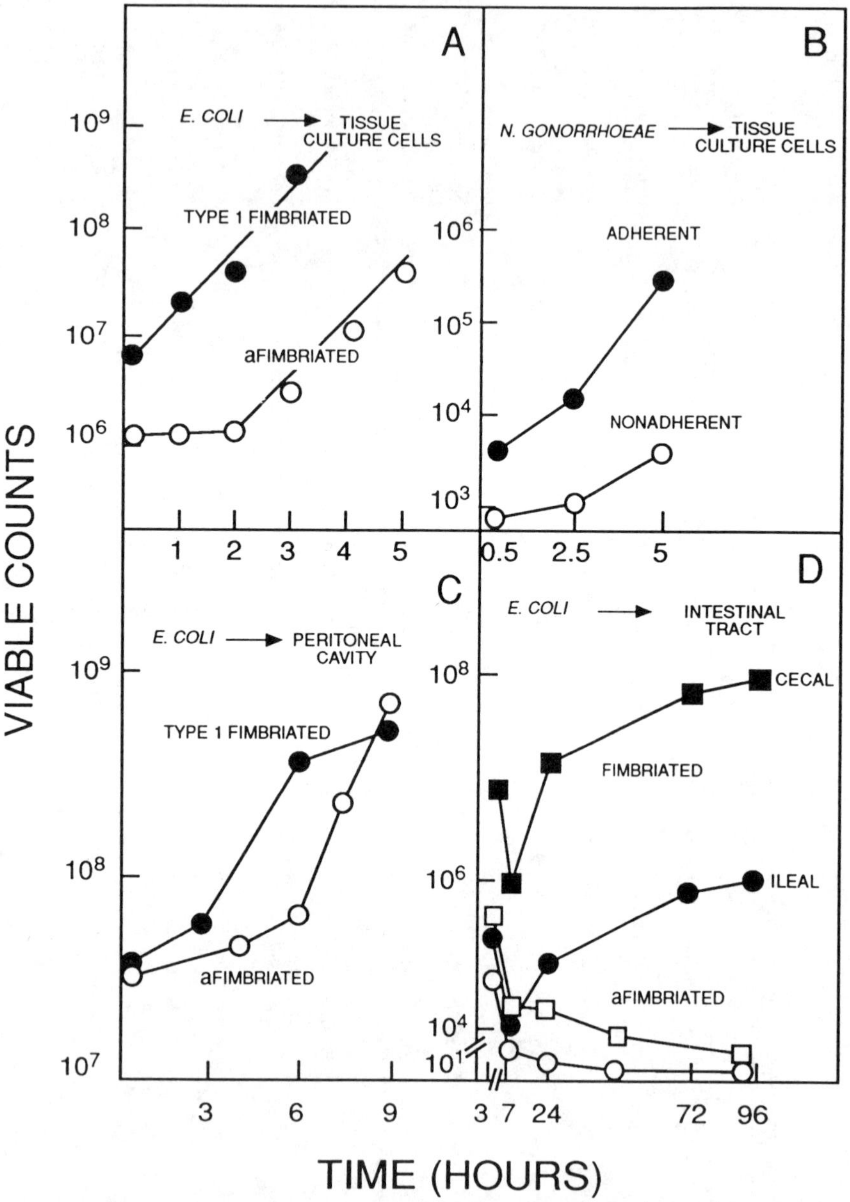

Figure 11–4. Growth advantages of bacteria adherent to tissue cells in vitro and in vivo. Derived from Zafriri et al., 1987, for *E. coli,* tissue culture; Bessen and Gotschlich, 1986 for *N. gonorrhoeae;* Alkan et al., 1986 for *E. coli*—peritoneal cavity; and Tanaka, 1982, for *E. coli*—intestinal tract.

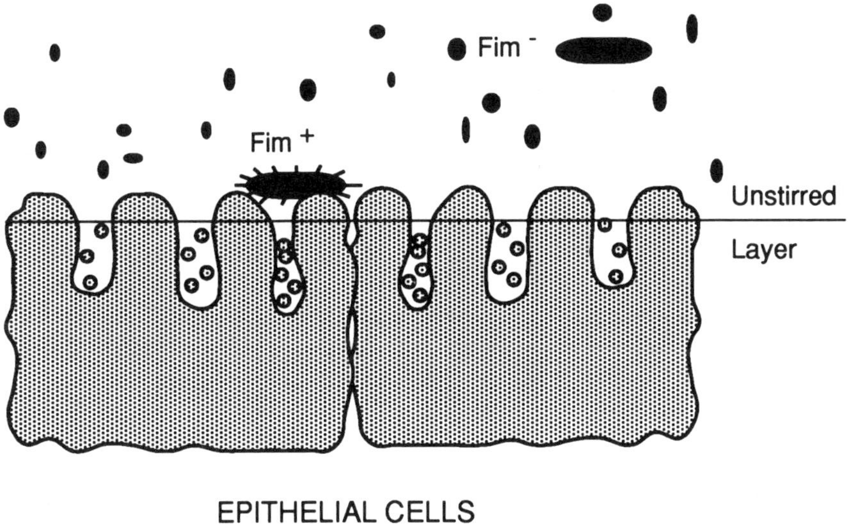

Figure 11–5. Schematic distribution of products secreted by epithelial cells (filled circle) and by bacteria (shaded oval) adherent to the cells, between the proximal and the distal milieu of the epithelial cell surfaces. The secretory products accumulate in pockets of the unstirred layer formed by the ruffle structure of the epithelial cell membrane and the lid of the adherent bacterium. (Modified from Zafriri et al., 1987).

Sometimes a growth advantage is gained even by transient association with the substratum. Adhesion processes in some bacteria have frequently been defined as involving two separable kinetic steps, one of which involves low-affinity interactions and is reversible, whereas the other is virtually irreversible due to a very low dissociation constants (see Chapter 2). The ability of *Vibrio cholerae* (Freter et al., 1981) and *E. coli* (Nevola et al., 1987) to colonize intestinal cells has been found to be dependent on the ability of the organisms to associate with and grow in the mucus layer before they reach, adhere to, and colonize mucosal cells of the gut. In some cases the chemical nature of the glycoprotein that binds the organisms has been defined (Cohen et al., 1985). Freter (1988) employed fecal mucus-coated glass, bathed by continuous flow as a substratum. Computer analysis of the growth of *E. coli* introduced into the system indicated the data could be explained only by assuming that there were at least two types of adhesion sites for the *E. coli* strain, mediating weak and strong adhesion, respectively. Multistep kinetics of adhesion may be explained in various ways. One explanation requires that the bacterium express at least two adhesins, one of which is low-affinity and the other of which is high-affinity. Another explanation may be related to phase transitions. A weakly adherent population may produce progeny that exhibit high-affinity adhesins for the same or different substrata. Another explanation is that initial adhesion may lead to cooperative bacteria–substratum interactions (Doyle et al., 1990).

In a hostile environment, however, bacterial multiplication requires a mechanism of escape from antimicrobials. Just as it is known that adherent bacteria undergo morphological and physiological changes, substrata can also be modified by adherent cells. For example, tissue culture cells are more readily "rounded" (a cytopathic effect specifically induced by the heat-labile enterotoxin) by adherent bacteria (Zafriri et al., 1987). When fimbriated and nonfimbriated isogenic mutants of *E. coli,* both secreting the same amount of labile toxin, were mixed with monolayers of mouse adrenal cells, the tissue culture cells tended to change morphology at relatively low densities of the *fim*$^+$ adherent mutant. The concentration of the toxin would be expected to be relatively high at the interface between the bound bacterium and the monolayer. The toxin secreted by the nonadherent *fim* mutant would be diluted into the total extracellular spaces, thereby rendering the Fim$^-$ organisms less toxic to the adrenal cells (Figure 11–6).

In another study, induction of lipopolysaccharide (LPS)-mediated inflammation in the urinary tract was found to be dependent on P-fimbriated bacteria adherent to the lining tissue (Linder et al., 1988). Furthermore, a complex between P fimbriae and LPS was much more toxic to the mouse bladder than LPS alone. The LPS is presumably delivered to the target tissue by the Galα-1,4Gal-specific fimbriae.

Toxins targeted to tissues by adherent bacteria may not be readily neutralized by soluble inhibitors (Ofek et al., 1990). For example, the enhanced effect due

Strain	No. of bacteria	% Rounded cells	Titer of toxin secreted in the growth medium
VL 645 (Fim+)	6.0×10^6	27 ± 4	1 : 160
	3.6×10^6	19 ± 3	
VL 647 (Fim-)	1.7×10^8	41 ± 5	1 : 160
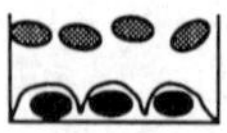	8.5×10^7	24 ± 6	
	4.2×10^7	15 ± 1	

Figure 11–6. Activity of labile toxin secreted by *Escherichia coli* nonadherent and adherent to Y-1 mouse adrenal cells. Fewer Fim$^+$ cells are required to cause the "rounding" of the tissue culture cells than the Fim$^-$ (nonadherent) bacteria. Note that the amount of toxin in the medium was the same for both bacterial strains. (From Zafriri et al., 1987. Courtesy of D. Zafriri.)

to labile toxin produced by adherent *E. coli* could not be neutralized by antitoxin that at a one-tenth lower concentration neutralized the toxic effect on nonadherent bacteria (Figure 11–7). Toxic effects of bacteria other than *E. coli* also seem to be neutralized more readily when the cells are in suspension, rather than adherent. Ofek et al. (1990) also found that trypan blue would neutralize streptolysin S at a lower dye concentration when the toxin was in a soluble form. These results may be partially explained by the fact that there is restricted diffusion near the cell–substratum interface from the bulk medium (see Figure 11–5).

Although bacterial films develop on animal surfaces can promote or impede the biodeterioration of the substratum, no evidence has been provided to show that this effect results directly from adhesive events (Little et al., 1990). Interaction of certain bacteria (e.g., *Shigella* spp., *Yersinia* spp., and *Neisseria gonorhoeae*) with target cells triggers cascades of reactions leading to endocytosis of the organisms. The relationships between adhesion and invasion have not been characterized. Logic states that invasion follows adhesion, but compelling evidence

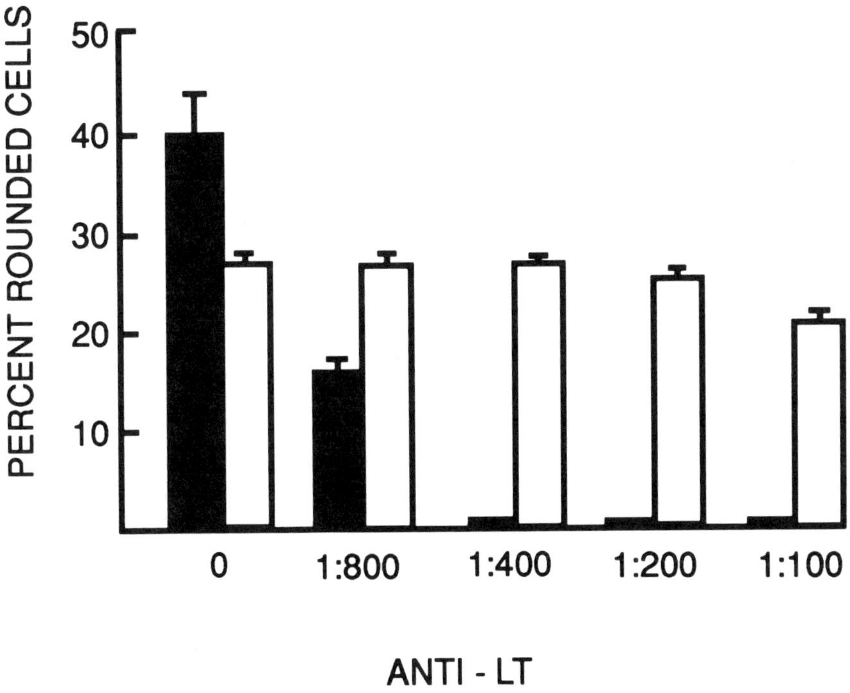

Figure 11–7. Inability of anti-labile toxin (anti-LT) to neutralize toxicity caused by LT-secreting bacteria. In the figure it is shown that the amount of antibody required to neutralize the toxicity caused by adherent *Escherichia coli* (open rectangle) is at least 10 times that needed to neutralize the toxicity caused by a nonadherent strain (darkened rectangle). (Derived from Ofek et al., 1990. Courtesy of D. Zafriri.)

for this exists only in the case of members of the genus *Yersinia*. *Yersinia* spp. produce a large polypeptide, termed invasin, that is expressed on their surfaces (Isberg and Falkow, 1985; Isberg et al., 1987). Invasin-bearing bacteria not only can adhere to but also penetrate target cells. In contrast, mutants deficient in *inv* gene (invasin gene) can neither adhere nor invade target cells. Furthermore, recombinant *E. coli* expressing the *Yersinia inv* gene product adheres and penetrates tissue culture cells. This is compelling evidence showing that invasion is dependent on adhesion. It appears, therefore, that the product of the *inv* gene, invasin, is a cell surface adhesin that binds the organisms to complementary receptors on target cells. In this regard, the *Yersinia* invasin receptors are different from receptors for adhesins of other bacteria, such as the fimbrial adhesins of enterobacteria. The receptors not only mediate binding of the bacteria but also serve to internalize the bound organisms, probably by triggering endocytosis. The receptor for the *Yersinia* invasin was found to be a membrane glycoprotein that belongs to a superfamily of glycoproteins called integrins (Ruoslahti and Pierschbacher, 1987). The integrins are transmembrane glycoproteins the carboxy terminii of which are cytoplasmically exposed and are bound to the cytoskeleton proteins responsible for cellular contractions and endocytosis. At present, it is not clear that the integrins mediate invasion of other invasive bacteria. It is interesting to point out, however, that phagocytosis of type 1 fimbriated *E. coli* involves interactions of the MS fimbrial adhesin with at least one member of the integrin superfamily (see Chapter 7).

It has been postulated that there is a coordinate regulation of multiple factors, such as adhesins and toxins, involved in bacterial pathogenesis by a positive regulatory element. This regulation is assumed to represent an adaptation of the pathogen to changing environments associated with a "pathogenic life-style" (Stibitz et al., 1988). It is tempting to speculate further that the phase variation phenomenon of pathogenic bacteria giving rise to multiple mechanisms of adhesion allows the emergence of phenotypes expressing a specific adhesin along with other determinants of pathogenicity by a coordinated positive regulatory mechanism (Hacker, 1990). Such potentially diverse phenotypes may be the best suited to adhere to, proliferate, and effectively exert toxic activity at a specific site during a specific stage of the infectious process.

It now seems clear that there are beneficial consequences to bacteria when they become adherent. They acquire growth advantages, resistance to antibodies and antibiotics, and exert enhanced toxic activities that escape neutralization. In some cases, invasion of target cells is tightly coupled with adhesion (Table 11–7).

Interference with Bacterial Adhesion to Substrata

A goal in adhesion research is to be able to inhibit or promote adhesion according to a particular need. On the one hand, inhibition of adhesion or dissociation of adherent bacteria may be critical in bacterial pathogenesis or in the development of biofilms or submerged objects. In contrast, it may be expedient to promote adhesion of a vaccine carrier strain or of a nonpathogen so as to exclude a potential pathogen. Evidence has been presented in this book that it is an advantage for a bacterium to adhere to a surface. Growth may be enhanced, cleansing mechanisms may be circumvented, and resistance to antimicrobial agents may be increased. Conditions have also been discussed whereby adhesion may not be beneficial for the bacteria. For example, attachment to phagocytes may lead to the elimination of the bacteria or deprivation of nutrients at the site of adhesion. In order for the bacteria to survive it is required sometimes that these adherent bacteria detach. Detachment would thus permit migration toward a more favorable environment (Figure 11–3). Evolutionary pressures have undoubtedly selected organisms the population of which contains adherent and nonadherent phenotypes. The emergence of these phenotypes may be due to complex genetic regulations of the expression of the adhesins or to naturally occurring inhibitors or promoters of adhesion. In fact, under natural conditions, the ability of the bacteria to adhere or not to adhere to a substratum is the net sum of modalities that promote and inhibit adhesion. The source of such modalities may be the substratum and its environment and/or the genome of the bacterium. Manipulation of bacterial adhesion on a particular substratum requires basic knowledge of the modalities that take place under natural conditions as well as of the nature of the molecules that mediate adhesion. Because very little is known about the molecular mechanisms of bacterial adhesion to environmental inert surfaces, it is not surprising that most of the approaches aimed at manipulating adhesion have been developed in systems involving adhesion to soft and hard tissues. In the following, the knowledge accumulated on naturally occurring inhibitors or promoters of adhesion is summarized. The various modalities of adhesion to mammalian soft tissues is discussed with the view of how colonization can be inhibited so as to prevent infection.

Numerous naturally occurring inhibitors or promoters of adhesion to soft tissue have been described. In principle, any constituent capable of binding to the bacterial surface can potentially act as an inhibitor or promoter of adhesion. Many bacterial pathogens have been found to interact with a wide variety of soluble macromolecules that occur naturally and may not be present at the site of adhesion. For only few of these macromolecules has a direct effect on adhesion been demonstrated. In some cases, the molecular basis of the macromolecule–bacterium interaction has been investigated. If a clear understanding of inhibition or promotion of adhesion is to be achieved, it will be necessary to define the origin of the inhibitor or promoter molecules. Some macromolecules are sparingly soluble and remain near their sites of secretion. Other molecules may be serum-borne and can affect sites far removed from their cellular origins.

Mucus constituents tend to reside near the tissues involved in their synthesis, whereas some constituents are found in virtually all anatomical sites. It follows that interaction of bacteria with mucus constituents may result in a dual effect: adhesion may be inhibited or may be enhanced, depending on the concentration of the mucus constituent and its degree of association with the cell surface where it can act as "bridge" between the bacterium and the substratum. Table 11–8 summarizes examples to illustrate the complexity of these effects. Whereas naturally occurring modulators of adhesion are hard to manipulate, several approaches employing various types of exogeneous modulators of adhesion have been developed.

One approach includes inhibition of adhesion by soluble inhibitors that structurally resemble the attachment site of the receptor molecule specific for binding the adhesin (Chapter 1). This so-called "receptor therapy" by receptor analogues has been demonstrated in several sets of experiments, each employing bacteria expressing lectins with different specificities whereby the inhibitor of the adhesin consists of derivatives of that carbohydrate (see also Table 5–20). It has been demonstrated that mannose and methyl-α-mannoside inhibit experimental infections or mucosal colonization by members of enterobacterial species expressing the MS fimbrial adhesin. Similarly, Galα-1,4Gal-containing saccharides have been shown to prevent urinary tract infection in both mice and monkeys by P-fimbriated *E. coli*. These findings provide some of the most convincing results for the central role of bacterial surface lectins in infection, and particularly in mucosal colonization. They also illustrate the great potential of simple carbohydrates in the prevention of infections caused by bacteria that express adhesins

Table 11–8. Modulation of bacterial adhesion by soluble macromolecular constituents of body fluids

Site of adhesion	Origin of constituent	Chemical nature of constituent	Interacting bacteria	Effect on adhesion	Reference(s)
Oral cavity	Saliva	Glycoprotein	*S. mutans*	Promotion	Kishimoto et al., 1989
		Glycoprotein	*S. sanguis*	Inhibition	Nesbitt et al., 1982a,b; Hogg and Embery, 1982
	Saliva	Fibronectin	*S. pyogenes*	Inhibition	Simpson and Beachey, 1983; Simpson et al., 1987
Gut	Goblet	Glycoprotein	*E. coli* (MR)	Inhibition	Drumm et al., 1988
	cells		*E. coli* (type 1)	Inhibition	Cohen et al., 1983, 1985
Urinary tract	Henle loop	Tamm–Horsfall	*E. coli*	Inhibition	Orskov et al., 1980; Chick et al., 1981
	Not defined	Mannooligo-saccharides	*E. coli* (types 1 and S)	Inhibition	Parkkinen et al., 1988
Inflammation	Serum	Albumin	*S. pyogenes*	Inhibition	See Chapter 6

with lectin activity. For example, studies on the carbohydrate specificity of type 1 fimbrial lectin of *E. coli* suggest that aromatic glycosides, such as 4-methylumbelliferyl-2-mannoside and *p*-nitro-*o*-chlorophenyl-2-mannoside, are potent inhibitors of this fimbrial adhesin. The results may provide a basis for the design of safe therapeutic agents that prevent adhesion in vivo and infection by enterobacterial strains that express mannose-specific fimbrial lectins (Firon et al., 1985, 1987).

Another experimental approach includes the use of animals that have been passively or actively immunized against the corresponding adhesins. This approach was exploited in studies of infections with *E. coli* expressing lectins or hemagglutinins that function as adhesins in a variety of experimental conditions. In all cases infection was prevented in the immunized animals challenged with strains expressing the fimbrial lectin (See Table 5–21). The only vaccine that is useful is a special case, in which the pregnant dam is immunized with the K88 fimbrial adhesin to induce antibodies in the colostrum which then protects suckling piglets from diarrhea caused by the K88-fimbriated bacteria. Both of the above approaches do, however, have problems of application. Receptor analogues composed either of simple carbohydrates or complex glycoconjugates would need to be continuously present on the mucosal surfaces because they act by the principle of competitive hapten inhibition. This may be difficult to achieve in face of the continuous self-cleansing function of mucosal surfaces. Although antibodies provide relatively long-lasting protection, their continuous presence on the mucosal surfaces is required and they need to be directed against the adhesin molecule rather than against the structure that carries the adhesin. This was not appreciated in early trials, and extensive serotype variations were indeed encountered among the fimbrial vaccines. It now seems more reasonable to plan vaccines composed of the adhesin molecule itself or of a structure common to all serotypes. Secretory IgA against the adhesin molecule can be induced best by oral immunization because parenteral immunization appears to stimulate local antibody production only weakly. Serum antibodies (both IgA and IgG) may also be found in mucosal secretions (Abraham et al., 1985). In this regard, the use of a live vaccine composed of attenuated *Salmonella typhi* (Levine et al., 1987) genetically modified to express the tip adhesin may circumvent some of the problems related to antigenic variation and induction of local immunity (Abraham et al., 1988; Madison et al., 1989) (Figure 11–8).

Another important problem is the issue of phase variation in the expression of adhesins as a strategy for survival of the organisms. Phase variation between several possible phenotypes, each of which expresses a different adhesin in a bacterial population of the same clone, may enable the organisms to avoid the inhibitory actions of antibodies or receptor analogues to one of the adhesive phenotypes. For example, whereas antibodies or mannooligosaccharides directed against the mannose-specific adhesin of *E. coli* would prevent the type 1 fimbriated phenotype from colonization, the other phenotype expressing an entirely

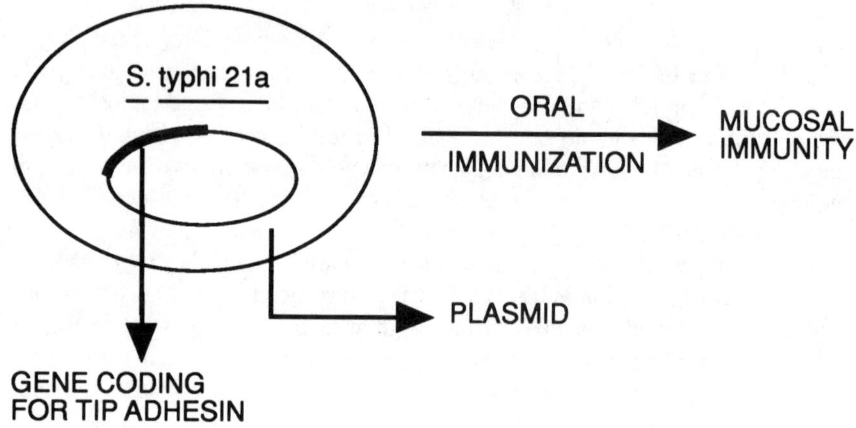

Figure 11–8. Hybrid vaccines that use carrier strains. The diagram shows the construction of a vaccine carrier in which the gene coding for the adhesin was mobilized into a plasmid caused by an avirulent strain of *S. typhi*21a. The carrier strain can now be used to induce mucosal immunity against the adhesin.

different adhesin and fimbrium (e.g., Galα-1,4Gal-specific adhesin of P fimbriae) and present in the same inoculum would not be inhibited. Future studies should take into consideration the various potential adhesin phenotypes in order to design an effective therapy for the prevention of serious bacterial infections (Figure 11–9).

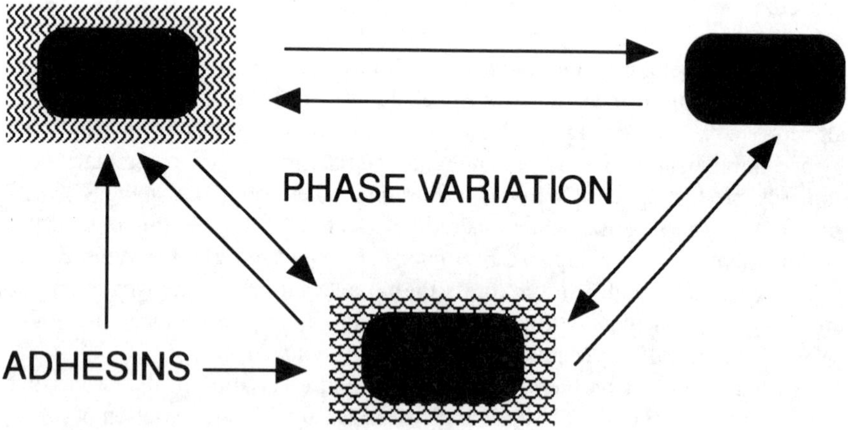

Figure 11–9. Diagram showing three phenotypes of a bacterial clone, one of which lacks a surface adhesin, whereas the other two express adhesins with distinct receptor specificities. The phenomenon of phase variation allows the emergence of daughter cells with adhesins different from that possessed by the mother cell. The figure emphasizes the notion that the protocol for the prevention of bacterial infections by inhibition of adhesin must include inhibitors against all potential adhesins.

Another approach is the use of metabolic inhibitors of the expression of adhesins. This approach is attractive in the sense that such metabolic inhibitors may be sublethal concentrations of antibiotics or other harmless agents. For example, alanine is an inhibitor of the expression of the K99 fimbrial adhesin (DeGraaf et al., 1980). As yet, this approach has not been exploited in experimental infections or in humans, although sublethal concentrations are bound to occur during the natural course of antibiotic treatment. Table 11–9 illustrates the types of antibiotics found to modulate adhesion of various species of bacteria. As can be seen the disadvantage of this approach is that the effects of sublethal concentrations of antibiotics are unpredictable. For example, the effect is dependent on the bacterial strain and may result in either enhancing or inhibiting adhesion.

An interesting and perhaps more practical approach is the use of dietary inhibitors of adhesion to prevent infection (Table 11–10). For example, several investigators have found that cranberry juice contains inhibitors of adhesion of *E. coli* to epithelial cells (Sobota, 1984; Schmidt and Sobota, 1988; Zafriri et al., 1989). More recently Ofek et al. (1991) found that of six fruit juices tested, those from the cranberry and blueberry (both belonging to the viccinia family) contained two distinct inhibitors of two types of adhesins produced by *E. coli*. One inhibitor is fructose which acts on the mannose-specific type 1 fimbriae. Fructose was found some years ago to inhibit type 1 fimbriae of *E. coli* (Salit and Gotschlich, 1977), *Salmonella typhimurium,* and *Shigella flexneri* (Old, 1972), albeit requiring relatively high concentrations as compared to D-mannose. The other inhibitor was a polymeric constituent of an as yet unknown structure and selectively inhibits mannose-resistant adhesins produced by urinary isolates of *E. coli,* including the P-fimbrial and nonfimbrial adhesins. Both inhibitors are found in the cranberry and blueberry juices at levels that are at least 10 times higher than those required to inhibit the corresponding adhesin in vitro. It is conceivable that inhibitory levels of the inhibitors are attained in the colon, where most uropathogenic *E. coli* reside. These findings support the possibility that the reported (Moen, 1962; Sternlieb, 1963; Papas et al., 1968) effects of cranberry juice on urinary tract infections are due to the presence of both MS and MR antiadhesive agents. If this is so, then blueberry juice should be as effective and it may also explain why juices other than cranberry or blueberry are not effective. These other juices contain only one inhibitor, fructose, which acts against only one of the adhesins (MS) of *E. coli*. These general results emphasize the importance of the phase variation phenomenon in the expression of multiple adhesins during the infectious process and the necessity for an approach aimed at inhibiting all of the potential adhesins (Figure 11–9).

Dietary lectins (Liener, 1968), such as those derived from legume or wheat germ, may also modulate adhesion. Gibbons and Dankers (1983) found that wheat germ agglutinin could form a part of acquired pellicle and then bind to *S. sanguis*. This mechanism of lectin-promoted adhesion of oral streptococci has

Table 11–9. Modulation of bacterial adhesion by subinhibitory concentrations of antibiotics

Pathogen	Adhesin	Substrata	Antibiotics	Reference(s)
A. Inhibition of adhesion				
Gram-negative bacteria				
Escherichia coli	Type 1 fimbriae	Human leukocytes, buccal epithelial cells, guinea pig erythrocytes, yeast cells	Benzylpenicillin, chloramphenicol, gentamicin, neomycin, novobiocin, oxolinic acid	Beachey et al., 1981, 1982; Eisenstein et al., 1981; Hammani et al., 1987; Ofek et al., 1979, 1989
	P-fimbriae	Human buccal	Ampicillin	Bassaris et al., 1984a,b; Dean and Kessler, 1988; Hammani et al., 1987; Sandberg et al., 1979; Vosbeck, 1982; Vosbeck et al., 1979
		Epithelial cells, intestinal cells, erythrocytes, uroepithelial cells, T 24 bladder carcinoma cells	Berberine sulfate, chloramphenicol, clindamycin, oxolinic acid streptomycin, sulfadiazin, sulfathiazole, tetracycline, trimethoprim	Stenqvist et al., 1982; Vaisanen et al., 1982; Sun et al., 1988a,b
		Epithelial cells, human buccal	Cefatoxime	Bassaris et al., 1984a
	K88 fimbriae	Pig erythrocytes	Colistin	Sogaard and Larsen, 1983
	K99 fimbriae	Pig and sheep	Chloramphenicol	Sogaard and Larsen, 1983; Ferrerios and Criado, 1983
		Erythrocytes	Colistin, gentamicin, tetracycline	Sogaard and Larsen, 1983

	CFA/I, II, III	Human and bovine erythrocytes	Benzylpenicillin, deoxycycline	Forestier et al., 1984
		Human intestinal epithelial cells	Minocycline, olandeomycin	
Neisseria meningitidis	ND	Human erythrocytes buccal epithelial cells	Ampicillin, benzylpenicillin, colistin, erythromycin, lincomycin, mystatin, rifampin, tetracycline, tobramyin	Kristiansen et al., 1983; Salit, 1983; Stephens et al., 1984
Neisseria gonorrhoeae	ND	Human buccal epithelial	Benzylpenicillin, tetracycline	Stephens et al., 1984
Pseudomonas aeruginosa	ND	Mucin	Cetrazidime	Vishwanath et al., 1987
Pseudomonas aeruginosa	ND	Epithelial cells	Clindamycin	Lianou et al., 1985
Vibrio proteolytica	ND	Polystyrene	Ampicillin, oxacillin, streptomycin	Paul, 1984
Gram-positive bacteria				
Staphylococcus aureus	Surface proteins LTA	Fibronectin	Chloramphenicol, erythromycin	Proctor et al., 1983
Staphylococcus aureus	ND	Epithelial cells, HeLa cells	Erythromycin, benzylpenicillin, cephalexin, mecillinam, pefloxacin	Mascellino et al., 1988; Miyake et al., 1989
Staphylococcus (coagulase negative)	ND	Plastic	Cephalothin, chloramphenicol, gentamicin, impenem, methicillin novobiocin, polymyxin B	Schadow et al., 1988
Streptoccocus pyogenes	LTA	Human buccal and pharyngeal cells	Benzylpenicillin, berberine sulfate, lincomycin, tetracycline	Alkan and Beachey, 1978; Sun et al., 1988b; Tylewska et al., 1981
Streptococcus pyogenes	ND	Epithelial cells	Erythromycin	Mascellino et al., 1988
Streptococcus sanguis	ND	Heart tissue	Benzylpenicillin, chloramphenicol, tetracycline, vancomycin	Bernard et al., 1981; Scheld et al., 1981; Lowy et al., 1983

continued

Table 11–9. Continued

Pathogen	Adhesin	Substrata	Antibiotics	Reference(s)
B. Promotion of adhesion				
Gram-negative bacteria				
Escherichia coli	ND	Human epithelial cells	Nalidixic acid	Vosbeck, 1982
Gram-positive bacteria				
Staphylococcus aureus	ND	Human buccal epithelial cells	Penicillin	Shibl, 1985
Staphylococcus aureus	ND	HeLa cells	Streptomycin, kanamycin, puromycin, gentamicin, dibekacin, astomycin, sisomycin	Miyake et al., 1989
Staphylococcus (coagulase negative)		Plastic	Rifampin	Schadow et al., 1988
Streptococcus mutans	ND	Hydroxylapatite	Penicillin	Peros and Gibbons, 1982
Streptococcus pyogenes	LTA	Buccal epithelial cells	Clindamycin	Alkan and Beachey, 1978
Streptococcus sanguis	ND	Heart tissue	Trimethoprim	Scheld et al., 1981

Compiled from reviews of Chabonen, 1987; Chopra and Linton, 1986; Schifferli and Beachey, 1988a,b; and Shibl, 1985.

ND, not determined.

Table 11–10. Examples of dietary products that modulate bacterial adhesion

Bacteria	Adhesin(s)	Dietary product	Active constituent	Effect on adhesion	Reference(s)
E. coli (uropathogen)	MS and MR	Cranberry and blueberry juices	Fructose, high molecular weight constituent	Inhibition	Ofek et al., 1991; Schmidt and Sobota, 1988; Sobota, 1984; Zafriri et al., 1989
S. pneumoniae	Gal-GlcNAc specific	Milk	Glycoconjugates	Inhibition	Andersson et al., 1986
H. influenzae	ND	Milk	Glycoconjugates	Inhibition	Andersson et al., 1986
S. sanguis	ND	Wheat bread	GlcNAc-specific	Enhancement	Gibbons and Dankers, 1983
S. mutans	ND	Table sugar	Sucrose	Enhancement	Germaine and Schachtele, 1976

ND, not defined.

Table 11–11. Strategies to prevent infections by inhibiting bacterial adhesion

Strategy	Advantages	Problems
Passive or active immunization with adhesin	Long-lasting protection.	Antigenic heterogeneity. Multiple adhesins. Mucosal immunity, e.g., S-IgA may be difficult to achieve.
Receptor therapy by receptor analogues	Directed against the functional group of the adhesin bypassing the antigenic variability. Can be achieved by simple low molecular weight inhibitors.	The analogue must be present constantly providing short-term protection. Toxicity of potent inhibitor. Multiple inhibitors must be present to account for multiple adhesins.
Metabolic inhibitors	Low doses of antibiotics eliminating toxicity and extra cost. Antibiotics may prevent expression of multiple adhesins.	Strains that are resistant to lethal doses are also resistant to sublethal doses.
Dietary inhibitors	Normal diet for many humans. A single dietary product may contain inhibitors of several adhesins.	Concentrations of inhibitor(s) are difficult to standardize.

not been studied in detail. Because the human diet contains lectins from various sources and lectins have been found to interact with various bacteria (Doyle and Keller, 1984), it would be expected that the lectins would interact with some of the microbiota. In the oral cavity, lectins may not only promote adhesion but they may also inhibit adhesion by aggregating the bacteria. Lectins may also resist digestion and may reach the alimentary canal in a functional form (Brady et al., 1978). Dietary inhibitors may be natural host products. This is the case in which milk was found to be rich in saccharides that specifically compete with the natural receptors of some bacterial adhesins (Andersson et al., 1986) (Table 11–10).

So far no adhesin vaccine has proven to be useful in humans for the prevention of infection by inhibiting adhesion. The data accumulated to date suggest that new approaches may ultimately prove useful in the future to modulate adhesion (Table 11–11 and Figure 11–9). The advantages of inhibiting adhesion to prevent bacterial colonization of a substratum rely on the assumption that the spread of strains with genotypic resistance to the method applied would be much slower than in the case of employing approaches aimed at killing the organisms.

References

Abraham, J.M., C.S. Freitag, R.M. Gander, J.R. Clements, V.L. Thomas, and B.I. Eisenstein. 1986. Fimbrial phase variation and DNA rearrangements in uropathogenic isolates of *Escherichia coli*. *Mol. Biol. Med.* **3**:495–508.

Abraham, S.N., J.P. Babu, C.S. Giampapa, D.L. Hasty, W.A. Simpson, and E.H. Beachey. 1985. Protection against *Escherichia coli*-induced urinary tract infections with hybridoma antibodies directed against type 1 fimbriae or complementary D-mannose receptors. *Infect. Immun.* **44:**625–628.

Abraham, S.N., D. Sun, J.B. Dale, and E.H. Beachey. 1988. Conservation of the D-mannose-adhesin protein among type 1 fimbriated members of the family *Enterobacteriaceae*. *Nature* **336:**682–684.

Adegbola, R.A. and D.C. Old. 1985. Fimbrial and non-fimbrial haemagglutinins in *Enterobacter aerogenes*. *J. Med. Microbiol.* **19:**35–43.

Alkan, M.L. and E.H. Beachey. 1978. Excretion of lipoteichoic acid by group A streptococci: influence of penicillin on excretion and loss of ability to adhere to human oral mucosal cells. *J. Clin. Invest.* **61:**671–677.

Alkan, M.L., L. Wong, and F.J. Silverblatt. 1986. Change in degree of type 1 piliation of *Escherichia coli* during experimental peritonitis in the mouse. *Infect. Immun.* **54:**549–554.

Aly, R. and S. Levit. 1987. Adherence of *Staphylococcus aureus* to squamous epithelium: role of fibronectin and teichoic acid. *Rev. Infect. Dis.* **9:**S341–S350.

Andersson, B., O. Ponnas, L.A. Hanson, T. Lagergard, and C. Svanborg-Eden. 1986. Inhibition of attachment of *Streptococcus pneumoniae* and *Haemophilus influenzae* by human milk and receptor oligosaccharides. *J. Infect. Dis.* **153:**232–237.

Aronson, M., O Medalia, D. Amichay, and O. Nativ. 1988. Endotoxin-induced shedding of viable uroepithelial cells is an antimicrobial defense mechanism. *Infect. Immun.* **56:**1615–1617.

Baddour, L.M., G.D. Christensen, W.A. Simpson, and E.H. Beachey. 1989. Microbial adherence. In: Mandel, G.L., R.G. Douglas, Jr., and J.F. Bennett (eds.), *Principles and Practice of Infectious Disease,* 3rd ed. John Wiley & Sons, New York, pp. 9–25.

Bagg, J., I.R. Poxton, D.M. Weir, and P.W. Ross. 1982. Binding of type-III group B streptococci to buccal epithelial cells. *J. Med. Microbiol.* **15:**363–372.

Baker, N., V. Minor, C. Deal, M.S. Ghahrabadi, D.A. Simpson, and D.E. Woods. 1991. *Pseudomonas aeruginosa* exoenzyme S is an adhesin. *Infect. Immun.* **59:**2859–2863.

Bar-Shavit, F., R. Goldman, I. Ofek, N. Sharon, and D. Mirelman. 1980. Mannose binding activity of *Escherichia coli,* a determinant of attachment and ingestion of the bacteria by macrophages. *Infect. Immun.* **29:**417–424.

Bartlett, D.H., M.E. Wright, and M. Silverman. 1988. Variable expression of extracellular polysaccharide in the marine bacterium *Pseudomonas atlantica* is controlled by genome rearrangement. *Proc. Natl. Acad. Sci. USA* **85:**3923–3927.

Bassaris, H.P., P.E. Lianou, E.G. Votta, and J.T. Papavassiliou. 1984a. Effects of subinhibitory concentrations of cefotaxime on adhesion and polymorphonuclear leukocyte function with gram negative bacteria. *J. Antimicrob. Chemother.* **14:**(*Suppl B*): 91–96.

Bassaris, H.P., P.E. Lianou, and J.T. Papavassiliou. 1984*b* Interactions of subminimal inhibitory concentrations of clindamycin and *Escherichia coli*: effects on adhesion and polymorphonuclear leukocyte function. *J. Antimicrob. Chemother.* **13:**361–367.

Beachey, E.H., B.I. Eisenstein, and I. Ofek. 1981. Sublethal antibiotics and bacterial adhesion. In: Elliott, K., M. O'Connor, and J. Whelan (eds.), *Adhesion and Microorganism Pathogenicity*. Pitman Medical Press, London, pp. 288–300.

Beachey, E.H., B.I. Eisenstein, and I. Ofek. 1982. Prevention of the adhesion of bacteria to mucosal surfaces: influence of antimicrobial agents. In: H.-U. Eickenberg, H. Hahn, and W. Opferkuch (eds.), *The Influence of Antibiotics on the Host–Parasite Relationship*. Springer-Verlag KG, Berlin, pp. 171–182.

Beachey, E.H., W.A. Simpson, I. Ofek, D.L. Hasty, J.B. Dale, and E. Whitnack. 1983. Attachment of *Streptococcus pyogenes* to mammalian cells. *Rev. Infect. Dis.* **5:**S670–S677.

Bernard, J.P., P. Francioli, and M.P. Glauser. 1981. Vancomycin prophylaxis of experimental *Streptococcus sanguis:* inhibition of bacterial adherence rather than bacterial killing. *J. Clin. Invest.* **68:**1113–1116.

Bessen, D. and E.C. Gotschlich. 1986. Interactions of gonococci with Hela cells: attachment, detachment, replication, penetration and the role of protein II. *Infect. Immun.* **54:**154–160.

Beuth, J., H.L. Ko, Y. Ohshima, A. Yassin, G. Uhlenbruck, and G. Pulverer. 1988. The role of lectins and lipoteichoic acid in adherence of *Staphylococcus saprophyticus*. *Zbl Bakt. A* **268:**357–361.

Bourgeau, G. and B.C. McBride. 1976. Dextran-mediated interbacterial aggregation between dextran-synthesizing streptococci and *Actinomyces viscosus*. *Infect. Immun.* **13:**1228–1234.

Bovre, K. and L.O. Froholm. 1972. Variation of colony morphology reflecting fimbriation in *Moraxella bovis* and two reference strains of *M. nonliquefaciens*. *Acta Pathol. Microbiol. Scand.* B **80:**629–640.

Brady, P.G., A.M. Vannier, and J.G. Banwell. 1978. Identification of dietary lectin, wheat-germ agglutinin in human intestinal contents. *Gastroenterology* **75:**236–239.

Buchanan, K., S. Falkow, R.A. Hull, and S.I. Hull. 1985. Frequency among *Enterobacteriaceae* of the DNA sequences encoding type 1 pili. *J. Bacteriol.* **162:**799–803.

Carruthers, M.M. and W.J. Kabat. 1983. Mediation of staphylococcal adherence to mucosal cells by lipoteichoic acid. *Infect. Immun.* **40:**444–446.

Chabanon, G. 1987. Bacterial adhesion: impact on antimicrobial therapy. *Path. Biol.* **35:**1365–1369.

Chick, S., M.J. Harber, R. MacKenzie, and Asscher, A.W. 1981. Modified method for studying bacterial adhesion to isolated uroepithelial cells and uromucoid. *Infect. Immun.* **34:**256–261.

Chopra, I. and A. Linton. 1986. The antibacterial effects of low concentrations of antibiotics. *Adv. Microbial Physiol.* **28:**211–259.

Christensen, G.D., W.A. Simpson, A.L. Bisno, and E.H. Beachey. 1982. Adherence of slime-producing strains of *Staphylococcus epidermidis* to smooth surfaces. *Infect. Immun.* **37:**318–326.

Christensen, G.D., L.M. Baddour, and W.A. Simpson. 1987. Phenotypic variation of *Staphylococcus epidermis* slime production *in vitro* and *in vivo*. *Infect. Immun.* **55:**2870–2877.

Chugh, T.D., G.J. Burns, H.J. Shuhaiber, and G.M. Bahr. 1990. Adherence of *Staphylococcus epidermidis* to fibrin-platelet clots *in vitro* mediated by lipoteichoic acid. *Infect. Immun.* **58:**315–319.

Ciardi, J.E., G. Rolla, W.H. Bowen, and J.A. Riley. 1977. Adsorption of *Streptococcus mutans* lipoteichoic acid to hydroxyapatite. *Scand. J. Dent. Res.* **85:**387–391.

Clerc, P. and P.J. Sansonetti, 1987. Entry of *Shigella flexneri* into HeLa cells: evidence for directed phagocytosis involving actin polymerization and myosin accumulation. *Infect. Immun.* **55:**2681–2688.

Cohen, P.S., R. Rossol, V.J. Cabelli, S.L. Yang, and D.C. Laux. 1983. Relationship between the mouse colonizing ability of a human fecal *Escherichia coli* strain and its ability to bind a specific mouse colonic mucous gel protein. *Infect. Immun.* **40:**62–69.

Cohen, P.S., J.C. Arruda, T.J. Williams, and D.C. Laux. 1985. Adhesion of human fecal *Escherichia coli* strain to mouse colonic mucus. *Infect. Immun.* **48:**139–145.

Costerton, J.W., T.J. Marrie, and K.J. Cheng. 1985. Phenomena of bacterial adhesion. In: Savage, D.C. and M. Fletcher (eds.), *Bacterial Adhesion*. Plenum Press, New York, pp. 3–44.

Cowan, M.M., K.G. Taylor, and R.J. Doyle. 1986. Kinetic analysis of *Streptococcus sanguis* adhesion to artificial pellicle. *J. Dent. Res.* **65:**1278–1283.

Cowan, M.M., K.G. Taylor, and R.J. Doyle. 1987a. Energetics of the initial phase of adhesion of *Streptococcus sanguis* to hydroxylapatite. *J. Bacteriol.* **169:**2995–3000.

Cowan, M.M., K.G. Taylor, and R.J. Doyle. 1987b. Role of sialic acid in the kinetics of *Streptococcus sanguis* adhesion to artificial pellicle. *Infect. Immun.* **55:**1552–1557.

Crichton, P.B. and D.C. Old. Differentiation of strains of *Escherichia coli:* multiple typing approach. *J. Clin. Microbiol.* **11:**635–640.

Dean, E.A. and R.E. Kessler. 1988. Quantitation of effects of subinhibitory concentrations of trimethoprim on P fimbria expression and *in vitro* adhesiveness of uropathogenic *Escherichia coli*. *J. Clin. Microbiol.* **26:**25–30.

DeGraaf, F.K. 1990. Genetics of adhesive fimbriae of intestinal *Escherichia coli*. *Curr. Top. Microbiol. Immunol.* **151:**29–53.

DeGraaf, F.K., P. Klaasen-Boor, and J.E. van Hees. 1980. Biosynthesis of the K99 surface antigen is repressed by alanine. *Infect. Immun.* **30:**125–128.

DeRienzo, J.M., J. Porter-Kaufman, J. Haller, and B. Rosan. 1985. Corncob formation: a morphological model for molecular studies of bacterial interactions. In: Mergenhagen, S.E. and B. Rosan (eds.), *Molecular Basis of Oral Microbial Adhesion*. American Society for Microbiology, Washington, pp. 172–176.

Dominick, M.A., M.J. F. Schmerr, and A.E. Jensen. 1985. Expression of type 1 pili of *Escherichia coli* strains of high and low virulence in the intestinal tract of gnotobiotic turkeys. *Am. J. Vet. Res.* **46:**270–275.

Doyle, R. and K. Keller. 1984. Lectins in diagnostic microbiology. *Eur. J. Clin. Microbiol.* **3:**4–9.

Doyle, R.J., M. Rosenberg, and D. Drake. 1990. Hydrophobicity of oral bacteria. In: Doyle, R.J., and M. Rosenberg (eds.), *Microbial Cell Surface Hydrophobicity*. American Society for Microbiology, Washington, pp. 387–419.

Drake, D., K.G. Taylor, A.S. Bleiweis, and R.J. Doyle. 1988. Specificity of the glucan-binding lectin of *Streptococcus cricetus*. *Infect. Immun.* **56:**1864–1872.

Drumm, B., A.M. Roberton, and P.M. Sherman. 1988. Inhibition of attachment of *Escherichia coli* RDEC-1 to intestinal microvillus membranes by rabbit ileal mucus and mucin *in vitro*. *Infect. Immun.* **56:**2437–2442.

Duguid, J.P. and D.C. Old. 1980. Adhesive properties of *Enterobacteriaceae*. In: Beachey E.H. (ed.), *Bacterial Adherence*. Chapman and Hall, New York, pp. 184–217.

Eisenstein, B.I. 1981. Phase variation of type 1 fimbriae in *Escherichia coli* is under transcriptional control. *Science* **214:**337–337.

Eisenstein, B.I. and D.C. Dodd. 1982. Pseudocatabolite repression of type 1 fimbriae of *Escherichia coli*. *J. Bacteriol.* **151:**1560–1567.

Eisenstein, B., I. Ofek, and E.H. Beachey. 1981. Loss of lectin-like activity in aberrant type 1 fimbriae of *Escherichia coli*. *Infect. Immun.* **31:**792–797.

Fachon-Kalweit, S., B.L. Elder, and P. Fives-Taylor. 1985. Antibodies that bind to fimbriae block adhesion of *Streptococcus sanguis* to saliva-coated hydroxyapatite. *Infect. Immun.* **48:**617–624.

Fein, J.E. 1981. Screening of uropathogenic *Escherichia coli* for expression of mannose-sensitive adhesins: importance of culture conditions. *J. Clin. Microbiol.* **13:**1088–1095.

Ferreiros, C.M. and M.T. Criado. 1983. Different expression and genetic control for the K99 antigen and its associated adhesin. *Curr. Microbiol.* **8:**221–224.

Finkelstein, R.A. and L.F. Hanne. 1982. Purification and characterization of the soluble hemagglutinin (cholera lectin) produced by *Vibrio cholerae*. *Infect. Immun.* **36:**1199–1208.

Finlay, B.B., F. Heffron, and S. Falkow. 1989. Epithelial cell surfaces induce *Salmonella* proteins required for bacterial adherence and invasion. *Science* **243:**940–943.

Firon, N., D. Duksin, and N. Sharon. 1985. Mannose-specific adherence of *Escherichia coli* to BHK cells that differ in their glycosylation patterns. *FEMS Microbiol. Lett.* **27:**161–165.

Firon, N., S. Ashkenazi, D. Mirelman, I. Ofek, and N. Sharon. 1987. Aromatic α-glycosides of mannose are powerful inhibitors of the adherence of type 1 fimbriated *Escherichia coli* to yeast and intestinal epithelial cells. *Infect. Immun.* **55:**472–476.

Fives-Taylor, P.M. and D.W. Thompson. 1985. Surface properties of *Streptococcus sanguis* FW213 mutants nonadherent of saliva-coated hydroxyapatite. *Infect. Immun.* **47:**752–759.

Fletcher, M. 1985. Effect of solid surfaces on the activity of attached bacteria. In: Savage,

D.C. and M. Fletcher (eds), *Bacterial Adhesion*. Plenum Press, New York, pp. 339–362.

Forestier, C., A. Darfeuille-Michaud, and B. Joly. 1984. Effect of antibiotics on adhesion of enterotoxigenic *Escherichia coli* strains. *Eur. J. Clin. Microbiol.* **3:**427–432.

Freter, R. 1988. Mechanisms of bacterial colonization of the mucosal surfaces of the gut. In: Roth, J.A. (ed.), *Virulence Mechanisms of Bacterial Pathogens*. American Society for Microbiology, Washington, pp. 45–60.

Freter, R. and P.C. M. O'Brien. 1981. Role of chemotaxis in the association of motile bacteria with intestinal mucosa: *in vivo* studies. *Infect. Immun.* **34:**234–240.

Gander, R.M. and V.L. Thomas. 1987. Distribution of type 1 and p pili on uropathogenic *Escherichia coli* 06. *Infect. Immun.* **55:**293–297.

Garber, N., N. Sharon, D. Shohet, J.S. Lam, and R.J. Doyle. 1985. Contribution of hydrophobicity to hemagglutination reactions of *Pseudomonas aeruginosa. Infect. Immun.* **50:**336–337.

Germaine, G.R. and C.F. Schachtele. 1976. *Streptococcus mutans* dextransucrase: mode of interaction with high-molecular-weight dextran and role in cellular aggregation. *Infect. Immun.* **13:**365–372.

Gibbons, R.J. and J. Dankers. 1983. Association of food lectins with human oral epithelial cells *in vivo*. *Arch. Oral Biol.* **28:**561–566.

Gibbons, R.J., E.C. Moreno, and I. Etherden. 1983. Concentration-dependent multiple binding sites on saliva-treated hydroxyapatite for *Streptococcus sanguis*. *Infect. Immun.* **39:**280–289.

Goldhar, J., R. Perry, and I. Ofek. 1984. Extraction and properties of nonfimbrial mannose-resistant hemagglutinin from a urinary isolate of *Escherichia coli*. *Curr. Microbiol.* **11:**49–54.

Goldhar, J., A. Zilberberg, and I. Ofek. 1986. Infant mouse model of adherence and colonization of intestinal tissues by enterotoxigenic strains of *Escherichia coli* isolated from humans. *Infect. Immun.* **52:**205–208.

Gunnarsson, A., P.A. Mardh, A. Lundblad, and S. Svensson. 1984. Oligosaccharide structures mediating agglutination of sheep erythrocytes by *Staphylococcus saprophyticus*. *Infect. Immun.* **45:**41–46.

Haahtela, K., E. Tarkka, and T.K. Korhonen. 1985. Type 1 fimbria-mediated adhesion of enteric bacteria to grass roots. *Appl. Environ. Microbiol.* **49:**1182–1185.

Hacker, J. 1990. Genetic determinants coding for fimbriae and adhesins of extraintestinal *Escherichia coli*. *Curr. Top. Microbiol. Immunol.* **151:**1–27.

Hammani, A., L. Aqueda, M. Archambaud, N. Marty, L. Lapchine, and G. Chabonon. 1987. Effects of sub-inhibitory concentrations of oxolinic acid on both hemagglutinating activity and adhesion to uroepithelia cells by *E. coli* strains isolated from urine: an *in vitro* study. *Path. Biol.* **35:**545–550.

Harper, M., A. Turvey, and A.J. Bramley. 1978. Adhesion of fimbriate *Escherichia coli* to bovine mammary-gland epithelial cell *in vitro*. *J. Med. Microbiol.* **11:**117–123.

Hasty, D., I. Ofek, H.S. Courtney, and R.J. Doyle. 1992. Multiple adhesins of streptococci. *Infect. Immun.* **60:**2147–2152.

Hattori, R. and T. Hattori. 1963. Effect of a liquid-solid interface on the life of microorganisms. *Ecol. Dev.* **16:**64–70.

Hattori, R. and T. Hattori. 1981. Growth rate and molar growth yield of *Escherichia coli* absorbed on an anion-exchange resin. *J. Gen. Appl. Microbiol.* **27:**287–298.

Hazlett, L.O., M. Moon, and R.S. Berk. 1986. *In vivo* identification of sialic acid as the ocular receptor for *Pseudomonas aeruginosa*. *Infect. Immun.* **51:**687–689.

Hazlett, L.O., M.M. Moon, M. Strejc, and R.S. Berk. 1987. Evidence for *N*-acetylmannosamine as an ocular receptor for *P. aeruginosa* adherence to scarified cornea. *Invest. Ophthalmol. Vis. Sci.* **28:**1978–1985.

Herzberg, M., K. Gong, G.D. MacFarlane, P.R. Erickson, A.H. Soberay, P.H. Krebsbach, G. Manjula, K. Schilling, and W.H. Bowen. 1990. Phenotypic characterization of *Streptococcus sanguis* virulence factors associated with bacterial endocarditis. *Infect. Immun.* **58:**515–522.

Hogg, S.D. and G. Embery. 1982. Blood-group reactive glycoprotein from human saliva interacts with lipoteichoic acid on the surface of *Streptococcus sanguis* cells. *Arch. Oral Biol.* **27:**261–268.

Hogt, A.H., J. Dankert, and J. Feijen. 1983. Encapsulation, slime production and surface hydrophobicity of coagulase-negative staphylococci. *FEMS Microbiol.* **18:**211–215.

Holderbaum, D., R.A. Spech, and L.A. Ehrhart. 1985. Specific binding of collagen to *Staphylococcus aureus*. *Collagen Rel. Res.* **5:**261–271.

Honda, T., M. Arita, and T. Miwatani. 1984. Characterization of new hydrophobic pili of human enterotoxigenic *Escherichia coli:* a possible new colonization factor. *Infect. Immun.* **43:**959–965.

Hultgren, S.J., W.R. Schwan, A.J. Schaeffer, and J.L. Duncan. 1986. Regulation of production of type 1 pili among urinary tract isolates of *Escherichia coli*. *Infect. Immun.* **54:**613–620.

Humphrey, B., S. Kjelleberg, and K.C. Marshall. 1983. Responses on marine bacteria under starvation conditions at a solid–water interface. *Appl. Environ. Microbiol.* **45:**43–47.

Isaacson, R.E., P.C. Fusco, C.C. Brinton, and H.W. Moon. 1978. *In vitro* adhesion of *Escherichia coli* to porcine small intestinal epithelial cells: pili as an adhesive factor. *Infect. Immun.* **21:**392–397.

Isberg, R.R. and S. Falkow. 1985. A single genetic locus encoded by *Yersinia pseudotuberculosis* permits invasion of cultured animal cells by *Escherichia coli* K–12. *Nature* **317:**262–264.

Isberg, R.R., D.L. Voorhis, and S. Falkow. 1987. Identification of invasin: a protein that allows enteric bacteria to penetrate cultured mammalian cells. *Cell* **50:**769–778.

Israele, V., A. Darabi, and G.H. McCracken, Jr. 1987. The role of bacterial virulence factors and Tamm–Horsfall protein in the pathogenesis of *Escherichia coli* urinary tract infection in infants. *Am. J. Dis. Child.* **141:**1230–1234.

Jannasch, H.W. and P.H. Pritchard. 1972. The role of inert particulate matter in the activity of aquatic microorganisms, *Mem. Ist Italian Idrobiol.* **29**(*Suppl*):289–308.

Javorsky, P., E. Rybosova, I. Havassy, K. Horsky, and V. Kmet. 1987. Urease activity of adherent bacteria and rumen fluid bacteria. *Physiol. Bohemosl.* **36:**75–81.

Karch, H.K., J. Heesemann, R. Laufs, A.D. O'Brien, C.O. Tacket, and M.M. Levine. 1987. A plasmid of enterohemorrhagic *Escherichia coli* 0157:H7 is required for expression of a new fimbrial antigen and for adhesion to epithelial cells. *Infect. Immun.* **55:**455–461.

Kishimoto, E., D.I. Hay, and R.J. Gibbons. 1989. A human salivary protein which promotes adhesion of *Streptococcus mutans* serotype C strains to hydroxylapatite. *Infect. Immun.* **57:**3702–3707.

Klotz, S.A. and R.L. Penn. 1987. Multiple mechanisms may contribute to the adherence of *Candida* yeasts to living cells. *Curr. Microbiol.* **16:**119–122.

Knutton, S., D.R. Lloyd, and A.S. McNeish. 1987. Identification of a new fimbrial structure in enterotoxigenic *Escherichia coli* (ETEC) serotype 0148: H28 which adheres to human intestinal mucosa: a potentially new human ETEC colonization factor. *Infect. Immun.* **55:**86–92.

Kolenbrander, P.E. 1982. Isolation and characterization of coaggregation defective mutants of *Actinomyces viscosus, Actinomyces naeslundii* and *Streptococcus sanguis. Infect. Immun.* **37:**1200–1208.

Kolenbrander, P.E. 1989. Surface recognition among oral bacteria: multigeneric coaggregations and their mediators. *CRC Crit. Rev. Microbiol.* **17:**137–159.

Korhonen, T.K., E. Tarkka, H. Ranta, and K. Haahtela. 1983. Type 3 fimbriae of *Klebsiella* sp.: molecular characterization and role in bacterial adhesion to plant roots. *J. Bacteriol.* **155:**860–865.

Kristiansen, B.E., L. Rustad, O. Spanne, and B. Bjorvatin. 1983. Effect of subminimal inhibitory concentrations of antimicrobial agents on the piliation and adherence of *Neisseria meningitidis. Antimicrob. Agents Chemother.* **24:**731–734.

Krivan, H.C., V. Ginsburg, and D.D. Roberts. 1988. *Pseudomonas aeruginosa* and *Pseudomonas cepacia* isolated from cystic fibrosis patients bind specifically to gangliotetraosylceramide (asialo GM1) and gangliotrioasylceramide (asialoGM2). *Arch. Biochem. Biophys.* **260:**493–496.

Kuriyama, S.M. and F.J. Silverblatt. 1986. Effect of Tamm–Horsfall urinary glycoprotein on phagocytosis and killing of type–1 fimbriated *Escherichia coli. Infect. Immun.* **51:**193–198.

Lambden, P.R., J.E. Heckels, L.T. James, and P.J. Watt. 1979. Variations in surface protein composition associated with virulence properties in opacity types of *Neisseria gonorrhoeae. J. Gen. Microbiol.* **114:**305–312.

Leffler, H. and C. Svanborg-Eden. 1981. Glycolipid receptors for uropathogenic *Escherichia coli* on human erythrocytes and uroepithelial cells. *Infect. Immun.* **34:**920–929.

Leffler, H., and C. Svanborg-Eden. 1986. Glycolipids as receptors for *Escherichia coli*, lectins or adhesins. In: Mirelman, D. (ed.), *Microbial Lectins and Agglutinins.* John Wiley & Sons, New York, pp. 83–111.

Levine, M.J., M.C. Herzberg, M.S. Levine, S.A. Ellison, M.W. Stinson, H.C. Li, and T. Van Dyke. 1978. Specificity of salivary-bacterial interactions: role of terminal sialic acid residues in the interaction of salivary glycoproteins with *Streptococcus sanguis* and *Streptococcus mutans*. *Infect. Immun.* **19:**107–115.

Levine, M.M., D. Herrington, J.R. Murphy, J.G. Morris, G. Losonsky, B. Tall, A.A. Lindberg, S. Svenson, S. Baqar, M.F. Edwards, and B. Stocker. 1987. Safety, infectivity, immunogenicity and *in vivo* stability of two attenuated auxotrophic mutant strains of *Salmonella* typhi, 541ty and 543ty as live oral vaccines in humans. *J. Clin. Invest.* **79:**888–902.

Lianou, P.E., H.P. Bassaris, E.G. Votta, and J.T. Pappvassiliou. 1985. Interaction of subminimal inhibitory concentrations of clindamycin and gram-negative aerobic organisms: effects on adhesion and polymorphonuclear leukocyte function. *J. Antimicrob. Chemother.* **15:**481–487.

Liener, I.E. 1986. Nutritional significance of lectins in the diet. In: Liener, I.E., N. Sharon, and I.J. Goldstein (eds.), *The Lectins: Properties, Functions, and Applications in Biology and Medicine*. Academic Press, Orlando, pp. 527–552.

Lillard, H.S. 1986. Role of fimbriae and flagella in the attachment of *Salmonella typhimurium* to poultry skin. *J. Food Sci.* **51:**54–56.

Linder, H., I. Engberg, I.M. Baltzer, K. Jann, and C. Svanborg-Eden. 1988. Induction of inflammation by *Escherichia coli* on the mucosal level: requirement for adherence and endotoxin. *Infect. Immun.* **56:**1309–1313.

Lindquist, B.L., E. Lebenthal, P.C. Lee, M.W. Stinson, and J.M. Merrick. 1987. Adherence of *Salmonella typhimurium* to small intestinal enterocytes of the rat. *Infect. Immun.* **55:**3044–3050.

Lis, H. and N. Sharon. 1986. Applications of Lectins. In: Liener, I.E., N. Sharon, and I.J. Goldstein (eds.), *The Lectins, Properties, Function and Applications in Biology and Medicine*. Academic Press, Orlando, pp. 294–374.

Little, B.J., P.A. Wagner, W.G. Characklis, and W. Lee. 1990. Microbial corrosion. In: Characklis, W.G. and K.C. Marshall (eds.), *Biofilms*. John Wiley & Sons, New York, pp. 635–670.

Lowy, F.D., S.D. Chang, E.G. Neuhaus, D.S. Horne, A. Tomasz, and N.H. Steigbigel. 1983. Effect of penicillin on the adherence of *Streptococcus sanguis in vitro* and in the rabbit model of endocarditis. *J. Clin. Invest.* **71:**668–675.

Maayan, M.C., I. Ofek, O. Medalia, and M. Aronson. 1985. Population shift in mannose-specific fimbriated phase of *Klebsiella pneumoniae* during experimental urinary tract infection in mice. *Infect. Immun.* **49:**785–789.

Madison, B., E.H. Beachey, and S.N. Abraham. 1989. Protection against *E. coli* induced urinary tract infection (UTI) with antibodies directed at the Fim H protein of type 1 fimbriae. Abstracts, Annual Meeting, American Society for Microbiology, p. 38.

Marrs, C.F., G. Schoolnik, J.M. Koomey, J. Hardy, J. Rothbard, and S. Falkow. 1985. Cloning and sequencing of a *Moraxella bovis* pilin gene. *J. Bacteriol.* **163:**132–139.

Marrs, C.F., W.W. Ruehl, G.K. Schoolnik, and S. Falkow. 1988. Pilin gene phase variation of *Moraxella bovis* is caused by an inversion of the pilin genes. *J. Bacteriol.* **170:**3032–3039.

Marshall, K.C. 1986. Adsorption and adhesion processes in microbial growth at interfaces. *Adv. Colloid Interface Sci.* **25:**59–86.

Mascellino, M.T., S. Catania, M.L. De Vito, and C. DeBac. Erythromycin, miocamycin and clindamycin towards adhesivity and phagocytosis of gram-positive bacteria. *Microbiologia* **11:**231–241.

Mattingly, S.J. and B.P. Johnston. 1987. Comparative analysis of the localization of lipoteichoic acid in *Streptococcus agalactiae* and *Streptococcus pyogenes*. *Infect. Immun.* **55:**2383–2386.

Maxe, I., C. Ryden, T. Wadström, and K. Rubin. 1986. Specific attachment of *Staphylococcus aureus* to immobilized fibronectin. *Infect. Immun.* **54:**695–704.

McBride, B.C., M. Song, B. Krasse, and J. Olsson. 1984. Biochemical and immunological differences between hydrophobic and hydrophilic strains of *Streptococcus mutans*. *Infect. Immun.* **44:**68–75.

McIntire, F.C., L.K. Crosby, J.J. Barlow, and K.L. Matta. 1983. Structural preferences of β galactoside-reactive lectins on *Actinomyces viscosus* T14V and *Actinomyces naeslundii*. *Infect. Immun.* **41:**848–850.

Melhem, R.A. and P.T. Loverde. 1984. Mechanism of interaction of *Salmonella* and *Schistosoma* species. *Infect. Immun.* **44:**274–281.

Minion, F.C., G.H. Cassell, S. Pnini, and I. Kahane. 1984. Multiphasic interactions of *Mycoplasma pulmonis* with erythrocytes defined by adherence and hemagglutination. *Infect. Immun.* **44:**394–400.

Mirelman, D. 1987. Ameba-bacterium relationship in amebiasis. *Microbiol. Rev.* **51:**272–284.

Mirelman, D. and I. Ofek. 1986. Introduction to microbial lectins and agglutinins. In: Mirelman, D. (ed.), *Microbial Lectins and Agglutinins*. John Wiley & Sons, New York, pp. 1–19.

Miyake, Y., A. Kohada, I. Fujii, M. Sugai, and H. Suginaka. 1989. Aminoglycosides enhance the adherence of *Staphylococcus aureus* to HeLa cells. *J. Antimicrob. Chemother.* **23:**79–86.

Moen, D.V. 1962. Observation on the effectiveness of cranberry juice in urinary infections. *Wis. Med. J.* **61:**282–283.

Morris, E.J. and B.C. McBride. 1984. Adherence of *Streptococcus sanguis* to saliva-coated hydroxyapatite: evidence for two binding sites. *Infect. Immun.* **43:**656–663.

Morris, E.J., N. Ganeshkumar, and B.C. McBride. 1985. Cell surface components of *Streptococcus sanguis:* relationship to aggregation, adherence, and hydrophobicity. *J. Bacteriol.* **164:**255–262.

Murray, P.A., M.J. Levine, L.A. Tabak, and M.S. Reddy. 1982. Specificity of salivary-bacterial interactions: II. Evidence for a lectin on *Streptococcus sanguis* with specificity for a NeuAcα2,3Galβ1,3GalNAc sequence. *Biochem. Biophy. Res. Commun.* **106:**390–396.

Nalbandian, J., M.L. Freedman, J.M. Tanzer, and S.M. Lovelace. 1974. Ultrastructure of mutants of *Streptococcus mutans* with reference to agglutination, adhesion and extracellular polysaccharide. *Infect. Immun.* **10**:1170–1179.

Nesbitt, W.E., R.J. Doyle, K.G. Taylor, R. Staat, and R.R. Arnold. 1982a. Positive cooperativity in the binding of *Streptococcus sanguis* to hydroxylapatite. *Infect. Immun.* **35**:157–165.

Nesbitt, W.E., K.G. Taylor, and R.J. Doyle. 1982b. Hydrophobic interactions and the adherence of *Streptococcus sanguis* to hydroxylapatite. *Infect. Immun.* **38:**637–644.

Nevola, J.J., D.C. Laux, and P.S. Cohen. 1987. *In vivo* colonization of the mouse large intestine and *in vitro* penetration of intestinal mucus by an avirulent smooth strain of *Salmonella typhimurium* and its lipopolysaccharide. *Infect. Immun.* **55:**2884–2890.

Nickel, J.C., I. Ruseska, J.B. Wright, and J.W. Costerton. 1985. Tobramycin resistance of *Pseudomonas aeruginosa* cells growing as a biofilm on urinary catheter material. *Antimicrob. Agents Chemother.* **27:**619–624.

Nishikawa, Y. and T. Baba. 1985. *In vitro* adherence of *Escherichia coli* to endometrial epithelial cells of rats and influence of estradiol. *Infect. Immun.* **50:**506–509.

Nowicki, B., M. Rhen, V. Vaisanen-Rhen, A. Pere, and T.K. Korhonen. 1984. Immunofluorescence study of fimbrial phase variation in *Escherichia coli*. KS71. J. Bacteriol. **160:**691–695.

Nowicki, B., M. Rhen, V. Vaisanen-Rhen, A. Pere, and T.K. Korhonen. 1985a. Fractionation of a bacterial cell population by adsorption to erythrocytes and yeast cells. *FEMS Microbiol. Lett.* **26:**35–40.

Nowicki, B., H. Holthofer, T. Saraneva, M. Rhen, V. Vaisenen-Rhen, and T.K. Korhonen. 1985b. Location of adhesion sites for P-fimbriated and for 075X-positive *Escherichia coli* in the human kidney. *Microb. Pathogen.* **1:**169–180.

Nowicki, B., J. Vuopio-Varkila, P. Viljanen, T.K. Korhonen, and P.H. Makela. 1986. Fimbrial phase variation and systemic *E. coli* infection studied in the mouse peritonitis model. *Microb. Pathogen.* **1:**335–347.

Ofek, I. and E.H. Beachey. 1978. Mannose binding and epithelial cell adherence of *Escherichia coli*. *Infect. Immun.* **22:**247–254.

Ofek, I. and E.H. Beachey. 1980. Bacterial adherence. *Adv. Int. Med.* **25:**505–532.

Ofek, I., E.H. Beachey, B.I. Eisenstein, M.L. Alkan, and N. Sharon. 1979. Suppression of bacterial adherence by subminimal inhibitory concentrations of beta-lactam and aminoglycoside antibiotics. *Rev. Infect. Dis.* **1:**832–837.

Ofek, I., J. Goldhar, Y. Eshdat, and N. Sharon. 1982. The importance of mannose specific adhesins (lectins) in infections caused by *Escherichia coli*. *Scand. J. Infect. Dis. Suppl.* **33:**61–67.

Ofek, I., S. Cohen, and R. Rachmani. 1989. Influence of antibiotics on adhesion of bacteria. *J. Chemother.* (*Suppl.* 4). **1:**6.

Ofek, I., D. Zafriri, J. Goldhar, and B.I. Eisenstein. 1990. Inability of toxin inhibitors to neutralize enhanced toxicity caused by bacteria adherent to tissue culture cells. *Infect. Immun.* **58:**3737–3742.

Ofek, I., J. Goldhar, D. Zafriri, H. Lis, R. Adar, and N. Sharon. 1991. Anti-adhesin activity in cranberry and blueberry juices. *N. Engl. J. Med.* **324:**1599.

Old, D.C. 1972. Inhibition of the interaction between fimbrial hemagglutinins and erythrocytes by D-mannose and other carbohydrates. *J. Gen. Microbiol.* **71:**149–157.

Old, D.C. and R.A. Adegbola. 1983. A new mannose-resistant hemagglutinin in *Klebsiella. J. Appl. Bacteriol.* **55:**165–172.

Ørskov, I., A. Ferencz, and P. Ørskov. 1980. Tamm–Horsfall protein or uromucoid is the normal urinary slime that traps type 1 fimbriated *Escherichia coli. Lancet* **1:**887.

Papas, P.N., C.A. Brusch, and G.C. Ceresia. 1968. Cranberry juice in the treatment of urinary tract infections. *Southwest Med. J.* **47:**17–20.

Parkkinen, J., R. Virkola, and T.K. Korhonen. 1988. Identification of factors in human urine that inhibit the binding of *Escherichia coli* adhesins. *Infect. Immun.* **56:**2623–2630.

Parry, S.H., S. Boonchai, S.N. Abraham, J.M. Salter, D.M. Rooke, J.M. Simpson, A.J. Bint, and M. Sussman. 1983. A comparative study of the mannose-resistant and mannose-sensitive hemagglutination of *Escherichia coli* isolated from urinary tract infections. *Infection* **11:**123–128.

Paul, J.H. 1984. Effects of antimetabolites on the adhesion of estuarine *Vibrio sp.* to polystyrene. *Appl. Environ. Microbiol.* **48:**924–929.

Pearl, H.W. 1985. Influence of attachment on microbial metabolism and growth in aquatic ecosystems. In: Savage, D.C. and M. Fletcher (eds.), *Bacterial Adhesion: Mechanisms and Physiological Significance*. Plenum Press, New York, pp. 363–400.

Pere, A., B. Nowicki, H. Saxen, A. Siitonen, and T.K. Korhonen. 1987. Expression of P, type-and type 1C fimbriae of *Escherichia coli* in the urine of patients with acute urinary tract infection. *J. Infect. Dis.* **156:**567–574.

Peros, W.J. and R.J. Gibbons. 1982. Influence of sublethal antibiotic concentrations on bacterial adherence to saliva-treated hydroxyapatite. *Infect. Immun.* **35:**326–334.

Ponniah, S., S.N. Abraham, M.E. Dockter, C.D. Wall, and R.D. Endres. 1989. Mitogenic stimulation of human lymphocytes by the mannose-specific adhesin on *Escherichia coli* type 1 fimbriae. *J. Immunol.* **142:**992–998.

Proctor, R.A., P.J. Olbrantz, and D.F. Mosher. 1983. Subinhibitory concentrations of antibiotics alter fibronectin binding to *Staphylococcus aureus. Antimicrob. Agents Chemother.* **24:**823–826.

Ramphal, R. and G.B. Pier. 1985. Role of *Pseudomonas aeruginosa* mucoid exopolysaccharide in adherence to tracheal cells. *Infect. Immun.* **47:**1–4.

Rhen, M., P. Klemm, and T.K. Korhonen. 1986. Identification of two new hemagglutinins of *Escherichia coli, N*-acetyl-D-glucosamine-specific fimbriae and a blood group M-specific agglutinin, by cloning the corresponding genes in *Escherichia coli* K–12. *J. Bacteriol.* **168:**1234–1242.

Rosenberg, M. 1984. Isolation of pigmented and nonpigmented mutants of *Serratia marcescens* with reduced cell surface hydrophobicity. *J. Bacteriol.* **160:**480–482.

Ruoslahti, E. and M.D. Pierschbacher. 1987. New perspectives in cell adhesion: RGD and integrins. *Science* **238:**491–497.

Salit, I.E. 1983. Effect of subinhibitory concentrations of antimicrobials on meningococcal adherence. *Can. J. Microbiol.* **29:**369–376.

Salit, I.E. and E.C. Gotschlich. 1977. Hemagglutination by purified type 1 *Escherichia coli* pili. *J. Exp. Med.* **146:**1169–1181.

Sandberg, T., K. Stenqvist, and C. Svanborg-Eden. 1979. Effects of subminimal inhibitory concentrations of ampicillin, chloramphenicol, and nitrofurantoin on the attachment of *Escherichia coli* to human uroepithelial cells *in vitro*. *Rev. Infect. Dis.* **1:**838–844.

Schadow, K.H., W.A. Simpson, and G.D. Christensen. 1988. Characteristics of adherence to plastic tissue culture plates of coagulase-negative staphylococci exposed to subinhibitory concentrations of antimicrobial agents. *J. Infect. Dis.* **157:**71–77.

Schaeffer, A.J., S.K. Amundsen, and L.N. Schmidt. 1979. Adherence of *Escherichia coli* to human urinary tract epithelial cells. *Infect. Immun.* **24:**753–759.

Schaeffer, A.J., W.R. Schwan, S.J. Hultgren, and J.L. Duncan. 1987. Relationship of type 1 pilus expression in *Escherichia coli* to ascending urinary tract infections in mice. *Infect. Immun.* **55:**373–380.

Scheld, W.M., O. Zak, K. Vosbeck, and M.A. Sande. 1981. Bacterial adhesion in the pathogenesis of infective endocarditis: effect of subinhibitory antibiotic concentrations on streptococcal adhesion *in vitro* and the development of endocarditis in rabbits. *J. Clin. Invest.* **68:**1381–1384.

Schifferli, D.M. and E.H. Beachey. 1988a. Bacterial adhesion: modulation by antibiotics which perturb protein synthesis. *Antimicrob. Agents Chemother.* **32:**1603–1608.

Schifferli, D.M. and E.H. Beachey. 1988b. Bacterial adhesion: Modulation by antibiotics with primary targets other than protein synthesis. *Antimicrob. Agents Chemother.* **32**:1609–1613.

Schmidt, D.R. and A.E. Sobota. 1988. An examination of the anti-adherence activity of cranberry juice on urinary and nonurinary bacterial isolates. *Microbios* **55:**173–181.

Sharon, N. and I. Ofek. 1986. Bacterial surface lectins specific for mannose. In: Mirelman, D. (ed.), *Microbial Lectins and Agglutinins*. John Wiley & Sons, New York, pp. 55–81.

Shaw, J.H. and S. Falkow. 1988. Model for invasion of human tissue culture cells by *Neisseria gonorrhoeae*. *Infect. Immun.* **56:**1625–1632.

Sherman, P.M. and E.C. Boedeker. 1987. Pilus-mediated interactions of the *Escherichia coli* strain RDEC-1 with mucosal glycoproteins in the small intestine of rabbits. *Gastroenterology* **93:**734–743.

Sheth, N.K., T.R. Franson, and P.G. Sohnle. 1985. Influence of bacterial adherence to intravascular catheters on *in vitro* antibiotic susceptibility. *Lancet* **ii:**1266–1268.

Shibl, A.M. 1985. Effect of antibiotics on adherence of microorganisms to epithelial cell surfaces. *Rev. Infect. Dis.* **7:**51–65.

Simon, M., J. Zeig, M. Silverman, G. Mandel, and R. Doolittle. 1980. Phase variation: evolution of a controlling element. *Science* **209:**1370–1374.

Simpson, W.A. and E.H. Beachey. 1983. Adherence to group A streptococci to fibronectin an oral epithelial cells. *Infect. Immun.* **39:**275–279.

Simpson, W.A., I. Ofek, and E.H. Beachey. 1980. Binding of streptococcal lipoteichoic acid to the fatty acid binding sites on serum albumin. *J. Biol. Chem.* **255:**6092–6097.

Simpson, W.A., H.S. Courtney, and I. Ofek. 1987. Interactions of fibronectin with streptococci: the role of fibronectin as a receptor for *Streptococcus pyogenes*. *Rev. Infect. Dis.* **9:**S351–S359.

Sobota, A.E. 1984. Inhibition of bacterial adherence by cranberry juice: potential use for the treatment of urinary tract infection. *J. Urol.* **131:**1013–1016.

Sogaard, H. and J.L. Larsen. 1983. The effect of antibiotics on mannose-resistant hemagglutination by K88-positive and K99-positive *Escherichia coli* strains. *J. Vet. Pharm. Ther.* **6:**187–193.

Speziale, P., M. Hook, L. Switalski, and T. Wadström. 1984. Fibronectin binding to a *Streptococcus pyogenes* strain. *J. Bacteriol.* **157:**420–427.

Speziale, P., M. Hook, L. Switalski, and T. Wadström. 1987. Binding of collagen to group A,B,C,D and G streptococci. *FEMS Microbiol. Lett.* **48:**47–51.

Stengvist, K., T. Sandberg, S. Ahlstedt, T.K. Korhonen, and C. Svanborg-Eden. 1982. Effects of subinhibitory concentrations of antibiotics and antibodies on the adherence of *Escherichia coli* to human uroepithelial cells *in vitro*. *Scand. J. Infect. Dis. Suppl.* **33:**104–107.

Stephens, D.S., J.W. Krebs, and Z.A. McGee. 1984. Loss of pili and decreased attachment to human cells by *Neisseria meningitidis* and *Neisseria gonorrhoeae* exposed to subinhibitory concentrations of antibiotics. *Infect. Immun.* **46:**507–513.

Sternlieb, P. 1963. Cranberry juice in renal disease. *N. Engl. J. Med.* **268:**57.

Stibitz, S., A.A. Weiss, and S. Falkow. 1988. Genetic analysis of a region of the *Bordetella pertussis* chromosome encoding filamentous hemagglutinin and the pleiotropic regulatory locus *vir*. *J. Bacteriol.* **170:**2904–2913.

Sun, D., S.N. Abraham, and E.H. Beachey. 1988a. Influence of berberine sulfate on synthesis and expression of Pap fimbrial adhesin in uropathogenic *Escherichia coli*. *Antimicrob. Agents Chemother.* **32:**1274–1277.

Sun, D., H.S. Courtney, and E.H. Beachey. 1988b. Berberine sulfate blocks adherence of *Streptococcus pyogenes* to epithelial cells, fibronectin and hexadecane. *Antimicrob. Agents Chemother.* **32:**1370–1374.

Sussman, M., S.N. Abraham, and S.H. Parry. 1983. Bacterial adhesion in the host-parasite relationship of urinary tract infection. In: Schulte-Wissermann, H. (ed.), *Clinical, Bacteriological Immunologic Aspects of Urinary Tract Infection in Children*. Thieme Stuttgart, pp. 103–112.

Swaney, L.M., Y.P. Liu, C.M. To, K. Ippen-Ihler, and C.C. Brinton, Jr. 1977. Isolation and characterization of *Escherichia coli* phase variants and mutants deficient in type 1 pilus production. *J. Bacteriol.* **130:**495–505.

Swanson, J. 1980. Adhesion and entry of bacteria into cells: Model of the pathogenesis of gonorrhea. In: Smith, H., J.J. Skehel, and M.J. Turner (eds.), *The Molecular Basis of Microbial Pathogenecity, Dahlem Konferenzen:* Verlag Chemi GMbh, Weinheim, pp. 17–40.

Swanson, J. and O. Barrera. 1983. Gonococcal pilus subunit size heterogeneity correlates with transitions in colony piliation phenotype, not with changes in colony opacity. *J. Exp. Med.* **158:**1459–1472.

Swanson, J. and J.M. Koomey. 1989. Mechanisms for variation of pili and outer membrane protein II in *Neisseria gonorrhoeae*. In: Berg, D.E. and M. Howe (eds.), *Mobile DNA*. American Society for Microbiology, Washington, pp. 743–761.

Tacket, C.O., D.R. Maneval, and M.M. Levine. 1987. Purification, morphology, and genetics of a new fimbrial putative colonization factor of enterotoxigenic *Escherichia coli* 0159:H4. *Infect. Immun.* **55:**1063–1069.

Tanaka, Y. 1982. Multiplication of fimbriate and nonfimbriate *Salmonella typhimurium* organisms in the intestinal mucosa of mice treated with antibiotics. *Jap. J. Vet. Sci.* **44:**523–527.

Teti, G., G. Orefici, F. Thomasello, R. Trifrletti, C. Fava, S. Recchia, and P. Mastrolni. 1985. Role of lipoteichoic acid in the adherence of group B streptococci to vaginal epithelial cells. In: Kimura, Y., S. Kotami, and Y. Shiokawa (eds.), *Recent Advances in Streptococci and Streptococcal Diseases*. Reed Books, Berkshire, UK, pp. 94–96.

Tylewska, S., S. Hjerten, and T. Wadström. 1981. Effect of subinhibitory concentrations of antibiotics on the adhesion of *Streptococcus pyogenes* to pharyngeal epithelial cells. *Antimicrob. Agents Chemother.* **20:**563–566.

Tylewska, S.K., V.A. Fischetti, and R.J. Gibbons. 1988. Binding selectivity of *Streptococcus pyogenes* and M-protein to epithelial cells differs from that of lipoteichoic acid. *Curr. Microbiol.* **16:**209–216.

Vaisanen, V., K. Lounatmaa, and T.K. Korhonen. 1982. Effects of sublethal concentrations of antimicrobial agents on the hemagglutination, adhesion, and ultrastructure of pyelonephritogenic *Escherichia coli* strains. *Antimicrob. Agents Chemother.* **22:**120–127.

Valentin-Weigand, P., J. Grulich-Henn, G.S. Chhatwal, G. Muller-Berghaus, H. Blobel, and K.T. Preissner. 1988. Mediation of adherence of streptococci to human endothelial cells by complement S protein (vitronectin). *Infect. Immun.* **56:**2851–2855.

Vishwanath, S., C-M. Guay, and R. Ramphal. 1987. Effects of sublethal inhibitory concentrations of antibiotics on the adherence of *Pseudomonas aeruginosa* to tracheobronchial mucin. *J. Antimicrob. Chemother.* **19:**579–583.

Vosbeck, K. 1982. In: Effects of low concentrations of antibiotics on *Escherichia coli* adhesion. In: Eickenberg, H.U., H. Hahn, and W. Opferkuch (eds.), *The Influence of Antibiotics on the Host–Parasite Relationship*. Springer-Verlag, Berlin and New York, pp. 183–193.

Vosbeck, K., H. Handschin, E.B. Menge, and O. Zak. 1979. Effects of subminimal inhibitory concentrations of antibiotics on adhesiveness of *Escherichia coli in vitro. Rev. Infect. Dis.* **1:**845–851.

Vosbeck, K., H. Mett, U. Huber, J. Bohn, and M. Petignat. 1982. Effects of low concentrations of antibiotics on *Escherichia coli* adhesion. *Antimicrob. Agents Chemother.* **21:**864–869.

Wadström, T., P. Speziale, F. Rozgonyi, A. Ljungh, I. Maxe, and C. Ryden. 1987. Interactions of coagulase-negative staphylococci with fibronectin and collagen as possible first stage of tissue colonization in wounds and other tissue trauma. *Zbl. Bakt.* (*Suppl. 16*) 83–91.

Weiss, A.A., and S. Falkow. 1984. Genetic analysis of phase change in *Bordetella pertussis*. *Infect. Immun.* **43:**263–264.

Weiss, E.I., J. London, P.E. Kolenbrander, A.S. Kagermeier, and R.N. Andersen. 1987. Characterization of lectin-like surface components on *Capnocytophage ochracea* ATCC 33596 that mediate coaggregation with gram-positive oral bacteria. *Infect. Immun.* **55:**1198–1202.

Wentworth, J.S., F.E. Austin, N. Garber, N. Gilboa-Garber, C.A. Paterson, and R.J. Doyle. 1991. Cytoplasmic lectins contribute to the adhesion of *Pseudomonas aeruginosa*. *Biofouling* **4:**99–104.

Wicken, A.J. 1980. Structure and cell membrane-binding properties of bacterial lipoteichoic acids and their possible role in adhesion of streptococci to eukaryotic cells. In: Beachey E. (ed.), *Bacterial Adherence*. Chapman and Hall, London, pp. 139–158.

Zafriri, D., Y. Oron, B.I. Eisenstein, and I. Ofek. 1987. Growth advantage and enhanced toxicity of *Escherichia coli* adherent to tissue culture cells due to restricted diffusion of products secreted by the cells. *J. Clin. Invest.* **79:**1210–1216.

Zafriri, D., I. Ofek, R. Adar, M. Pocino, and N. Sharon. 1989. Inhibitory activity of cranberry juice on adherence of type 1 and type P fimbriated *Escherichia coli* to eukaryotic cells. *Antimicrob. Agents Chemother.* **33:**92–98.

Zhang, X-h., M. Rosenberg, and R.J. Doyle. 1990. Inhibition of the cooperative adhesion of *Streptococcus sanguis* to hydroxylapatite. *FEMS Microbiol. Lett.* **71:**315–318.

Index